COMMERCIAL METHODS OF ANALYSIS

COMMERCIAL METHODS OF ANALYSIS

by Foster Dee Snell, A.M., Ph.D.

and

Frank M. Biffen, B.Sc., F.R.I.C.

WITH PHOTOMICROGRAPHS BY
George Lord

REVISED EDITION

CHEMICAL PUBLISHING COMPANY, INC.

NEW YORK

1964

CHEMICAL PUBLISHING CO., INC.
NEW YORK

Printed in the United States of America

PREFACE TO SECOND EDITION

The favorable reception accorded the first edition of *Commercial Methods of Analysis* has led us to prepare this revised edition. In the intervening twenty years, there have been introduced many methods of carrying out special determinations often characterized as the "sophisticated" methods because of their complexity. Typical examples are infrared analysis and gas chromatography. They are so specialized that each would require a book for any comprehensive treatment. Therefore, it has been the judgment of the authors not to try to introduce them.

So changes are surprisingly few. Analysis of minerals has not changed over these past twenty years and it is improbable that it will change over the next twenty. The same statement applies to many, many other determinations. The refractometer shown looks different in this edition from the model of twenty years ago but it operates in the same way. Our thanks go to the manufacturers and distributors for many illustrations of new apparatus and equipment.

Even the comments in the preface to the first edition are equally cogent today, so we suggest that you review it.

FOSTER DEE SNELL,
FRANK M. BIFFEN.

NEW YORK, N.Y.,
March, 1964.

PREFACE TO FIRST EDITION

Any new book in the field of analytical chemistry should present new material, and preferably a new point of view, in order to justify its existence. Many meritorious and deservedly well-known textbooks have been published on theoretical analytical chemistry. This book has another purpose—to show the method of approach to analysis of the innumerable complex commercial products existing on the market today. Many of these are colloids, some are emulsions that may contain half a dozen ingredients besides water and oil, a majority contain organic as well as inorganic ingredients, and as unknown samples most of them would present problems extremely difficult of solution by the inexperienced chemist.

In order to make the book of more universal use, standard methods, such as those given in the "Methods of Analysis of the American Society for Testing Materials" and the "Methods of Analysis of the Association of Official Agricultural Chemists," have in many cases been introduced, though often not in the exact words of these methods. In line with the general scheme of the book, simplifications have been adopted in some parts, and, what is more important, explanations of steps, the reason for which is not sufficiently clear, have been given. Standard methods are not given in full, only such determinations as seem necessary to the general analyst. If a sample must be analyzed in detail, strictly in conformance with these official methods, the original methods should be referred to. Other, usually considerably shorter, methods are sometimes introduced.

Particular care has been taken to set forth in logical sequence the preliminary steps that necessarily precede the actual analysis of samples of unknown composition. Both in this and in subsequent procedure, an endeavor is made to clarify the meaning of each step and often to give the reason for it. This is to develop the inexperienced analyst into an experienced one by logical methods, to show the importance of what may at first seem to be unimportant details of procedure, details which, if omitted, will in many cases lead to incorrect results. Methods are as simple and direct as possible, compatible with accuracy. Where data are available, limits of accuracy are discussed as to what is desirable and what is attainable for a particular type of sample, or sometimes for a particular application. The practical point of view of the commercial laboratory has been emphasized by methods that aim at general economy of time and materials without sacrificing reliability of results.

A feature of the book is an attempt to set forth succinctly, frequently by stating the reactions involved, the methods of calculating the results of determinations in order to show reason for the formula given.

As a matter of organization, after such routine and widely used determinations as those for nitrogen, sulfur, and halogen compounds, only commercial types of samples are considered. Condensation has been accomplished by profuse cross reference to methods in other parts of the book, in nearly every case to specific portions rather than to general paragraphs or chapters.

Special methods have been included that have been found to be saving of both time and labor. Many modern methods of analysis and methods for analyzing of substances of recent importance, such as synthetic plastics and synthetic elastomers, have been included, even when only qualitative information is available. Emphasis is also laid on the growing importance of the determination of minute quantities.

The useful new branch of qualitative analysis by spot tests is condensed to one chapter (Chap. 5). The Precision Scientific Company of Chicago has designed a micro kit (Fig. 69) for use with this book. It contains only the reagents required in Chap. 5.

It is hoped that this volume will prove a useful text to the student who has already become familiar with the tools of quantitative analysis and the routine methods, and who desires further training in this field by a study of commercial products. It may also serve as a manual to the industrial analyst whose work is not confined to simple control operations.

Many of the methods and ideas presented here have been developed and used over a period of two decades in commercial laboratory operations. As with most things new in the field of chemistry, much time, effort, and research have gone into their making.

FOSTER DEE SNELL,
FRANK M. BIFFEN.

NEW YORK, N.Y.,
July, 1944.

CONTENTS

CHAPTER		PAGE
PREFACE		v
CHAPTER		
1.	GENERAL	1
2.	TOOLS OF THE ANALYST	7
3.	GENERAL PROCEDURES	37
4.	PRELIMINARIES TO THE ANALYSIS	75
5.	QUALITATIVE ANALYSIS BY MICRO AND SPOT TESTS	81
6.	HYDROGEN-ION DETERMINATION AND ELECTROMETRIC TITRATION	112
7.	HALOGENS	127
8.	COMBINED NITROGEN	140
9.	SULFUR AND ITS COMPOUNDS	157
10.	MINERALS	176
11.	WATER ANALYSIS	226
12.	FIXED ALKALIES AND ALKALINE SALTS	254
13.	CEMENT AND MORTAR	275
14.	FERROUS METALS	282
15.	ALUMINUM AND LIGHT ALUMINUM ALLOYS	311
16.	COPPER-BASE ALLOYS	323
17.	SAPONIFIABLE FATS AND OILS	337
18.	SULFATED OIL AND RELATED PRODUCTS	367
19.	SOAP AND SOAP PRODUCTS	381
20.	HEAVY-METAL SOAPS	405
21.	LUBRICATING GREASES	408
22.	WAXY SUBSTANCES	423
23.	SOLVENTS AND THINNERS	434
24.	ALCOHOLS	457
25.	PAINTS, VARNISHES, LACQUERS, AND RESINS	468
26.	SYNTHETIC RESINS	487
27.	WHITE PIGMENTS, EXTENDERS, AND MINERAL FILLERS	496
28.	MIXED WHITE PIGMENTS, EXTENDERS, AND MINERAL FILLERS	509
29.	COLORED PIGMENTS AND MINERAL COLORS	522
30.	OIL AND WAX EMULSIONS	540
31.	RUBBER	559
32.	MISCELLANEOUS RUBBER ANALYSES AND SYNTHETIC ELASTOMERS	581
33.	ANALYSIS OF COAL	596
34.	GAS ANALYSIS	603
35.	PAPER, PAPER FIBERS, TEXTILE FIBERS, AND STARCHES	614
36.	SUGAR AND SUGAR PRODUCTS	654
37.	PETROLEUM PRODUCTS	662
38.	PYRETHRINS AND ROTENONE	710
39.	INDICATOR, REAGENT, AND VOLUMETRIC SOLUTIONS	717
INDEX		733

COMMERCIAL METHODS OF ANALYSIS

CHAPTER 1

GENERAL

If a chemist were given a commercial material of a specific type and asked to determine its composition, he might well be in a quandary as to how to approach the task unless he had much prior experience with that specific type. A sample is rarely analyzed without some general preliminary information as to its probable composition; but, equally definitely, such preliminary information is rarely sufficiently comprehensive to indicate all the materials present.

The ultimate purpose to be served by the analysis is important. Lack of this essential preliminary information may lead to much lost time in doing unnecessary work. How accurate need the results be? Time should not be spent in the use of highly accurate methods when a briefer and less accurate procedure will give all the information needed. Conversely, and more commonly, work that is not of adequate accuracy may vitiate all the results obtained.

Is it essential to find out everything possible about the sample, to do a complete analysis? In many instances, proper advance information will indicate that only part of the analysis will ever be used. How much of the sample is available? Considerable advance planning must often be done to get the necessary results from a small sample. Can two or three or more determinations be made on the same portion of the sample? This must often be known in advance to permit such use. What size of sample is desirable for each determination? Naturally this depends on the amount of the substance being determined, which should be known within some general limits before proceeding. What methods are to be used for each determination? These and many other questions must be answered by the analyst before he weighs out the first part of the sample.

Purpose of the Analysis. Analyses are not made merely to be decorated with the signature of the analyst and filed. Each has one or more purposes. If the purpose is to present a comparison with a competitor's product, the approach is not the same as it is if the analysis is to serve as the basis of determining whether the product is of good quality, and that in turn differs from an analysis for the purpose of permitting manufacture of a similar product. It is necessary to find out as precisely as possible the purpose of the analysis.

It is well at the beginning to stress what will always be tacitly assumed throughout, that in commercial practice economy of operation is paramount. The ability to do a thing well and efficiently is ever desirable and one of the marks of a well-trained and practical chemist; more especially is this so in applied chemistry. Business, to be profitable, must be run on sound economic principles. If the making of chemical analyses is a minor or major part of that business, they, as much as any other part of the business, must be performed with a view to a favorable balance sheet in terms of time and materials.

Accuracy. To the worker who takes pride in his work—and one who does not is not worth his salt—raising the question of accuracy may seem almost sacrilegious. Reflection, however, will indicate that results with accuracy to parts per million are only occasionally needed. In determining the amount of abrasive in a tooth paste, for example, accuracy within 1 per cent will often be sufficient; a tube of that brand from another batch may vary by more than that. In determining the amount of fatty acid in a fatty oil, an error of 0.1 per cent may make the difference between the product's being classifiable as an edible oil or as semicrude for further refining. An error of a few parts per million in the arsenic content of a dyestuff may mean the difference between a quality suitable for certification and an industrial dyestuff at a fraction of the price per pound.

In a check against specifications, will the person receiving a report be grateful for a result reading to 20.86 per cent when the requirement is that an ingredient be between 20 and 22 per cent? Clearly this will take longer than to obtain a result of 20.8 or 20.9 per cent, if only in longer time for weighing or longer to get the end point in titration, and longer to calculate. So, the accuracy—and the method—should be suited to the requirements.

It is equally true, and perhaps this is the more common fault, that lack of sufficient accuracy nullifies work not only on that determination but on others. A short method that for other purposes may be excellent may become but waste of time for the sample on hand.

Poor, slipshod work will always give inaccurate results, and it may, though it should not, be necessary to warn against this. On no account should such work be tolerated, whether results are being expressed as one part in ten or one part in ten million. There are times when, to obtain any results at all, in the nature of the case a higher degree of accuracy is obtained than is necessary. When this is unavoidable, it should occasion no worry.

In general, it may be said that the determination of the amounts of an element in parts per million involves special methods that are usually far more lengthy than less accurate procedures.

Size of Sample. This is governed by several factors. If only a small original sample is available, more thought and ingenuity must be used. It is remarkable how much information can be extracted from very small amounts of a sample by careful forethought. Much real satisfaction is experienced by the worker who is able to get all the information and accuracy required from what would normally be called an inadequate size of sample by judiciously arranging the order of the determinations. A certain sense of what is best and fitting becomes part of the analyst's intuition, an intuition that, however, is based on logical chemical reasoning based in turn on sound chemical knowledge. Analysis is the application of known chemical facts in a logical sequential manner.

Even if the sample is sufficient, preliminary thought as to how much to use will be amply repaid. Large samples that give bulky precipitates and are usually unwieldy all through the analysis should be avoided. It is well to aim at obtaining a precipitate weighing less than 1 gram, preferably about half of that; a colored solution that can easily be matched; or a titration value of, conveniently, between 20 and 40 ml. Such figures often cannot be obtained, and the above must be looked upon not as hard and fast rules but rather as indicative of what may be termed optimum conditions. Experience is the best teacher, and even a short period can enable one to become quite expert at judging what is required.

At the same time, too small a sample should be avoided if possible. The experimental error becomes proportionately greater; and, theoretically at any rate, a limit may be reached at which this error is equal to the actual determination. In these cases, and also when any blank that may be made is nearly as great as the determination, the degree of accuracy is seriously impaired. Experimental error varies both with the worker and with the method. In the first case, the experimental error will, with a reasonably good worker, not be of great consequence. In the second case, a large experimental error may be inherent in the method used. If this is so, it is essential to become aware of the fact before employing such a method.

Choice of Method. The many methods presented in a textbook as available for a certain determination may well be a source of worry. What method is best under the circumstances is a question that can be answered only by careful attention to these methods. First of all, if possible, find out what other substances are in the sample. Then it may be found necessary to eliminate one or more of the given methods as this or that other substance interferes or may have to be removed before the estimation can be made.

If the approximate amount of the required element is known, this may also limit the methods that can be employed. Choice may also be made advantageously with a view to further work that may be made on the same portion of sample. Ease of manipulation and availability of apparatus may be determining factors, and the method chosen may be decided by the required accuracy of the result.

In any event, the methods given should be read carefully with particular attention to the effect of interfering substances. The whole determination may become invalid if the worker fails to pay sufficient attention to any such particulars. This source of error is very difficult to trace if it is supposed that a given method has been properly followed; and not only is an inaccurate result derived for the element under consideration, but the whole analysis may be wrongly interpreted.

That the method selected should be thoroughly read, understood, and followed might seem a most obvious fact. Nevertheless, it is emphasized here because experience has shown that many errors and incomprehensible results are traceable to careless reading and interpretation. For the most part, there is a sound theoretical reason for each step given, and no deviation should be made without an equally sound reason. Unfortunately, little explanation is, as a general rule, given as to why certain procedures are to be followed. Even this failure may sometimes be turned to good account in that the worker is enabled to find out the reason for himself from theoretical considerations. With beginners, however, this is frequently not possible, and even experienced chemists may have real difficulty in this respect. To sum up, for this matter is of real importance, read the method carefully, find out as well as possible the reason for each step, and follow it faithfully. Let it be said that most so-called short cuts turn out to be otherwise. This does not preclude the possibility that an experienced person can sometimes improve the method and modify it to suit his particular requirements. But that experienced person will not take the short cut unless his experience shows it to be satisfactory, and that means that at some time he blazed the trail.

Calculation of Results. Another matter that seems self-evident should be mentioned. Good work needs equally good mathematical interpretation. A reliable analyst is necessarily a reliable arithmetician. One little error in calculation has wasted many a day's work and many a firm's money. In working out results, four-place logarithms are usually the preferred tool. A 20-inch slide rule possesses nearly the same accuracy and is not unduly expensive considering the time it saves. It is acceptable for most commercial calculations. But even if calculated on the slide rule, all figures used in the computation should be set down so that they can be checked. Results should ordinarily be checked by an associate. For this purpose a 10-inch slide rule has adequate accuracy and saves time. Here too, the greater ease of reading the 20-inch rule justifies the higher cost. All work sheets should be considered a part of the analysis and filed with the original data. A worker may have to check back on his results tomorrow or next week or next year.

Results are usually reported as per cent and calculated to the first or second place of decimals. To go beyond the second place in dealing with whole percentages is a waste of effort and quite out of keeping with the accuracy of usual methods. When the result is entirely in a fraction of a percentage, normal accuracy to one or two parts per thousand dictates that results be expressed as three significant figures. Often the details of the method will indicate a lower accuracy, and two significant places are all that are justified.

Consistency and order should be aimed at in calculating and presenting results. Unless there is sound reason for doing otherwise, such as that one substance is present in a comparatively large quantity and another in a very small amount, all results of a particular analysis should be given calculated to the same decimal place.

As a final check on himself, a worker should make sure that the correct conversion factor for the weight of precipitate to its derivative or the appropriate normality factor of acid, alkali, or other titrating medium, the proper weight of sample, and the proper aliquot of that sample have been employed.

CHAPTER 2

TOOLS OF THE ANALYST

Although the analyst is a professional man, he has his tools just as do men in other professions, for example, the surgeon, and an intimate knowledge of their virtues and limitations is essential. Only some of the more important typical tools can be discussed. Numerous illustrations are used to save space even though the item shown may be of a well-known type.

Beakers and Flasks. The majority of operations that involve heating are carried out in beakers or flasks. The glass of which they are made may be divided into borosilicate and all other kinds.

Pyrex is a brand of borosilicate glass with a sufficiently low coefficient of thermal expansion, $0.0_{5}32$ per degree centigrade, so that it can be and is made of substantial thickness. Breakage due to heat strain is so rare as to be properly attributable to an overlooked crack or other defect. It is reasonably but not entirely resistant to alkalies. It is glass, however, and will crack or shatter on violent impact with hard surfaces.

A special low-alkali type of Pyrex is rarely referred to because of its limited availability. A newer type, Vycor, is used as crucibles and otherwise as a substitute for fused silica. Its coefficient of expansion is about one-fourth that of Pyrex, for all practical purposes negligible. Because of cost, it is not apt to be much used as beakers or flasks.

Some laboratories stock only Pyrex-ware beakers and flasks; in others a type of soft soda glass is also used. With a coefficient of thermal expansion about three times that of Pyrex, soda glass is best used for operations in which heating is not performed and attack on the glass is of no great consequence. When soda glass is used as beakers and flasks, it should be rigorously kept separate from Pyrex; failing that, the chemist should look at the brand mark before heating the contents. Failure to do one or the other may result in lost time and patience.

Availability of a generous supply of glassware represents an economy in laboratory operation. The majority of operations in beakers are carried out in 250- or 400-ml. low form sizes with a spout (Fig. 1). This is also known as a Griffin beaker. Similarly the 250-ml. conical flask with wide or narrow mouth (Figs. 2 and 3) possesses a wide utility. This is also known as an Erlenmeyer flask. For titrations, such a flask possesses definite elements of superiority to beakers. Thus the titrating medium is run in from the buret with one hand by the experienced operator while the flask is swirled with the other hand. Prompt mixing without danger of splashing gives real economy of time.

Such apparently minor points as these are illustrative of many that must receive consideration for economical operation. It will be found that the employment of methods that are as simple as permissible for the purpose will in the aggregate make for both speed

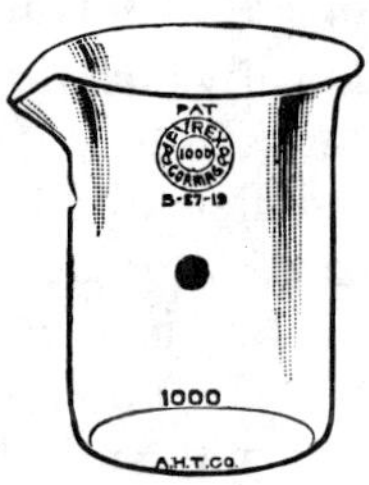

FIG. 1.—Low form beaker with spout.

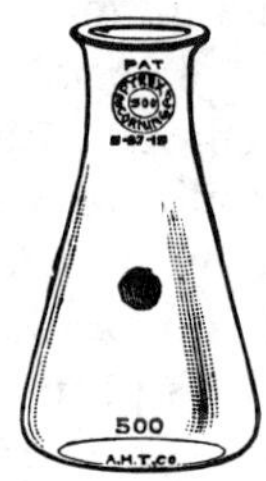

FIG. 2.—Widemouthed conical flask.

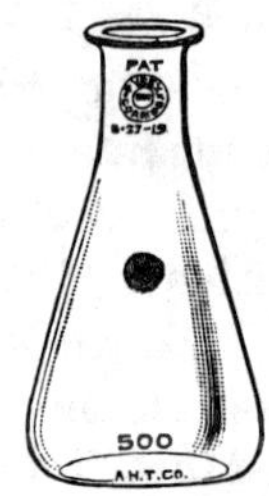

FIG. 3.—Conical flask with narrow mouth.

and skill in performance. Good manipulative ability coupled with the choice of equipment best suited to the job in hand is a big step forward to success as an analyst. It is to attain this end that many points that may appear obvious but that, as experience shows, are often neglected are mentioned from time to time.

Volumetric Glassware. This is normally of soft glass. Little advantage is gained by use of the initially more expensive Pyrex.

Two main classes of volumetric ware are on the market. A certificate of accuracy within certain close limits, issued by the National Bureau of Standards, is supplied with one class. The other class is stated to conform to wider but still reasonably close limits of accuracy. The former costs appreciably more than the latter and may, especially in the case of burets and pipets, be worth the extra expenditure. It is good practice if time and facilities permit to purchase the latter type and calibrate it. For the majority of purposes no further degree of accuracy than that already present is required.

A word as to keeping volumetric apparatus clean is in order. The tips of pipets and burets are liable to become clogged, particularly if solutions immiscible with water are allowed to dry in them. Much labor is needed to clean such equipment, and often during the process the piece may be damaged. It is time well spent to clean such glassware just after it is used. A very fine wire, judiciously

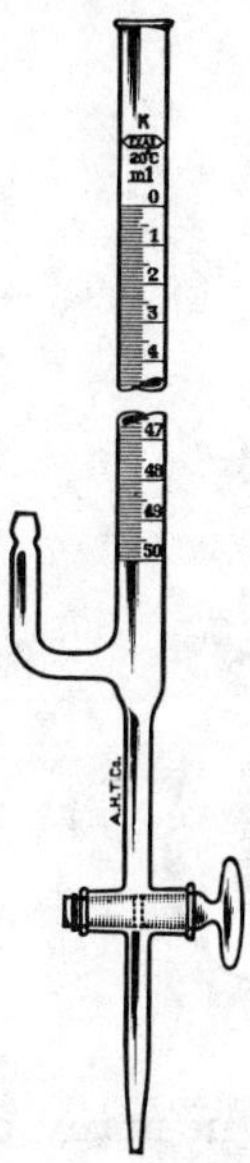

FIG. 4.—Standard buret with side arm for refilling from reservoir.

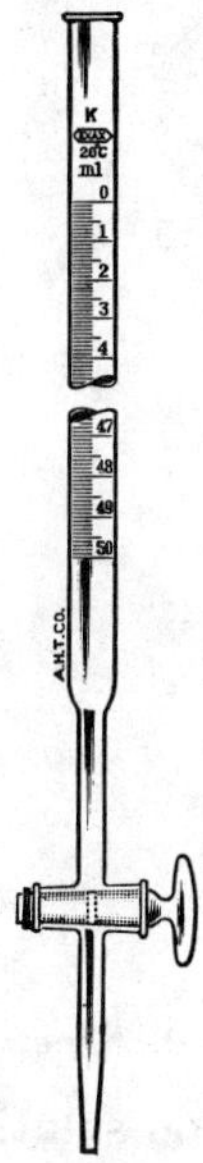

FIG. 5.—Standard buret.

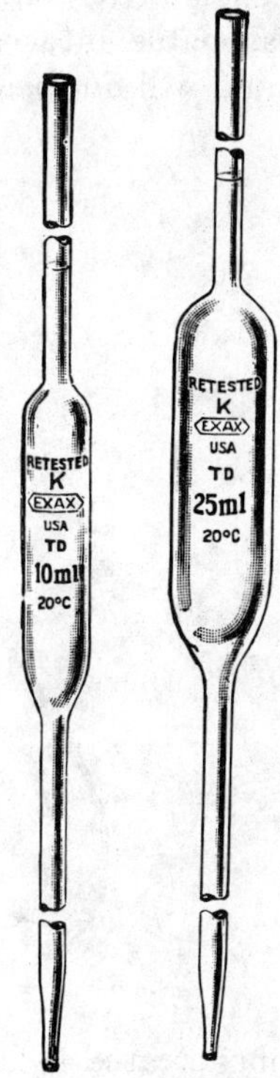

FIG. 6.—Pipets.

applied, helps in cleaning out clogged tips. Stopcock grease or the petroleum jelly commonly used for the purpose of lubrication should not be applied too generously.

Normally soap and water, preferably with the addition of a little trisodium phosphate, sodium metasilicate, or soda ash, will serve to clean glassware. When any substantial number of chemists are working in the same laboratory, it is good economy to accumulate soiled glassware and have the junior of the group, if no laboratory boy is available, do a general cleaning at regular intervals, not less frequently than daily.

Chromic acid or sodium bichromate in concentrated sulfuric acid, often simply called "cleaning solution," is a very efficient cleaner, particularly if it is hot, for removal of organic contamination. Burets (Figs. 4 and 5) and pipets (Fig. 6) have a tendency to become greasy in continuous use and should therefore be cleaned at reasonable intervals. If this is not done, the correct volume of liquid will not be delivered, because of surface tension effects. Hot

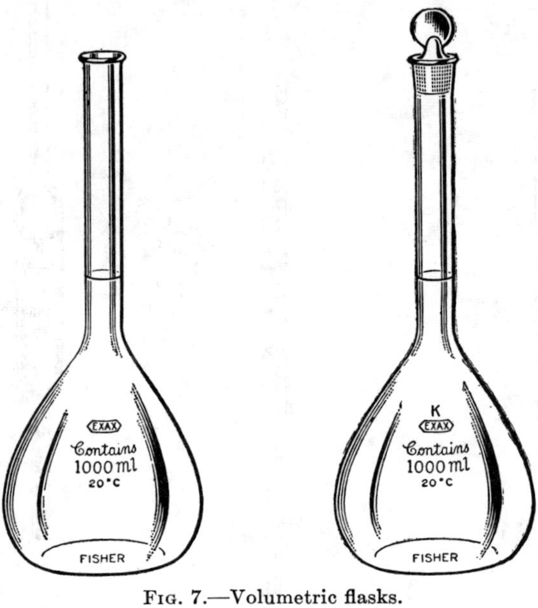

FIG. 7.—Volumetric flasks.

concentrated sodium hydroxide solution is effective in many cases for greasy glass or when material is burned on. It should not be used on graduated or calibrated glassware (Fig. 7), for it will attack glass to some extent. Care must be exercised in use of either of these cleaning solutions, because they are very corrosive to the skin.

Glass Joints. Laboratories use relatively few pieces of equipment with ground-glass joints, other than burets and separatory funnels. The most common is a stopcock. However, nearly every laboratory has some. By agreement among manufacturers, such joints of more complex apparatus are of standard size and taper so that units are interchangeable and replaceable. For such types of equipment as are so provided, it is good economy to get the replaceable type. Glassware will get broken, and the purchase of a complete replacement unit is thus avoided.

Watch glasses in large numbers are an essential item of laboratory equipment. They are of ordinary soda glass or of Pyrex. Necessarily large and small sizes are used to fit the varying sizes of beakers, but the major use is to cover 150- , 250- and 400-ml. sizes. When used to prevent contamination by dust and dirt, they may well fit closely, with steam escape provided at the spout of the beaker. But often a beaker is covered during evaporations, to prevent contamination and to avoid loss of material by spattering. Then close fitting is a defect. Special glasses for this use are ridged to hold them from contact with the lip of the beaker other than at a few points (Fig. 8).

FIG. 8.—Ridged watch glass in use.

Equal effectiveness but less convenience is provided by supporting the watch glass on a glass triangle laid on the top of the beaker or by placing three to four small glass hooks around the edge of the beaker to support the watch glass. In such cases a somewhat oversized watch glass is preferable, for it furnishes better protection from adventitious contamination by dust. Depending on the use, the watch glass may require rinsing into the beaker, or even cleaning with a policeman if spattering of a precipitate has occurred.

Stirring Rods and Ebullators. Stirring rods are a necessary adjunct of the chemical laboratory, even if they are used too often in titrating. One of the most common practices is to round off each end in a Bunsen flame. Experience, and this is backed by theory, indicates the desirability of leaving one end rough as an aid to even boiling. Bubbles tend to form on these sharp edges. A very useful ebullator may be made by drawing out a narrow tube to a near capillary size and sealing about 2 to 3 mm. of it to the end of a thin rod. When it is placed in a liquid and the liquid is heated, just below the boiling point a fine stream of bubbles appears to issue from the capillary and continues to issue throughout the boiling. This is a great aid in the prevention of bumping. The ebullator should be withdrawn while the liquid is still boiling. If it is allowed to cool in the liquid, the vapor in the capillary condenses and it becomes filled with liquid. Thus before the ebullator can be useful again, the liquid must be removed. Practically, a new one must be made.

Separatory Funnels. These are of three distinct shapes, a pear shape or conical, a tubular and a globe type (Figs. 9, 10, and 11). The pear shape is preferable when a small amount of lower layer is to be drawn off; otherwise there is little choice. Many standard methods specify one shape when another can be used without introduction of error. In general a short broad stem is preferable to a long thin one because the short stem usually drains promptly and dripping of the liquid during handling is avoided. There is seldom occasion to use sizes other than 250 and 500 ml.

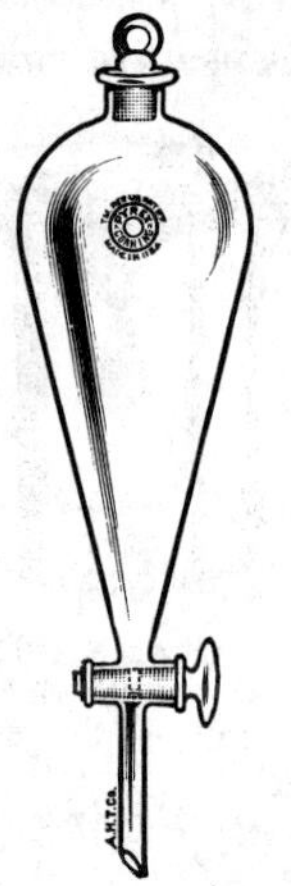

FIG. 9.—Pear-shaped separatory funnel.

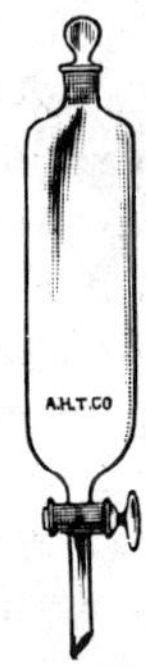

FIG. 10 —Tubular separatory funnel.

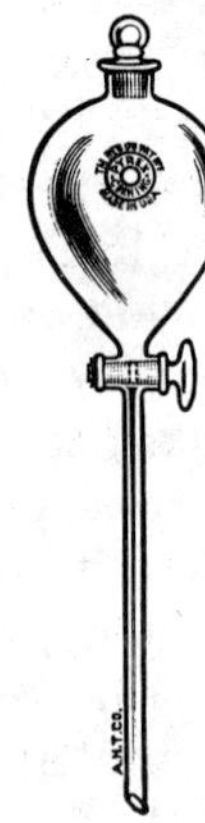

FIG. 11.—Globe-type separatory funnel.

Wash Bottles. The wash bottle is an essential unit in analytical chemistry. Figure 12 shows four types. The rubber tubing connecting the jet to the inner tube adds great flexibility in washing out beakers. The inner tube is bent at the end so as to enable the last portion of liquid to be blown out. Though a wicker collar is a great convenience for use with hot-water wash bottles, strong tightly wound cord is a satisfactory alternative. For solvents, such as ether, that attack rubber, an all-glass wash bottle is indicated. A number of fancy wash bottles with little gadgets attached are available. Polyethylene squeeze bottles are very useful for unheated solutions. In a busy laboratory, at least one wash bottle is kept on a hot plate to supply hot water. It is also a great convenience to have a small wash bottle containing alcohol and another containing ether, each labeled to save time in identification.

Jones Reductor. This unit permits reduction of a solution by retaining it in a column of amalgamated zinc for the required time. The item is shown in Fig. 13. To fill the unit, place a few glass

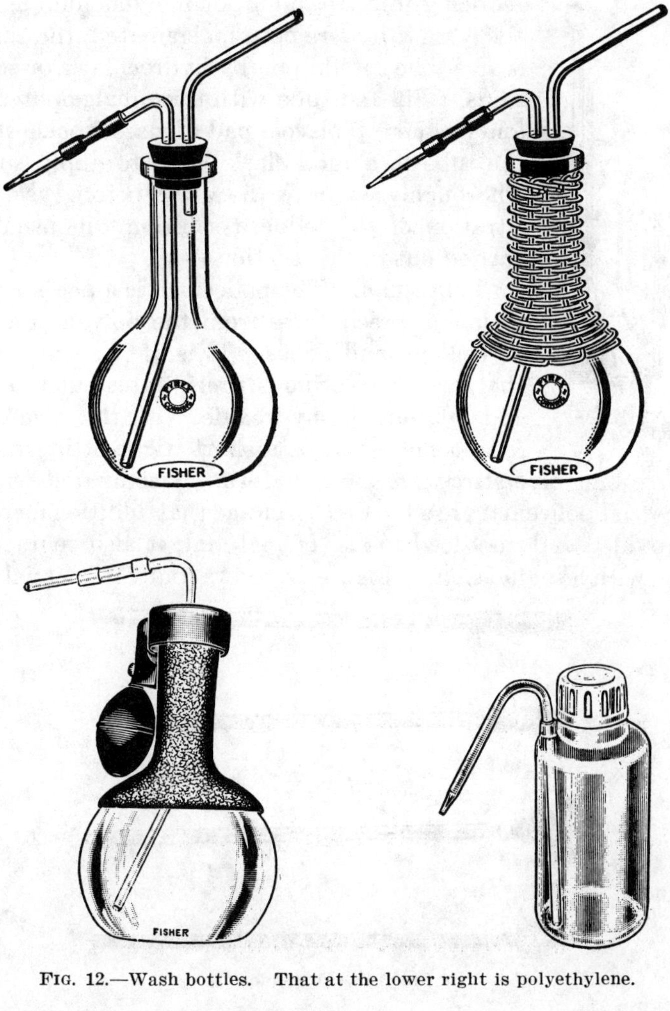

FIG. 12.—Wash bottles. That at the lower right is polyethylene.

beads in the bottom of a tube having a stopcock at its lower end. The tube should be about 30 cm. long and have an inside diameter of about 18 mm. On the beads place a layer of glass wool and then a thin layer of asbestos prepared for use in Gooch crucibles. Amalgamate 20- to 30-mesh zinc by first cleaning it with hydrochloric acid and adding mercuric chloride to the mixture with stirring until the zinc is well covered with mercury. This is evidenced by no hydrogen being evolved. While capable in this condition of reducing iron, titanium, or molybdenum as effectively as if it were not amalgamated, the zinc will scarcely be acted upon by hydrochloric or sulfuric acids. Fill the tube with this amalgamated zinc, and on the top place a pad of glass wool. Attach the tube to a filter flask so as to apply suction. Thoroughly wash with water before use. The titration of the reduced solution will usually be carried out in this suction flask.

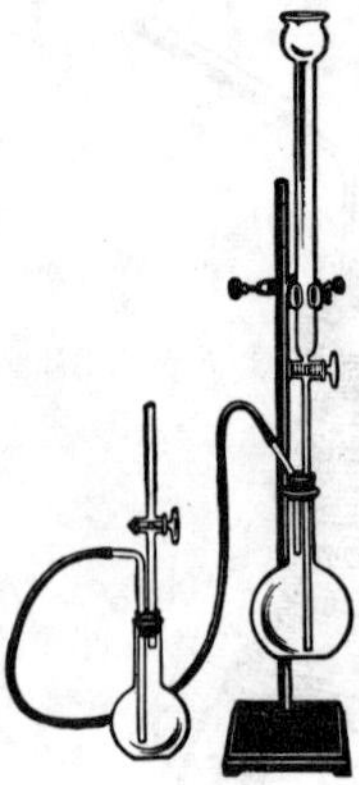

FIG. 13.—Jones reductor.

Policeman. The policeman is a necessary tool to loosen precipitates from the bottom and sides of beakers and flasks. It is of particular use in the case of gelatinous precipitates such as silica and aluminum hydroxide. In its elementary form a policeman is a short, tight-fitting piece of rubber tubing on a stirring rod. This form, as compared with the commercial policeman, has the disadvantage that additional crevices are provided and may lead to loss of material; it also contacts the surface with less efficiency. Figure 14 shows four commercial types

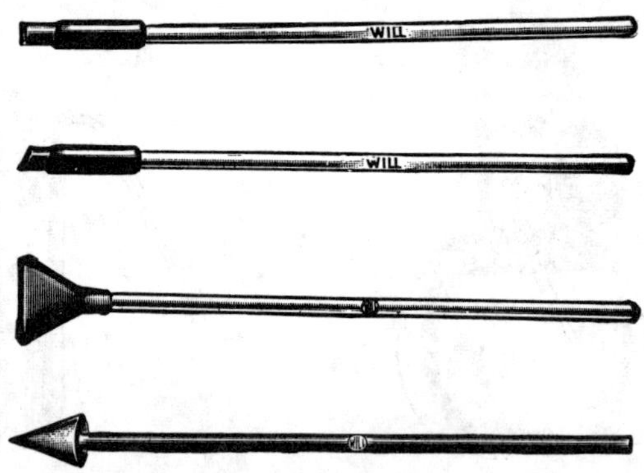

FIG. 14.—Policemen.

of which the long, flat-ended type is the most generally useful. When used, make sure that any precipitate adhering to the policeman is washed off with a strong stream of water from a wash bottle.

Chemicals. Pure chemicals are another tool of the chemist. If possible, use reagents from the original containers in which the chemicals are received, thus avoiding possible contamination or error in transfer to service bottles. Avoid service bottles with cork stoppers. The majority of manufacturers of analytical-grade chemicals now supply them in 1-pound bottles with plastic screw-on tops or ground-glass stoppers. In a few instances, purchase in larger containers and transfer to service bottles is justified by the saving in cost. But always bear in mind that a single error may be dangerous, will surely be expensive in terms of wasted time, and may be expensive in many other ways.

After use, immediately replace the cap on a bottle to minimize contamination, dehydration, or absorption of moisture. And replace the bottle in its place on the shelf to save time not only for others who may want to use it but also for yourself the next time you want it. Alphabetical arrangement on stock shelves is usual and logical.

For analytical work chemically pure (C.P.) grades of chemicals are generally used. Some are sold as conforming to the published standards of the American Chemical Society. C.P. is a standard that is set by the individual manufacturer. Tolerances as to impurities are shown on the bottle labels and must be consulted and taken into account when traces are to be determined. In such cases it may be necessary to run blank determinations on even the most highly purified chemicals.

Some familiarity with other grades of chemicals is desirable. U.S.P. grades are as specified in the United States Pharmacopoeia and N.F. as in the National Formulary. They are either equivalent to C.P. or lower in grade, too often the latter. The cost of such grades of chemicals is often less than for C.P. They should be used with caution and only when it is known that results will not be vitiated by their use.

Technical-grade chemicals usually cost appreciably less than the purer chemicals. They are normally not suitable in analytical procedures. However they often are used as an adjunct to analytical procedures. In a desiccator, technical calcium chloride is used; for freezing baths, technical sodium chloride is suitable; technical grades of solvents such as chloroform, ether, and carbon tetrachloride are satisfactory, provided that they leave no residue.

Every chemical necessitates precautions depending on the material. No analytical chemical should come in contact with the hands. And above all bear in mind that pure is a relative term. Depending on the grade, a specified impurity may be expressible in the first decimal place of percentage or in the third or the fifth. It may not be detectable. But all analytical reagents contain some impurities. Chemically pure is a convenient expression that should mean best quality but cannot mean exactly what it says.

Alcohol. This is one specific chemical that merits individual consideration. Although educational institutions obtain it without tax at under one dollar a gallon, all others must pay the tax, which is approximately $20 per gallon of 95% alcohol. A United States permit for purchase and storage is required; in addition many states also require a permit. Practically all analytical procedures can be as satisfactorily conducted with specially denatured alcohol 3A, usually called SD3A, which is tax-free. This contains 5 gallons of synthetic methanol to 100 gallons of ethyl alcohol. Its purchase and storage requires another type of permit, but a bond is not required unless purchases exceed 100 gallons per year.

This is not the place to go into great detail. Each chemical laboratory must maintain familiarity with Federal and state laws, submit at least one report a month, and comply with a host of regulations. There is no way of evading the burden. Completely denatured grades of alcohol can be purchased without license or regulations but are not suitable for analytical purposes.

Distilled Water. Although one is inclined to think of water as just water, it is either a tool or a chemical according to the particular use and in part according to the point of view. Ample supplies of good quality water and of distilled water are essential to any laboratory. If the local water supply is high in dissolved solids, over 300 parts per million (ppm.), cleaned glassware should be rinsed with distilled water before it is set aside to dry. Very few laboratories are small enough to be able to supply their full requirements for distilled water by purchase. Units for producing distilled water in the laboratory are available of as low capacity as ½ gallon per hour. The difference in original cost is small enough so that a still of ample capacity with some margin for expansion should be provided. This is particularly true because the best quality of distillate is obtained by a considerable degree of incompleteness of condensation.

There is no definite choice between the three trade types of still, Barnstead (Fig. 15), Precision (Fig. 16), and Stokes (Fig. 17). All are available in designs heated by gas, electricity, or steam. The latter is suitable only for a laboratory in an industrial plant that has steam the year round.

Many laboratories use *deionized water*, which for most analytical purposes is as good as distilled water. Laboratory scale apparatus for producing this is available, including disposable units in which the ion-exchange resin is not regenerated.

Gas Burners. Various types of burners derived from the Bunsen burner are standard equipment in every laboratory. In the most elementary form, gas is regulated at the stopcock and air at the burner. Such designs find little use in the analytical laboratory. The next step in development provides both gas and air adjustment

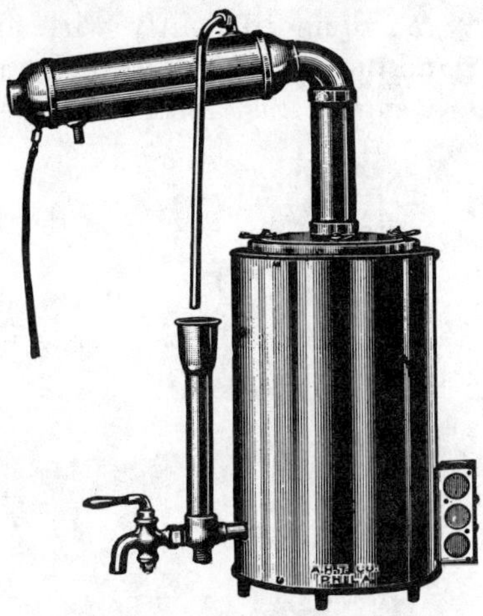

Fig. 15.—Barnstead still.

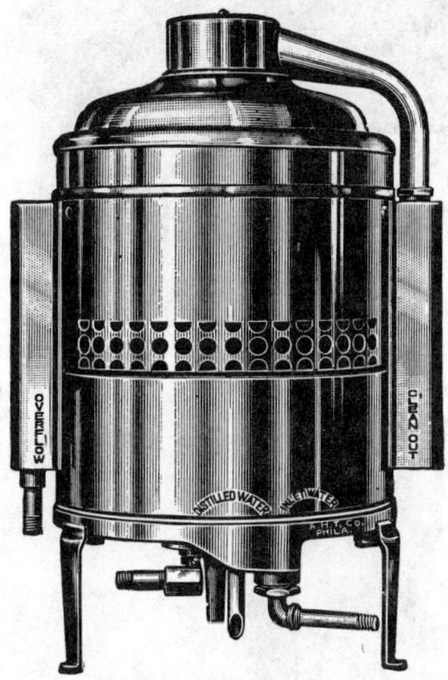

Fig. 16.—Precision still.

on the burner. These burners, which are assigned different names by different manufacturers, are about as simple a design as is useful to the analyst (Fig. 18). They usually do not have a grid over the top. When intense local heat is desired, the Meker burner is used (Fig. 19). This has a fine-mesh grid over the top, and in effect each opening in the grid acts as a small burner. Proper regulation of the relative gas and air supply is necessary with all these types to avoid flashing back, the burning of gas inside the tube. The Meker burner is often incorrectly called a blast burner. It fulfills most of the functions formerly performed by a blast burner. Correctly the latter is a special design of burner fed with both air and gas under pressure, thus getting more concentrated combustion and higher temperature (Fig. 20). For the maximum temperature, such as for working Pyrex or hard potash glass, oxygen under pressure is supplied. In the true blast burner, control of the form of the flame is also provided, making possible, for example, a thin hot flame for use in some types of glassworking. Although minor glasswork is done in the laboratory, major work is usually supplied by professionals, because it is more accurate and, everything considered, more economical.

A useful form of burner not in such common use is the luminous flame, or Argand burner (Fig. 21). This type is most useful when low-temperature ashing is required. With careful use, complete ashing may be done efficiently with an Argand burner far below red heat. More evenly distributed heat is obtained, and combustions of many samples may be performed without fusion, which might otherwise cause trouble.

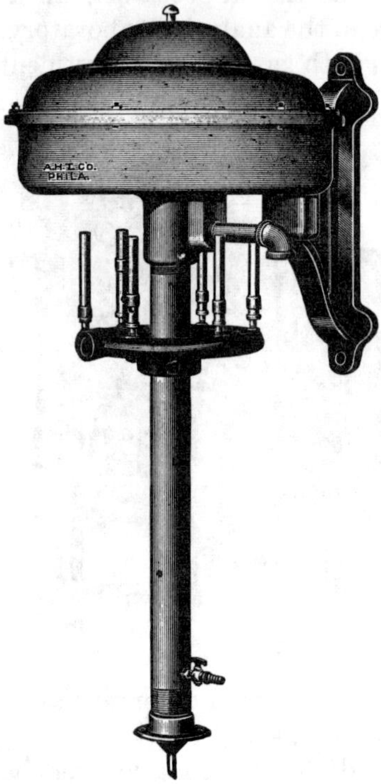

Fig. 17.—Stokes still.

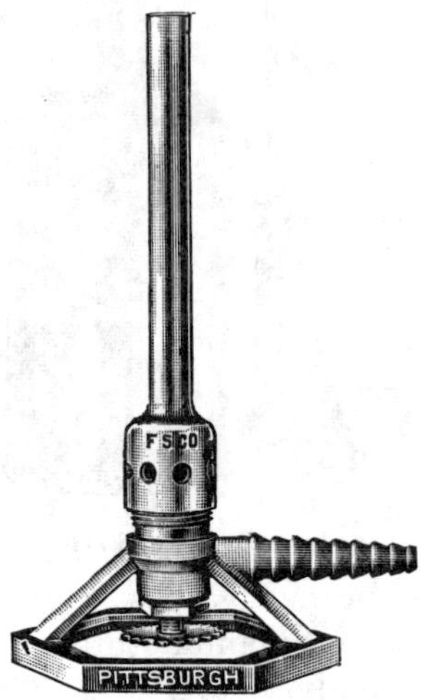

Fig. 18.—Bunsen burner with gas and air control.

In all these comments, it is assumed that usual city gas supplies are available such as carbureted water gas, coke-oven gas, or natural gas diluted with inert gas. These all have about the same heating value, around 500 B.t.u. per cubic foot. In many regions undiluted natural gas or gas from pentane in steel bottles is used. Adjustment to provide more air relative to the gas is required. Special burners for such gases are available, but the usual types are adjustable to fit their use.

When work is being concentrated in a limited area of workbench, the number of gas outlets is often less than the number of burners required. This is easily provided for by connecting two or more burners to a single outlet with a T or Y in the tubing. They are then individually controlled by the gas control on the burner. Safety dictates that all be turned off at the gas cock when the last person leaves at night and that each burner on a line be inspected when the cock is open, to avoid escaping gas.

Triangles. In elementary chemistry, triangles in many cases consist of iron wire with unglazed porcelain protectors on the three sides. For continuous use these are uneconomical. Also the

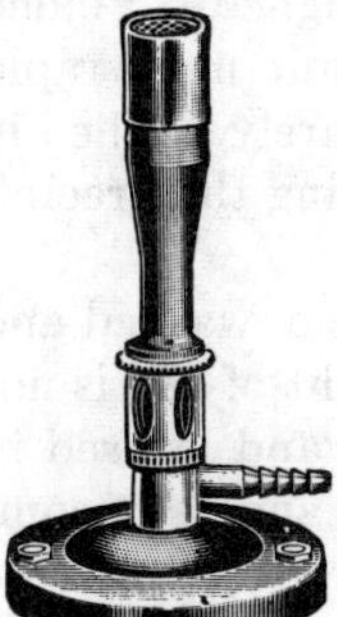

FIG. 19.—Meker burner.

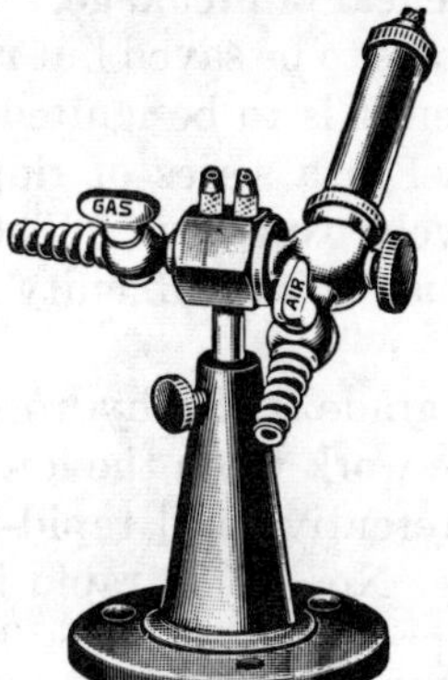

FIG. 20.—Blast burner.

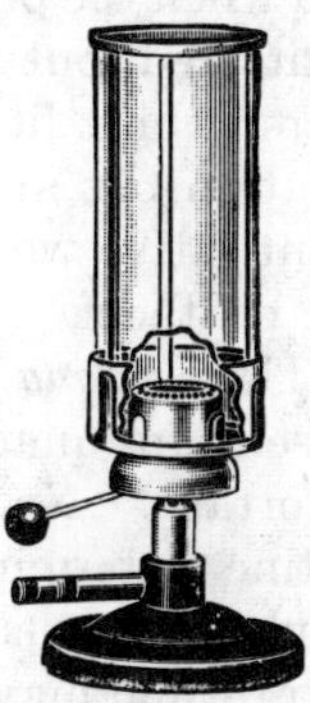

FIG. 21.—Argand burner.

porous porcelain absorbs anything spilled on it and thereafter will leave a mark and often a weighable deposit on porcelain or platinumware. For utility, triangles are preferably made of Nichrome wire. Very resistant to oxidization, this wire is comparatively thin, but strong, and allows the gas flame to reach practically every part of the crucible or dish.

Wire Gauzes. These similarly are often of iron, with or without a round asbestos central area. But wire gauzes, like triangles, are preferably made of Nichrome wire. A cost of six to eight times that of plain iron gauze is offset by a life exceeding that ratio many times. When distribution of heat is required, a piece of asbestos paper on such a gauze is both inexpensive and effective.

Filter Paper. The filter paper is a necessary adjunct in filtration. Both time and expense are economized by selection of the grade best suited to the purpose. The best known are the Whatman grades, of English manufacture, which will be used for illustration. Three general broad qualities are purchasable: qualitative, semiquantitative, and quantitative grades.

The *qualitative* grade is hydrochloric acid–washed and is used for work when the ash weight is unimportant. No. 1 is for filtering precipitates of average fineness. No. 2 is thicker and therefore not so rapid-filtering. Both are recommended for all-round qualitative use. No. 3 is thicker still and is used for very fine precipitates and for Büchner funnels. It is fairly resistant to alkalies. No. 4 is soft, open, and rapid-filtering and is recommended for gelatinous precipitates. No. 5 is tough, hard, and of close texture and may be used with a vacuum with little fear of breakage. All these grades may be used when the precipitate is to be saved but redissolved, when it is to be titrated, but not when it is to be ignited and weighed. Folded filters, which fit a funnel in a series of ridges, permit more rapid filtration but are relatively expensive. They are rarely applied in quantitative work because of the difficulty of washing the precipitate on them.

The *semiquantitative* grade is also hydrochloric acid–washed and is useful for quantitative work when the lowest weight of ash is not important. No. 30 is retentive and rapid-filtering and is used in routine determinations. No. 31 is rapid-filtering and is recommended for gelatinous precipitates. No. 32 is retentive, of close texture, and may be used for the filtration of fine precipitates. The ash of an 11-cm. circle in each case weighs just over 0.0003 gram.

Quantitative filter paper is of practically pure cellulose fiber. The pulp has been washed with hydrochloric and hydrofluoric acid until it is substantially free from silica and other inorganic sources of ash. Such papers are always properly stored in the box in which they come, which carries on the back a statement of the ash content per unit. They should be used in exact quantitative work when the ash weight is important and should usually give a weight of ash in the fourth decimal place, frequently in the fifth. Correction for the ash introduced by the paper is therefore rarely applied, the cases being those of unavoidable ignition of amounts of precipitate of the order of less than 10 mg.

Whatman No. 40 is probably the most used general paper. It will retain well-digested barium sulfate precipitates. For gelatinous precipitates such as iron and aluminum hydroxide or silicic acid, a more open texture is provided in No. 41. It must be handled carefully to avoid breaking and is entirely unsuited to fine precipitates such as barium sulfate. For finely divided precipitates, a close-texture grade is provided as No. 42. A corresponding acid-hardened grade is No. 50, used mainly with vacuum or for other procedures that put abnormal strain on the paper. A rapid filtering hardened grade is No. 41H. The hardened grades are resistant to alkalies and will stand up under stronger acids than the usual grades. Occasionally in control work by titration, these grades are used more than once. Precipitates can be dissolved or washed off, owing to the hard, smooth surface. Occasionally No. 44, which is very thin and retentive, is used in exact work when an absolute minimum of ash and sorption is essential. Other grades for special purposes are available, such as an ether-extracted grade. The ash of the grades 40 to 44 varies from 0.00015 to 0.00007 gram for an 11-cm. circle.

There are other well-known papers. Munktells, a Swedish paper, is produced in a full line as is Carl Schleicher and Schull, a German paper. The latter is now produced in the United States also. Suitable qualitative papers are produced by many American manufacturers, but as yet there is no well-known American quantitative paper.

For all-round laboratory use, Nos. 40 and 41 Whatman filter papers for quantitative work and Nos. 1 and 4 qualitative papers or corresponding qualities of other makes will be found sufficient to cover the majority of requirements. Filter papers can be purchased in sizes from 4.25 to 50 cm. As a rule, use the smallest filter paper compatible with the size of precipitate, rate of filtration, and volume of liquid to be filtered. Correspondingly, select the size of sample to give an amount of precipitate that can be conveniently and accurately handled.

Filter Pulp. The speed of filtration of any filter paper diminishes as its pores become clogged with precipitate. It is therefore advantageous to allow the insoluble matter to settle before filtration is commenced. This settling and the subsequent filtration are facilitated by adding a slurry of filter paper in the solution after precipitation. Gelatinous silica, finely divided barium sulfate, sulfides, sulfur, slimy ferrocyanides, colloidal hydrated oxides, etc., are advantageously filtered using filter pulp. The incorporation of filter pulp with a precipitate not only ensures its retention on the filter but also facilitates the cleaning of the beaker. Ignition of the precipitate is more likely to give a finely divided powder which rapidly acquires constant weight and is more readily attacked by fluxes in a subsequent fusion. Owing to absorption of solution during filtration, it is especially important to wash thoroughly before ignition.

Filter pulp is sold commercially in qualitative and quantitative grades in the form of pellets as well as in moist form for immediate use. If it is not available in the laboratory, filter paper can be pulped for the purpose.

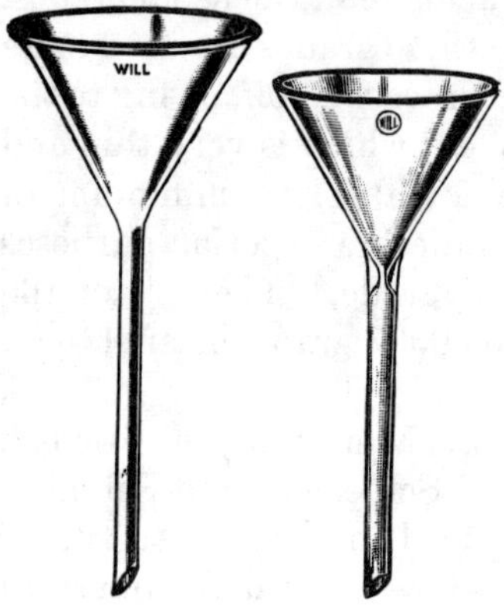

FIG. 22.—Funnels with and without constriction.

Funnels must team up in performance with the filter paper. In general the maximum speed of filtration is obtained by fitting the paper to adhere closely to the funnel. This is made easier if the stem of the funnel is full of filtrate, producing a very mild suction, promoted by a constriction at the top of the stem (Fig. 22). If bubbles progressively appear in the stem of the funnel, the paper is not properly folded and fitted. In general, it is well to fit the paper and wet it to see that it performs properly before the solution to be filtered is added. Place the stem of the funnel against the side of the receiving beaker to prevent splashing. To save time in filtration, wash by decantation as much as possible. Limited use is made of vacuum to speed filtration through a filter paper on a funnel, a perforated cone of gold or

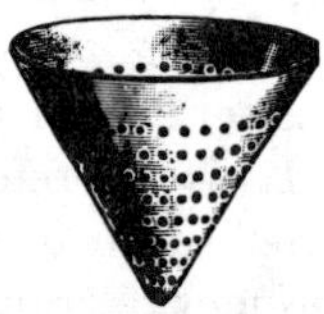

FIG. 23.—Filter cone.

FIG. 24.—Gooch crucible with integral bottom.

FIG. 25.—Caldwell crucible with separate bottom.

platinum being used to prevent failure of the tip of the filter paper (Fig. 23).

Filter Crucibles. Direct filtration of a precipitate can be performed under vacuum with three types of crucibles. Gooch crucibles have a perforated bottom, either integral or as a separate disk

(Figs. 24 and 25). The filtering medium is provided as a thin layer of acid-and-alkali-washed asbestos put in place with vacuum, then dried in the oven or ignited according to whether the precipitate is to be dried or ignited.

Glass filter crucibles have glass sides, usually transparent, and a porous fritted glass bottom (Fig. 26). The bottoms are of varying

FIG. 26.—Typical glass crucible with fritted bottom.

FIG. 27.—Alundum crucible.

porosity according to the type of precipitate. Complete removal of the precipitate is often difficult. Porcelain crucibles with fritted porcelain bottoms are also available.

Alundum crucibles are of aluminum oxide, fritted together (Fig. 27). The principal objections to their use are a tendency to

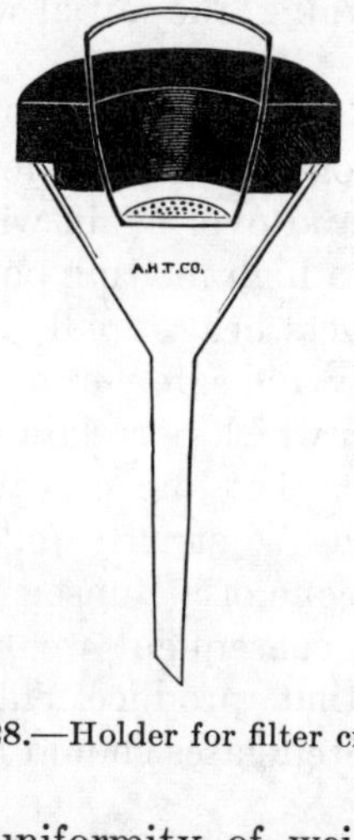

FIG. 28.—Holder for filter crucibles.

FIG. 29.—Gooch crucible holder with glass crucible.

lack uniformity of weight on repeated ignition and difficulty in removal of one type of precipitate to use for another type. When repeated determinations of the same type of material are being conducted, as in control work, Alundum crucibles often permit economy of time, successive precipitates even being collected on top of each other without reweighing. The same technic is sometimes employed with Gooch or fritted-glass crucibles. The relatively high price of Alundum and fritted crucibles is a limitation on their use.

These types of crucibles are useful only for granular as distinguished from gelatinous precipitates. Attempts to use them for the latter type are futile. Usually precipitates in filter crucibles are finished up by washing with alcohol or acetone and then with ether to complete removal of moisture. The ether must be evaporated before either drying in the oven or ignition.

Gooch, Alundum, and fritted-bottom crucibles are fitted either to a glass funnel with a rubber ring (Fig. 28) or to a holder usually known as a Gooch-crucible holder (Fig. 29).

Plain Crucibles. The best crucibles are of platinum; second choice is silica or Vycor Pyrex. A poor third choice is porcelain, but it is often necessarily used for student work. A good initial quality of porcelain is essential to avoid failure due to heat shock. Crucibles must be discarded when the glaze is destroyed unless they are used continuously with precipitates of the same composition. For economy of time, crucibles should be numbered and that number burned in. Subsequent to the first weighing, the initial weight can quickly be approximated closely.

Platinumware. Contradictory as it may appear on the face of it, in proper use platinum apparatus is economical. Crucibles and dishes made of platinum can be used over and over again without being damaged. Because it is a metal with a high melting point, is unattacked by many substances, and conducts heat rapidly, platinum may be used to ignite substances rapidly and safely at a saving of much time and equipment in operations in which porcelain would corrode or crack. It is essential, however, that the precautions issued by the makers of platinum apparatus be strictly followed. Any reducing flame, such as the blue inner cone of a Bunsen flame, will tend to crystallize the platinum, with consequent weakening and cracking of the unit. Substances that produce chlorine, ammonia, sulfur dioxide, and highly carbureted gases should not be heated in platinumware.

Compounds containing lead, silver, zinc, tin, bismuth, antimony, and arsenic should not be heated in platinum, for they may be reduced to metals and alloy with the platinum. Such an alloy always weakens the metal and a hole, possibly at first just a pinhole or a crack, is soon in evidence.

Fusions with alkali hydroxides or nitrates should not be performed in platinum vessels.

Phosphates, in the presence of reducing matter, filter paper, for example, should be ignited with great care, for reduction to phosphorus with subsequent attack on the platinum may result.

Sometimes a sample that is not suitable for the use of platinum may be so processed as to permit its use. For example, when reducing substances such as sulfides are present, heat a weighed sample in a porcelain crucible until the reducing matter has been volatilized or oxidized by the air. In the absence of unsuitable organic matter, this may be by heating with nitric acid. Similarly, extract lead from a sample with hot nitric or hydrochloric acid of a concentration depending on the nature of the sample. The sample is then suitable for handling in platinum. These are only illustrative examples. But remember that each error may cost several dollars for smelting and re-forming the platinum.

To clean platinum crucibles and dishes, fusion with technical potassium acid sulfate is usually efficacious. The melt is readily soluble in hot water. Should the adhering substance be acidic, it may be necessary to follow with a sodium carbonate fusion. This melt is not nearly so easily soluble in hot water. It is usually helpful to be able to detach the melt from the vessel. This may be accomplished by plunging the bottom of the crucible or dish into running cold water while it is still at a dull red heat. Some experience is needed to judge the correct temperature at which to heat the melt before immersing in water. The melt will separate sharply from the crucible or dish, owing to difference in rate of expansion. This method may also be applied with care to determinations calling for sodium carbonate fusions in platinum. After cleaning by fusion, the platinum vessel may be polished by rubbing with wet fine sea sand; 80 to 100 mesh is convenient.

Wooden molds supplied, if desired, with the platinum apparatus are useful to keep the vessel in proper shape. With the wooden mold inside, the platinum may be carefully, but with some force, rolled against a wooden surface such as a bench. To smooth out small dents, the rounded wooden handle of a spatula may conveniently be applied inside the crucible or dish, which may be gradually rotated between the hand and the spatula handle.

Platinum that has become unusable owing to continual wear or to maltreatment may be turned in for about 80 or 90 per cent of the then current price of a similar item of equal weight. Mathematically one finds that only a few porcelain crucibles can be purchased for the resulting net cost of a platinum crucible.

When properly used, these vessels can be employed almost daily and still be in excellent condition after the lapse of two or more years. Practically speaking, a busy analytical laboratory can hardly operate without platinum crucibles and dishes. When silica is determined by volatilization with hydrofluoric acid, a platinum vessel must be used; no other readily available metal or substance will remain unattacked by this acid at elevated temperatures.

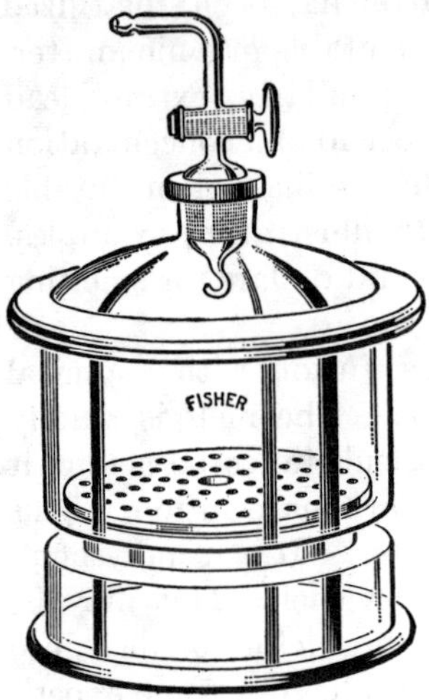

Fig. 30.—Vacuum desiccator.

Although so useful, the fact remains that the platinum articles themselves are intrinsically valuable and should be treated as such and carefully locked up when not in use. Crucible tongs tipped with platinum are highly desirable though not essential for handling platinum apparatus.

Desiccators and Desiccants. An ample number of both large and small desiccators, supplied with fresh desiccant, should be readily available. Some types combine the use of desiccant with a vacuum (Fig. 30) and, though primarily designed to dry substances that may be affected by air or that necessitate low pressure to dry thoroughly, are a great convenience for everyday use. Heated articles placed in a desiccator warm the air, which contracts on cooling to produce a partial vacuum that makes removal of the lid very difficult with an ordinary desiccator. Further, when the lid is opened in the usual manner by sliding, the sudden influx of air may occasionally blow precipitates from a crucible. With a vacuum desiccator, the difficulty is avoided by slowly letting in air at the valve before attempting to remove the lid.

By far the most common drying agent is anhydrous calcium chloride of about 8 mesh. Solid sodium hydroxide is sometimes used, preferably as pellets. It tends to coat with saturated solution. Porosity and efficiency are promoted by use of soda lime, a mixture of sodium hydroxide and calcium oxide. When an extremely low aqueous tension is required, phosphorus pentoxide is employed. It quickly becomes coated with a sirupy coating of phosphoric acid and thus loses its efficiency. Except for special reasons it is undesirable. Concentrated sulfuric acid is sometimes used, presumably because of ready availability. Any liquid desiccant is undesirable.

Newer desiccants offered under brand names with their literature usually indicating their composition are anhydrous magnesium perchlorate, anhydrous calcium sulfate, and anhydrous alumina. Some have an indicator to show when efficiency has been lost; usually they can be regenerated. But in the last analysis calcium chloride is cheap and meets over 99 per cent of the requirements of a desiccant.

Rough Balance. Many approximate weighings are required without a high degree of accuracy. For this purpose a laboratory is normally equipped with one or more rough balances (Fig. 31). Of

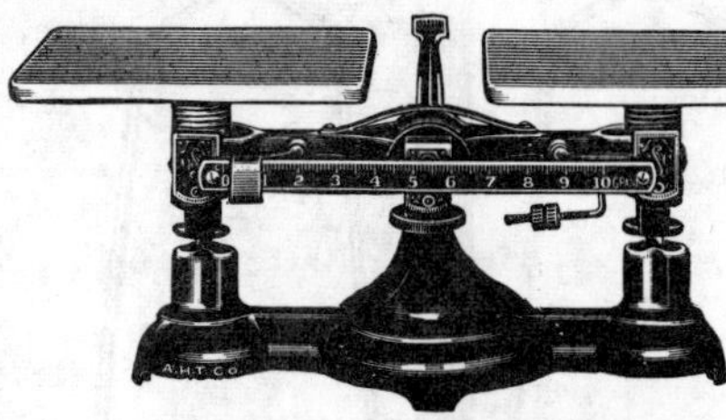

FIG. 31.—Rough balance.

these varying types one may have a notched beam equivalent to 200 grams, another a beam graduated for a lesser weight and additional capacity provided by metallic weights. Analytical weights should never be used with the rough balance; a corresponding set of rough weights costs but little.

Analytical Balance. For routine work, the analytical balance selected will weigh up to 200 grams and is normally sensitive to 0.1 mg. Assuming that the minimum weight of sample or precipitate is 0.1 gram, this places a limitation on accuracy of 1 part per 1,000, or 0.1 per cent. The majority of weighings are of larger amounts and correspondingly more accurate. Occasionally it is necessary because the available sample is small or for some other reason to weigh amounts of sample or precipitate less than 0.1 gram. The analyst must correspondingly bear in mind the equivalent limitation on accuracy, as an illustration, that, in weighing 0.01 gram, accuracy to only 1 per cent is to be expected.

The standard type of analytical balance with a graduated beam and rider is widely used (Fig. 32). Its use represents poor economy considering the number of times daily on the average that the analyst uses the balance. An improvement is the chainomatic balance (Fig. 33). For the simplest model, weights are used as usual down to 100 mg. Instead of fractional weights, the length of a gold chain hanging from the beam is varied by a rack and pinion. Accuracy of reading on a vertical scale is promoted by a vernier. Another model has in addition a notched beam that is calibrated for 0.1 to 1 gram in

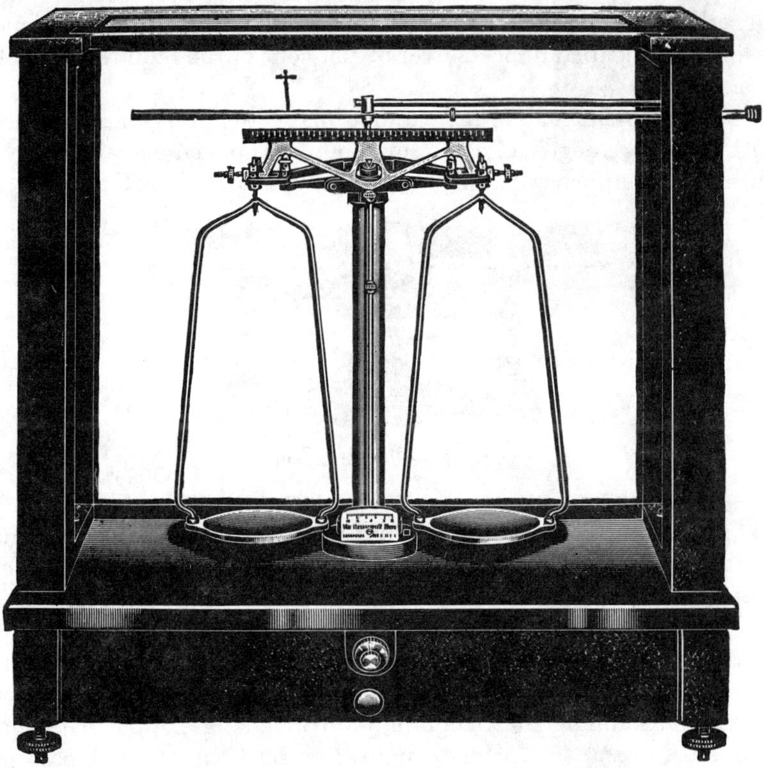

FIG. 32.—Plain analytical balance.

0.1-gram units (Fig. 34). The rider provided can best be described as a roller with flanges to keep it in the notch selected. Control for moving it from notch to notch is provided extending through the case of the balance. Thus for its use the case is closed after weighing to 1 gram, and the rest of the weighing is completed by outside controls. These types promote accuracy and economize on time. Other controllable devices for applying weights to the pan are available.

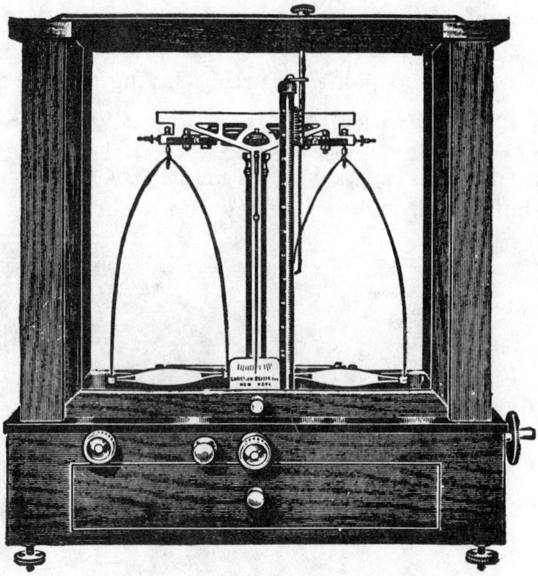

FIG. 33.—Chainomatic analytical balance.

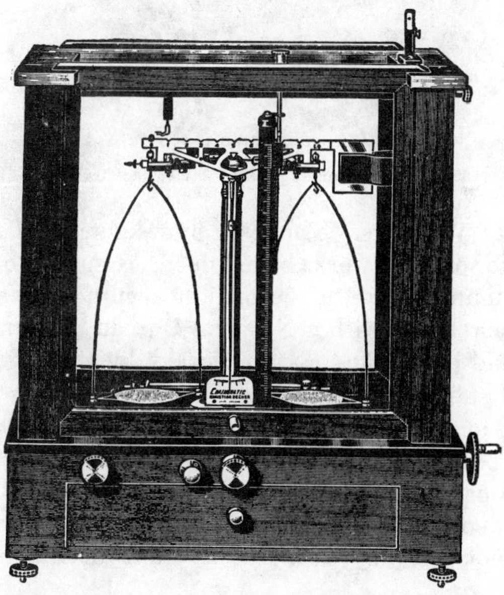

FIG. 34.—Chainomatic balance with notched beam.

Further saving of time is provided by a magnetic damping device that quickly stops the swing of the balance without interfering with the accuracy of the end point. Further saving of time is provided by the use of a single pan balance as shown in Figure 35. The speed and accuracy obtainable frequently offsets the higher initial cost.

A balance lamp is a good investment, since concentrating the light in the balance promotes rapidity and accuracy.

FIG. 35.—Sartorius Selecta Analytical Balance.

Analytical Weights. The type of weights used may to some extent depend on the work being done. Usually brass weights, gold- or platinum-plated, of the highest accuracy are used. These need not be purchased with a National Bureau of Standards certificate but should be checked as to accuracy by the user, by methods available in detail elsewhere.

Provision for keeping the atmosphere in the analytical balance dry to avoid adsorption of moisture films on glass is usual. A crystallizing dish or small Petri dish of calcium chloride is common but may promote corrosion. A watch glass holds so shallow a layer as to become moist unduly soon. Soda lime or other desiccant is preferable.

Drying Ovens. Routine practice in a laboratory usually requires a drying oven operated at 110°C., which may be altered for special purposes, or others may be available. Gas-heated ovens are nearly obsolete, except for occasional use at relatively high temperatures. Normally an electrically heated and thermostatically controlled type is used (Fig. 36). These are adjustable for a range such as 30 to 200°C. Thus they may be used for incubation at body temperature or for nonquantitative drying at higher temperatures than 110°C. They should be designed so that the temperature will be reasonably uniform throughout and convection currents will provide for removal

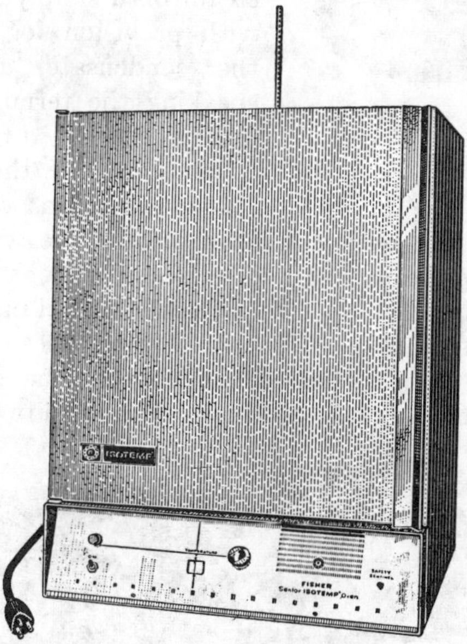

FIG. 36.—Electric drying oven.

of moisture and possibly of fumes. For some special purposes, such as in food analysis, a vacuum oven is required. Except that it operates under a vacuum it is similar to other ovens in design. Sometimes a vacuum desiccator can be successfully used in the usual electric oven, thus avoiding purchase of another expensive unit for occasional use.

Water Bath or Steam Bath. Essentially water baths and steam baths are equivalents, a provision for heating at not over 100°C. for gradual evaporation without danger of bumping or for driving off inflammable solvents such as ether.

Basically a beaker filled with water will do but is hardly economical. Single water baths in the small laboratory are consolidated for larger groups into a large shallow bath with multiple openings (Figs. 37 and 38). The bath should be fitted with a series of rings to fit various sizes of apparatus. Individual baths are heated with a Bunsen burner, larger ones by built-in gas burners or occasionally by electricity. In many industrial plants, a variable amount of steam is bled into the water in the water bath to provide the desired temperature. The steam bath simply provides steam with provision for draining away the condensate, and practically speaking the terminology is commonly applied to either indiscriminately. If the bath is used for evaporation of inflammable solvents, contact of vapor with free flame must be avoided. If built in, the water or steam bath should be located under a hood or in some other well-ventilated place. Water baths should preferably have a constant-level device and be connected to a source of water supply. This prevents their being heated to dryness. The device should be given attention to keep it in good operating condition.

FIG. 37.—Unit water bath with constant-level device.

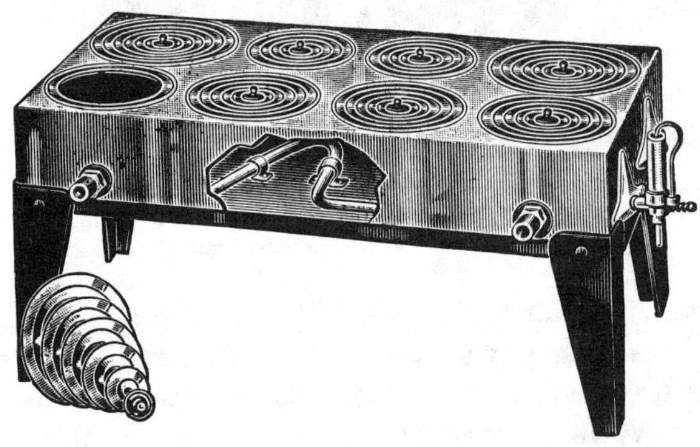

FIG. 38.—Typical multiple-unit water bath.

Electric Hot Plates. By far the most convenient method of heating in the laboratory is with an electric hot plate. This should have fully enclosed elements to avoid oxidation by the laboratory atmosphere, contamination from occasional spillage, and possible short circuit.

Two useful types are shown in Fig. 39. Such units are expensive in first cost but with even reasonable care can be used almost indefinitely. It is usual to provide three levels of heating so that such a plate may be used to keep solutions warm, just simmering, or boiling vigorously. The units are of especial value when a flameless source of heat is desired, as in evaporating or refluxing of inflammable liquids. There again the necessity of enclosed heating elements becomes evident. Many extraction units are heated by multiple electric heaters. It is a good plan to cut a red light into the line so that such a plate cannot inadvertently be left on when not in use.

FIG. 39.—Convenient types of hot plates.

Centrifuge. Although it is actually indispensable in only a few types of work a centrifuge often seems essential. A small centrifuge operated by hand will take test tubes but is a makeshift. A large electrically operated centrifuge with replaceable heads is really needed (Fig. 40). These permit the centrifuging of test tubes, bottles, special glass units for throwing the precipitate down into a graduated capillary, etc. Many filtrations can be expedited by centrifuging before filtration, decanting, and washing the precipitate one or more times in a centrifuge bottle. Many extractions of solids by liquid when time does not permit the use of the Soxhlet extractor are most efficiently performed by stirring with solvent, centrifuging, and repeating three or more times. Cautions in use include attention to balancing the load in the centrifuge, starting up, and stopping reasonably slowly. If only a single sample is run, use a counterbalance. In putting material in a centrifuge, bear in mind it will have a tremendous force applied; and, if the container is not properly packed and seated in a rubber cushion in the holders, it will break.

FIG. 40.—Electric centrifuge illustrating various types of tube holders.

The centrifuge, properly used, will often separate troublesome emulsions and precipitates that can hardly be separated by gravity or filtration. The types of tubes used for special purposes with a centrifuge are too numerous to illustrate. The majority of operations are carried out in test tubes or bottles.

Electrodeposition Apparatus. This is an essential in laboratories doing much nonferrous metal analysis. In other laboratories the use is only occasional and can usually be avoided. For intermittent use, storage batteries charged through a rectifier for local alternating current will serve. When used more frequently, a motor generator is desirable.

Stopcocks and Stoppers. Stopcocks and stoppers should be kept lightly greased to avoid sticking, particularly when equipment is put away for a brief time. When equipment is put away for a considerable time, the stopcocks and stoppers should be withdrawn, reinserted with a small strip of paper, and held in place by a rubber band. But in spite of cautions, the "other fellow" will let them get frozen in place.

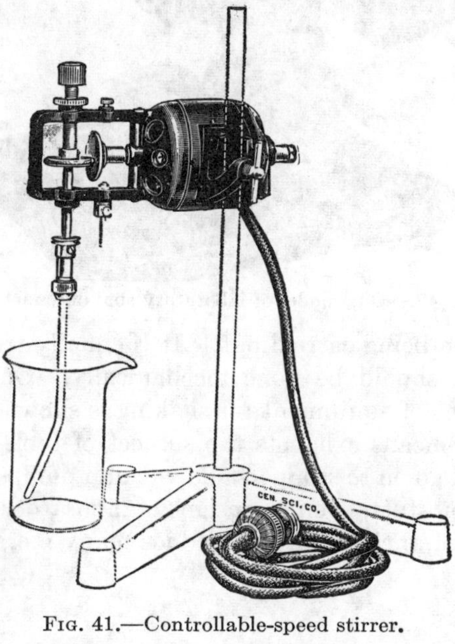

Fig. 41.—Controllable-speed stirrer.

The first trial for removal is preferably by blowing with hot water from a wash bottle and applying gentle pressure. If the part will not loosen in that way, the frozen joint may be gently warmed over a free flame, after it is certain that no inflammable solvents are present. Judiciously rapping the narrower end of the stopper or stopcock helps. If the joint will not loosen in these ways, one of the commercial types of appliances should be used. In any laboratory of substantial size, removers for glass stopcocks and for glass stoppers will soon repay their cost in convenience.

Mixing Equipment. When instructions call for stirring for 10 minutes, few chemists expect to stand and patiently stir with a stirring rod. Various simple types of mixers are available, varying in type from a modified malted-milk mixer to those designed especially for the laboratory. A single example is shown in Fig. 41, that particular model being driven by a flexible shaft and at any desired speed. The stirrers are made of glass by the operator to suit the

Fig. 42.—One model of laboratory shaking machine.

size of operation being carried out. It is equally true that shaking for 10 minutes should be done mechanically. One of the many available designs of equipment for shaking is shown in Fig. 42.

This by no means exhausts the subject of tools of the analyst. Rather it could go on to stoppers and vacuum pumps, humidity controls and thermostats, even to designs of laboratory benches. The next chapter will, incidentally, introduce many more tools.

CHAPTER 3

GENERAL PROCEDURES

Some operations are conducted in nearly every analysis, others in many of them. To avoid repetition, many of these general procedures as set forth here are referred to from more specific chapters by cross references.

Sampling. GENERAL. Before an analysis can be conducted, a sample must be taken. One cannot select one lump from a pile of coal, scoop out a handful of seed from the top of a bag or ladle out a half pint of milk from the top of a can and expect it to be representative of a carload of these commodities. Sampling of commercial products that are mixtures is even more difficult. Proper sampling takes time, and since methods cannot be prescribed for every possible material it also requires discretion. Obviously the most competent analysis of a nonrepresentative sample will be of little value. Improper sampling is probably the most common error of inexperienced chemists.

The sampling procedures are largely presented here in a general sense. Thus for cutting down the size of a sample of solids, one might use a shovel, a trowel, or a spatula according to whether the original sample was a ton, a few pounds, or a few hundred grams. The size of individual particles must correspond.

In general, from multiple units, sample every unit if not over 10, 1 in every 10 if there are 100 in the shipment, and not less than 1 in every 100 no matter how large the shipment. For bulk shipments such as grain and coal, take a few pounds representative of the shipment at random from different parts during unloading. Fortunately for the chemist, professional samplers usually handle this when large shipments are involved.

In sampling, the purpose must be considered and judgment used. If moisture is to be determined, protect the material from evaporation as far as possible during sampling. Small samples are normally transferred to tightly sealed bottles or cans for protection. As an extreme example, if dissolved oxygen is to be determined in water, the sample container must be completely filled so that oxygen cannot be gained or lost from the air over the liquid; and in such case the sample container is fitted with inlet and outlet tubes for withdrawal of sample without its coming into contact with air.

COARSE SOLIDS. The usual method is known as mixing and quartering. If in lumps, crush the larger ones so that roughly uniform size is obtained. Place in a flattened pile, divide into four quarters, and discard the opposite quarters. Mix the remainder well by scooping from the outside toward the center, again flatten, and discard the quarters. Repeat this as many times as necessary. For minerals or coal, this quickly comes down to a size at which error can occur. In that case crush the entire sample to smaller size, and

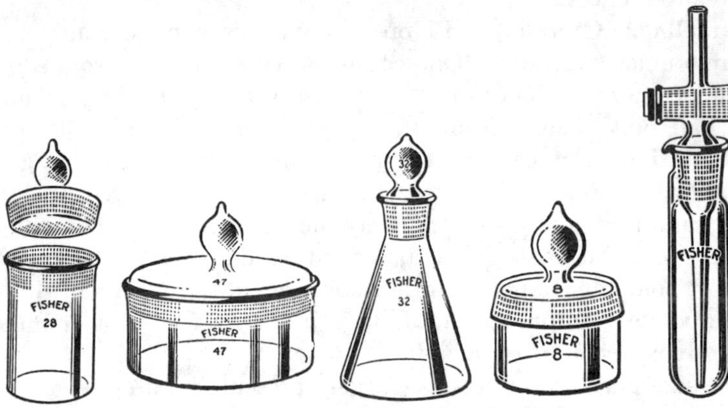

FIG. 43.—Typical weighing bottles.

continue to quarter. Eventually from a large sample 1 to 5 pounds should result, and this will require further crushing and quartering in the laboratory to obtain it in a condition in which 1 gram or so may be weighed out with assurance that it is representative.

For a more detailed description of the procedure as applied to coal, an example of a difficult material to sample, consult the American Society for Testing Materials Specification D271-58.

Laboratory samples are either weighed out directly or taken from a weighing bottle (Fig. 43).

FINE SOLIDS. These are sampled in the same way as liquids.

LIQUIDS. Homogeneous liquids are sampled by taking out a portion at regular intervals during processing or unloading. If the sample is in drums, a thin tube, usually referred to as a thief, is inserted to the bottom and removed with the finger over the end. Thus a column of the liquid from top to bottom is removed. The various samples are bulked, well mixed, and portions of a pound or so that appear to be representative are taken. If volatile, change in composition by loss of the volatile portion must be avoided. When carried out on a large scale, automatic sampling devices are provided.

In the laboratory relatively large portions from this sample are often pipeted out and the weight calculated from the specific gravity. This is not feasible unless the viscosity is similar to that of water; otherwise the volume delivered by the pipet is not accurate. The method is particularly desirable for volatile liquids, which will normally have low viscosity. For high-viscosity liquids, taking the sample from a weighing bottle is usual.

Nonhomogeneous liquids present other problems. They may separate into two layers. Often the volume of each liquid is measured, and each is sampled and analyzed separately. This may double the work, since many materials distribute between the two layers. There are two alternatives. It may be possible to take samples of each layer that are in the ratio of occurrence of the two layers and combine them for analysis. This is most feasible when both layers are a substantial percentage of the whole. In some cases, the layers can be well mixed and a representative sample quickly taken, as by pouring into a preweighed bottle or flask.

GASES. These are relatively unusual and involve collection by displacement of a liquid in which the gas is insoluble. Special gas sample tubes are used, usually filled with water saturated with salt. Occasionally mercury is used. In exceptional cases, the tube is evacuated and use of a displacement liquid avoided. This is discussed in more detail in Chap. 34.

Weighing. Often direct weighing of samples must be accomplished expeditiously. As a matter of economy of time, speed is always desirable. A majority of original commercial samples will contain moisture or other volatile substance. Calcium oxide produced by ignition absorbs both moisture and carbon dioxide rapidly during weighing. The accuracy of such weighings must be balanced against possible change during weighing. In general, accuracy to 1 part in 1,000 is adequate. Thus for a 1-gram sample the weight in tenths of a milligram is only approximated. Similarly for a 10-gram sample strict accuracy in milligrams is not essential. One usually tries to get somewhat higher accuracy than is essential but balances this against undue loss of time in nonessential accuracy. For example, weighing a 10-gram sample accurately in tenths of milligrams, which represents parts per 100,000, has no significance.

Generally samples that lose moisture or other volatile substance are taken from a weighing bottle by difference. Loss of volatile matter is thus unimportant. But of course this would not apply on a sample to be used for isolation of the volatile matter.

It is rarely essential to weigh while hot, or even warm. If necessary, only an approximate weight can be obtained. Normally anything to be weighed is first cooled in a desiccator to avoid producing convection currents by warming the surrounding air while weighing.

Any container to be weighed should as a general practice be heated either in the oven at 110° or by ignition, and cooled in a desiccator. This avoids such errors as sorption of water vapor on the surface of a weighing bottle, which can be substantial. This is also the reason why weighing bottles of samples are often stored in a desiccator.

Determination of Volatile Matter. OVEN DRYING. The most common method for determination of volatile matter is drying in an oven. Volatile matter is usually moisture, but there are numerous exceptions.

Weigh a suitable sample into a tared dish. One of the most suitable units is a shallow flat-bottomed glass moisture dish approximately 6 to 7 cm. in diameter and 4 cm. deep. Crystallizing dishes or Petri dishes are suitable, but round porcelain evaporating dishes are not satisfactory. Shallow light aluminum dishes are convenient when reaction with the sample will not occur. Aluminum milk-bottle hoods are an inexpensive form, discarded after a single use. If a weighing bottle is used for the purpose, it should be shallow and is cooled with the cover off. Spread the sample uniformly over the bottom of the dish. For some substances, such as sirup, milk, or soap solutions, the evaporating surface is increased by putting sand in the dish before weighing. Sand also prevents the formation of glassy surfaces that imprison volatile matter.

When repeated determinations of the same kind are made, the time for complete drying becomes established. Commonly, however, the sample is removed from the oven, cooled in a desiccator and weighed, then replaced in the oven for 1 hour. On the next weighing, it is established whether constant weight has been obtained. This is far from as simple as would appear. In many instances, a substance is present that is difficultly volatile, and small losses continue. This will be true if glycerin or a light lubricating oil is present. Oxalic acid volatilizes slowly at 110°. In other cases, oxygen is absorbed by the sample, and loss of volatile matter is paralleled by oxidation reactions. It is therefore not unusual to find that after a time the sample gains in weight. With such materials present, an approximation is obtained that is helpful although not accurate.

Samples containing more than a trace of volatile matter that has a flash point of 110° or lower should never be placed in the 110° oven.

VACUUM DRYING. When oxidation of the sample being dried can occur, it is often prevented by use of a vacuum oven, thus removing the oxygen that could cause oxidation. This type of drying is quite commonly applied to food products.

Weigh a sample into a suitable dish as in the previous method. The temperature to be used must be correlated with the vacuum obtainable. Dry to a constant weight in a vacuum oven at a uniform temperature not less than 20° or more than 25° above the boiling point of water at the working pressure, which should not exceed 100 mm. of mercury. Cool in a desiccator and weigh. Constant weight is attained when successive dryings for 1-hour periods show an additional loss of not more than 0.05 per cent.

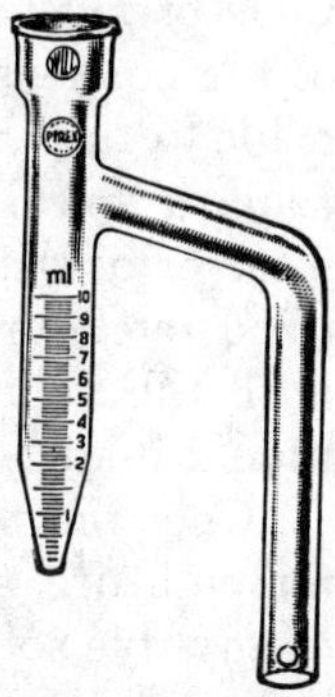

FIG. 44.—Moisture distillation tube.

SOLVENT DISTILLATION. When an appreciable amount of moisture is present and enough of the sample is available, solvent distillation is convenient. Weigh a sample that will yield several grams of water into a 500-ml. short-necked round-bottom flask. Add 100 to 150 ml. of toluene or xylene, and connect with the suitable calibrated tube illustrated (Fig. 44), which is in turn connected to a vertical condenser. The tubes are usually known as Dean-Stark tubes and are calibrated to 5 or 10 ml. in 0.1-ml. units. Cork stoppers must be used for connections for these solvents attack rubber. It is better to use all-glass apparatus. Before use, the solvent may have been shaken with water and allowed to stand for it to separate. The solubility of water in these hydrocarbons is so small, however, that normally the commercial grades can be used without previous treatment. Methods often prescribe one or the other of these solvents; usually there is little choice. Toluene has the advantage of a lower boiling point, about 110°; in some cases it is used because the higher boiling point of xylene, 137 to 140°, will cause decomposition of the sample.

To prevent the water from sticking to the sides of the cleaned graduated receiver, see that it is scrupulously clean, and fill it with the saturated xylene through the condenser. Wrap the flask and tube leading to the receiver with asbestos paper. Heat the flask in an oil bath at a temperature such that the solvent distills and condenses at approximately 100 drops per minute. The moisture evaporates and is carried over by the solvent vapor. Both are condensed, but water is heavier than the solvent and substantially insoluble in it. Therefore, water collects in the calibrated tube. Purge the reflux condenser during the distillation with 5-ml. portions of xylene to wash down any moisture adhering to the walls of the condenser. The water in the receiver or condenser may be made to separate, when globules of water stick to the sides, by using a spiral copper wire. Move the wire up and down in the receiver occasionally. When the greater part of the water has distilled over, increase the distillation rate to 200 drops per minute until the volume of water fails to increase, but distill for not less than 2 hours.

Cool the tube to 20°, and read the volume of water. At this temperature, the density of water is 0.998. Placing a single crystal of any dye soluble in water and insoluble in the solvent in the tube before setting up simplifies reading the volume of water but is not essential.

Foaming can be minimized or prevented by adding a lump of rosin, unless alkalies or alkaline salts are present. Fused sodium acetate and potassium bisulfate are sometimes used for the same purpose. If the liquid over the water in the receiver is cloudy, let the tube stand overnight for separation to take place.

Alcohol, acetone, and other water-soluble solvents, if present, will distill over with the xylene or toluene. Provided that a ternary system is not formed by solution of xylene, the volume can be determined, then the specific gravity, and from that the amounts of water and water-miscible solvent can be calculated. If ammonia is present in the sample, sometimes it can be acidified with a drop or two of concentrated sulfuric acid to avoid distillation of the ammonia. If that device is not applicable, the volume of aqueous ammonia distilled must be corrected by taking the specific gravity and calculating the water content from specific-gravity tables, or alternatively calculating the total weight of distillate and subtracting the weight of ammonia as shown by titration. The first of these alternatives is the simpler.

If the sample contains a hydrated salt, due allowance must be made in some cases for water of hydration not liberated at the temperature of boiling toluene or xylene, whichever is used. For example, borax decahydrate loses its water of hydration at 200°; sodium metasilicate dehydrates to $3H_2O$ at 100°.

Drying in a Desiccator. Occasionally a weighed sample is dried in a desiccator without heating. Checking until constant weight is reached is essential, and the method is desirable only in special cases. Usually it requires many days and is expedited by use of a vacuum desiccator.

Ignition. Inorganic substances may often be ignited at a low temperature, usually meaning below red heat, without decomposition. In that case all moisture is driven off, whether free or combined as water of hydration.

Conductimetric Method. Under controlled conditions, it is possible to determine moisture contents with a low degree of accuracy by conductivity. The most common application is for moisture in wood.

Gases. Moisture in gases is determined by passing a known volume through a weighed tube containing a suitable absorbent. It must neither absorb any other gas present nor react with the main gas. Thus calcium chloride is suitable for absorbing moisture from air; soda lime is unsuited because it also takes up carbon dioxide. Calcium chloride is not suited for determining moisture in hydrogen sulfide gas. because it reacts to liberate hydrogen chloride.

Specific Gravity. The terms specific gravity and density are usually used interchangeably. In the case of pure liquids, those referred to as chemical individuals, it serves as a means of identification. With mixtures such as alcohol and water, it is often the easiest means of quantitative analysis after it has been established that only the specified solvents are present. It is rarely important for solids.

Specific-gravity Bottle. The specific-gravity bottle, often known as the pycnometer, is the simplest and most commonly used method for determining specific gravity (Fig. 45). A laboratory should have these available in various sizes from 1 to 50 ml. When enough of the sample is available, use the larger sizes. But if one is determining the specific gravity of a liquid fraction isolated from a sample, often only 1 ml. is available. Greater precision in handling them becomes essential to obtain accuracy.

The bottle must first be calibrated; and, since this calibration will probably serve for hundreds of determinations, it is essential to do it as carefully as possible. Clean both the bottle and its stopper with hot chromic acid–sulfuric acid cleaning solution, and wash thoroughly. Place in the oven until completely dry, and cool in a desiccator with active moisture absorbent. When cool, insert the stopper and weigh to the nearest 0.1 mg. Strictly speaking, a correction has to be applied for the weight of air in the bottle, but this is negligible, both because it is so small and because the same correction would be applied to standard and sample. Error caused by the expansion or contraction of the glass bottle due to change in temperature is negligible under ordinary conditions, since the coefficient of expansion of glass is relatively small compared with that of most liquids.

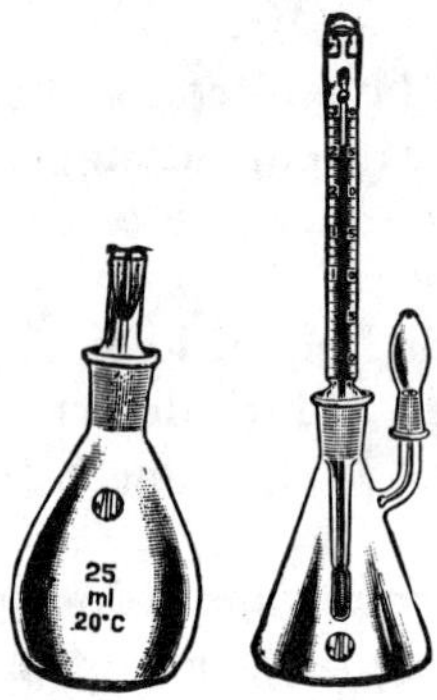

FIG. 45.—Types of specific-gravity bottles.

Fill the bottle with pure recently boiled distilled water at a stated temperature, such as 25°. With bottles that are provided with a thermometer, the adjustment is comparatively simple. For others, immerse the bottle with the neck just above the surface in a beaker containing a relatively large volume of water at 25°. Leave in this bath, which is kept constant at 25°, for at least 15 minutes. Remove by the neck, see that the bottle is full, and quickly but carefully insert the stopper with a slight twisting motion so that no air bubbles are formed in the neck. The liquid in the bottle is forced upward through the capillary in the stopper and overflows. Immediately remove excess of liquid from the top of the bottle but leave the capillary full to the brim. An experienced worker does this by a quick flick of the finger. Do not use an absorbent cloth or paper for this purpose, for almost invariably liquid will be removed from the capillary. Dry the outside of the bottle carefully with a lintless cloth. Be careful not to handle the bottle more than is absolutely essential. The hands will warm up the contents and cause liquid to exude from the capillary. Weigh the whole.

The difference between this weight and the weight of the bottle and stopper alone gives the weight of the water contained by the bottle at 25°, less the weight of the same volume of air. Repeat this procedure at various other temperatures at which the bottle will be used, conveniently 4°, 15.6°, 20°, 30°, and perhaps 35°. The one weight of bottle and stopper will suffice for all temperatures.

For determination of specific gravity, weigh the bottle filled in a similar manner with the sample liquid adjusted to a temperature for which the bottle has been calibrated. Sometimes, particularly with volatile liquids, it is helpful after adjusting the volume of contents and drying to squirt a little ether on the sides of the bottle. This evaporates quickly and absorbs heat from the liquid contents, which accordingly contract. Wipe again and weigh before the liquid warms up and starts oozing out of the capillary in the stopper.

Specific gravity is the weight of a volume of liquid at a specified temperature compared with that of water at a specified temperature. These temperatures are usually but not necessarily the same. Comparison is sometimes with water at the temperature of maximum density, 4°. With a series of calibrations, as discussed, it is a simple matter to obtain the density of the liquid at more than one reference standard. One may compare with water at the same temperature by dividing the determined value by the weight of water contained in the pycnometer at the given temperature. One may compare with water at 4° by dividing the determined value by the weight of water contained in the pycnometer at that temperature.

When many specific-gravity determinations are being made, it is at the then current room temperature, which may easily be a different temperature from those at which the bottle was calibrated. For this purpose, plot at least three calibrations of the weight of water contained in the pycnometer at different temperatures. This furnishes a check on the accuracy of calibration, for they should fall on a straight line. From this graph, the weight of water contained by the bottle at any experimental temperature used can be read off.

When dealing with very viscous liquids, it is difficult to remove entrapped bubbles of air. If not otherwise removable, they are separated by packing the bottle, without stopper, in sand in a centrifugal unit. The neck of the bottle should protrude from the sand, and centrifuging should be continued until all bubbles rise to the surface. They may often be broken with a pointed glass rod or fine platinum wire after they come to the surface.

Specific-gravity Hydrometer. The hydrometer, or specific-gravity spindle (Fig. 46), does not permit of accuracy approaching that of the specific-gravity bottle. If large amounts of sample are available and the hydrometer is calibrated over a limited range, sufficient accuracy for rough analysis is obtained. Essentially it is a weighted bulk attached to a calibrated stem. When immersed in a liquid, the hydrometer sinks to a level at which it displaces its own weight, which depends on the density of the liquid. The hydrometer stem is so calibrated that the specific gravity may be obtained directly by reading the stem at the level of the liquid. Anyone inexperienced in its use is likely to make a serious error in reading the level of the meniscus. Cleanliness of the hydrometer is essential.

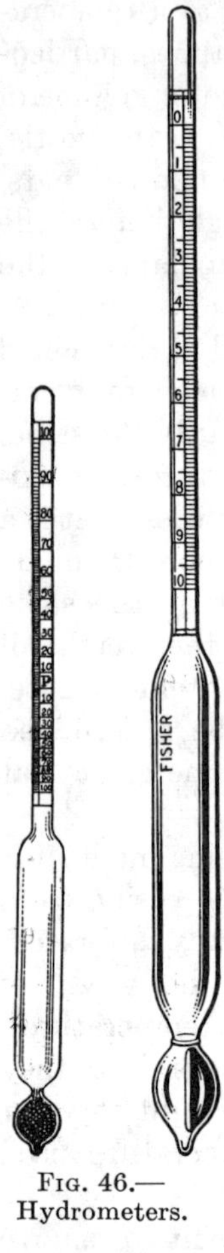

FIG. 46.—Hydrometers.

For some industries hydrometers are available that read directly in the desired substance, as for alcohol in water. This is necessarily calibrated for a specific temperature, and the water-alcohol mixture must either be at that temperature or correction tables must be applied. Liquids heavier than water and those lighter than water are read on different hydrometers or sets of hydrometers.

BAUMÉ HYDROMETER. Density in degrees Baumé is widely used in many industries. There are separate scales for liquids heavier than water and liquids lighter than water; the density in Baumé degrees does not *per se* specify which is meant. Tables readily available in every handbook permit translation to or from specific gravity. In the event of an inadequate sample or unavailability of a Baumé hydrometer, density can be obtained with the specific-gravity bottle and translated from the table or by calculation from the following formulas.

For liquids lighter than water:

$$\text{Degrees Baumé} = \frac{140}{\text{specific gravity } 60°\text{F.}/60°\text{F.}} - 130$$

For liquids heavier than water:

$$\text{Degrees Baumé} = 145 - \frac{145}{\text{specific gravity } 60°\text{F.}/60°\text{F.}}$$

Use of the tables is simpler and therefore preferable.

API HYDROMETER. The American Petroleum Institute has a special density scale. Correction tables and conversion tables are available in Tagliabue's "Manual for Inspectors of Petroleum."[1] Conversion can also be made by formula.

$$\text{Degrees API} = \frac{141.5}{\text{specific gravity } 60°\text{F.}/60°\text{F.}} - 131.5$$

BRIX HYDROMETER. This type of hydrometer is also often called a saccharometer because the scale is used exclusively in the

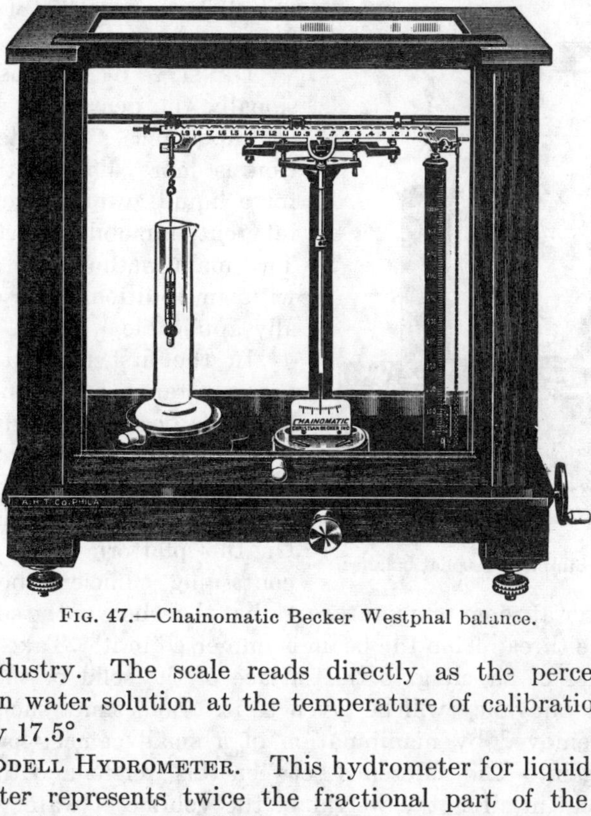

FIG. 47.—Chainomatic Becker Westphal balance.

sugar industry. The scale reads directly as the percentage of sucrose in water solution at the temperature of calibration, which is usually 17.5°.

TWADDELL HYDROMETER. This hydrometer for liquids heavier than water represents twice the fractional part of the density. Thus a density of 1.06 is 12° Twaddell, density of 1.30 is 60° Twaddell. It is an English system used mainly in the textile industry in the United States.

[1] "Tag Manual for Inspectors of Petroleum," C. J. Tagliabue Manufacturing Co., Brooklyn, N.Y., 1939.

Other specific-gravity scales are occasionally encountered: barkometer in the tanning industry, lactometer or Quevenne degrees in the milk industry, etc. Many can be identified by reference to an unabridged dictionary.

WESTPHAL BALANCE. This apparatus consists essentially of a balance arranged to weigh a plummet that carries a thermometer. That illustrated in Fig. 47 is of chainomatic type, but simpler ones like that in Fig. 48 are available. The plummet hangs in a jar of the sample liquid, care being taken that it hangs centrally and does not touch the side of the jar. This balance is desirable only if large numbers of densities are to be obtained.

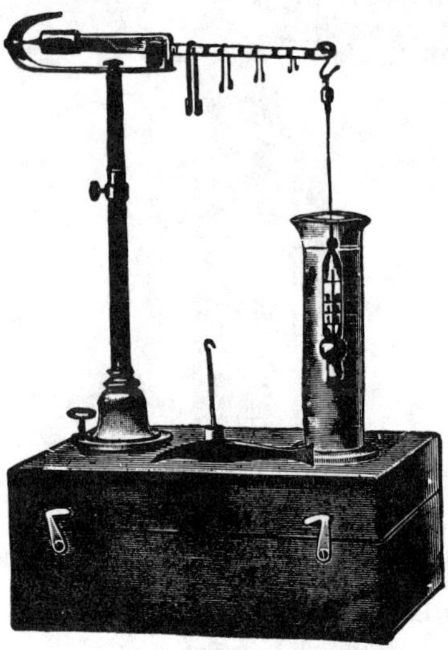

FIG. 48.—Simple Westphal balance.

DENSITY OF SOLIDS. Occasionally the density of a solid is of importance. The determination is more difficult than that for a liquid, owing largely to the inherent immobility of the solid. Two main methods are available with an additional one occasionally applicable.

In the first method suspend the substance by a fine thread from the beam of a balance and weigh at room temperature. Then place a small platform over, but not touching, the balance pan. On this platform place a beaker containing sufficient boiled distilled water at room temperature so that the whole of the solid hanging by the thread from the beam is immersed in it. Take care that no bubbles of air cling to the surface of the solid or the thread; otherwise buoyancy will be given to it. Sometimes such bubbles can be removed by manipulation of a small camel's-hair brush. Weigh again. The difference between this weight and the weight of the solid in air is the weight of the volume of water displaced by the solid. Consequently the original weight of the solid divided by this difference in weight is the specific gravity of the solid at that temperature relative to water at the same temperature. This method is convenient for substances that can readily be suspended in water. It is applicable to solids with a gravity less than unity by using a liquid with a gravity less than that of the solid and in which it is not soluble. The specific gravity of the liquid must be known or determined. From this the weight of an equal volume of water at that temperature is easily calculated by dividing the weight of the liquid displaced by the specific gravity of the liquid at the temperature of the determination.

For powders that cannot be treated in the above way, the second method is useful. Weigh a suitable amount into a calibrated specific-gravity bottle. Fill the bottle containing the powder with a liquid that has a density lower than that of the powder and that does not dissolve it. It is advisable first to wet the sample thoroughly with this liquid to avoid air bubbles. These are sometimes removable by tamping or centrifuging.

Finally fill the bottle, insert the stopper, and weigh by the usual technic. The specific gravity of the powder is calculated as follows:

$$\text{Sp. gr. of powder} = \frac{\text{wt. of bottle and powder} - \text{wt. of bottle}}{\left\{\begin{array}{l}\text{wt. } H_2O\\ \text{contained}\\ \text{by bottle}\\ \text{when full}\end{array}\right\} - \dfrac{\left\{\begin{array}{l}\text{wt. of}\\ \text{bottle, powder,}\\ \text{and liquid}\end{array}\right\} - \left\{\begin{array}{l}\text{wt.}\\ \text{of bottle}\\ \text{and powder}\end{array}\right\}}{\text{sp. gr. of liquid}}}$$

It will be seen that this is the conventional formula, dividing the weight of the sample by the weight of an equal volume of water.

A method that is very convenient is to add lumps of the solid to liquids of various densities until one is found in which it just remains suspended. It must not dissolve in the liquid. The method is rarely applicable, for liquids in the range of densities desired are not ordinarily available.

Melting Point. The melting point of a chemical individual is a constant often useful in its identification. Frequently one must work with materials such as resins, solid fats, and asphalts which are by nature mixtures rather than chemical individuals. The melting point sometimes determines or assists in determining commercial grades. In such cases, it is often defined by a combination of judgment and the method employed, as will be clear from some of the methods given.

Approximate or Preliminary Determination. When very accurate results are not necessary and speed is essential, the following method is useful:[1] Place 2 mg. or less of the finely powdered

[1] Stahl, Glenn W., *Ind. Eng. Chem., Anal. Ed.*, **13**, 545 (1941).

sample upon the bulb of a 150 or 360° thermometer having a thin glass mercury reservoir. The position of the solid should be such that the temperature may be read by a direct glance at the thermometer graduations. Hold the thermometer horizontally with the bulb approximately 8 inches above a 2- or 3-inch nonluminous Bunsen-burner flame, the solid resting on the top surface of the bulb. Focus attention on the thermometer bulb and slowly lower it about 1 inch every 10 seconds. When the solid fuses at any point on the upper surface of the bulb, quickly read the temperature without withdrawing the bulb from above the flame. Accuracy may be increased by spreading the molten solid in a thin film along the bulb by means of a clean spatula or glass rod, if the solid has not decomposed. Allow to cool when the crystals become clearly visible on the shiny bulb. Repeat the heating procedure, and take the melting point as that at which the film disappears. In the neighborhood of 175°, the heat exchange is so variable that the individual melting points may vary by 5°.

Closed-end Capillary. This is the classic method for determination of melting point of a pure solid. It is seldom applicable other than to chemical individuals. Prepare a fine capillary, usually with a bore of not over 1 mm., seal one end, and insert fine crystals of the sample by tamping. These are often most conveniently worked down by rubbing the tube with a dull file to cause vibration. Fasten the capillary tube by means of a small rubber band to a thermometer in such a way that the solid matter in the tube is next to the bulb of the thermometer. With sufficient experience the rubber band becomes unnecessary, and the capillary clings to the thermometer by being moistened with the bath liquid.

Suspend the thermometer and capillary tube by any convenient means in a beaker containing water, glycerin, concentrated sulfuric acid, or other suitable noninflammable liquid that is transparent and boils above the expected melting point of the sample being tested. The size of this beaker depends somewhat on the experience of the operator or the accuracy required. When the worker has limited experience or when very accurate work is required, it is preferable to use 300 ml. of liquid in a 400-ml. beaker. If the worker has more experience or approximations are required, a 100-ml. beaker will suffice. Another form of tube for containing the heating bath is shown in Fig. 49. Heating at the proper rate promotes circulation and therefore uniformity of temperature in the bath.

Heat the liquid fairly rapidly with constant stirring until it is within about 25° of the melting point and then more slowly so that when the melting point is being approached the rise is not over 1° a minute. At the melting point, the solid liquefies more or less sharply, depending on the type of substance and its purity. Opaque materials frequently become clear at this point. The method assumes that the bulb of the thermometer is in equilibrium with the liquid. The thermometer is desirably selected so that there is no substantial amount of exposed column of mercury above the liquid; otherwise in very accurate work a correction is necessary.

For the usual range of melting points, sirupy phosphoric acid is a convenient medium for heating instead of a liquid that boils above the expected melting point. The evaporation of water provides the necessary stirring. The liquid must cool and have a few milliliters of water added before reuse.

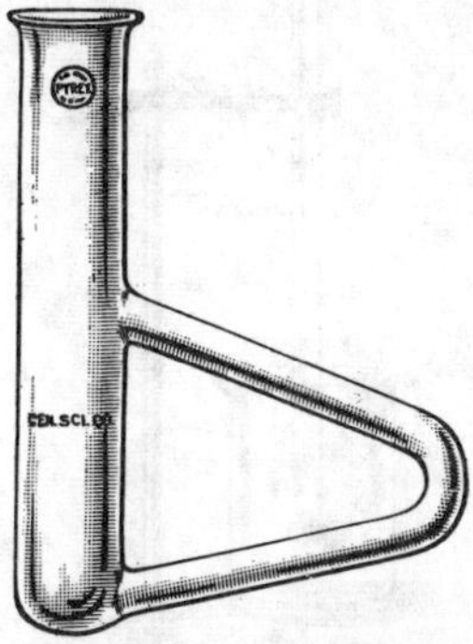

FIG. 49.—Thiele tube for use in determination of melting point.

The purity of a substance may sometimes be judged from the character of the melting point. When a sharp point is obtained, the substance will in general be pure. An indefinite point indicates the presence of impurities.

OPEN-END CAPILLARY. The melting points of many substances, when they are insoluble in water, glycerin, or other suitable liquid, may be determined by using a capillary tube open at both ends. Fill one end with the substance. Often it is melted, inserted in the tube, and chilled in the refrigerator for hours before the melting point is determined. Attach to a thermometer as with a closed-end capillary and so suspend in a beaker that the thermometer bulb is 3 cm. from the bottom of the beaker and the liquid in the beaker is 5 cm. above the mercury bulb. Heat gradually as for the closed-end capillary. At the melting point, the weight of the liquid pushes the melted sample up to the level of the liquid. This happens more or less sharply, depending on the nature of the substance. The method is particularly useful for soft solids, for materials that would give unsatisfactory melting points by the closed-end capillary method, and for nonhomogeneous samples.

The melting points of some substances such as greases that do not give a sharp melting point, owing to their heterogeneous composition, are determined by using in place of the capillary a tube 8 cm. long and 4 mm. in diameter open at both ends. The large amount of sample in this tube takes longer to heat evenly than if in a capillary, and consequently the heating should be conducted especially slowly with thorough stirring.

BALL-AND-RING METHOD. This is method E28-58T of the ASTM as applicable to asphalts, tars, pitches, rosins, and natural and synthetic resins. Its applicability to other purposes is readily visualized by one who has used the equipment. In general it is applied to materials that do not soften at a definite temperature. For this reason, the determination must be made by a fixed, arbitrary, and closely defined method if the results obtained are to be comparable.

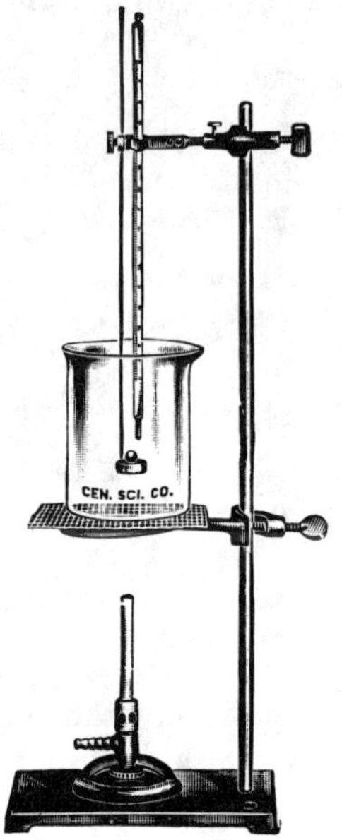

FIG. 50.—Equipment for melting point by ball-and-ring method.

The apparatus is illustrated in Fig. 50, the essential dimensions being as follows: ring—depth 6.35 mm., inside diameter at top 17.46 mm. and at bottom 15.88 mm., outside diameter 20.6 mm., all accurate to 1 mm; ball—9.53 mm. in diameter weighing 3.45 to 3.55 grams. In case of question of conformity of dimensions of other parts, consult the original specification. The specification requires that one heat at least twice the quantity of sample to fill the desired number of rings, for several checks may be run, and in no case less than 40 grams, avoiding use of the surface layer of the sample. When working with fractions isolated from commercial samples, it is often impossible to heat such a large quantity. This represents one of the differences between specification testing and the exigencies of commercial analysis. The sample should be heated in a clean container in an oven, or on a hot plate, sand or oil bath. Do not heat above the temperature necessary to pour the material readily without inclusion of air bubbles or for a total period of more than 15 minutes.

Preheat the ring to approximately the temperature at which the material is to be poured, rest the ring on an amalgamated brass plate, and pour in the sample. When cool, trim flush with the top of the ring, and make a small indentation in the center for the ball. If a repeat determination is necessary, use a fresh sample and a clean container.

Fill the beaker to a depth of not less than 9 cm. with freshly boiled distilled water to avoid formation of air bubbles on heating, or with glycerin precooled to not less than 45° (113°F.) below the anticipated softening point but in no case lower than 5° (41°F.) for materials melting below 80° (176°F.) and in no case lower than 35° (95°F.) for materials melting above 80° (176°F.)

Suspend the filled ring so that the lower surface is 2.5 cm. above the bottom of the beaker and so that its upper surface is at least 5.1 cm. below the surface of the water. Suspend a thermometer of suitable range and accuracy so that the bottom of the bulb is level with the bottom of the ring and within 1.0 cm. of it but not touching. Maintain the initial temperature, with the ball in the liquid but not on the specimen, for 15 minutes. Using forceps, place the ball in the center of the material in the ring. Heat uniformly in such a manner that the bath temperature after the first 3 minutes rises at a rate of 5° (9°F.) per minute within an accuracy of 10 per cent. Tests in which the rates of heating vary are unreliable. The softening point is taken as the temperature at the instant the sample touches the bottom of the beaker or an amalgamated plate or piece of filter paper placed on the bottom to protect the vessel and save time and trouble in cleaning. With care and proper attention to details, duplicate determinations should not differ by more than 1.0°.

MERCURY CUP. Place a thin sliver of material on the surface of mercury in a crucible in which is immersed the bulb of a thermometer. Heat the mercury very slowly with a small flame until the sliver suddenly collapses. The temperature read on the thermometer is the melting point. This method is applicable to some waxes, resins, and fats.

Refractive Index. The physical constant known as the refractive index depends on the fact that a ray of light passing at an angle from one transparent medium to another is always bent from its original course. The angle from the perpendicular of the incident ray is the angle of incidence, and the corresponding angle of the ray after passing into the second medium is the angle of refraction. The ratio of the sine of the angle of incidence to the sine of the angle of refraction is the index of refraction. Practically, one need only consider it as a factor varying, in general, inversely with the density of the medium. As used, the reference medium is air and the value is a whole number. For the determination of the refractive index of solids, see page 60.

Abbe Refractometer. Although other instruments are used, that commonly employed for liquids is the Abbe refractometer shown in Fig. 51. An Abbe refractometer can be employed for substances that liquefy below 150°, but by far the major use is with saponifiable fats and oils and with essential oils. The determination is rapidly made with only a couple of drops of liquid and furnishes a constant for identification of a fat or oil more easily obtained than chemical data. When only a distinction between two oils is desired, the refractive index is often sufficient.

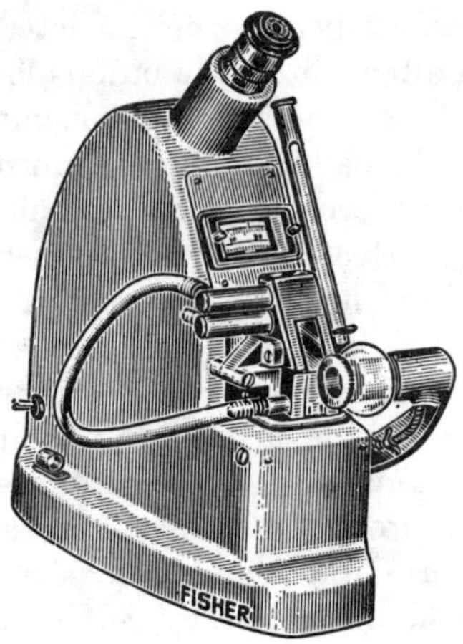

Fig. 51.—Bausch and Lomb design of Abbe refractometer for refractive index of oils.

The physical apparatus consists of a double prism that receives the fluid and can be rotated about a horizontal axis by means of a rack and pinion and radial arm, a telescope for viewing the critical line of total refraction, and a sector rigidly connected to the telescope upon which a scale is engraved in terms of refractive indices. This scale is usually supplemented by a vernier that permits reading the fourth decimal place with higher accuracy. The double prism is encased in a water jacket through which preheated water is allowed to flow so that the refractive index may be determined at any required temperature, normally at 25° or at 40°.

To operate an Abbe refractometer, wet the surface of the lower prism thoroughly with a few drops of the sample, taking great care not to scratch the prism. A glass rod well rounded at the end is suitable to apply the material. Close the prisms firmly by use of the screw head. Pass water at the temperature required through the prisms. A copper spiral tube in the water-supply line is conveniently jacketed for temperature control by a can or beaker of water, suitably heated above or cooled below the desired temperature. The temperature of the water flowing through the prisms is then controlled by varying the rate of passage of water from the supply through the coil to the prisms.

The standard temperature for fats is 40°, and for oils and other liquids it may be 15, 20, or 25°. Often the temperature selected depends on that at which data are available for comparison. Allow the material to reach the temperature of the prisms as measured

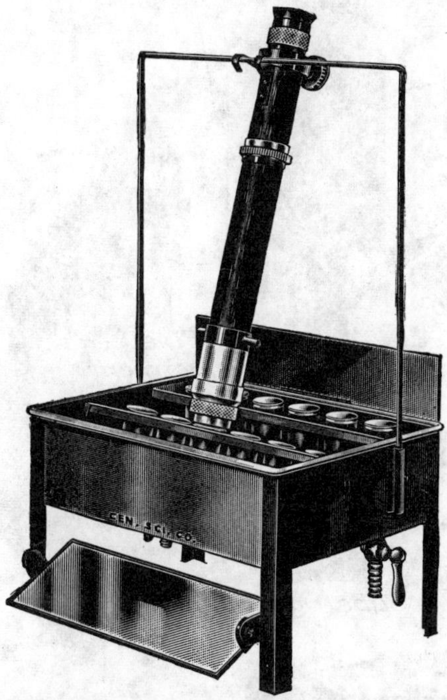

FIG. 52.—Immersion refractometer.

by the thermometer. Use artificial or diffused daylight as illumination. Move the alidade on the side scale back and forth until the field of vision shows light and dark portions. Adjust the compensator screw at the side until the line of separation is as clear and sharp as possible, and adjust the alidade so that this line crosses the intersection of the two cross hairs in the eyepiece, which should be in focus. Read the refractive index to the fourth decimal place, and report at the temperature at which it was taken.

From time to time check the accuracy of the instrument by using liquids of known refractive index. Monobromonaphthalene, which has a value of 1.6580 at 20°, and water, which reads 1.3330 at 20°, are suitable. Sometimes a quartz plate supplied with the

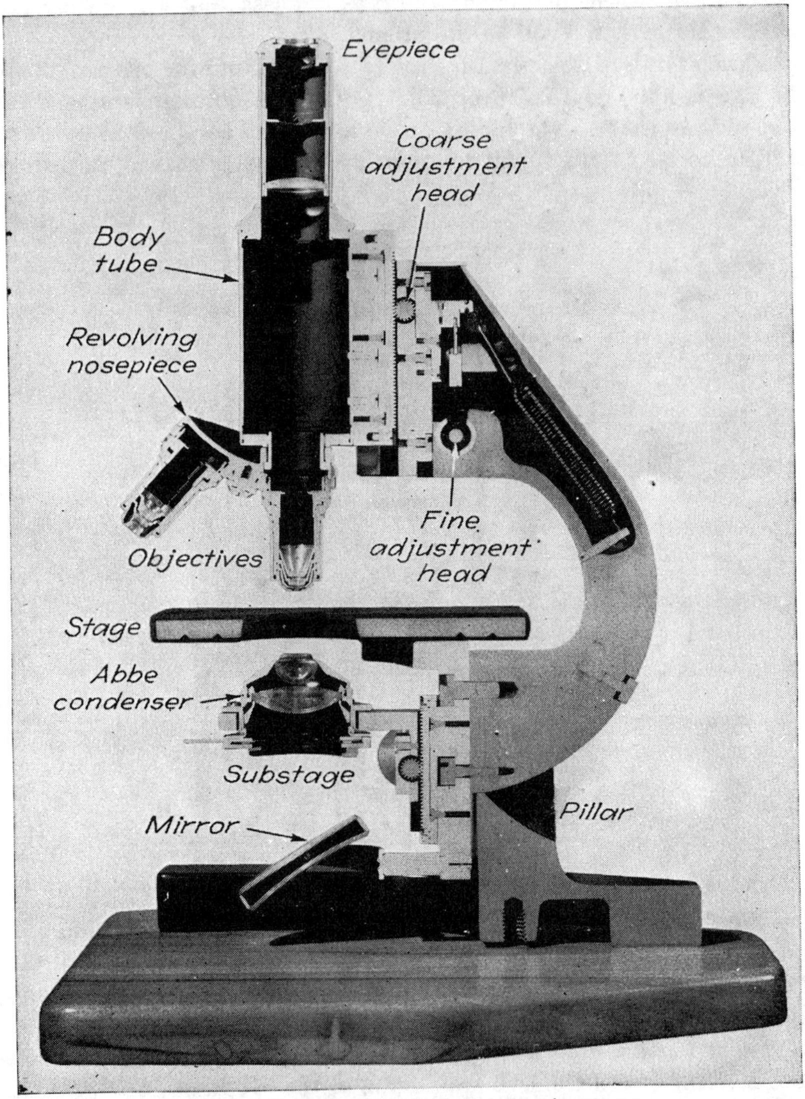

FIG. 53.—Cross section of laboratory microscope.

instrument is used for the purpose. Other convenient reference standards are given in many chemical handbooks.

If the temperature at which a refractive index of a fat or oil was taken varies from that of available data, a temperature correction can be applied with satisfactory accuracy. Add 0.00038 for each degree the temperature is to be reduced; subtract that amount for each that it is to be raised. Thus a value of 1.4608 determined at 20° would be corrected to 1.4627 at 15° or to 1.4589 at 25°.

Immersion Refractometer. This instrument, as shown in Fig. 52, is particularly designed to cover a small index range with a higher accuracy than is possible with the Abbe type. It consists essentially of a telescope with interchangeable measuring prisms and a compensator contained in the telescope for achromatizing the shadow line. Auxiliary prisms for the examination of small quantities are also provided. The sample is placed in a constant-temperature bath. The prism of the instrument is immersed in the sample. The reading is made directly.

This instrument is most commonly used for measuring the refractive indices of solutions of alcohol, sugar, and dilute acids. Different ranges are made available by changing the prisms. The strength of solutions may readily be obtained by comparing the readings with those of known similar solutions.

The Microscope. It is often necessary to study details of a specimen more closely than is possible with the naked eye. This requirement may be met by examination through a hand glass or a simple type of magnifier such as a thread counter. If sufficient information is not obtainable in this way, the microscope is resorted to.

There are numerous levels or grades of microscopes, differing in accuracy of definition, range of magnification of the available eyepieces and lenses, etc. Space limitations prevent discussion here of details of manipulation of the instrument. For further study of that subject, consult Volume I of "Handbook of Chemical Microscopy" by E. W. Chamot and C. W. Mason, John Wiley & Sons, Inc., New York, 1938.

The general principle of the microscope is shown by the accompanying photographic cross section (Fig. 53). The degree of magnification is varied by changing either the eyepiece or the objective. Any but the most elementary microscope has two or more of each. The magnifications obtained by different combinations of eyepiece and objective are shown in Table 1.

A completely equipped laboratory will have a research microscope equipped with revolving stage, mechanical stage, polarizer, analyzer, substage condenser, measuring accessories, and various illuminators and camera attachments. This is of real utility and will pay for the rather high initial outlay. Such equipment can be purchased with a few of the accessories, and others can be added from time to time. A complex type is shown in Fig. 54.

TABLE 1.—RELATION OF LENS, OBJECTIVE, AND MAGNIFICATION

Objective magnification No.	Eyepiece					
	5 ×	6.4 ×	7.5 ×	10 ×	12.5 ×	15 ×
2	10	12.8	15.0	20	25.0	30.0
2.6	13	16.6	19.5	26	32.5	39.0
3.2	16	20.5	24.0	32	40.0	48.0
4	20	25.6	30.0	40	50.0	60.0
6	30	38.4	45.0	60	75.0	90.0
10	50	64.0	75.0	100	125.0	150.0
21	105	134.4	157.5	210	262.5	315.0
31	155	198.4	232.5	310	387.5	465.0
40	200	256.0	300.0	400	500.0	600.0
43	215	275.2	322.5	430	537.5	645.0
45	225	288.0	337.5	450	562.5	675.0
60	300	384.0	450.0	600	750.0	900.0
80	400	512.0	600.0	800	1,000.0	1,200.0
97	485	620.8	727.5	970	1,212.5	1,455.0
98	490	627.2	735.0	980	1,225.0	1,470.0

However, much good work can be performed with an elementary or student microscope costing very much less. It is important that the microscope should have good objectives and eyepieces. These must be treated with great care and cleaned with special tissue or the softest nonlint cloth. Do not allow the objective to rest upon the specimen, for this is liable to injure the lens, which is made of comparatively soft glass. If possible, cover the substance being examined with a microscope cover glass. Separation of the particles of a fine powder is best obtained by adding a drop of liquid. What liquid to use depends on the type and solubility in that liquid of the specimen and also the results desired. For many purposes, a 50 per cent mixture of water and glycerin is very useful.

Do not use too much of a powdered sample, or the field will be crowded and accurate observation will become impossible. At high magnifications especially it will be noticed that the particles, being in different planes, are not all in focus at once.

Care in adjusting the lighting is well repaid. Observe the effects with different amounts of light obtained by manipulating the light diaphragm, raising or lowering the substage condenser, or changing the position of the mirror. Details that cannot be seen with one arrangement of the lighting will often show up when

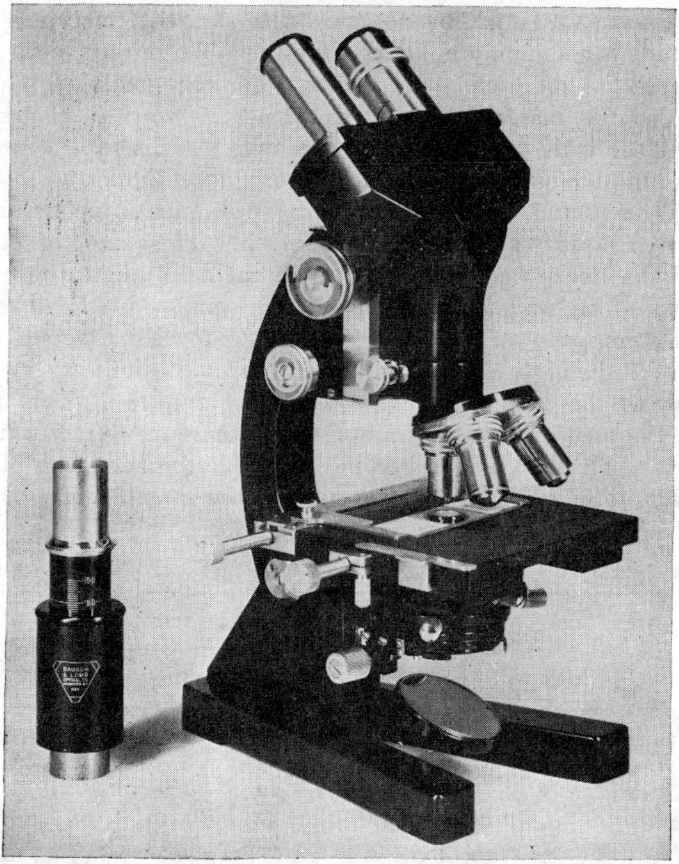

FIG. 54.—Research microscope with alternative monocular to replace the binocular.

illuminated at a different angle. If feasible, observe by both transmitted and direct light. It is usually best to observe the specimen with the lowest magnification that will give the desired results. The higher the magnification the less the depth of focus. However, viewing under a high magnification, although the sample may not be fully in focus, may often give unsuspected or important information.

Always keep the microscope and accessories clean, away from heat, damp, and dust and, when not in use, housed in the case provided for them. Frequently a preliminary observation under the microscope will save much chemical analysis. Substances that give similar chemical reactions may often be readily distinguished when so examined.

Refractive Index of Solids. Although the refractive index of liquids is commonly obtained with the Abbe refractometer, that instrument is not applicable to solids. For very small-size particles and also for materials bounded by nearly vertical planes, the Becke test with the microscope is simple and useful. Immersed in a suitable mounting medium, a thin band of light is visible outlining the particle when a narrow axial illuminating cone and an objective of low aperture are employed. This band of light is called the Becke line. The narrow bright halo moves toward the medium of higher refractive index if the focus is raised and toward the medium of lower refractive index if the focus is lowered.

Use media of widely different known refractive indices, gradually narrowing the ranges until the Becke line vanishes. This occurs when the medium and the sample are of the same refractive index. At this point the particle becomes invisible, because rays of light are then refracted through it at exactly the same angle as that at which they are refracted through the medium.

Table 2.—Liquids for Determination of Refractive Index of Solids

	Vapor pressure		Refractive index at 20°
	°	mm.	
α-Bromonaphthalene	110	3.5	1.6582
n-Butyl phthalate	110	3.8	1.4932
Heptylic acid	90	1.9	1.4234
Mesitylene	20	27.15	1.4981
Ethyl propionate	20	27.75	1.3841

Various liquids for the purpose have been recommended by Chamot and Mason in their "Handbook of Chemical Microscopy." The set[1] shown in Table 2 is useful, since five readily available substances cover the whole range of refractive indices normally encountered. The range of refractive index between 1.6582 and 1.4234 is covered by mixtures of α-bromonaphthalene and heptylic

[1] Kunz, A. H., and Spulnik, J., *Ind. Eng. Chem., Anal. Ed.*, **8**, 485 (1936).

acid. Above 1.4932, normal butyl phthalate can replace the more expensive heptylic acid. Mixtures of mesitylene and ethyl propionate cover the range between 1.4981 and 1.3841.

Technical-grade chemicals are suitable. Use an Abbe refractometer to obtain the refractive indices of the chemicals themselves and of their mixtures. Relative vapor pressures of the reagents are indicated in the table as a guide in their use. Such figures are an indication of the amount of change that may reasonably be expected to occur on brief exposure to the air.

The Colorimeter.[1] Although the term may imply that this is as definite an instrument as a microscope, actually many forms are available. Any instrument which will compare the intensity of color of a gas, liquid or solid, with that of a standard is a colorimeter. The simplest form is a pair of glass tubes such as Nessler tubes (Fig. 55). In an elaboration of that, such tubes are enclosed, producing a type similar to a Wesson colorimeter. Further refinements and more complex mechanism give instruments, of which the Duboscq colorimeter (Fig. 56), is an example. Each refinement of detail adds somewhat to the precision of the instrument. It is probable that 90 per cent of the colorimetric results are obtained in chemical laboratories with Nessler tubes. There are three different technics commonly employed. For the purpose of the description that follows, it will be assumed that two colored solutions are to be compared, the colors having been developed from any of the well-known procedures. The color of a solution is often recorded in arbitrary units of standard glasses, not usually so applied as to constitute a fourth method.

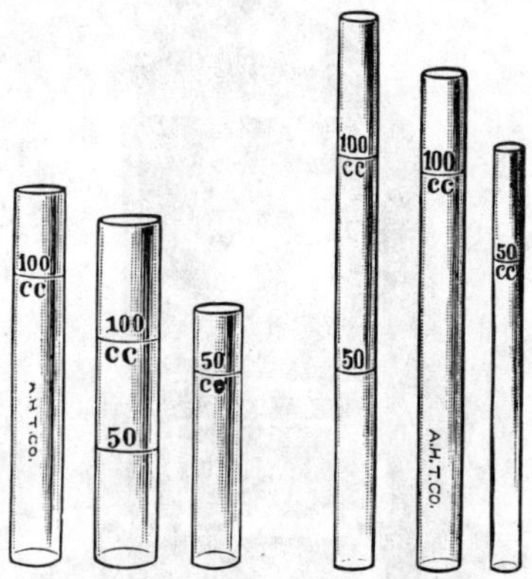

FIG. 55.—Typical Nessler tubes.

DILUTION METHOD. This is carried out with a pair of graduated tubes which for this purpose must be clear, free from flaws, and identical as to thickness of glass, internal diameter, and graduation. Matched pairs of graduated Nessler tubes are suitable.

[1] This subject is covered in more detail in Foster Dee Snell and Cornelia T. Snell, "Colorimetric Methods of Analysis," Vol. 1, D. Van Nostrand Company, Inc., New York, 1936.

Place the standard and the sample in similar graduated tubes. The colors of the two solutions should be nearly alike. The darker is then to be diluted to match the lighter. Although water is usually a suitable diluent, the method of development of the color enters in; if the solution is strongly acid or high in salts a similar diluent free from color is required. Add the diluent to the darker solution, mixing after each addition, until the colors of the two solutions, observed horizontally, appear to be identical. The amounts of colored substance present are then proportional to the

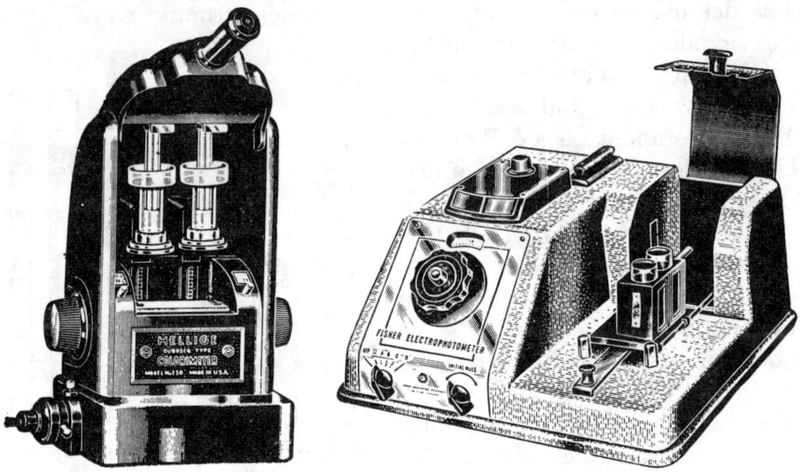

Fig. 56.—Colorimeters. Left one is Duboscq type. Right one is an electric filter photometer.

volumes present in the two tubes. This is true only if Beer's law holds, that the color is proportional to concentration. This is assumed to be correct over a small range of color change, even though it may not be correct over a longer range.

Duplication Method. This method is occasionally referred to as a colorimetric titration because a standard solution is used to duplicate the color in the sample and is usually added from a buret. The method is usually carried out in a pair of calibrated Nessler tubes. They need not be graduated.

Place a known volume of the sample in which the color has been developed in one tube, or dilute the entire developed sample to a known volume in one tube. To the other tube add the same reagents as were added to the sample to a volume of water amounting to nearly the same volume as the sample. This statement as to the same reagents must go back over the history of the sample. For example, if at an early stage the sample was ashed with sulfuric acid and subsequently that acid was neutralized to sodium sulfate, the standard should contain a weight of sodium sulfate equal to that in the volume of sample that is to be compared. The degree of dilution of the blank varies according to the concentration of the standard solution to be used.

To the blank containing the same reagents as those used for the sample, add a standard solution, carefully mixing after each addition, until the color of the sample is duplicated by that produced in the blank. The color of the blank having been made to duplicate that of the sample, they may differ in volume. It is often simple to calculate or estimate the amount of standard that would be necessary if the volumes were equal.

To eliminate error and to correct for the variation in salt concentration, duplicate the volume as well as the color of the standard by adding water and standard alternately until the two solutions are identical in both color and volume. The volume of standard required to duplicate the sample is then a direct measure of the amount of test substance in the sample, or in the volume matched if an aliquot was used.

Balancing Method. In essence this consists in taking a definite depth of sample in one tube and matching it by adding an amount of standard to another tube such that when viewed through the depths of the solutions the color intensities will be identical. Although this may be carried out in Nessler tubes, it is quite commonly done with more complex instruments such as the Duboscq colorimeter shown diagrammatically in Fig. 57. For purpose of illustration, the Duboscq colorimeter will be described and its application to this method given.

The two independent tubes *A* and *B* are of the same size and hold the solutions of the unknown and the standard. Each is mounted in a holder *M*, *N*, which slides up and down in a slit cut in the backboard of the instrument and is operated by a rack and pinion. Light is reflected upward through the tubes by a mirror *G*. Directly over tubes *A* and *B*, which contain the solutions to be compared, are two glass plungers *O*, *P* of a diameter less than that of *A* and *B*. The bottoms of these plungers are finely ground and in the best instruments are fused on or the plungers are of solid glass. In the less expensive instruments, the bottoms are fastened with an adhesive.

The telescope K, for observation of the colors, is perpendicular to the base so that the operator looks downward into the instrument. The light reflected upward through the solutions in A and B is so

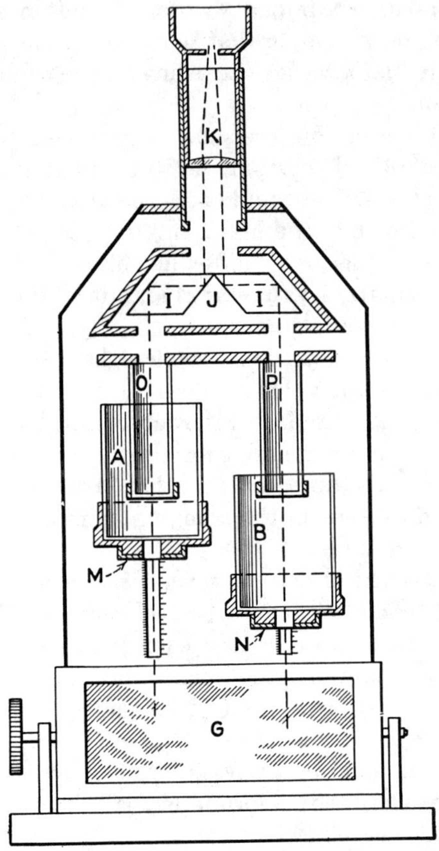

FIG. 57.—Diagrammatic illustration of Duboscq colorimeter.

reflected by the prisms I, I in the box J that two fields appear side by side, one from A and one from B. The arrangements of the prisms is such that the images observed in the field of the telescope are those of the bottoms of the plungers O, P rather than of the entire depth of liquid in A and B. By means of suitable reduction of the aperture by screens, reflection from the sides of the tubes is cut off.

For use, place the instrument to face a source of light, and adjust mirror *G* at the proper angle to reflect skylight upward through *A* and *B*. Fill both cups about half full of clear water and bring the plungers into contact with the water. When this is observed through the eyepiece, a circle with a fairly indistinct line through the middle will be seen. There should be no detectable difference between the intensity of the two halves of the field. If there is a difference, adjust the location of the instrument so that they are equally bright.

Place the cups in position, and turn them up until they just touch the bottom of the plungers. Set the scale so that the zero mark corresponds with this setting of the cups, or adjust the cup carriers so that the scales are at the zero marks when the cups just touch the bottoms of the plungers. Be sure that the cups are not interchanged after setting the zero point, for the bottoms may not be of equal thickness.

Place in one cup a solution of the standard and in the other the developed sample. Move the cup having the lighter color upward until the plunger just touches the surface of the liquid, or comes to a definite reading standardized for the method. Then move the other cup upward, observing its movement through the eyepiece, until the image of the base of the plunger in that liquid appears to be of the same intensity as that of the image observed from the other field. The instrument is then balanced, and the depths of liquids underneath the plungers are inversely proportional to the concentration of color present in each. The slits in which the holders of *A* and *B* move are calibrated so that the depths of liquid may be read directly and errors in reading depths of liquid in glass are eliminated.

As a further method of elimination of side light, it is desirable to surround the cups with a suitable shield, furnished with many instruments, or to use cups with opaque sides. Modifications of this apparatus differing in details of construction are available from various manufacturers.

With this type of instrument, the nature of the color is not important, provided that it is apparently the same in sample and standard, for only the intensities are compared. By simple modification the same instrument is suitable for colorimetry, nephelometry, and bichromatic pH work.

Color with Standard Glasses. Once the depth is specified, the color of any solution can be read in terms of the three component

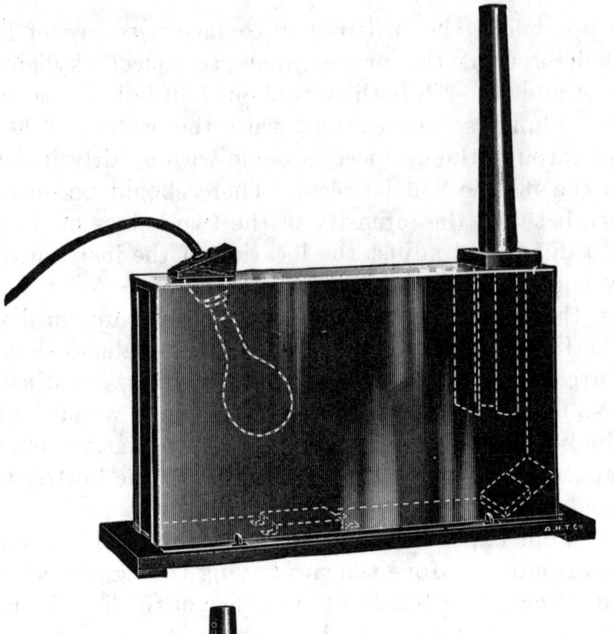

FIG. 58.—Two types of Wesson colorimeter.

colors red, blue, and yellow. The best recognized standard for this is Lovibond glasses. A typical determination is reading the color of a vegetable oil. A specified depth is placed in one tube, such as that of the Wesson colorimeter (Fig. 58). A similar depth of water in the other tube serves in part to match the optical effect of the depth of oil. Standard glasses, which cannot be taken up here, are then placed over the standard tube until the color of the oil is duplicated.

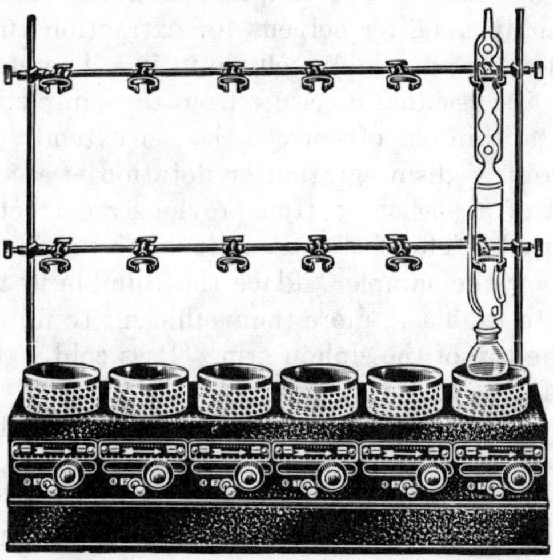

FIG. 59.—Multiple Soxhlet extraction unit.

FIG. 60.—Soxhlet extractor.

Extraction. The operation of extraction in general consists in leaching a solvent-soluble material from one that is insoluble. Practically it may consist of soaking a material in solvent in a beaker, and decanting the extract, of repetitions of this operation in centrifuge bottles with the precipitate thrown down in each case between additions of solvent, of Soxhlet extraction, or of operation of a rubber extractor. The first is inefficient and hardly quantitative.

SOXHLET EXTRACTION. Figure 59 shows a complete Soxhlet extraction apparatus with ground-glass joints. Unless used continuously, they are often assembled by joining the three units of the fat-extraction flask, the Soxhlet extractor, and a condenser with cork stoppers (Fig. 60). Rubber is attacked by many of the solvents used. Figure 59 shows a typical multiple-unit heater for carrying out such extractions.

Fat and oil extractions from seeds, food products, and impregnated substances are commonly made with this apparatus. In the case of seeds, crush as well as possible in a mortar before extracting. Further crushing may be necessary after most of the extractable matter is removed. The sample is then reextracted. Experience or stated methods give the necessary time to extract a sample completely.

Most samples need to be water-free before extraction; otherwise complete extraction may not occur. For samples that are not bone-dry, ethyl ether is normally a better solvent for extraction than petroleum ether, for water is appreciably soluble in it. Therefore the solvent serves to dry the residual moisture from the sample.

Place the sample in a thimble of correct size to extend just above the siphon arm; and, if disintegration or flotation is apt to occur, cover with a wad of absorbent cotton previously extracted with the solvent being used. The cotton also serves to distribute the condensed solvent over the sample. Place the thimble in the extractor. Add solvent to the flask, more than sufficient to fill the empty extractor up to the top of the siphon arm. Pass cold water through the condenser jacket.

Heat the flask with a suitable source of heat according to the solvent. For light solvents such as ethyl ether and petroleum ether, a water bath is convenient and lessens fire hazard. For higher boiling solvents such as benzene, alcohol, and chlorinated hydrocarbons, use either a Bunsen flame under an asbestos gauze or, preferably, an electric heater.

The solvent, the vapors of which ascend through the side arm of the extractor, will condense and drip into the thimble. The rate of extraction can be varied at will by adjusting the heat. When the extractor is filled to the top of the side arm, the solvent, now containing soluble matter from the sample, siphons over into the flask. Thus the device serves to make a series of periodic extractions with fresh solvent, draining the solvent automatically. Rapid extraction tends to keep the extractor at a temperature little less than the boiling point of the solvent. Methods often specify the rate of extraction and thereby control the temperature in the extractor.

Occasionally, extracted fatty or waxy material may clog the siphon arm. In that case, heat the arm carefully to liquefy the gelled extract.

When extraction is complete, the solvent must be removed from the extracted material. This is conveniently, safely, and economically performed by removing the thimble and sample and collecting the solvent in the empty extractor. When the extractor is nearly full, disconnect it momentarily from the flask and condenser and by tipping it cause the contents to siphon into a supply of recovered solvent. This may in some cases be reused directly; in others it requires redistillation or drying.

When the major part of the solvent has been removed, dry the last traces from the extract in the oven. Any water in the extract, usually heavier than the extract, tends to sink to the bottom, and spattering may occur. To prevent this, add a few milliliters of absolute alcohol to the extract and evaporate to dryness on the water bath before placing in the oven. The alcohol in so evaporating carries the water with it as vapor.

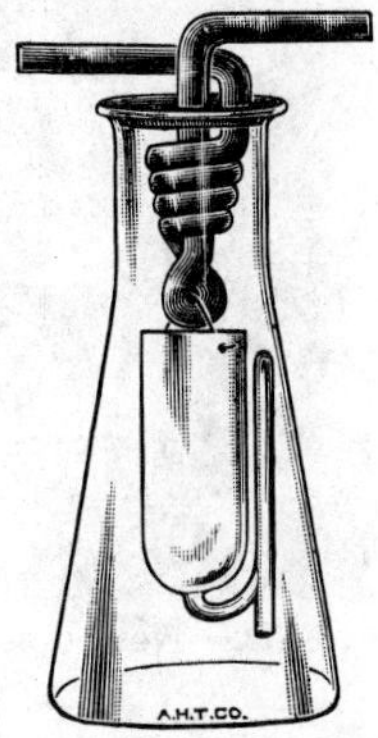

FIG. 61.—Rubber extractor.

Many substances extracted, oils, for example, are appreciably volatile at 110°. Therefore the extraction is most commonly made with low-boiling solvent such as ether or low-boiling petroleum ether. The last trace of these is quickly evaporated at 110° and minimizes loss of the extract.

Although extractor, flask, and condenser with replaceable ground-glass joints are desirable they are seldom essential. For a laboratory running continuously on miscellaneous work, it is desirable to have a series of condensers for Soxhlet extractors set up as a permanent installation. Extractors can then be put into place at will. For condensation of the solvent, a bulbed condenser as illustrated is most satisfactory; a coil type will occasionally fill up with solvent if operated rapidly.

RUBBER EXTRACTOR. Though designed for extraction of rubber, this unit has many other uses. It is illustrated in Fig. 61. The specific dimensions for conformity to ASTM specifications appear in Fig. 108. The rubber extractor performs like the Soxhlet extractor with one essential difference: the solvent extracting the sample is at its boiling point.

Particle Size. The properties of many industrial products depend on maximum particle size and distribution of the particles over a range of sizes. Paint pigments, sand, fine abrasives, and rubber fillers are only a few examples. The various methods apply under different sets of circumstances.

DRY-MESH ANALYSIS. This is the most common and generally the simplest method of determining particle size. For laboratory use, a nest of screens 5 inches in diameter that fit one into the other with a lid at the top and a tray at the bottom, as illustrated in Fig. 62, is convenient. The screens in the nest vary according to the sample. For sand, a set with 20, 40, 60, 80, and 100 openings per linear inch is used. For finer abrasives, sieves at 120, 140, 170, and 200 mesh may be introduced. Rarely would all be used; usually separation into three to five fractions is desired. Mechanical shaking is desirable and is essential for extreme accuracy. Take a known weight such as 20 grams, place on the top screen, and cover with the lid. Shake, using a sideways as well as a moderate up-and-down motion. Continue the shaking until it is clear that no further separation occurs. Weigh the amounts left on each screen and in the pan. Calculate as percentages of original sample, and report. For a distribution in the range of 20 to 100 mesh, this would show the following:

FIG. 62.—Nested set of standard sieves.

Retained on 20 mesh	=	per cent
Passing 20, retained on 40 mesh	=	per cent
Passing 40, retained on 60 mesh	=	per cent
Passing 60, retained on 80 mesh	=	per cent
Passing 80, retained on 100 mesh	=	per cent
Passing 100 mesh	=	per cent

With very fine materials, the more commonly used screens are 200, 230, 270, and 325 mesh. Since these screens are very fine and the corresponding sizes of particles tend to cling together, dry meshing cannot be carried out. It is usual to employ water to wet and separate the particles. The screens cannot be nested as they can with coarser particles.

Place the sample on the finest sieve to be used. Preferably wet the screen first with alcohol and then with water. Fine screens, owing to surface-tension effects, are difficult to wet with water without first using alcohol. Immerse in a container of water so that the water rises in the screen, but do not cover the whole screen with water. By careful up-and-down movements, allow the water to wash the abrasive through. With some finely powdered substances, it is necessary to brush gently, using a soft camel's-hair brush. This should break up flocculates but not break down individual particles. Be sure the brush is finally thoroughly washed out into the sieve. Continue until no more goes through. This is observable as indicated by the turbidity or clarity of the water. Gently running water is advantageous. Apply a towel to the bottom of the sieve to absorb most of the moisture left. Dry in the oven, brush the residue onto a tared watch glass, and weigh. Place this residue on the wetted screen of the next coarser size, and repeat.

Each fraction is weighed, and the amounts passing one sieve and retained on the next highest are obtained by subtraction. Report as in the case illustrated. Screens of 200 mesh and more are very fragile, and extreme care must be used in handling them. Those with 500 meshes to the linear inch are available, but practical limitations indicate that 325 mesh is about as fine as can be used with any approach to accuracy. That is a mesh dimension of about 44 microns. With paint pigments, the procedure is often varied by washing through with kerosene instead of water.

THE MICROSCOPE. This method is of limited application, since only very small quantities of the material can be examined at a time under the microscope. Qualitative observation is rapid and often useful. Quantitative determination is tedious and should be used only when insufficient sample is available for a screen or other analysis or when more precise measurements are required.

Standard marked microscope slides, or inserts for the microscope eyepiece, are available that are of definite distance between the markings at stated magnifications.

Place a very small amount of sample on a slide wet with water, 50 per cent aqueous glycerin, or other suitable mounting medium and cover with a thin microscope glass. Focus on the particles, and read their dimensions in terms of divisions on the standardized scale. Count the number of those of approximately the same size in one field. Do the same for the other sizes present. Repeat the whole, using as many different fields as possible, and average the results.

SPECIAL EQUIPMENT METHODS. Various types of equipment are available especially for grading of particle size, often with accuracy down to 5 or 2 microns. They will merely be mentioned, for they are available only in specialized laboratories.

Mechanical separation is accomplished by air with equipment based on the cyclone air separator. The fractions are collected and weighed and are available for further work.

Another method depends on the fact that the rate of fall of a particle through a liquid of known viscosity is related to the size of the particle. When the specific gravity of the particles and the time of fall through a chosen distance are known, the size of the particles can be calculated with the aid of Stokes's law. An approximation is obtained by collecting the particles in a graduated tube. The volume collected within a specified time corresponds to a limited range of particle size.

A third method depends on the same principle as the previous method. A photometric instrument reads the amount of light that passes through specified sections of a cylinder in which the material is suspended. These readings are taken at definite time intervals during settling, and results are read by comparing with calibration curves obtained with materials of known density and particle size.

Fractionation. Various types of distilling heads are available, a few of which are illustrated in Fig. 63. Each has some advantage over the others. The object in all is to cause the higher boiling fractions to condense and return to the distillation flask while the lowest boiling fraction is vaporized and carried over into the condenser. Considerable surface for condensation is necessary, and most columns allow for this.

The Hempel distilling head is the simplest form illustrated and for general use is about 315 mm. body length, 95 mm. reduced end, 22 mm. outside diameter of body tube, 13 mm. inside diameter of body tube, and 7 mm. outside diameter of side tube. The body is filled with glass beads or short pieces of tubing.

Distilling heads are readily made in the laboratories, the size depending on the amount of sample available. Make the bottom of the body tube as wide as is consistent with serving the purpose of supporting the beads or glass tubing. This allows the vapors and the condensed liquid to flow uninterruptedly so that the column will not become full of condensed liquid. Use beads of such size, or glass tubing of such length and breadth that the vapors and the condensed liquid are enabled to flow freely.

Short pieces of glass tubing about 1 cm. in length and 4 mm. outside diameter give ample condensing surface and are not so liable to cause clogging as are beads. Fill the column with these nearly up to the side arm. Insert a cork containing a thermometer at the top and adjust the bulb of the thermometer so that it is just below the side arm.

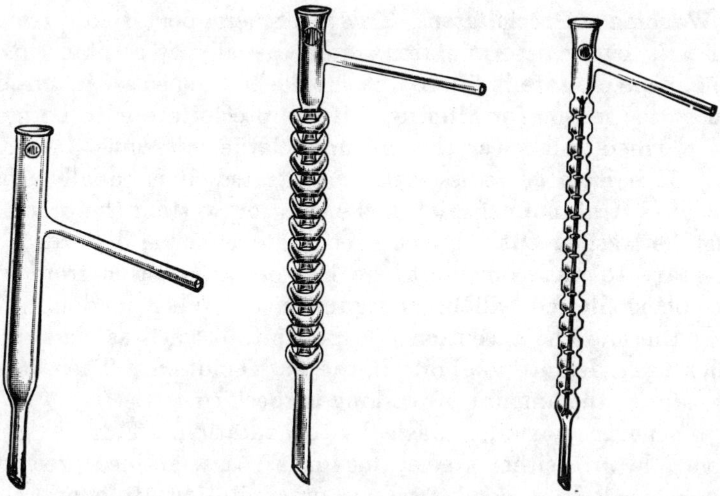

FIG. 63.—Various forms of distilling heads.

Shield the column and the flask containing the liquid to be fractionated from drafts. The source of heat should likewise be shielded to avoid uneven heating. Raise the temperature of the flask gradually until the vapors of the liquid condense in the column. Continue the heating until the vapor starts to condense in the end of the side arm that is in the condenser. Maintain the heat so that a steady stream of drops but not a continuous stream of liquid passes into the receiver. The initial boiling point is usually taken as that reading on the thermometer at which the first drop comes from the condenser. The short time lag between the time when it first condenses and that when it issued from the condenser is assumed to be sufficient to heat the bulb of the thermometer to the correct temperature.

Smaller columns may be made and used when only limited amounts of liquid are available. In these it is more important than ever to maintain as free a passage as possible for the vapor and condensed liquid, to keep the lower end as wide as conveniently possible.

Change the receiver whenever definite breaks occur in the distillation temperature. Increase the heat gradually as the temperature rises. Record the volume distilled over each definite range of temperature. It is then frequently desirable to obtain the specific gravity of each fraction in order to estimate the composition of the distillate.

Washing of Precipitates. This is a general operation carried out in nearly every determination, often casually as a minor matter. When a precipitate is filtered, it should be washed with consideration of the reason for filtering. If the precipitate is to be ignited and weighed, it is clear that all nonvolatile salts must be washed out. If it is to be redissolved and titrated, it is equally evident that any titratable material in the solution wetting the precipitate must be washed out. If the precipitate is to be discarded, it is necessary to wash only until no loss of the solution from which it is being filtered will be of significance. When in doubt, wash more times. The exception is a precipitate, such as lead sulfate, which has a definite solubility in the wash solution. Then, in some procedures, the amount of washing is specified.

Whenever possible, wash by decantation. This is usually applicable with either a gelatinous or a coarse-grained precipitate. Follow prescribed conditions for precipitation; frequently they affect the form of the precipitate. Washing gelatinous precipitates on the filter is helped along by detaching the gel from the paper by a strong stream of water from the wash bottle.

CHAPTER 4

PRELIMINARIES TO THE ANALYSIS

In the type of analysis under discussion one does not have a sample with specific instructions as to the determinations to be made; one has a sample. Therefore, the preliminary to any quantitative work is to decide what determinations are to be made.

Appearance. Often the type of product is indicated by physical examination. That the product is a colorless liquid eliminates a great many possibilities; a colored liquid introduces others. If the liquid is viscous, that tells another story, eliminates some ingredients, and suggests others. If a sample is a soft amorphous solid, it suggests some procedures; a hard crystalline mass offers others, a finely ground solid yet others. If the sample is a liquid, does it foam when shaken? If it does, a surface-active agent is present, perhaps soap; if it does not, one knows at least that soap is not present.

Commercial products are usually submitted in the original package. The label showing what the product is used for is a clue to one of experience and also often to one with imagination. If it is used to polish silver, one may expect a mild abrasive and other things; if used as a hand lotion, then softening agents such as glycerin or glycols are probably present; if an opaque protective coating a pigment will be present, etc.

Pause a few minutes to consider. Many of the quantitative determinations you might do would take hours; even the qualitative determinations to follow shortly will take a few minutes each. If you are directing your own work, sit down, consider what you should tell yourself to look for. You may save time later by spending some now.

Odor. Characteristic odors often furnish a clue, although here experience is more often necessary. But nearly everyone knows the odor of phenol and of alcohol. True, odor may be a clue to only one of five or even ten materials present. But it is a step forward to have a clue to even one material. Often the substance that gives the odor is added as a masking odor, a cover odor. This is particularly true if it is of a perfume type. One can sometimes identify by odor more than one substance in a product. Perhaps it contains a little ammonia and pine oil. Each can be detected quite independently in the presence of the other.

If the original substance is a liquid, the odor is often brought out by heating in a test tube. In general it is best to smell it, warm gently, smell again, etc., until it is nearly at boiling. An ammoniacal odor in the vapors may be ammonia or triethanolamine, or morpholine or perhaps some other substance. If alcohol is present, the vapors can often be ignited even though the solution itself may be too dilute to burn. Other odors brought out by so heating are those from dilute solutions of formic or acetic acid, formaldehyde, etc.

If the original substance is an odorless solid, occasionally heating will be helpful. Warm a small amount on a crucible cover without contact with the flame; a broken one will do as well as a new one, but be sure it is clean.

Inflammability. If nothing more, a suitable test for inflammability will determine whether organic matter is present or whether the sample is purely inorganic. For this, carefully heat a portion of solid on the tip of a spatula blade, some of a liquid in a spoon. Does the solid catch fire on heating? Does it entirely burn up? Does the liquid burn? If it does not burn, when heat is continued to complete evaporation to solids, the liquid was presumably either water or chlorhydrocarbon. Observe the behavior of the deposited solid as though the sample were originally solid.

If the sample burns, observe the type of flame. Aromatic compounds usually burn with very smoky flames. Aliphatic hydrocarbons usually give luminous but not smoky flames; alcohols and glycols give nonluminous bluish flames. Of course in a mixture this is only a guess at one ingredient.

If the sample burns, after observing the flame blow it out and note the odor of the vapors. General types of odors characteristic of fats or oils and of waxes are often distinguishable. The glycerides give an acrid odor from a trace of acrolein. The odor from waxes is describable only as waxy. It is often a help to burn in a similar way a known sample of the suspected material. One often burns a sample in several successive stages and notes the odor at each stage. Sometimes materials are successively burned away. This may give one or more clues.

If an ash is left that has significance, either there is an inorganic material present or the sample contains a salt of an organic material.

If the ash is dissolved in water and the solution is alkaline, a soap may have been present. Of course, also, the original solution or solid may have contained an inorganic alkali or alkaline salt. In this type of detective work, no clue is final but all must be correlated with other evidence to be conclusive.

Solubility. If the sample is a liquid, is it miscible with water? If so, this limits the vehicle to a few such as water, lower alcohols, and ketones. If not miscible, does it show a limited solubility in water? If insoluble in water, is the sample soluble in or miscible with petroleum ether, with alcohol, or with benzene? Those are further clues.

If the sample is a solid, is it soluble in cold water? If it dissolves in water only on heating, then cool the solution to see if it crystallizes out. If it does, it was soluble in hot water but much less so in cold; if not, a reaction occurred on heating.

And of course all these tests are only approximate, only relative. Often roughly quantitative solubilities need to be determined, by adding the substance 1 gram at a time to 100 ml. of water until it will no longer dissolve.

Flame Reactions. Use the familiar flame test of qualitative analysis. Dip a clean platinum wire in the solution and hold in the hottest part of the Bunsen flame. Often the flame test from a liquid will be improved by adding hydrochloric acid. A few particles of a solid substance may be made into a paste with a drop of hydrochloric acid for test. Be sure to clean the wire after every test.

The yellow flame of sodium is so familiar as to prove hardly anything. But a trace or a brief sodium flame is probably contaminant; only a small amount is required to give a strong flame test. Substantial amounts of other ions may be present but masked by the intense yellow flame. Observe through the blue of a cobalt glass for potassium, which shows red under that test. Better check against a known potassium compound both as to color and permanence. Many other flames are confusing but do not show as red through the cobalt glass.

In reasonable absence of sodium, calcium will give a brick red, lithium a bright crimson, strontium a less brilliant crimson. Barium will give a green flame but so will copper and borates under proper conditions.

It is best to check each flame against knowns simply because one cannot well expect to have every flame test in mind unless they are used frequently. Unusual ones, such as the green flame of a halogenated hydrocarbon when tested on a copper wire, may be applied occasionally. But hydrochloric acid would give the same test.

Indicators. This test is naturally limited in applicability to liquid samples and solutions of solids. Liquid samples may be turbid or contain much suspended matter. Perhaps to get the indicator reaction it may be well to add a pinch of diatomaceous earth to a few milliliters and filter; perhaps the precipitate is white or settles readily, and this is not necessary. Perhaps the sample is an emulsion, in which case filtration is out of the question.

Try the reaction in a few drops with phenolphthalein solution, and if it does not turn pink add methyl orange to see if it turns red. Pink with phenolphthalein indicates a pH above about 8.3. If such a color is present, it may indicate the presence of a water-soluble alkali such as sodium hydroxide, potassium hydroxide, barium hydroxide, or calcium hydroxide. Qualitative addition of sodium bicarbonate solution will show whether it is one of the latter two. But it may also indicate the presence of soap, trisodium phosphate, any alkali silicate even the most siliceous, sodium carbonate or commercial sodium bicarbonate, which always contains a trace of carbonate. There are many other possibilities, few other probabilities. A suspension of magnesium oxide or of calcium carbonate will give the reaction. One can often form a judgment from the intensity of the pink color, considered in connection with the amount of substance known to be dissolved.

A reaction with methyl orange to give a definite red color shows a pH below 4.0 and may indicate an acid, organic or inorganic, or an acid salt such as sodium acid sulfate, or a salt that hydrolyzes in solution, sodium silicofluoride, for example.

Tests with these two simple indicators should be only a preliminary to one more test for pH. Every laboratory should have one of the several mixed indicators called Universal indicators. This means that the pH of the solution over a range of at least 3 to 11 can be estimated with an accuracy to one unit or less by addition of the indicator to the solution. Papers or sets of papers for the range are also sold. The subject is discussed in detail in Chap. 6 starting on page 114.

Information about the pH of the solution will often be useful in interpretation of later data. As an example, if soap is found present and the pH is taken, if around 11 there is a more alkaline substance present, if around 9 the soap is not a straight soap, that would give a pH of about 10, almost regardless of the concentration.

Effect of Acid. Add 1:4 hydrochloric acid in the cold to the liquid product, a solution of it or an aqueous suspension of the solid substance. If there is no indication of reaction, gradually heat to boiling with due attention to what is happening. An odorless noncombustible gas indicates that carbonates were present. This is readily verified by leading the gas into barium hydroxide or calcium hydroxide solution.

Many metals react to evolve hydrogen gas, which is odorless when pure. Practically, impurities usually give a characteristic odor, and the gas may be burned with a blue flame at the mouth of the test tube. The ease of reaction taken in conjunction with the electromotive series of metals from any general chemistry book is a clue to the metal probably present. Odors such as those of hydrogen sulfide and sulfur dioxide are readily recognizable.

During the heating of solids with acid, note whether the solid partly or completely dissolves. Suitable qualitative tests are in order at this point, after filtering if necessary. Test for iron and aluminum by precipitation with excess ammonium hydroxide. Filter this off, and test for calcium with ammonium oxalate. After removing it, if present, test for magnesium with disodium phosphate and ammonium hydroxide. The suggestions are only examples; naturally one is guided by what is to be expected in the particular type of sample. Where possible, apply confirmatory qualitative tests.

Acidification gives other clues. Emulsions are usually broken down. Solutions that foam before acidifying but not afterward are probably soap; if they continue to foam after acidifying rather than separating fatty acids, then other types of surface-active agents such as sulfated alcohols are probably present. Saponin is a possibility.

Effect of Alkali. Render the solution or suspension of sample strongly alkaline with 10 per cent sodium hydroxide solution. Note whether substantial amounts of heat are evolved. An odor of ammonia readily noted indicates appreciable amounts of ammonium salts. Small amounts may be detected only with moist red litmus paper on heating. Such small amounts of ammonia may be due to amines or proteins rather than ammonium salts. Evolution of hydrogen is due to reaction of the alkali with metallic aluminum, zinc, or magnesium.

Tests for Negative Radicals. In general these are applied on the clear original solution or on a solution of an insoluble sample in dilute nitric acid. Typical tests are silver nitrate solution followed by excess nitric acid for chlorides, barium chloride added to a solution slightly acid with hydrochloric acid for sulfates. Note the amounts of precipitate formed as a guide to the amounts of sample to be taken later for quantitative analysis. Many others are applicable. Spot tests, discussed in detail in Chap. 5, are rapid and generally reliable.

Selection of Quantitative Methods. Even one with considerable experience does not blindly proceed to an analysis of the sample after these preliminaries. Chart your course. A fairly good idea of the composition of the sample will have been obtained. Now devise a scheme of analysis. If at all possible, plan this so that a series of determinations can be carried out in succession, thus saving weighings and total time consumed. Often the shortest method does not represent an economy of time if a more involved procedure will permit carrying out other determinations at the same time. It is a sound maxim to avoid short cuts unless previous experience with the methods has established that they give the results with adequate accuracy.

One should in general apply accepted methods of analysis, those which have become standardized by use. But bear in mind that the methods must be fitted together, modified as to sequence or technic to fit the particular sample. In a preliminary reading of the methods, give attention to every step and the reason for it. This is a necessary preliminary to any modification needed to fit the circumstances.

And finally, all too often, standard methods are not available. Then you must consider what properties of the substances being determined can be applied quantitatively. And results must be cross-checked in some way to establish accuracy of such non-standard method.

CHAPTER 5

QUALITATIVE ANALYSIS BY MICRO AND SPOT TESTS

The conventional scheme of qualitative analysis is applicable on a microscale, with some modifications. Supplementing that, a very useful and comparatively new branch of analytical chemistry is based on use of small amounts, usually drops or small particles of sample. Many of the tests are the same as those which may be employed in a test tube with larger amounts of sample and reagent. Therefore, aside from furnishing the newer technic, this chapter may be convenient to recall reactions for macro qualitative use. Some reactions that are not specific in macro work are made so in spot testing by use of the capillary properties of filter paper. Thus a filter paper is impregnated with the reagent and a drop of sample added. This frequently localizes interferences. The development of the technic is largely due to Feigl.[1] Prepared papers for use with some spot-test reactions are sold commercially in coarse and fine texture. Some are prepared with a chemically inert water-repelling barrier which restricts the spot-test reaction to a uniform area, usually 100 sq. mm. The intensity of the resultant colored spot serves as a roughly quantitative measure of the abundance of the constituent in the test drop by comparison with similarly prepared standards.

The tests are generally quite simple to carry out. Their very simplicity is, however, a source of danger, and it is necessary that very careful attention be paid to interfering ions. Owing to the sensitivity of the tests, the reagents employed must be pure. The apparatus used should be very clean and kept away from drafts or even the open laboratory, so that contamination is not present when the tests are made.

The amount of apparatus required is surprisingly small and consists of smaller pieces of apparatus than usual in order to handle the drops. Such apparatus is available from the usual supply houses, typical items being shown in Figs. 64 to 68. Fig. 69 shows an electrobalance for weighing up to 100 mg. with precision to ±0.02%, accuracy to ±0.05%. On the right is a miniature carbon and hydrogen combustion furnace.

[1] FEIGL, FRITZ, "Spot Tests" Vol. I Inorganic applications, Vol. II Organic applications. Translated by Ralph E. Oesper, Fourth Edition, Elsevier Publishing Company, 1954.

Because the tests use such small quantities, they are of particular utility in dealing with commercial products when only small amounts

FIG. 64.—Micro stirring rod.

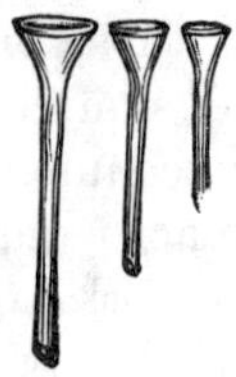

FIG. 65.—Micro funnels.

FIG. 66.—Micro dropping bottle.

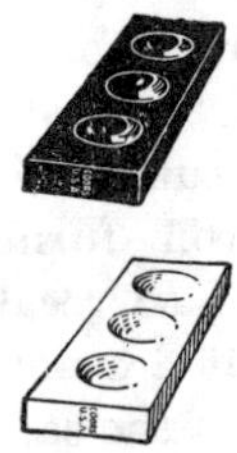

FIG. 67.—Micro spot plate.

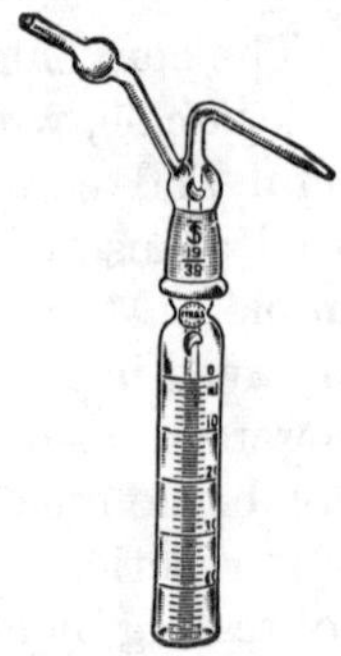

FIG. 68.—Micro wash bottle.

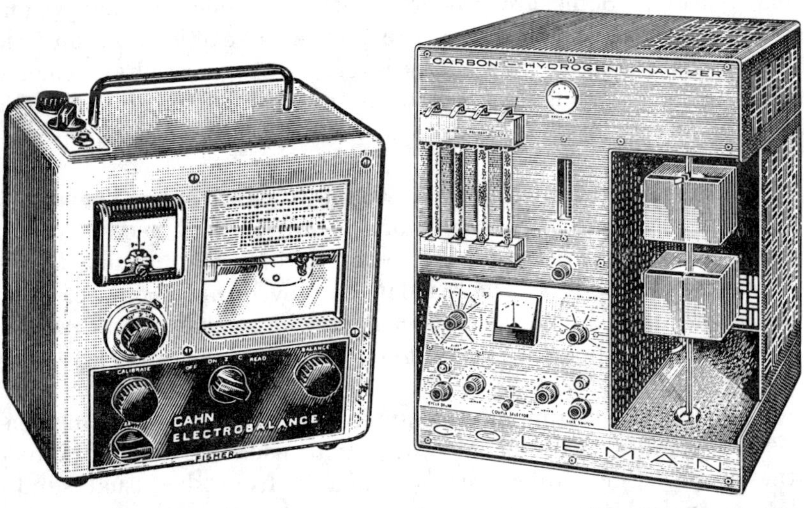

FIG. 69.—Left an electric microbalance. Right a microcombustion apparatus for determining carbon and hydrogen.

of material are available, as is so often the case. Even when large amounts are on hand, the tests frequently commend themselves by their simplicity and distinctive character.

When traces of a substance are being looked for in the presence of large amounts of foreign substance, it is important to remember that the foreign substance may well reduce the sensitivity of the test. It is always best to confirm positive results with one reagent by the use of another reagent. Occasionally a test is absolutely specific for an element, but even so it must usually be performed under carefully observed specific conditions. There are radicals, of which tin and antimony are examples, that normally cause much difficulty but can be identified with ease and certainty by spot tests.

MICRO QUALITATIVE SCHEME

A system for carrying out a complete schematic analysis using not more than 0.3 gram of sample has been developed.[1] As it is presented here, reference is needed to later parts of the chapter for identifying tests in many cases and for confirmatory tests in all cases.

Sample. A sample already in solution presents no problem. If not so available, dissolve by the usual technics, with hydrochloric acid, nitric acid, *aqua regia,* potassium pyrophosphate fusion, sodium carbonate fusion, etc. If sodium carbonate fusion is used, render just acid with 1:1 nitric acid.

A neutral solution of sample requires no further preparation. If strongly acid, depending on the elements expected to be present, the technic will vary. Often it is necessary only to neutralize approximately with ammonium hydroxide. When ammonium salts are objectionable, remove excess acid by evaporation. If this is hydrochloric or nitric acid, evaporate just to dryness, taking care not to decompose the salt; then cool and take up with water. If excess sulfuric acid is present, greater care is required, for the excess must be fumed off without decomposing the residual salts. Often after such evaporation it is necessary to clarify the aqueous solution by adding a single drop of 1:1 nitric acid.

Group 1: Silver, Mercurous Mercury, and Lead. Heat the solution and add a drop of 1:3 hydrochloric acid. If precipitation occurs, continue such additions until precipitation is complete. Filter while still hot, and cool the filtrate well. Lead is shown by fine white crystals. Reserve this liquid for further testing. The residue on the filter is silver chloride, mercurous chloride, or both. Wash with hot water and then with 1:5 ammonium hydroxide. A black residue indicates that mercurous ion was present. Acidify the ammoniacal solution with hydrochloric acid. A white precipitate indicates that silver chloride was present.

Group 2: Mercuric Mercury, Lead, Copper, Cadmium, Bismuth, Arsenic, Antimony, and Tin. If lead was present, filter to remove

[1] Davies, W. C., *J. Chem. Education,* **17,** 231–234 (1940).

the lead chloride. Add bromine water and boil to oxidize the ions present. Allowing for the amount of hydrochloric acid already added, render the solution about $2N$. This requires addition of a total volume of 1:3 acid equal to the volume of the solution before testing for Group 1. Pass in hydrogen sulfide gas for about 5 minutes. Dilute to about ten times the volume, at which the acidity will be about $0.2N$. Pass in hydrogen sulfide gas for 5 minutes longer and filter. The mixed sulfides on the filter paper may contain all the members of this group. Reserve the filtrate for testing for further groups.

First wash the precipitate on the filter with approximately 0.1 per cent ammonium nitrate solution, rejecting the washings. Then wash with hot 8 per cent sodium hydroxide solution, reserving the precipitate. To a portion of the alkaline solution add an equal volume of concentrated hydrochloric acid, and warm gently for a few minutes. If arsenic is present, yellow arsenious sulfide is precipitated. Filter from this sulfide, and boil the filtrate until hydrogen sulfide gas is no longer given off. Test a portion of this filtrate for antimony. To about 1 ml. of filtrate add a few particles of aluminum to reduce tin, and test for that element.

Subject the residual sulfides, insoluble in sodium hydroxide solution, to a few milliliters of hot 1:1 nitric acid. If a black residue remains, it is mercuric sulfide. Dissolve this in bromine water, and test to confirm. Dilute a portion of the acid filtrate, and render alkaline with 1:1 ammonium hydroxide. A blue coloration indicates the presence of copper. If a precipitate is thrown down at the same time, filter it and discard the filtrate. Test a portion of this residue for bismuth. Wash another portion of the residue with 1:10 sulfuric acid, then with water. A white residue at this point is lead sulfate. Dilute another portion of the nitric acid filtrate and make strongly alkaline with hot 8 per cent sodium hydroxide solution. A white precipitate indicates the presence of cadmium.

Group 3: Iron, Aluminum, Chromium, and Manganese. Boil the acid filtrate from which sulfides were precipitated, both to remove hydrogen sulfide and to concentrate to a suitable small volume. Add an excess of bromine water and boil off the excess. Add 1 gram of solid ammonium chloride and a slight excess of 1:1 ammonium hydroxide. Just bring to boiling and filter. Reserve this filtrate for testing for other groups.

Treat the residue on the filter with hot 8 per cent sodium hydroxide solution. The precipitate that remains may contain ferric hydroxide, chromium hydroxide, and manganese hydroxide. Test small portions for the three elements. The alkaline filtrate may contain aluminum. Some chromium may dissolve but will not interfere with the test for aluminum. Any chromium so dissolved can be reprecipitated by further boiling of the alkaline washings.

Group 4: Nickel, Cobalt, Manganese, Zinc, and Chromium. Pass hydrogen sulfide gas into the hot ammoniacal filtrate from the Group 3 hydroxides. Filter and reserve the filtrate for later tests. Dissolve the residue in a few milliliters of concentrated hydrochloric acid. It may be necessary to add 1 ml. of 3 per cent hydrogen peroxide solution to dissolve the precipitate completely. Boil off most of the acid, and then add a little water. Test portions of this solution for each of the metals of this group.

Group 5: Barium, Calcium, and Strontium. Acidify the filtrate from the Group 4 sulfides with acetic acid, and boil to concentrate and remove the hydrogen sulfide. If any precipitate forms during this treatment, it should be manganese incompletely removed in an earlier stage. Add 1:1 ammonium hydroxide to render the solution alkaline and a drop of 3 per cent hydrogen peroxide solution. Filter from the resulting manganese precipitate, if any. Add an excess of saturated aqueous ammonium carbonate solution, usually a couple of drops, to the filtrate. Heat to boiling and filter, reserving the filtrate for later tests.

Dissolve the precipitate from the filter with a few milliliters of 10 per cent acetic acid. Heat this solution to boiling to drive off carbon dioxide, and cool. To a portion add a fragment of solid sodium acetate, and when dissolved add a drop of dilute potassium bichromate solution. A yellow precipitate is barium chromate and indicates the presence of barium. Warm the balance of the solution, and add saturated sodium carbonate solution or a few particles of solid. Filter, wash well, and discard the filtrate and washings. Dissolve the residue from the filter with a few milliliters of dilute acetic acid, and test this solution for calcium and strontium.

Group 6: Magnesium, Sodium, and Potassium. Divide the filtrate from precipitation of the previous group into two parts. To one add a few crystals of an ammonium phosphate, and when dissolved make strongly ammoniacal. A white, crystalline precipitate of magnesium ammonium phosphate indicates the presence of magnesium. Scratching the inside of the tube with a glass stirring rod often promotes precipitation. Filter this precipitate, wash briefly, and dissolve in a couple of milliliters of hot 8 per cent sodium hydroxide solution. Test this solution for magnesium by confirmatory methods. Evaporate the other portion to dryness, and carefully ignite at a low temperature to remove ammonium salts. Add water and test for sodium and potassium, provided that neither was introduced in preparation of the sample.

Ammonia. A very convenient method of testing is to place 5 to 10 ml. of sample in a 150- by 20-mm. test tube. Stretch a red litmus paper across the top of the test tube, holding it in place with a spring-type test-tube holder. Moisten the litmus paper with a couple of drops of water. Add a substantial excess of 8 per cent sodium hydroxide solution down the side of the test tube from a pipet, avoiding touching the litmus paper. Heat cautiously, shaking to prevent bumping or spattering until vapors issue from the tube. This will not only detect very small amounts of volatile alkalinity by a slight development of blue color but will enable an experienced person to estimate roughly the amount to be expected. If trouble is encountered with bumping, render the sample distinctly acid with mineral acid and filter, using the filtrate for this test.

TESTS FOR INDIVIDUAL IONS

Generally speaking, when a single test is described, its application will involve three actual tests. One of these will be a blank to make sure that the reagents do not contain the test substance as impurity, or in a few cases to give a comparative color for judging the result of the test. The second will be a test with a known small concentration of the test substance, usually but slightly more than the limit of identification. Such a parallel test often permits roughly judging the concentration of the test substance. The third is of the solution under test. For brevity the necessity for such detail is mentioned in the tests that follow only when it is of especial importance.

At least two specific tests are given if two are available. If known, the sensitivity of the test is stated. A rough limit of the concentration in which it is applicable is expressed in parts per million, abbreviated ppm. That is the same as milligrams per liter. To express in terms of per cent, multiply by 0.0001.

Cations. ALUMINUM. *By Alizarin.* Prepare alizarin paper by impregnating filter paper with a 1 per cent solution of alizarin in alcohol, and dry. If chromium, iron, manganese, thorium, and magnesium are absent, place a drop of the solution under test on a strip of this paper. Expose the strip to ammonium hydroxide fumes. A red coloration indicates aluminum.

If the interfering ions mentioned may be present, make the solution under test alkaline with a drop of 8 per cent sodium hydroxide solution. Place a drop of this solution on a piece of plain filter paper. Apply a drop of a solution containing 3 grams of barium nitrate in 97 grams of a saturated solution of ammonium thiocyanate. Let stand for ½ minute, and add a drop of 1 per cent alcoholic alizarin solution. Dry in a current of warm air and extract the excess alizarin with alcohol. Dry and compare the red color with that of a blank similarly treated.

As an alternative procedure for the removal of interfering substances, make the solution neutral or very slightly acid and then place a drop on a strip of filter paper impregnated with a 5 per cent solution of potassium ferrocyanide. The solution spreads out, carrying the aluminum with it and leaving the ferrocyanide precipitate of the interfering metals. Add a drop of water to the center of the spot and a drop of 1 per cent alcoholic alizarin solution near the damp outer circle. The two damp areas will spread into each other. Expose the paper to ammonia vapor and dry. A red zone appears where the two spots meet if aluminum is present.

The test will detect 0.00015 mg. of aluminum in a concentration of 3 ppm.

By Aluminon, the Ammonium Salt of Aurin Tricarboxylic Acid. The red coloration of aluminum with aluminon is a very sensitive qualitative test. Provision is necessary to avoid interference by antimony, beryllium, bismuth, iron, lead, mercury, large amounts of phosphates, reducing agents, silicates, tin, and titanium. Reducing agents, if present, are eliminated by boiling with a few drops of 3 per cent hydrogen peroxide. To the neutral solution under test add a few particles of solid ammonium carbonate to precipitate interfering positive ions. Filter and wash any precipitate. Evaporate the combined filtrate and washings to 5 ml. or less. Add equal volumes of 1:10 hydrochloric acid and 25 per cent ammonium acetate solution. To 3 drops of this solution add 1 drop of a 0.1 per cent aqueous solution of aluminon. A bright red coloration indicates aluminum.

If the pH of the final test solution is adjusted to 6.3 to 7.3 by the slow addition of ammonium carbonate solution, the test becomes more sensitive. The ammonium carbonate will decompose the red lake otherwise formed when chromium is present. When phosphate is present in large excess, precipitate the aluminum from the original sample solution as the basic phosphate by the addition of sufficient 1:1 ammonium hydroxide to make the solution distinctly alkaline, and heat just to boiling. Filter and dissolve in a few milliliters of 1:10 hydrochloric acid. Then test as described, providing for possible interference by positive ions.

The procedure described will detect about 0.0001 mg. of aluminum in a concentration of 0.1 ppm.

AMMONIA. *By Nessler's Reagent.* To a drop of the solution under test add a drop of Nessler's reagent, page 720. A yellow color indicates ammonia. For testing solid substances, heat with 8 per cent sodium hydroxide solution, or stronger, and place a piece of filter paper moistened with Nessler's reagent over the mouth of the test tube.

By Silver Nitrate and Formaldehyde. To prepare the reagent solution, treat 10 ml. of a 20 per cent silver nitrate solution with 5 drops of 40 per cent formaldehyde solution and a drop of 8 per cent sodium hydroxide solution. Filter off any metallic silver that forms.

Place a drop of the solution under test and 1 to 2 drops of 8 per cent solution of sodium hydroxide in a micro test tube. Carefully place a drop of the silver nitrate–formaldehyde reagent on the end of a rod, and suspend over the solution in the test tube. The most convenient technic is to have the rod project from a small cork, which has a vent cut at one side to avoid development of pressure. Warm gently for 2 to 3 minutes. According to the amount of ammonia evolved, a black or gray separation of metallic silver is visible in the suspended drop. This can be seen more distinctly when transferred to a filter paper.

The limit of sensitivity is 0.00005 mg. of ammonia in a concentration of 1 ppm.

ANTIMONY. *By Rhodamine B.* Antimony can be detected by rhodamine B in the absence of bismuth, gold, mercury, thallium, and titanium chlorides or of relatively large amounts of bromide, dichromate, iodide, iron, molybdate, perchlorate, permanganate, persulfate, thiocyanate, thiosulfate, or tungstate. Place a drop of the solution under test on a spot plate, add a crystal of sodium nitrite, and acidify with a drop of 1:1 hydrochloric acid. When effervescence ceases, add 2 to 3 drops of a 0.1 per cent aqueous solution of rhodamine B. A change from red to violet occurs when antimony is present.

The method is sensitive to 0.0005 mg. of antimony in a concentration of 10 ppm.

By 2-Methylbenzothiazole. For this test large amounts of bisulfite, cyanide, iodide, and thiosulfate must not be present. If lead, mercury, or silver are present add a drop of 1 per cent potassium iodide, more if necessary to complete precipitation, and filter.

To a drop of the solution under test, add a drop of a 15 per cent solution of potassium iodide and a drop of a 1 per cent solution of 2-methylbenzothiazole in alcohol. Antimony gives a yellow turbidity. In the presence of bismuth the turbidity is yellow-red in color. When cupric and ferric ions are present, destroy the free iodine that is liberated, by addition of a drop of 5 per cent solution of sodium bisulfite.

The method will detect 0.02 mg. of antimony in a concentration of 20 ppm.

ARSENIC. *By Mercuric Bromide.* The coloration developed in the Gutzeit method for determination of arsenic (page 208) is qualitatively applicable. Reduce a drop or two of the solution under test in a micro test tube by gently warming with aluminum powder and a drop of 8 per cent sodium hydroxide solution. Halfway down the tube place a plug of absorbent cotton moistened with 10 per cent lead acetate solution to remove any hydrogen sulfide gas. Place a strip of filter paper moistened with 5 per cent alcoholic mercuric bromide solution over the mouth of the tube. A yellow color finally becoming brown indicates arsenic.

The test will detect 0.001 mg. of arsenic in a concentration of 1 ppm.

As Arsenate by Silver Nitrate. Place a drop of the solution or a few milligrams of the solid sulfide precipitate in a micro crucible. Warm with a few drops of 1:1 ammonium hydroxide and 3 per cent hydrogen peroxide solution. This will oxidize any arsenic present to arsenate. Sulfides must be completely decomposed, and the final solution must be colorless. Acidify the solution with 10 per cent acetic acid, and add 1 to 2 drops of 1 per cent silver nitrate solution. A red-brown precipitate or coloration indicates arsenic.

The test is sensitive to 0.006 mg. of arsenious acid in a concentration of about 125 ppm.

BARIUM. *By Rhodizonic Acid.* Prepare a test paper by impregnating filter paper with a 0.1 per cent aqueous solution of rhodizonic acid in alcohol, and dry. Spot a freshly made piece of this paper with the neutral or slightly ammoniacal solution under test. A red color indicates barium. If strontium is present, place a drop of 1 per cent acetic acid on the red spot, and any red color due to strontium will disappear. Lead will also cause a red coloration, which can be made to disappear by exposing the test paper to ammonia fumes.

The reagent will detect 0.001 mg. of barium in a concentration of 20 ppm.

By Sodium Rhodizonate. Place a drop of the neutral or slightly acid solution to be tested on a strip of filter paper. Add a drop of a 5 per cent aqueous solution of sodium rhodizonate. A red color indicates barium, provided that other heavy metals are absent. Add a drop of 1:4 hydrochloric acid, which will dissolve any interfering color due to strontium.

The limit of sensitivity is 0.0005 mg. of barium in a concentration of 10 ppm.

BERYLLIUM. *By Quinalizarin.* If ammonium salts are present, render the solution distinctly alkaline by dropwise addition of 8 per cent sodium hydroxide solution, and boil until the ammonia has been driven off. Then render the solution faintly acid with 1:1 hydrochloric acid. If both iron and magnesium are present, they will give an interfering red color; either alone does not give this reaction. To remove them, evaporate the solution of sample to dryness, and fuse in a nickel crucible with a small piece of sodium hydroxide. Take up the fusion with a few milliliters of water. Filter the insoluble residue, which contains the iron and magnesium.

Prepare a fresh saturated solution of quinalizarin in alcohol, and add a drop of this solution to a drop of the solution under test on a spot plate. Add a drop of 1:10 ammonium hydroxide or of 1 per cent sodium hydroxide solution. Subject to some further tests for interfering ions, a blue color indicates the presence of beryllium. Copper and nickel interfere, but their color is removed by adding a drop of 1 per cent potassium cyanide solution. A yellow color will remain after addition of cyanide if cobalt is present but is not sufficiently intense to mask the reaction. If ferric salts are present, add a drop of 1 per cent sodium tartrate solution to decolorize. If magnesium has not been removed, it gives a blue color with the reagent. Add a drop or two of saturated bromine water, and such color due to magnesium will be destroyed.

The limit of sensitivity is 0.00014 mg. of beryllium in a solution containing 3 ppm.

BISMUTH. *By Thiourea.* Add a few drops of 1:4 hydrochloric acid to the solution of the sample, and, if an insoluble chloride is precipitated, filter. If chromate or ferric ions are present, add a 10 per cent sodium sulfite solution dropwise to reduce the solution. If antimony is present, remove by hydrogen sulfide precipitation. These precautions provide for the ions that would interfere.

Place a small crystal of thiourea on a strip of filter paper, and add a drop of the slightly acid solution under test. If bismuth is present, the crystal will at once turn yellow, and the near-by paper will be stained yellow to orange.

The limit of sensitivity is 0.001 mg. of bismuth in a solution containing 35 ppm.

By 8-Hydroxyquinoline. Saturate filter paper with a mixture of 5 parts by volume of a 1 per cent solution of 8-hydroxyquinoline in alcohol and one part by volume of a 25 per cent solution of potassium iodide. If copper or ferric ion are present in the sample add a drop of 10 per cent sodium bisulfite to the test paper. Place a drop of the solution under test on the paper. A red-orange color indicates bismuth.

The limit of sensitivity is 0.001 mg. of bismuth in a concentration of 20 ppm.

CADMIUM. *By p-Nitrobenzenediazoaminoazobenzene, Cadion.* If chromium, cobalt, copper, iron, magnesium, or nickel ions are present, add a few drops of a saturated solution of Rochelle salt, sodium potassium tartrate. If silver is present, precipitate by dropwise addition of a 1 per cent potassium iodide solution, and filter. If mercury is present, precipitate with hydrogen sulfide, and filter. Cyanides must be boiled off from acid solution.

Dissolve 0.02 gram of Cadion in 100 ml. of 0.02*N* alcoholic potassium hydroxide solution. Place a drop of this solution upon a strip of filter paper resting upon a thick piece of blotting paper. Add a drop of the solution under test, made slightly acid with acetic acid. Add a drop of 10 per cent potassium hydroxide solution. A bright pink spot surrounded by a violet-blue ring indicates cadmium.

The limit of sensitivity is 0.000025 mg. of cadmium in a concentration of 0.5 ppm.

By Dinitrodiphenylcarbazide. If other elements of Group 2 are present, render the test solution definitely alkaline with 1:1 ammonium hydroxide, and filter. All except copper are so eliminated, and that is provided for by later addition of sodium hydroxide.

Place a drop of the acid, neutral, or ammoniacal solution under test on a spot plate, and add a drop each of 8 per cent sodium hydroxide solution and of 10 per cent potassium cyanide. Next add a drop of a 0.1 per cent alcoholic solution of dinitrodiphenylcarbazide and two drops of a 40 per cent solution of formaldehyde, and stir. A blue-green coloration or precipitate indicates the presence of cadmium. The reagent alone is red in alkaline solution. This is altered to violet by formaldehyde.

The limit of sensitivity is 0.0001 mg. of cadmium in a concentration of 20 ppm.

CALCIUM. *By Ammonium Ferrocyanide.* Magnesium reacts in the same way as calcium. If magnesium is present, render the solution slightly alkaline with 1:1 ammonium hydroxide. Add a few drops of saturated aqueous ammonium carbonate, heat to boiling, filter, and wash. Dissolve the precipitate in 10 per cent acetic acid. Strontium does not interfere, but judgment must be used in interpretation in the presence of barium, since barium also precipitates from concentrated solutions.

To a drop of the solution under test add 3 drops of a saturated aqueous solution of ammonium ferrocyanide and a drop of alcohol. A white crystalline precipitate or cloudiness, best observed against a black background, indicates calcium.

The limit of detection by this technic is 0.025 mg. of calcium in a concentration of 500 ppm. If the test is carried out with somewhat larger quantities in a test tube, small traces of cloudiness are more easily observed than in a shallow drop. The limit of sensitivity may reach 1 ppm.

As the Osazone of Sodium Dihydroxytartrate. To prepare the reagent, dissolve 0.3 gram of phenylhydrazine hydrochloride in 5 ml. of water and add 0.2 gram of sodium dihydroxytartrate in 0.8 ml. of 1:1 hydrochloric acid. Heat on the steam bath for 30 minutes. Filter the precipitated osazone, wash with alcohol and dry. Weigh; then digest in a flask with a little alcohol. Add 17 per cent of the weight of the osazone in terms of sodium hydroxide in the form of 10 per cent alcoholic solution. The sodium salt of the compound soon precipitates. Filter, wash with alcohol, and dry.

All metals interfere except ammonium and alkali ions and magnesium when present in more than ten times the amount of calcium. Transfer a drop of the neutral or slightly alkaline solution under test to a watch glass with a black background. Add a small crystal of the reagent. In the absence of calcium, the crystal dissolves. When calcium is present, first a fine skin, and then a more or less dense precipitate is formed, depending on the concentration of calcium. When ammonia and ammonium salts are present, do not warm, or a condensation product of the reagent will separate out.

The limit of sensitivity is 0.00001 mg. of calcium in a concentration of 0.2 ppm.

CHROMIUM. *As Silver or Lead Chromate.* Chromium is to be tested as the chromate. If suspected to be present in the reduced form as chromic ion, it must first be oxidized. To do this place a drop of the solution under test on a filter paper, add a drop of 20 per cent sodium hydroxide solution, and hold in bromine vapor. Spot the paper with a drop of 5 per cent silver nitrate solution strongly acidified with acetic acid. A red coloration indicates chromium. Spot another part of the paper with a drop of 5 per cent lead acetate solution strongly acidified with acetic acid. A yellow coloration confirms the presence of chromium.

This method will detect 0.006 mg. of chromium in a concentration of 125 ppm.

By Diphenylcarbazide. As in the preceding method, the chromium must be converted to chromate. Render the solution under test definitely but not strongly acid to litmus with 1:1 sulfuric acid. To a drop of the solution add a drop of saturated bromine water and then 2 to 3 drops of 10 per cent potassium hydroxide solution, enough to make it definitely alkaline to litmus paper. Mix well; add a crystal of phenol and a drop of a 1 per cent alcoholic solution of diphenylcarbazide. Add 10 per cent sulfuric acid drop by drop until the red color of the alkaline diphenylcarbazide disappears. A residual blue-violet color indicates chromium. The phenol converts the excess bromine, after oxidation, to inactive tribromophenol.

Cobalt, manganese, copper, and nickel ions give colored precipitates. In case they are present, allow the precipitate to settle, and examine the color of the solution. These ions do not present any insuperable interference unless the amount is over 300 times the concentration of chromium.

The sensitivity in the absence of interfering ions is 0.00025 mg. of chromium in a concentration of 5 ppm.

COBALT. *By Ammonium Thiocyanate.* Copper must be absent. To a drop of the slightly acid solution under test, add 3 drops of acetone or of a mixture of equal parts of amyl alcohol and ethyl ether. Then add a drop of a saturated solution of ammonium thiocyanate in the same solvent. A blue coloration indicates cobalt. Ferric iron gives a red color that interferes. To destroy the red color developed with ferric iron, add a small crystal of sodium fluoride. Copper salts give a red-brown coloration or precipitate that interferes. To test nickel salts for cobalt, the strength of the solution should not exceed 0.2 per cent of nickel.

The limit of sensitivity is 0.0005 mg. of cobalt in a concentration of 10 ppm.

COPPER. *By α-Benzoinoxime.* Ions precipitated by ammonium hydroxide will interfere unless precautions are taken. Add a drop of a 1 per cent solution of Rochelle salt, sodium potassium tartrate, to the test spot before exposure to ammonia fumes to prevent this interference. Large amounts of tartrate reduce the sensitivity of the test. Appreciable amounts of ammonium salts prevent the precipitation of the oxime. Evaporate strongly acid solutions and those containing much ammonium salt to dryness, and heat to volatilize the acid or ammonium salt. Take up the residue in a drop of 1:100 hydrochloric acid.

Impregnate filter paper with a 2 per cent alcoholic solution of α-benzoinoxime. Place a drop of the slightly acid solution under test on a strip of the reagent paper, and expose the paper to ammonia vapors. A green coloration indicates copper.

The limit of sensitivity is 0.0001 mg. of copper in a concentration of 2 ppm.

By Salicylaldoxime. Palladium and gold interfere. Dissolve 1 gram of the reagent in 5 ml. of alcohol, and dilute to 100 ml. with water. To a drop of the neutral solution under test add a drop of 10 per cent acetic acid and a drop of the reagent. Observe using a black background. A dirty green precipitate or turbidity indicates the presence of copper. The presence of the acetic acid prevents interferences other than those mentioned. Most of the metallic ions give precipitates with the reagent in neutral or slightly alkaline solution.

The limit of sensitivity is 0.0005 mg. of copper in a concentration of 10 ppm.

FERRIC IRON. *By Potassium Ferrocyanide.* To a drop of the neutral or slightly acid solution under test, on a filter paper, add a drop of a 10 per cent solution of the reagent. The blue color of ferric ferrocyanide, also known as Prussian blue, indicates the presence of ferric iron.

The limit of sensitivity is 0.0001 mg. of ferric iron in a concentration of 2 ppm.

By Potassium Thiocyanate. Phosphates, arsenates, fluorides, oxalates, tartrates and other organic hydroxyl compounds, and many compounds that give stable complex ferric salts, can so reduce the concentration of ferric ion that the reaction is inhibited. Considerable amounts of mercury salts use up the thiocyanate so that more reagent than usual may be required. Cobalt, chromium, copper, and nickel salts reduce the sensitivity of the test owing to the color of their ions or their reaction with the thiocyanate. Nitrites must be oxidized to nitrates, as by dropwise addition of 0.1 per cent potassium permanganate; otherwise they give a red color with the reagent.

To a drop of the acid solution under test add a drop of a 1 per cent solution of potassium thiocyanate. A deep red color indicates the presence of ferric iron.

The limit of sensitivity is 0.0005 mg. of ferric iron in a concentration of 5 ppm.

FERROUS IRON. *By α,α′-Dipyridyl.* Ferrous iron gives this reaction, or ferric iron may be reduced to ferrous iron for the purpose. To reduce ferric iron, add an equal volume of a saturated solution of hydrogen sulfide to the acid sample, and boil off the excess of reagent. Impregnate filter paper with a 2 per cent alcoholic solution of the reagent. Spot a strip of the reagent paper with a drop of the solution under test, first rendering slightly acid. A red or pink color indicates ferrous iron. Large amounts of halides or sulfates precipitate the red complex.

To detect ferrous iron in the presence of a large excess of ferric iron, transfer a drop of the slightly acid solution under test to a micro crucible coated with paraffin wax, add a few small crystals of sodium fluoride, and then a drop of the α,α′-dipyridyl reagent.

The limit of sensitivity is 0.00003 mg. of ferrous iron in a concentration of 0.6 ppm.

By Potassium Ferricyanide. To a drop of the neutral or slightly acid solution under test, on a filter paper, add a drop of a 10 per cent solution of the reagent. The blue color of ferrous ferricyanide, also known as Turnbull's blue, indicates that ferrous iron is present.

The limit of sensitivity is 0.0001 mg. of ferrous iron in a concentration of 2 ppm.

GOLD. *By Benzidine.* Platinum and ferric salts give similar colors that interfere with detection of that due to gold. Prepare a 0.05 per cent solution of benzidine hydrochloride in 10 per cent acetic acid. To a drop of the solution under test placed on a filter paper add a drop of the reagent solution. A blue color indicates the presence of gold.

The limit of sensitivity is 0.00002 mg. of gold in a concentration of 20 ppm.

By p-Dimethylaminobenzylidenerhodanine. Palladium gives a similar color that interferes. Impregnate filter paper with a 1 per cent solution of the reagent in alcohol. If cuprous, mercurous, or silver ions are present, add a drop of a 1 per cent potassium bromide solution. Now to the test paper add a drop of the neutral or slightly acid solution under test. A red-violet coloration or precipitate indicates gold.

The limit of sensitivity of the method is 0.004 mg. of gold in a concentration of 0.2 ppm.

LEAD. *By Benzidine or Tetramethyldiaminodiphenylmethane.* Chromates, permanganates, and other oxidizing agents interfere but will be absent if the metals of the second group have been isolated by hydrogen sulfide precipitation. The test with benzidine is specific for lead as distinguished from other metals precipitated by hydrogen sulfide, except bismuth. Spot a piece of filter paper with a drop of a 1:3 ammonium hydroxide solution, a drop of 3 per cent hydrogen peroxide, and a drop of the solution under test. Hold over steam for a minute to oxidize the lead to lead dioxide and to decompose the excess of hydrogen peroxide. Add a drop of a 2 per cent solution of benzidine hydrochloride in 1 per cent hydrochloric acid. A blue color indicates lead if bismuth is absent.

If the amount of bismuth present is not more than ten times the amount of lead the second reagent, tetramethyldiaminodiphenylmethane, can be used. For the purpose, dissolve 0.05 gram of the reagent in 10 ml. of glacial acetic acid, and dilute to 100 ml. with water. The method of carrying out the test, the sensitivity, and the color produced are the same as with benzidine.

The limit of sensitivity is 0.0015 mg. of lead in a concentration of 30 ppm.

By Rhodizonic Acid. This reagent is specific in the absence of barium and strontium. Impregnate a filter paper with a 0.1 per cent solution of rhodizonic acid in ethyl alcohol, and dry. Add a drop of the solution under test. If lead is present, a red color appears.

The limit of sensitivity is 0.01 mg. of lead in a concentration of about 200 ppm.

MAGNESIUM. *By Quinalizarin.* Lanthanum and beryllium salts interfere, although the beryllium-quinalizarin compound is differentiated by its stability toward oxidizing agents, as described under beryllium. It is advisable to remove all metals precipitated by ammonia hydroxide and ammonium carbonate. Large amounts of ammonium salt and the presence of phosphate decrease the sensitivity of the test.

To 2 drops of the neutral solution under test add 2 drops of a 0.02 per cent solution of quinalizarin. Then add 3 drops of 8 per cent sodium hydroxide solution. A blue precipitate or coloration, which is easily discernible when compared with the blue-violet of a blank run simultaneously, indicates the presence of magnesium.

The limit of sensitivity is 0.00025 mg. of magnesium in a concentration of 5 ppm.

By Thiazole Yellow. If only cobalt, nickel, and zinc are present, prevent their interference by adding a drop of a 1 per cent potassium cyanide solution to a drop of the solution under test before adding the reagent. Alternatively, precipitate cobalt, nickel, and manganese by adding ammonium sulfide, filter and boil off the excess. If aluminum, tin, and cadmium are present, precipitate them as the hydroxides with ammonium hydroxide. Mere traces of magnesium are apt to be precipitated with substantial amounts of these hydroxides and so not give the test.

To a drop of the solution under test add a drop of a 0.1 per cent solution of the reagent and a drop of $0.1N$ sodium hydroxide solution. An orange or red color appears if magnesium is present.

The limit of sensitivity is 0.0015 mg. of magnesium in a concentration of 30 ppm.

MANGANESE. *By Tetramethyldiaminodiphenylmethane.* In this test manganese is converted to permanganate by periodate, then serves to oxidize the reagent. In a sense the manganese acts catalytically since it will be oxidized to permanganate so long as excess periodate is present. Chromium salts will also be oxidized to chromates by the periodate, and this also oxidizes the reagent; therefore chromium must be absent.

To a drop of the solution under test, add a drop of a saturated solution of potassium periodate, a drop of 10 per cent acetic acid and 2 drops of a 1 per cent solution of the reagent in chloroform. An intense blue color indicates the presence of manganese.

The limit of sensitivity is 0.000001 mg. of manganese in a concentration of 0.02 ppm.

By Catalytic Oxidation with Ammonium Persulfate. Mix a drop of the solution under test with a drop of concentrated sulfuric acid in a micro crucible. Add a drop of a 0.1 per cent solution of silver nitrate and stir. Drop in a few milligrams of ammonium persulfate, and gently heat the mixture. A violet coloration indicates manganese. In this the silver ions serve as catalyst to permit oxidation of manganese to permanganate by persulfate. Since chlorides, bromides, and iodides would precipitate silver, the silver salt must be in excess over such amount of these halides as is present.

The limit of sensitivity is 0.0001 mg. of manganese in a concentration of 2 ppm.

MERCURY. *By Aniline and Stannous Chloride.* Silver ion must not be present, for it will react in much the same way as mercury. Place a drop of the solution under test on a filter paper, add a drop of a freshly prepared 20 per cent solution of stannous chloride and then a drop of aniline. A brown to black color due to the reduced metal indicates mercury.

The limit of sensitivity is 0.001 mg. of mercury in a concentration of 20 ppm.

By p-Dimethylaminobenzylidenerhodanine. If silver is present, precipitate it as the chloride. Then to a drop of the solution under test, add a drop of a saturated solution of the reagent in alcohol or acetone. A change from a colorless or faint yellow to violet-red or a precipitate indicates mercury. If the solution contains excess of chlorides or is more than 0.1N with nitric or sulfuric acid, add a drop of saturated solution of sodium acetate as a buffer. The color developed then is usually pink. If sodium acetate is used, the blank will be yellow to orange.

In the presence of copper, add 5 drops of a 10 per cent trisodium phosphate solution to a drop of the solution under test, which must not be more acid than N. Then add a drop of the reagent solution. A violet color indicates mercury. Green copper phosphate is present in the blank under these conditions.

The limit of sensitivity is 0.00033 mg. of mercury in a concentration of 7 ppm. In the presence of copper the test is only about one-third as sensitive.

MOLYBDENUM. *By Potassium Thiocyanate.* Phosphoric acid, phosphates, and organic acids form complex molybdates that may hinder or prevent this reaction. Mercury salts and nitrites react with the thiocyanate and may make the reaction less sensitive or even prevent it.

Soak a strip of filter paper in 2:3 hydrochloric acid; add a drop of the solution under test and on this a drop of a 5 per cent solution of potassium thiocyanate. A red color appears if iron is present. Add a drop of a 5 per cent solution of stannous chloride or sodium thiosulfate to reduce the iron. A vermilion color remaining indicates the presence of molybdate.

The normal limit of sensitivity is 0.0001 mg. of molybdenum in a concentration of 2 ppm.

By Potassium Ethyl Xanthogenate. To a drop of the neutral or slightly acid solution under test add a grain of potassium ethyl xanthogenate and 2 drops of 1:4 hydrochloric acid. A deep violet to purple color indicates the presence of molybdate in very small amounts. A black, oily drop soluble in benzene, chloroform, and carbon disulfide indicates large amounts of molybdate. This sensitive reaction is specific for molybdenum. Oxalic, tartaric and citric acids, which form complexes with molybdates, decrease the sensitivity when present in large amounts.

The limit of sensitivity is 0.00004 mg. of molybdenum in a concentration of 0.8 ppm.

NICKEL. *By Dimethylglyoxime.* To a drop of the neutral solution under test, add a drop of a 1 per cent alcoholic solution of dimethyl glyoxime followed by a drop of 1:3 ammonium hydroxide. A red spot or ring indicates nickel. When iron is present, to a drop of the solution under test, add a drop of a saturated sodium tartrate solution and finally a drop of the reagent. A red spot or ring indicates nickel.

In the presence of cobalt, copper, and manganese, use an oxidation procedure. To a drop of the acid solution under test add a drop of 3 per cent hydrogen peroxide and a drop of a saturated solution of sodium carbonate. When considerable cobalt is present, a green precipitate forms and slowly darkens. With smaller amounts of cobalt, a green coloration is formed. Add a drop of the reagent. A red spot or ring indicates nickel.

Large amounts of oxidizing agents make the reaction more sensitive. To a drop of the solution under test add 1 to 2 drops of bromine water and make alkaline with 1:3 ammonium hydroxide to the disappearance of the bromine color. The oxidation by the hypobromite formed is complete in 1 to 2 minutes. Then add a drop of the reagent. A red- or orange-colored solution indicates nickel.

The limit of sensitivity is 0.00016 mg. of nickel in a concentration of 3.5 ppm. The sensitivity is reduced to about one-fourth of this when iron, cobalt, copper, or manganese are present.

By Rubianic Acid. Copper or cobalt interfere with this test. The presence of much ammonium salt reduces the sensitivity. To a drop of the solution under test on a filter paper, add a drop of 1 per cent alcoholic solution of rubianic acid. Develop in ammonia fumes. A blue to violet color indicates nickel.

The limit of sensitivity is 0.000012 mg. of nickel in a concentration of 0.8 ppm.

POTASSIUM. *By Sodium Dipicrylamine.* Prepare the reagent solution by dissolving 0.2 gram of dipicrylamine in 2 ml. of hot 5 per cent sodium carbonate solution and diluting to 20 ml. If ammonium salts are present, evaporate the solution under test to dryness, and ignite gently to volatilize ammonium compounds. Take up with water for use. Cesium and rubidium compounds must be absent. Place a drop of the neutral solution under test on a filter paper. At once add a drop of the reagent solution and then a drop of 1:5 hydrochloric acid. An orange-red spot indicates potassium.

The limit of sensitivity is 0.003 mg. of potassium in a concentration of 100 ppm.

By Sodium Cobaltinitrite and Silver Nitrate. If ammonium salts are present remove as in the preceding method. Thallium and lithium compounds must be absent. Place a drop of the neutral or acetic acid solution under test on a black spot plate. Add a drop of 0.05 per cent silver nitrate solution. This addition of silver compounds increases the sensitivity fourfold. Now add a small crystal of sodium cobaltinitrite. A yellow precipitate or cloudiness indicates the presence of potassium.

The limit of sensitivity is 0.001 mg. of potassium in a concentration of 20 ppm.

SELENIUM. *By Thiourea.* Nitrates and large amounts of copper interfere. Tellurium and bismuth give yellow precipitates that are distinguishable only by comparison with a blank. To a few crystals of thiourea on a filter paper add a drop of the solution under test. An orange-red precipitate of selenium indicates the presence of a selenite.

The limit of sensitivity is 0.0001 mg. of selenium in a concentration of 2 ppm.

By Hydriodic Acid. Place a drop of concentrated hydriodic acid, or a small crystal of potassium iodide and a drop of concentrated hydrochloric acid, on a filter paper. Add a drop of the acid solution under test. The brown-black color that develops is completely decolorized by a drop of a 5 per cent solution of sodium thiosulfate if selenium is absent. If a red-brown color remains, a selenite is indicated. Tellurous acid gives a similar color originally but is decolorized with sodium thiosulfate unless present in large quantities.

The limit of sensitivity is 0.001 mg. of selenium in a concentration of 40 ppm.

SILVER. *By Potassium Chromate.* Add a drop of a 10 per cent ammonium carbonate solution to a drop of the solution under test on a watch glass. Remove a drop of the clear solution from any precipitate by means of a filter paper and add a drop of a 1 per cent solution of potassium chromate in *N* acetic acid. A red circle of silver chromate indicates silver.

The limit of sensitivity is 0.002 mg. of silver in a concentration of 40 ppm.

By p-Dimethylaminobenzylidenerhodanine. Impregnate filter paper with a saturated solution of the reagent in acetone. If the solution under test is neutral, mix with an equal volume of 1:10 nitric acid. If already acid, add a tiny crystal of potassium nitrate. Add a drop of this test solution to the test paper. A red-violet stain or precipitate indicates silver. When cuprous or mercury ions are present, add to the colored precipitate a few drops of 1:4 hydrochloric acid or 10 per cent ammonium chloride solution. The organic silver compound remains, being less soluble than silver chloride, but the cuprous or mercury compounds are decomposed to form chlorides.

In the presence of gold, platinum, and palladium, to a drop of the acid solution under test add a drop of a 10 per cent potassium cyanide solution and then a drop of the reagent solution. Finally stir in a drop of 1:4 nitric acid. A red color indicates silver.

The limit of sensitivity is 0.00002 mg. of silver in a concentration of 0.4 ppm. In the presence of cuprous, mercury, gold, platinum, and palladium ions, the sensitivity is considerably reduced.

Sodium. *By Zinc Uranyl Acetate.* The presence of 0.5 per cent of salts of aluminum, ammonium, barium, calcium, cadmium, cobalt, copper, magnesium, mercury, manganese, nickel, potassium, strontium, and zinc interferes. More than 0.1 per cent of lithium salts interferes. To prepare the reagent solution, mix 5 grams of uranyl acetate with 3 ml. of warm 30 per cent acetic acid, and dilute to 25 ml. Mix 15 grams of zinc acetate with 3 ml. of 30 per cent acetic acid, and dilute to 25 ml. Mix these two solutions, and warm. Add approximately 0.05 gram of sodium chloride, allow to stand for 24 hours, and filter. To a drop of the neutral solution under test on a dark spot plate add 8 drops of zinc uranyl acetate reagent, and stir. A yellow cloud or precipitate indicates the presence of sodium.

The limit of sensitivity is 0.012 mg. of sodium in a concentration of 250 ppm.

Strontium. *By Sodium Rhodizonate.* Neutralize the solution under test, using litmus as indicator. Impregnate filter paper with a saturated solution of potassium chromate, and dry. Spot with the solution under test. Barium and strontium in neutral solution give chromates under these conditions. After a minute add a drop of a 1 per cent solution of sodium rhodizonate. A brown-red color or circle indicates the presence of strontium. In this test strontium chromate is sufficiently soluble to react with the sodium rhodizonate while barium chromate is not.

The limit of sensitivity is 0.004 mg. of strontium in a concentration of 80 ppm.

Thallium. *By Potassium Iodide.* To a drop of the weakly acid solution under test placed on a watch glass with a black background add a drop of a 10 per cent solution of potassium iodide. If a precipitate appears, add a drop of a 2 per cent solution of sodium thiosulfate. Warm and stir. Any mercuric iodide dissolves in the excess potassium iodide, while any silver or lead iodide dissolves in the sodium thiosulfate. A yellow precipitate remaining after this treatment indicates thallium.

The limit of sensitivity is 0.0006 mg. of thallium in a concentration of 12 ppm.

By Benzidine. Before this test, oxidizing agents must be reduced. Silver must be precipitated by addition of chlorides. Cerium, cobalt, and manganese must be absent. If thallous ions are suspected, first oxidize by heating with *aqua regia*, evaporate to dryness, and take up in water. Place a drop of the solution under test on a spot plate, and add a drop of concentrated ammonium hydroxide, followed by a drop of a 0.1 per cent solution of benzidine hydrochloride in 10 per cent acetic acid. A blue color or precipitate indicates thallic ions.

The limit of sensitivity is 0.0003 mg. of thallium in a concentration of 10 ppm.

TIN. *By Mercuric Chloride and Aniline.* If tin is present in the tetravalent form, add a small piece of magnesium metal to the acid solution to reduce it. After this reaction, the solution must be distinctly acid or antimony will interfere. Place a drop of the solution to be treated on a piece of test paper made by impregnating filter paper with a 2 per cent solution of mercuric chloride and drying. Follow with a drop of aniline to produce mild alkalinity. Too great alkalinity causes antimonious ions to reduce mercuric chloride. A black or brown precipitate or coloration indicates tin.

The limit of sensitivity is 0.0006 mg. of tin in a concentration of 12 ppm.

By Cacotheline. Antimonious ions, ferrous ions in the presence of fluorides and phosphates, sulfites, thiosulfates, and colored ions interfere with this test. Reduce the tin, if necessary, as in the previous method. Impregnate a filter paper with a saturated solution of cacotheline; and, before it is quite dry, spot it with a drop of the solution under test, made slightly acid with 1:10 hydrochloric acid. The formation of a red ring surrounded by a colorless zone on the yellow paper indicates tin.

The limit of sensitivity is 0.0002 mg. of tin in a concentration of 4 ppm.

TITANIUM. *By Hydrogen Peroxide.* Titanium in a solution acidified with hydrochloric or sulfuric acid gives a characteristic reaction, subject to the absence of interfering ions. Fluorides prevent the reaction. Chromates, vanadates, molybdates, and cerium salts give colored reaction products with hydrogen peroxide. The color due to iron can be suppressed by the addition of phosphoric acid. Large amounts of acetates, nitrates, chlorides, bromides and colored ions reduce the sensitivity. To a drop of the acid solution under test add a drop of 3 per cent hydrogen peroxide. A yellow color indicates titanium.

The limit of sensitivity is 0.002 mg. of titanium in a concentration of 40 ppm.

By Chromotropic Acid, 1.8-*Dihydroxynaphthalene*-3.6-*Disulfonic Acid.* No substantial amount of nitric acid must be present in the test solution. To remove this interference, add one-third the volume of 10 per cent sulfuric acid, and evaporate to sulfur trioxide fumes. Let cool and take up with water. Hydrochloric acid need not be replaced by sulfuric. To a drop of the acid solution under test add a drop of a 5 per cent solution of chromotropic acid. A clear red-brown precipitate or coloration indicates the presence of titanium. This is modified by interfering ions but can usually be distinguished, even in their presence. Uranyl salts give a brown color that is destroyed by a drop of stannous chloride in hydrochloric acid. Ferric ions give a deep green, mercury a yellow, silver salts black.

The limit of sensitivity is 0.005 mg. of titanium in a concentration of 100 ppm.

TUNGSTEN. *By Potassium Thiocyanate.* Spot a filter paper with a drop of 20 per cent hydrochloric acid. Add a drop of the solution under test and then a drop of 10 per cent potassium thiocyanate solution followed by a drop of 10 per cent stannous chloride solution. A blue color indicates tungsten. If molybdenum is present, a red color is produced that is destroyed by the addition of a drop of concentrated hydrochloric acid.

The limit of sensitivity is 0.004 mg. of tungsten in a concentration of 80 ppm.

URANIUM. *By Potassium Ferrocyanide.* A neutral or acetic acid solution is suitable for testing directly. If acid with mineral acid, add a crystal of sodium acetate to serve as buffer. Copper or iron interferes, but this can be eliminated. In their presence, add a drop of the test solution to a drop of a 10 per cent solution of potassium iodide on a filter paper. Then add a drop of a 10 per cent sodium thiosulfate solution to decolorize the liberated iodine. In the absence of copper or iron, simply add a drop of the solution under test to a filter paper. Now add a drop of 10 per cent potassium ferrocyanide solution. A brown stain indicates uranium.

The limit of sensitivity is 0.001 mg. of uranium in a concentration of 20 ppm.

VANADIUM. *By Hydrogen Peroxide.* If ferric ions are present, add a drop of 10 per cent phosphoric acid to a drop of the solution under test. If titanium is present, add a tiny crystal of sodium fluoride. Cerium ions, molybdates, chromates, iodides, and bromides, as well as large amounts of colored metallic ions, reduce the sensitivity of the test. Now to the drop of test solution, with or without additions, add a drop of 1:4 sulfuric acid, mix, and allow to stand a few minutes. Then add a drop of 1 per cent hydrogen peroxide solution. A further drop may be necessary, but excess reagent decreases the sensitivity. A red-brown to blood-red, or, in very dilute solutions, a pale brown to pink color indicates vanadium.

The limit of sensitivity is 0.0025 mg. of vanadium in a concentration of 50 ppm.

By 8-Hydroxyquinoline. If titanium or iron is present, neutralize 2 ml. of the solution under test, add 0.5 ml. excess of 33 per cent sodium hydroxide solution, and filter. If molybdenum or tungsten is present, render distinctly acid with 30 per cent acetic acid, add a drop of 10 per cent barium acetate solution, and filter. Now to a drop of the filtrate in a micro test tube add 1 to 2 drops of 10 per cent acetic acid and a drop of the reagent prepared by dissolving 2.5 grams of 8-hydroxyquinoline in 100 ml. of 5 per cent acetic acid. Add 2 drops of chloroform, stopper, and shake the test tube. A red to brown-violet color in the chloroform layer indicates vanadium. A blank test is especially important in testing for very small amounts of vanadium.

The limit of sensitivity is 0.0001 mg. of vanadium in a concentration of 2 ppm.

ZINC. *By Cobalt-mercury Thiocyanate.* Prepare the mercury thiocyanate reagent by dissolving 9 grams of mercuric chloride and 9 grams of ammonium thiocyanate in 100 ml. of water. Allow to stand for 4 days and filter. To a drop of the solution under test add a drop of a 0.02 per cent solution of cobalt chloride in 0.5*N* hydrochloric acid and a drop of the mercury thiocyanate solution. Rub the spot plate with a glass rod for 15 seconds. According to the amount of zinc present, either at once, or at the longest after 2 minutes, a blue precipitate is formed. In the absence of zinc, the precipitation first begins after 2 to 3 minutes. Minute quantities of zinc catalyze the formation of the cobalt compound, which, by itself, is always slow in forming. If a red color due to ferric thiocyanate interferes, remove this by adding a few milligrams of sodium fluoride 1 to 2 minutes after the addition of the reagent.

The limit of sensitivity is 0.0005 mg. of zinc in a concentration of 10 ppm.

By Diethylaniline and Ferricyanide. Traces of iron do not interfere with this test. In the procedure, ferricyanide is reduced to ferrocyanide, which forms precipitates with cobalt, nickel, manganese, and copper. Those elements must therefore be absent.

For the test, add to 15 drops of a 2 per cent potassium ferricyanide solution, 10 drops of a 0.25 per cent solution of diethylaniline, monoethylaniline free, in 50 per cent sulfuric acid. Add a drop of the solution under test. If the yellow color of the solution turns brown to reddish brown, zinc is indicated.

The limit of sensitivity is 0.002 mg. of zinc in a concentration of 20 ppm.

Zirconium. *By Alizarin.* If the solution under test contains sulfates, add 10 per cent barium chloride solution dropwise until precipitation is complete. Digest and filter. Fluorides, molybdates, organic hydroxyacids, phosphates and tungstates would interfere and must therefore be absent. The final solution must be substantially neutral. Prepare the reagent by dissolving 5 grams of alizarin in 100 ml. of alcohol, adding 1:3 hydrochloric acid, dropwise, until a yellow color is formed. Then add an equal volume of alcohol, and filter.

To a drop of the solution under test in a micro crucible, add a drop of the reagent solution. Boil the contents of the crucible. A red to violet coloration indicates zirconium, beryllium, aluminum, titanium, or thorium. Now add a drop of *N* hydrochloric acid. Only the zirconium compound is unaffected. Much zirconium gives a red-violet color and precipitate; a little gives a trace of red-brown coloration.

The limit of sensitivity is 0.0005 mg. of zirconium in a concentration of 10 ppm. When aluminum or thorium is present, 0.001 mg. of zirconium may be detected in the presence of 0.5 mg. of these metals.

By β-Nitroso-α-naphthol. Fluorides must be absent. Remove sulfates as in the preceding method. Impregnate filter paper with a 2 per cent alcoholic solution of β-nitroso-α-naphthol, and dry. Add a drop of the solution under test. A more or less deep, red coloration on the yellow-brown paper indicates zirconium. The addition of a drop of dilute acid makes the color more vivid.

The limit of sensitivity is 0.0002 mg. of zirconium in a concentration of 4 ppm.

Anions. Since most anions are fairly readily identified by macroscopic methods, usually little advantage is gained by using micro or spot test procedures. Therefore only a few tests for anions are given.

Borate. *By Methyl Alcohol.* Add a drop of sulfuric acid to a drop of the solution under test in a micro crucible. Add 5 drops of methyl alcohol, mix, and ignite. A green flame indicates a borate.

The limit of sensitivity is 0.00001 mg. of boron in a concentration of 0.2 ppm.

By Curcumin. Curcumin, present in turmeric, also known as curcuma, gives a sensitive test for borates. If the sample contains organic matter, destroy it by boiling in acid solution with a few drops of 3 per cent hydrogen peroxide until the peroxide is completely destroyed. If necessary, evaporate the sample solution to dryness, and ignite gently.

Prepare turmeric test paper by soaking filter paper in a solution made by boiling 20 grams of curcuma in 50 ml. of ethyl alcohol, filtering, and diluting the filtrate with 50 ml. of distilled water. Dry the impregnated paper and spot with a drop of the solution under test which has been made acid by adding one-fifth of its volume of 1:3 hydrochloric acid solution. When boric acid or borates are present, the spot turns red on drying. A red color is also given by chlorates, chromates, hydrogen peroxide, nitrates, copper, ferric iron, molybdenum, or nickel. To eliminate this interference, spot the red area with a drop of a 1 per cent solution of sodium hydroxide. Any red color due to borate or boric acid will then turn to a light blue or green.

The limit of sensitivity is 0.00002 mg. of boron in a concentration of 0.4 ppm.

BROMATE. *By Manganous Sulfate, Sulfuric Acid, and Benzidine.* To a drop of the solution under test in a micro test tube, add a drop of a 2 per cent manganous sulfate solution, slightly acid with sulfuric acid. Warm for 2 to 3 minutes on a water bath, and cool. Add a few drops of a solution prepared by dissolving 0.05 gram of benzidine hydrochloride in 10 ml. of glacial acetic acid, diluting to 100 ml. with water, and filtering. Now add a few small crystals of sodium acetate. A more or less deep blue color indicates the presence of bromate. Chlorates and iodates do not interfere under these conditions.

The limit of sensitivity is 0.04 mg. of potassium bromate in a concentration of 400 ppm.

CHLORATE. *By Manganous Sulfate, Phosphoric Acid, and Diphenylcarbazide.* If persulfates are present in the solution under test, add a drop each of 1 per cent silver nitrate solution and 1:10 sulfuric acid to a small volume. Evaporate to sulfur trioxide fumes, cool, and dilute to the original volume. Periodates must be absent. To a drop of the solution in a small porcelain crucible add a drop of a solution prepared by mixing equal volumes of saturated manganous sulfate solution and sirupy phosphoric acid. Warm rapidly and allow to cool. A more or less deep violet coloration indicates chlorate. Very pale colors may be intensified by adding a drop of a 1 per cent alcoholic solution of diphenylcarbazide.

The limit of sensitivity is 0.00008 mg. of potassium chlorate in a concentration of 1 ppm.

FERRICYANIDE. *By Benzidine.* Oxidizing agents such as chromates and molybdates must be absent. To a drop of the solution under test add a drop of a 1 per cent lead nitrate solution and a drop of a saturated solution of benzidine hydrochloride in 10 per cent acetic acid. A blue color or precipitate indicates the presence of ferricyanide. When ferrocyanide is present, a white benzidine precipitate is formed, and more of the benzidine solution is required.

The limit of sensitivity is 0.001 mg. of potassium ferricyanide in a concentration of 20 ppm.

FERROCYANIDE. *By Uranyl Acetate.* Moisten a filter paper with 20 per cent uranyl acetate solution. Now apply a drop of the solution under test. A brown precipitate or coloration formed within 2 minutes indicates the presence of ferrocyanide. After that time ferricyanide commences to react with the reagent.

The limit of sensitivity is 0.05 mg. of potassium ferrocyanide in a concentration of 20 ppm.

FLUORIDE. *By Zirconium-alizarin.* If the solution under test contains oxalates, evaporate to dryness and ignite gently to decompose them. Then redissolve the residue in water, adding a drop of 50 per cent acetic acid if necessary. If sulfates are present add 2 per cent benzidine hydrochloride solution dropwise until precipitation is complete. If the solution is acid with mineral acid, add a crystal of sodium acetate. Arsenates, phosphates, and thiosulfates must be absent. Mix 3 volumes of a 1 per cent alcoholic alizarin solution with 2 volumes of a 0.4 per cent aqueous solution of zirconium chloride. Impregnate filter paper with this solution, and dry.

Place a drop of 50 per cent acetic acid solution on a piece of the test paper and add a drop of the solution under test, which must be neutral or acid with acetic acid. A yellow color on the red paper indicates the presence of fluoride.

The limit of sensitivity is 0.001 mg. of fluorine in a concentration of 20 ppm.

IODATE. *By Hypophosphorous Acid.* To a drop of the neutral solution under test add a drop of starch solution and a drop of an approximately 1 per cent solution of hypophosphorous acid. A transitory blue color indicates the presence of iodate. Iodine is first formed rapidly, and this slowly reacts to form hydriodic acid. Bromates and chlorates do not react under these conditions.

The limit of sensitivity is 0.001 mg. of iodic acid in a concentration of 20 ppm.

NITRATE. *By Brucine.* If nitrites are present, they must be removed. To do this, acidify the solution under test, add a crystal of sodium azide to decompose the nitrite, and allow to stand for 10 minutes. Heat to boiling and boil for 5 minutes. Cool and use as the sample. Chlorates must be absent. To a drop of concentrated sulfuric acid add 3 mg. of brucine and stir to dissolve. Add a drop of the solution under test. A bright red color turning yellowish red on standing indicates the presence of nitrate.

The limit of sensitivity is 0.00006 mg. of nitric acid in a concentration of about 1 ppm.

NITRITE OR NITRATE. *By Sulfanilic Acid and α-Naphthylamine.* This reaction is primarily for detection of nitrites. In the absence of nitrites, nitrates may be detected by reduction to nitrite. For this, acidify a neutral sample solution with acetic acid, or add a crystal of sodium acetate to one containing mineral acid. Then add zinc dust and heat gently. In any case the final solution should be neutral or acid with acetic acid.

As reagents first dissolve 1 gram of sulfanilic acid in 100 ml. of 30 per cent acetic acid with the aid of heat. Separately dissolve 1 gram of α-naphthylamine in 230 ml. of boiling water and cool. Decant the colorless supernatant liquid from the latter and mix with 100 ml. of glacial acetic acid. To a drop of the solution under test add a drop of the sulfanilic acid reagent and a drop of the α-naphthylamine reagent. The gradual development of a red color indicates the presence of a nitrite.

The limit of sensitivity is 0.00001 mg. of nitrous acid in a concentration of 0.2 ppm.

PERIODATE. *By Manganous Sulfate and Phosphoric Acid.* In the absence of chlorates, test for periodates by the reaction with this reagent already given for chlorates. The reaction and sensitivity are the same.

PERSULFATE. *By Benzidine.* Chromates, permanganates, ferricyanides, and hypohalogenites must be absent. Chlorates, perchlorates, bromates, iodates, and nitrates do not react but reduce the sensitivity of the reaction. Periodates give a brown coloration that interferes. If the solution is acid with mineral acid, dissolve a crystal or two of sodium acetate in it. A neutral solution should not be acidified. To a drop of the solution under test add a drop of a 2 per cent solution of benzidine hydrochloride in 4 per cent acetic acid. A blue color indicates the presence of persulfate.

The limit of sensitivity in neutral solution is 0.00025 mg. of potassium persulfate in a concentration of 5 ppm. In acid solution it is only about one-fourth as sensitive.

PHOSPHATE. *By Ammonium Molybdate and Benzidine.* If hydrogen peroxide is present, heat the solution under test to boiling until it is volatilized or decomposed. When silica is present, use a special procedure.

In the absence of silica, spot filter paper with a drop of ammonium molybdate solution made by mixing 100 ml. of a 5 per cent solution of ammonium molybdate and 35 ml. of 30 per cent nitric acid. Dry the filter paper, and add a drop of the solution under test, then a drop of a 0.05 per cent solution of benzidine hydrochloride in 10 per cent acetic acid. Finally add a drop of a saturated sodium acetate solution. A blue spot or ring indicates the presence of phosphate.

If soluble silica is present, spot the filter paper with the solution under test. To each drop add a drop of the ammonium molybdate solution previously described, in every 100 ml. of which 15 grams of tartaric acid have been dissolved. Place the paper in warm air for a few minutes to accelerate the reaction. Then add a drop of the benzidine hydrochloride solution, and hold the paper over ammonia fumes. A blue color indicates the presence of phosphate.

In the absence of silica, the limit of sensitivity is 0.00005 mg. of phosphorus pentoxide in a concentration of 1 ppm.; in the presence of silica, 0.0015 mg. of phosphorus pentoxide in a concentration of 30 ppm.

SILICATE. *By Ammonium Molybdate and Benzidine.* If necessary, partly neutralize the solution under test so that it is no more than 0.5*N* acid. If phosphate is present, a modified procedure, the second of those which follow, is required. In the absence of phosphate, transfer a drop to a filter paper, and add a drop of the ammonium molybdate solution used in the phosphate test. Place this in warm air for a few minutes. Add a drop of the benzidine solution used in the phosphate test, and hold the paper over ammonia fumes. A blue color indicates the presence of silicate. If the reaction is carried out in a porcelain crucible, the sensitivity of the test is appreciably increased. A blank must be run on the crucible to make sure that no silica is dissolved from it.

If the solution under test contains no more than 0.2 per cent of phosphate, to a drop in a micro test tube add 2 drops of the ammonium molybdate solution described for the phosphate test and centrifuge. Using a capillary tube, decant the supernatant liquid to a micro crucible. Gently warm, then cool, and add 2 drops of a 1 per cent oxalic acid solution to destroy the last traces of ammonium phosphomolybdate. Add a drop of the benzidine solution described for the phosphate test and 2 to 3 drops of saturated sodium acetate solution. A blue color indicates the presence of silicate.

When phosphate is absent, the sensitivity is about 0.001 mg. of silica in a concentration of 20 ppm.; when phosphate is present, it is about 0.006 mg. in 125 ppm.

SULFATE. *By Barium Rhodizonate.* Place a drop of a 5 per cent barium chloride solution on a filter paper and add a drop of fresh 5 per cent sodium rhodizonate solution. A red color appears. Now add a drop of the solution under test which may be either acid or alkaline. If the red color disappears completely or even in part, sulfate is present.

The limit of sensitivity is 0.001 mg. of sulfuric acid in a concentration of 5 ppm.

SULFIDE. *By Sodium Nitroprusside.* Render the solution under test definitely alkaline with 1:1 ammonium hydroxide. Then to a drop of it add a drop of a 1 per cent solution of sodium nitroprusside. A more or less intense violet color indicates the presence of sulfide. The test may also be performed with filter paper impregnated with a 1 per cent solution of sodium nitroprusside in a 1:5 solution of ammonium hydroxide.

The limit of sensitivity is 0.001 mg. of sodium sulfide in a concentration of 20 ppm.

SULFITE. *By Sodium Nitroprusside.* Carefully neutralize the solution under test. Mix a drop of a 10 per cent potassium ferrocyanide solution, a drop of a cold saturated solution of zinc sulfate or nitrate, and a drop of a 1 per cent sodium nitroprusside solution. A white precipitate of zinc ferrocyanide is formed. Now add a drop of the sample solution. If the precipitate develops a red color, sulfite is present.

The limit of sensitivity is 0.003 mg. of sodium sulfite in a concentration of 60 ppm.

CHAPTER 6

HYDROGEN-ION DETERMINATION AND ELECTROMETRIC TITRATION

The titration of an acid is a measure of the total acidity present but does not express the activity of the acid. The concentration of hydrogen ion, C_H, in a solution is a function of the concentration of acid and degree of ionization. These definitions apply equally readily to bases, and, by suitable consideration of the equilibria involved, to hydrolysis accompanied by ionization of a product of hydrolysis to give hydrogen ion or hydroxyl ion.

The values of C_H may vary so greatly that charting on a linear scale is not practical. The values of C_H are, therefore, commonly expressed as the pH value, which is the logarithmic value, $\log_{10} 1/C_H$.

The dissociation of pure water is 10^{-7} mole per liter. The concentration of hydrogen ion is, therefore, 10^{-7}, and the concentration of hydroxyl ion is the same. If an acid is added, the amount of hydrogen ion is increased and the amount of hydroxyl ion correspondingly reduced. According to the law of mass action

$$\frac{C_H \times C_{OH}}{C_{H_2O}} = \frac{10^{-7} \times 10^{-7}}{1} = 10^{-14}$$

If the amount of acid were such as to increase the concentration of hydrogen ion to 0.01 mole per liter, which obtains in about $0.01M$ hydrochloric acid, the hydroxyl ion would be correspondingly reduced. The very simple expression $C_H \times C_{OH} = 10^{-14}$ defines the concentration of either when the other is known.

The pH scale is by definition the negative exponent of 10, which will give the concentration of hydrogen ion. The concentration of hydroxyl ion is simultaneously defined. It follows that for neutrality, when $C_H = C_{OH}$, the hydrogen-ion concentration of a solution is expressed as pH 7. Since $C_H \times C_{OH}$ always equals 10^{-14}, when C_H is high, its inverse expression is low. Hence, the lower the pH is, the higher the acidity present. Conversely the higher the pH is, the higher the alkalinity present.

Since the pH is a logarithmic function, it follows that a solution with a pH one unit above another contains one-tenth the concentration of hydrogen ion and ten times the concentration of hydroxyl ion.

Table 3 gives the approximate pH of 0.1*N* solutions of some common acids, bases, and salts as illustrations of the variation of pH among solutions of like concentration. The subject is of great importance because of the effect of pH on many industrial processes and because the experienced analyst can often use the pH of a solution as a guide. To illustrate, if a solution has a pH of 8.0, the analyst can rule out determinable amounts of many salts that react alkaline by hydrolysis.

TABLE 3.—pH VALUES FOR 0.1*N* SOLUTIONS, AT 20°, ROUNDED TO THE FIRST DECIMAL

	pH		pH
Hydrochloric acid	1.0	Sodium bicarbonate	8.4
Sulfuric acid	1.2	Borax	9.2
Phosphoric acid	1.5	Ammonia	11.3
Acetic acid	2.9	Sodium carbonate	11.6
Alum	3.2	Trisodium phosphate	12.0
Carbonic acid	3.8	Sodium metasilicate	12.2
Boric acid	5.2	Sodium hydroxide	13.0

METHODS OF MEASURING pH

The several methods of measuring pH include the hydrogen electrode and antimony electrode, colorimetry, the quinhydrone electrode, and the glass electrode.

The hydrogen electrode consists of platinum coated with platinum black in contact with a stream of hydrogen. It is easily poisoned and requires great skill in its manipulation. The latter comment applies to a lesser degree to the antimony electrode. Neither will be described, because they are rarely used.

Colorimetric methods are not applicable to very turbid and colored solutions, although rough pH values may be obtained in those with indicator papers. Colorimetry is the simplest and most widely used method of determining pH and will be described in some detail because little or no special equipment is necessary.

The quinhydrone electrode cannot be used for pH values over 9 or for oxidizing and reducing solutions. It has no real advantage over the glass electrode and will be touched on only briefly.

Although colorimetry is most convenient for occasional determinations or for rough tests, the glass electrode is the most useful and accurate for frequent tests. It has a working range of pH 1 to

9.5 with the usual electrode and to pH 11 and even higher with special electrodes. It is not easily poisoned or affected by oxidizing and reducing conditions, but corrections must be applied for alkali-ion effects at the higher pH values.

Colorimetry. The colorimetric determination of pH of a solution varies in sensitivity and accuracy from mere approximation with an indicator or indicator paper, through high accuracy by comparison with a series of buffers containing the indicator, to methods in slightly buffered solutions where the buffer effect of the indicator is taken into account. This last type will not be described. As the hydrogen-ion concentration of a solution containing a colorimetric pH indicator is changed, the hue as well as the intensity of color is usually changed. There are exceptions, the so-called monochromatic pH indicators with which only the intensity of color is altered. In all cases the method is essentially comparison of the color developed in the sample with that developed in known standards.

UNIVERSAL INDICATORS. For rough estimation, a mixed indicator, called a universal indicator, has a place. One such indicator has the composition, 70 mg. tropaeolin 00, 100 mg. methyl orange, 80 mg. methyl red, 400 mg. bromothymol blue, 500 mg. phenolphthalein, and 100 mg. alizarin yellow R, in 100 ml. of 50 per cent alcohol. One drop is used in 10 ml. of solution to give the colors shown in the table below. The commercial availability of other universal indicators with accompanying color scales makes it rarely necessary for the laboratory to prepare its own mixed indicator.

If accuracy to about one unit is not sufficient, it serves to spot the area in which to apply more sensitive methods.

pH	Color
2.0	Orange red
3.0	Red orange
4.0	Orange
5.0	Yellow orange
6.0	Orange yellow
6.5	Yellow
7.0	Green yellow
8.0	Green
9.0	Green blue
10.0	Violet blue to blue violet
11.0	Violet
12.0	Violet to violet red
> 12.0	Violet red

INDICATOR PAPERS. The well-known blue and red litmus strips are an elementary illustration of this type of method for measuring hydrogen-ion concentrations. More versatile papers are available, of which the Hydrion shown in Fig. 70 is a good example. Paper is impregnated with a universal indicator. Colors printed on the dispenser correspond to different pH values. Such papers have the same general use and accuracy as a liquid universal indicator. For use with turbid or colored solutions, always observe the color of the paper where capillary action has drawn up the solution, thus avoiding any discoloration on the paper from the solution.

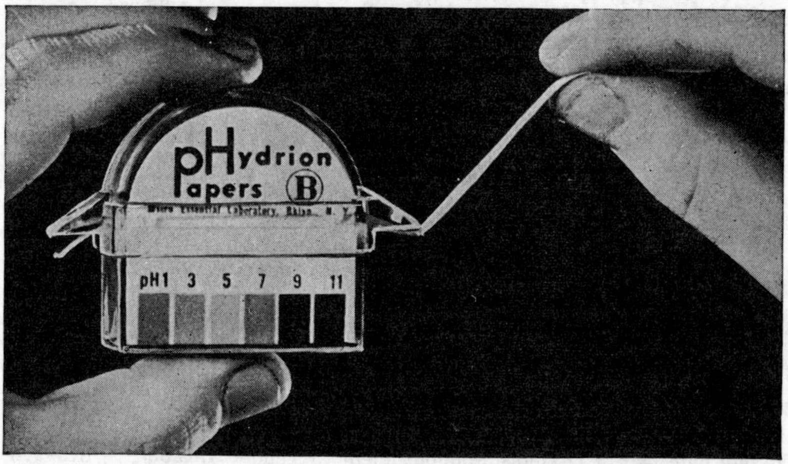

FIG. 70.—Handy unit for determination of pH by universal indicator papers.

COMPARISON WITH INDICATORS IN BUFFERS. When a colorless sample solution is used, the color developed with a properly selected indicator can be compared with the same indicator in a series of known buffer solutions. This is usually accurate to 0.1 unit of pH when conducted by an analyst with some experience.

INDICATOR SOLUTIONS. There are four important classes of indicators: phthaleins, sulfonephthaleins, nitrophenols, and azo compounds. Some of the more important are shown in Table 4. Table 5 gives a selection of suitable indicators together with the usual strength and the amount of alkali required to dissolve them. To prepare, grind the indicator in an agate mortar with the sodium hydroxide solution, wash into a calibrated flask, and dilute to the mark. These are available commercially in solutions of the specified concentrations.

TABLE 4.—COLOR CHANGES AND pH RANGE OF COMMON INDICATORS

Trade name	Chemical name	Color		pH range
		Acid	Basic	
Methyl violet	Benzylpentamethyl-*p*-rosaniline and analogues	Yellow	Violet	0–2.0
Thymol blue	Thymolsulfonephthalein	Red	Yellow	1.2–2.8
Methyl yellow	*p*-Dimethylaminoazo-benzene	Red	Yellow	2.9–4.0
Methyl orange	Sodium dimethylamino-azobenzene-sulfonate	Red	Orange-yellow	3.1–4.4
Bromophenol blue	Tetrabromophenol-sulfonephthalein	Yellow	Blue-violet	3.0–4.6
Bromocresol green	Tetrabromo-*m*-cresol-sulfonephthalein	Yellow	Blue	3.8–5.4
Methyl red	Dimethylaminobenzene-*o*-carboxylic acid	Red	Yellow	4.4–6.2
Litmus	Azolitmin	Red	Blue	4.5–8.3
Chlorophenol red	Dichlorophenol-sulfonephthalein	Yellow	Red	4.8–6.4
Bromophenol red	Dibromophenol-sulfonephthalein	Yellow	Red	5.4–7.0
Bromothymol blue	Dibromothymol-sulfonephthalein	Yellow	Blue	6.0–7.6
Phenol red	Phenolsulfonephthalein	Yellow	Red	6.4–8.0
Cresol red	*o*-Cresolsulfonephthalein	Yellow	Red	7.2–8.8
Metacresol purple	*m*-Cresolsulfonephthalein	Yellow	Purple	7.4–9.0
Thymol blue	Thymolsulfonephthalein	Yellow	Blue	8.0–9.6
Phenolphthalein	Phenolphthalein	Colorless	Red	8.0–9.8
Thymolphthalein	Thymolphthalein	Yellow	Blue	9.3–10.5
Alizarin yellow	Sodium *p*-nitroaniline-azosalicylate	Yellow	Red	10.1–12.0

TABLE 5.—STANDARD INDICATOR SOLUTIONS

Indicator	pH range	Weight, grams	Ml. 0.01*N* sodium hydroxide	Dilute to
Bromophenol blue	3.10–5.00	0.02	3.28	250 ml.
Methyl red	4.05–5.95	0.02	....	250 ml.
Bromocresol purple	5.30–7.20	0.03	8.34	250 ml.
Bromothymol blue	6.15–8.05	0.02	3.54	250 ml.
Phenol red	6.75–8.65	0.01	3.10	250 ml.
Cresol red	7.15–9.05	0.02	5.76	250 ml.
Thymol blue	7.85–9.75	0.02	4.76	250 ml.

Buffer Solutions. As standards it is necessary to have aqueous mixtures of salts which on dilution with water will, to all practical purposes, not change. There are numerous series of such standard buffers. The quoted pH of the buffer solution depends primarily on determination by electrometric methods.

Such buffer solutions can be purchased in sealed bottles. They are also available in the form of tablets of mixed salts ready to dissolve in a specified volume of water. Although these are convenient, it is sometimes necessary for the laboratory worker to prepare his own. Therefore, there follows a series of solutions for preparation of mixed buffers, and Tables 6 and 7 show the mixtures of solutions of C.P. chemicals necessary for various pH values. As these tables overlap, buffers from one set can be compared with another to ensure accuracy.

0.2*M* Potassium Chloride. Dissolve 14.912 grams of the dry salt in warm water and when cool dilute to exactly 1 liter.

0.2*M* Hydrochloric Acid. Dilute the concentrated acid to approximately 0.2*M*, and standardize against 0.2*M* sodium hydroxide solution.

0.2*M* Acid Potassium Phthalate. Dissolve 40.836 grams in water and dilute to 1 liter. If the purity of the salt is subject to any question, titrate a portion of this solution to check its concentration.

0.2*M* Sodium Hydroxide. Dissolve 100 grams of sodium hydroxide in 100 ml. of water in a Pyrex flask. Stopper the flask, and allow to stand at least overnight for any carbonate to settle out. It is essential that substantial amounts of carbonate be absent. Decant carefully, or filter through a hardened filter paper, and dilute to approximately 0.2*M*. Standardize against standard acid or against several portions of acid potassium phthalate. Store in a paraffin-lined bottle.

0.2*M* Monopotassium Phosphate. Dissolve 27.312 grams of the dry salt in water and dilute to 1 liter. Check the solution to ensure that it gives a distinct red with methyl red and a distinct blue with bromophenol blue.

0.2*M* Boric Acid. Dissolve 12.369 grams of dry boric acid and 14.912 grams of dry potassium chloride in water, and dilute to 1 liter.

0.05*M* Sodium Carbonate. Dry the sodium carbonate to constant weight at 160°. Dissolve 5.3 grams in water, and dilute to 1 liter.

TABLE 6.—SERIES OF BUFFER MIXTURES

pH	Ml. of 0.2*M* potassium chloride	Ml. of 0.2*M* hydrochloric acid	Dilute to
1.0	0.00	59.50	100 ml.
1.1	2.72	47.28	100 ml.
1.2	12.45	37.55	100 ml.
1.3	20.16	29.84	100 ml.
1.4	26.30	23.70	100 ml.
1.5	31.18	18.82	100 ml.
1.6	35.03	14.95	100 ml.
1.7	38.12	11.88	100 ml.
1.8	40.57	9.43	100 ml.
1.9	42.51	7.49	100 ml.
2.0	44.05	5.95	100 ml.
2.1	45.27	4.73	100 ml.
2.2	46.24	3.76	100 ml.

pH	Ml. of 0.2*M* potassium acid phthalate	Ml. of 0.2*M* hydrochloric acid	Dilute to
2.2	50	46.60	200 ml.
2.4	50	39.60	200 ml.
2.6	50	33.00	200 ml.
2.8	50	26.50	200 ml.
3.0	50	20.40	200 ml.
3.2	50	14.80	200 ml.
3.4	50	9.95	200 ml.
3.6	50	6.00	200 ml.
3.8	50	2.65	200 ml.

pH	Ml. of 0.2*M* potassium acid phthalate	Ml. of 0.2*M* sodium hydroxide	Dilute to
4.0	50	0.40	200 ml.
4.2	50	3.65	200 ml.
4.4	50	7.35	200 ml.
4.6	50	12.00	200 ml.
4.8	50	17.50	200 ml.
5.0	50	23.65	200 ml.
5.2	50	29.75	200 ml.
5.4	50	35.25	200 ml.
5.6	50	39.70	200 ml.
5.8	50	43.10	200 ml.
6.0	50	45.40	200 ml.
6.2	50	47.00	200 ml.

TABLE 6.—SERIES OF BUFFER MIXTURES.—(*Continued*)

pH	Ml. of 0.2*M* mono-potassium phosphate	Ml. of 0.2*M* sodium hydroxide	Dilute to
5.8	50	3.66	200 ml.
6.0	50	5.64	200 ml.
6.2	50	8.55	200 ml.
6.4	50	12.60	200 ml.
6.6	50	17.74	200 ml.
6.8	50	23.60	200 ml.
7.0	50	29.54	200 ml.
7.2	50	34.90	200 ml.
7.4	50	39.34	200 ml.
7.6	50	42.74	200 ml.
7.8	50	45.17	200 ml.
8.0	50	46.85	200 ml.

pH	Ml. of 0.2*M* boric acid, 0.2*M* potassium chloride	Ml. of 0.2*M* sodium hydroxide	Dilute to
7.8	50	2.65	200 ml.
8.0	50	4.00	200 ml.
8.2	50	5.90	200 ml.
8.4	50	8.55	200 ml.
8.6	50	12.00	200 ml.
8.8	50	16.40	200 ml.
9.0	50	21.40	200 ml.
9.2	50	26.70	200 ml.
9.4	50	32.00	200 ml.
9.6	50	36.85	200 ml.
9.8	50	40.80	200 ml.
10.0	50	43.90	200 ml.

pH	Ml. of 0.05*M* sodium carbonate	Ml. of 0.05*M* borax
9.2	0.00	100.00
9.4	35.70	64.30
9.6	55.50	44.50
9.8	66.70	33.30
10.0	75.40	24.60
10.2	82.15	17.85
10.4	86.90	13.10
10.6	91.50	8.50
10.8	94.75	5.25
11.0	97.30	2.70

TABLE 6.—SERIES OF BUFFER MIXTURES.—(*Continued*)

pH	Ml. of 0.1M disodium phosphate	Ml. of 0.1M sodium hydroxide	Dilute to
11.0	25	4.13	50 ml.
11.2	25	6.00	50 ml.
11.4	25	8.67	50 ml.
11.6	25	12.25	50 ml.
11.8	25	16.65	50 ml.
12.0	25	21.60	50 ml.

TABLE 7.—SIMPLE BROAD-RANGE BUFFER MIXTURE

pH	Ml. of 0.2M disodium phosphate	Ml. of 0.1M citric acid
2.2	0.40	19.60
2.4	1.24	18.76
2.6	2.18	17.82
2.8	3.17	16.83
3.0	4.11	15.89
3.2	4.94	15.06
3.4	5.70	14.30
3.6	6.44	13.56
3.8	7.10	12.90
4.0	7.71	12.29
4.2	8.28	11.72
4.4	8.82	11.18
4.6	9.35	10.65
4.8	9.86	10.14
5.0	10.30	9.70
5.2	10.72	9.28
5.4	11.15	8.85
5.6	11.60	8.40
5.8	12.09	7.91
6.0	12.63	7.37
6.2	13.22	6.78
6.4	13.85	6.15
6.6	14.55	5.45
6.8	15.45	4.55
7.0	16.47	3.53
7.2	17.39	2.61
7.4	18.17	1.83
7.6	18.73	1.27
7.8	19.15	0.85
8.0	19.45	0.55

0.05*M* Sodium Borate. To ensure the proper water of crystallization, recrystallize the salt from water and dry over hydrated sodium bromide, $NaBr.2H_2O$, to constant weight. Dissolve 19.10 grams in water, and dilute to 1 liter.

0.1*M* Disodium Phosphate. Prepare the dihydrate by exposing the dodecahydrate crystals to the atmosphere for 2 weeks. To check the composition, dry at 100° and 20 to 30 mm. pressure for 1 day, and then carefully ignite to constant weight. This should result in 25.28 ± 0.1 per cent loss. Dissolve 17.805 grams of the salt in water, and dilute to 1 liter.

0.1*M* Citric Acid. To ensure the proper water of crystallization, recrystallize the salt from water, and dry over hydrated sodium bromide, $NaBr.2H_2O$. To check the amount of water of hydration, dry at 20 to 30 mm. pressure at 70°. The acid shall remain colorless and lose 8.58 ± 0.1 per cent of moisture. Dissolve 21.008 grams in water and dilute to 1 liter. As a further check on the concentration, it is advisable to titrate a portion of the solution with 0.2*N* barium hydroxide solution to a distinct red with phenolphthalein.

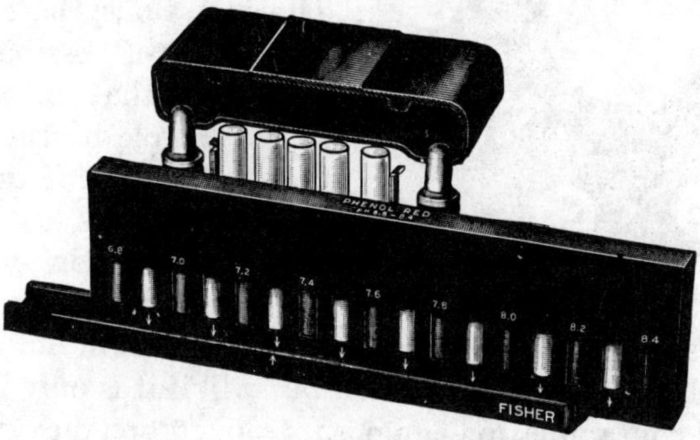

Fig. 71.—Series of standard buffers and comparison with sample.

Procedure. Use uniform tubes, for which test tubes without lips, 1.5 by 15 cm., are suitable. In a series of such tubes, place 10-ml. portions of the various buffer solutions to cover the desired range. To each add 1 ml. of standard indicator solution, and mix well. If the buffer standards are sealed, they will keep for months.

To 10 ml. of the sample add 1 ml. of the appropriate indicator, mix well, and compare with the series of standard buffers. This is conveniently done in a comparator such as that shown in Fig. 71, but such equipment is not essential. View against a background of diffused light. When the sample solution is at all turbid, at the back of the tube containing the sample and the indicator place a similar test tube containing some of the sample without any indicator. By this means comparable turbidities are obtained in the sample and standard.

Instead of buffer solutions with added indicator, glass color standards such as the set shown in Fig. 72 may be used. They are usually considered to offer somewhat lower accuracy. The color is developed in the sample as usual.

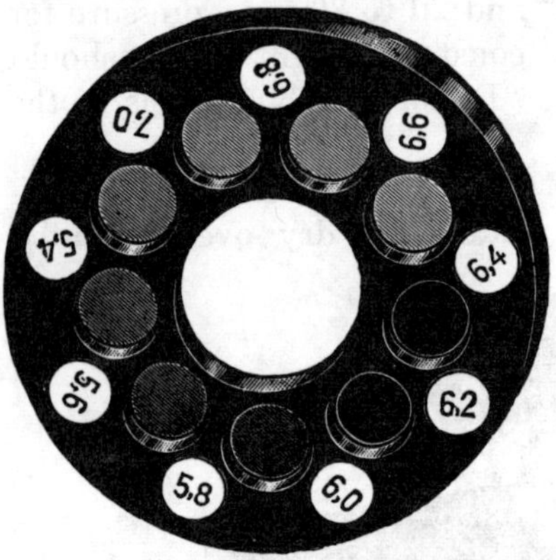

Fig. 72.—Hellige glass color standards.

Sources of Error in Colorimetric pH. As in so many other determinations, there are numerous sources of possible error. It is essential that the temperature of the sample be the same as of the standards. If the sample is turbid and has to be filtered, selective sorption on the filter paper may alter the pH. Dilution of the sample with water to reduce its color will not appreciably alter the pH if it is highly buffered but will if it is only lightly buffered. Acid contamination can occur from carbon dioxide in the air. Alkali contamination can arise from reaction with the alkali of a glass container in which the sample is stored. Salt error can occur when comparing solutions containing different concentrations of salts but the same concentration of indicator. The cause of this is unknown. Neither the acid nor the alkaline colors of bromophenol blue, bromocresol green, bromothymol blue, phenol red, cresol red, or thymol blue are affected. The acid colors of methyl orange, tropaeolin 00 and methyl red, and the alkaline colors of various nitrophenols and salicyl yellow are intensified in the presence of salts. The same types of indicators do not always behave alike. Methyl red and methyl orange show very small salt errors under different conditions, probably owing to their amphoteric character. Protein error is similar to the salt error; colloidal error is probably a better term for it. This type of error occurs in solutions containing protein and solutions containing soap.

The Glass Electrode. This electrode depends upon electric potentials developed on glass surfaces, varying with the type of glass and with the pH in the solution in contact with the glass. The glass electrode is not affected by oxidation-reduction potentials in the solution. It has no salt or protein error and usually comes to equilibrium immediately. This electrode can be used in colored, turbid, and colloidal solutions and lends itself readily to the measurement of pH in even a few milliliters of solution.

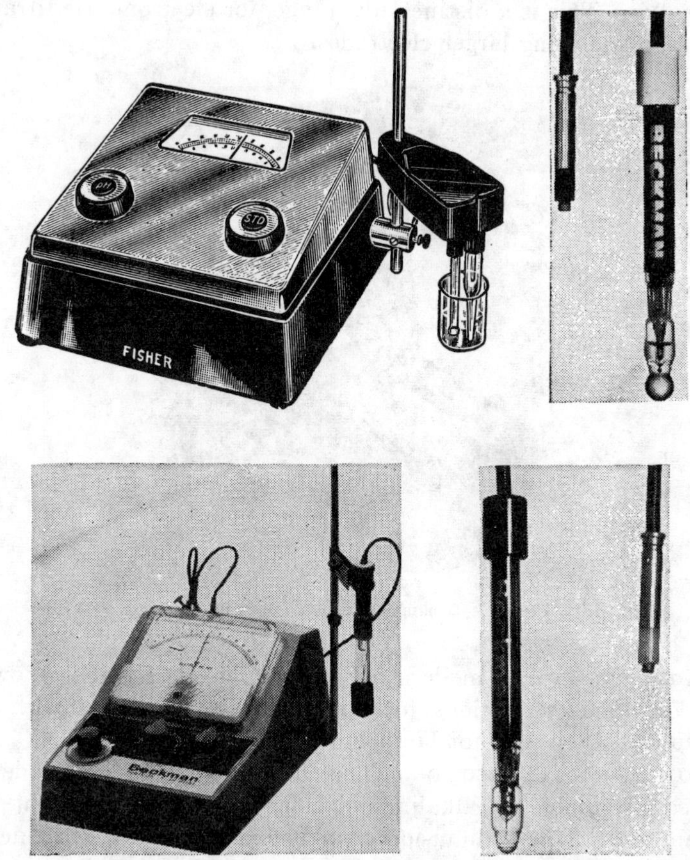

FIG. 73.—pH meters with glass electrodes.

Modern glass electrodes are strong enough to withstand moderately severe treatment. Some can be used throughout the range from low to high pH values, and do not require correction for sodium-ion concentration. Many types are available, some being shown in Figs. 73 and 74.

The assembly is usually sold in a convenient unit, housing the glass electrode, a small calomel half cell, dry batteries, a vacuum tube, galvanometer, and scale. Most instruments have temperature compensators incorporated and a scale reading both in pH and in millivolts. Many types of glass electrodes and of reference electrodes may be used interchangeably with those in the original apparatus. This is a distinct advantage, for electrometric titrations may be made using larger electrodes.

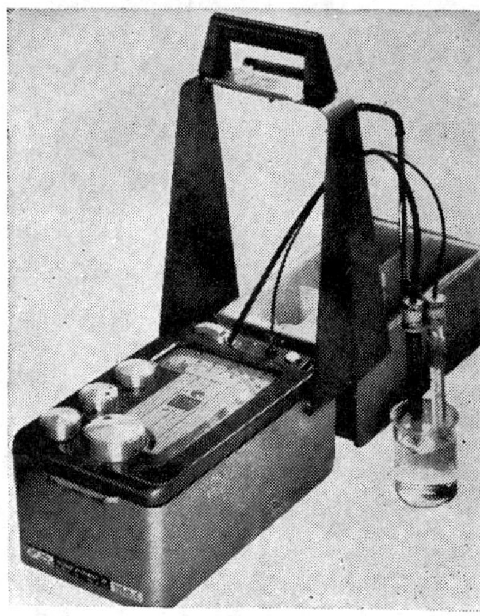

FIG. 74.—Beckman pH meter with dual scale.

A typical form of modern pH glass electrode meter is shown in Fig. 74. Since directions for operation are furnished with each instrument, they will not be included here.

SODIUM-ION CORRECTION. The ordinary glass electrode has limited usefulness in alkaline solutions above pH 9.5 containing sodium ions. The sodium-ion error increases rapidly with increase of pH and of sodium-ion concentration. Although this error may be estimated when the sodium-ion concentration is known, irregularities in electrode behavior and lack of knowledge of the exact sodium-ion concentration frequently make accurate pH values above 10 difficult to obtain. As an example, with the normal glass electrode in a solution *N* in sodium ion, a low value of 0.1 unit occurs at pH 9.5, and of 0.55 unit at pH 11. Charts are supplied by the manufacturers giving the corrections for different sodium-ion concentrations at different pH levels. Nevertheless, the ordinary glass electrode should not be relied on when a correction of 0.2 pH or greater is required.

For more alkaline solutions, special glass electrodes are available. With such an electrode, an error of 0.1 unit is reached only at pH 13.

Electrometric Titration. The glass electrode lends itself readily to electrometric titration. The quinhydrone electrode can be used for this purpose but has a lesser range and is not so convenient in that quinhydrone must be added to the solution under test.

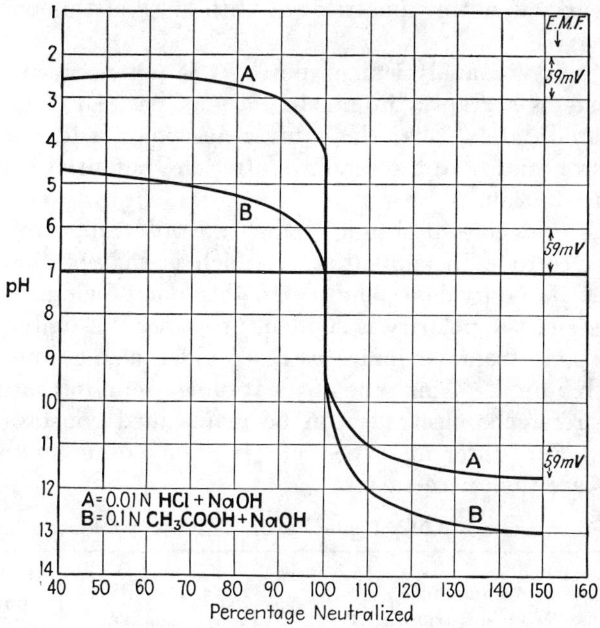

FIG. 75.—Typical electrometric titrations.

PROCEDURE. Insert the electrodes toward the side of the beaker containing the solution under test. It is convenient to have a variable-speed stirrer so set that the stirring arm is well clear of the electrodes. Suspend a thermometer in the solution. Start up the stirrer slowly, and read the pH of the solution directly on the meter. Add the acid or alkali solution in small amounts from a buret clamped above the beaker, taking pH readings after each addition. Plot the volume of the titrating medium against the pH. This method of titration is particularly suitable for turbid or colored solutions with which it would be impossible to obtain a distinct end point with any colorimetric indicator.

Such curves are shown in Fig. 75. If desired and the electrometer used is so calibrated, one may plot pH and the millivolts on the same graph. The reading in millivolts will be the difference between the reading of the solution and that of the reference electrode. Whatever kind of reference electrode is used, one unit in pH change in the solution will always correspond to a change in the potential of 59.1 millivolts at 25°. Table 8 shows that the greatest change in potential relative to change in concentration is between 99.9 and 100.1 or at 100 per cent neutralization. This point of greatest change is the equivalence point, or, if the titration curve is not symmetrical at both sides of this point, very close to it.

When used for analytical purposes, it is not necessary to plot the readings as a graph; the maximum may be read directly from the tabulated data. Near the equivalence point, add the reagent drop by drop and note the readings after the potential is constant after each addition.

It is also possible to obtain the neutralization point by using a reference electrode the potential of which is that of the titrated solution at its equivalent point. On this point being reached in such a system, the polarity is suddenly reversed. A galvanometer or capillary electrometer indicates this sudden change; no potentiometer is required. This type of titration is useful in routine work when the reference electrode can be maintained constant at one potential. For different types of titration, different reference electrodes are then necessary.

TABLE 8.—TITRATION OF 0.01*N* HYDROCHLORIC ACID WITH SODIUM HYDROXIDE

% neutralized	Change in concentration, %	H^+	pH	Potential change, millivolts	$\frac{\text{Change in potential}}{\text{Change in concentration}}$
0		10^{-2}	2		
90		10^{-3}	3		
	9			59.1	6.6
99		10^{-4}	4		
	0.9			59.1	65.7
99.9		10^{-5}	5		
	0.1			2×59.1	1182
100		10^{-7}	7		
	0.1			2×59.1	1182
100.1		10^{-9}	9		
	0.9			59.1	65.7
101		10^{-10}	10		
110		10^{-11}	11		

CHAPTER 7

HALOGENS

The quantitative methods that follow are primarily directed to chlorine compounds but generally are adaptable to other halides by suitable choice of factors. Thus the silver chloride precipitation is equally applicable for silver bromide or iodide; the method for chlorate may be applied to bromate or iodate. In the qualitative tests, all three halides are considered.

QUALITATIVE

Elements. Free chlorine or bromine displace iodine from potassium iodide solution. Chlorine vapors are greenish yellow, bromine vapors red, and iodine vapors violet. They bleach many colored substances when moist. All are absorbed in dilute alkali to form a mixture of the halide and hypohalite.

Silver Nitrate. In the absence of bromides and iodides, silver nitrate precipitates white curdy silver chloride from nitric acid solutions of soluble chlorides. This precipitate is distinguished from the insoluble silver bromide, which is pale yellow in color, and from the insoluble silver iodide, which is yellow in color, by the fact that it is readily soluble in ammonium hydroxide solution. Silver bromide is slowly soluble and silver iodide difficultly soluble. Concentrated ammonium carbonate solution dissolves silver chloride, whereas silver bromide is very slightly soluble and silver iodide insoluble.

Potassium Permanganate. When strong potassium permanganate solution is added to an acidified, concentrated solution of sample and warmed, the free halide is liberated. This is difficult of detection unless enough sample is available. When so liberated, it may be further identified as the element. Manganese dioxide and concentrated sulfuric acid give the same liberation of the element.

Chlorate. Careful addition of crystals of solid chlorate to warm concentrated sulfuric acid results in explosive liberation of chlorine peroxide. Acidified, concentrated chlorate solutions liberate chlorine from chlorides.

Perchlorate. Determine by elimination of other chlorine compounds. Strongly acidify with hydrochloric acid and boil vigorously to decompose chlorate, hypochlorite, and chlorite. Precipitate the remaining chloride with silver nitrate. Filter, evaporate to dryness with sodium carbonate, and fuse. The perchlorate is converted to chloride and is detected with silver nitrate.

Hypochlorite. Hypochlorite exists only in alkaline solution. In acid solution the equivalent is chlorine water. The alkaline solutions unless very dilute have a detectable odor resembling free chlorine. Such solutions decolorize indigo suspension but not permanganate solution.

Chlorite. Chlorites decolorize both indigo suspension and permanganate solution.

QUANTITATIVE

In the procedures that follow, the preparation of sample and methods of analysis are given in sufficient detail so that they should be applicable to any compound or mixture containing a substantial amount of halogen. Determine minute amounts nephelometrically, as described in Chap. 11.

Silver Chloride. GRAVIMETRIC METHOD. In this determination, the chloride is quantitatively precipitated as silver chloride. Other halides if present will precipitate at the same time. Although silver sulfate is somewhat insoluble in water, interference by sulfate is avoided if the technic is carefully followed.

SAMPLE. The desired weight of sample is one that will contain about 0.2 to 0.25 gram of chloride. For this purpose the amount of original sample varies greatly according to the chloride content. To provide for duplicates, if an adequate sample is available, it is preferable to make up 250 ml. and use 50-ml. aliquots. Thus, as an illustration, about 2 grams of sodium chloride would constitute an appropriate sample.

If soluble in water, dissolve the sample, and dilute to 250 ml. If the sample is insoluble in water and consists of a mineral halide, with or without silica present, or consists of or contains silver halide, fuse with five times its weight of potassium carbonate. This may be done in a platinum crucible unless lead or mercury compounds are present; if they are, use a nickel crucible. Extract the fused and cooled melt with hot water, and filter if there are insolubles present. Render the filtrate just acid with 1:1 nitric acid, then definitely alkaline with 1:1 ammonium hydroxide. Let the solution stand for a few hours, filter off any precipitate, and dilute to 250 ml.

This technic is applicable to samples containing lead chloride, but an alternative is to weigh out individual samples and dissolve them in hot water, in which lead chloride is sufficiently soluble for carrying out the determination. If the sample contains mercurous chloride, render the solution or suspension of sample alkaline with 10 per cent sodium hydroxide solution, and bring to a boil. Filter off the insoluble mercury compounds, neutralize with 1:1 nitric acid, cool, and dilute to 250 ml.

PROCEDURE. Prepare a 4.8 per cent solution of silver nitrate as reagent. Each milliliter of this will precipitate 0.01 gram of chloride. Transfer the aliquot of sample to a 250-ml. beaker, and dilute to about 100 ml. Add 5 ml. of 1:4 nitric acid to the approximately neutral solution. Then add the 4.8 per cent silver nitrate solution slowly with constant stirring. When practically all the chloride has been precipitated, the silver chloride will begin to coagulate.

Cover with a watch glass and bring to a boil to coagulate the precipitate completely. Vigorous stirring may be necessary. Allow the beaker and contents to cool to room temperature, for silver chloride is slightly soluble in hot water. Silver chloride decomposes slowly in light to the subchloride, Ag_2Cl, which is purple; therefore it should cool in a dark cabinet.

When cold, filter through paper, washing by decantation. Test the filtrate for the presence of chloride, and, if positive, repeat the determination with a fresh aliquot. Finally transfer the precipitate to the paper, and wash with cold 1:50 nitric acid until the increments of wash solution show no test for silver on adding a chloride solution. Dry the filter paper and contents in an air oven. Remove as much as possible of the precipitate to a glazed paper, and carefully burn the filter paper in a weighed porcelain crucible. Add a drop of each of 1:1 nitric and hydrochloric acids to the residue to oxidize any reduced silver chloride. Now add the precipitate from the glazed paper, carefully drive off all moisture and excess acid from the crucible and gently heat to about 150°. If the precipitate is dark, a further treatment with hydrochloric and nitric acids is required to complete the conversion to silver chloride. Reheat, cool in a desiccator, and weigh.

Alternatively, filter the precipitate by suction through a prepared and weighed Gooch crucible that has been heated to 150°. Wash as before with 1:50 nitric acid. Dry in the oven, finally at 150°, cool, and weigh.

The accuracy of this determination is primarily limited only by the accuracy of weighing or volumetric measurement of the original sample and the accuracy of weighing of the final precipitate. Normally that leads to accuracy better than 1 per cent of the absolute amount present and with care even to 0.1 per cent.

By suitable choice of factors, the same method is applicable to precipitation and determination of bromides and iodides.

CALCULATION.

$$\text{Weight of AgCl} \times 0.4078 \times \frac{100}{\text{weight of sample}} = \text{per cent sodium chloride}$$

Corresponding factors for other forms are NH_4Cl, 0.3732; KCl, 0.5201; Cl, 0.2474.

Volumetric Method with Sodium Chromate Indicator. In this method, often known as the Mohr method, advantage is taken of the fact that silver chloride is more insoluble than silver chromate; alkali chromate added as indicator does not react until all the chloride has combined with silver.

Sample. Prepare the solution of the sample as for the gravimetric determination. If the chlorides present are such as to give a strongly acid reaction, such as antimony chloride, for example, add 25 ml. of a 10 per cent solution of sodium acetate as buffer before dilution to volume. If colored ions are present, iron, for example, carefully precipitate with a small excess of 1:1 ammonium hydroxide. In that case filter and render the filtrate faintly acid with acetic acid before diluting to volume.

The 50-ml. aliquot will require a titration of 40 to 50 ml. of 0.1*N* silver nitrate solution for maximum accuracy. The method is also applicable to bromides and iodides.

Procedure. Transfer the aliquot of sample solution to a 250-ml. conical flask. Add 0.5 ml. of a 10 per cent solution of potassium chromate as indicator. As a comparison solution, add the same amount of indicator to 50 ml. of water.

From a buret carefully add 0.1*N* silver nitrate solution, swirling the flask vigorously all the while. Toward the end of the titration the precipitate will show signs of coagulating. Add the titrating medium more slowly now, finally drop by drop. When the color is perceptibly darker than that of the comparison solution, owing to the formation of red silver chromate, the end point is attained. With practice this point can readily be seen on the addition of 1 drop excess of the standard. The accuracy is of the same order as in gravimetric determination of chloride, provided that the volume of standard silver nitrate used in titration approaches 50 ml.

Calculation.

$$\text{Ml. silver nitrate solution} \times \text{normality} \times 0.05845 \times \frac{100}{\text{weight of sample}} = \text{per cent sodium chloride}$$

Corresponding factors for other forms are NH_4Cl, 0.0535; KCl, 0.07456; Cl, 0.03546.

Volumetric Method with Dichlorofluorescein Indicator. This sorption indicator may be used in neutral or weakly acid solutions as an alternative to the chromate indicator.

Procedure. To the 50-ml. aliquot of sample for titration add 0.5 ml. of 0.2 per cent alcoholic solution of fluorescein, 0.2 per cent aqueous solution of sodium fluoresceinate, or a 0.1 per cent aqueous solution of the sodium salt of dichlorofluorescein. Titrate with 0.1*N* silver nitrate solution. When near the end point the silver chloride coagulates, and at the end point the precipitate suddenly turns a deep pink color. In very dilute solutions, a distinct change to an orange shade prior to the pink tint is taken as the end point. The accuracy attainable is about the same as with the chromate indicator. The method is also applicable to bromides and iodides. The calculation is identical with that using the chromate indicator.

Volumetric Back-titration with Thiocyanate Indicator. In many cases a sample cannot be titrated in neutral solution. This method, often known as the Volhard method, is based on the fact that, in the presence of silver and ferric ions, alkali thiocyanates combine selectively with silver. When excess is present, it reacts with the ferric ions to form ferric thiocyanate. Arsenic, antimony, bismuth, cadmium, cobalt, copper, iron, lead, manganese, nickel, and zinc do not interfere.

Procedure. Prepare the sample as for other chloride determinations except that colored ions may be present and that adjustment to neutrality by addition of buffer is unnecessary. Transfer 50 ml. to a 250-ml. conical flask, add 2 to 3 ml. of concentrated nitric acid, and mix. From a buret add 0.1*N* silver nitrate solution in definite excess. At this point, the silver chloride is well coagulated. Let the precipitate settle, and add 1 ml. more of silver nitrate to the upper layer. If more chloride precipitates, continue the addition until excess is proved to be present. Filter the precipitate, and wash it with cold water until free of silver nitrate, combining the washings with the filtrate.

Now add to the filtrate and washings 5 ml. of a saturated solution of ferric ammonium alum or of a 10 per cent solution of ferric sulfate. Titrate the solution with 0.1*N* ammonium or potassium thiocyanate solution (page 725). Swirl the flask continually, and take as end point when a permanent reddish-brown color is produced. This titration is equivalent to the excess silver nitrate added. Subtract the thiocyanate back-titration from the silver nitrate titration. The difference is equivalent to the chloride. The accuracy attainable is about the same as with the chromate indicator. The method is also applicable to bromides or iodides.

Calculate as for direct titration with silver nitrate with the chromate indicator.

Chlorine in Organic Compounds. Carius Method. This method is also suitable for other combined halogens. The substance is decomposed by fuming nitric acid in a sealed, heavy-walled combustion tube, such as those shown in Fig. 76, under pressure. Silver nitrate present precipitates silver halide. This may be weighed, or, using a known amount of silver, the excess silver may be titrated. It is usual to determine the halide gravimetrically. Any metals present go to their nitrates. Arsenic, phosphorus, and sulfur are oxidized to arsenic, phosphoric, and sulfuric acids. An element of danger due to explosion is present, and the method must be performed with great care throughout, using a tube furnace designed to resist explosion and so located that persons do not pass

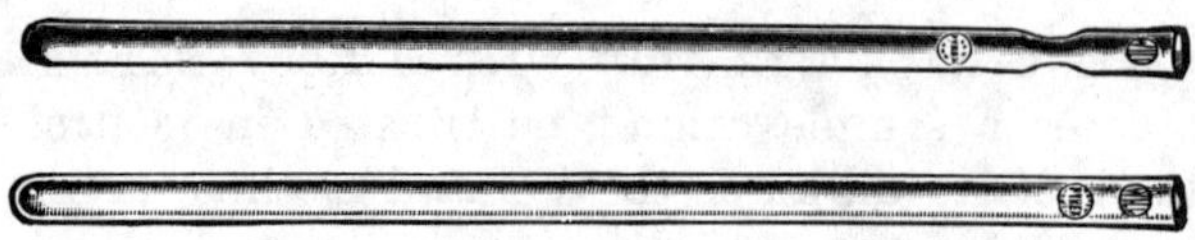

FIG. 76.—Tubes for Carius determinations.

in front of it. Figure 77 shows a furnace designed for gas heating. Such furnaces are also made for heating electrically.

PROCEDURE. The tube is usually about 50 cm. long, 2 cm. in diameter, and about 2 mm. thick, closed at one end. It is kept vertical until it has been sealed. By means of a funnel, add 0.5 to 1 gram of powdered silver nitrate to the tube. Add 3 ml. of chlorine-free, fuming nitric acid by means of a long-stemmed funnel. Avoid wetting the open end of the tube.

FIG. 77.—Gas-heated furnace for Carius determinations.

Weigh 0.1 gram of the sample into a narrow thin-walled glass tube closed at one end. Introduce this tube into the bomb so that it floats on the nitric acid. Be careful that the acid does not come in contact with it before the open end of the tube is sealed. If the tube is already constricted, heat it at the constriction, otherwise a few millimeters from the open end. A blast burner with controllable shape of flame is desirable for this purpose. Finally pull out into a thick-walled capillary and fuse the end. This capillary is an intentional point of weakness in case too great pressure builds up in the tube. To prevent fragility due to quick cooling, coat the capillary with a heavy layer of soot by adjusting the burner to give a luminous flame. When the capillary has cooled, with the tube still vertical, wrap the tube in asbestos paper. Insert into the removable iron tube of a bomb furnace, and then carefully turn it to horizontal with the capillary directed toward a blank wall or a window, in case of failure. The capillary should just extend outside the tube. Now insert the loaded iron tube into the cold bomb furnace. Heat the furnace, so regulating it that a temperature of 200° is attained after 3 hours. Higher heating may be necessary. If so, gradually raise the temperature 50° every 3 hours. Aliphatic substances normally decompose completely after 4 hours at 200°. Aromatic substances usually require 8 to 10 hours at 250 to 300°. Do not heat above 300°.

Let the bomb cool in the furnace. Relieve the pressure with the tube still in the furnace by heating the capillary with a burner until it "blows." Then remove the tube, file, and break off a piece of the capillary. If any drops of oil or organic matter are present, the determination should be discarded and a new one started and heated to a higher temperature, but not above 300°.

Wash the contents of the bomb quantitatively with water into a beaker, avoiding transfer of any pieces of broken glass. If decomposition is complete, this is a sample for further processing for halide. If the silver nitrate was not weighed, it can be determined only gravimetrically. In that case, dilute to about 100 ml., and filter through a paper, washing by decantation. Then proceed as usual, starting (page 129) at "Finally transfer the precipitate to the paper, and . . . ".

If a precipitate of silver chloride or silver bromide becomes contaminated with fragments of broken glass from the capillary, dilute the suspension to about 50 ml., and add 1:1 ammonium hydroxide with vigorous stirring until the silver halide is dissolved. Filter and reprecipitate the halide by careful addition of 1:1 nitric acid dropwise with vigorous stirring. Then dilute to about 100 ml., and filter the silver halide through a paper, washing by decantation. Complete, starting (page 129) with "Finally transfer the precipitate to the paper, and . . . ".

In case silver iodide becomes contaminated with glass, a different procedure must be used. Complete as though no glass were present. Moisten the ignited residue with 1:10 sulfuric acid, and add a few pieces of pure zinc. The reaction decomposes silver iodide. When reaction is complete, filter, and wash free of iodine and water-soluble materials. Dry, weigh, and ignite. Subtract the weight of glass from that of the total precipitate.

For volumetric determination, dilute the decomposed sample to about 100 ml. To determine the equivalent of the silver nitrate added, divide the weight in grams by 0.017 to obtain the equivalent in milliliters of 0.1N silver nitrate. Then complete by the thiocyanate method, starting (page

131) at "Filter the precipitate, and wash it with cold water until free of silver nitrate . . . ". Contaminating glass will not affect this result.

The accuracy to be expected in this determination is limited by the amount of sample. Absolute accuracy to 0.5 per cent of the amount of halide is as good as can be expected.

Sodium Peroxide Fusion Method. Substances that do not readily volatilize on heating may be decomposed by fusion with sodium peroxide. The method is easier to operate with organic samples than the Carius method but more limited in applicability and often of lower accuracy.

Procedure. To 0.2 gram of the substance in a small steel crucible add 16 to 18 times its weight of sodium peroxide. For a substance of which the carbon and oxygen content is as low as 25 to 50 per cent, add half its weight of pure granulated sugar. Add an equal weight of sugar if this percentage is less than 25. The crucible should not be more than two-thirds full. Place it in a larger porcelain crucible containing water sufficient to cover the steel crucible to within 1 to 2 cm. of its top. Close the steel crucible with a cover containing a small hole through which an iron wire can be inserted. Heat such a wire to redness, and fire the charge by inserting it through the hole. The combustion may be quite violent. As soon as it is complete, upset the steel crucible and contents into the water in the porcelain crucible. Cover the latter with a watch glass and heat gently until the fusion is dissolved. This is recognized when the evolution of bubbles ceases.

Remove the steel crucible, wash carefully, and filter the solution. To neutralize the alkalinity and reduce chlorates, bromates, or iodates, either add a saturated aqueous solution of sulfur dioxide or pass sulfur dioxide gas through the solution until an excess is present. Then boil off the excess. The resulting solution is a sample for gravimetric determination of halogen, starting on page 129 at "Add 5 ml. of 1:4 nitric acid to the approximately neutral solution.

The accuracy is limited by the possible losses due to volatilization, incomplete combustion, and sorption during filtration.

Chlorates. These are determined by reducing to chloride and are then estimated in the usual manner. The methods are also applicable to bromates and iodates.

Sample. Chlorates in any sample not soluble in water may be decomposed and lost in dissolving. Therefore precautions not described here will have to be developed as applying to the specific sample. For water-soluble samples, use an amount that will contain 0.2 to 0.25 gram of chloride. Weight out about five times that amount, dissolve in water, and dilute to 250 ml. Pipet out 50 ml. for each determination.

PROCEDURE. *Reduction with Ferrous Sulfate.* Dilute the sample to about 100 ml., and add 50 ml. of a 10 per cent solution of ferrous sulfate. Heat to boiling with constant stirring, and boil for 15 minutes. Cool and add dropwise sufficient 1:1 nitric acid to dissolve the precipitated basic ferric salt. Then determine the chloride gravimetrically, starting (page 129) at "Add 5 ml. of 1:4 nitric acid to the approximately neutral solution."

Reduction with Sulfurous Acid. Either add a saturated aqueous solution of sulfur dioxide or pass sulfur dioxide gas through the aliquot of the sample until excess is present. Boil off the excess gas and determine the halogen gravimetrically, starting (page 129) at "Add 5 ml. of 1:4 nitric acid to the approximately neutral solution."

Reduction with Zinc. To the neutral aliquot of chlorate sample solution add about 5 ml. of 10 per cent acetic acid. Then add about 1 gram of powdered zinc and boil for about 1 hour. The zinc should have been largely used up. Determine the chloride gravimetrically, starting (page 129) at "Add 5 ml. of 1:4 nitric acid to the approximately neutral solution," checking to be sure that the zinc dissolves completely and readily in the added nitric acid.

Calculation is the same as for chlorides, with substitution of the following factors: AgCl to $NaClO_3$, 0.7426; to $KClO_3$, 0.8550; to NH_4ClO_3, 0.7081.

Alkali-metal Perchlorates. These are reduced to chlorides on heating with the ammonium chloride.

PROCEDURE. Intimately mix 0.2 to 0.5 gram of the perchlorate in a platinum crucible with 2 grams of ammonium chloride. Platinum is essential to catalyze the reaction. Cover with a watch glass, and heat gently for 90 to 120 minutes. The following reaction takes place:

$$3NaClO_4 + 8NH_4Cl = 3NaCl + 4N_2 + 8HCl + 12H_2O.$$

Keep the temperature below the melting point of the chloride formed by the reaction, or the platinum may be attacked by the melt. Add more ammonium chloride, mix, and reheat as before.

Excess ammonium chloride decomposes at 350°, which is well below the fusion points of sodium or potassium chloride. When the decomposition of excess ammonium chloride is complete, dissolve in water and dilute to about 100 ml. Determine the chloride gravimetrically, starting (page 129) at "Add 5 ml. of 1:4 nitric acid to the approximately neutral solution." Factors for calculation are: AgCl to $NaClO_4$, 0.8543; to $KClO_4$, 0.9666.

Mixed Chlorates and Perchlorates. Determine the chlorate on one portion of sample as though perchlorate were absent. The result will be chlorate without perchlorate being affected. On another portion of sample, determine the total chlorine present as chlorate and perchlorate by heating with ammonium chloride in the method described for perchlorate. The difference between the silver chloride from the two methods is that derived from perchlorate.

Mixed Chlorides, Chlorates, and Perchlorates. Determine chloride on aliquots of the sample solution diluted to about 100 ml. Use the gravimetric procedure, starting (page 129) at "Add 5 ml. of 1.4 nitric acid to the approximately neutral solution." The chlorate and perchlorate are unaffected. On other aliquots, apply the procedure for chlorates. The result is the sum of chloride and chlorate and will yield chlorate by difference from the preceding determination. To obtain perchlorate, weight out a sample, and apply the method of reduction of perchlorates by heating with ammonium chloride. By calculation of the difference from the chloride and chlorate determinations, perchlorate is obtained.

Sodium Hypochlorite Bleach. This bleaching solution is a common article of commerce and contains sodium hypochlorite, NaOCl, sodium chlorate, sodium chloride, sodium carbonate, and either sodium hydroxide or sodium bicarbonate. Since it is relatively unstable, samples must be kept in dark bottles in a cool place.

Since sodium hypochlorite is used for its oxidizing properties, the usual form of reporting the analysis is as the amount of available chlorine. The molecular weight of sodium hypochlorite is 74.45, and each molecule is equivalent to a molecule of free chlorine. Since Cl_2 equals 70.91, it happens that the figure for available chlorine is close to the actual sodium hypochlorite content. In any case results are only approximate, for some loss of chlorine during transfer is inevitable.

PROCEDURE. *Available Chlorine by Sodium Arsenite.* A diluted solution of the sample is desirable unless it contains less than 0.25 per cent of available chlorine. For bleach solution expected to contain 10 per cent or more of available chlorine, pipet 10 ml. into a 250-ml. calibrated flask containing about 100 ml. of water, keeping the tip of the pipet well below the surface of the water. Dilute to the mark, and mix well. Use correspondingly larger volumes of weaker solutions for dilution to 250 ml.

Introduce a 50-ml. aliquot of the diluted sample below the surface of 100 ml. of water in a conical flask. Swirl the contents of the flask, and run in 0.1*N* sodium arsenite solution (page 730) until nearly the full quantity expected to be required has been added. Place a drop of the solution on a piece of starch-iodide paper. If a brown stain is formed, appreciable available chlorine is unreacted. A light blue color indicates that the reaction is nearly complete. Continue to add the sodium arsenite solution, repeating the spot test until only a light blue color is formed.

Repeat the titration, adding the sodium arsenite solution to within a few drops of the end point indicated by the first titration. Add 5 ml. of 10 per cent potassium iodide solution and 1 ml. of a 0.5 per cent solution of soluble starch, and continue the titration until the blue color disappears. Chlorates, which are usually present in a greater or lesser quantity, do not interfere.

CALCULATION.

$$2NaOCl + As_2O_3 = As_2O_5 + 2NaCl$$

$$\text{Ml. sodium arsenite solution} \times \text{normality} \times 0.03546 \times \text{aliquot} \times \frac{100}{\text{weight of sample}} = \text{per cent available chlorine}$$

The factor for calculation of the result as NaOCl is 0.3726.

To convert available chlorine to sodium hypochlorite multiply by 1.05.

Total Chlorine. To 30 ml. of 3 per cent hydrogen peroxide in a 250-ml. calibrated flask add 10 ml. of the undiluted sample solution, delivering from a pipet with the tip well below the surface of the liquid. For dilute bleach solutions, increase the volume added, that specified being based on the usual commercial bleach with 10 to 13 per cent of available chlorine. This changes all forms of chlorine present to chloride. Swirl the flask vigorously for a minute, dilute to the mark, and mix. Pipet an aliquot of 25 ml., or more if the flask does not contain about 1 gram of chloride, into a conical flask. Neutralize the aliquot by careful addition of 1:1 nitric acid until the sample solution is faintly acid. Then determine the total chloride gravimetrically, starting at the procedure on page 129.

If chlorate is substantially absent, as is the case with a well-made bleach, the total chlorine can be taken in conjunction with data on available chlorine to calculate chloride originally present as sodium chloride. If the chlorate is appreciable, it must be separately determined by a method that will only be outlined here. Reduce the available chlorine in a substantial sample to chloride, using sodium arsenite or hydrogen peroxide. Then determine chloride in an aliquot by the gravimetric method. Determine chlorate in another aliquot by one of the methods that reduces chlorate to chloride. The difference between the two determinations will be chlorine present as chlorate.

CALCULATION.

$$(\text{Ml. } 0.1N \text{ silver nitrate solution} - \text{ml. } 0.1N \text{ potassium thiocyanate solution}) \times \text{aliquot} \times 0.003546 \times \frac{100}{\text{weight of sample}} = \text{per cent total chlorine}$$

In the absence of sodium chlorate, calculate sodium chloride as follows:

$$(\text{Per cent total chlorine} - \text{one-half the per cent of available chlorine}) \times 1.6486 = \text{per cent sodium chloride}$$

When sodium chlorate is present, subtract its equivalent as sodium chloride from the sodium chloride calculated to give the correct sodium chloride figure.

Total Alkalinity as Sodium Oxide. This will usually be carried out with 10 ml. of strong bleach. If the alkalinity is low, it is desirable to use a larger sample in order to get a titration of proper magnitude. Measure a volume of 3 per cent hydrogen peroxide which is three times the volume of sample to be used into a 250-ml. conical flask. Neutralize this to a methyl-orange end point with 0.1N sodium hydroxide. Pipet the volume of sample for titration into the neutralized peroxide. Swirl vigorously for about 1 minute, and titrate with 0.1N hydrochloric acid, using methyl orange as indicator.

CALCULATION.

$$\text{Ml. hydrochloric acid} \times \text{normality} \times \text{aliquot} \times 0.031 \times \frac{100}{\text{weight of sample}} = \text{per cent sodium oxide}$$

Sodium Hydroxide. The volume of sample is determined as in the preceding method. To 50 ml. of 10 per cent barium chloride solution and a volume of 3 per cent hydrogen peroxide solution three times the sample to be used in a 250-ml. conical flask add 0.5 ml. of phenolphthalein indicator. Neutralize with 0.1N sodium hydroxide solution. Then pipet the sample solution into the flask, and swirl vigorously for about 1 minute. In addition to reducing hypochlorite to chloride, this mixture precipitates the carbonate present. Titrate the hydroxide alkalinity with 0.1N hydrochloric acid to the disappearance of the pink color.

CALCULATION.

$$\text{Ml. hydrochloric acid} \times \text{normality} \times \text{aliquot} \times 0.040 \times \frac{100}{\text{weight of sample}} = \text{per cent sodium hydroxide}$$

Sodium Bicarbonate. If no free sodium hydroxide is found, sodium bicarbonate can be expected to be present in a greater or lesser amount. Prepare neutralized barium chloride–hydrogen peroxide mixture as directed in the determination of sodium hydroxide. Separately pipet the same amount of sample as used in the determination of total alkalinity into a volume of 0.1N sodium hydroxide solution equivalent to the acid used in that determination, and mix well. Now mix these two solutions, and swirl vigorously for 1 minute. Titrate the excess sodium hydroxide with 0.1N hydrochloric acid to the disappearance of the pink color.

Calculation.

$$\text{Ml. sodium hydroxide solution} \times \text{normality} \times \text{aliquot} \times 0.084 \times \frac{100}{\text{weight of sample}} = \text{per cent sodium bicarbonate}$$

Sodium Carbonate. When sodium hydroxide is present, calculate as follows:

$$[\text{Per cent sodium oxide} - (\text{per cent sodium hydroxide} \times 0.775)] \times 1.709 = \text{per cent sodium carbonate}$$

When sodium bicarbonate is present, calculate as follows:

$$[\text{Per cent sodium oxide} - (\text{per cent sodium bicarbonate} \times 0.369)] \times 1.709 = \text{per cent sodium carbonate}$$

CHAPTER 8

COMBINED NITROGEN

Aside from its presence in elemental form in the air, nitrogen must be considered in the forms of nitrites, nitrates, ammonia, proteins, and other organic combinations. By far the most common method of determination is its reduction to ammonia by the Kjeldahl method, followed by distillation and titration.

QUALITATIVE

Ammonia and Volatile Amines. Free ammonia is readily detected by its characteristic odor. Some aliphatic amines have similar odors that can confuse a worker who has limited experience. If the ammonia is in combined form, the same statements hold true after the sample has been rendered alkaline with sodium hydroxide. Details of a convenient method of testing are given on page 86. A trace of ammonia or ammonium compound is detected as well as estimated by reaction with Nessler reagent, as described on page 88.

Nitrates. By Ferrous Sulfate. Mix 1 ml. of the unknown solution with 10 ml. of concentrated sulfuric acid in a test tube. Cool and carefully flow down the side of the tube 10 ml. of a saturated solution of ferrous sulfate. Do not mix the solutions, but allow one to rest upon the other. At the junction of the solutions, a brown ring, varying in intensity with the concentration, forms when nitrate is present. Bromides, bromates, chromates, chlorates, ferro- and ferricyanides, iodides, iodates, and permanganates interfere.

By Diphenylamine. Dissolve 0.5 gram of diphenylamine, $(C_6H_5)_2NH$, in 100 ml. of concentrated sulfuric acid diluted with 20 ml. of water, and place 2 ml. of this solution in a test tube. Cover with 2 ml. of the solution under test. A deep blue color is produced at the interface when nitrates are present. Many oxidizing reagents interfere and give the same test.

Nitrites. Add a few crystals of potassium iodide to a few milliliters of the test solution, and acidify with acetic acid. A yellow coloration due to free iodine, detectable in small quantities by the addition of starch, may indicate nitrite. Some oxidizing substances will also give this test. Nitrous oxide is evolved on acidifying a solution of a nitrite.

To detect nitrous acid in nitric acid, add potassium permanganate solution acidified with 1:3 sulfuric acid. Nitrites and many reducing agents decolorize the permanganate.

Amines. All primary amines are alkaline in aqueous solution, the aliphatic compounds being stronger bases than ammonia while the aromatic compounds are only weakly basic. In many cases the aromatic amines have only a limited solubility in water, do not turn red litmus blue, and are very little ionized. Secondary amines are more basic than primary amines; tertiary amines are still more basic. All form salts with acids, the salts of aromatic amines giving acid reactions in water owing to hydrolysis.

ALIPHATIC AMINES. Most aliphatic amines when heated give an odor resembling that of ammonia. The accompanying vapors are usually alkaline to litmus paper. When boiled with sodium hydroxide, ammonia, in small or large quantities, is often given off. A series of qualitative reactions is briefly indicated.

Nitrous Oxide Test. To the cold dilute hydrochloric acid solution of the test sample add a solution of sodium nitrite. *Primary amines* yield alcohols with a copious evolution of nitrogen. $RNH_2 + HNO_2 = ROH + N_2 + H_2O$. *Secondary amines* yield nitrosoamines, the lower members of which are yellowish liquids of characteristic odor, only slightly soluble in water.

$$R_2NH + HNO_2 = R_2NNO + H_2O$$

Concentrated hydrochloric acid reconverts the nitrosoamine to the secondary amine. *Tertiary amines* are either unaffected or are oxidized.

PRIMARY AMINES IN GENERAL. Warm 1 ml. of the solution with 2 drops of chloroform and 2 ml. of a strong alcoholic solution of potassium hydroxide. A very disagreeable and characteristic odor of the carbylamine results.

$$RNH_2 + CHCl_3 + 3KOH = 3KCl + 3H_2O + RNC$$

This odor is so pungent and penetrating that the reaction should be carried out in a hood and the whole mixture immediately acidified and washed down the drain as soon as the reaction is completed.

AROMATIC AMINES. *Benzenesulfonyl Chloride Test.* *Primary amines* in the presence of aqueous alkali on treatment with benzenesulfonyl chloride yield primary substituted benzenesulfonamides, soluble in alkali.

$$C_6H_5SO_2Cl + RNH_2 + NaOH = C_6H_5SO_2NHR + NaCl + H_2O$$

On extracting the alkaline solution with ether, the free sulfonamide is obtained. This may be isolated and its melting point obtained. *Secondary amines* under the above conditions yield secondary substituted benzenesulfonamides which, having no replaceable hydrogen atom, are insoluble in alkali.

$$C_6H_5SO_2Cl + RNHR' + NaOH = RNR'SO_2C_6H_5 + NaCl + H_2O$$

Tertiary amines, on treatment of the dilute hydrochloric acid solution with a solution of sodium nitrite, yield the corresponding *para* nitroso derivatives when the *para* position is not already occupied.

$$C_6H_5NRR' + HNO_2 = NOC_6H_4NRR'(1{:}4) + H_2O$$

The hydrochlorides of these substances are usually yellow and are soluble in water. Alkali liberates the free base as a greenish solid. Reduction with zinc dust and acetic or hydrochloric acid, yields the corresponding derivatives of the *para* diamines. On passing hydrogen sulfide into a solution of such a diamine in dilute hydrochloric acid and then oxidizing the mixture with a few drops of ferric chloride solution, a deep blue or violet coloration is produced, owing to the formation of a thionine dye of the methylene-blue type.

QUANTITATIVE

Kjeldahl Method. This most useful method is employed to determine the nitrogen content of many organic substances. The organic matter is destroyed with sulfuric acid, and the nitrogen is changed to ammonia, which is held by the acid as sulfate. Nitrates must first be reduced by zinc dust, salicylic acid, or other suitable means. Nitrogen in the form of oxides and as cyanides must also be reduced first. Some compounds containing nitrogen do not fully decompose when treated by this method. This can always be ascertained by running a determination on the type of material known to be present.

The method is applied on a very large scale, as indicated by such equipment as that shown in Figs. 78 and 79. The units used for single determinations are assembled from standard laboratory equipment in much the same form.

KJELDAHL-GUNNING-ARNOLD METHOD. *Nitrates Absent.* The amount of sample will depend on its nitrogen content. For fertilizers, use 0.7 to

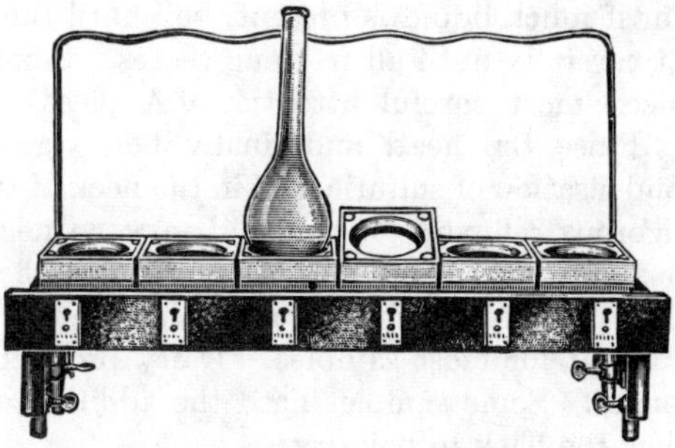

FIG. 78.—Multiple Kjeldahl digestion unit.

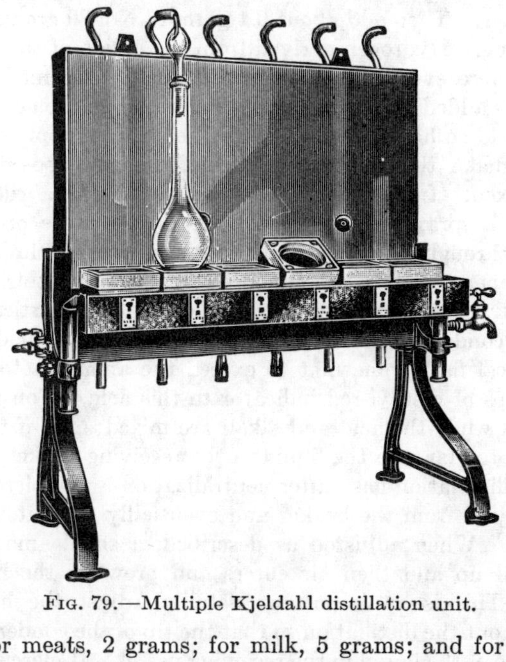

FIG. 79.—Multiple Kjeldahl distillation unit.

3.5 grams; for meats, 2 grams; for milk, 5 grams; and for soils, 7 to 14 grams. Introduce the sample into a 500-ml. Kjeldahl flask. Add 0.5 gram of copper sulfate crystals as a catalyst. Mercuric oxide, 0.78 gram, or an equivalent amount of mercury may replace the copper sulfate, but in this case sodium sulfide must be added later to remove the mercury before distillation, thus preventing the formation of mercury-amido compounds, which are not completely decomposed by sodium hydroxide. Copper selenide, or selenium itself, are even more efficient catalysts of the reaction.

Next add 10 to 15 grams of nitrogen-free sodium or potassium sulfate and 25 ml. of concentrated sulfuric acid. Place the flask in an inclined position, resting on an asbestos board with a hole to fit the flash. Heat the flask gently, or if much liquid is present, boil until the contents begin to char, then heat carefully until all frothing ceases. Some samples foam vigorously and need most careful attention. A piece of paraffin may reduce foaming. Raise the heat, and finally boil vigorously, but not enough so that condensation of sulfuric acid in the neck of the flask will not be complete. Vigorous refluxing of the acid may be necessary to wash down carbonaceous matter deposited on the neck of the flask by foaming. Fifteen to thirty minutes of brisk boiling after the solution becomes a pale straw color is sufficient for most samples. Dairy products, leather, etc., require a little longer. Some samples need the addition of more sulfuric acid. Do not allow the flask to boil dry.

Remove the flask, and allow to cool. Dilute to 200 ml. by careful addition of water. Then add about 0.1 gram of No. 80 granulated zinc to prevent bumping. Hydrogen derived from the action of the alkali on the zinc effects a more even boiling. Conventionally, sodium hydroxide for neutralization is added as a separate layer underneath the acid layer. For this purpose, add sufficient of a saturated solution of commercial sodium hydroxide to make the solution definitely, but not too alkaline when thoroughly mixed. Immediately close the flask with a rubber stopper through which a spray trap is inserted. An alternative procedure is to add the alkali through a small separatory funnel passing through a second hole in the stopper. No loss of ammonia can result with this method.

The spray trap has previously been connected to a vertical condenser. The tip of this condenser just dips into sufficient standard acid in a 400-ml. beaker or conical flask somewhat to exceed the ammonia to be distilled. Add a few drops of methyl-red indicator to this acid. Considerable heat will be evolved when the acid and alkali are mixed; and, if the tip of the condenser dips too far into the liquid in the receiving beaker, atmospheric cooling of the distillation flask after neutralization is complete will tend to suck up the liquid from the beaker and eventually draw it back into the distilling flask. When adjusted as described, a small amount of liquid may be sucked up and then air enters and prevents the liquid's being drawn over. This is an important detail. Adjust the height of the beaker throughout the distillation so that the tip of the condenser is always near the surface of the liquid in the receiving vessel. If mercury is present, immediately after adding the alkali add 10 ml. of a 10 per cent solution of sodium sulfide. When copper is present, the intense blue color formed with excess alkali acts as an indicator. On boiling later, a brown copper hydroxide is precipitated.

Make sure all the connections of the apparatus are tight. Mix the contents of the flask by swirling, and commence heating. Distill slowly at first, and finally boil rather vigorously. Boil for at least 20 minutes, preferably longer. Should the methyl red start to turn yellow do not reject the determination, but add more standard acid solution. Ammonia is so soluble in water that a small excess in solution will not be lost on brief exposure to the air. When distillation is complete, turn out the flame, and immediately lower the receiving vessel. Wash down the condenser tube and its tip into the receiver. Titrate the excess acid with standard alkali.

A saturated solution of boric acid is occasionally used to absorb the ammonia. This fixes the ammonia, but since boric acid is neutral to methyl red or methyl orange, the ammonia may be directly titrated with standard acid.

CALCULATION.

$$[(\text{Ml. of acid} \times \text{normality}) - (\text{ml. of alkali} \times \text{normality})] \times 0.01401 \times \frac{100}{\text{weight of sample}} = \text{per cent nitrogen}$$

For ammonia the factor is 0.01703. When results are to be reported as protein, multiply the nitrogen by 6.25 or the ammonia by 5.7. Occasionally, for specific proteins other factors are applied, such as 5.7 for milk proteins. For amines, substitute the appropriate equivalent weight.

Nitrates Present. To the sample in the flask add 30 ml. of concentrated sulfuric acid and 1 gram of salicylic acid. Shake, and allow to stand for 5 to 10 minutes with frequent agitation. The salicyclic acid forms a nitro compound. Add 5 grams of sodium thiosulfate, and heat the solution for 5 minutes. Cool, and continue as for the usual Kjeldahl method, starting (page 143), at "Add 0.5 gram of copper sulfate crystals as a catalyst."

Ammonia. BY NESSLER REAGENT. In reaction with ammonia, Nessler reagent, potassium mercuric iodide, produces a yellow to brown color that is proportional in intensity to the concentration of ammonia present.

The method is of particular importance in water analysis. Direct nesslerization is described here. With sewage and industrial wastes, in order to liberate all the ammonia but minimize hydrolysis of organic nitrogen compounds, the sample is predistilled at pH 7.4 (page 240). By methods not given here, small amounts of other nitrogen compounds may be converted to ammonia and determined by this procedure. Kjeldahl distillation is typical of the methods of conversion to ammonia. The methods given are substantially those of the American Public Health Service.[1]

[1] "Standard Methods for the Examination of Water and Sewage," 11th ed., American Public Health Association, New York, 1960.

AMMONIA- AND MAGNESIUM-FREE WATER. Treat distilled water with a slight excess of bromine water and let stand overnight. Then distill, discarding the first 20 per cent and the last 20 per cent.

Alternatively remove ammonia by shaking for 5 minutes with 3 grams of 60- to 80-mesh Permutit per liter, and decant. The Permutit may be used repeatedly. If traces of magnesium remain after the latter treatment, they may cause cloudy tubes with the reagent. To remove magnesium, add 25 ml. of 33 per cent sodium hydroxide solution and 8 ml. of 10 per cent copper sulfate solution per liter. Mix well, let settle overnight, and decant the clear upper layer.

PROCEDURE. Add 1 ml. of Nessler reagent (page 720) to 50 ml. of original or diluted sample solution containing not over 0.07 mg. of ammonia. Mix well and allow to stand for 10 minutes. Compare with natural or artificial standards if satisfactory in clarity and intensity of color. If the intensity of color is too great, repeat, using an appropriate dilution of the sample with ammonia- and magnesium-free distilled water.

If the reagent gives a precipitate with the sample, the latter must be clarified. For this, take 100 ml. of sample in a bottle, and add 1 ml. of 10 per cent copper sulfate solution. Shake, and add 1 ml. of 33 per cent sodium hydroxide solution. Again shake, and let stand to clarify, or centrifuge. If the sample contains hydrogen sulfide, substitute 1 ml. of 10 per cent lead acetate solution or 1 ml. of saturated zinc sulfate solution for the copper sulfate solution. If these clarifying agents fail to be effective, try adding them in the reverse order. No correction need be applied for these amounts of clarifying agents.

After treatment with copper or lead solution for clarification, add 2 to 4 drops to 33 per cent Rochelle salt solution to prevent clouding. Use 50 ml. as sample, and proceed as above, starting with "Add 1 ml. of Nessler reagent . . . ".

Permanent Standards. Dissolve 2 grams of potassium chloroplatinate, K_2PtCl_6, in a small volume of water, and add 100 ml. of concentrated hydrochloric acid. Dilute to 1 liter with water. Dissolve 12 grams of cobaltous chloride, $CoCl_2.6H_2O$, using dry crystals, in a small volume of water, and add 100 ml. of concentrated hydrochloric acid. Dilute to 1 liter. Measure into 50-ml. Nessler tubes the volumes of these solutions, indicated in Table 9, and dilute to volume. The values given in the table are approximate; actual equivalents of the standards thus prepared will differ with the quality of the Nessler reagent and the color sensitiveness of the analyst's eye. They should be compared with nesslerized natural standards and the record of the value of each tube modified as necessary. Such comparisons should be made for each newly prepared Nessler solution and checked by each analyst. These artificial standards keep satisfactorily for several months if protected from dust.

Natural Standards. Dissolve 3.818 grams of pure ammonium chloride in ammonia-free water, and dilute to 1 liter. Dilute 10 ml. of this stock

TABLE 9.—COMPOSITION OF PERMANENT STANDARDS FOR DETERMINATION OF AMMONIA BY NESSLER'S REAGENT

Value in ammonia nitrogen, mg.	Volume of platinum solution, ml.	Volume of cobalt solution, ml.
0.000	1.2	0.0
0.001	1.8	0.0
0.002	2.8	0.0
0.004	4.7	0.1
0.007	5.9	0.2
0.010	7.7	0.5
0.014	9.9	1.1
0.017	11.4	1.7
0.020	12.7	2.2
0.025	15.0	3.3
0.030	17.3	4.5
0.035	19.0	5.7
0.040	19.7	7.1
0.045	19.9	8.7
0.050	20.0	10.4
0.060	20.0	15.0
0.070	20.0	22.0

solution to 1 liter with ammonia-free water. Each milliliter contains 0.01 mg. of nitrogen or 0.01216 mg. of ammonia. Make up a series of standards by diluting 0.2, 0.4, 0.7, 1.0, 1.4, 1.7, 2.0, and 2.5 ml. of the diluted standard ammonium chloride solution to 50 ml. with ammonia-free water. Add 1 ml. of Nessler reagent to each at the same time the sample is treated, mix well, and compare after 10 minutes. Such natural standards are used for checking the artificial standards more commonly used in routine work.

CALCULATION.

$$\text{Ml. standard} \times \frac{0.01216}{1{,}000} \times \frac{10^6}{\text{ml. of sample}} = \text{ppm. as ammonia}$$

For the result as nitrogen, substitute the factor 0.01.

Free Ammonia and Amines by Titration. OTHER ALKALIES ABSENT. For free ammonia or amines in solution when no other alkaline materials are present, direct titration is feasible. The technic is applicable to so many free amines and alkaloids that the applications are too numerous to list. The alkalinity and, therefore, the size of sample desirable will vary with the structure.

The sample must be selected so as to permit of a titration with 40 to 50 ml. of standard solution. Such standard solution is usually 0.1N or 0.5N, but in exceptional cases and with resulting lower accuracy it may be 0.01N, 0.05N or N. The selection of sample is not particularly difficult. If it cannot be precalculated from the expected concentration, a preliminary sample is roughly titrated, and from the results the desired size of sample and concentration of standard solution are selected. For the description that follows, it will be assumed that the titration is with 0.5N acid.

PROCEDURE. WATER-SOLUBLE SAMPLES. If enough of the sample is available, weigh out an amount five times that which is to be titrated. Transfer to a 250-ml. calibrated flask and dilute to volume.

If the sample gives off fumes, weigh in a tared, glass-stoppered weighing bottle. In that case, transfer to a beaker containing about 200 ml. of water. Remove the stopper of the weighing bottle underneath the solution by manipulation with a glass rod. Mix well and transfer to a 250-ml. calibrated flask. Wash the beaker and weighing bottle, transfer the washings to the flask, and dilute to volume.

Pipet a 50-ml. aliquot of sample into a 500-ml. conical flask containing about 200 ml. of water. Add sufficient 0.5N standard acid to give a small excess. Mix well, and titrate back to a methyl-orange or methyl-red end point with 0.1N alkali. Calculate as in the Kjeldahl method, but substitute the proper factor if an amine is being titrated.

WATER-INSOLUBLE SAMPLES. If the sample does not dissolve in water readily, as is the case with some amines, weigh out five times the amount to be titrated. Transfer to a 250-ml. volumetric flask, with water if desired. Then add a known excess of standard acid solution to combine with the compound, most of the acid-amino compounds being water-soluble. Dilute to volume, and mix well.

Pipet a 50-ml. aliquot into 150 to 200 ml. of water. Add sufficient standard alkali to combine with the standard acid used in dissolving and a few milliliters excess. Back-titrate at once with the standard acid solution. Although this seems roundabout, it often gives a far better end point than straight titration of the excess acid would give. Calculate as in the Kjeldahl method, but substitute the factor for the proper amine.

Free Ammonia. OTHER ALKALIES PRESENT. When ammonia is present in the sample, it may be separated by distillation into standard acid. Many amines will partly distill under similar conditions and must, therefore, be absent. The technic is substantially that of the last steps of the Kjeldahl determination on page 143.

PROCEDURE. Select the sample for distillation and strength of alkali for absorption so that the titration will be 40 to 50 ml. Transfer the sample to a Kjeldahl flask. Set up the apparatus there described. Immediately close the flask with a rubber stopper through which a spray trap is inserted. Continue, starting (page 144) with "The spray trap has previously been connected to . . . ".

Combined Ammonia by Distillation. This determination gives the total free and combined ammonia present.

PROCEDURE. Weigh into a 500-ml. Kjeldahl flask sufficient of the solid or liquid sample to give an amount of ammonia equivalent to about 40 ml. of 0.1*N* or 0.5*N* acid. Add about 200 ml. of water to dissolve the sample. Continue as for the Kjeldahl determination (page 144) starting with "Then add about 0.1 gram of No. 80 granulated zinc to prevent bumping . . . ". To obtain combined ammonia, subtract the free ammonia obtained by direct titration or distillation without addition of sodium hydroxide from the total ammonia. Calculate the excess to the ammonium salt present.

Ammonia in the Presence of Organic Amines. In commercial practice, frequently organic compounds are present that partly distill or give off ammonia on distillation with strong alkali, more or less slowly. Thus, triethanolamine, mono- and diethanolamine, morpholine, urea, proteins, glue, and gelatin will all give alkaline vapors under these conditions. An accurate determination of the ammonia present is not then possible. A rough estimate may be obtained by very quickly distilling off the ammonia. The small amount of ammonia from any organic compound present will be roughly balanced by any small amount of residual ammonia left undistilled. With mono- and diethanolamine, urea, morpholine, and other substances that readily, though not quantitatively, give off alkalinity, this method is not applicable. In all such circumstances, determine total nitrogen by the Kjeldahl method. Other methods, depending on solubilities, may be employed to determine the amino compound.

If the acid-amino compound, as is usually the case, is water-soluble, a useful method is to just acidify the solution with hydrochloric acid, and, in the absence of, or in the presence of a known amount of other soluble matter, evaporate to dryness on a steam bath and then in an air oven. Heat sufficiently long to drive off any excess hydrochloric acid, but not so long that the amino compound volatilizes. From the weight of the hydrochlorides of ammonia and the amino compound and the amount of total nitrogen present, the relative amounts of each may be calculated.

As an example, consider a mixture containing ammonia and triethanolamine, the molecular weight of the commercial triethanolamine being about 140. Such a combination is not at all infrequent. The following data are from actual determinations. The ammonia was quickly distilled and titrated. Triethanolamine distills, but slowly. A total nitrogen figure was obtained by the Kjeldahl method. The mixed hydrochlorides were obtained as described above.

	Per Cent
Ammonia by direct distillation, calculated as nitrogen..............	0.139
Total nitrogen by Kjeldahl method..............................	0.224
Total water soluble after acidification with hydrochloric acid, less the hydrochloride of fixed alkali present, determined in the ash.......	1.82

If x = proportion of total nitrogen due to triethanolamine, then, since the molecular weight of triethanolamine hydrochloride is 140 + 36.5 = 176.5, and that of ammonium chloride is 53.5, the total weight per cent of the mixed hydrochlorides is

$$\left(0.224x \times \frac{176.5}{14}\right) + \left[0.224(1 - x)\frac{53.5}{14}\right] = 1.82$$

From this

$$x = 0.490$$

Hence

$$0.490 \times 0.224 = 0.110 \text{ per cent nitrogen due to triethanolamine}$$

and

$$0.510 \times 0.224 = 0.114 \text{ per cent nitrogen due to ammonia}$$

$$0.110 \times {}^{140}\!/_{14} = 1.10 \text{ per cent triethanolamine present}$$

$$0.114 \times {}^{17}\!/_{14} = 0.138 \text{ per cent ammonia present}$$

Nitrate. By Reduction to Ammonia. Powdered Devarda alloy, containing 50 parts of copper, 45 parts of aluminum, and 5 parts of zinc, when acted upon by sodium hydroxide solution, produces hydrogen, which readily reduces nitrates in solution to ammonia.

Procedure. Dissolve 0.5 gram of the nitrate in 100 ml. of water, and transfer to a special Devarda flask (Fig. 80) with a ground-in joint containing a tube reaching nearly to the bottom of the flask. For very accurate work, connect this flask to a scrubbing flask to absorb carbon dioxide and that in turn to one containing concentrated sulfuric acid to absorb ammonia fumes from the laboratory. The Devarda flask has another side arm that connects to a wide glass tube, containing glass wool and heated by an electric coil or by steam passed through a surrounding jacket.

When this special apparatus is not available, use a 500-ml. flask with a two-hole rubber stopper, one hole connecting to the acid and alkali scrubbers and the other hole used to hold a spray trap. The purpose of the heated scrubber or the spray trap is to prevent alkaline spray's being carried over into a 150-ml. conical receiving flask containing 25 ml. of 0.5*N* sulfuric acid with which it is connected by glass tubing. A further 150-ml. conical flask containing 2 ml. of the standard acid is connected to the receiving flask. Make provision so that a current of air may be pulled through the apparatus by connecting this last flask to a water pump.

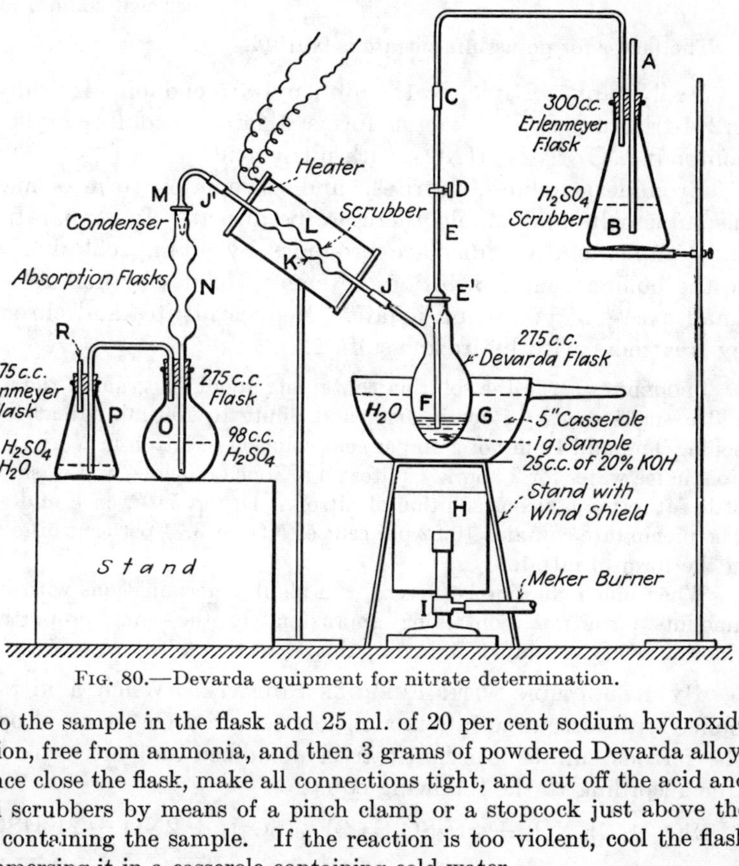

FIG. 80.—Devarda equipment for nitrate determination.

To the sample in the flask add 25 ml. of 20 per cent sodium hydroxide solution, free from ammonia, and then 3 grams of powdered Devarda alloy. At once close the flask, make all connections tight, and cut off the acid and alkali scrubbers by means of a pinch clamp or a stopcock just above the flask containing the sample. If the reaction is too violent, cool the flask by immersing it in a casserole containing cold water.

Allow the reaction to continue for 20 minutes, during which time the scrubber containing the glass wool is heated. Now heat the Devarda flask on a water bath, starting with cold water. Open the stopcock or pinch clamp to the acid and alkali scrubbers, and apply gentle suction to the receiving flask containing the 2 ml. of acid. Allow 5 to 6 bubbles a second to pass. Heat the flask, and aspirate air through the entire apparatus for 20 minutes.

Disconnect the scrubber or spray trap from the receiving flask, washing down the tube in the acid, inside and out. Add the acid from the second receiving flask, washing out thoroughly, and titrate the excess acid with standard alkali.

CALCULATION.

$$(\text{Ml. } 0.1N \text{ acid} - \text{ml. } 0.1N \text{ alkali}) \times 0.008501 \times \frac{100}{\text{weight of sample}} = \text{per cent sodium nitrate}$$

The factor for potassium nitrate is 0.01011.

BY NITRON. Nitron is 1:4-diphenyl-3:5-endanilo-4:5 dihydro-1:2:4-triazole, $C_{20}H_{16}N_4$. It forms a fairly insoluble crystalline compound, $C_{20}H_{16}N_4.HNO_3$, with nitric acid.

Bromides, iodides, nitrites, and chromates form somewhat insoluble salts and should therefore be absent. In preparation of the sample solution, displace bromides by adding chlorine water to the boiling neutral solution. Remove iodides by boiling with a small excess of potassium iodate. Reduce nitrites and chromates by treatment with hydrazine sulfate.

PROCEDURE. To the solution containing about 0.1 gram of the nitrate add a slight excess of acetic acid, and dilute to 100 ml. Heat to near boiling, and add 10 ml. of a 10 per cent solution of nitron in N acetic acid. Cool in ice water for 2 hours. Filter on a Gooch crucible, and wash with a little saturated aqueous solution of nitron. Dry at 110°, cool, and weigh. The precipitate contains 16.52 per cent of NO_3 or 3.74 per cent of nitrogen in the form of nitrate.

When much chloride is present, run blank determinations with known amounts of nitrate containing approximately the same proportion of chloride as in the sample, and apply this correction to the sample.

BY REDUCTION WITH FERROUS SULFATE. When a nitrate is boiled with ferrous sulfate in acid solution, in the absence of oxygen the ferrous sulfate is quantitatively oxidized and nitric oxide set free according to the following equation

$$6FeSO_4 + 2KNO_3 + 5H_2SO_4 = 3Fe_2(SO_4)_3 + 2NO + 2KHSO_4 + 4H_2O$$

After distillation of the nitrogen oxide, the amount of ferrous salt can be titrated quantitatively.

PROCEDURE. To a solution containing 0.1 gram of the nitrate in a 250-ml. conical flask add 50 ml. of $0.1N$ ferrous ammonium sulfate solution (page 720) and 70 ml. of concentrated hydrochloric acid. Fit the flask with a two-hole rubber stopper. One hole carries a tube leading to a flask containing a suspension of 50 grams of sodium bicarbonate in 100 ml. of water. The second hole in the stopper carries a medicine dropper containing 3 ml. of a 1 per cent solution of ammonium molybdate to be later used as a catalyst for the reaction.

To displace air from the flask while cold and thus prevent the side reaction of NO to NO_2, remove the stopper from time to time, and add 4 grams of sodium bicarbonate, a little at a time. Quickly close the mouth of the flask after each addition, thus displacing all the air.

Bring the solution in the flask to a boil for 2 to 3 minutes. Add the catalyst by pinching the bulb of the medicine dropper. Boil gently for about 10 minutes, and then replace the bicarbonate suspension by a fresh saturated solution. Remove the flame, and cool the flask in cold water. This will result in drawing in saturated bicarbonate solution to replace condensed steam. When room temperature is attained, remove the stopper, and add 35 ml. of a 40 per cent solution of ammonium acetate for every 50 ml. of solution to be titrated and then 4 ml. of 85 per cent phosphoric acid. The acetate, by buffer action, reduces the concentration of the strong acid to 1 to $2N$.

Titrate slowly with $0.1N$ potassium dichromate solution (page 728), using 7 drops of a 1 per cent solution of diphenylamine in concentrated sulfuric acid as indicator. Near the end of the titration, the green color of the solution deepens to a blue-green or, when much iron is present, to a grayish blue. Continue to add the dichromate solution drop by drop until the color changes to an intense violet-blue. The ferrous sulfate solution should have been standardized under the same conditions.

Calculation.

$$(\text{Ml. } 0.1N \text{ ferrous sulfate solution} - \text{ml. } 0.1N \text{ dichromate solution}) \times 0.00337 \times \frac{100}{\text{weight of sample}} = \text{per cent potassium nitrate}$$

The factor for sodium nitrate is 0.002834.

Nitrite. By Potassium Permanganate Oxidation. In acid solution, potassium permanganate oxidizes nitrite to nitrate quantitatively. All other materials oxidizable by potassium permanganate must be absent.

Procedure. Dissolve a preliminary sample containing 0.1 gram of the nitrite in 100 ml. of water, and add 10 ml. of 1 : 1 sulfuric acid. Run in $0.1N$ potassium permanganate solution (page 728) until a permanent pink is obtained. To a fresh sample similarly dissolved in water, add this amount of potassium permanganate solution and 10 ml. in excess. Now add 10 ml.

of 1:1 sulfuric acid. If the acid is added before the permanganate, there is danger of loss of nitrous acid by volatilization. Boil for a few minutes and titrate the excess permanganate with 0.1*N* oxalic acid solution (page 727).

CALCULATION.

$$2KMnO_4 + 5KNO_2 + 3H_2SO_4 = K_2SO_4 + 2MnSO_4 + 3H_2O + 3H_2O + 5KNO_3$$

$$(\text{Ml. } 0.1N \text{ permanganate solution} - \text{ml. } 0.1N \text{ oxalic acid solution}) \times 0.004255 \times \frac{100}{\text{weight of sample}} = \text{per cent potassium nitrite}$$

The factor for sodium nitrite is 0.00345.

Hydroxylamine. RASCHIG METHOD. In the absence of other reducing agents for ferrous compounds, hydroxylamine can be quantitatively determined by such reduction. The reaction is $2NH_2OH + 2Fe_2(SO_4)_3 = 4FeSO_4 + 2H_2SO_4 + N_2O + H_2O$.

PROCEDURE. Dissolve a sample containing about 0.1 gram of the hydroxylamine salt in 5 ml. of water in a conical flask, add 30 ml. of cold saturated ferric-ammonium alum solution, and mix well. Then add 10 ml. of 1:4 sulfuric acid, and mix again. Heat to boiling for 5 minutes. Dilute to 300 ml., and titrate the reduced iron immediately with 0.1*N* potassium permanganate solution (page 728).

CALCULATION.

$$\text{Ml. potassium permanganate titration} \times \text{normality} \times 0.01652 \times \frac{100}{\text{weight of sample}} = \text{per cent hydroxylamine}$$

Cyanide. VOLHARD METHOD. Cyanides are determined in the absence of halogens or other materials that react with silver nitrate to reduce or precipitate it. Excess silver nitrate is added and the excess back-titrated with a thiocyanate.

PROCEDURE. Dissolve a sample containing about 0.1 to 0.2 gram of cyanide in 75 ml. of water. Just neutralize with 1:3 nitric acid. Add 0.1*N* silver nitrate solution (page 729) until no more precipitate of silver cyanide is formed, and then add a slight excess. Make definitely acid with 1:3 nitric acid, and dilute to about 250 ml. Add 5 ml. of ferric nitrate indicator solution (page 717), and titrate with 0.1*N* potassium thiocyanate (page 729) or ammonium thiocyanate (page 725) until a permanent red color just appears in this solution. This end point is best seen in a porcelain casserole and in artificial light.

CALCULATION.

$$(\text{Ml. } 0.1N \text{ silver nitrate solution added} - \text{ml. } 0.1N \text{ thiocyanate used}) \times 0.006511 \times \text{aliquot} \times \frac{100}{\text{weight of sample}} = \text{per cent potassium cyanide}$$

The factor for sodium cyanide is 0.0049.

LIEBIG METHOD. Cyanide is also directly titrated with silver nitrate in alkaline solution. Halides and other materials that react with silver must be absent.

PROCEDURE. Dissolve sufficient sample to contain about 0.2 gram of the alkali cyanide in 75 ml. of water. Add 5 ml. of 10 per cent potassium hydroxide solution, free from chlorides. Titrate with 0.1*N* silver nitrate solution (page 729) with constant stirring, until a faint permanent turbidity forms. Potassium silver cyanide first forms. This is soluble in water, but when all the free cyanide is combined as this salt, further addition of silver nitrate decomposes it, precipitating silver cyanide. Use a dark background to view the end point.

CALCULATION.

$$\text{Ml. silver nitrate solution used} \times \text{normality} \times 0.1302 \times \frac{100}{\text{weight of sample}} = \text{per cent potassium cyanide}$$

The factor for sodium cyanide is 0.0980.

Ferrocyanide. In the absence of other substances that reduce permanganate, alkali ferrocyanide is directly determined by its reducing value.

PROCEDURE. Dissolve a sample of sufficient size to contain about 1 gram of the alkali ferrocyanide in 100 ml. of water, and acidify with 10 ml. of 1:1 sulfuric acid. Titrate in a casserole with 0.1*N* potassium permanganate solution (page 728) to a permanent pink color. The end point is difficult to detect accurately. The acid ferrocyanide solution is milky with a bluish tinge, and on the addition of the permanganate a yellow shade is first obtained. This turns to green. Finally, on the addition of excess permanganate, it becomes pink. Restandardize the permanganate solution against C.P. potassium ferrocyanide, $K_4Fe(CN)_6.3H_2O$, to minimize the error in reading the end point.

CALCULATION.

$$\text{Ml. permanganate used} \times \text{normality} \times 0.3683 \times \frac{100}{\text{weight of sample}} = \text{per cent potassium ferrocyanide}$$

The factor for sodium ferrocyanide is 0.3039.

Ferricyanide. The method for ferricyanide is related to that for ferrocyanide. The ferricyanide is quantitatively reduced in alkaline solution to ferrocyanide and the latter determined.

PROCEDURE. Dissolve a sample of such size that it will contain about 5 grams of the alkali ferricyanide in 75 ml. of water. Add 10 per cent potassium hydroxide to this until it is definitely alkaline to phenolphthalein. Heat the solution to boiling, and add an excess of a saturated solution of ferrous sulfate. At first, light brown ferric hydroxide is precipitated. When black ferrous-ferric hydroxide begins to form, the reaction is complete, and excess ferrous sulfate is being added. Cool, and dilute to 250 ml. in a calibrated flask. Filter through a dry paper. Titrate a 50-ml. aliquot with potassium permanganate solution, as in the case of ferrocyanide.

CALCULATION.

$$\text{Ml. permanganate used} \times \text{normality} \times \text{aliquot} \times 0.3292 \times \frac{100}{\text{weight of sample}} = \text{per cent potassium ferricyanide}$$

The factor for sodium ferricyanide is 0.2809.

CHAPTER 9

SULFUR AND ITS COMPOUNDS

The form of sulfur that must receive primary consideration is the sulfate. Others of lesser importance are sulfite, sulfide, thiosulfate, elementary sulfur and organically combined sulfur, in addition to the acids, sulfur dioxide and hydrogen sulfide. There are other forms, but they are encountered infrequently. Organic sulfur may be present as sulfonates, organic sulfates, or in lower stages of oxidation.

QUALITATIVE

Soluble sulfates are usually identified by precipitation of the insoluble barium salt. If solid, dissolve a sample in water, and filter out any insoluble matter. Make the solution distinctly acid to litmus with 1:1 hydrochloric acid. Heat to boiling, add a few drops of 10 per cent barium chloride solution, and mix. In the presence of substantial amounts of sulfate, an immediate precipitation of finely divided barium sulfate occurs. If only a small amount of sulfate is present, the precipitate may be slow in becoming visible. In such cases let the solution stand for an hour or two. Sometimes it is desirable to centrifuge, using a similar portion to which barium chloride has not been added as a blank. A fine white precipitate in the bottom, when compared with the blank, is then to be taken as a positive test. The presence of a small amount of sulfate in products containing sulfites is normal and due to atmospheric oxidation.

Insoluble sulfates must first be converted to soluble sulfates. For this purpose, fuse a portion of the sample with anhydrous sodium carbonate, using a platinum crucible if no substance injurious to platinum is present, otherwise a base-metal crucible. Let the melt cool, and dissolve in hot water so far as practicable. The element originally present as sulfate will now be present as the carbonate. Filter to obtain a clear solution of the sodium sulfate, and proceed as for soluble sulfates above, starting with "Make the solution distinctly acid to litmus . . . ". If silicates are present, filter off any precipitated silica before adding the barium chloride.

Sulfites when acidified with 1:1 hydrochloric acid evolve the characteristic pungent sulfur dioxide. The gas evolved may be led into acidified potassium permanganate solution, which it will decolorize. Also sulfite solutions may be treated with an excess of bromine water and then carefully acidified with 1:1 hydrochloric acid. The sulfite is thereby converted to sulfate. Boil off the excess of bromine water, and proceed as with soluble sulfates (page 157) starting at "Heat to boiling, add a few drops . . .".

Soluble sulfides when acidified with 1:1 hydrochloric acid evolve the characteristic odor of hydrogen sulfide. A piece of filter paper moistened with lead acetate solution will turn black when placed at the mouth of a test tube in which this reaction occurs. This test is exceedingly sensitive.

Insoluble sulfides can often be converted to soluble sulfides by fusion with sodium carbonate, but this should not be carried out in platinum. The aqueous extract on filtration when acidified will then liberate hydrogen sulfide.

Thiosulfates when acidified with 1:1 hydrochloric acid evolve sulfur dioxide and give a white opalescence of precipitated sulfur. They are strong reducing agents and readily decolorize free iodine in neutral or acid solution.

Organic sulfur compounds must be oxidized to decompose them and convert the sulfur to sulfate. It is most convenient to do this in an oxygen bomb (page 161) and test the washings for sulfur. If such a bomb is not available, the Eschka method (page 162) for sulfur can be applied as a qualitative test. Volatile sulfur compounds are not effectively converted by the Eschka method.

GRAVIMETRIC

Determination of Total Sulfur as Sulfate. SAMPLE. *Silicates, Rocks, and Insoluble Sulfates Other Than Lead Sulfate.* For silicates and rocks that may contain sulfide and sulfate, fuse the finely powdered sample in a platinum crucible with six times its weight of sodium carbonate and 0.5 to 1.0 gram of potassium nitrate as an oxidizing agent. If the sample is known to consist of barium sulfate, omit the potassium nitrate. Mix the sample intimately with the fusion mixture. To prevent sulfurous gases from the flame from coming in contact with the sample, insert the crucible in a hole in an asbestos board. To prevent absorption of sulfur gases from the laboratory place a platinum cover on the crucible. This also serves to retain the heat. First heat over a low flame until the reaction ceases, then increase the heat to a strong blast. While still red hot, dip the bottom half of the crucible in cold water to facilitate detachment of the melt from the crucible. When cool, dissolve in hot water. Filter the alkaline solution from insoluble matter such as the carbonates derived from insoluble sulfates, and wash the residue on the filter thoroughly with approximately 0.1 per cent sodium carbonate solution.

Carefully acidify the filtrate with 1:1 hydrochloric acid, and heat to boiling to drive off the carbon dioxide. Filter to remove gelatinous silica, if any is present, and wash well with 1:20 hydrochloric acid.

If silica was present, the resulting filtrate contains a small amount of silica that will contaminate a barium sulfate precipitate. For very accurate work, therefore, evaporate the filtrate to dryness, and bake at 150° for a few minutes. Take up the residue with a suitable volume of warm 1:20 hydrochloric acid. Bring just to the boil, immediately filter out the silica, and wash well with 1:20 hydrochloric acid. The filtrate is the sample for sulfate determination. Many of these steps can be by-passed if silica is absent. If a high degree of accuracy is not essential, do not carry out the second stage of silica removal, because provision for its removal is made later in the method.

Lead Sulfate. This would ruin a platinum crucible in the fusion method. Boil the sample, usually of about 1 gram, with 50 ml. of a 20 per cent solution of sodium carbonate for 30 minutes. Pass carbon dioxide through the solution while boiling. Filter off the lead carbonate, and wash the residue on the paper with 0.1 per cent sodium carbonate solution. Carefully acidify the filtrate and washings with hydrochloric acid, and boil to free from carbon dioxide. This is then ready for precipitation of sulfate.

Calcium and Strontium Sulfates. These can be processed as described for lead sulfate. It is more convenient to suspend a sample of 0.5 to 1.0 gram in about 20 ml. of hot water and add 1:1 hydrochloric acid with vigorous stirring until the sulfate has dissolved. If other, insoluble matter is present, it must then be filtered off before precipitation of sulfate.

Samples Containing Chlorates and Nitrates. If the original sample contains substantial amounts of chlorates or nitrates or they are introduced in processing, they must be removed, or the amount of barium sulfate precipitate will be inaccurate. If it is necessary to remove these radicals, add to the sample solution half its volume of concentrated hydrochloric acid, and evaporate to dryness. Take up the residue with 25 ml. of 1:20 hydrochloric acid, and use as sample.

PROCEDURE. Neutralize the solution of sulfate, unless it contains calcium or strontium sulfate, with 1:1 ammonium hydroxide, and add a slight excess, as is usual in precipitation of iron and aluminum. If calcium or strontium is present, this precaution must be omitted. Warm, but do not boil, to coagulate iron and aluminum hydroxides, if present. If precipitation occurs at this point, filter and wash the filter paper well with hot water. For work of the highest acuracy, use 1:20 ammonium hydroxide as the wash solution.

Neutralize the filtrate with 1:1 hydrochloric acid, dilute to about 250 ml., and add 4 ml. more of 1:1 hydrochloric acid. Bring to a boil, and add with constant stirring sufficient boiling 10 per cent barium chloride solution to precipitate all the sulfate, and a small excess. This is conveniently heated in and added from a test tube. Ten milliliters of a 10 per cent barium chloride solution will precipitate the equivalent of 0.13 gram of sulfur. The barium sulfate as originally precipitated is in too finely divided form for filtration. Therefore, boil the solution for 30 minutes and then allow the precipitate to settle. Alternatively, digest on a steam bath for at least 1 hour. Filter through a No. 40 Whatman filter paper or equivalent. An ignited and weighed Gooch crucible may be substituted if silica is known to be absent. Wash at least ten times with hot water. The final washings should give no test for chloride with silver nitrate solution.

Transfer the filter paper and precipitate to a weighed platinum or porcelain crucible. Dry with a low flame; then carefully burn off the filter paper. Finally heat strongly, but do not blast. Alternatively, place the crucible in a muffle furnace, burn off the paper at the mouth of the furnace, and gradually advance the crucible into the center of the muffle.

If silica has not been completely removed at an earlier stage, it should be removed here. The precaution is a desirable one in all accurate work and requires but a few moments. Add a couple of drops of 1:4 sulfuric acid and a drop of 48 per cent hydrofluoric acid to the precipitate in the cooled crucible. Evaporate off the water on a steam bath to avoid spattering. Heat gently over a flame to volatilize the sulfuric acid, and then heat strongly. This treatment not only volatilizes any silica present but also reconverts any barium sulfate reduced by the burning filter paper. Cool in a desiccator, and weigh.

Some workers prefer to add the barium chloride solution to the dilute solution of the sample while cold. This is then slowly heated to boiling with stirring and is claimed to give larger crystals of barium sulfate with less contamination. For the usual type of work, the precaution does not justify the extra time required.

CALCULATION.

$$\text{Weight of } BaSO_4 \times 0.1373 \times \frac{100}{\text{weight of sample}} = \text{per cent sulfur}$$

The following factors are useful:

Sulfur dioxide, SO_2	0.2744
Sulfur trioxide, SO_3	0.3430
Sulfuric acid, H_2SO_4	0.4202

Sulfate radical, SO_4	0.4115
Sodium sulfate, Na_2SO_4	0.6086
Potassium sulfate, K_2SO_4	0.7465
Calcium sulfate, $CaSO_4.2H_2O$	0.7376

Total Sulfur in Organic Materials. OXYGEN BOMB METHOD. This is the most accurate means of determination of organic sulfur, since the sample is burned in an atmosphere of oxygen, the sulfur being converted to sulfuric acid. The entire operation takes place in a hermetically sealed bomb.

Apparatus. The bomb should have a capacity of not less than 300 ml. and should be constructed so that liquids may be easily drained out and so that it will not leak during the test. The inner surface may be of platinum, gold, porcelain enamel, stainless steel, or similar material that is unaffected by the combustion process or product. The head gasket and lead wire insulation must be resistant to heat and chemical action. The sample cup may be made of platinum, glazed silica, or other suitable material with a capacity of not less than 2.5 or more than 5.0 ml. With a platinum cup, use platinum wire as a firing wire. Platinum or iron wire of about 34 Brown and Sharpe gauge may be used with a silica cup. Practically, this means that the usual combustion bomb for pressure combustion of coal, oil, etc., for calorific value is applicable.

SAMPLE. Normally not more than 2 grams of sample can be burned at one time. Therefore this method is not recommended when very small amounts of sulfur are present in an organic sample. If the sample will burn readily, weigh out 2 grams; if not, weigh out 1 gram. If the sample will not burn readily, add an equal quantity of pure, sulfur-free benzoic acid or naphthalene, and mix intimately.

PROCEDURE. Put the sample in the cup, and in turn place this in the bomb. Put the firing wire in place. This must be clean and so arranged that it touches the sample. When iron wire is used, cut a piece about 5 cm. but not more than 10 cm. in length and coil in a small spiral. To connect the terminals, it is permissible to use a piece of platinum wire of somewhat larger diameter than the iron wire, to complete the connection. Place about 5 ml. of distilled water in the bomb to saturate with water vapor the oxygen used for combustion. When corrosive compounds are produced on burning, as when organic chlorine substances are burned, add 0.5 gram of sulfur-free sodium carbonate to this water in the bomb. Close the bomb tightly, and admit pure oxygen under pressure slowly, in order to avoid blowing the sample from the cup. The desirable minimum pressure is indicated in the following table:

Capacity of Bomb, ml.	Minimum Gauge Pressure, Atmospheres
300–350	40
350–400	35
400–450	30
450–500	27.5
Over 500	25

Place the bomb in a container of cold water, and ignite the sample by making the appropriate electrical connection. After 10 minutes, remove the bomb, and release the pressure at a uniform rate such that the operation requires not less than 1 minute. Open the bomb, wash down the cover, and observe the inside for traces of unburnt sample or sooty deposit. If either is found, repeat the determination.

Transfer the contents of the bomb and the cup to a beaker, washing well with water, but do not exceed a total volume of 350 ml. Preferably the volume is 200 to 250 ml. If necessary, filter through a washed qualitative paper, and wash the filter paper. To the solution and washings, add 2 ml. of concentrated hydrochloric acid and 10 ml. of saturated bromine water to oxidize any sulfites to sulfates. Evaporate in a steam bath or on a hot plate to about 75 ml. Determine sulfate by precipitation, starting (page 160) at "Bring to a boil, and add with constant stirring sufficient boiling 10 per cent barium chloride solution . . . ".

Eschka Method. This method not only is standard for determination of sulfur in coal but is applicable to any organic compound that will not volatilize before burning.

Procedure. To 1 gram of the finely ground sample, add 3 grams of Eschka mixture, composed of 2 parts of C.P. porous calcined magnesium oxide and 1 part of anhydrous sodium carbonate, both sulfur-free. Intimately mix in a platinum crucible, and cover with 2 grams more of the Eschka mixture. Place the open crucible in a hole in an asbestos sheet, and heat very gradually to carry out the combustion. Charring normally takes place. Stir frequently with a platinum wire to allow access of air. After 30 minutes, increase the heat to a dull redness. The carbon content is burned out when the sample becomes yellow or light brown in color. Remove the heat, and cool the crucible.

Place the crucible and charge in 100 ml. of hot water, and boil gently for 30 minutes. Filter by decantation, and finally transfer the insoluble matter to the filter. Wash well with boiling water so that the filtrate and washings amount to about 200 ml. Add 3 to 5 ml. of bromine to oxidize any sulfites, make just acid with 1:1 hydrochloric acid, add 4 ml. excess, and boil the solution until the excess of bromine has been boiled off.

Precipitate the sulfate by the usual gravimetric method (page 160), starting with "Bring to a boil, and add with constant stirring sufficient boiling 10 per cent barium chloride solution . . . ".

Always run a blank, using the same amount of Eschka mixture and bromine. Subtract the barium sulfate so obtained from the amount obtained from the sample before calculating.

CARIUS METHOD. This is essentially the same as the corresponding method for chlorine determinations in organic compounds (page 131). Because of its insolubility in nitric acid, barium nitrate cannot be used in the way silver nitrate is in the determination of halogens. Aliphatic sulfides cannot be analyzed by this method, for the sulfones so formed by the action of the nitric acid are not decomposable in the Carius tube.

PROCEDURE. Follow the procedure for chlorine from the beginning (page 132), using the same weight of sample, through the step "Wash the contents of the bomb quantitatively with water into a beaker, avoiding transfer of any pieces of broken glass . . . ". To this solution add about 2 grams of sodium chloride and half its volume of concentrated hydrochloric acid. Evaporate to dryness to remove nitrates. Take up the residue with 50 ml. of 1:20 hydrochloric acid. Continue as usual for sulfate (page 160), starting with "Bring to a boil, and add with constant stirring sufficient boiling 10 per cent barium chloride solution . . . ".

SODIUM PEROXIDE FUSION METHOD. This is essentially the same as the corresponding method for chlorine determination in organic compounds (page 134).

PROCEDURE. Follow as there given through "Remove the steel crucible, wash carefully and filter the solution . . . ". Then dilute to about 100 ml., carefully neutralize with 1:1 hydrochloric acid and add 2 ml. excess. Precipitate the sulfate as usual (page 160), starting at "Bring to a boil, and add with constant stirring sufficient boiling 10 per cent barium chloride solution . . . ".

Sulfide Sulfur. There are numerous methods of determination of sulfide sulfur by evolution as hydrogen sulfide and absorption of the sulfide in alkaline or neutral solutions of lead, cadmium, silver, etc. Those which follow are representative.

SULFIDE AS LEAD SULFATE. *Apparatus.* Prepare a unit for evolution of hydrogen sulfide. The 250-ml. flask for the sample contains 10 grams of mossy zinc and 50 ml. of water. Its two-hole rubber stopper carries a separatory funnel and an outlet tube. The outlet tube leads to two 250-ml. absorption flasks with the usual inlets and outlets connected in series. As the absorption liquid, prepare an alkaline lead nitrate solution by pouring a saturated solution of lead nitrate into 100 ml. of a solution of 56 grams of potassium hydroxide in 140 ml. of water until the precipitate no longer dissolves on vigorous stirring. This requires about 3 volumes of the lead nitrate solution to 1 of the alkaline solution. Put 100 ml. of this in the first absorption flask and 50 ml. in the second.

PROCEDURE. Transfer about 0.5 gram of sample to the first flask, more if the sulfide content is low. For example, use 5 grams of iron or steel drillings, and even then the amount of sulfide being handled may be undesirably low. Add about 50 ml. of concentrated hydrochloric acid from the separatory funnel. Heat slowly, and finally bring to a boil until steam appears in the first absorption flask. Disconnect the flask while still heating so that the absorption liquid will not suck back. The sulfide is now in the absorption flasks as lead sulfide, usually all in the first. Let the precipitated lead sulfide settle, and filter. Wash first with cold water, then with hot, until the washings no longer give a test for lead. Dissolve the lead sulfide from the paper with hot 1:4 nitric acid, and add 2 ml. of concentrated sulfuric acid to the lead solution. Heat this to sulfur trioxide fumes to drive off the nitric acid and convert the lead completely to the sulfate. Take up after cooling, using about 10 ml. of cold water. Filter on a Gooch crucible, and wash with several portions of 50 per cent alcohol, finally with ether. Dry and weigh the lead sulfate.

CALCULATION.

$$\text{Weight of lead sulfate} \times 0.1057 \frac{100}{\text{weight of sample}} = \text{per cent sulfur as sulfide}$$

The factor for zinc sulfide is 0.3213.

SULFIDE AS CADMIUM SULFIDE. As absorption solution, dissolve 3.2 grams of cadmium chloride in 100 ml. of water and 60 ml. of concentrated ammonium hydroxide. Place this in the absorption system described for the lead sulfate method. Use the same weight of sample and method of evolution as given there.

Filter the precipitated cadmium sulfide on a Gooch crucible, and wash with 1:20 ammonium hydroxide until the washings no longer shows a test for cadmium. Dry at 100° and weigh.

CALCULATION.

$$\text{Weight of cadmium sulfide} \times 0.2219 \times \frac{100}{\text{weight of sample}} = \text{per cent sulfur as sulfide}$$

The factor for zinc sulfide is 0.6743.

SULFIDE AS BARIUM SULFATE. This is an alternative procedure superimposed on the cadmium sulfide method. Filter the cadmium sulfide precipitate on a small filter paper, and wash with 1:20 ammonium hydroxide until the washings no longer show a test for cadmium. Place the filter paper with the sulfide in a beaker, and cover with a watch glass. Mix with 5 ml. of water and 1 gram of potassium chlorate. Add 5 ml. of concentrated hydrochloric acid. Allow to stand for about 10 minutes to oxidize the sulfide to sulfate. Warming at this time might cause a loss of sulfide. Finally warm to be sure the reaction is complete, add about 50 ml. of water, and filter. Wash the residue on the filter paper well with water.

Complete as for the usual sulfate precipitation (page 160), starting at "Neutralize the filtrate with 1:1 hydrochloric acid, dilute to about 250 ml., and add 4 ml. more of 1:1concen trated hydrochloric acid."

CALCULATION.

$$\text{Weight of barium sulfate} \times 0.1374 \times \frac{100}{\text{weight of sample}} = \text{per cent sulfur as sulfide}$$

The factor for zinc sulfide is 0.4174.

Sulfide and Polysulfide Sulfur. Samples such as iron pyrites and others that are not simple sulfides will not give their entire sulfide content by evolution as hydrogen sulfide. In those cases, either a determination of total sulfur is necessary or the sulfur can all be reduced to sulfide sulfur. For the latter, mix a suitable sample according to the sulfur content, such as a 5-gram sample of iron ore, with an equal weight of a mixture containing equal amounts of aluminum powder, carbon, and sodium bicarbonate. Wrap in a filter paper, place in a nickel crucible, and cover with a lid. Heat gently at first, gradually increasing the heat so that it is finally blasted for 30 minutes. Cool, and transfer to the evolution flask. Proceed to determine sulfides by one of the several methods immediately preceding.

Sulfate and Sulfite in the Same Sample. This is carried out stepwise. The sulfite is decomposed and the resulting sulfur dioxide displaced by carbon dioxide in the absence of air. Such a sample gives the true sulfate. In another sample, sulfite is oxidized to sulfate and the total determined as barium sulfate. An alternative method of determining sulfite in the absence of other reducing substances is by direct titration (page 172). Yet another, if the reducing substances are not volatile, is by distillation of the sulfur dioxide and its absorption and determination as sulfate (page 168). Select the combination that appears simplest.

SULFATE. Measure out a suitable sample of solution and dilute to 250 ml., or suspend a 1- to 5-gram sample of solid in 250 ml. of water in a 500-ml. conical flask. Pass carbon dioxide through the solution to displace air and protect from air exposure. Heat the solution to boiling, and add 1:1 hydrochloric acid until it is definitely acid, then 4 ml. in excess. Continue to boil and pass the carbon dioxide through until decomposition of sulfite is complete. Filter if necessary. Then precipitate the sulfate by the usual method given on page 160, starting at "Bring to a boil, and add with constant stirring sufficient boiling 10 per cent barium chloride solution . . . ". This gives the amount of sulfate.

Sulfite. To another similar sample dissolved or suspended in water add 2 ml. of bromine; and, when the solution is saturated, carefully acidify by addition of 1:1 hydrochloric acid, continuing until 4 ml. in excess have been added. Boil this solution until the excess of bromine has been removed. The sulfite and sulfate are now both present as sulfate. Precipitate this as in the previous case. These two results should be calculated in terms of sulfur. The difference between them is the sulfur present as sulfite.

Sulfate and Sulfide in the Same Sample. The problem and solution are the same here as for sulfate and sulfite. The sulfide may be volatilized in the absence of air to give the true sulfur as sulfate; in the absence of other reducing substances it may be titrated directly (cf. page 171), or it may be volatilized and titrated with iodine (page 171).

Sulfate. Proceed as for sulfate in the presence of sulfite as previously described, the only difference in the first operation being that hydrogen sulfide rather than sulfur dioxide is being removed.

Sulfide. Proceed as for sulfite in the presence of sulfate. In the second operation the sulfide is oxidized with bromine to sulfate just as the sulfite was. The result in terms of sulfide sulfur is then obtained by difference.

VOLUMETRIC

Sulfate. Benzidine Method. Benzidine sulfate is practically insoluble in dilute hydrochloric acid solution, although benzidine hydrochloride is soluble. Benzidine itself is a weak base and is neutral to phenolphthalein. Therefore, the acidity of the sulfate in benzidine sulfate can be titrated directly with alkali.

Any sample containing soluble sulfate in neutral solution is suitable. Methods of conversion of insoluble sulfate to soluble sulfate for gravimetric use (page 157) may also be applied here, provided that precipitation does not occur on neutralizing. The final sample solution should be one containing about 0.5 gram of sulfate radical in 250 ml. in a calibrated flask.

Procedure. Prepare a benzidine hydrochloride solution by mixing 6.7 grams of the base or 9.57 grams of benzidine hydrochloride, with 20 ml.

of water. Add 20 ml. of concentrated hydrochloric acid, and dilute the mixture to 1 liter. The solution is brown. A slight precipitate may settle out on standing, but the solution is unimpaired. One milliliter of this solution is equivalent to 0.00357 gram of sulfuric acid, H_2SO_4.

Transfer a 50-ml. aliquot of sample to a beaker, and render nearly neutral with 0.5*N* sodium hydroxide. Stir in vigorously an equal volume of the reagent solution, and allow the precipitate to settle. Filter on a large Gooch crucible through a double filter paper, applying only gentle suction. Use some of the clear filtrate to wash the last traces of precipitate from the beaker into the Gooch crucible. Finally wash the crucible with 20 ml. of cold water in small portions. This procedure is necessary to minimize the loss due to the slight solubility of the precipitate in water.

Transfer the precipitate and filter paper to a conical flask, add 50 ml. of water, and shake thoroughly. Warm to 50°, add phenolphthalein indicator, and titrate with 0.1*N* sodium hydroxide solution. When near the end point, boil the liquid for 5 minutes, cool somewhat, and finish the titration.

CALCULATION.

$$\text{Ml. alkali} \times 0.1 \times 0.04904 \times \frac{100}{\text{weight of sample}} = \text{per cent sulfuric acid}$$

ORGANIC AND INORGANIC SULFATE BY BENZIDINE. Both organic and inorganic sulfate are precipitated by benzidine. The organic sulfate compound is soluble in alcohol; the inorganic is insoluble. The reaction of the organic sulfate is

$$2C_nH_{2n+1}SO_4Na + C_{12}H_8(NH_2.HCl)_2 = (C_nH_{2n+1}SO_4HNH_2)_2C_{12}H_8 + 2NaCl$$

ORGANIC SULFATE. Weigh 6.426 grams of benzidine hydrochloride into about 50 ml. of water. Heat until dissolved, filter if necessary, and dilute to 1 liter.

Transfer 10 ml. of the benzidine hydrochloride solution to a 250-ml. conical flask. If the sample is alkaline, neutralize with *N* hydrochloric acid. If 0.5 to 2.0 gram per liter of the sulfated compound is present, heat the reagent and 10 ml. of sample to about 40°, and add the sample to the reagent with constant stirring. If less than 0.5 gram per liter is present, increase the volume of sample solution to 50 ml. or even 100 ml. If over 2 grams per liter is present, increase the volume of reagent to 20 ml. and, if necessary, reduce the volume of sample solution to 15 or even 10 ml.

Rotate the flask vigorously until the precipitate flocculates, and allow to stand for 5 to 10 minutes. Filter through a qualitative paper, rinse the flask three times with 15-ml. portions of water, and pour each through the filter paper. Wash the filter and precipitate three times more with water, and allow the wash water to drain well each time. Remove the remaining water in the neck of the funnel. Now wash the precipitate with boiling neutral alcohol until 40 to 60 ml. of filtrate are collected in a 50-ml. graduated cylinder. Pour this alcohol solution into the original precipitation flask, add 3 to 4 drops of a 0.04 per cent alcoholic solution of bromocresol purple, and warm nearly to boiling for 5 minutes on a water bath.

Titrate the hot solution with $0.01N$ sodium or potassium hydroxide until the yellow color turns blue. Always run a blank on the filter paper, which is normally slightly acidic, by pouring 50 ml. of hot alcohol through the folded paper and titrating as before.

CALCULATION.

$$(\text{Ml. alkali} - \text{blank}) \times 0.01 \times 0.2883 \times \frac{100}{\text{weight of sample}} = \text{per cent sodium lauryl sulfate}$$

Other factors are:

Sodium myristyl sulfate, $C_{14}H_{29}SO_4Na$	0.3163
Sodium palmityl sulfate, $C_{16}H_{33}SO_4Na$	0.3443
Sodium stearyl sulfate, $C_{18}H_{37}SO_4Na$	0.3724

The above factors are for pure products. For a technical compound containing inorganic salts and mixtures of different homologues, determine an empirical factor, using 25 ml. of a 0.10 per cent solution of the compound by proceeding exactly as in the method.

Inorganic Sulfate. The precipitate remaining after washing with alcohol in the previous method is the inorganic sulfate-benzidine compound. This can be titrated and the sodium sulfate calculated from it. Since washing with water always dissolves a little benzidine sulfate, it is preferable to start with a fresh sample and minimize water washing.

PROCEDURE. Proceed as for organic sulfate, but, after all the solution has been removed by filtration, immediately wash, not with water but with approximately 50 ml. of hot neutral alcohol. Transfer the filter paper and residue to a flask, and add 50 ml. of boiled water. Warm to incipient boiling, add 5 to 6 drops of a 0.02 per cent alcoholic solution of phenol red, and titrate while hot with $0.01N$ sodium or potassium hydroxide solution until a red color appears. No blank is necessary, for the alcohol washes the filter paper free from acidity.

CALCULATION.

$$\text{Ml. alkali} \times 0.01 \times 0.071 \times \frac{100}{\text{weight of sample}} = \text{per cent sodium sulfate}$$

Sulfur Dioxide. MONIER-WILLIAMS METHOD. Sulfur dioxide in solution or formed by the action of acid upon sulfites is distilled into a solution of hydrogen peroxide containing a known amount of hydrochloric acid. The increase in acidity may be titrated, or the sulfuric acid may be precipitated and weighed. This method is suitable for small quantities of sulfur dioxide in food and other products where it is used as a preservative. It is applicable in the presence of volatile sulfur compounds other than sulfur dioxide.

PROCEDURE. The apparatus is shown in Fig. 81. Connect a 750-ml. round-bottomed flask *B* to an upward-sloping reflux condenser *D*, the lower

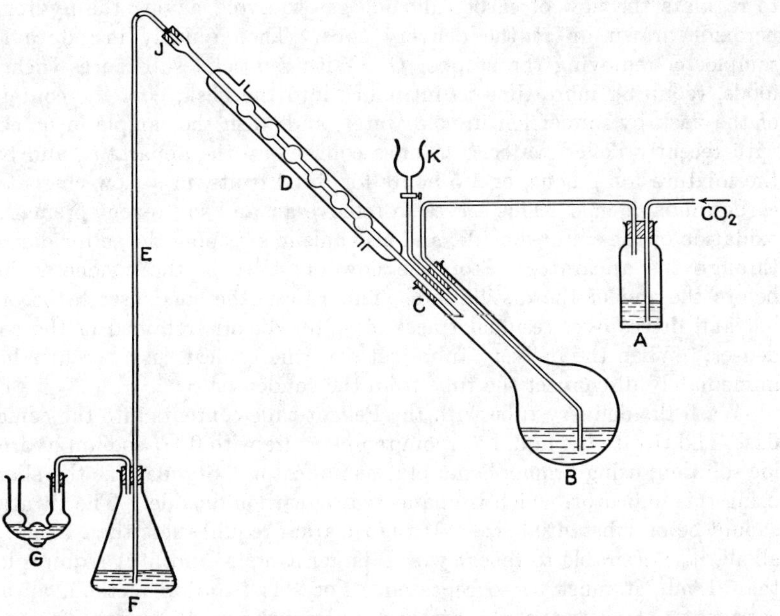

FIG. 81.—Apparatus for quantitative determination of sulfur dioxide by distillation.

end of which is cut off at an angle. Pass carbon dioxide gas through a 10 per cent solution of sodium carbonate in the flask *A* to remove any chlorine. Also connect a dropping funnel *K* to *B* through a three-hole stopper *C*. Connect the upper end of the condenser *J* to a 200-ml. conical flask *F* by a long piece of glass tubing *E*. To the conical flask connect a Peligot tube *G*, which is used to catch any traces of sulfurous acid not absorbed in the conical flask. Use rubber stoppers throughout. Add 15 ml. of neutral 3 per cent hydrogen peroxide solution to the conical flask and 5 ml. to the Peligot tube. Since hydrogen peroxide usually contains sulfuric acid, purify it as follows: Dilute 30 per cent hydrogen peroxide solution with its own volume of water, and neutralize with a solution of barium hydroxide, using bromophenol blue solution (page 116) as indicator. Allow the precipitate to settle in the cold. Filter from the barium sulfate, and determine the strength of the peroxide solution by permanganate titration (page 719). Finally adjust to 3 per cent strength. The bromophenol-blue indicator in the hydrogen peroxide solution remains unaffected for some time.

Add 300 ml. of water and 20 ml. of concentrated hydrochloric acid to the flask *B*, connect up the apparatus, pass a current of carbon dioxide gas, and boil for a few minutes. The sample is now to be added to flask *B*. If the sample is liquid, add through the dropping funnel *K*. If it is an easily transferable solid, first cool the contents of the flask somewhat, taking care to regulate the flow of carbon dioxide gas to avoid having the hydrogen peroxide drawn up in the delivery tube. Then quickly introduce the sample by removing the stopper *C*. With semisolid substances such as foods, requiring more time to introduce into the flask, cool the contents of the flask by immersion in cold water, and wash the sample in quickly with recently boiled water. At once connect up the apparatus, and boil the mixture for 1 hour, or 1.5 hours for dried fruits, in a slow current of carbon dioxide gas. This serves to remove air and so prevent premature oxidation of the sulfur dioxide, and also aids in sweeping the sulfur dioxide through the apparatus. Stop the flow of water in the condenser just before the end of the distillation. This causes the condenser to become hot and drives over residual traces of sulfur dioxide retained in the condenser. When the delivery tube just over the conical flask becomes hot, immediately disconnect the tube from the condenser.

Wash the delivery tube with the Peligot tube contents into the conical flask, and titrate the liquid at room temperature with $0.1N$ sodium hydroxide solution, using bromophenol blue as indicator. Standardize the alkali, using this indicator, which is unaffected by carbon dioxide. The titration should be of substantial size. If the titration requires less than 10 ml. of alkali, it is desirable to repeat with a larger sample; and, if it requires less than 1 ml., it must be so repeated. For this titration method, always carry out a blank on the reagents, and subtract the result from the titration.

As an alternative or as a check on the titration, the sulfur can be determined gravimetrically. For this, dilute the neutralized solution to about 100 ml., and add 1 ml. of concentrated hydrochloric acid. Then proceed as for sulfate on page 160, starting with "Bring to a boil, and add with constant stirring sufficient boiling 10 per cent barium chloride solution . . . ". Subtract the blank obtained on the reagents if a high degree of accuracy is required; usually it is unnecessary to do so.

Calculation.

$$\text{Ml. of alkali used} \times \text{normality} \times 0.03203 \times \frac{100}{\text{weight of sample}} = \text{per cent sulfur dioxide}$$

$$\text{Weight of BaSO}_4 \times 0.2744 \times \frac{100}{\text{weight of sample}} = \text{per cent sulfur dioxide}$$

Sulfide. BY IODINE. For rapid work with possibly lower accuracy, the sulfide distilled and precipitated as cadmium sulfide (page 164) may be titrated.

Transfer the contents of the absorption flasks, or the first if no color shows in the second, to a liter beaker. Dilute to nearly 1 liter with cold water. Carefully add 1:3 hydrochloric acid until distinctly acid, avoiding vigorous stirring because it would promote loss of hydrogen sulfide. The cadmium sulfide will completely dissolve. Titrate the sulfide with standard 0.1*N* iodine solution.

CALCULATION.

$$\text{Ml. titration} \times \text{normality of iodine solution} \times 0.01603 \times \frac{100}{\text{weight of sample}} = \text{per cent sulfide as sulfur}$$

Sulfides or Hydrosulfides. BY IODINE. The amount of sample required will depend on the sulfide or hydrosulfide content. Weigh out or measure a volume of sample equivalent to 0.05 to 0.075 gram of sodium sulfide. Dissolve in 50 ml. of water or dilute to about 50 ml. Add an excess of 0.1*N* iodine solution (page 726), usually 50 ml. Render the solution distinctly acid with 1:1 hydrochloric acid or 30 per cent acetic acid. Then titrate the excess iodine solution with 0.1*N* sodium thiosulfate solution (page 730), using starch indicator toward the end of the titration.

CALCULATION.

$$(\text{Ml. } 0.1N \text{ iodine solution} - \text{ml. } 0.1N \text{ thiosulfate titration}) \times 0.001703 \times \frac{100}{\text{weight of sample}} = \text{per cent sulfide calculated as hydrogen sulfide}$$

The corresponding factor as sodium hydrosulfide is 0.002804 and as potassium hydrosulfide 0.003608.

Sulfide and Hydrosulfide. BY IODINE. The following equations show that twice as much acidity is present when iodine acts on sulfides as when it acts on hydrosulfides.

$$H_2S + I_2 = 2HI + S$$
$$NaHS + I_2 = NaI + HI + S$$

In the latter case, the total acidity is unaltered, although the pH level at which it is available is altered. The original reactions with hydrochloric acid are as follows:

$$Na_2S + 2HCl = 2NaCl + H_2S$$
$$NaHS + HCl = NaCl + H_2S$$
$$H_2S + I_2 = 2HI + S$$

It follows that the total hydrochloric acid required to decompose an alkali sulfide is equivalent to the hydriodic acid later liberated by the oxidation reaction. In the case of a hydrosulfide, on the other hand, twice as much acidity is liberated as was required to decompose the sample. In the absence of other alkaline materials, therefore, the amounts of sulfide and hydrosulfide can be determined acidimetrically, but with only a fair degree of accuracy.

PROCEDURE. Weigh out or measure out a sample the same as for sulfide or hydrosulfide, and dissolve in 50 ml. of water, or dilute to 50 ml. Add a known excess of 0.1*N* iodine solution, then a known excess of 0.1*N* hydrochloric acid. Then titrate the excess of iodine with 0.1*N* sodium thiosulfate solution. Finally titrate the excess acidity with 0.1*N* sodium hydroxide solution, using methyl orange as indicator. Subtract from this latter titration the milliliters of 0.1*N* acid added.

CALCULATION.

$$(\text{Ml. } 0.1N \text{ iodine solution} - \text{twice } 0.1N \text{ alkali solution}) \times 0.003903 \times \frac{100}{\text{weight of sample}} = \text{per cent sodium sulfide}$$

$$\text{Ml. } 0.1N \text{ alkali solution} \times 0.005607 \times \frac{100}{\text{weight of sample}} = \text{per cent sodium hydrosulfide}$$

Sulfites and Acid Sulfites. The titration of sulfurous acid by an alkali to a phenolphthalein end point is twice the titration necessary to obtain an end point using methyl orange as indicator. Sodium sulfite is neutral to phenolphthalein and alkaline to methyl orange, while sodium bisulfite is neutral to methyl orange but acid to phenolphthalein.

SODIUM SULFITE. Weigh out a sample equivalent to about 0.25 to 0.4 gram of anhydrous sodium sulfite, and dissolve in about 50 ml. of water, or measure out a corresponding amount of liquid sample. Titrate with 0.1*N* acid to a methyl-orange end point.

CALCULATION.

$$\text{Ml. acid} \times \text{normality} \times 0.1261 \times \frac{100}{\text{weight of sample}} = \text{per cent sodium sulfite}$$

The factor for $Na_2SO_3.7H_2O$ is 0.2522.

SODIUM BISULFITE. Weigh out a sample equivalent to about 0.3 to 0.5 gram of anhydrous sodium bisulfite, and dissolve in about 50 ml. of water, or measure out a corresponding amount of liquid sample. Titrate with 0.1*N* alkali to a phenolphthalein end point.

CALCULATION.

$$\text{Ml. alkali used} \times \text{normality} \times 0.1041 \times \frac{100}{\text{weight of sample}} = \text{per cent sodium bisulfite}$$

SODIUM SULFITE AND BISULFITE TOGETHER. Weigh out a sample equivalent to about 0.5 to 0.6 gram of sodium sulfite and bisulfite, and dissolve in about 50 ml. of water, or measure out a corresponding amount of liquid sample. First titrate with 0.1*N* alkali to the phenolphthalein end point to convert bisulfite to sulfite. Calculate the titration as bisulfite. Then titrate with standard acid to the methyl-orange end point. Subtract from this last titration the alkali titration to the phenolphthalein end point, and calculate the difference to sulfite. The mechanism will be clear from the following equations:

$$NaHSO_3 + NaOH = Na_2SO_3 + H_2O$$
$$Na_2SO_3 + HCl = NaHSO_3 + NaCl$$

SULFITES IODOMETRICALLY. Sulfurous acid, either combined or free, reacts with iodine in solution, forming sulfuric and hydriodic acids.

$$SO_2 + I_2 + 2H_2O = H_2SO_4 + 2HI$$

Always titrate the iodine solution with the sulfurous acid or sulfite. Low results are obtained by the reverse procedure, unless the solution contains less than 0.04 per cent of sulfur dioxide.

PROCEDURE. Dissolve 2.5 grams of the sample or an equivalent amount of sulfurous acid, and dilute to exactly 250 ml. Place 50 ml. or less of 0.1*N* iodine solution, depending on the strength of the sample solution, in a beaker, and add 0.5 ml. of concentrated hydrochloric acid. Now run in the solution of the sample from a buret, stirring constantly. When the yellow color is almost gone, add 2 ml. of starch solution, and titrate to the disappearance of the blue color. Calculate the weight of sample used for the titration from the volume of sample solution added.

CALCULATION.

$$\text{Ml. iodine solution} \times \text{normality} \times 0.03201 \times \frac{100}{\text{weight of sample}} = \text{per cent sulfur dioxide}$$

The following factors are useful:

Na_2SO_3	0.06303
$Na_2SO_3.7H_2O$	0.1261
$NaHSO_3$	0.05203
$Na_2S_2O_5$	0.04743

Sodium metabisulfate, $Na_2S_2O_5$, is the anhydride of $NaHSO_3$.

Sodium Thiosulfate. Sodium thiosulfate and iodine react in solution to give sodium iodide and sodium tetrathionate.

$$2Na_2S_2O_3 + I_2 = 2NaI + Na_2S_4O_6$$

Procedure. Weigh out a sample equivalent to 0.5 to 1.0 gram of hydrated sodium thiosulfate or 0.03 to 0.5 gram of anhydrous sodium thiosulfate, and dissolve in about 100 ml. of water. For liquid samples, measure out an equivalent amount, and dilute if necessary. Add 2 ml. of a 0.5 per cent starch solution, and titrate directly with the 0.1*N* iodine solution (page 726) to a definite blue coloration.

Calculation.

$$\text{Ml. iodine} \times \text{normality} \times 0.1581 \times \frac{100}{\text{weight of sample}} = \text{per cent sodium thiosulfate}$$

The factor for $Na_2S_2O_3.5H_2O$ is 0.2481.

Sodium Thiosulfate and Sodium Sulfite Together. The sodium sulfite can be fixed as an insoluble formaldehyde complex if the solution is kept cold. The thiosulfate can then be titrated. By difference from the titration of the two together, the sulfite is determined.

Procedure. A suitable sample is 0.5 to 1.0 gram. Dissolve, and dilute to about 50 ml. To this solution add 10 ml. of 40 per cent formaldehyde solution and 10 ml. of 2*N* acetic acid per gram of sample. The formaldehyde-bisulfite compound separates on cooling. Keep the solution below 15° to prevent dissociation of this complex. Add 2 ml. of a 0.5 per cent starch solution, and titrate directly with 0.1*N* iodine solution (page 726) to a definite blue coloration. Calculate this as the thiosulfate present.

Repeat the titration with another sample of identical amount, but do not add the formaldehyde. Subtract the first titration from the second, and calculate the difference to bisulfite in the usual way.

Sodium Bisulfate. Weigh out or measure out a sample equivalent to 0.25 to 0.5 gram of sodium bisulfate, and dissolve in water to give a volume of about 50 ml. Add a few drops of methyl-orange indicator solution, and titrate with 0.1*N* alkali.

CALCULATION.

$$\text{Ml. alkali} \times \text{normality} \times 0.1201 \times \frac{100}{\text{weight of sample}} = \text{per cent sodium bisulfate}$$

Persulfates. OXALIC ACID METHOD. In the presence of silver sulfate as catalyst, oxalic acid reduces persulfates to sulfates.

$$H_2C_2O_4 + K_2S_2O_8 = K_2SO_4 + H_2SO_4 + 2CO_2.$$

PROCEDURE. To about 0.5 gram of the persulfate dissolved in water in a conical flask, add 50 ml. of 0.1*N* oxalic acid (page 727) and 0.2 gram of silver sulfate dissolved in 20 ml. of 10 per cent sulfuric acid. Heat the flask for 30 minutes on a steam bath to expel the carbon dioxide formed. Dilute the liquid to 100 ml., and titrate the excess oxalic acid at about 40° with 0.1*N* potassium permanganate solution (page 728). Subtract this titration from the 50 ml. of oxalic acid originally added to obtain the amount of oxalic acid used.

CALCULATION.

$$\text{Ml. oxalic} \times \text{normality} \times 0.1352 \times \frac{100}{\text{weight of sample}} = \text{per cent potassium persulfate}$$

The factor for $(NH_4)_2S_2O_8$ is 0.1141.

FERROUS SULFATE METHOD. In the cold, ferrous salts are quantitatively oxidized to ferric salts by persulfates,

$$2FeSO_4 + K_2S_2O_8 = Fe_2(SO_4)_3 + K_2SO_4$$

PROCEDURE. To 0.25 gram of the persulfate dissolved in water in a conical flask add a considerable excess of 0.1*N* ferrous ammonium sulfate solution (page 726). A suitable volume is 100 ml. Dilute the solution with 150 ml. of hot water, and titrate the excess ferrous salt with 0.1*N* potassium permanganate solution (page 728). Deduct this titration from the permanganate equivalent of the ferrous sulfate solution added. The difference is due to that oxidized by the persulfate.

The calculations and factors are identical with those in the oxalic acid method, replacing the milliliters of oxalic acid used by the milliliters of ferrous sulfate that reacted with the sample.

CHAPTER 10

MINERALS

The routine scheme for mineral analysis is generally used for an inorganic material containing common elements such as silica, iron, aluminum, and calcium. It may be applied to total solids from evaporation of a boiler water or scale from a boiler, to a sample of ash from a synthetic resin or the ash of a sludge of unknown composition, as well as to foundry slag and the use implied in the title, analysis of minerals. In short, in general laboratory work this chapter may well apply in so many types of samples as to receive more use than any other.

Since the presence of a very small amount of an element is sometimes of great importance, some methods of determination of traces are given. The majority of such determinations, including all those given here, are carried out colorimetrically. Some not included are by careful titration with reagents that are 0.01*N* or 0.001*N*.

Sample. If the sample is in massive form, it must be crushed down in a jaw crusher. That is rarely the job of the analytical laboratory. Usually samples delivered have been crushed to not over barley-size grains. For such a sample, quarter and grind (page 38) until a representative sample of not over 100 grams is available. Grinding should be in an agate mortar, porcelain being a poor second choice. During such sampling and subsequent use of the sample, exercise every precaution to see that the inevitable coarser grains do not "float" to the surface.

If the sample is wet, it is desirable to determine moisture on one sample and prepare another dried at 110° for analysis on a moisture-free basis.

QUALITATIVE

If the sample is soluble in concentrated hydrochloric acid, dissolve in that acid, by boiling if necessary. If the sample is like siliceous materials or slags and is not soluble in concentrated hydrochloric acid, mix about 0.5 gram of finely ground sample with

3 to 5 grams of anhydrous sodium carbonate in a 15- to 20-ml. platinum crucible, and fuse (page 188), starting with "Place a platinum cover on the crucible, and heat the latter with a small Bunsen flame . . . ". The worker must know from the nature of the sample that it will not injure platinum (page 24) and take preliminary precautions against the presence of reducing substances such as sulfur, and against lead, mercury, etc. In this chapter, substances injurious to platinum are assumed to be absent.

If the sample contains substantial amounts of silica it is most satisfactory to determine the silica quantitatively and dilute the filtrate to a known volume. Use small portions of this silica-free solution for qualitative examination and aliquots for the balance of the analysis. The subject of qualitative examination has been covered in Chap. 5.

MICROSCOPIC

One of the most characteristic properties of a mineral is its refractive index. Those considered here are silicas, including diatomaceous earth, pumice, talc, pyrophyllite, feldspar, clay, and asbestos. Analytical and physical data on others can be found in any good textbook on mineralogy.

Silicas. Minerals in this class are used as abrasives for cutting, for polishing, and as fillers. For various purposes, quite different types are applied. For microscopic examination, occasionally 100 magnifications is sufficient. More often 250 or 450 magnifications is desirable.

For general identification, the refractive index is simple and easily applied. It falls between 1.540 and 1.560, even for the somewhat impure grades. Place the specimen under the microscope in benzyl alcohol, which has a refractive index of 1.540. As the focus is raised, the Becke line (page 60) will move toward the inside of the grains. Similarly, in benzyl benzoate with refractive index 1.560, on raising the focus the Becke line will move toward the medium. Other media, single substances or mixtures, may be used in place of those mentioned.

Crystalline Silica. This is derived from crushed quartz or sand. A representative specimen is shown in Fig. 82. Analytically it is almost entirely silica. The color is usually white. Under the microscope, it shows up as irregular, transparent, glasslike pieces usually with no definite observable structure. It is normally quite a harsh abrasive and feels gritty when rubbed between the finger and thumb. Hexagonal prisms and pyramids are sometimes observable. This type of silica is in general a cutting abrasive or a filler.

Semidecomposed Silica. Long weathering will leach out soluble salts from a silicate rock, leaving an impure silica which under the microscope appears as semitransparent, irregular particles with no definite structure. Different rocks weather differently, and various types of semidecomposed silica of various colors are available. These show little difference under the microscope, although they are highly variable in analytical composition. Rottenstone is the residue from leaching of siliceous limestone and falls in this class. Tripoli, shown in Fig. 83, is another member of the family. Experience is necessary to become familiar with structural characteristics of each so that they can be identified

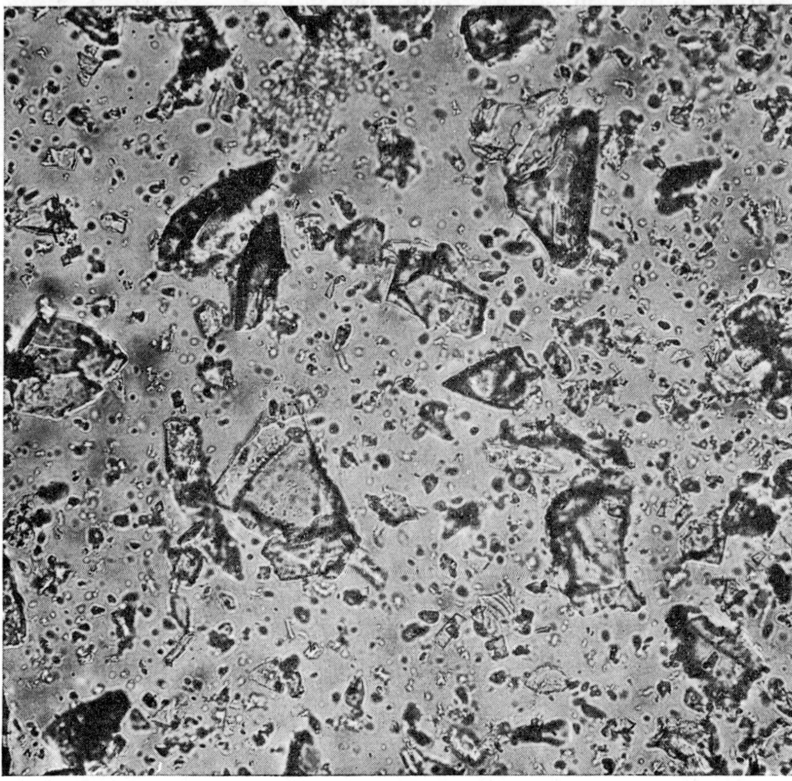

Fig. 82.—Crystalline silica, 250×.

promptly. Silicas of this type vary in hardness, generally depending upon how much they have been decomposed. They are definitely softer than the crystalline silicas and shade without any clear line of demarcation into the decomposed silicas. They are cutting abrasives for soft surfaces, polishing abrasives for hard surfaces, and fillers.

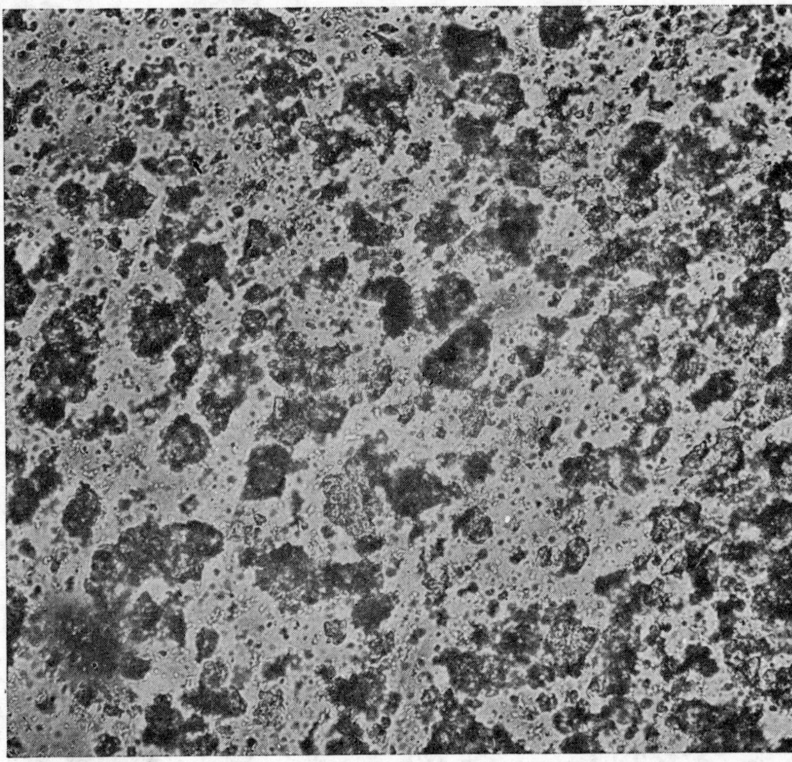

Fig. 83.—Tripoli, 250×.

Decomposed Silica. The term is purely a relative one, for analytically they often show the presence of several per cent of iron and aluminum. They are white or grayish white. Examined microscopically, they appear as fully opaque amorphous particles. As a rule they are quite soft, as is evidenced by the feel of them when rubbed between two fingernails. They are polishing or very mild cutting abrasives for soft metal surfaces, probably because of their porous rather than solid structure.

The polishing properties of an abrasive depend not only upon the type and size of the particles but also to a considerable extent upon the gradation of size of the particles. Some industrial silicas owe their advantages over others of similar type entirely to differences in this property. The variation in particle size is estimated qualitatively by examining several fields under the microscope but then, like so many other procedures, must be interpreted in terms of experience, the memory of how other specimens appeared.

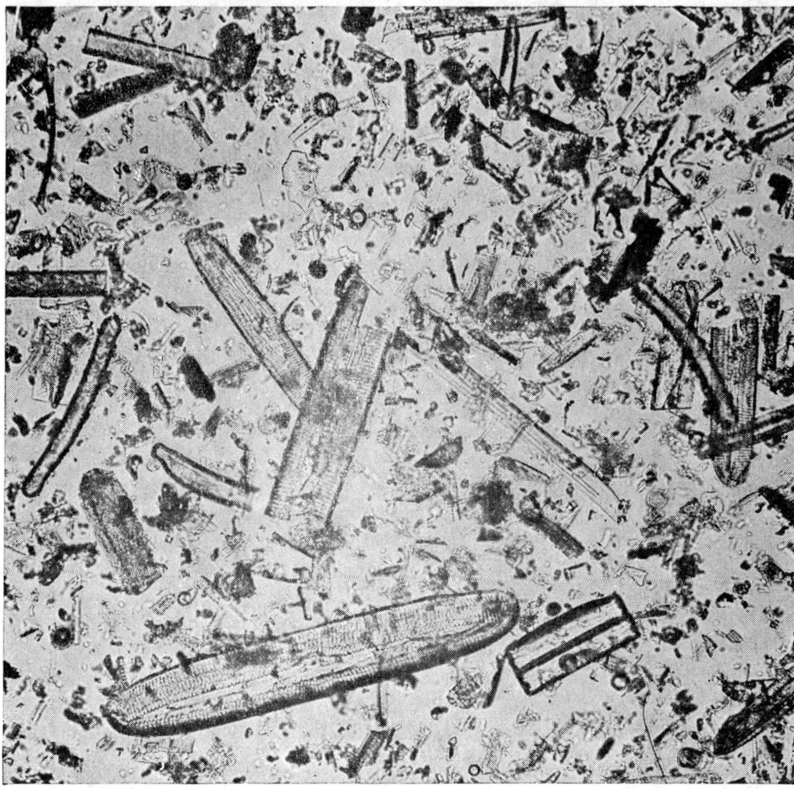

FIG. 84.—Fresh-water diatomaceous earth, 250 ×.

Diatomaceous Earth. This form of silica is composed of the skeletons of tiny aquatic organisms and finds use as a polish for soft surfaces, as a filler, as a filter medium, and as an insulator for heat. Colors are white, gray, yellow, and pink, often due to heat-treatment before grading. Because of its structure, diatomaceous earth is easily broken down during mechanical use, forming many minute broken edges that are sufficiently fragile to give a smooth abrasive action. Other names for the same material are diatomite and kieselguhr, the latter a German name applied prior to development of domestic deposits. Many grades are best known by trade names.

Diatomaceous earth is further classifiable in terms of the source, whether of fresh water or salt water origin. In general, the fresh water deposits such as those of Canada show elongated, bladelike structures (Fig. 84); salt water deposits show roughly circular and

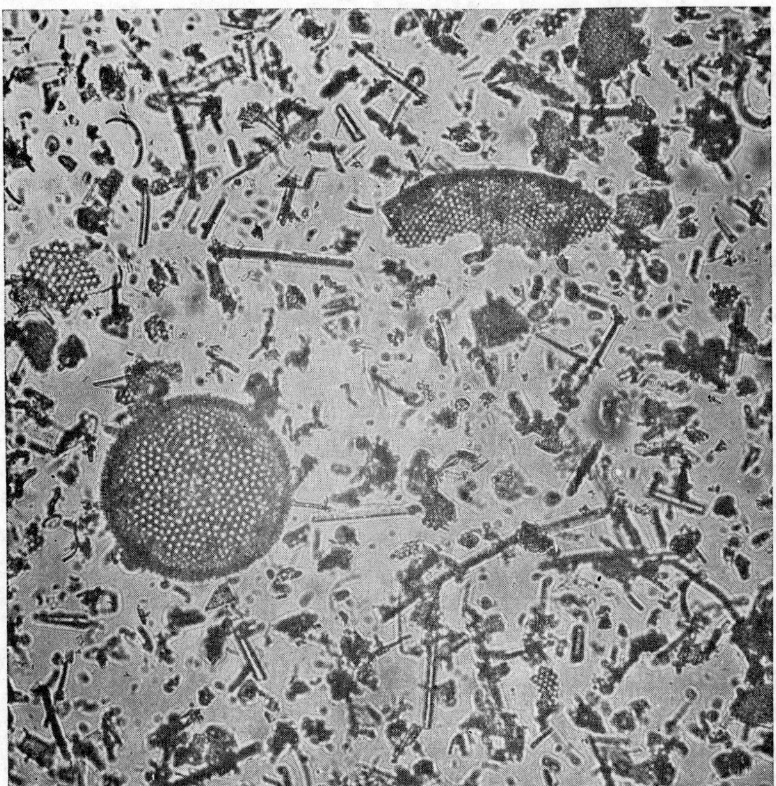

FIG. 85.—Salt-water diatomaceous earth, 250 ×.

honeycomb-like forms (Fig. 85), readily distinguishable by one with even small experience. In general it is necessary to use a magnification of at least 450 times in order to see the individual diatoms clearly. Many of those in a field will have been crushed and will be present as fragments. Some air-separated samples need even higher magnification for clear observation.

Pumice or Pumicite. These are grades of light, foamy, amorphous glasses such as is shown in Fig. 86, obtained by crushing of lava deposits. Analytically they are variable silicates of aluminum and the alkali metals. Both are normally grouped as pumice, the term pumicite having no clear significance. Pumice is white, gray, yellow, brown, or even reddish. Microscopically pumice is often glasslike in appearance with striations. These striations are not always present, and it is at times difficult to differentiate pumice

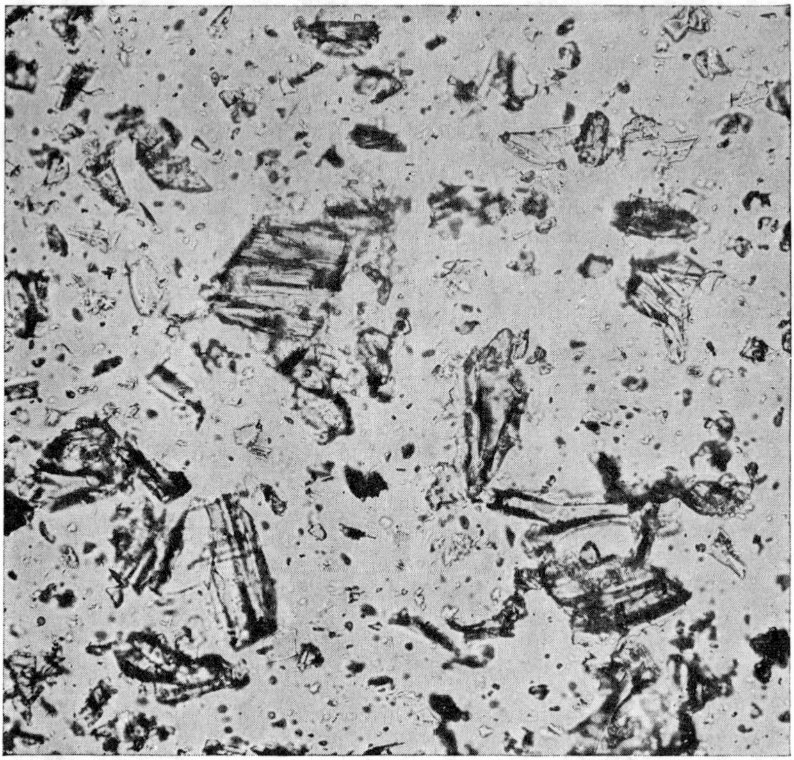

FIG. 86.—Pumice, 250 ×.

from crystalline silica by direct observation. Since pumice is an amorphous glass, it is distinguished from silica by the fact that glasses are completely dark no matter in which direction they may be turned when placed between crossed Nicol prisms. The refractive index of pumice is somewhat below that of clove oil, 1.526. Pumice may therefore be identified by observing the Becke line when immersed in this oil or another medium of similar refractive index. The Becke line will move toward the medium when the focus is raised if the specimen is pumice.

Talc. This form of magnesium silicate is relatively soft and is usually white. It has a slippery feel when rubbed between the fingers. Under the microscope, it shows a platelike structure, sometimes occurring as flat tablets and sometimes as masses of plates lying side by side (Fig. 87).

The refractive index of talc varies from 1.57 to 1.59 and is higher than that of either silica or pumice. When benzyl benzoate is used

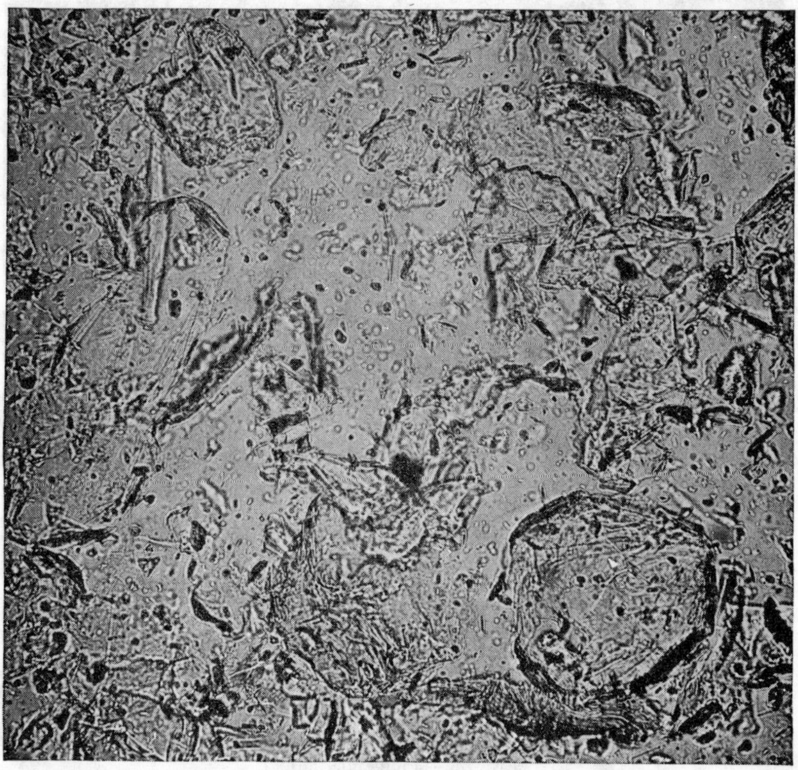

Fig. 87.—Talc, 250×.

as medium, the Becke line will move toward the inside of the grain of talc when the focus is raised. The crystal structure may be either orthorhombic or monoclinic. It is widely used as a dusting powder and filler.

Pyrophyllite. This hydrated aluminum silicate is soft and has a slippery feel similar to talc, for which it is sometimes mistaken. Microscopic identification is not easy, because the crystal form is orthorhombic and the refractive index 1.588, practically the same as that of talc. The two are readily differentiated by chemical analysis.

Feldspar. This hard mineral is a potassium or sodium aluminum silicate. Good grades are white, although an off tone to gray is sometimes found. Microscopically, although in the form

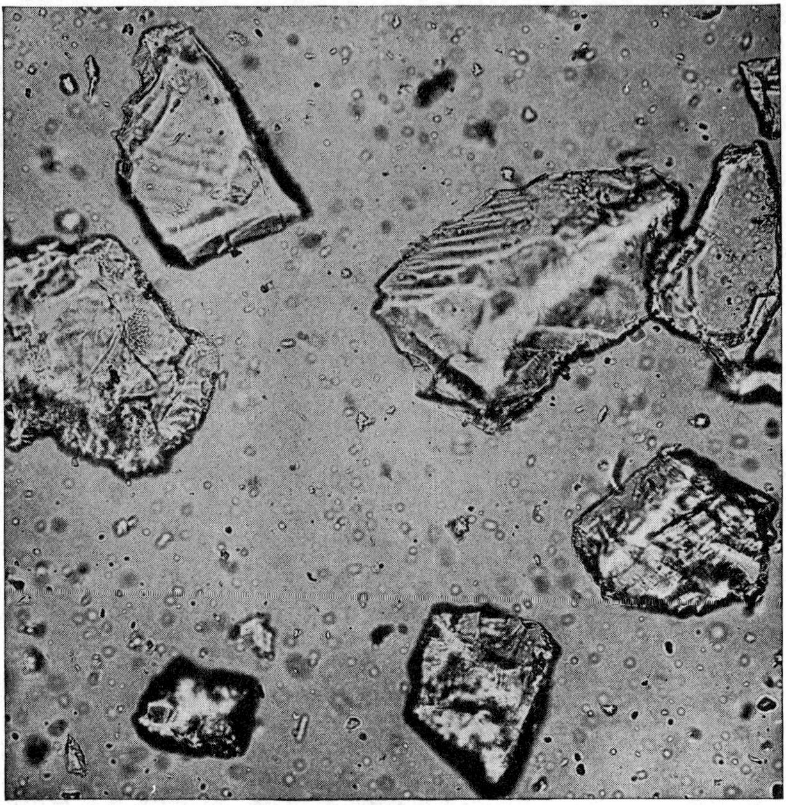

Fig. 88.—Feldspar, 250×.

of plates, it resembles pumice and crystalline silica, as may be noted from Fig. 88. Feldspar may often be distinguished from pumice by the absence of striations and from crystalline silica by the absence of hexagonal particles. Chemical analysis for confirmation of microscopic observation is highly desirable. The refractive index of all types of feldspar is below 1.540. When benzyl alcohol is used as medium, the Becke line will move toward the medium when the focus is raised.

Orthoclase is another common form of potassium aluminum silicate. It occurs as monoclinic crystals with a refractive index of 1.524.

Clay. China clay, or kaolin, is a soft white, amorphous, hydrated aluminum silicate that shows no definite structure under the microscope. This negative property is illustrated in Fig. 89.

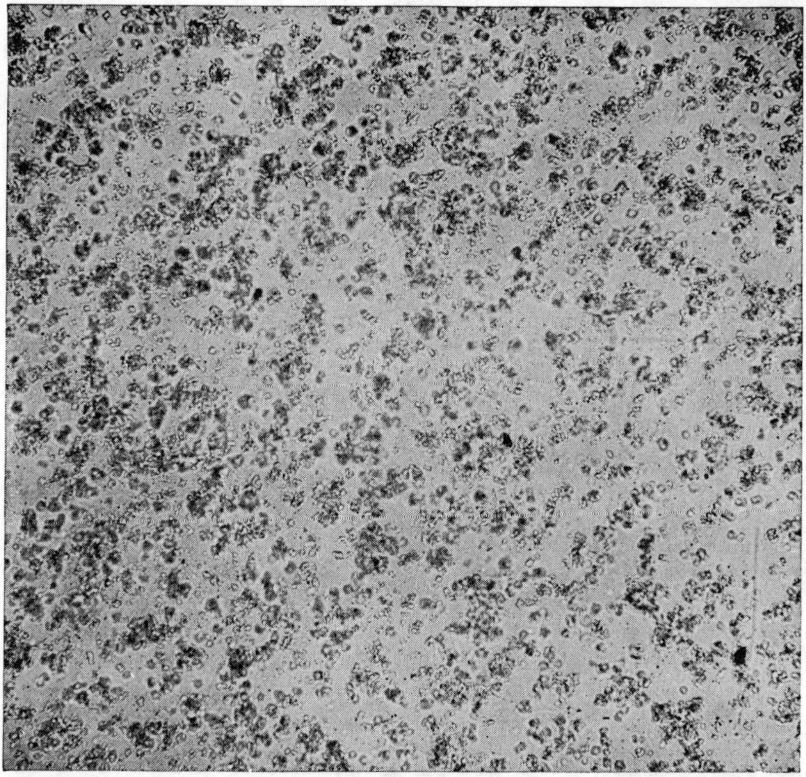

Fig. 89.—Clay, 250×.

It has a refractive index of 1.540 and consequently is not visible microscopically using benzyl alcohol as medium. The Becke line is not easily observable in examining amorphous powders, for particles with perpendicular sides cannot be located. Other grades of clays are used for varied purposes and may be colored owing to impurities.

Bentonite. This hydrated aluminum silicate is white to yellow in color, distinguishable from kaolin in that it swells in water. Apart from this property, it appears under the microscope to be very similar to clay, showing an amorphous structure. The refractive index is of the order of 1.55.

Asbestos. The most common form of asbestos is serpentine, which is a hydrated magnesium silicate. Amphibole asbestos contains calcium silicate with the magnesium silicate and is structurally weaker. The microscope shows a characteristic fibrous

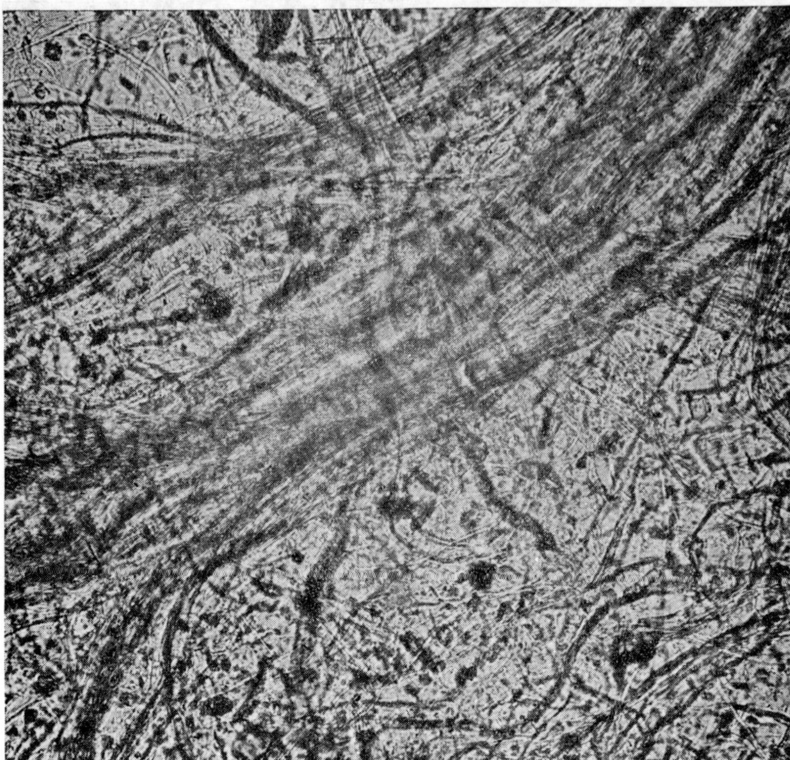

Fig. 90.—Asbestos, 250 ×.

structure composed of bundles of silky fibrous rods, as illustrated in Fig. 90. The refractive index varies appreciably with the type and ranges from 1.490 to 1.571. Although serpentine is monoclinic, the crystal structure is not readily apparent. Asbestine is a trade name for short-fiber talc.

QUANTITATIVE

Loss at 110°. Weigh out 0.5 or 1.0 gram of sample into a weighed platinum crucible and place in an oven at 110° until constant weight is attained. An apparently dry sample will usually show several tenths of a per cent of loss at 110° and often several per cent. The smaller sample is satisfactory if silica and carbonates are low; a larger one is needed if silica exceeds 50 per cent or if losses on drying and on low- and high-temperature ignition will be large.

Loss on Low-temperature Ignition. Cover the crucible containing the oven-dried sample, and heat carefully in such a manner that the crucible does not get red hot in any portion. The temperature will usually not exceed 500°. Most water of hydration, if so firmly bound as not to be lost at 110°, will be driven off. Few carbonates decompose at this temperature. Carbon burns off more effectively than at a higher temperature; if present, the heating should be continued until all black or gray color due to it is gone. Occasionally agitate the contents of the crucible by gently shaking or stirring the contents with a platinum wire so as to allow air to penetrate to all parts of the sample. Samples will rarely fuse at this temperature. If they do, they are probably high in alkali-metal salts. It is sometimes desirable to crush the ashed and partly cooled sample against the bottom of the crucible set on a hard inorganic surface, if fritting but not actual fusion has occurred. For the purpose, use a small agate pestle, and brush off the pestle with a small camel's-hair brush. Oxidation can be expedited by placing a small Nichrome triangle on the crucible and a porcelain lid on it, thus reflecting heat downward, at the same time leaving room for air access. When such ignition is completed, cool, and weigh.

Loss on High-temperature Ignition. Heat the crucible and contents after low-temperature ignition over a Meker-type burner. Note any evidence of reaction or fusion. Some water of hydration of minerals is driven off only at this relatively high temperature. Most carbonates, notably calcium carbonate, are decomposed to the oxide. Some sulfates decompose, and a few compounds will volatilize under these conditions. When blasting is completed, cool, and weigh.

Samples Soluble in Hydrochloric Acid. Use the sample from high-temperature ignition if no volatilization of inorganic material has taken place. If volatilization has taken place in high-temperature ignition, use an original sample unless organic matter is present; if it is, use a sample after low-temperature ignition.

If the sample is soluble in hydrochloric acid brush it with a camel's-hair brush into a 250-ml. beaker, or transfer the platinum crucible with the sample to the beaker, and add 25 ml. of 1:4 hydrochloric acid. Warm until solution of the sample is complete, and remove the crucible with a glass rod or platinum-tipped tongs, rinsing it down carefully with a wash bottle. Be sure that the resulting solution is free from suspended matter; if any is found that does not go into solution on boiling, prepare a fresh sample and use the procedure for samples insoluble in hydrochloric acid.

Samples Insoluble in Hydrochloric Acid. Follow the same precautions in selecting the sample as for acid-soluble samples. The following method is applicable to most samples, except that, in the presence of barium sulfate, a modification, which follows immediately after this one, is necessary.

Add five to ten times its weight of anhydrous C.P. sodium carbonate to the sample in the platinum crucible, and mix intimately. Alternatively, use a mixture of equal parts of sodium and potassium carbonates, which will fuse at a lower temperature. Place a platinum cover on the crucible, and heat the latter with a small Bunsen flame, gradually increasing the heat until the reaction and evolution of gases cease. Reaction may not commence until the whole is bright red. Different samples will require different treatment to avoid spattering or frothing over. Acid oxides in reacting will liberate carbon dioxide from the carbonate melt. Finally heat for 15 to 30 minutes at the full bright red heat produced by a Meker-type burner, after apparent reaction has ceased. Seize the crucible at the top edge with tongs, preferably platinum-tipped, and carefully swirl the liquid melt to form as large a surface as possible around the sides of the crucible. A worker with sufficient experience can then so cool the melt that it can be removed by slight tapping. This is done by chilling the bottom of the crucible in cold water while still slightly red, taking care not to immerse to the lip. Such quick chilling does not injure the platinum.

If possible, transfer the solid melt to a 250-ml. beaker, rinsing the crucible and the cover with hot water from a wash bottle. Otherwise transfer the crucible to the beaker, and rinse off the cover with hot water. Add about 75 ml. of water, and heat to boiling. Fused sodium carbonate in massive form is not readily soluble, hence the advantage of swirling in the crucible while hot and removal from the crucible so as to expose both sides of the melt for solution. An experienced worker can promote the solution of the melt by covering the beaker with a watch glass and carefully adding 1:4 hydrochloric acid to the hot solution in amounts sufficient to reduce the carbonate concentration but insufficient to acidify. Too rapid addition will cause violent frothing; too much acid will decompose the sodium silicate present and coat the melt with silicic acid. Too much acid is particularly undesirable if the crucible is in the beaker. A flattened end of a stirring rod will be effective to crush the residual portions of melt, particularly if it is not in the crucible.

When solution is complete, let the contents of the beaker cool somewhat, and if the crucible is present remove it with a glass rod or platinum-tipped tongs. In that case rinse the crucible with hot water, and if necessary use a policeman. Examine the solution carefully to be sure that everything has been dissolved.

Occasionally a sample will not react completely with carbonates. In that case carefully acidify the carbonate solution, filter on quantitative paper, and wash the residue with hot water. The further procedure is a matter of judgment. If the sample is siliceous, then the residue may be incompletely reacted silicate. In that event, ash the paper and fuse again with sodium carbonate, starting as above at "Add five to ten times its weight of anhydrous C.P. sodium carbonate . . . ". If the sample contains oxides that will not dissolve in carbonate but will dissolve in pyrosulfate, proceed as follows:

Transfer to the same crucible used for the carbonate fusion, and burn off the paper. Add 3 to 5 grams of C.P. potassium bisulfate or potassium pyrosulfate, and mix well. Heat gradually, for the bisulfate will froth, until all visible reaction ceases. This is a tedious job requiring care. Then place the lid on the crucible and heat to a bright red, occasionally swirling the melt to be sure that no undissolved particles remain. Finally in cooling, use the procedure recommended for a carbonate melt, and dissolve in 75 ml. of water. Addition of 10 to 20 ml. of 1:4 hydrochloric acid may promote solution. Combine the solution with that of the acidified carbonate melt as the sample solution.

Sample Modified for Determination of Barium Sulfate. When barium sulfate is present in the carbonate melt, the melt will dissolve incompletely in water, leaving a residue of barium carbonate. Proceed as usual up to solution of the melt. Acid must not be added. When the melt is well disintegrated, filter and wash the precipitate of barium carbonate until the wash waters are no longer alkaline. Remove the filtrate and wash waters, and acidify with 1:4 hydrochloric acid. This is the sample solution free from barium. Separately dissolve the precipitate on the filter with 25 ml. of hot 1:4 hydrochloric acid, washing the paper until the washings are no longer acid. Any detectable undissolved residue on the paper indicates that the fusion was incomplete, and it must be started again with a fresh sample. Use the acidified filtrate as a sample solution for determination of barium as the sulfate.

Barium as Sulfate. Heat the solution nearly to boiling, and carefully add 5 ml. of 1:10 sulfuric acid dropwise with vigorous stirring. Heat to boiling, and digest on a steam bath until the precipitate settles readily. Let settle, and filter on a retentive paper, washing well with small volumes of 1:100 sulfuric acid. Ignite, and weigh.

Calculation.

$$\text{Weight of barium sulfate} \times \frac{100}{\text{weight of sample}} = \text{per cent barium sulfate}$$

Silica. Hydrochloric Acid Method. Add 20 ml. of concentrated hydrochloric acid to the solution of sample, and evaporate to dryness, preferably immersing the beaker in a steam bath. If it is necessary to expedite the operation, evaporate to separation of solids or incipient bumping over a Bunsen burner, and finish on a hot plate. In the latter operation, bumping will often start, in which case set the beaker on a Nichrome triangle on the hot plate. The beaker should be covered with a loosely fitting watch glass. Evaporate completely to dryness, and heat to at least

160° by careful flaming, but do not decompose iron or aluminum salts. Undue baking will convert these to difficultly soluble oxides.

When the beaker has cooled sufficiently, add about 5 ml. of concentrated hydrochloric acid, and dilute to about 100 ml. Warm and filter through a coarse filter paper. In warming, endeavor to dissolve soluble salts, but do not boil for any extended period, or the silica will become gelatinous and therefore difficult to filter. Carefully use a policeman on the beaker, and rinse into the filter. Wash the precipitate on the paper well, preferably with a stream from the wash bottle, which separates the precipitate from the paper and agitates it well. Unless the silica content is under 2 to 3 per cent, a second evaporation to remove silica is necessary. If the silica content is sufficiently low and if extreme accuracy is not essential, the second evaporation can be omitted. For the most accurate procedure, evaporate this filtrate to dryness, and bake, as in the case of the acidified original sample. Filter the resulting small precipitate of silica on another paper to expedite filtration and washing. Wash each precipitate until the wash water no longer shows a positive test for chloride with silver nitrate. Combine the filtrates and wash waters, and set aside as the solution for several subsequent determinations.

Ignite the two papers together in a platinum crucible, cool, and weigh. Add to the weighed precipitate a few drops of concentrated sulfuric acid and 2 to 3 ml. of 48 per cent hydrofluoric acid. Evaporate carefully to avoid spattering. This volatilizes silica as silicon tetrafluoride. The sulfuric acid prevents any iron and aluminum volatilizing as fluorides. If a large amount of silica was present, let cool, add another 2 to 3 ml. of 48 per cent hydrofluoric acid, and repeat the evaporation. Finally raise the temperature to volatilize the sulfuric acid, and ignite at a red heat to decompose any traces of metallic sulfates. Cool, and weigh. The loss in weight is the silica. If the technic has been well carried out, little nonvolatile residue will be found present.

If the residue is substantial and further determinations are to be made on the solution from which silica has been removed, fuse the residual oxides with a little potassium bisulfate or pyrosulfate, dissolve in water, and add to the filtrate from the silica determination. Another and less accurate alternative with fairly small residues is to assume that it consists of iron and aluminum oxides and count the residue as part of those oxides when later precipitated. If this is done, be sure to correct for the aliquot used in iron and aluminum determination before adding this weight.

Perchloric Acid Method. This differs from the hydrochloric acid method only in elimination of heating to dryness to insolubilize silica. Efficient dehydration of silica is obtained in one step. If small amounts of organic matter, or even easily oxidizable inorganic matter are present, add 40 ml. of 1:1 nitric acid to the sample solution, and heat to boiling. Perchloric acid must be used only with great caution and by prescribed methods; otherwise explosions are liable to occur.

To the solution of the sample add 5 ml. of 60 to 70 per cent perchloric acid. Cover with a watch glass, and heat on a hot plate until the water, and nitric acid if present, have been evaporated and dense fumes of perchloric acid fill the beaker, condensing on and draining down the sides of the beaker. Little loss of perchloric acid should occur. Continue so to heat under these conditions for 15 minutes, let cool, and dilute with 50 ml. of water. Heat to boiling, filter the silica, and wash on the paper with 25 ml. of 1:5 hydrochloric acid. Wash with water until the wash water no longer shows a positive test for chloride with silver nitrate. Ignite in a platinum crucible, cool, and weigh. Complete as for the hydrochloric acid method (page 190), starting at "Add to the weighed precipitate a few drops of concentrated sulfuric acid and 2 to 3 ml. of 48 per cent hydrofluoric acid."

Modification for Commercial Silicas. With commercial silicas when the silica content exceeds 90 per cent and the balance rarely contains significant amounts of anything other than iron and aluminum, the fusion can be by-passed.

Weigh 0.5 gram of the sample into a weighed platinum crucible. Ignite with a Bunsen flame for a few minutes, cool, and weigh. The loss in weight is the sum of loss at 110° and low- and high-temperature ignition, rarely more than 2 per cent, even with amorphous silicas. Add 2 to 3 ml. of 48 per cent hydrofluoric acid and 2 to 3 drops of 1:4 sulfuric acid. Heat carefully, to avoid spattering, just to fumes. Let cool, and add 1 ml. more of the hydrofluoric acid. Evaporate to complete dryness, and ignite at a red heat. Cool, and reweigh. The loss in weight is due to silica.

Fuse the residue with five to ten times its weight of potassium bisulfate or pyrosulfate, dissolve in dilute hydrochloric or sulfuric acid, and use as sample for determination of iron and aluminum.

Iron and Aluminum Oxides. Phosphorus Absent. In the absence of more than a trace of phosphates, the iron and aluminum are precipitated as the hydroxides. Such a trace of phosphate means an amount that will barely show a qualitative test with ammonium molybdate (page 258) on the solution for determination of iron and aluminum. The usual sample is an aliquot of the solution from the silica determination.

Procedure. Evaporate, if necessary to reduce the total volume, cool, and dilute to 250 ml. in a calibrated flask. Occasionally, as with filtrates from commercial silicas or samples low in iron and aluminum, the entire solution is to be used as a single sample. At this stage, one must consider the relative amounts of elements probably present. The aliquot taken is also to be used for calcium and magnesium. It may be further aliquoted for these determinations; but, if calcium and magnesium are low and aluminum and associated metals high, it may be necessary to take an unduly large aliquot for iron and aluminum to provide appreciable amounts of calcium and magnesium.

To 50 ml. of the sample solution, or other aliquot, add sufficient concentrated hydrochloric acid to make the total present about 10 ml. per 100 ml. of solution. This is designed to furnish a concentration of ammonium chloride of about 5 grams per 100 ml. of solution after neutralization. Failure to have sufficient ammonium chloride present may permit calcium and magnesium to precipitate here or magnesium during the calcium determination. If manganese is present, add excess of bromine water, and boil off the excess. This precaution can do no harm if manganese is absent.

Just neutralize the solution to methyl-orange or methyl-red indicator with 1:4 ammonium hydroxide, and add a slight excess. Do not add a substantial excess and boil, for the precipitate will assume a gelatinous and very difficultly filterable condition. Boil for 2 to 3 minutes, and at once filter through a free-filtering paper such as an 11-cm. Whatman No. 41. Use a policeman on the beaker, and rinse into the filter paper. Wash thoroughly with hot water.

For very accurate work or if the precipitate is voluminous, place another beaker under the funnel, and dissolve the precipitated hydroxides with 20 ml. of hot 1:4 hydrochloric acid. Finally wash the paper with hot water until a drop of wash water gives no test with silver nitrate. Reprecipitate as above, starting at "Just neutralize the solution . . . ". Filter on the paper previously used, and collect the filtrate with that previously obtained, as sample solution for further determinations.

Ignite the precipitate and paper in a platinum or porcelain crucible, platinum if iron is to be determined later. First heat at a low temperature to dry and burn off the paper, finally blasting for several minutes. Ignition in an electric muffle is very convenient. In the latter case, first place the crucible at the mouth of the muffle to dry the paper, and then gradually advance it, placing it in the hottest part when the paper is burned off.

Cool, weigh the mixed oxides, and calculate as per cent of the original sample. These oxides contain the iron and aluminum. If titanium was present, it is also present as the dioxide. The same is true of manganese. Any traces of phosphorus present in the original sample, will be in this precipitate. If the titanium and manganese are present to only a minor extent, they are often ignored.

In the absence of significant amounts of interfering elements, which is the usual case commercially, this combined weight of aluminum and iron is sufficient for many purposes, particularly if the precipitate is nearly white, indicating no more than a trace of iron content. The combined oxides are often reported as R_2O_3, signifying that iron and aluminum have not been separated. Otherwise the methods that follow are applied for determination of iron and the interfering elements.

CALCULATION.

$$\text{Weight of } R_2O_3 \times \text{aliquot} \times \frac{100}{\text{weight of sample}} = \text{per cent } R_2O_3$$

IRON AND ALUMINUM AS PHOSPHATES. Iron and aluminum can be quantitatively precipitated as the phosphates in the presence of ammonium acetate and acetic acid. Calcium does not precipitate as phosphate under these conditions; it does when the precipitation is made in ammoniacal solution. The method is a convenient one to apply when some phosphate is present, since an attempt to precipitate iron and aluminum as hydroxides would give a precipitate more or less contaminated with phosphate. For a reasonably accurate precipitation of the phosphates, the weight of precipitate should not exceed 0.1 gram.

PROCEDURE. Transfer the filtrate from which silica has been eliminated to a 250-ml. calibrated flask, and dilute to volume. Pipet out a 50-ml. aliquot as sample, even less if the iron and aluminum content is high. Add about 10 ml. of 2 per cent disodium phosphate solution unless phosphate is known to be in excess. Carefully add 1:4 ammonium hydroxide until a slight permanent precipitate persists, and then 1:4 hydrochloric acid dropwise, until the precipitate is just dissolved. Boil the solution for 3 minutes, and add about 10 ml. of 50 per cent ammonium acetate solution. Add 3.2 ml. of glacial acetic acid, continue to boil until precipitation is complete, and allow the precipitate of iron and aluminum phosphates to settle. Filter at once, and wash the precipitate well with hot water. Ignite at a low temperature in a platinum crucible until the paper is charred and burned off, thus avoiding reduction of the phosphate. Finally ignite, cool, and weigh as mixed iron and aluminum phosphates.

In case of doubt as to this precipitate's being all in the form of phosphates, redissolve in 1:10 nitric acid and use as a sample for determination of phosphate (page 201 *et seq.*). Calculate the phosphate as phosphorus pentoxide, and take the iron and aluminum oxides by difference.

One method of calculation is on the assumption that the precipitate is all aluminum phosphate, a situation usually approached commercially. Otherwise, determine the iron content of the mixed phosphates. For this, fuse the mixed phosphate precipitate, in the crucible in which it was ignited, with ten times the weight of sodium carbonate. Dissolve the cooled melt in 1:6 sulfuric acid solution. This is a sample for determination of iron by any of the methods that follow.

CALCULATION. If iron is substantially absent,

$$\text{Weight of phosphates} \times 0.4178 \times \text{aliquot} \times \frac{100}{\text{weight of sample}} = \text{per cent aluminum oxide}$$

In case the precipitate is mainly iron phosphate, the corresponding factor for iron oxide is 0.5294.

If substantial amounts of both iron and aluminum are present

$$[\text{Weight of mixed phosphates} - (Fe_2O_3 \times 1.8889)] \times \text{aliquot} \times 0.4178 \times \frac{100}{\text{weight of sample}} = \text{per cent aluminum oxide}$$

Iron Oxide. If the iron and aluminum and possibly other interfering elements are to be reported separately, add 2 to 3 grams of C.P. potassium bisulfate or pyrosulfate to the platinum crucible containing the mixed oxides, mix well, and fuse. The use of platinum ensures the high temperature requisite for solution of ignited iron oxide. When the melt is clear, cool and dissolve in about 50 ml. of 1:10 sulfuric acid. If only iron is to be determined, use this as the sample. If iron and titanium are to be determined, dilute this to 100 ml. with 1:10 sulfuric acid, and pipet out 50 ml. for determination of iron. If other elements are to be determined, dilute to 250 ml. with 1:10 sulfuric acid, and take aliquots suitable for the determinations. If the iron content is substantial, use one of the titration methods; if small, do it colorimetrically.

Permanganate Method. For the iron determination a Jones reductor (page 13) is convenient. Wash out the reductor with 1:20 sulfuric acid and then with water, leaving sufficient water in the column to cover the zinc. These washings should show a pink color with 1 to 2 drops of 0.1*N* potassium permanganate solution. Run 50 ml. of 5 per cent sulfuric acid through the reductor while regulating the rate to about 100 ml. per minute. Next pass the iron solution through the reductor to convert all the iron to the ferrous state. Follow it with 200 ml. of 1:20 sulfuric acid and that with 50 ml. of water, leaving sufficient water in the reductor to cover the zinc.

Alternatively take a 50-ml. aliquot of the sample in a beaker, and add 50 ml. of 1:10 sulfuric acid. Put three to four 1-cm. sticks of zinc in the solution, and let stand at about 40° until a test on a spot plate shows no color with potassium ferrocyanide. At once pour off the solution from the zinc into another beaker, and rinse the zinc with a few milliliters of hot water. Any zinc metal transferred to the beaker in which the sample is to be titrated will invalidate the result.

Titrate the ferrous iron with 0.1*N* potassium permanganate solution (page 728) to a faint pink color that persists for 30 seconds and, for very accurate work, deduct a blank.

The titration as outlined must not be in a solution containing hydrochloric acid, or more than the equivalent amount of permanganate will be required. Addition of 1 ml. of a solution containing 6.7 grams of hydrated manganous sulfate, 13.8 grams of sirupy phosphoric acid, and 13 grams of concentrated sulfuric acid per 100 ml. for each milliliter of concentrated hydrochloric acid in the solution being titrated will overcome this and permit titration in hydrochloric acid solution.

CALCULATION.

$$10FeSO_4 + 2KMnO_4 + 8H_2SO_4 = 5Fe_2(SO_4)_3 + K_2SO_4 + 2MnSO_4 + 8H_2O$$

$$(\text{Ml. titration of sample} - \text{ml. titration of blank}) \times \text{normality of permanganate} \times 0.07985 \times \text{aliquot} \times \frac{100}{\text{weight of sample}} = \text{per cent ferric oxide}$$

The corresponding factor for iron as Fe is 0.05585.

DICHROMATE METHOD. Prepare the reduced iron solution as for the permanganate method. Titrate with a $0.1N$ solution of potassium dichromate (page 728). Determine the end point with a 2 per cent solution of potassium ferricyanide on a spot plate. The end point is shown by the absence of blue color from a drop of titrated solution and a drop of indicator. If the iron content is not known approximately, it is best to reserve about 20 per cent of the solution and add it after an end point has been obtained. Greater accuracy will be obtained by use of a blank.

This procedure is slightly less accurate than the permanganate method, owing to a less satisfactory end point. Its advantages are: the oxidizing solution is a primary standard, the titration can be carried out in hydrochloric acid solution, and turbid solutions can be titrated.

CALCULATION.

$$K_2Cr_2O_7 + 6FeSO_4 + 8H_2SO_4 = 2KHSO_4 + Cr_2(SO_4)_3 + 3Fe_2(SO_4)_3 + 7H_2O$$

$$(\text{Ml. titration of sample} - \text{ml. titration of blank}) \times \text{normality of dichromate} \times 0.07985 \times \text{aliquot} \times \frac{100}{\text{weight of sample}} = \text{per cent ferric oxide}$$

The factor for iron as Fe is 0.05585.

COLORIMETRICALLY BY THIOCYANATE. This is the most sensitive and convenient method for determining ferric iron in solutions containing as little as 0.02 ppm. of iron. Many organic compounds, copper, fluorides, nitrous oxides, phosphates, silver, and titanium interfere. As presented it provides not only for aliquots from mineral analysis but for many other samples.

PROCEDURE. If the sample is organic, ignite at as low a temperature as is possible, and dissolve in hot 1:20 sulfuric acid. When the ash is refractory, fuse with potassium bisulfate or pyrosulfate, and dissolve the melt in 1:20 sulfuric acid. When interfering substances are present, precipitate the iron as the hydroxide with 1:1 ammonium hydroxide or as sulfide in approximately neutral solution. Wash the precipitate, and dissolve in 1:20 sulfuric acid. Dilute the solution to 100 ml. with the same strength acid, and test qualitatively for the approximate amount of this solution to be used as an aliquot. For analysis of minerals use an aliquot of the solution free from silica (page 190).

To the aliquot in a 50-ml. Nessler tube add a few drops of 0.1*N* potassium permanganate solution (page 728) until a slight pink color persists, to oxidize the iron and any reducing substance present. Now add 5 ml. of a 10 per cent ammonium or potassium thiocyanate solution. Dilute to the mark with 1:20 sulfuric acid, and mix. At the same time, to another 50-ml. Nessler tube add 35 ml. of the 1:20 sulfuric acid and 5 ml. of the thiocyanate solution. From a buret add a standard iron solution containing 0.1 mg. of iron per ml. until a near match is obtained. Carefully dilute to the mark with 1:20 sulfuric acid, adding more standard solution if necessary. A series of standards containing varying amounts of iron may be used as an alternative to the duplication method outlined.

It is essential that the strength and amount of acid and of thiocyanate solution be the same in the blank and in the sample, for the color intensity of the complex ion varies with these factors and is not directly proportional to the amount of iron present.

The determination can be made more delicate by extracting the iron compound with ether or amyl alcohol or preferably with a mixture of 5 volumes of amyl alcohol to 2 volumes of ether. The greater accuracy is due to the undissociated salt's not ionizing in the nonpolar solvent. Such extraction is advisable for amounts of 5 ppm. or less. Titanium produces yellow thiocyanate compounds extractable by amyl alcohol–ether mixtures but not by ether alone. The presence of sodium sulfate in the solution lowers the color, but hydrochloric acid restores it. Chlorides and nitrates do not affect the color. Phosphates do not seriously affect the color up to several hundred times the amount of iron present.

STANDARD. Dissolve 0.8634 gram of hydrated ferric ammonium sulfate or 0.2376 gram of hydrated ferric sulfate in 1:20 sulfuric acid, and dilute nearly to 1 liter with the same strength of acid. Add a few drops of 0.1*N* potassium permanganate solution until a faint pink blush persists for 1 minute, and dilute to volume with 1:20 sulfuric acid. This standard solution contains 0.1 mg. of iron per ml. In case of doubt as to the amount of iron present in the standard, as when salts available are partly dehydrated, determine iron in the standard by permanganate titration (page 194).

CALCULATION.

$$\text{Ml. standard iron solution} \times 0.1 \times \text{aliquot} \times \frac{1{,}000}{\text{weight of sample}} = \text{ppm. of iron}$$

To report as ferric oxide, Fe_2O_3, multiply by 1.43.

Titanium. TITRIMETRICALLY. As sample use a 50-ml. aliquot of the solution provided for determination of iron and aluminum. Wash out a Jones reductor (page 13) with 1:20 sulfuric acid and then with water, leaving sufficient water in the column to cover all the zinc. These washings should not require more than 1 or 2 drops of $0.1N$ potassium permanganate solution to obtain a pink color.

Dissolve 345 grams of hydrated C.P. ferric ammonium sulfate in water, and dilute to 1 liter. To ensure the absence of ferrous iron, add $0.1N$ potassium permanganate solution until a pink color persists for 5 minutes. Measure about 25 ml. of this in a graduated cylinder, and place in the receiver.

Regulate the suction on the receiver while running 50 ml. of 5 per cent sulfuric acid through the reductor so that the rate is about 100 ml. per minute. Be sure that the lower end of the reductor is under the surface of this solution in the receiver, for the reduced titanium solution must be transferred to this without coming into contact with the air. Now pass the titanium solution through the reductor at 100 ml. per minute, and follow it with 100 ml. of 1:20 sulfuric acid, then with 100 ml. of water. The zinc in the reductor must remain covered at all times.

Gradually release the suction, wash the tube immersed in the solution, and immediately titrate the reduced iron with $0.1N$ potassium permanganate solution. Run a blank determination, using the same reagents and washing the reductor in exactly the same way. Subtract this blank from the titration of the determination. The reduced titanium solution quantitatively reduces ferric iron, the resulting ferrous iron being then determined.

CALCULATION.

$$(\text{Ml. titration} - \text{ml. blank}) \times \text{normality of permanganate} \times 0.0799 \times \frac{100}{\text{weight of sample}} = \text{per cent titanium dioxide}$$

The factor for titanium is 0.0479.

This figure will include any iron present as well as any metal reducible by zinc such as chromium and arsenic. Determine iron separately, and apply as a correction to the percentage of titanium dioxide.

COLORIMETRICALLY. The usual method of determining small amounts of titanium colorimetrically is by oxidation with hydrogen peroxide in acid solution to develop a yellow color. Vanadium, chromium, tungsten, and molybdenum interfere. Fluorides bleach the color. Fusion with sodium carbonate and a small amount of sodium nitrate, followed by leaching with hydrogen peroxide solution, eliminates chromium, vanadium, and molybdenum. Titanium remains in the residue and is soluble by fusion with equal weights of ammonium sulfate and sulfuric acid.

Ferric salts that produce a yellow color in acid solution may be allowed for by adding a similar quantity of iron to the standard. A correction may also be applied, since 0.1 gram of ferric oxide in 100 ml. of 1:40 sulfuric acid solution is equal in color intensity to 0.2 mg. of titanium oxide when oxidized by hydrogen peroxide. Phosphoric acid causes a slight fading of the yellow color produced by the reaction. At least 1:40 sulfuric acid is necessary to prevent hydrolysis to metatitanic acid or to a basic sulfate. The presence of large amounts of alkali sulfates causes bleaching unless a large excess of sulfuric acid is present.

The range of maximum sensitivity is 1.5 to 20 mg. of titanium oxide per 100 ml. In this range the minimum perceptible difference is 6.5 per cent. If several readings are taken and averaged, the error should not exceed 2 per cent. When less than 0.3 mg. of titanium dioxide is present per 100 ml., dissociation occurs, and the color is not strictly proportional to the concentration.

Procedure. Transfer 25 ml. of the sample solution to a 50-ml. Nessler tube. To a similar tube add 20 ml. of water and salts and acid similar to the amounts present in the sample. To each add 20 ml. of 1:5 sulfuric acid.

Add 5 ml. of 3 per cent commercial hydrogen peroxide solution to both the sample and the standard. To the standard add sufficient standard titanium solution to match the color of the sample. Adjust the volume of the standard to equal that of the sample by careful addition of water and standard solution if necessary. The color will remain constant for several days.

Standard. To prepare the standard titanium solution, fuse 1 gram of pure titanium dioxide with 10 grams of ammonium bisulfate made by mixing approximately equal weights of ammonium sulfate and sulfuric acid. Take up the melt with 200 ml. of 1:4 sulfuric acid, and dilute to 1 liter with water. If necessary this can be standardized by evaporation of a measured portion to dryness in platinum and ignition to constant weight. If necessary, add ammonium carbonate to assist in volatilizing sulfuric acid in this determination.

Manganese. The amounts of manganese found in minerals are small and usually determined colorimetrically. The sample is an aliquot of the solution prepared for determination of iron and aluminum.

Three methods are in general use. Oxidation with potassium or ammonium persulfate, using silver as a catalyst, occasionally gives a color slightly different from that of true permanganate, and sometimes oxidation may not be complete. The method is usually sufficiently accurate. The use of potassium periodate as oxidizing agent avoids the necessity of precipitating chlorides, which are removed during the process. At times very small amounts of manganese do not readily oxidize by this reagent. The sodium bismuthate method necessitates filtration or the use of the centrifuge to remove excess of the reagent. When properly performed, this method is reliable for very small amounts of manganese.

More than 2 mg. of manganese per 100 ml. will give too deep a color for comparison. This is overcome by accurate dilution. Phosphates may be added to prevent interference by iron. Large amounts of chromium are oxidized to chromates and interfere.

PROCEDURE. *Persulfate Method.* Moisten the potassium or ammonium persulfate a day or two before use, for persulfate oxidation does not proceed smoothly when the added salt is dry. This is probably due to the small solubility of the salt.

To a suitable aliquot of a sample in acid solution, prepared as described (page 198), add 1 ml. of 2 per cent silver nitrate solution. Filter or decant from any silver chloride. Be sure that excess silver nitrate is present. At the same time prepare a blank containing the same acid concentration and salts as are present in the sample, which will vary somewhat according to the previous history of the sample but is usually approximately 1:10 sulfuric acid.

Add 1 gram of moist potassium or ammonium persulfate per milligram of manganese to the sample and the same amount to the blank. Heat both in a water bath until the permanganate color is fully developed, usually about 10 minutes. Cool, and transfer to comparison tubes. Dilute the sample to 50 or 100 ml. and the blank to 45 or 90 ml. It is convenient to use as the standard $0.01N$ or even $0.001N$ potassium permanganate solution made from the pure reagent standardized in solution at $0.1N$ strength (page 728) and diluted with boiled distilled water. Add this permanganate solution from a buret, mixing thoroughly after each addition, until a match is obtained, as determined by looking down the tubes against a white background. Finally adjust the volume of the standard to match that of the blank, and check that the colors are the same.

So long as excess persulfate is present, the color remains permanent. Thus if turbidity is present, the developed sample may be filtered or allowed to stand for the turbidity to settle, after which the solution can be decanted into another tube for comparison.

CALCULATION.

$$\text{Ml. permanganate} \times \text{normality} \times 0.011 \times \text{aliquot} \times \frac{10^6}{\text{weight of sample}} = \text{ppm. of manganese}$$

Periodate Method. Adjust the concentration of acid so that about 90 ml. of solution contains not less than 5 to 6 ml. of concentrated sulfuric acid and 20 ml. of concentrated nitric acid or 5 to 10 ml. of sirupy phosphoric acid. Add either one or both of these last two acids to the sulfuric acid solution. If the sample contains chlorides, evaporate with sulfuric acid to fumes of sulfur trioxide, take up with water, and add acid, as above. Prepare a blank containing the same acids and salts as the sample, and dilute to about 90 ml.

Add 0.2 to 0.4 gram of potassium or sodium periodate to the sample and blank, and boil for 1 minute. If even traces of oxidizable matter have not been removed, the color will not develop until a large excess of periodate is present. If the hydrated salt, $Na_3IO_4.3H_2O$, is used, take about 1.5 times as much. Keep hot for 5 to 10 minutes, let cool, and dilute the sample to 100 ml. Match the color by addition of a standard permanganate solution to the blank, as in the persulfate procedure (page 199) starting at "It is convenient to use as the standard 0.01*N* or . . . ". The calculation is the same.

The high concentration of acid is necessary to prevent precipitation of manganese, as well as other metals such as silver, lead, mercury, or bismuth, as iodate or periodate. A higher concentration of acid than given does no harm if it is present in both sample and standard. Iron, if present in considerable amounts, is kept in solution and rendered colorless, or relatively so, by phosphoric acid. A known amount of ferric nitrate may be added to the standard if too much iron is present in the sample, but usually phosphoric acid is sufficient to remove the color.

For routine work, a permanganate standard containing a slight excess of periodate, kept stoppered, will be stable for at least 3 months. If excess periodate is present, the standards are also stable for several months. More than a trace of chlorides must be absent from the sample.

Bismuthate Method. Dilute an aliquot of the sample and a blank of similar composition except for the manganese content to about 40 ml. Add 0.1 gram of sodium bismuthate per mg. of manganese to the sample. Boil, if necessary, to obtain the true permanganate color, stir, and let settle. Filter through a Gooch crucible, containing a mat of ignited asbestos that has been washed with permanganate solution and then with boiled distilled water. Dilute the sample to 50 ml. and the blank to 45 ml. Match the sample by addition of permanganate to the blank as in the persulfate procedure (page 199), starting at "It is convenient to use as the standard 0.01*N* or . . . ".

The addition of a few crystals of periodate after oxidation with the bismuthate tends to keep the color from fading, as otherwise often occurs after the bismuthate has been removed. As alternatives to filtering when somewhat lower accuracy is permissible, centrifuge, or let the excess of sodium bismuthate settle from the developed sample in the dark for 30 minutes. Then pipet an aliquot into another tube, and match the color in a similar blank.

Phosphorus. The amount of phosphorus present will determine whether a gravimetric or titrimetric method for substantial amounts or a colorimetric method for traces is to be applied. For an accurate estimate of the amount present in the original sample, start with a fresh sample, and remove the silica.

Magnesium Phosphate Precipitation. This method depends on precipitation of magnesium ammonium phosphate and its ignition to magnesium pyrophosphate. It is applied for determination either of phosphate radical or of magnesium. Supplementary evidence is necessary if the phosphate is to be calculated as trisodium phosphate. The method is official with the American Society for Testing Materials, Association of Official Agricultural Chemists, and American Oil Chemists Society.

Procedure. Take an aliquot to contain about 0.5 gram of sample. Dilute to about 90 ml., and add 5 grams of ammonium chloride. Add a few drops of methyl-red or methyl-orange indicator and 1:1 hydrochloric acid dropwise until slightly acid. If considerable alkalinity has been neutralized, the solution has warmed up, in which case cool in running water. Add 25 ml. of magnesia mixture (page 719), and mix well. Slowly add 1:1 ammonium hydroxide with constant stirring until a precipitate of magnesium ammonium phosphate begins to form. Cease addition of ammonium hydroxide, but continue stirring until precipitation appears to be complete. Add a few drops of phenolphthalein solution, and continue addition of 1:1 ammonium hydroxide dropwise until the solution is alkaline to that indicator. Add 15 ml. excess of 1:1 ammonium hydroxide, and set aside for 4 hours in an ice bath or preferably overnight at room temperature.

Decant the supernatant liquid through a quantitative filter paper, and wash the precipitate in the beaker several times with 1:20 ammonium hydroxide, pouring the washings through the filter paper in such a way as to wash thoroughly the deposit on the filter as well as the residue in the beaker. Let the paper drain, and discard the filtrates.

Dissolve the precipitate from the filter with 25 ml. of 1:1 hydrochloric acid, and catch the solution in the beaker containing the bulk of the precipitate. Wash the paper thoroughly with 1:20 hydrochloric acid. Dilute the filtrate and washings to about 100 ml. The precipitate should all be dissolved at this point. Add 2 ml. of magnesia mixture to the solution, and precipitate magnesium ammonium phosphate as previously with 1:1 ammonium hydroxide but adding 10 ml. excess in this case.

After precipitation for at least 2 hours in an ice bath or preferably overnight at room temperature, filter on a quantitative paper. Wash the precipitate on the filter with 1:20 ammonium hydroxide until the wash waters show no trace of chlorides with silver nitrate. Ignite the paper gently in a platinum or porcelain crucible so that the paper is destroyed without flaming. Gradually raise the temperature, after carbon has been destroyed, to the full temperature of a Meker burner, or alternatively ignite in a muffle furnace at a bright red heat. Cool in a desiccator, and weigh. Repeat the ignition until constant weight is established.

CALCULATION. The precipitation reaction is

$$H_3PO_4 + MgCl_2 + 3NH_4OH = MgNH_4PO_4 + 2NH_4Cl + 3H_2O$$

On ignition the decomposition is

$$2MgNH_4PO_4 = Mg_2P_2O_7 + 2NH_3 + H_2O$$

$$\text{Weight of } Mg_2P_2O_7 \times 0.6377 \times \text{aliquot} \times \frac{100}{\text{weight of sample}} = \text{per cent phosphorus pentoxide}$$

Other factors that may be substituted are trisodium phosphate dodecahydrate, 3.4149; trisodium phosphate monohydrate, 1.6350; anhydrous trisodium phosphate, 1.4731.

PHOSPHOMOLYBDATE-PYROPHOSPHATE METHOD. This is an AOAC method primarily intended for phosphate in fertilizers but applicable to total phosphate in any solution. The phosphate is first separated from contaminants by precipitation as the phosphomolybdate, instead of as magnesium ammonium phosphate. Reprecipitation is then as in the preceding method.

PROCEDURE. Transfer an aliquot of sample equivalent to 0.25 to 0.5 gram of anhydrous trisodium phosphate or to 0.1 to 0.2 gram of phosphorus pentoxide to a 250-ml. beaker. Add 1:1 nitric acid until acid to litmus, than 1:1 ammonium hydroxide until alkaline to litmus, then render just acid dropwise with 1:1 nitric acid. Dissolve about 15 grams of ammonium nitrate in the solution, or a lesser amount estimated to allow for the amount formed in the solution during neutralization. Heat to about 90°, and add about 70 ml. of ammonium molybdate solution (page 718) for every 0.1 gram of phosphorus pentoxide present. Be sure that sufficient is added. Digest with occasional stirring at 65° for 1 hour to complete precipitation.

Let the precipitate of ammonium phosphomolybdate settle, and add a few milliliters of ammonium molybdate solution to the supernatant liquor. If further precipitation occurs, discard and start with a new sample, adding more ammonium molybdate solution so that the precipitation is completed at one time.

Filter on a quantitative paper, and wash the beaker and precipitate on the filter with cold 1 per cent ammonium nitrate solution. Discard the wash waters, and filtrate. Dissolve precipitate adhering to the original precipitation beaker in 25 ml. of 1:1 ammonium hydroxide. Pour this over the precipitate on the paper to dissolve it, and catch the solution in a fresh beaker. Rinse the precipitation beaker and paper with 10 ml. of 1:1 ammonium hydroxide, then with 15 ml. of hot water. Add a piece of litmus paper, and neutralize with 1:1 hydrochloric acid. Alternatively add bromothymol blue directly to the solution, and neutralize. Add magnesia mixture (page 719) at the rate of about 1 drop per second with vigorous stirring to a total of 15 ml. for every 0.1 gram of phosphorus pentoxide or every 0.25 gram of anhydrous trisodium phosphate present. This is best added from a buret or, with sufficient experience, from a pipet. Mix well during the addition, and let stand for 15 minutes. Continue as for magnesium phosphate precipitation (page 201), starting at "Add 15 ml. excess of 1:1 ammonium hydroxide, and set aside for 4 hours . . . ".

Phosphomolybdate Volumetrically. This official method of the AOAC for phosphate in fertilizers is simpler than the preceding. In the complex reaction, each molecule of phosphorus pentoxide is equivalent to 46 molecules of sodium hydroxide. It follows that this method is particularly applicable to small samples. Careful attention to details of acidity, time, and temperature of precipitation are essential to accuracy.

Procedure. Transfer an aliquot of sample equivalent to about 0.04 to 0.1 gram of phosphorus pentoxide to a 250-ml. beaker. Less sample can be used with suitable modifications. Add about 7 ml. of concentrated nitric acid, less an allowance for any present in the solution. Add 1:1 ammonium hydroxide until the solution is approximately neutral. This may be adjusted by testing with litmus paper. If any precipitate is formed on addition of the ammonium hydroxide, it will slowly redissolve at the end point on stirring. Dilute to 75 to 100 ml., and warm or cool as necessary to 25 to 30°.

To 100 ml. of the usual ammonium molybdate reagent (page 718) add 5 ml. of concentrated nitric acid, and filter if necessary. If not over 0.1 gram of phosphorus pentoxide is in the sample, add 20 to 25 ml. of this reagent; if more is present, proportionally increase the amount. Place in a shaker or insert a mechanical stirrer, and shake or stir at room temperature for 30 minutes. A somewhat less satisfactory alternative is to heat to 45 to 50° and add the reagent. Maintain at that temperature on a corner of the hot plate or over a low flame with occasional stirring for 30 minutes.

Let the precipitate settle momentarily, and at once decant the upper layer through a quantitative filter paper. Add a few drops of ammonium molybdate solution to the filtrate, and if further precipitation occurs start with a new sample. Wash the precipitate in the beaker with two 25-ml.

portions of 1 per cent ammonium nitrate solution, let settle, and decant the washings through the filter paper. Transfer the precipitate to the filter paper, and discard the filtrate and accumulated washings. Wash the precipitate on the paper until the washings from two fillings of the funnel give a pink color with a drop of phenolophthalein solution and a drop of the standard sodium hydroxide solution used in subsequent solution of the precipitate.

Transfer the filter paper with the precipitate to a 250-ml. conical flask. Add about 25 ml. of water, and disperse the precipitate and paper in it. Add standard sodium hydroxide solution until the precipitate is completely dissolved, as shown by an absence of yellow color. Add a few drops of phenolphthalein solution, and back-titrate with standard acid to disappearance of the color of the indicator.

CALCULATION. The complex reaction of precipitation is

$$H_3PO_4 + 12(NH_4)_2MoO_4 + 21HNO_3 = (NH_4)_3PO_4.12MoO_3 + 21NH_4NO_3 + 12H_2O$$

The subsequent titration is

$$2(NH_3)_3PO_4.12MoO_3 + 46NaOH + H_2O = 2(NH_4)_2HPO_4 + (NH_4)_2MoO_4 + 23Na_2MoO_4 + 23H_2O$$

$$(\text{Alkali titration in ml.} \times \text{normality}) - (\text{acid titration in ml.} \times \text{normality}) = \text{ml. of } N \text{ alkali used}$$

$$N \text{ alkali in ml.} \times \frac{0.0710}{23} \times \text{aliquot} \times \frac{100}{\text{weight of sample}} = \text{per cent phosphorus pentoxide}$$

The corresponding factors are: for trisodium phosphate dodecahydrate, 0.3802; for trisodium phosphate monohydrate, 0.1820; and for anhydrous trisodium phosphate, 0.1640.

COLORIMETRIC. Reduction of a phosphomolybdate gives a blue color that is used to estimate phosphorus in very small amounts, much smaller than are possible by the yellow of the phosphomolybdate. Many reducing agents have been proposed and used, such as hydrazine sulfate, stannous chloride, hydroquinone, reduced molybdic acid, benzidine, and aminonaphthol sulfonic acid. Different forms of sample require different treatments, which cannot be adequately dealt with here.

The method that follows is a modification of that of the American Public Health Association for the determination of small amounts of phosphates in water. By rough comparison in tubes, it is accurate to about 3 ppm. in its range of maximum sensitivity.

Applied with a colorimeter, it will be accurate to 0.3 ppm. Excessive color or turbidity interfere, and these interfering substances are assumed to be absent.

PROCEDURE. Transfer a sample containing about 4 to 6 mg. of phosphorus calculated as phosphate radical to a 100-ml. tube, and dilute to about 50 ml. Add a drop of phenolphthalein indicator solution, and if the solution is alkaline then add 1:2 sulfuric acid until the solution is colorless. With only a small amount of alkalinity and no silica in the sample, this neutralization is unnecessary.

Into a second tube containing about 50 ml. of water add 3 ml. of a standard phosphate solution containing 0.2865 grams of thoroughly dry C.P. monopotassium phosphate per liter or 0.2 mg. of PO_4 radical per ml. Dilute the sample and standard to about 70 ml. To each add 1 ml. of 1:2 sulfuric acid, and mix. Then add exactly 10 ml. of a 2.5 per cent solution of ammonium molybdate in 1:4 sulfuric acid, prepared by first dissolving 2.5 grams of the salt in 50 ml. of water and later adding 1:2 acid to dilute to volume. Mix well.

As reagent, grind 0.5 gram of 1-amino-2-naphthol-4-sulfonic acid, dissolve in 195 ml. of 15 per cent sodium bisulfite solution, and dilute to 200 ml. with 20 per cent sodium sulfite solution. Add 4 ml. of this to the sample and standard, and mix well. Dilute each to 100 ml., and mix. After 5 minutes, preferably at 37°, compare the standard with the sample by pouring the standard into another tube until the depth required to match the sample is obtained. Alternatively balance in a colorimeter. The color is stable for about 1 hour.

CALCULATION. The standard contains 0.6 mg. of phosphate radical in 100 ml. Therefore, by balancing, the content in the sample is an inverse ratio of the depths that match.

$$\frac{0.6 \times \text{depth of standard}}{\text{Depth of sample}} = \text{milligrams of phosphate radical in sample}$$

If the tubes are of the same diameter, this ratio may also be taken by volume. To convert results to phosphorus pentoxide multiply the result by 0.7474, to phosphorus by 0.3265.

Aluminum. BY DIFFERENCE. From the combined oxides containing aluminum, ordinarily the subtraction of iron as ferric oxide gives the aluminum by difference. However if titanium, manganese, and phosphorus, or any of them are present they must also be subtracted as TiO_2, Mn_2O_3, and P_2O_5, respectively. The more such subtractions are made, the more errors are accumulated in the aluminum determination and the less the degree of accuracy. This is the main reason for the inaccuracy of determinations of aluminum. In the presence of interfering elements, aluminum is not determinable directly, as are so many other metals.

Copper. COLORIMETRICALLY. The amount of copper in a silicate mineral may be significant if it is to be used in compounding rubber. The amount in commercial inorganic residues analyzed by mineral-analysis methods may be important. Unless the amount is large, determine it colorimetrically on a suitable aliquot of the solution after the precipitation of iron and aluminum oxides.

Diethyldithiocarbamate gives a brown coloration with copper in neutral, acid, or ammoniacal solution. This will detect 0.01 ppm. in the final solution and is most accurate in the range of 0.1 to 1.0 ppm.

SAMPLE. *By Dry Ashing.* Ash as large a quantity of material as is convenient in a platinum dish, previously cleaned with hot nitric acid. Use as low a temperature as possible, preferably heating, after charring, in a muffle at a temperature not greater than 600°. Place a clean Nichrome triangle over the dish containing the charred sample and a porcelain cover on top of this to reflect the heat down if the carbon does not readily burn off. Take up the ash in a few milliliters of concentrated nitric acid, and heat. Transfer the solution to a beaker, dilute to 50 ml., and heat again. Filter into a 100-ml. calibrated flask, and dilute the cooled filtrate and washings to volume.

By Wet Ashing. A more precise determination is obtained by acid digestion and preliminary separation of the copper as the sulfide. The following technic for separation may also be applied to solutions of ash. In it most of the interfering metals are eliminated by precipitation of the copper, but it is more tedious as well as more accurate. For many purposes, the greater accuracy is not essential.

Digest the sample in a 500-ml. Kjeldahl flask with an appropriate volume of concentrated sulfuric acid, occasionally adding a few milliliters of concentrated nitric acid until all organic matter is destroyed. Fume off most of the sulfuric acid, and cool. Dilute to a suitable volume, and neutralize to 1 per cent acid. Pass in hydrogen sulfide gas for 10 to 15 minutes. Filter or centrifuge to separate the precipitated copper sulfide, and wash with 1 per cent sulfuric acid saturated with hydrogen sulfide.

Dissolve the precipitated copper sulfide in 5 ml. of hot 1:1 nitric acid, and heat to boil off the hydrogen sulfide. Then transfer this solution to a 50 or 100-ml. calibrated flask, and dilute to volume.

Minerals. Use an aliquot of the solution from which silica has been removed (page 190).

PROCEDURE. Test a few milliliters of the solution qualitatively with a 0.1 per cent aqueous solution of sodium diethyldithiocarbamate. This will give the experienced worker an idea as to the size of aliquot to use in the determination.

Measure out an aliquot containing not over 1 mg. of copper. If iron and aluminum are not already absent, add 1:1 ammonium hydroxide until it is just alkaline. Heat to boiling, filter, and wash the precipitate. Dissolve the precipitated hydroxide, even if slight in amount, in a suitable volume of hot 1:10 nitric acid, and reprecipitate with 1:1 ammonium hydroxide solution as before. Filter into the first filtrate. This removes iron, aluminum, etc., which might interfere by precipitation later. Occasionally it is necessary to centrifuge to precipitate a cloud not removable by filtering.

Reduce the volume to not over 30 ml. by boiling. Cool, and transfer to a 50-ml. Nessler tube. Wash the beaker with 5 to 10 ml. of water, and transfer to the tube. Add 5 ml. of concentrated ammonium hydroxide to keep any zinc present in solution, and finally add 5 ml. of 0.1 per cent sodium diethyldithiocarbamate solution. Dilute to 50 ml., and mix well. At the same time, to 30 ml. of distilled water, add similar amounts of ammonium hydroxide solution and of reagent solution. Prepare a copper standard by dissolving 0.0393 gram of hydrated copper sulfate, $CuSO_4.5H_2O$, in water, and dilute to 1 liter. This contains 0.01 mg. of copper per ml. Add this standard solution from a buret to the blank, mixing well after each addition. When the blank matches the sample, which should require not more than 10 ml. or the color will be too dark to match, carefully adjust to the volume of the sample by adding distilled water and reagent in suitable proportions. Finally confirm by observing through the tubes endwise, onto a white paper before a window or suitably diffused light, and make any further adjustments necessary. Always observe and then reverse the tubes and observe again, for it is not always possible to get completely diffused light. If larger numbers of samples are being run, it is then more convenient to prepare a series of color standards that are stable for at least 1 hour.

A very slight turbidity due to an interfering ion may be present without introduction of serious error. The addition, both to the sample and to the blank, of 1 to 2 ml. of a 1 per cent solution of gum tragacanth before addition of the reagent adds to the brilliancy of the solution.

Even with these precautions, turbidity may still result. In such a case, after approximately duplicating the sample with the standards, extract the copper diethyldithiocarbamate compound from the solutions in the tubes with carbon tetrachloride, using two to three 10-ml. portions for each tube. Compare the extracts in dry Nessler tubes. If not equal in color, pipet out known amounts of the sample extract or of the blank extract until they match, and calculate the copper content from the relative volumes used.

CALCULATION.

$$\text{Ml. standard copper solution} \times 0.01 \times \text{aliquot} \times \frac{1{,}000}{\text{weight of sample}} = \text{ppm. of copper}$$

Arsenic. GUTZEIT METHOD. No substantial amounts of arsenic are found in common minerals. On the other hand, it is not unusual to determine arsenic in ppm. in a mineral, in which industrially such a content would be significant. Since the arsenic may be present in a mineral ash, the method as outlined goes beyond strictly mineral analysis to provide for the separation of arsenic from organic substances. Thus it is also applicable to a food product.

In this method, arsenic is first reduced to arsenious acid and then, in a hydrogen generator, to arsine. The arsine is led over paper treated with mercuric chloride or bromide, preferably the latter. It produces a stain that is compared with stains produced by standard amounts of arsenic. Normally it is not applied for amounts under 0.001 mg. of arsenic.

SAMPLE. *Inorganic.* If large amounts of inorganic material are present from which the arsenic must be separated, a distillation as the chloride with absorption in dilute alkali is advisable. Place an original sample or its ash containing 0.02 to 5 mg. of arsenic in a 500-ml. distilling flash with a low side arm. Close the flask, and connect with a condenser by rubber stoppers previously soaked in hot 1:3 hydrochloric acid. Connect the exit tube from the condenser to a wash bottle to deliver the distillate into 25 ml. of approximately $0.1N$ sodium hydroxide. Place another wash bottle with a second 25-ml. portion of $0.1N$ sodium hydroxide, in series with the first. To the sample add 2 grams of ferrous sulfate, 2 grams of potassium chloride, and 0.2 gram of potassium bromide.

Add 50 ml. of concentrated sulfuric acid, and distill. The distillation should be so carried out that the solution does not boil, but hydrogen chloride is evolved and carries over the arsenic chloride with it. The temperature should rise to 90° in about 10 minutes when distillation will be complete. Sulphur dioxide fumes passing over do not interfere. Dilute the combined alkaline solutions and washings from the wash bottles to a suitable volume. Use an aliquot as sample.

Organic by Wet Ashing. Weigh a suitable amount of sample to contain 0.02 to 5 mg. of arsenic into an 800-ml. Pyrex Kjeldahl flask, add 25 to 50 ml. of concentrated nitric acid, then cautiously add 20 ml. of concentrated sulfuric acid. Place the flask on an asbestos mat with a 2-inch hole. Warm slightly, but discontinue heating if foaming becomes excessive. When the reaction has quieted, heat cautiously, and rotate the flask occasionally to prevent caking of the sample upon the glass exposed to the flame. Maintain oxidizing conditions in the flask at all times during digestion by adding small quantities of concentrated nitric acid carefully whenever the mixture turns brown or darkens. If heated to vigorous fuming while reducing material is present, substantial loss of arsenic may occur. This is avoided by keeping the temperature low until oxidation is complete. Continue the digestion until organic matter is destroyed, and heat sufficiently to evolve copious fumes of sulfur trioxide. The final solution should be water-white, or, at most, a light straw color. Cool, and add 75 ml. of water and 25 ml. of saturated ammonium oxalate solution to assist in expelling oxides of nitrogen from the solution. Evaporate again until fumes appear. Cool, and dilute with water to a suitable volume. Use an aliquot as sample.

Some samples containing stable organic arsenic compounds, such as oils, shrimp, and tobacco, require special treatment to complete the oxidation of organic arsenic to inorganic arsenic or to destroy organic interferences. Treat these first with nitric and sulfuric acids, and dilute. Then to an aliquot of the solution add 5 ml. of 20 per cent magnesium nitrate solution, and evaporate to dryness. Carefully ignite until carbon-free. Dissolve the residue in 20 ml. of 10 per cent hydrochloric acid, and dilute to a known volume. Use an aliquot as sample.

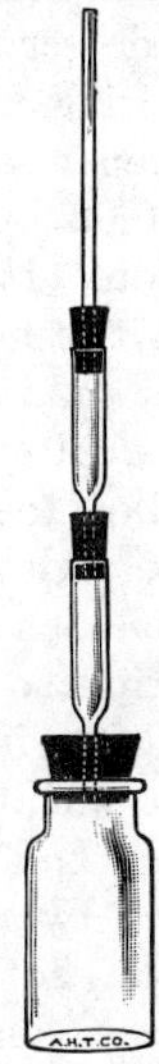

FIG. 91.—Gutzeit apparatus for determination of arsenic.

MERCURIC BROMIDE PAPER. In preparing the mercuric bromide paper, use commercial arsenic papers or cut from paper of uniform weight and texture into strips exactly 2.5 mm. wide and about 12 cm. long. Uniformity in width and texture of the paper is of great importance. Irregular texture produces irregular impregnation. To sensitize, soak the strips 1 hour or longer in a 5 per cent filtered solution of mercuric bromide in alcohol. When the quantity and character of the zinc used produces rapid evolution of arsine, the concentration of the mercuric bromide solution is increased to 6 per cent. With slow evolution of arsine, use paper soaked in a 3 to 5 per cent solution. If the strips are in sheets, cut off the two sides before soaking, and leave the strips attached at the ends. After sensitizing, remove the strips and dry individual ones on glass rods or groups by waving them in the air. When nearly dry, place the strips between clean sheets of paper; subject them to pressure long enough to take out bends or curls. Store in a dry, dark place. Aging of impregnated strips usually results in markedly fainter and longer stains. The best stains are obtained with papers not over 2 days old. When ready for use, cut individual strips off squarely ½ inch from one end, and insert this end into the narrow tube of the apparatus. Handle by the portion to be cut off.

PROCEDURE. The sample may have been prepared as above or may be a simple acid solution from other sources. The acidity must be known, if necessary by titration of an aliquot. Place an amount of sample solution not larger than 30 ml. in the 2-ounce widemouthed bottle of the apparatus shown in Fig. 91. This should contain 0.01 to 0.03 mg. of arsenic expressed as arsenious oxide. If the sample solution contains only hydrochloric acid, add sufficient to make the total amount equivalent to 5 ml. of the concentrated acid. If the sample contains sulfuric acid, this may be supplemented by sufficient concentrated hydrochloric acid to be equivalent to a total of 5 ml. Alternatively, exactly neutralize with 25 per cent arsenic-free sodium hydroxide, and add 5 ml. of concentrated hydrochloric acid.

Add 5 ml. of a 15 per cent aqueous solution of potassium iodide. Now add 4 drops of stannous chloride solution prepared by dissolving 40 grams of arsenic-free stannous chloride, $SnCl_2.2H_2O$, in concentrated hydrochloric acid and diluting to 100 ml. with the same strength acid. The stannous chloride serves to absorb any iodine liberated and to sensitize the zinc, promoting smooth evolution of hydrogen.

Prepare standards unless they are already available, corresponding to 0.01, 0.02, and 0.03 mg. of arsenious oxide, As_2O_3, from a standard arsenic solution. It is essential that the standards contain the same kind and amount of acid as the samples. Accordingly, add this amount of acid to each standard. If sulfuric acid in the sample has been neutralized, add an equivalent quantity of arsenic-free sodium sulfate to the standards. Mix, and allow to stand for 30 minutes at not less than 25° or for 5 minutes at 90°. Dilute with water to 40 ml.

Fit each bottle for the samples and standards, which should be of uniform size and design, with a one-hole rubber stopper containing a glass tube 1 cm. in diameter and 6 to 7 cm. long, with an additional constricted end to facilitate connection. Place a small wad of glass wool in the constricted end of the tube, and add 3.5 to 4 grams of 30-mesh dry sand, cleaned by washing successively with hot 10 per cent sodium hydroxide solution, hot concentrated nitric acid, and hot distilled water. Take care to have the same quantity in each tube. Moisten the sand with 10 per cent lead acetate solution, and remove the excess by light suction. The lead acetate serves to remove any hydrogen sulfide that may be evolved during the reaction. The combination in Fig. 91 provides for a second tube with cotton and sand, but this is not essential or usual.

Connect the tubes by means of rubber stoppers with narrow glass tubes of 2.6 to 2.7 mm. internal diameter and 10 to 12 cm. long. Introduce a strip of mercuric bromide paper centrally in each. A bore as great as 3 mm. will allow the strip to curl, which results in an uneven stain and a poor end point.

To each bottle add 10 to 15 grams of stick zinc in lengths of about 1 cm., or 2 to 5 grams of 30-mesh granular zinc. Add equal amounts to each bottle, and as far as possible equalize the surface area exposed. Immerse the apparatus to within 1 inch of the top of the narrow tube in a water bath kept at 20 to 25°, and allow the evolution to proceed for 90 minutes. Remove the strip, and average the length of the stain on both sides in millimeters. Compare with standards prepared at the same time or carefully preserved permanent standards prepared by a similar technic. Frequent blanks should be run. These should not show more than 0.001 mg. of arsenious oxide.

Stains are stable for 6 months. When impregnated by dipping in melted paraffin, they will keep even longer. Continued preparation of standards may be avoided if, once carefully made, they are so preserved. These may be compared visually with the sample. Alternatively plot a graph using length of the stain in millimeters as ordinates and milligrams of arsenious oxide as abscissas. This is only strictly valid while using the same lots of reagents and exactly the same technic.

STANDARDS. Dissolve 1 gram of C.P. arsenious oxide in 25 ml. of a 20 per cent solution of arsenic-free sodium hydroxide. Saturate with carbon dioxide gas to prevent oxidation, and dilute to one liter with freshly boiled water. Each milliliter of this solution contains 1 mg. of arsenious oxide. Dilute 40 ml. of this solution to 1 liter. Make 50 ml. of the diluted solution to 1 liter. Each milliliter of this latter solution contains 0.002 mg. of arsenious oxide. Thus 5, 10, and 15 ml. of this solution furnish the usual standards. Prepare such diluted solutions at frequent intervals. The concentrated standard keeps indefinitely.

CALCULATION.

$$\text{MgAs}_2\text{O}_3 \times \text{aliquot} \times \frac{1{,}000}{\text{weight of sample}} = \text{ppm. of As}_2\text{O}_3$$

Other useful factors are: arsenious oxide × 0.757 = arsenic, grains per pound × 143 = ppm., ppm. × 0.007 = grains per pound.

Lead. Although lead is not a normal ingredient of minerals, very small amounts may be significant, in mineral fillers for cosmetics, for example, in the talc of a face powder specifically. The method as presented goes further to include the lead in ash from organic materials.

DITHIZONE METHOD. Diphenylthiocarbazone, more commonly referred to as dithizone, gives colored precipitates that are soluble in organic reagents such as chloroform and carbon tetrachloride with many metals, including lead. The method is more specific in the presence of potassium cyanide. Under these conditions, bismuth, cobalt, phosphates, thallium, stannous tin, and more than traces of copper and ferric iron interfere. Most of these are eliminated by precipitating the lead as sulfide in acid solution. If bismuth, which would also precipitate, is present, it is removed from the carbon tetrachloride extract by extracting with a solution of potassium cyanide, or with the reagent at an earlier stage. The resulting lead dithizone solution is cherry-red.

Since the method is very sensitive, it is essential that all apparatus used be first treated with hot nitric acid and rinsed with distilled water. Always run a blank under the same conditions as the sample, using the same amounts of reagents and the same apparatus.

SAMPLE. *Organic.* For such samples as have to be ashed, peelings of fruit, bones, fish and organs, for example, dry in a cleaned platinum or porcelain dish, and ash in a muffle furnace at a temperature not higher than 450°. An Argand burner with the flame kept reasonably low and the dish protected from extraneous matter is suitable. If arsenic is being determined in the same sample, the residue from distillation of arsenic trichloride may be used. Consider the dry ash as an inorganic sample or the residue from arsenic distillation as solution of such a sample.

Inorganic. Dissolve the ash in 1:9 hydrochloric acid, and dilute to several times the volume. Modify the kind or concentration of acid used in special cases. Boil, and filter if necessary. Make the cooled filtrate slightly alkaline to litmus with 1:1 ammonium hydroxide. Allow to stand for 2 to 5 minutes. If a precipitate forms, redissolve by warming with a few drops of 1:1 hydrochloric acid. Heat the solution nearly to boiling, and pass hydrogen sulfide gas into the hot solution. If no precipitate forms, render distinctly acid with 1:1 hydrochloric acid. Neutralize the solution with 1:16 ammonium hydroxide using bromothymol blue as indicator. If a black sulfide forms, allow to stand for 15 minutes, and add a small amount of paper pulp to entrain the precipitate. Centrifuge or filter, and dissolve the precipitated sulfide in 10 ml. of 1:9 hydrochloric acid solution. Cool, and dilute this solution to a known volume. Use portions of this to determine the intensity of coloration with the reagent and therefore the size of aliquot to be used.

If bismuth is present, adjust the lead solution to about pH 2.0 with 1:1 nitric acid or 1:1 ammonium hydroxide, using metacresol purple as indicator. Extract the solution with several 25-ml. portions of the dithizone reagent in chloroform, until the last portion remains unchanged in color. The lead remains in the acid solution. Finally neutralize to bromothymol-blue indicator with 1:16 ammonium hydroxide solution, and dilute to volume.

PROCEDURE. Transfer a quantity of the solution containing 0.005 to 0.015 mg. of lead to a separatory funnel. Add 15 ml. of a cyanide citrate solution made by mixing 30 ml. of a 5 per cent potassium cyanide solution, 15 ml. of a 5 per cent ammonium citrate solution made slightly alkaline with ammonium hydroxide, and 5 ml. of concentrated ammonium hydroxide solution, with 450 ml. of water. Add an excess of a solution of dithizone in chloroform from a buret, each ml. containing 0.025 mg. of the reagent, 1 ml. at a time. Shake vigorously after each addition. Excess dithizone is evidenced by that fact that the red lead–dithizone compound takes on a purple tint due to free dithizone itself, which will be green in chloroform solution. As the next step, add chloroform from a buret to make the insoluble layer total 10 ml. in volume. Make an alkaline potassium cyanide solution by adding 10 ml. of 5 per cent potassium cyanide solution and 5 ml. of concentrated ammonium hydroxide solution to 500 ml. of water. Shake 20 ml. of this with the developed sample solution in chloroform, using a separatory funnel. Allow to separate thoroughly, and draw off the chloroform solution. Repeat the extraction with 20-ml. portions of the alkaline potassium cyanide solution until the aqueous layer is colorless, indicating the complete removal of the excess of free dithizone. Transfer the chloroform solution of the lead dithizone compound to a small test tube, stopper, and allow to stand to clarify the slight turbidity. Compare the cherry-red color of the lead-dithizone solution visually or in a colorimeter with standard solutions prepared similarly and simultaneously. Also run a blank at the same time.

The intensity of the red color is not strictly proportional to the amount of lead present. Hence it is very important to match the sample against a standard of almost the same intensity. When using a colorimeter, plot a curve representing the colorimetric readings of standard solutions containing 0.001 to 0.40 mg. of lead. This may then be referred to when other samples are run, and no standards need be made. However, always run and subtract the value of a blank.

Modified Procedure. *Excess Dithizone.* This method is official with the AOAC and is given in more detail in their methods. The sample and standards are compared in the presence of excess dithizone. The resulting shades are better differentiated than slight differences in the red lead–dithizone compound alone, and the method is thus more sensitive.

Proceed up to the sentence above, commencing "As the next step add chloroform from a buret . . . ". Then run the chloroform solution containing the lead dithizone compound and excess dithizone into a small separatory funnel containing 25 ml. of 1:50 nitric acid free from nitrous acid. Shake to extract, and draw off the green dithizone layer into another separatory funnel containing a further 25-ml. portion of the nitric acid. Shake, allow layers to separate, and discard the chloroform fraction. Filter the acid extracts containing the lead, in succession through a small pledget of wet cotton inserted in the stem of a small funnel, into a 50-ml. flask or glass-stoppered cylinder, using the second acid extract to wash out the funnel in which the first extraction was made. This procedure removes chloroform globules. Make up any slight deficiency in volume with the 1:50 nitric acid, and mix.

Place an aliquot part, or the entire amount, of the 50 ml. of nitric acid solution of the lead in a separatory funnel. If an aliquot is taken, make to 50 ml. with the 1:50 nitric acid. Add 10 ml. of an ammonia-cyanide mixture. Make this by adding to 100 ml. of 10 per cent recrystallized, phosphate-free potassium or sodium cyanide in a 500-ml. volumetric flask, sufficient redistilled ammonium hydroxide to introduce 19.1 grams of ammonia, and complete to volume with redistilled water. Determine the strength of the redistilled ammonia by titration or by specific gravity.

Mix, and immediately develop the color by shaking for one minute with the proper amount of standard dithizone solution, as shown in Table 10.

Draw off the lower layer into a tube, and compare with standards, prepared similarly, commencing at the nitric acid extraction from chloroform solution. If the range is exceeded, either repeat with a smaller aliquot, or reextract with excess dithizone before draining from the funnel, and isolate once more in 50 ml. of the nitric acid reagent. Develop the color, and compare with standards covering a higher range. Interpolation between steps of the various ranges should be easily made. If an aliquot of the 50 ml. of the 1:50 nitric acid solution in which the lead had been isolated is taken, subtract only a corresponding amount of the total reagent blank from the amount of lead found.

STANDARD. Dissolve 1.5985 grams of lead nitrate, $Pb(NO_3)_2$, in 100 ml. of 1:500 nitric acid. Dilute 1 ml. of this solution to 1 liter with acid of the same strength. The dilution contains 0.01 mg. of lead in 1 ml.

TABLE 10.—STANDARD VOLUMES FOR LEAD BY DITHIZONE

Pb ranges micrograms (0.001 mg.)	Concentration, mg./liter	Volume of standard dithizone solution
0–5	4	5
0–10	4	10
0–20	8	10
0–50	8	25
0–100	10	30
0–200	20	30

Zinc. Like arsenic and lead, the presence of a trace of zinc, an amount to be expressed in ppm., may often be significant in an inorganic residue.

The well-known potassium ferrocyanide precipitation of zinc (page 504) can be adapted to nephelometric estimation of such small amounts by careful control of acidity, salt concentration, and time of standing. Iron and copper must be removed. Accuracy to 0.05 mg. can be obtained. The most accurate range is 0.20 to 0.25 mg. of zinc in the sample.

SAMPLE. If the sample contains organic matter, ash 1 to 5 grams in a muffle furnace or at a low red heat, avoiding fusion of the ash. Zinc is not volatilized under these conditions. Cool, and add 15 ml. of water and 7 ml. of 1:4 hydrochloric acid to the ash. Heat in a steam bath until half the liquid has evaporated. If much silica is present, dehydrate, redissolve in 1:20 hydrochloric acid, and filter.

With dilute aqueous solutions such as water, acidify 1 liter with concentrated hydrochloric acid, and evaporate to dryness in a platinum or a porcelain dish. Extract the residue with 1:20 hydrochloric acid, and filter.

In either case, dilute the filtrate to about 95 ml., and add 5 grams of C.P. sodium citrate, $Na_3C_6H_5O_7.2H_2O$. Add 2 ml. of a solution containing 3.939 grams of copper sulfate, $CuSO_4.5H_2O$, per liter, and a drop of a 0.04 per cent solution of thymol-blue indicator. Add 1 per cent potassium hydroxide solution until the solution becomes yellow, and then add a drop of 0.04 per cent bromophenol-blue indicator. If the solution is bluish at this point, add dilute acid until the yellow color is just restored, giving a pH of 3.0 to 3.4. Saturate the cold solution with hydrogen sulfide. Filter the zinc and copper sulfides, washing well with hydrogen sulfide solution to free from iron salts. Dissolve the combined sulfides in a few milliliters of a mixture of equal parts of concentrated nitric and hydrochloric acids. Filter or decant and dissolve the residue in a few milliliters of concentrated hydrochloric acid. Dilute about ten times with water, and repeat the precipitation as sulfide, omitting the addition of sodium citrate.

In the absence of citrate buffer, the adjustment of acidity is much sharper and requires more care. Dissolve the reprecipitated sulfides with mixed acids as before. Add 5 ml. of 1:4 hydrochloric acid to the residue and 20 ml. of water. Zinc and copper are now free from iron. Dilute the solution with 5 volumes of water, and slowly saturate the cold solution with hydrogen sulfide. Filter to separate the precipitated copper sulfide from the zinc, which does not precipitate in this strongly acid solution and is in the filtrate. Evaporate the filtrate to dryness, and dissolve the residue in 4 to 5 drops of 1:4 hydrochloric acid and a little water. The solution of the residue at this point is occasionally refractory, and it is desirable to see that it is moistened throughout with acid and warmed slightly before the addition of water. Transfer to a 25-ml. volumetric flask, and dilute to volume. Test a portion to estimate the proper aliquot to use.

PROCEDURE. To an aliquot in a 50-ml. Nessler tube, believed to contain about 0.20 mg. of zinc, add 14 ml. of $0.1N$ potassium hydroxide solution, and carefully neutralize the excess potassium hydroxide to phenolphthalein with $0.1N$ hydrochloric acid, adding 1 ml. of acid in excess. In similar tubes, take 5 ml. of 2 per cent potassium chloride solution and 1 ml. of $0.1N$ hydrochloric acid, and dilute to about 40 ml. To these add suitable amounts of standard to be equivalent to 0.05, 0.10, 0.20, 0.30, 0.40, and 0.50 mg. of zinc. Dilute the sample and each standard to 49 ml., add 1 ml. of 2 per cent potassium ferrocyanide solution, and mix at once. Compare the turbidity of the developed sample with that of the standards against a black background. Readings are difficult in artificial light. Masking the meniscus helps.

STANDARD. Prepare a standard zinc solution by dissolving 0.1000 gram of pure zinc in *N* hydrochloric acid, and dilute to 1 liter with the same acid. Each milliliter of this contains 0.1 mg. of zinc. For use dilute 10 ml. of this stock solution to 100 ml. Each milliliter now contains 0.01 mg. of zinc.

Calcium as the Oxalate. PHOSPHORUS ABSENT. For this determination, take the solution from which iron and aluminum and possibly other oxides have been precipitated. Either use the entire solution or dilute to a known volume and take an aliquot equivalent to 0.05 to 0.1 gram of calcium oxide.

As a variation, if small amounts of magnesium but large amounts of calcium are present, it may be necessary to take a sample or aliquot with a larger calcium content; conversely, if large amounts of magnesium and small amounts of calcium are present, either the proper size of calcium sample may be determined and the filtrate aliquoted, or the size of the calcium precipitate may be reduced. Either of these changes in the amount of calcium determined lowers the accuracy of the determination.

PROCEDURE. *By Titration.* Heat the sample solution to boiling, and add 20 ml. of boiling ammonium oxalate solution saturated at room temperature. Boil for 1 minute, and allow to stand at least 20 minutes or until the precipitate has settled. Filter, first by decantation, through a retentive paper. Very small amounts of calcium oxalate may require long standing to settle out. Make sure excess reagent was present by adding more to the filtrate. If precipitation occurs, refilter, or preferably use another aliquot with a larger volume of reagent. Save the filtrate for determination of magnesium.

For work of the highest accuracy, or if substantial amounts of magnesium are present, redissolve and reprecipitate. For this dissolve the precipitate from the paper by adding about 40 ml. of hot 1:4 hydrochloric acid. Wash the paper free of acid with hot water, collecting the washings with the calcium solution. Carefully add 1:4 ammonium hydroxide until the solution is just alkaline to methyl orange. Heat to boiling, add 5 ml. of boiling ammonium oxalate solution, and proceed as before. Filter on the same paper as at first, and combine the filtrate with that from the first precipitation for determination of magnesium.

When the reprecipitation is complete, or after the first precipitation if a second is unnecessary, wash the precipitate well with hot water, always bearing in mind, however, that 0.0014 gram of calcium oxalate dissolves in 100 ml. of boiling water. Make a hole in the apex of the filter paper, and wash the precipitate with a stream of hot water, into a 250-ml. conical flash. Be careful to wash any precipitate from the folds of the filter paper. Pour a solution containing 5 ml. of concentrated sulfuric acid in 100 ml. of water over the filter paper, and wash thoroughly with hot water. Bring the contents of the conical flask to a boil, and at once titrate with 0.1*N* potassium permanganate solution (page 728), swirling the flash well, until a faint permanent pink color persists for 30 seconds. An oxalate radical is oxidized to carbon dioxide and water for each calcium present.

$$5CaC_2O_4 + 2KMnO_4 + 8H_2SO_4 = 5CaSO_4 + K_2SO_4 + 2MnSO_4 + 10CO_2 + 8H_2O$$

The accuracy can be improved if only small amounts of calcium are present by running a blank, washing similar amounts of acid but no ammonium oxalate through a similar filter paper. This blank is negligible with appreciable amounts of calcium.

CALCULATION.

$$\text{Ml. titration} \times \text{normality of permanganate} \times \text{aliquot} \times 0.02804 \times \frac{100}{\text{weight of sample}} = \text{per cent calcium oxide}$$

The factor for calcium carbonate is 0.05004; that for calcium sulfate is 0.06807.

By Ignition. Precipitate, and reprecipitate if necessary, as for titration. After washing, transfer the paper and precipitate to a platinum or porcelain crucible. Dry, and char over a low flame or at the front of a muffle furnace, and finally ignite at full heat. Cool in an efficient desiccator, and weigh while just warm, to minimize absorption of water by the ignited calcium oxide. The crucible must previously be weighed under the same conditions.

This method is preferable for very small amounts of calcium oxide, of the order of a few milligrams or less, for the titration blank in the previous method is liable to be of a similar order of magnitude.

CALCULATION.

$$\text{Weight of calcium oxide} \times \text{aliquot} \times \frac{100}{\text{weight of sample}} = \text{per cent calcium oxide}$$

To report as calcium carbonate, multiply by the factor 1.785.

PHOSPHORUS PRESENT. The usual calcium oxalate precipitation in ammoniacal solution would give a mixed oxalate-phosphate of indefinite composition. In the ammonium acetate solution from which iron and aluminum phosphates are precipitated (page 193), only calcium oxalate precipitates.

PROCEDURE. Heat the combined filtrate and washings from the aluminum and iron phosphate precipitate, or an aliquot, to 50°, and add 10 ml. of warm saturated ammonium oxalate solution. Allow to stand in a warm place until the precipitate has settled. Filter, wash well, and determine either volumetrically or gravimetrically.

Magnesium as the Phosphate. If necessary, reduce the volume of the combined filtrates from the calcium determinations to about 200 to 300 ml. by boiling. If the amount of magnesium in the sample is too great, cool, dilute to a known volume, and use an aliquot.

PROCEDURE. Add 10 ml. of 10 per cent disodium hydrogen phosphate solution, cool to room temperature, and add with stirring a volume of concentrated ammonium hydroxide equivalent to one-ninth of the final volume. Appreciable magnesium will at once produce a crystalline precipitate of magnesium ammonium phosphate. Let stand for at least 4 hours. The smaller the precipitate is, the longer the time of standing that is necessary. Very small amounts of precipitate must stand overnight to settle out thoroughly. Scratching the side of the beaker will help to start minute crystals and hasten the precipitation.

Filter cold by decantation through a retentive paper. Wash the precipitate from the beaker onto the filter paper, using 1:16 ammonium hydroxide, and wash on the paper with the same solution. Complete the transfer after loosening the precipitate, which invariably adheres to the sides and bottom of the beaker, by using a policeman. Thoroughly wash out all the soluble salts. Transfer the precipitate and filter paper to a weighed platinum or porcelain crucible, and heat gently to dry and char the filter paper. Gradually increase the heat to the full blast of a Meker-type burner. Heat to constant weight, and weigh as magnesium pyrophosphate.

CALCULATION.

$$\text{Weight of ignited precipitate} \times 0.3623 \times \text{aliquot} \times \frac{100}{\text{weight of sample}} = \text{per cent magnesium oxide}$$

To report as magnesium carbonate, the factor is 0.7572.

Sodium and Potassium. AS SULFATES. Precipitate all metals other than the alkali metals by the procedures given before this. If alkali metals are to be determined, fusion with sodium carbonate or potassium bisulfate is not permissible. With appreciable amounts of alkali metals in a sample, it should be soluble in acid with reasonable ease. If not, use the method for insoluble silicates (page 187).

If phosphate is present, as when magnesium has been determined, add saturated barium hydroxide solution in slight excess to precipitate it, and evaporate to 50 ml. without filtration. Add sufficient ammonium carbonate solution to precipitate the excess barium hydroxide completely. Filter, and wash the precipitate with hot water. Concentrate the filtrate and washings in a platinum dish. Carefully add an amount of 1:1 sulfuric acid slightly in excess of the carbonate present so that it is decomposed with evolution of carbon dioxide. Evaporate on a water bath and then in an oven in order to avoid spattering. Finally, carefully heat to white fumes over a small flame. Heat to a dull red until no more fumes are evolved. When cooled, add about 1 gram of solid ammonium carbonate to react with the excess sulfur trioxide, and carefully ignite at a low red heat until all the ammonium carbonate and ammonium sulfate are driven off. Cool, and weigh the combined sodium and potassium sulfates. If the salts are mainly sodium salts, it is often conventional to assume that the combined sulfates are all sodium sulfate and calculate to sodium oxide, Na_2O.

This procedure is also applicable to sodium or potassium salts of organic acids. Add sulfuric acid to the solid in a platinum dish, and evaporate, using caution until carbon is all burned out.

CALCULATION.

$$\text{Weight of sulfate} \times 0.4364 \times \text{aliquot} \times \frac{100}{\text{weight of sample}} = \text{per cent sodium oxide}$$

AS CHLORIDES. Treat as in the determination as sulfates with a slight excess of barium hydroxide solution. This precipitates sulfates as well as phosphates. Evaporate, and add sufficient ammonium carbonate solution to precipitate the excess barium hydroxide completely. Filter, wash with hot water, and concentrate the filtrate and washings in a platinum dish. Carefully add 2 to 3 ml. of concentrated hydrochloric acid a drop at a time to avoid spattering. Evaporate on a water bath, then in the oven at 110°, and finally ignite very carefully over a small flame to volatilize ammonium chloride until constant weight is obtained. Do not heat at too high a temperature, for potassium chloride is appreciably volatile when strongly ignited. Cool, and weigh as mixed sodium and potassium chlorides. Comments about mixed sulfates also apply to mixed chlorides. If the potassium is to be separately determined to get a value for sodium, obtain the mixed alkalies as the chlorides by this method.

CALCULATION.

$$\text{Weight of chloride} \times 0.5303 \times \text{aliquot} \times \frac{100}{\text{weight of sample}} = \text{per cent sodium oxide}$$

IN INSOLUBLE SILICATES. In this method, the substance is heated with a mixture of ammonium chloride and calcium carbonate. The alkalies are obtained as chlorides, while the silica is converted to calcium silicate, and the other metals are chiefly left as oxides. A typical equation follows.

$$K_2O.Al_2O_3.6SiO_2 + 6CaCO_3 + 2NH_4Cl = 2KCl + 6CaSiO_3 + Al_2O_3 + 2NH_3 + 6CO_2 + H_2O$$

Since even the purest of reagents usually contain a small amount of alkali salt, always run a blank using the same quantity of reagents.

Thoroughly mix 0.5 gram of the silicate with 0.5 gram of C.P. ammonium chloride in an agate mortar. Add 3 grams of C.P. calcium carbonate, and again mix well. Transfer to a finger-shaped platinum crucible about 8 cm. long and with a diameter of about 2 cm. at the top and 1.5 cm. at the bottom. Larger quantities require larger crucibles. Complete the transfer by rinsing the mortar with 1 gram more of the calcium carbonate, which is added to the crucible.

Cover the crucible, incline, and heat gently for about 15 minutes. At the end of this period, no more ammonia is evolved. Stronger heating volatilizes ammonium chloride, which, in turn, decomposes into ammonia and hydrochloric acid. The latter will react with the calcium carbonate, forming the soluble chloride.

Raise the temperature until no more than three-fourths of the crucible is brought to a dull red heat. Maintain at this temperature for 1 hour. After cooling, remove the sintered mass if possible by gently tapping the inverted crucible. If this does not serve, add water, and carefully warm the crucible and contents. Transfer to a platinum dish, using a policeman if necessary, and heat with 50 to 75 ml. of water for 30 minutes, disintegrating any lumps with an agate pestle. Pour the clear solution through a filter paper, and wash several times by decantation. Finally transfer the residue to the paper, and wash free of chlorides.

To make sure that the silicate has been completely decomposed, dissolve the residue in 1:1 hydrochloric acid. No insoluble matter should remain. If there is an insoluble residue, repeat the determination with a fresh sample.

To the filtrate add sufficient 1:1 ammonium hydroxide and 10 per cent ammonium carbonate solution to precipitate any soluble calcium. Heat, and filter. Dissolve the precipitate in hydrochloric acid, and reprecipitate with 1:1 ammonium hydroxide and 10 per cent ammonium carbonate. Again filter, and wash. Evaporate the combined filtrates to dryness. To avoid decrepitation, place in the oven to drive off the last few milliliters of moisture. Remove ammonium salts by heating carefully over a moving flame. Cool, dissolve the residue in a little water, and precipitate the last traces of calcium with 1:1 ammonium hydroxide and hot 10 per cent ammonium oxalate. Let stand as long as possible, not less than 2 hours and preferably 12 hours. Filter, and concentrate the filtrate by evaporation. Determine the mixed alkalies as chlorides, as previously described. Start at the beginning of the method. Subtract the blank determination from the mixed sodium and potassium chlorides.

Sodium and Potassium by Calculation from Sulfates and Chlorides. Determine the sum of the two alkalies both as sulfate and as chloride. Calculate the percentages of sodium (x) and of potassium (y) as follows: Na = 23.0; K = 39.1; Na_2SO_4 = 142.05; K_2SO_4 = 174.25; NaCl = 58.45; KCl = 74.56.

$$\frac{142.05}{23 \times 2}x + \frac{174.25}{39.1 \times 2}y = \text{per cent total sulfates}$$

$$\frac{58.45}{23}x + \frac{74.56}{39.1}y = \text{per cent total chlorides}$$

Then

$$3.088x + 2.228y = \text{per cent total sulfates}$$

$$2.541x + 1.907y = \text{per cent total chlorides}$$

Solve these two equations for x and y.

The total alkali as sulfate may be determined on the total chlorides by the simple procedure of fuming down the chlorides with sulfuric acid.

This method is convenient and reliable when appreciable quantities of both the alkalies are present. When only a small amount of one is present, the accuracy is less satisfactory but is sufficiently good to warrant the use of this simple method in all but very accurate analyses.

Potassium. As the Chloroplatinate. Potassium chloroplatinate is practically insoluble in 80 per cent alcohol, whereas the corresponding sodium salt is soluble. Sodium chloride is not readily soluble in 80 per cent alcohol. It is therefore essential to convert both sodium and potassium to the form of their chloroplatinates. Lithium, magnesium, strontium, and calcium, if present, will not interfere.

Procedure. Dissolve the mixed chlorides in water in a porcelain dish, and add a slight excess of chloroplatinic acid solution containing 10 per cent of platinum. Since two molecules of alkali chloride are equivalent to one of platinum, calculate the approximate number of milliliters of 10 per cent solution required to give chloroplatinic acid equal to 1.7 times the weight of mixed chlorides.

If any precipitate forms when hot, add sufficient water to dissolve, thus preventing any occlusion of mother liquor in a mass of crystals suddenly formed. Evaporate until the residue solidifies on cooling. Thoroughly drench with 80 per cent alcohol, SD3A containing methanol, for example, and stir well.

Filter through a small paper by decantation, and wash by decantation with 80 per cent alcohol. Dry both the filter paper and the dish to remove the alcohol, and transfer the contents of the dish to a weighed platinum crucible. Wash the dish with a few milliliters of hot water, and pour through the filter paper to dissolve any precipitate, catching the filtrate in the platinum crucible. Evaporate the water from the crucible on a steam bath, and then heat for a few minutes to constant weight in an oven at 135°. Loosely cover the crucible at first in the oven, for decrepitation is liable to occur.

CALCULATION. The factor for potassium chloroplatinate to potassium chloride is 0.3067, but experience indicates that a more satisfactory average factor to use is 0.3058.

$$\text{Weight of chloroplatinate} \times 0.3058 \times \text{aliquot} \times \frac{100}{\text{weight of sample}} = \text{per cent potassium chloride}$$

AS THE PERCHLORATE. Potassium perchlorate is insoluble in a butyl alcohol–ethyl acetate mixture but the sodium salt is soluble. Dissolve the mixed alkali chlorides in water in a 150-ml. beaker, and add two to three times the equivalent quantity of 60 to 70 per cent C.P. perchloric acid. Practically, this means about 2.5 times the weight of mixed alkali chlorides. Evaporate to fumes and finally to dryness. Any liquid condensed on the sides of the beaker must be removed by brushing with a free flame. Cool the beaker and contents, dissolve in a few milliliters of hot water, and again evaporate to dryness.

Add 10 to 20 ml. of a mixture of equal parts by volume of normal butyl alcohol and ethyl acetate to the cooled beaker, and digest at near the boiling point for 2 to 3 minutes. Cool to room temperature, and decant through an ignited and weighed Gooch crucible. Wash the precipitate three times by decantation with the same solvent to remove most of the soluble perchlorates.

Dissolve the residue in a minimum of hot water, and again evaporate to dryness. Extract with 10 ml. of the mixed solvents as before, digest, cool, and filter through the original crucible. Transfer the precipitate to the crucible, using a wash bottle containing the mixed solvents. Wash the crucible ten to fifteen times with 0.5- to 1-ml. portions of the solvent. Dry the crucible and precipitate at 110° for a few minutes and finally for 15 minutes at 350°. Cool, and weigh the potassium perchlorate. Sodium and lithium present are converted to their perchlorates by this method and may be recovered and weighed by evaporation of the solvent.

CALCULATION.

$$\text{Weight of potassium perchlorate} \times 0.5381 \times \text{aliquot} \times \frac{100}{\text{weight of sample}} = \text{per cent potassium chloride}$$

Sodium. This metal is not usually determined directly but is calculated by difference after the potassium and sodium have been determined and the potassium separately determined.

BY ZINC URANYL ACETATE. This method is used to determine small amounts of sodium in the presence of moderate amounts of potassium, ammonium, barium, calcium, magnesium, and small amounts of lithium and strontium. Phosphates, silicates, oxalates, and tartrates interfere. It is most satisfactory when applied to the alkali metal chlorides.

PROCEDURE. Prepare zinc uranyl acetate solution as follows: To 77 grams of uranyl acetate, $UO_2(C_2H_3O_2)_2.2H_2O$, add 13.3 ml. of glacial acetic acid and 410 ml. of water. Warm to 70°, and stir to dissolve. To 231 grams of zinc acetate, $Zn(C_2H_3O_2)_2.3H_2O$, add 6.6 ml. of glacial acetic acid and 262 ml. of water. Warm to 70°, and stir to dissolve. Mix the two solutions while still warm. Cool to 25°, and allow to stand for 24 hours.

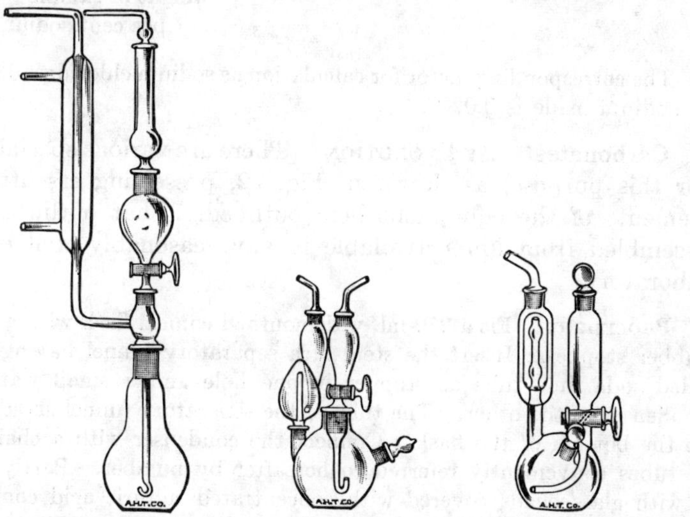

FIG. 92.—Typical apparatus for determination of carbon dioxide by evolution.

Filter into a clean, dry bottle. This solution is practically permanent but may require filtering from time to time.

Separate the alkali metal chlorides from a sample. Dissolve in water, and take an aliquot if necessary; not more than 20 mg. of sodium chloride or more than 50 mg. of potassium chloride should be present. Evaporate the sample solution to be used in a small beaker until salts are about to separate.

To the evaporated sample solution add at least 10 ml. of the zinc uranyl acetate reagent for each milliliter of the concentrated solution, and stir vigorously. Keep in a water bath at 20° ± 1° for 30 to 45 minutes, and stir occasionally. Filter through a weighed Gooch crucible, using gentle suction.

Pass all the solution through the crucible, and suck it dry. Wash all the precipitate onto the crucible with 2-ml. portions of the reagent solution, sucking the crucible dry after each portion. Now wash with five 2-ml. portions of the reagent solution, sucking dry as before. Next wash the precipitate with five 2-ml. portions of 95 per cent alcohol saturated with sodium zinc uranyl acetate at 20°. Finally wash with ether to remove the alcohol, and suck the filter dry of ether. Wipe the outside of the crucible clean with a moist and then with a dry towel. Place in the balance case, preferably one containing a drying agent, for 10 minutes, and weigh as $NaC_2H_3O_2.Zn(C_2H_3O_2)_2.3UO_2(C_2H_3O_2)_2.6H_2O$.

CALCULATION.

$$\text{Weight of precipitate} \times 0.01495 \times \text{aliquot} \times \frac{100}{\text{weight of sample}} = \text{per cent sodium present}$$

The corresponding factor for calculation as sodium chloride is 0.0381 and as sodium oxide is 0.0202.

Carbonates. BY EVOLUTION. There are various special devices for this purpose, as shown in Fig. 92, possessing essentially the elements of the equipment here outlined. That outlined can be assembled from units available in any reasonably well equipped laboratory.

PROCEDURE. Fit a 250-ml. widemouthed conical flask with a two-hole rubber stopper. Insert the stem of a separatory funnel having a small, filled soda–lime tube as stopper in one hole and a small water-cooled condenser in the other. The tube of the separatory funnel should extend to the bottom of the flask. Connect the condenser with a chain of five U tubes conveniently referred to hereafter by number. Partly fill tube 1 with glass beads covered with concentrated sulfuric acid containing a little silver sulfate. Keep the amount used at a minimum to lessen back pressure. This dries the evolved gases and absorbs traces of hydrochloric acid gas. Fill tube 2 with granular anhydrous calcium chloride. This also serves to dry the evolved gas. Fill tube 3 with soda lime to absorb the carbon dioxide. It may also be filled with special absorbents, Ascarite, for example.

Fill the first half of tube 4 with the same reagent as tube 3 but the last half with granular calcium chloride to make sure that no moisture liberated by the reaction $2NaOH + CO_2 = Na_2CO_3 + H_2O$ escapes the soda lime. Tube 5 contains granular calcium chloride and is a precaution to protect against moisture entering. A bottle containing concentrated sulfuric acid completes the train. It is a further protection against moisture and also serves to detect leaks in the apparatus. Compare the size and rate of the bubbles that pass through this with those passing through tube 1. Weigh a sample expected to contain about 0.2 to 0.25 gram of carbon dioxide into the flask, and assemble the equipment. With the separatory funnel open, slowly draw air through the apparatus with a pump. Tube 1 serves to show the rate at which it is passing. Continue this until tubes 3 and 4 reach constant weight.

Turn off the suction, and close the stopcock of the separatory funnel. Add a small volume of dilute sulfuric acid or phosphoric acid, including some excess, to this funnel, the amount depending on the size and nature of the sample. Release portions of this acid into the flask containing the sample from time to time, letting the pressure of the gas evolved serve for the present to carry the gas into tubes 1 to 4. When evolution of gas has ceased, gently boil the contents of the flask to drive off dissolved carbon dioxide. At the time this heating is stopped, open the stopcock of the separatory funnel, and turn on the suction to draw air through the apparatus. Continue this for about 30 minutes, and remove tubes 3 and 4 for weighing. Replace, again draw air through for 30 minutes, and reweigh to ensure constant weight. During handling, close the stopcocks of the U tubes; or, if they are not provided with stopcocks, close them with talc-free rubber policemen, and weigh them with the tubes before and after absorption. If the procedure is handled with good technic, the absorption will be practically entirely in tube 3. The total increase in weight is due to carbon dioxide.

CHAPTER 11

WATER ANALYSIS

This chapter describes suitable methods for the analysis of the most common industrial solvent, water. The analyses are necessarily for small amounts of dissolved substances, amounts that may vary by many hundreds of per cent. Water analysis is usually expressed, in milligrams per liter, as parts per million, usually written as ppm.

The wide divergence in quality of water is more evident if its various uses are considered. Distilled water often contains traces of ammonia and has a total solids content, although this may be only a fraction of a ppm. Potable waters and those used in industrial processes vary widely, from 20 to 30 ppm. to several hundred ppm. Boiler waters may have a relatively high total solids content, provided that it is not such as to give undesirable adherent deposits in the boiler. Cooling waters may have a large solids content, provided that they are not corrosive and do not deposit to an undue extent on the condenser surfaces. Perhaps it is best to visualize waters for analysis as varying from distilled water to sea water.

The main reference is "Standard Methods of Analysis of Water and Sewage," published by the American Public Health Association (APHA). In this, a strictly chemical text, bacteriological methods are not touched on.

The selection of determinations to carry out on a water sample depends primarily on the purpose for which the water is to be used. This will now be illustrated for a few of the most common types.

Boiler Feed Water. Suspended solids, dissolved solids, total alkalinity, total hardness, chloride, sulfate, and pH are a minimum. Mineral analysis, carbonate, hydroxide, and dissolved oxygen are frequently needed.

Boiler Contents and Blowdown. This is a concentrated form of the boiler feed water. Suspended solids, dissolved solids, total alkalinity, carbonate, hydroxide, chloride, sulfate, and pH are a minimum. If sulfite and/or phosphate are being used in treatment, they should be included. Mineral analysis and oil are frequently needed.

Municipal Water Supplies. The bacteriological examination is of the greatest importance. In addition, color, turbidity, dissolved solids, hardness, alkalinity or acidity, iron, manganese, fluoride, free chlorine, and pH value are determined. Occasionally a complete mineral analysis is run.

Swimming Pools. Apart from bacteriological examination, only turbidity, residual and free chlorine, and pH are significant.

Collection of Samples. It cannot be too strongly stated that considerable trouble should be taken to ensure that a really representative sample of the water, under the conditions in which it is in use, is obtained. For most purposes, this includes attention to the dissolved gases. Therefore, a sample should normally be taken in a clean bottle, which must be filled to overflowing and tightly stoppered without introduction of air. Usually the sample must be not less than 1 liter; 1 gallon is better. If the same sample is to be used for bacteriological examination, the bottle must be not only clean but also sterile.

Suspended Solids. Measure 500 ml. of well-mixed sample into a volumetric flask, and filter through a dried and weighed Gooch crucible containing an asbestos mat. Wash free from chloride with small portions of hot water. Dry, cool, and weigh. The increase in weight is due to the suspended solids.

Alternatively, determine the total solids content of a 500-ml. sample by evaporating to dryness on a steam bath and drying at 103° for 1 hour. From this, subtract the dissolved solids to determine the suspended solids.

Dissolved Solids. Either evaporate the filtrate obtained in the determination of suspended matter, or filter 500 ml. of sample to free it from suspended matter. This may be in a 600- or an 800-ml. beaker or in a smaller beaker to which repeated additions of the sample are made. Take great care that any deposit on the sides of the beaker due to evaporation of the water is not touched by the flame from the burner. When it is evaporated to about 50 ml., transfer the liquid to a weighed platinum dish. Use a policeman on the beaker, and transfer to the dish with distilled water. Evaporate the remaining water on a steam bath, and dry for at least 1 hour in an oven at 103°. Cool in a desiccator, and weigh.

Calculation.

$$\text{Weight of solids} \times \frac{10^6}{500} = \text{ppm. dissolved solids}$$

Ash on Dissolved Solids. This is sometimes determined as a rough indication of the amount of organic matter present. Bicarbonates are also decomposed, and some other salts are altered. Ignite the platinum dish containing the dissolved solids to constant weight. The total solids are sometimes similarly ignited to ash.

Analysis of Total Solids, Dissolved Solids, or Ash. Any of these may be analyzed for mineral content. This is usually the dissolved solids or their ash. It is rarely necessary to analyze both dissolved solids and total solids. Make a qualitative test for phosphate on a separate sample of water to determine whether the procedure must be modified. Use as sample either the solids from the previous determinations or those from evaporation of a larger amount of water. This selection will depend on the amount of solids present. The accuracy is improved by having a substantial sample, 0.5 gram or more.

This analysis has all the possibilities that exist for analysis of a mineral (see Chap. 10). The usual series of determinations is as follows: Dissolve in hydrochloric acid (page 187) or by fusion (page 187). Determine silica by the hydrochloric acid method (page 189). If the water is in its natural state, significant amounts of phosphorus are rare, but often samples are of phosphate-treated waters. Determine iron and aluminum oxides by the appropriate method according to whether appreciable phosphorus is absent (page 191) or present (page 193). Determine iron oxide colorimetrically with thiocyanate (page 195). If appreciable amounts of manganese are present, determine colorimetrically by any one of the three methods (pages 198 to 201).

If the water contains appreciable phosphorus, determine it colorimetrically (page 204). Determine larger amounts titrimetrically (page 203). Determine calcium as the oxalate by titration (page 216), allowing for the phosphorus if present (page 217). Determine magnesium gravimetrically as the phosphate (page 218).

It is rarely necessary to determine sodium and potassium. When determined, they usually are not separated but are reported together as total alkali metals (pages 218, 219). In the rare cases when they have to be separately determined, either the perchlorate (page 222) or chloroplatinate (page 221) method is proper.

Chloride. For this, use a freshly filtered sample and determine either titrimetrically (page 130) or nephelometrically.

Nephelometrically. The classic method for the determination of traces of chloride is nephelometrically as silver chloride. Other ions forming insoluble silver salts, such as bromides and iodides, must be absent. Other sources of error are possible sorption of ions by the silver chloride during peptization and coagulation, differences in the coagulating action of the two precipitating agents, and peptizing and coagulating effects of other compounds present.

The rate of mixing is very important, reproducibility being better with slow mixing. Over the usual range, accuracy to 2 to 3 per cent is obtained.

Procedure. *By Balancing.* Select a sample of solution containing about 0.2 mg. of chloride, and dilute to 20 ml. with chloride-free distilled water. As standard, take 10 ml. of a solution containing 0.01 mg. of chloride per ml., add any other reagents known to be present in the sample, and dilute to 20 ml. with chloride-free distilled water.

To sample and standard, add 10 ml. of $0.1N$ nitric acid and 10 ml. of a $0.005N$ silver nitrate solution. Mix, and place in a water bath at 40° for 30 minutes or longer. Cool rapidly to room temperature, and compare within 30 minutes in a nephelometer. It is advisable to rinse out the nephelometer tubes first with distilled water and then with the solutions to be introduced. The eyepiece of the instrument should be kept in exactly the same position throughout a series of comparisons.

If developed at 20°, the opalescence increases slowly to an almost constant value in 1 hour. Heating to 60° gives full development in less time but results in a greater opalescence, which falls off on continued heating. After 30 minutes at 60°, coagulation begins. Heated at 40°, this coagulation does not occur, and almost constant opalescence is obtained after 30 minutes. Further heating at this temperature has no effect. Exposure to diffused daylight has a perceptible effect that is due to coloration of the white opalescence. If greater sensitivity is desired, dilute the developed sample, and standard to 100 ml. with chloride-free alcohol.

By Duplication. Transfer 20 ml. of sample containing about 0.2 mg. of chloride to one 50-ml. Nessler tube and an equal volume of water containing the same known impurities to another. Dilute each to about 30 ml., and add 5 ml. of concentrated nitric acid and 2 ml. of 0.1 per cent gelatin solution. Dilute the sample solution to 49 ml., add 1 ml. of $0.1N$ silver nitrate solution, and mix. Dilute the standard to about 40 ml., add 1 ml. of $0.1N$ silver nitrate solution, and mix. Add standard sodium chloride solution containing 0.02 mg. of chloride per ml. to this standard, with mixing, until the opalescence of the standard matches that of the sample, comparing them against a black background. Adjust the volumes by adding water to make up the volume of the lesser before final comparison.

STANDARD. Dissolve 0.3297 gram of sodium chloride in water, and dilute to 100 ml. As a standard containing 0.02 mg. of chloride per ml., dilute 10 ml. of this solution to 1 liter.

Sulfate. If substantial amounts of sulfate are present, they may be determined gravimetrically; lesser amounts may be determined by titration by a special method.

GRAVIMETRIC. Use 100 to 2,000 ml. according to the content of sulfate, evaporating the larger samples in the beaker in which sulfate is to be precipitated. Finally, at a volume a little over 100 ml., add 2 ml. of concentrated hydrochloric acid. Complete as on page 160, starting at "Bring to a boil, and add with constant stirring sufficient boiling 10 per cent barium chloride solution to precipitate . . . ".

TITRIMETRICALLY BY TETRAHYDROXYQUINONE. Use the clear or filtered sample. For waters with sulfate content in terms of SO_4 of under 100 ppm., use 100 ml., and evaporate to 25 ml. For sulfate content of 100 to 800 ppm., use 25 ml.; for 800 to 2,000 ppm., use 10 ml., and dilute to 25 ml. Between 2,000 and 4,000 ppm. of sulfate, use 5 ml., and dilute to 25 ml.

PROCEDURE. Carry out the procedure at 20 to 25°. Add a drop of 0.5 per cent alcoholic solution of phenolphthalein to the prepared sample in a 150-ml. conical flask, and carefully titrate with 0.05*N* hydrochloric acid solution until colorless. Add 25 ml. of isopropyl alcohol and 0.20 to 0.22 gram of disodium tetrahydroxyquinone indicator made by grinding 1 gram of the reagent with 30 grams of dry C.P. potassium chloride. The solution will develop a deep yellow color.

Prepare 0.025*N* barium chloride solution by dissolving 3.054 grams of barium chloride, $BaCl_2.2H_2O$, in water and making up to volume in a 1-liter calibrated flask. Determine its actual normality by precipitating barium sulfate from 50 ml. of the boiling solution, following the gravimetric method for barium (page 189). To the solution add 1 ml. of concentrated hydrochloric acid, and dilute to about 125 ml. Then continue, "Heat the solution nearly to boiling, and carefully add 5 ml. of 1:10 sulfuric acid dropwise with vigorous stirring."

When the indicator is dissolved, start the titration by addition of the 0.025*N* barium chloride solution. Add this steadily, constantly swirling the flask, until a brown color begins to form near the end point. Continue the titration a drop at a time, still swirling the flask, until a salmon-pink color appears.

This method is not wholly satisfactory in the presence of phosphates. Corrections can be made when the amount of phosphate present is known, by using prepared tables shown in the methods of the APHA.

CALCULATION.

$$\text{Ml. of barium chloride solution} \times \text{normality} \times 0.04803 \times \frac{10^6}{\text{weight of sample}} = \text{ppm. of sulfate as } SO_4$$

The factor for sodium sulfate is 0.07103.

Alkalinity. Add 4 drops of 0.5 per cent alcoholic phenolphthalein indicator solution to 100 ml. of the filtered sample in a conical flask or a porcelain casserole. Titrate with 0.02*N* hydrochloric acid until colorless. If the sample is colorless after the phenolphthalein indicator is added, add 4 drops of a 0.05 per cent aqueous methyl orange solution, and titrate to an orange-pink end point.

Processed waters that have been treated with phosphate, such as those softened with orthophosphate, will not conform to the following rules, but otherwise the nature of the alkalinity can be predicted from the titration. If the titration to the phenolphthalein end point is zero, the alkalinity is due solely to the bicarbonate. When there is no further titration to the methyl orange end point after the phenolphthalein end point, only hydroxide alkalinity is present. When the phenolphthalein end-point titration is half the total titration, the alkalinity is due to carbonate alone. When the phenolphthalein end-point titration is greater than half the total titration, the alkalinity is due to both carbonate and hydroxide; when it is less than half the total titration, the alkalinity is due to both carbonate and bicarbonate. All kinds of alkalinity are frequently expressed in terms of calcium carbonate.

CALCULATION.

$$\text{Ml. total acid titration} \times \text{normality} \times 0.053 \times \frac{10^6}{\text{volume of sample in ml.}} = \text{ppm. total alkalinity as sodium carbonate}$$

To calculate as calcium carbonate, use the factor 0.05005.

$$[(2 \times \text{ml. acid titration to phenolphthalein}) - \text{ml. total acid titration}] \times \text{normality} \times 0.01701 \times \frac{10^6}{\text{volume of sample in ml.}} = \text{ppm. hydroxide alkalinity as hydroxide radical}$$

To calculate as sodium hydroxide, use the factor 0.040.

To calculate as calcium carbonate, use the factor 0.05005.

$$[\text{Ml. total acid titration} - 2\ (\text{ml. acid titration to phenolphthalein})] \times \text{normality} \times 0.06102 \times \frac{10^6}{\text{volume of sample in ml.}}$$
$$= \text{ppm. bicarbonate alkalinity as bicarbonate radical}$$

To calculate as sodium bicarbonate, use the factor 0.084.

To calculate as calcium carbonate, use the factor 0.05005.

$$2\ (\text{ml. total alkalinity titration} - \text{ml. acid titration to phenolphthalein}) \times \text{normality} \times 0.030 \times \frac{10^6}{\text{volume of sample in ml.}}$$
$$= \text{ppm. carbonate alkalinity as } CO_3 \text{ in the presence of hydroxide}$$

To calculate as sodium carbonate, use the factor 0.053.

To calculate as calcium carbonate, use the factor 0.05005.

$$2\ (\text{ml. acid titration to phenolphthalein}) \times \text{normality} \times 0.030 \times \frac{10^6}{\text{volume of sample in ml.}}$$
$$= \text{ppm. carbonate alkalinity as } CO_3 \text{ in the absence of hydroxide}$$

Barium Chloride Method. Because the end points of the titrations, especially that of phenolphthalein, are not always very exact and any error may therefore be doubled, the titration method, although frequently sufficiently accurate, is replaced by the barium chloride method when more precise results are required.

Procedure. First determine total alkalinity exactly as in the double titration method. To determine hydroxide alkalinity, add 25 ml. of a neutral 10 per cent barium chloride solution to 100 ml. of the clear or filtered sample in a conical flask or a porcelain casserole. Mix thoroughly to precipitate any carbonate present. Add 4 drops of a 0.5 per cent alcoholic phenolphthalein indicator solution. If a pink coloration due to free hydroxide appears, titrate with 0.02*N* hydrochloric acid solution to a colorless end point. Do not use standard sulfuric acid, for the barium chloride will use it up in precipitation of barium sulfate.

When hydroxide alkalinity is present, no bicarbonate ions can be in solution. If hydroxide is absent, add a small measured excess of 0.02*N* sodium hydroxide solution—usually 10 ml. is sufficient—to 100 ml. of a clear or filtered sample in a conical flask or porcelain casserole. Free hydroxide ions will combine with bicarbonate ions to produce carbonate ions, and the amount of hydroxide consumed is a measure of the bicarbonate present. Now precipitate the total carbonate in the solution by the addition of 25 ml. of a neutral 10 per cent solution of barium chloride. Titrate the excess sodium hydroxide to a phenolphthalein end point exactly as in the hydroxide alkalinity determination (pages 231 to 232), starting with

"Add 4 drops of a 0.5 per cent alcoholic phenolphthalein indicator solution . . . ".

CALCULATION.

$$\text{Ml. acid titration} \times \text{normality} \times 0.01701 \times \frac{10^6}{\text{volume of sample in ml.}}$$
$$= \text{ppm. hydroxide alkalinity as hydroxide radical}$$

To calculate as sodium hydroxide, use the factor 0.040.
To calculate as calcium carbonate, use the factor 0.05005.

$$(\text{Ml. } 0.02N \text{ alkali added} - \text{ml. } 0.02N \text{ acid titration}) \times 0.02 \times 0.06102$$
$$\times \frac{10^6}{\text{volume of sample in ml.}} = \text{ppm. bicarbonate alkalinity as } HCO_3$$

To calculate as sodium bicarbonate, use the factor 0.084.
To calculate as calcium carbonate, use the factor 0.05005.

$$(\text{Ml. total alkalinity titration} - \text{ml. acid titration due to hydroxide})$$
$$\times \text{normality} \times 0.030 \times \frac{10^6}{\text{volume of sample in ml.}}$$
$$= \text{ppm. carbonate alkalinity as } CO_3 \text{ in the presence of hydroxide}$$

To calculate as sodium carbonate, use the factor 0.053.
To calculate as calcium carbonate, use the factor 0.05005.

$$(\text{Ml. total alkalinity titration} - \text{ml. alkali used up by bicarbonate})$$
$$\times \text{normality} \times 0.030 \times \frac{10^6}{\text{volume of sample in ml.}}$$
$$= \text{ppm. carbonate alkalinity as } CO_3 \text{ in the presence of bicarbonate}$$

Acidity. TOTAL ACIDITY. For this determination, use a tall cylinder to contain the sample and thus lessen the surface of the sample and minimize loss of dissolved carbonic acid during titration. To 100 ml. of the sample, add 10 drops of a 0.5 per cent alcoholic phenolphthalein indicator solution, and rapidly titrate with $0.02N$ sodium hydroxide solution. Stir gently while adding the alkali solution until a faint, permanent pink color appears.

CALCULATION.

$$\text{Ml. alkali titration} \times \text{normality} \times 0.05 \times \frac{10^6}{\text{volume of sample}}$$
$$= \text{ppm. total acidity expressed as calcium carbonate}$$

FREE MINERAL ACIDS. To 100 ml. of the sample in a conical flask or porcelain casserole add 4 drops of a 0.05 per cent aqueous methyl orange solution, and titrate to an orange-yellow color with $0.02N$ sodium hydroxide solution. The calculation is exactly the same as for total acidity. If the nature of the mineral acid is known, calculate to that acid, using its milliequivalent factor.

Free Carbon Dioxide. CALCULATION.

$$(\text{Ml. alkali used by total acidity} - \text{ml. alkali used by free mineral acid}) \times \text{normality} \times 0.022 \times \frac{10^6}{\text{volume of sample in ml.}} = \text{ppm. free } CO_2$$

Any weak organic acids, if present, will be included in this figure.

Hardness. TOTAL HARDNESS. *Calculation Method.* The most accurate method of determining hardness is by calculation from the mineral analysis. It is always expressed as calcium carbonate. Calculate the calcium present as carbonate, and add to this the equivalent magnesium and any iron, aluminum, and manganese, calculated as calcium carbonate.

CaO to $CaCO_3$ = 1:1.785
MgO to $CaCO_3$ = 1:2.482
Fe_2O_3 to $3CaCO_3$ = 1:1.880
Al_2O_3 to $3CaCO_3$ = 1:2.946
Mn to $CaCO_3$ = 1:1.822

Express the sum as the total hardness in ppm. of calcium carbonate. Hardness is most frequently expressed as grains per gallon. Divide ppm. by 17.15 to obtain grains per gallon.

Carbonate and Noncarbonate Hardness. When the total hardness is greater than the carbonate and bicarbonate alkalinity, both calculated as calcium carbonate, the amount of hardness equivalent to the alkalinity is called carbonate hardness, and the amount of hardness in excess of this is called noncarbonate hardness. When the total hardness is equal to or less than the sum of the carbonate and bicarbonate alkalinity, there is no noncarbonate hardness; the total hardness is all carbonate hardness. Carbonate hardness is sometimes termed temporary hardness, and noncarbonate hardness is called permanent hardness.

Soap Method. The soap method gives approximate results in estimating hardness. It is the most widely used method for determining this value. Water will not lather when soap is added until all the calcium, magnesium, iron, and aluminum in solution have been precipitated as the corresponding soaps. A pseudo hardness is caused in brines by their higher sodium content, which prevents solution of sodium soaps. A series of standard solutions is required to produce the soap solution.

Preparation of a Standard Calcium Chloride Solution. To 0.5 gram of pure dry calcium carbonate in a 150-ml. conical flask add from a pipet 3 ml. of 1:1 hydrochloric acid, avoiding spattering, and allow to stand with occasional shaking until the calcium carbonate has completely dissolved. Add 25 ml. of water, and boil gently to expel carbon dioxide. Cool, and add 1:10 ammonium hydroxide solution dropwise until the solution is neutral to litmus. Transfer the solution to a 500-ml. volumetric flask, dilute to the mark with carbon dioxide-free distilled water, and mix well. One milliliter of this solution equals 1 mg. of calcium carbonate.

Preparation of a Stock Soap Solution. Shake vigorously approximately 100 grams of pure powdered Castile soap in 1 liter of 80 per cent alcohol. Allow this solution to stand at least overnight, and decant off the clear solution for standardization. This solution is approximately nine to ten times as strong as the standard soap solution desired.

Place 120 ml. of the stock solution in a liter volumetric flask, dilute to the mark with 80 per cent alcohol, and mix well. Pipet exactly 5 ml. of the standard calcium chloride solution into a 250-ml. glass-stoppered bottle. A 250-ml. centrifuge bottle with a rubber stopper is an acceptable substitute. Dilute with exactly 50 ml. of carbon dioxide-free distilled water. Then add the soap solution to be standardized, from a buret, 0.2 ml. at a time. Shake the bottle vigorously after each addition. The end point is reached when a lather is formed over the entire surface of the water and remains continuous over the surface for 5 minutes after the bottle is laid on its side. Repeat, adding the soap 0.1 ml. at a time. Then, the number of milliliters of the diluted soap solution that must be diluted to 1 liter equals the milliliters of diluted soap solution required in titration times 1,000/5. Transfer this amount to a 1-liter volumetric flask, and make up to the mark with 80 per cent alcohol.

For further correction, a lather factor is required. This factor is defined as the amount of standard soap solution required to produce a permanent lather in 50 ml. of distilled water. Determine this exactly as in the sample, adding the soap solution 0.1 ml. at a time. The lather factor will be 0.5 to 1.4 ml. according to the kind of soap.

The standard soap solution must now be corrected for the lather factor. The number of milliliters of this soap solution that must be diluted to 1 liter to obtain the final soap solution is therefore

$$(5 - \text{lather factor}) \times \frac{1{,}000}{5}.$$

Prepare the soap solution, so adjusted. Then check the strength of the resulting soap solution by titration against the standard calcium chloride solution. If necessary, adjust again, and repeat the standardization until 5 ml. of standard calcium chloride solution requires exactly 5 ml. plus the lather factor of the standard soap solution.

One milliliter of this solution prepared after correcting for lather factor is equivalent to 1 mg. of calcium carbonate.

PROCEDURE. Measure 50 ml. of the water sample into a 250-ml. glass-stoppered bottle. A 250-ml. centrifuge bottle with a rubber stopper is a convenient substitute. If the water is acid, neutralize to methyl orange indicator by titration with 0.02N sodium hydroxide solution. Then add 0.2 ml. of standard soap solution at a time from a buret, shaking the bottle vigorously after each addition. The end point is reached when a lather is formed over the entire surface of the water and remains continuous over the surface for 5 minutes after the bottle is laid on its side. For more accurate work, repeat, adding 0.1 ml. at a time toward the end of the titration.

If more than 15 ml. of standard soap solution is required for the titration, repeat, using a 25-ml. or smaller sample, and dilute to 50 ml. with carbon dioxide-free distilled water. Titrate as above, and multiply the milliliters of soap solution used by the appropriate factor.

If magnesium is present, a false end point, in which a transient lather forms that generally remains less than 5 minutes, is observed. This is at an approximate dividing line between the precipitation of calcium and that of magnesium soaps. The difference between the true end point and the false end point may be considered as representing magnesium. This value when calculated to hardness gives a rough estimate of the magnesium hardness. The hardness due to calcium is obtained by subtracting the magnesium hardness from the total hardness. To avoid mistaking the false end point for the true one, read the buret after the titration is apparently finished, and add about 0.5 ml. more of the soap solution. If the end point was a false one due to magnesium, the lather will disappear. Soap solution must then be added until the true end point is reached.

CALCULATION.

$$(\text{Ml. of standard soap solution} - \text{lather factor}) \times \frac{1{,}000}{50}$$
$$= \text{ppm. total hardness as calcium carbonate}$$

Nitrite. SAMPLE. If the sample is colored, shake with aluminum hydroxide suspension (page 718) to decolorize, and filter. Use a 50-ml. sample if the nitrite content is less than 0.025 ppm. in terms of nitrogen; otherwise, use a proportionately smaller amount.

REAGENTS. *Sulfanilic Acid Solution.* Dilute 302 grams or 288 ml. of glacial acetic acid to 1,000 ml. in a volumetric flask, using distilled water. Mix well, and add 8 grams of sulfanilic acid, shaking until dissolved.

Naphthylamine Acetate Solution. Dilute 302 grams or 288 ml. of glacial acetic acid to 1 liter in a volumetric flask, using distilled water. Mix well, and add 5 grams of α-naphthylamine, $C_{10}H_7NH_2$. Shake well, and filter through washed absorbent cotton.

Sodium Nitrite Standard Solutions. Dissolve 1.1 grams of silver nitrite in 500 ml. of distilled water. Add slowly, with constant stirring, 0.45 gram of C.P. sodium chloride dissolved in about 100 ml. of distilled water. This precipitates the silver as chloride. Allow to settle, and filter into a 1-liter volumetric flask. Dilute to the mark, and mix well. This is the stock solution containing 100 ppm. of nitrite expressed as nitrogen.

Dilute 100 ml. of this stock solution to 1 liter with distilled water in a volumetric flask. It now contains 10 ppm. of nitrogen as nitrite. Mix well, and dilute 50 ml. of this solution to 1 liter in a volumetric flask with recently boiled, nitrite-free, distilled water. To preserve, add 1 ml. of chloroform, and keep in a bottle that has been sterilized in boiling water. Each milliliter of this solution equals 0.0005 mg. or 0.5 ppm. of nitrogen.

Procedure. Place the clear colorless sample in a 50-ml. Nessler tube. If the volume of sample is less than 50 ml., dilute to the mark with nitrite-free water. At the same time prepare eleven 50-ml. Nessler tubes containing 0.0, 0.1, 0.2, 0.4, 0.6, 0.8, 1.0, 1.2, 1.6, 2.0, and 2.5 ml. of standard sodium nitrite solution containing 0.5 ppm. of nitrogen as nitrite. Dilute each to 50 ml. They will then correspond to 0, 0.001, 0.002, 0.004, 0.006, 0.008, 0.010, 0.012, 0.016, 0.020, and 0.025 ppm. of nitrogen as nitrite if a 50-ml. sample was used. If less than 50 ml. is used, the reading must be multiplied by a suitable factor. Add 1 ml. of sulfanilic acid solution and 1 ml. of naphthylamine solution to the sample and to each standard. Mix well. Allow to stand for 10 minutes, and compare the color of the sample with that of the standards. This comparison must be made within 30 minutes of the time of developing the colors.

Calculation.

$$\text{Mg. of nitrogen in matched standard} \times \frac{1{,}000}{\text{volume of sample in ml.}} = \text{ppm. nitrogen}$$

To express the result as nitrite radical, multiply by 3.284.

Nitrate. Sample. For this determination, it is necessary previously to have determined color, chloride, total alkalinity to the methyl orange end point, and nitrite. If the nitrate content, determined roughly by qualitatively testing by this method, is below 4 ppm., use 100 ml. of sample. If more than 4 ppm. of nitrate is present, use proportionately less sample. If the nitrite content is below 1 ppm., the color below 10 ppm., and chloride as chlorine below 30 ppm., no preliminary treatment of the sample is required. If the nitrite is above 1 ppm., oxidize it to nitrate by titrating with 0.02N potassium permanganate solution until a faint pink color persists.

If the color is above 10 ppm., add 3 ml. of aluminum hydroxide suspension (page 718), and thoroughly mix. After letting it stand a few minutes, filter, and wash with distilled water. Use the filtrate and washings for the nitrate test, or treat for the removal of chlorides, if necessary.

When the chloride content of the sample is more than 3 mg., which corresponds to 30 ppm. if a 100-ml. sample is being used, first neutralize to methyl orange indicator with 0.02*N* sulfuric acid. Then add to the cold solution, with stirring, enough silver sulfate solution to precipitate all but about 0.1 mg. of chloride. Do not use an excess of silver sulfate solution. Filter, wash the precipitate, and use the filtrate and washings for the test.

REAGENTS. *Silver Sulfate Solution.* Dissolve 4.397 grams of C.P. silver sulfate, Ag_2SO_4, in distilled water in a 1-liter volumetric flask, dilute to the mark, and mix well. Keep in an amber-colored bottle. One milliliter equals 1 mg. of chloride. The solution is 0.0282*N*.

Phenoldisulfonic Acid. To 25 grams of pure white phenol in a 400-ml. beaker add 150 ml. of concentrated sulfuric acid. Stir until solution is complete. Then add 75 ml. of fuming sulfuric acid containing 15 per cent free sulfur trioxide, and stir well. Heat at 100° for 2 hours, cool, and keep in a covered glass jar.

Potassium Hydroxide Solution. Dissolve 67.5 grams of C.P. potassium hydroxide in a small amount of water. Transfer with distilled water to a 100-ml. volumetric flask. Allow to cool, dilute to the mark, and mix well. One milliliter equals 0.675 gram of C.P. potassium hydroxide, approximately 12*N*.

Standard Potassium Nitrate Solution. Dissolve 0.7216 gram of C.P. potassium nitrate, KNO_3, in distilled water, and dilute to the mark in a 1-liter volumetric flask. Mix well, and accurately measure 50 ml. of this solution into a small porcelain dish. Evaporate to complete dryness on a steam bath. Add 2 ml. of phenoldisulfonic acid, and rub with the flat end of a glass rod to wet the residue thoroughly. Dissolve in distilled water, and transfer to a 500-ml. volumetric flask. Dilute to the mark, and mix well. One milliliter of this solution equals 0.01 mg., which is 10 ppm. of nitrogen or 0.04426 mg. of nitrate radical.

PROCEDURE. Place the sample, previously treated if necessary, in a small porcelain dish. If alkaline, neutralize to the methyl orange end point with 0.02*N* sulfuric acid. Calculate the necessary acid from the total alkalinity determination; do not add indicator to the sample, or the color will invalidate the determination. Evaporate to complete dryness on a steam bath. Cool, add 2 ml. of phenoldisulfonic acid reagent, and rub with the flat end of a glass rod in order to wet the dried solids thoroughly. Dilute with about 20 ml. of distilled water, and add dropwise, with stirring, 12*N* potassium hydroxide solution until the maximum color is attained. Not more than 6 ml. should be required. Filter into a 50-ml. Nessler tube, wash the filter free from color, and finally dilute to the mark. Compare with standards.

STANDARDS. Prepare nitrate standards over the range suspected. Convenient amounts of standard nitrate solution are 0.0, 0.1, 0.2, 0.4, 0.6, 0.8, 1.0, 2.0, 4.0, 6.0, 8.0, 10.0, 20.0, 30.0, and 40.0 ml., corresponding to one-tenth these figures in ppm. as nitrogen on a 100-ml. sample. Add these to 50-ml. Nessler tubes. To each add 2 ml. of 12*N* potassium hydroxide solution, and dilute the mixture to the mark with nitrate-free distilled water. Stopper tightly to prevent evaporation and contamination. Standards so prepared will keep for several weeks without deterioration.

CALCULATION.

$$\text{Ml. of nitrogen in matched standard} \times \frac{1{,}000}{\text{volume of sample in ml.}} = \text{ppm. of nitrogen}$$

To express the results as nitrate radical, multiply by 4.426. If nitrite was present and was oxidized to nitrate, deduct the equivalent nitrite nitrogen from this result to obtain the true value for nitrate.

Reporting of Results. Sometimes it is desired to report the results of the mineral analysis of water in the form of hypothetical combinations. No definite statement as to actual content can be made, owing to ionization in solution. It is usual to combine the ions in the inverse order of the solubility of the resulting combinations. Bicarbonates of calcium and magnesium are changed to carbonates on evaporation, and it is the general practice to include the bicarbonate radical with the carbonate. The following plan is suggested:

$CaCO_3$ = the smaller of Ca or $(CO_3 + HCO_3)$
$Mg(OH)_2$ = the smaller of Mg or OH
$MgCO_3$ = the smaller of Mg-$Mg(OH)_2$ or $(CO_3 + HCO_3) - [Mg(OH)_2 + CaCO_3]$
$CaSO_4$ = the smaller of (Ca-$CaCO_3$) or SO_4
$CaCl_2$ = Ca-$(CaCO_3 + CaSO_4)$
$MgSO_4$ = the smaller of Mg $- [Mg(OH)_2 + MgCO_3]$ or (SO_4-$CaSO_4$)
$MgCl_2$ = Mg-$[Mg(OH)_2 + MgCO_3 + MgSO_4]$
Na_2SO_4 = SO_4-$(CaSO_4 + MgSO_4)$
KCl = K
NaCl = the average of Cl-$(CaCl_2 + MgCl_2 + KCl)$ and Na-$(Na_2SO_4 + NaNO_3 + Na_2CO_3 + NaOH)$
$NaNO_3$ = NO_3
Na_2CO_3 = $(CO_3 + HCO_3) - (CaCO_3 + MgCO_3)$
NaOH = OH-$Mg(OH)_2$

SiO_2, Fe_2O_3, Al_2O_3 are reported as such. Phosphate is rarely present in significant amounts except in treated boiler waters. If present, calculate in order to calcium, magnesium, and sodium.

When sodium and potassium are obtained by difference,

$$NaCl = Na - (Na_2SO_4 + NaNO_3 + Na_2CO_3 + NaOH)$$

pH Value. Obtain this by any convenient method. The most accurate means is by the glass electrode (page 123).

Waters more alkaline than the range of this instrument are rarely encountered. Buffer solutions (page 117) and indicators (page 115) are useful. For field work, a pH paper (page 115), giving different colors at different pH values, may be used for rough estimation.

It frequently happens that the sample of water is so unbuffered that definite pH values are difficult to obtain. Always make the pH determination as soon as the sample bottle is opened, since gases in solution, particularly carbon dioxide, may readily be given off on exposure or when the sample is shaken. Conversely, with highly unbuffered samples, carbon dioxide may be absorbed from the surrounding atmosphere. This is indicated by the pH declining.

Ammonia. Free ammonia is recovered quantitatively only when the distillation mixture is maintained at a pH of 7.4 or above. Waters with high ammonia content are nesslerized directly. Others require the addition of a phosphate buffer for pH 7.4 before the sample is distilled. Always use a fresh sample for this determination.

PROCEDURE. Set up a distilling flask and condenser. Put distilled water in the flask, and distill until the distillate shows no test with Nessler reagent (page 720), indicating that the flask and condenser have become free of ammonia. Empty the distilling flask, and put in it 500 ml. of the sample water or a smaller volume diluted to 500 ml. with ammonia-free distilled water. Add 10 ml. of phosphate buffer containing 14.3 grams of anhydrous monopotassium phosphate (KH_2PO_4) and 90.15 grams of dibasic potassium phosphate trihydrate ($K_2HPO_4.3H_2O$) per liter. Distill at a rate of 6 to 10 ml. per minute collecting four 50-ml. portions in Nessler tubes. Alternatively, distill 250 ml. into a 250-ml. calibrated flask, and use a 50-ml. portion as sample. Save the contents in the flask for albuminoid nitrogen.

Determine the nitrogen by the Nessler method on page 145, starting at Procedure. If the series of distillates was used, add the results to obtain the total ammonia. Unless the amount in the last tube distilled is substantially zero, repeat with a diluted sample. Calculate as nitrogen.

Albuminoid Nitrogen. Albuminoid nitrogen is the nitrogen equivalent of ammonia formed or liberated from nitrogenous matter by the action of alkaline potassium permanganate in water after first distilling off any ammonia. It bears a variable ratio to organic nitrogen in highly polluted waters. In waters of apparently slight pollution, the values obtained by this method are related to color and amount of colloidal matter.

Alkaline Permanganate Solution. Boil 1,200 ml. of distilled water in a 2,500-ml. porcelain dish for 10 minutes to drive off ammonia. Add 16 grams of C.P. potassium permanganate, and dissolve by stirring. To prepare the $9N$ alkali, dissolve 505 grams of pure potassium hydroxide or 360 grams of pure sodium hydroxide in a little water, and dilute to 1 liter with ammonia-free water. Allow to cool and settle, and filter through asbestos. Add 800 ml. of clear $9N$ potassium or sodium hydroxide solution to the aqueous permanganate solution, and dilute to 2,500 ml. with ammonia-free water. Evaporate to 2,000 ml. Determine the ammonia in 50 ml. of this reagent by nesslerization (page 145), and use as a correction in subsequent determinations.

Procedure. To the cool contents remaining in the distilling flask after removal of ammonia, add 50 ml. or more, depending largely on the degree of pollution of the water, of the alkaline permanganate solution. Complete as for ammonia (page 240), starting at "Distill at a rate of 6 to 10 ml. per minute . . .".

Dissolved Oxygen. Winkler Method. *Rideal-Stewart Modification.* Dissolved oxygen quickly erodes and pits boilers and boiler tubes; hence its determination is of importance in boiler waters.

Collection of Sample. When collecting a sample of cold water from a faucet or valve connection, open the faucet or valve and insert a glass tube or one of pure gum rubber. This will not take all the water flowing from the opening; a major part should be wasted. Collect a sample for each determination in a 250-ml. narrow-necked glass-stoppered bottle. To avoid entrainment or absorption of atmospheric oxygen, the glass or gum-rubber tube extends to the bottom of the bottle. The water being so led in will therefore be delivered without entrained air below the surface of the liquid in the bottle. Allow the bottle to overflow for at least 10 minutes at the rate of 1 or 2 liters per minute. Immediately stopper without enclosing an air bubble below the stopper.

If the sample is hot, it must be cooled to room temperature before it is collected. To perform this, lead it from the point of collection through a copper coil immersed in running water or, preferably, an ice bath. Then lead into the sample bottle, as described above for cold water.

To obtain a sample from a tank or pond, use both a gallon bottle and the 250-ml. glass-stoppered bottle. Provide each with a two-hole rubber stopper carrying a glass inlet and outlet tube, the inlet tubes extending to the bottom of the bottles and the outlet tubes extending only to the bottom of the stoppers. Connect the outlet tube of the sample bottle with the inlet tube of the gallon bottle. Immerse the bottles in the water to the desired depth, and apply suction to the outlet of the gallon bottle.

PROCEDURE. The first steps should be carried out in the field. Remove the stopper from the bottle, and immediately proceed with the analysis in the bottle in which the sample was collected. All reagents must be added from a pipet dipping well under the surface of the sample. Some small overflowing of the bottle will occur when the stopper is inserted after each reagent is added.

Conditions are first provided for oxidation of nitrites, sulfides and sulfites, organic matter, and ferrous compounds. Add 0.7 ml. of concentrated sulfuric acid and then 1 ml. of potassium permanganate solution containing 6.32 grams per liter. This is approximately 0.2*N*. Insert the stopper, and mix by inverting the bottle several times. If a noticeable excess of potassium permanganate is not present after 20 minutes, again add 1 ml. of the permanganate solution. If this is still insufficient, add more. It is permissible to use a stronger solution. After the potassium permanganate color has persisted for 20 minutes, destroy the excess by adding 1 ml. of approximately 2.17*N* potassium oxalate solution. Prepare this by dissolving 20 grams of C.P. potassium oxalate, $K_2C_2O_4,H_2O$, in water and diluting to 1 liter.

The next steps provide for rapid absorption of the dissolved oxygen by manganous hydroxide. For this, prepare a filtered manganous sulfate solution containing 480 grams of $MnSO_4.4H_2O$ or 400 grams of $MnSO_4.2H_2O$ per liter. Also prepare an alkaline potassium iodide solution by dissolving 700 grams of potassium hydroxide and 150 grams of potassium iodide in water, cooling, and diluting to 1 liter. Add 1 ml. of the manganous sulfate solution and 5 ml. of the alkaline potassium iodide solution to the sample, and mix well. Manganous hydroxide is precipitated and will settle on standing.

After the precipitate has settled, add 1 ml. of concentrated sulfuric acid, and mix by shaking. The dissolved oxygen now liberates free iodine from the potassium iodide present. After the acid has been added and the stopper replaced at this step, there is no further change, and the rest of the determination may be performed in the laboratory within a few hours.

Transfer 200 ml. of the contents of the bottle to a 500-ml. conical flask. Titrate with 0.025*N* sodium thiosulfate solution. Add 0.5 ml. of a 0.5 per cent starch solution after the color of iodine is nearly gone, and then complete the titration to the disappearance of the blue color.

In some highly accurate work, a correction for the loss of dissolved oxygen by displacement with the reagents is applied. This is outside the limit of accuracy of anyone except a highly experienced operator.

Calculation.

$$\text{Ml. sodium thiosulfate} \times \text{normality} \times 0.008 \times \frac{10^6}{\text{volume of sample in ml.}} = \text{ppm. dissolved oxygen}$$

Using $0.025N$ sodium thiosulfate and titrating a 200-ml. aliquot of sample, this is

$$\text{Ml. sodium thiosulfate} = \text{ppm. dissolved oxygen}$$

Color. This test is usually applicable to municipal and potable waters by comparison with known color standards. Remove any turbidity by settling or by centrifuging. Do not filter, for color may be sorbed by the filter paper.

Procedure. Fill a 100-ml. Nessler tube with the clear water sample to a height equal to that of standards. Compare the sample and standards by looking vertically downward through the tubes upon a white or mirrored surface placed at such an angle that light is reflected up through the liquid. If the color of the water is greater than 70 ppm., dilute a known quantity of water, and correct the reading accordingly.

Preparation of Standards. To 1.245 grams of C.P. potassium chloroplatinate, K_2PtCl_6, and 1.000 gram of C.P. crystallized cobaltous chloride, $CoCl_2.6H_2O$, in a 1-liter volumetric flask, add 200 ml. of distilled water and 100 ml. of concentrated hydrochloric acid. Shake until dissolved, dilute to the mark, and mix well. This solution contains 500 ppm. of platinum and has a color value assigned of 500. To prepare standards of 5, 10, 15, 20, 25, 30, 35, 40, 50, 60, and 75 ppm., transfer 1, 2, 3, 4, 5, 6, 7, 8, 10, 12, and 15 ml. to 100-ml. uniform Nessler tubes, and dilute to volume with distilled water. If the tubes are well stoppered, this color is substantially permanent. For routine checking on the same water, closer cut standards over a narrow range are often prepared.

Turbidity. The standard unit of turbidity is considered as that produced by 1 ppm. of silica in the form of diatomaceous earth, fuller's earth, or bentonite, in distilled water. Comparison can be made either directly with standards or somewhat more conveniently by measurement of the obscuring power against a light. The silica standards are calibrated in terms of their obscuring power.

Jackson Turbidimeter Method. This apparatus, shown in Fig. 93, consists of a graduated glass tube and a standard candle made of beeswax and spermaceti gauged to burn 7.39 to 8.16 grams per hour. The apparatus is supported in a vertical position so that the center line of the tube passes through the center line of the candle. A spring device keeps the upper surface of the candle firmly pressed against the top of the support. The latter is exactly 3 inches below the bottom of the glass tube. This tube has a flat polished bottom. When readings are made, most of the tube is enclosed in a metal case in order to avoid nephelometric effects. Tubes are available reading either in ppm. or in centimeters of depth. Similar tubes using an electric filament as the light are also available and are used for other turbidimetric determinations but are not APHA standard.

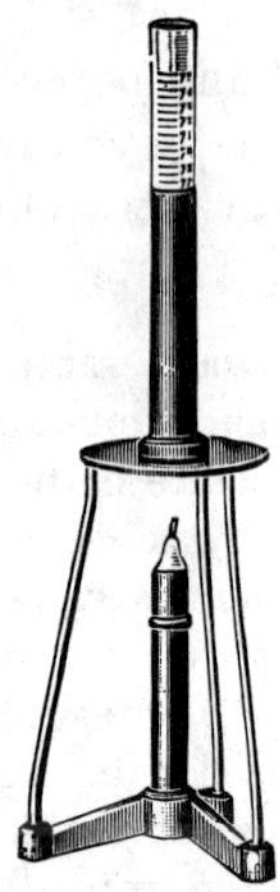

FIG. 93.—Jackson turbidimeter.

PROCEDURE. Clean the tube both inside and out, be sure that the outside bottom surface is dry, and place in the apparatus. Trim the candle, removing any charred portion of the wick that is easily broken off with the fingers, and see that the candle presses against the top of the candleholder.

Light the candle, and at once pour the turbid sample into the graduated glass tube until the image of the candle flame just disappears from view. Pour in the sample very slowly when the candle flame is only faintly visible, and make the observation in a darkened room or with a black cloth over the head. The candle should be kept lighted for only a few minutes at a time in order to avoid heat effects.

Remove the graduated glass tube, and read the depth of liquid either in ppm. of turbidity or to the nearest 0.1 cm. Readings in centimeters are converted to ppm. according to Table 11.

TURBIDITY STANDARDS METHOD. Standards of suspended siliceous materials flocculate less and keep better if prepared with a buffer solution having a pH of about 7.8. For this, mix 50 ml. of a solution of 12.404 grams of boric acid per liter with 50 ml. of a solution of 14.912 grams of potassium chloride per liter and with 3.97 ml. of a solution of 8.000 grams of sodium hydroxide per liter. Dilute the resulting mixture to 2 liters with freshly boiled distilled water.

Then to 5 grams of high-quality bentonite or fuller's earth in a liter volumetric flask add 100 ml. of distilled water or buffer solution. It is desirable to let bentonite soak overnight. Then shake until the bentonite or fuller's earth is thoroughly wetted and swelled and all lumps are broken up. Dilute to the mark with water or buffer solution, and mix well. Allow the suspension to stand about 24 hours, during which the heavier particles will have settled out and the turbidity will have been reduced to slightly more than 1,000 ppm. Decant carefully, and test the suspension with the candle turbidimeter. Adjust to 1,000 ppm. by adding water or buffer solution.

A suitable series of standards is prepared in 250-ml. uniform bottles. The series varies according to the degree of turbidity of water being compared. An appropriate general series consists of 250, 100, 50, 25, 15, 10, 5, 2.5, and 1 ml. diluted to 250 ml. and corresponding to 1,000, 400. 200, 100, 60, 40, 20, 10, and 4 ppm. Usually a more closely cut series is prepared for the specific range involved. Be very careful to keep the 1,000-ppm. solution agitated during the measuring of an accurate aliquot.

TABLE 11.—CONVERSION OF JACKSON TURBIDIMETER DEPTHS TO PPM.

Depth of liquid, centimeters	Turbidity, ppm.	Depth of liquid, centimeters	Turbidity, ppm.
2.3	1,000	11.4	190
2.6	900	12.0	180
2.9	800	12.7	170
3.2	700	13.5	160
3.5	650	14.4	150
3.8	600	15.4	140
4.1	550	16.6	130
4.5	500	18.0	120
4.9	450	19.6	110
5.5	400	21.5	100
5.6	390	22.6	95
5.8	380	23.8	90
5.9	370	25.1	85
6.1	360	26.5	80
6.3	350	28.1	75
6.4	340	29.8	70
6.6	330	31.8	65
6.8	320	34.1	60
7.0	310	36.7	55
7.3	300	39.8	50
7.5	290	43.5	45
7.8	280	48.1	40
8.1	270	54.0	35
8.4	260	61.8	30
8.7	250	72.9	25
9.1	240		
9.5	230		
9.9	220		
10.3	210		
10.8	200		

Compare the sample in a similar bottle with the standards by viewing through the bottles; these are placed before a frosted screen behind which are two lamps, the whole preferably enclosed in a box. The matching may also be made by looking transversely through the solutions at a dark object, such as a crosspiece on a window frame or against a light background. Always shake the bottles immediately before making the observation.

Turbidities below 5 are determined instrumentally. For details the APHA methods should be consulted.

Free Chlorine. This determination is required on waters that have been treated with chlorine or hypochlorites, such as water to be used as potable supplies, for swimming pools, or as condenser water. Chloramines, if present, as well as the free chlorine will be measured by this method.

At a pH less than 4, chlorine develops a color with the reagent, varying from pale yellow at low concentrations to an orange at higher concentrations.

Oxidizing agents such as ferric compounds, manganic compounds, nitrites, algae, organic iron compounds, and lignocellulose interfere, producing colors similar to or affecting the intensity of the color produced by chlorine. The method is valid in waters containing less than 0.3 ppm. of iron, 0.01 ppm. of manganic manganese, or 0.3 ppm. of nitrite nitrogen.

Orthotolidine Reagent. To 1 gram of orthotolidine, melting point 129°, in a mortar add 5 ml. of 1:4 hydrochloric acid, and grind to a thin paste. Dissolve this in 200 ml. of water, and transfer to a liter volumetric flask. Dilute to 505 ml. with water; then make up volume with 1:4 hydrochloric acid. Mix well, and store in an amber-colored bottle at room temperature. Prepare fresh every 6 months.

Procedure. Remove any turbidity by centrifuging the sample. Adjust the temperature to be between 20 and 40°. Place 100 ml. of the clear sample in a 100-ml. Nessler tube having the graduation mark 300 $\pm$ 6 mm. from the bottom. Add 1 ml. of orthotolidine reagent, and mix with a plunger-type stirrer.

If chloramine is absent, nitrites and iron are each less than 0.3 ppm., and manganic manganese is not over 0.01 ppm., it is necessary only to set the sample in the dark for 5 to 15 minutes for the color to develop. If chloramine is present, be sure to let the sample stand for the full 15 minutes, but no more. A preliminary test to observe the approximate time for maximum color development is often made.

If highly alkaline waters give blue-green colors with the reagent, add sufficient 1:4 hydrochloric acid to reduce the pH to 4 or less. Usually 1 ml. will be sufficient. If the iron content is greater than 0.3 ppm. and less than 1 ppm., also add 1 ml. of 1:4 hydrochloric acid after the reagent before diluting to 100 ml.

If manganic manganese is present in excess of 0.1 ppm. or if ferric iron is present in amounts greater than 1 ppm., proceed to obtain a reading due to chlorine plus manganic salts and/or iron. Then to a second sample add 2 or 3 drops of 6*N* sulfuric acid and 10 or 12 drops of 0.02 per cent hydrogen peroxide solution. Mix well and allow to stand for 40 minutes.

TABLE 12.—PERMANENT COLOR STANDARDS FOR FREE CHLORINE TEST

Free chlorine, ppm.	Copper sulfate solution, ml.	Potassium dichromate solution, ml.
0.01	0.0	0.8
0.02	0.0	2.1
0.03	0.0	3.2
0.04	0.0	4.3
0.05	0.4	5.5
0.06	0.8	6.6
0.07	1.2	7.5
0.08	1.5	8.2
0.09	1.7	9.0
0.10	1.8	10.0
0.15	1.8	15.0
0.20	1.9	20.0
0.25	1.9	25.0
0.30	1.9	30.0
0.35	1.9	34.0
0.40	2.0	38.0
0.50	2.0	45.0
0.60	2.0	51.0
0.70	2.0	58.0
0.80	2.0	63.0
0.90	2.0	67.0
1.00	2.0	72.0

Add 1 ml. of the orthotolidine reagent, and read the tube after 5 minutes. This gives the true residual chlorine in the water. Prepare a hydrogen peroxide solution fresh at least every week.

Compare the color produced with that of standards, looking down through the tubes against a white background illuminated with good daylight or with diffused light from a daylight bulb. Avoid direct sunlight. Report as ppm. of chlorine.

PERMANENT COLOR STANDARDS. Prepare standard copper sulfate solution as follows: Dissolve 1.5 grams of copper sulfate, $CuSO_4.5H_2O$, in 50 ml. of water in a 100-ml. volumetric flask. Add 1 ml. of concentrated sulfuric acid, dilute to the mark with water, and mix well. This solution is approximately 0.12*N*, and each milliliter contains approximately 0.0096 gram of anhydrous copper sulfate.

Similarly prepare standard potassium dichromate solution as follows: Dissolve 0.25 gram of potassium dichromate, $K_2Cr_2O_7$, in 900 ml. of water in a 1-liter volumetric flask, and add 1 ml. of concentrated sulfuric acid. Dilute to the mark with water, and mix well.

Prepare standards in Nessler tubes similar to that to be used for the sample. Add the exact quantities of the copper sulfate and potassium dichromate solutions shown in Table 12. Dilute to the 100-ml. mark in each case with distilled water, and mix. Seal by cementing micro cover glasses over the tops of the tubes, using collodion, Canada balsam, or other suitable cement. If not exposed to sunlight, these are stable for 6 months.

TEMPORARY STANDARDS. These must be made up fresh for each test. First prepare zero-chlorine-demand water. For this, add to distilled, ammonia-free water sufficient chlorine to produce approximately 0.5 ppm. excess. Allow to stand for 30 minutes, and then boil for at least 30 minutes. Cool to 25° before using, and make up fresh daily.

Next prepare standard chlorine water. Rapidly bubble chlorine gas for 30 minutes or longer through 500 ml. of water in an amber-colored bottle placed in a hood. This should produce a solution containing rather more than 1 gram of chlorine per liter. Standardize this solution each time it is used as follows: Dissolve 1 gram of C.P. potassium iodide in 100 ml. of water in a 250-ml. conical flask. Add 10 ml. of glacial acetic acid and then from a pipet exactly 5 ml. of the chlorine water. Shake well, and after a minute titrate with 0.025*N* sodium thiosulfate solution (page 730), using starch indicator (page 717) toward the end of the titration. The chlorine concentration in milligrams per milliliter is ml. 0.025*N* thiosulfate $\times$ 0.886/5.

As standards, add the solution of chlorine water to zero-chlorine-demand water in a series of 100-ml. Nessler tubes in amounts necessary to give the desired range of ppm. Fill to the mark with zero-chlorine-demand water, and develop the color in the same way and at the same time as that of the sample.

Chlorine Demand. Polluted waters such as sewage, effluents, trade wastes, and waters highly contaminated from organic sources frequently require chlorination to control odor and bacterial growth. A measure of the chlorine demand may be required in order to estimate the quantity of chlorine necessary to add to a given volume of water. Chlorine demand is defined as the quantity of chlorine required to produce a residual chlorine content between a trace and 0.1 ppm. after 10-minute contact.

PROCEDURE. To 250 ml. of the sample in an 8-ounce reagent bottle or equivalent, add from a buret standard chlorine water containing 0.5 mg. of chlorine, and agitate the mixture well. Test for free chlorine by withdrawing 0.25 ml. of the solution with a pipet and mixing with a drop of orthotolidine reagent (page 246) on a spot plate. Continue adding the standard chlorine water in 0.5-ml. portions, agitating and testing as before until only a slight excess of chlorine is present, indicated by a light yellow coloration. The quantity of chlorine required is designated the immediate chlorine demand.

To a second similar sample add standard chlorine water in an amount equivalent to the immediate chlorine demand plus 0.5 ml., agitate gently, and allow to stand for exactly 10 minutes. Immediately test a portion for residual chlorine as directed. Repeat the procedure, adjusting the chlorine added, until the results conform to the definition of chlorine demand.

CALCULATION.

$$\text{Ml. standard chlorine water} \times \text{grams chlorine per liter in standard chlorine water} \times \frac{1{,}000}{250} = \text{ppm. chlorine demand}$$

Residual Chlorine. This method is for use with highly alkaline sewage and is not satisfactory for public or normal water supplies. The stronger acid used with the reagent reduces the interference of nitrites.

PROCEDURE. Prepare orthotolidine reagent exactly as with the more dilute acid (page 246) to the point at which the transfer is made to a 1-liter volumetric flask. Dilute to 425 ml. with water. Add 80 ml. of concentrated hydrochloric acid, dilute to the mark with 1:4 hydrochloric acid, and mix well. Preserve in an amber-colored bottle at room temperature, and prepare fresh every 6 months.

Mix 10 ml. of the sample at 20° or slightly above, with 1 ml. of this orthotolidine reagent. Allow to stand for 5 minutes, and compare with color standards prepared at the same time. Report the results in ppm. of chlorine.

COLOR STANDARDS. Restandardize the chlorine solution (page 248), and immediately dilute a volume containing 10 mg. of chlorine to 100 ml. in a volumetric flask with zero-chlorine-demand water (page 248). Mix well. Dilute 1 ml. of the latter solution to 100 ml. with zero-chlorine-demand water, and mix well to obtain a solution containing 1 ppm. of chlorine.

From this dilute chlorine solution and zero-chlorine-demand water, prepare 10-ml. portions containing 0, 0.2, 0.4, 0.6, 0.8, and 1.0 ppm. of chlorine. Treat in the same way as the sample and at the same time. Permanent standards may be purchased from supply houses, but these must be checked against standard chlorine solutions.

Chloramine. Chloramine is widely used instead of chlorine for the sterilization of water. It is measured by the orthotolidine test for residual chlorine.

The pH of the solution determines the type of chloramine present. At pH 8.5 or higher, monochloramine alone is formed; below pH 4.4, nitrogen trichloride is produced. Between these two pH values mono- and dichloramine coexist in a ratio fixed by the actual pH. Sufficient acidification converts monochloramine to nitrogen trichloride according to the equation

$$3NH_2Cl + 2H^+ = 2NH_4^+ + NCl_3$$

In dilute solutions, less than 1 minute suffices to complete this reaction at 15°.

Below pH 4.4, two-thirds of the nitrogen present as chloramine is converted to ammonium ion. The chloramine content is then obtained by multiplying the observed increase in ammonia nitrogen by 3/2 times 35.46/14, or 3.8. Using 50-ml. Nessler tubes, the factor is 3.8/5, or 0.76, which, for convenience, may be considered 3/4. The Nessler tube readings correspond to ppm. of nitrogen when 10-ml. samples are read.

PROCEDURE. Fill two 50-ml. Nessler tubes with the sample solution. To one add 0.2 ml. of 1:10 sulfuric acid to lower the pH below 4.4, mix by inverting the tube, and allow to stand for 1 minute. If the pH of the sample solution was below 8.6, add 0.1 ml. of a 5 per cent solution of sodium carbonate to the other Nessler tube, and mix the contents. Now add 2 ml. of Nessler reagent (page 720) to each tube, and compare with standards not later than 1 minute after the addition of the reagent (pages 146 to 147). Approximately 90 per cent of the full color is developed in 1 minute at 20°. The difference in the readings of the two tubes multiplied by the factor is a measure of the chloramine present.

This procedure is subject to the limitations common to direct nesslerization of hard waters. Also, the usual clarification procedures may affect the chloramine determination.

Fluoride. BY ZIRCONIUM-ALIZARIN. The fluoride content of waters has been receiving increasing attention, with respect to municipal water, because of the contention that an adequate supply results in better teeth, though too much causes mottling. The determination is affected by the presence of other ions, particularly sulfate.

REAGENT. Dissolve 0.5 gram of zirconium oxychloride, $ZrOCl_2.8H_2O$, in distilled water in a 100-ml. volumetric flask, dilute to the mark, and mix well. Dissolve 0.1 gram of alizarin sodium monosulfonate in 100 ml. of distilled water, and store in an amber-colored bottle in a cool place.

To 25 ml. of the zirconium oxychloride solution add dropwise, with agitation, 25 ml. of the alizarin monosulfonate solution. If a turbidity persists, allow the mixture to stand until it settles, usually overnight. Finally add 100 ml. of distilled water, and mix well. Store this diluted reagent in an amber-colored bottle in a cool place, and prepare fresh every few days.

DILUTION WATER. The sulfate content of the sample must be known. As a minimum, prepare distilled water to which sufficient sodium sulfate has been added to match that in the sample. In general, that will be satisfactory in waters containing less than 500 ppm. of total solids. If a complete mineral analysis is available, it is preferable to match it in fluoride-free form. For this purpose, prepare the dilution water by adding suitable volumes of the solutions in Table 13 to distilled water.

FLUORIDE STANDARD. Dissolve 0.2210 gram of C.P. sodium fluoride in distilled water in a 100-ml. volumetric flask, dilute to volume, and mix well. This contains 0.1 per cent of fluoride, or 1,000 ppm. Dilute 5 ml. of this solution to 100 ml. in a volumetric flask, and mix well. It now contains 50 ppm.

PROCEDURE. Place 50 ml. of the clear sample in a 50-ml. Nessler tube. Prepare a series of nine 50-ml. Nessler tubes containing 0, 0.2, 0.4, 0.6, 0.8, 1.0, 1.2, 1.4, and 1.6 ml. of the standard sodium fluoride solution, and dilute to 50 ml. with water made up from salt solution to correspond to the analysis of the sample water. These standards so made correspond to 0, 0.2, 0.4, etc., ppm. of fluorine, respectively. Add from a pipet exactly 5 ml. of 5*N* hydrochloric acid to each of the 50-ml. Nessler tubes containing the standards and the samples. Mix with a plunger type of stirrer, such as a glass rod flattened at one end. Then add 1 ml. of the zirconium-alizarin indicator, and again mix well. Allow the standards and sample to stand at room temperature about 18 hours, mix again, and compare.

If the fluoride content is greater than 1.6 ppm., repeat, using a sample diluted to 50 ml. with distilled water. Note that, in this case, the salt solution for dilution of the standards is to be similarly diluted.

Should some slight amount of the indicator be thrown down as a reddish precipitate due to phosphate or organic matter, disperse by rapid stirring, and make the color comparison before separation can again take place.

TABLE 13.—SALT SOLUTIONS FOR USE IN FLUORIDE TEST

Salt	Formula	Grams per liter	Milli-equivalents per liter	Milligrams per milliliter	
				Basic ion	Acidic ion
Calcium sulfate......	$CaSO_4$	1.36	20	0.4	0.96
Calcium chloride.....	$CaCl_2$	11.1	200	4.0	7.1
Magnesium chloride..	$MgCl_2$	9.5	200	2.4	7.1
Sodium sulfate.......	Na_2SO_4	14.2	200	4.6	9.6
Sodium chloride......	NaCl	11.7	200	4.6	7.1
Sodium bicarbonate..	$NaHCO_3$	16.8	200	4.6	12.2
Sodium silicate.......	Na_2SiO_3	12.2	200	4.6	7.6
Ferric chloride.......	$FeCl_3$	10.8	200	3.7	7.1
Aluminum chloride...	$AlCl_3$	8.9	200	1.8	7.1

Sulfite. This method is used in the routine control of feed water or boiler water to which sulfites are added for the purpose of absorbing dissolved oxygen. The sample used for this test must be cooled as drawn from the boiler and at once stoppered. Allow to settle, but do not filter. Test immediately after settling.

PROCEDURE. To 100 ml. of the sample in a 250-ml. conical flask add 10 ml. of glacial acetic acid, and 0.5 ml. of starch solution (page 717). Titrate with 0.2*N* iodine solution (page 726) until the solution turns just blue. This oxidizes sulfite to sulfate. For somewhat more accurate results, acidify the sample with 10 ml. of glacial acetic acid, add a measured excess of 0.2*N* iodine solution, and titrate the excess iodine with standard 0.02*N* sodium thiosulfate solution (page 730). Perform a blank determination, using distilled or tap water in place of the sample. Subtract the milliliters of sodium thiosulfate solution used from the milliliters of iodine solution added in both cases, and subtract the blank from the sample determination to give the milliliters of iodine solution consumed by the sulfite.

CALCULATION.

$$\text{Ml. standard iodine solution used} \times \text{normality} \times 0.06303 \times \frac{10^6}{100} = \text{ppm. sodium sulfite}$$

Oil. Oil is undesirable in boiler water because it may form thin films of relatively great heat-insulating properties and may modify scale formation in an undesirable manner. It may also have an effect on priming and foaming of the boiler.

Short Method. Thoroughly mix the sample. At once measure 2 liters into a beaker or porcelain dish, and evaporate to dryness on a steam bath, taking care to prevent contamination. Cool the container, add 100 ml. of petroleum ether, and rub up the mixture thoroughly with a glass rod. Ethyl ether or benzene may also be used as solvent. Filter into a weighed fat-extraction flask. Wash the filter with a little of the solvent, and evaporate the contents of the flask on a steam bath. Place the flask in the oven until the odor of petroleum ether is gone, but no longer. Cool, and weigh the extracted oil.

Calculation.

$$\text{Weight of oil} \times \frac{10^6}{2{,}000} = \text{ppm. of oil}$$

Accurate Method. Dissolve 10 grams of iron in 200 ml. of concentrated hydrochloric acid. Oxidize to the ferric state with just sufficient 1:1 nitric acid, and dilute the mixture to 1 liter.

To 500 or 1,000 ml. of the well-mixed sample add 5 ml. of the ferric chloride solution. Bring nearly to boiling, and add enough 1:1 ammonium hydroxide to precipitate all the iron. This precipitate will occlude the oil in the sample. Allow the precipitate to settle for a few minutes, and filter through a coarse, fat-free filter paper. Wash the filter free of ammonia; then dry the paper and precipitate at 100°. Extract the dried precipitate and paper in a Soxhlet extractor with ethyl ether until the extract is colorless. Evaporate the solvent, and dry the flask in the oven for the minimum time to remove the ether. Cool, and weigh the oil directly.

CHAPTER 12

FIXED ALKALIES AND ALKALINE SALTS

This major class of industrial products finds wide application either as such or in compounded products. The terminology is confusing industrially, for salts that hydrolyze to give an alkaline reaction are referred to as alkalies in many industries. Here the term alkali will be limited to free caustic alkalies, others being referred to as alkaline salts. The term fixed alkalies is used to exclude volatile alkalies such as ammonia and amines.

In respect to total use, the sodium compounds are the important class. In compounded products, potassium compounds have some minor importance because of special properties. For brevity, only sodium compounds are considered. In every case, by suitable modification of factors, the methods are directly applicable to potassium compounds.

Not only are sodium compounds less expensive normally, but the relative weights of the sodium and potassium radicals, 23 and 39, result in more molecules or more active material per pound. Frequently the potassium compounds are more soluble.

The important members of the class under discussion are the following:

1. Sodium hydroxide, $NaOH$.
2. Sodium orthosilicate, Na_4SiO_4.
3. Sodium sesquisilicate, $Na_6Si_2O_7$.
4. Sodium metasilicate, Na_2SiO_3.
5. Trisodium phosphate, Na_3PO_4.
6. Sodium carbonate, Na_2CO_3.
7. Sodium sesquicarbonate, $Na_2CO_3(NaHCO_3)_{1.25}$.
8. Tetrasodium pyrophosphate, $Na_4P_2O_7$.
9. Disodium phosphate, Na_2HPO_4.
10. Sodium tetraborate, $Na_2B_4O_7$.
11. Sodium bicarbonate, $NaHCO_3$.
12. Sodium metaphosphate, $NaPO_3$.

This incomplete list is in the approximate order of the degree of alkalinity of their aqueous solutions of similar concentration by weight. The ratio of carbonate to bicarbonate varies in sodium sesquicarbonate from different manufacturers. Numerous commercial silicates are omitted because they differ in composition from one manufacturer to another. Practically speaking, all commercial silicates are more alkaline than sodium sesquicarbonate.

All are listed as the anhydrous salts, because practically all occur in that form. Some are sold as hydrates and are generally uneconomical in cost per pound relative to the anhydrous salt content. Generally the practice in commercial analysis in dealing with such mixtures is to calculate the composition in the form of anhydrous salts unless there is evidence that a specific hydrate is present and, when the data are all assembled, to identify the specific hydrates by applying judgment and trial calculations.

This list of 12 materials, which includes by no means all the possibilities but all the probabilities, would seem to offer a tremendous number of possible combinations, 66 binary combinations alone. Actually, although practically all could be mixed as dry materials, many are incompatible in aqueous mixture. To illustrate, sodium hydroxide and sodium bicarbonate will react to form sodium carbonate, with probable excess of one material. It seems well to indicate briefly by numbers pairs that are incompatible by reason of the difference in pH level at which they are stable, as illustrated by the example just cited: 1,3; 1,4; 1,7; 1,9; 1,10; 1,11; 2,7; 2,9; 2,10; 2,11; 3,7; 3,9; 3,10; 3,11; 4,7; 4,9; 4,10; 4,11; 5,7; 5,9; 5,10; 5,11; 6,9. Although this leaves 43 pairs, actually only a fraction of those as well as of the ternary and quaternary mixtures are encountered, fortunately for the analyst.

Although sodium hexametaphosphate, $NaPO_3$, is not alkaline, it is commonly sold as an "adjusted" compound in which sodium carbonate has been used to raise the alkalinity. Manufacturers' restrictions on its sale and use limit the occasions when it is encountered.

Soap is an alkaline salt and could properly appear in this list. It has been excluded because a separate chapter is devoted to it. It would fall just below sodium carbonate in the listing.

The diversity of products in which these materials are used is so great as to preclude more than an incomplete list. Alone or as mixtures they are used in metal cleaning, metal etching, glass cleaning, dishwashing, laundering, dairy cleansing and sterilizing, building of soaps, etc.

The presentation must necessarily possess elements of vagueness to allow for application, in one case to a water-soluble salt or mixture of salts, in another to a complex commercial mixture such as a soap built with alkaline salts, and in other cases to various liquid compound specialties.

QUALITATIVE

Indicators. With many of these materials, addition of a drop of phenolphthalein solution will produce a pink color. To an experienced worker the intensity is a rough colorimetric indication of the degree of alkalinity. A series of indicators for different pH levels are often used. For this purpose, universal indicators or indicator paper is useful (pages 114 and 115). One must bear in mind that soap also gives an alkaline reaction of about pH 10.2. Sodium sulfite also is alkaline to phenolphthalein.

Some members of the class fall below the level at which phenolphthalein reacts, pH 8.3. Sodium bicarbonate theoretically does so, although commercially it contains a trace of sodium carbonate, which gives the reaction. Sodium metaphosphate, unless adjusted, is definitely below. Often a quick titration to phenolphthalein and methyl orange end points gives qualitative information needed and is useful quantitatively later.

Acid. CARBONATES. Acidify a solution of sample with 1:5 nitric acid, using this acid in order that the same solution may later be tested for phosphate. A good deal of acid may be required to neutralize the sample, and the solution may heat up considerably in neutralization if substantial amounts of alkalinity are present. Evolution of an odorless gas normally indicates the presence of a carbonate or bicarbonate. In case of doubt, use a test tube fitted with an exit tube, and pass the evolved gas into a solution of barium hydroxide. In using this technic it is essential that the sample solution be heated steadily and continuously until the exit tube is withdrawn; otherwise, the barium hydroxide solution will be sucked back into the sample tube. A white precipitate of barium carbonate indicates a positive test. Sulfites or bisulfites correspondingly evolve sulfur dioxide, but the odor is easily recognized.

SILICATES. When silicates are present in appreciable amounts, the addition of acid will produce a gelatinous precipitate of silica. The absence of a precipitate does not preclude the possibility that a silicate may be present, for moderate amounts of silicic acid may remain in colloidal dispersion upon acidification and are thus not visible. When no precipitate is produced upon the acidification of a product suspected to contain a silicate, add 1:1 ammonium hydroxide. This often brings down a flocculent precipitate of silicic acid, which is insoluble on reacidification.

Very small amounts of silicates often cannot be identified qualitatively and are detected only on later quantitative analysis of the ash. They then show up as the usual residue volatile with hydrofluoric acid. Such small amounts of silicates may have substantial importance, as for inhibition of corrosion in a special cleaning compound or to prevent corrosion of aluminum tubes by a product sold in such containers. When small amounts of silica are found, usually judgment and knowledge of the contaminants present in commercial-quality materials are necessary to decide whether or not this is adventitious.

Addition of acid precipitates some water-soluble gums. They are differentiated from silica by filtering on ashless paper and ignition or by qualitative tests for gums (Table 35, page 545).

Alcohol. Of the alkaline materials under discussion only sodium hydroxide and potassium hydroxide are completely soluble in alcohol, and these hydroxides are usually accompanied by traces of carbonates that are confusing.

If the sample is a solution, concentrate as much as possible, even by drying to a solid if feasible. Add a large excess of alcohol; warm, and, depending on conditions, let settle, centrifuge, or filter. Test the clear alcoholic solution with phenolphthalein. Substantial alkalinity usually indicates the presence of free sodium hydroxide, occasionally of potassium hydroxide.

Alcohol will extract minor amounts of free alkali from the higher silicates and commercial grades of trisodium phosphate; thus, a small amount of alkalinity is not positive evidence of the presence of uncombined alkali hydroxide. This is increased by any dilution of the alcohol by sample solution. Filter paper will sorb appreciable amounts of free alkali; therefore, if the alcoholic solution has been filtered rather than centrifuged, absence of a test with phenolphthalein does not preclude the presence of a trace of free alkali, usually unimportant in this stage of the determination.

Silver Nitrate. For this test carefully add 1:5 nitric acid from a buret or pipet to a solution of the sample until a small piece of litmus paper floating in the solution indicates approximate neutrality. Add an excess of 5 per cent silver nitrate solution.

A white precipitate in the absence of chlorides is normally due to the presence of metaphosphate or pyrophosphate radical, confirmable by the molybdate test. A yellow precipitate, soluble in

1:1 nitric acid, indicates the presence of orthophosphate radical, confirmable by the molybdate test. A brown precipitate of silver oxide indicates that the solution was not properly neutralized and may mask the other colors. This is soluble in excess of ammonium hydroxide.

Ammonium Molybdate. Add a considerable excess of 1:1 nitric acid to a solution of the sample. If carbonates are present, they are liberated by the acid. If large amounts of silicates are present, silicic acid is precipitated by acidification and the silica removed by centrifuging or filtering. Removal of small amounts of silica is not essential in testing qualitatively. Nearly neutralize by careful addition of 1:1 ammonium hydroxide.

Add an excess of ammonium molybdate reagent (page 718), and warm. A dense yellow precipitate shows the presence of orthophosphate. A small amount of yellow precipitate may indicate a small amount of orthophosphate or may be due to hydrolysis of complex phosphates. A white precipitate under these conditions is molybdic acid and indicates that the solution was too acid.

Strongly acidify an equal portion of sample solution with 1:1 nitric acid, and boil gently for several minutes, the longer the better. This at least partly hydrolyzes metaphosphate and pyrophosphate to orthophosphate. Cool, and nearly neutralize by addition of 1:1 ammonium hydroxide. Test with ammonium molybdate reagent as before. A much denser yellow precipitate under these conditions than from a sample of similar size before boiling with acid indicates that pyrophosphate or metaphosphate was present. The trace of yellow precipitate shown in the original test for orthophosphate will increase gradually on standing or more rapidly on heating if complex phosphates are present, for they hydrolyze in the solution containing the reagent.

Albumen. Add about a 1 per cent aqueous solution of albumen, egg white, for example, to the solution of the sample, acidified with glacial acetic acid. A white precipitate serves to distinguish metaphosphate from other phosphates.

Sulfuric Acid and Methanol. Transfer a portion of solid sample or concentrated solution to a porcelain dish or crucible. Add concentrated sulfuric acid in substantial excess, and mix well. Cautiously mix in a few milliliters of methanol, and light. Borates will react to form methyl borate boiling at 65°, which burns with a green flame, best observed in semidarkness. Occasionally, depending on the sample, it is necessary to heat. Ethyl alcohol gives a similar reaction but is less sensitive.

For greater sensitivity mix the sample and acid in a test tube previously fitted with a stopper carrying an exit tube drawn down to a capillary. Cool the mixture well, and add an equal volume of methanol. Mix well with cooling. Replace the stopper and exit tube, and warm gently. Ignite the vapor from the capillary, and observe in semidarkness against a white background.

QUANTITATIVE

Sample. If available as a solid, weigh out a sample of 10 to 20 grams, both to ensure that it is representative and to provide aliquots for various determinations without the necessity of making further weighings. Dissolve in water, recently boiled to free it from carbon dioxide, and cooled; dilute to 500 ml.; and keep well stoppered. Use aliquots of 10 to 100 ml., usually 25 or 50 ml., according to the concentration of the substance being determined. If only a limited sample is available, a smaller weight must necessarily be taken and diluted to a lesser volume. This further implies that often the same aliquot must be used for two or more determinations. Often the strength of standard acid or alkali used for titration can be adapted to the sample available. Thus, where standard alkali is called for, one might titrate a fair-sized aliquot with $0.5N$ or a much smaller one with $0.1N$. If the sample is a liquid, it may need to be diluted or concentrated and then made up to a known volume.

In calculation these steps must be allowed for in terms of the aliquot used. Thus, if 10 ml. were used for a determination from a dilution to 500 ml., the aliquot factor would be 50; if 50 ml. were used, the aliquot factor would be 10.

Moisture. Unless the sample is known to be anhydrous, moisture is determined for later use in obtaining a material balance. If the sample is a dilute solution, evaporate a suitable sample to constant weight in the oven at 110°. If the sample is a dry solid and does not fuse at 110°, it may be dried in the oven. Some carbonation may be unavoidable. If bicarbonate is present, loss of carbon dioxide will occur. Results are interpreted in the light of other results obtained. Some samples require determination of moisture by the solvent-distillation method (page 41). Many hydrates do not lose all their water of hydration at 110° or the approximately 140° of boiling xylene. An example is sodium metasilicate, which loses its water of hydration at various temperature levels. Calculate and report as loss at 110°; if determination is by the xylene method, so mark the result.

Ash. As a routine matter, weigh out 1 to 2 grams of solid or of residue from evaporation of a solution to dryness. If platinum is to be used, be sure that free sodium hydroxide or potassium hydroxide is absent. Ignite just below redness to drive off carbon dioxide from bicarbonates, and burn out organic matter. Cool, and weigh. Ignite at a full red heat to confirm the accuracy of low-temperature ashing or to detect any substance decomposable at a higher temperature. Cool, and weigh. Calculate each result as per cent loss on ignition under the specific conditions.

Sodium Hydroxide. If no alkaline salt giving an alkalinity above pH 8.0 is present, the amount of hydroxide may be titrated directly. Up to 1 to 2 per cent of carbonate in the free alkali may be due to carbonation by air exposure. Titrate a 25- or 50-ml. aliquot of the sample solution with standard acid, using phenolphthalein as indicator.

CALCULATION.

$$\text{Titration in ml.} \times \text{normality of acid} \times 0.0400 \times \text{aliquot} \times \frac{100}{\text{weight of sample}} = \text{per cent sodium hydroxide}$$

To express the alkalinity as sodium oxide, Na_2O, replace the factor 0.0400 by 0.0310. The corresponding factors for potassium hydroxide and potassium oxide are 0.0561 and 0.0471.

Sodium Carbonate. COLD TITRATION. Measure an aliquot that is expected to give a total titration of 20 to 50 ml. of the standard acid to be used. Add phenolphthalein as indicator, and titrate. The end point with this indicator represents conversion of all the carbonate to bicarbonate. Add methyl red as indicator, and continue to a new end point. If methyl red is not available, use methyl orange with slightly lower accuracy. This end point represents complete conversion of the carbonate through the bicarbonate stage to carbonic acid. The total titration to methyl red should be double that to phenolphthalein and serves as a confirmation of the absence of uncombined alkali or of bicarbonate. Calculation is based on the total titration to the methyl red end point.

CALCULATION.

$$\text{Titration in ml.} \times \text{normality of acid} \times 0.0530 \times \text{aliquot} \times \frac{100}{\text{weight of sample}} = \text{per cent sodium carbonate}$$

For calculation as potassium carbonate, use 0.0691 as factor in place of 0.0530.

HOT TITRATION. Since methyl red and methyl orange give somewhat indistinct end points in the presence of carbonic acid, this procedure gives greater accuracy.

Measure an aliquot of the sample, and add standard sulfuric acid to approximately the methyl red end point and 5 ml. in excess. Heat to gentle boiling for about 2 minutes, and cool. Titrate to the methyl red end point with standard alkali. Methyl orange may be substituted for methyl red as indicator with slightly lower accuracy. Standard hydrochloric acid may be substituted for standard sulfuric acid, provided that care is exercised not to boil vigorously.

CALCULATION.

$$(\text{Acid titration in ml.} \times \text{normality of acid}) - (\text{alkali titration in ml.} \times \text{normality of alkali}) = \text{ml. of } N \text{ acid used}$$

When later methods refer back to this procedure, it is this corrected volume of N acid that is used for substitution in the method of calculation indicated for the specific mixture.

$$\text{Ml. of } N \text{ acid} \times 0.053 \times \text{aliquot} \times \frac{100}{\text{weight of sample}} = \text{per cent sodium carbonate}$$

EVOLUTION OF CARBON DIOXIDE. In the presence of other alkaline substances the absolute method of determination of the amount of carbonate present is by evolution and absorption as carbon dioxide. This does not differentiate between the amount present as carbonate and that present as bicarbonate, or in some cases as both. It also does not differentiate between the material present as a soluble compound and that present as an insoluble one such as calcium carbonate.

The method is described in detail under Mineral Analysis (page 224). Either an aliquot of sample solution or a weighed sample may be used.

CALCULATION.

$$\text{Weight of carbon dioxide} \times 2.4086 \times \text{aliquot} \times \frac{100}{\text{weight of sample}} = \text{per cent sodium carbonate}$$

Other factors are 1.9090 for sodium bicarbonate, and 2.2743 for calcium carbonate.

Sodium Bicarbonate. ACID TITRATION. Carbon dioxide is lost from sodium bicarbonate solution on boiling. A solution of pure sodium bicarbonate gives no color with phenolphthalein. Determine by either of the methods for titration of sodium carbonate.

CALCULATION.

$$\text{Titration in ml.} \times \text{normality of acid} \times 0.0840 \times \text{aliquot} \times \frac{100}{\text{weight of sample}} = \text{per cent sodium bicarbonate}$$

ALKALINE TITRATION. The methods of the American Society for Testing Materials (ASTM) provide a titration of the acidity of sodium bicarbonate by standard alkali. Measure out a suitable aliquot, and titrate with standard sodium hydroxide, using 10 per cent silver nitrate solution on a spot plate as outside indicator. This gives a white precipitate of silver carbonate when tested until the end point is reached, when a dark brown precipitate of silver oxide is obtained.

CALCULATION.

$$\text{Titration in ml.} \times \text{normality of alkali} \times 0.0840 \times \text{aliquot} \times \frac{100}{\text{weight of sample}} = \text{per cent sodium bicarbonate}$$

Sodium Hydroxide and Sodium Carbonate. Such mixtures are frequently encountered, usually as detergents. The term "special alkali" is applied to them commercially and sometimes also to other mixtures. Three methods of determination are offered in order that the analyst may select the one best suited to the condition of his sample, available equipment, and the accuracy desired.

ALCOHOL EXTRACTION. Sodium hydroxide is soluble in alcohol. Sodium carbonate is insoluble. To apply this method, the sample is preferably available as a dry pulverulent solid.

PROCEDURE. Weigh out a sample according to the amount available, 5 or 10 grams if available. Add about 100 ml. of neutral alcohol, and heat to boiling. Centrifuge or filter, and wash the residue with neutral alcohol until the washings show no alkaline reaction to phenolphthalein. The sodium hydroxide is now in the filtrate; the sodium carbonate remains as a solid residue. If the sample is a solution, a technic using absolute alcohol can be employed. Evaporate a sample that contains 5 to 10 grams of solids so far as feasible. Usually this will be to a paste containing 50 to 75 per cent of water. Roughly estimating the water content, add neutral absolute alcohol in sufficient amount so that as diluted it is 95 per cent alcohol. Heat, and proceed as for solid samples.

Dilute the solution of sodium hydroxide to a suitable volume, usually 250 ml., and titrate a suitable aliquot, usually 50 ml., with standard acid. Necessarily this is flexible, depending on the variable composition being analyzed. Dissolve the residue of sodium carbonate in water, and analyze according to one of the methods previously given.

CALCULATION.

$$\text{Titration in ml.} \times \text{normality of acid} \times \text{aliquot} \times 0.0400 \times \frac{100}{\text{weight of sample}} = \text{per cent sodium hydroxide}$$

No accurate check of the composition so determined against gravimetric total solids of liquid products is possible. In determination of total solids, absorption of carbon dioxide from the air will have occurred, introducing an error so that the sum of sodium hydroxide and sodium carbonate determined is substantially less than the total solids. Thus the sum of the titrated values and loss at 110° will be somewhat less than 100 per cent.

Direct Titration. This is not inherently so accurate as the preceding method but is applicable to solutions of these two materials. In titration with acid, all the alkalinity of sodium hydroxide and half the alkalinity of sodium carbonate are titrated before the phenolphthalein end point is reached. Thus sodium carbonate has been titrated to sodium bicarbonate. Titration from the phenolphthalein end point to the methyl red or methyl orange end point then titrates the remaining half of the alkalinity derived from the sodium carbonate.

Procedure. Pipet out a suitable aliquot of the sample solution for titration. Add a few drops of phenolphthalein solution, and titrate cold with standard acid, usually hydrochloric acid. The end point is one that requires considerable experience to determine with a high degree of accuracy. This will be referred to in calculation as the phenolphthalein titration. Add methyl red or methyl orange, and continue the titration to its end point. The titration from the phenolphthalein end point to the methyl red end point will be referred to as the methyl red titration.

Calculation.

$$\text{Methyl red titration in ml.} \times \text{normality of acid} \times 0.106 \times \text{aliquot} \times \frac{100}{\text{weight of sample}} = \text{per cent sodium carbonate}$$

In this, note that instead of multiplying the methyl red titration by 2 and using 53 as the equivalent weight of sodium carbonate it is less involved to take 106 as the equivalent weight, which it is so far as this particular titration is concerned.

$$(\text{Phenolphthalein titration in ml.} - \text{methyl red titration in ml.}) \times \text{normality of acid} \times 0.040 \times \text{aliquot} \times \frac{100}{\text{weight of sample}} = \text{per cent sodium hydroxide}$$

The determination of sodium hydroxide combines two errors, a condition inherent in any such difference method. The comment at the end of the previous calculation also applies here.

Evolution Method. Determine the sodium carbonate as carbon dioxide by evolution, as outlined on page 224. Calculate this not only as sodium carbonate in the sample, by the method previously outlined, but also to the equivalent per cent of sodium hydroxide as below. Titrate an aliquot of the sample to the methyl red or methyl orange end point, as outlined in the previous method, but note only the total titration.

CALCULATION.

$$\left(\text{Total titration in ml.} \times N \times \text{aliquot} \times 0.040 \times \frac{100}{\text{weight of sample}}\right)$$
$$- \left(\text{weight of } CO_2 \times 0.9090 \times \text{aliquot} \times \frac{100}{\text{weight of sample}}\right) = \text{per cent NaOH}$$

This calculation will be found on inspection to combine all the steps of conversion of the actual data of the two analytical processes on which the per cent of sodium hydroxide depends. This method is susceptible of relatively high accuracy.

Sodium Carbonate and Sodium Bicarbonate. A mixture of these compounds is sold under the designation of sodium sesquicarbonate. The composition is apt to range from 1 to 1.5 sodium bicarbonate to 1 sodium carbonate. It is also designated commercially as modified soda. While a mixture within this range is most probable, mechanical mixtures may be encountered in any ratio. Sodium carbonate normally contains a trace of sodium bicarbonate. The two methods available are modifications of those previously described.

DIRECT TITRATION. Pipet out an aliquot of the sample solution for titration. Add phenolphthalein solution, and titrate to the end point with standard acid. This titrates half the alkalinity of the sodium carbonate, converting it to the bicarbonate. Then add methyl red or methyl orange indicator, and titrate to that end point. The principle is a modification of that discussed in more detail as applied to mixtures of sodium hydroxide and sodium carbonate.

CALCULATION.

$$\text{Phenolphthalein titration in ml.} \times \text{normality of acid} \times 0.1060 \times \text{aliquot} \times \frac{100}{\text{weight of sample}} = \text{per cent sodium carbonate}$$

$$(\text{Methyl red titration in ml.} - \text{phenolphthalein titration in ml.}) \times N \times 0.0840 \times \text{aliquot} \times \frac{100}{\text{weight of sample}} = \text{per cent } NaHCO_3$$

The sum of these two determinations together with the moisture present, will approach 100 per cent. Moisture by the xylene method is necessary in analysis of mixtures of carbonate and bicarbonate.

ACID AND ALKALINE TITRATION. Determine the total alkalinity as for sodium carbonate by hot titration (page 260). Titrate sodium bicarbonate by alkaline titration as previously outlined (page 261).

CALCULATION.

$$(N \text{ acid in ml.} \times 0.053 \times \text{aliquot}) - (\text{bicarbonate titration in ml.} \times N \text{ alkali} \times 0.053 \times \text{aliquot}) \times \frac{100}{\text{weight of sample}} = \text{per cent } Na_2CO_3$$

$$\text{Titration in ml.} \times \text{normality of alkali} \times 0.0840 \times \text{aliquot} \times \frac{100}{\text{weight of sample}} = \text{per cent sodium bicarbonate}$$

Trisodium Phosphate. This is a salt of the tribasic acid, phosphoric acid, H_3PO_4. Fortunately, for analytical purposes the point of change from phosphoric acid to monosodium phosphate is at the methyl red or methyl orange end point, about pH 4.0; that of change from monosodium phosphate to disodium phosphate is at the phenolphthalein end point, about pH 8.3. Practically, this means that one equivalent of alkalinity of trisodium phosphate is quantitatively titratable to the phenolphthalein end point and that one equivalent of alkalinity of disodium phosphate is titratable between the phenolphthalein and methyl orange end points. The behavior in titration will be recognized as identical with that of sodium carbonate, the end product of titration being monosodium phosphate rather than carbonic acid.

COLD TITRATION. Follow the procedure for sodium carbonate, the difference being only in calculation. Titration to phenolphthalein and methyl orange end points is desirable as a check to ensure the latter is twice the former. Owing to commercial variations, the titration to phenolphthalein is usually a little more than half the total titration, the deviation varying in the products of different manufacturers but never amounting to more than a couple of per cent. If the sample were disodium phosphate, it would titrate like sodium bicarbonate. Thus the reactions are as follows:

To the phenolphthalein end point,

$$Na_3PO_4 + HCl = Na_2HPO_4 + NaCl$$

To the methyl orange end point,

$$Na_2HPO_4 + HCl = NaH_2PO_4 + NaCl$$

Below the methyl orange end point,

$$NaH_2PO_4 + HCl \rightleftarrows NaCl + H_3PO_4$$

The reverse of this sequence occurs in attempts to titrate phosphoric acid; in this case, no titration to trisodium phosphate as an end point is possible.

CALCULATION.

$$\text{Total titration in ml.} \times \text{normality of acid} \times 0.0820 \times \text{aliquot} \times \frac{100}{\text{weight of sample}} = \text{per cent trisodium phosphate}$$

This gives the result in terms of anhydrous trisodium phosphate. To calculate as the monohydrate, use the factor 0.0910 in place of 0.0820. Commercial crystallized material approximates $Na_3PO_4.12H_2O$, for which the factor is 0.1901. For calculation in terms of P_2O_5, the factor is 0.0355.

GRAVIMETRIC. A more reliable method is to determine the phosphorus gravimetrically as magnesium pyrophosphate (page 201) or by precipitation of the phosphomolybdate and conversion to the pyrophosphate (page 202). The total phosphorus present is then calculated as trisodium phosphate.

VOLUMETRIC. When the sample is limited in amount or the phosphate in the sample is low, the determination may be done volumetrically. This titration method (page 203) is rapid and shows good precision in experienced hands.

COLORIMETRIC. Reduction to molybenum blue is a sensitive but lengthy method for determination of traces (page 204).

Disodium Phosphate. If present alone, determine by any of the methods specified for trisodium phosphate, substituting the appropriate conversion factor in the calculations. For conversion to the anhydrous salt from magnesium pyrophosphate, this is 1.2756, as the dodecahydrate 3.2173. In the phosphomolybdate titration the respective factors are 0.1422 and 0.3582. Disodium phosphate is seldom used in compounded alkaline products because of the relatively low pH value of its solutions as compared with the more common alkaline salts.

Sodium Pyrophosphate or Sodium Metaphosphate. These salts do not react as orthophosphate but are converted to it for determination as magnesium pyrophosphate (pages 201 and 202) or phosphomolybdate (page 203). To the aliquot of sample solution, add one-third its volume of 1:1 nitric acid, and boil gently for 20 minutes. The reactions that occur from hydrolysis in acid solution are $H_4P_2O_7 + H_2O = 2H_3PO_4$ and $HPO_3 + H_2O = H_3PO_4$. In application of the procedures for orthophosphate, make due allowance for the nitric acid so added, or ammonium nitrate if it is neutralized with ammonium hydroxide.

For conversion from magnesium pyrophosphate, the conversion factor for anhydrous tetrasodium pyrophosphate is 1.1947, for the decahydrate 2.004, and for the metaphosphate, which is always anhydrous, 0.9164. Similar factors in the phosphomolybdate method are 0.1330, 0.2231, and 0.1020, respectively.

An electrometric titration method for sodium pyrophosphate is also available, based on the acidity liberated by the action of zinc sulfate on an acid pyrophosphate in accordance with the following reactions:

$$Na_4P_2O_7 + 2HCl = Na_2H_2P_2O_7 + 2NaCl$$
$$Na_2H_2P_2O_7 + 2ZnSO_4 = Zn_2P_2O_7 + Na_2SO_4 + H_2SO_4$$

The method is inaccurate in the presence of polyphosphates.

PROCEDURE. Weigh a sample containing the equivalent of 1 gram of tetrasodium pyrophosphate into a 250-ml. beaker. Dissolve in sufficient water, cold so as to inhibit hydrolysis to orthophosphate, so that the resulting solution just covers the electrodes of a glass electrode assembly (page 123). Adjust the pH of the solution to exactly 3.8 with $0.2N$ hydrochloric acid, using the technic described on page 125. If carbonates are present pass carbon dioxide-free air or nitrogen through the liquid until free of carbon dioxide. Readjust, if necessary, to pH 3.8. Prepare a zinc sulfate solution by dissolving 125 grams of hydrated zinc sulfate, $ZnSO_4.7H_2O$, in water and diluting to 1 liter, filtering and adjusting the pH to 3.8.

Add 50 ml. of this solution to the sample solution, mix carefully, and allow 5 minutes for the reaction to become complete, as shown by the pH becoming constant. Titrate the liberated acid with $0.2N$ sodium hydroxide until a pH of 3.8 is again attained.

CALCULATION.

$$\text{Ml. of alkali to readjust to pH 3.8} \times \text{factor} \times \frac{100}{\text{weight of sample}} = \text{per cent tetrasodium pyrophosphate}$$

The factor above is the fraction of a gram of tetrasodium pyrophosphate, $Na_4P_2O_7$, equivalent to 1 ml. of $0.2N$ sodium hydroxide solution used for titration as calculated by standardization against pure tetrasodium pyrophosphate.

Sodium Carbonate and Trisodium Phosphate. Titrate an aliquot of the sample to the methyl red or methyl orange end point as described for titration of sodium carbonate alone, using the procedure for either cold titration or hot titration (page 260). Determine the phosphate present in another aliquot by the magnesium pyrophosphate (page 201), phosphomolybdate-pyrophosphate (page 202), or volumetric phosphomolybdate method (page 203); and calculate the results by either of the first two methods as anhydrous trisodium phosphate. In case of shortage of sample, that used for titration may then be used for phosphate determination, usually by taking an aliquot equivalent to about 0.1 gram of phosphorus pentoxide. The most convenient combination of methods is to determine the phosphorus volumetrically.

CALCULATION.

Phosphate determined gravimetrically:

$$\left(\text{Total titration in ml.} \times N \text{ acid} \times 0.053 \times \text{aliquot} \times \frac{100}{\text{weight of sample}}\right) - \left(\text{per cent } Na_3PO_4 \times \frac{Na_2CO_3}{Na_3PO_4}\right) = \text{per cent } Na_2CO_3$$

Phosphate determined volumetrically:

$$[(\text{Total acid titration} \times N \times \text{aliquot}) - (P_2O_5 \text{ titration} \times N \times \text{aliquot} \times 2/23)] \times 0.053 \times \frac{100}{\text{weight of sample}} = \text{per cent } Na_2CO_3$$

The factor 2/23 introduced is due to the titration of trisodium phosphate requiring 2 equivalents of acid while titration of the phosphomolybdate precipitate from the same phosphate requires 23 equivalents of alkali.

Sodium Silicates. Numerous sodium silicates are used in industry. Their composition is commonly expressed by the ratio of sodium oxide to silicon dioxide. Thus the most commonly used solid silicate, sodium metasilicate, Na_2SiO_3, is expressed as $Na_2O:SiO_2 = 1:1$. Many silicates have indefinite ratios, usually more siliceous than the metasilicate, such as 1:2.5, 1:3.25, and 1:3.86. These are usually liquids or difficultly soluble solids, although the 1:2 ratio solid dissolves rather readily in water. More alkaline solid silicates than the metasilicate are a sesquisilicate, 1.5:1, $Na_6Si_2O_7$, and an orthosilicate, 2:1, Na_4SiO_4. The strongly alkaline solid silicates are usually sold in hydrated form, the most common being sodium metasilicate pentahydrate, $Na_2SiO_3.5H_2O$. The pH values of aqueous solutions of the silicates are strikingly high. In general, metasilicate and those more alkaline are termed alkaline silicates, and those less alkaline than metasilicate are termed siliceous silicates. It will be noted that, except for sodium metasilicate, this corresponds to the ingredient which is higher in the mathematical ratio.

The methods of analysis of silicates are in general the same whatever may be the ratio of sodium oxide to silicon dioxide.

The total alkalinity is determined by titration and the silica by isolation. It is not feasible to obtain moisture by evaporation, for the alkaline silicates readily carbonate on drying, and it is very difficult to drive off the last traces of water from the siliceous silicates owing to the glassy nature of the solids.

Alkalinity. Dilute a suitable aliquot of sample solution to about 150 ml., and titrate to the methyl red or methyl orange end point according to the hot titration method for carbonates (page 260).

Calculation.

$$\text{Titration in ml.} \times \text{normality of acid} \times 0.031 \times \text{aliquot} \times \frac{100}{\text{weight of sample}} = \text{per cent sodium oxide}$$

If all the alkalinity is due to sodium oxide derived from sodium silicate and the ratio of the silicate is known, as occasionally by microscopic examination of the crystals of hydrated sodium metasilicate, the factor for sodium metasilicate can be substituted in the preceding and sodium metasilicate calculated directly with somewhat questionable accuracy. Both solid silicates and solutions have usually absorbed some carbon dioxide from the air. Therefore, careful analysis will often show a mixture with a small amount of carbonate and a large amount of metasilicate, and it becomes a matter of judgment to decide whether this is due to exposure or intentional admixture. Some clue to this is given by first titrating to the phenolphthalein end point and then going on to that of methyl red or methyl orange.

Silica. The methods of determination of silica are well established for mineral analysis. Measure out an aliquot of sample that will yield 0.2 to 0.4 gram of silica; a larger amount will be bulky and therefore difficult to wash and ignite. Proceed by the hydrochloric acid method (page 189), using it in full.

Calculation of Ratio of Sodium Oxide to Silicon Dioxide. The ratio will be roughly evident on inspection, for the molecular weights are sodium oxide = 61.99, silicon dioxide = 60.06. Quantitatively it is obtained as follows, whichever is smaller being expressed as 1:

$$\text{Sodium oxide factor} = \frac{\text{per cent sodium oxide}}{61.99}$$

$$\text{Silicon dioxide factor} = \frac{\text{per cent silicon dioxide}}{60.06}$$

Judgment must then be applied to identification of the ratio since analytical methods will not give values that correspond exactly with commercial ratios, particularly when some carbonation may have occurred.

Sodium Carbonate, Trisodium Phosphate, and Sodium Metasilicate. Determine the total alkalinity of the sample to a methyl red or methyl orange end point as for a mixture of sodium hydroxide and sodium carbonate, using either hot or cold titration.

Determine silica by the hydrochloric acid method (page 189), reserving the filtrate for determination of phosphate. Silica must be removed before phosphate can be determined.

Determine the phosphate in the filtrate by the magnesium pyrophosphate (page 201), phosphomolybdate-pyrophosphate (page 202), or phosphomolybdate titration (page 203) methods. The latter is simplest and quickest and will be assumed in the calculations that follow. Frequently the most desirable sample for this determination is only an aliquot of the aliquot used for the other two determinations.

CALCULATION.

$$\left(\text{Total titration in ml.} \times \text{normality of acid} \times \text{aliquot} \times 0.053 \times \frac{100}{\text{weight of sample}}\right) - \left(\text{per cent } Na_2SiO_3 \times \frac{Na_2CO_3}{Na_2SiO_3}\right) - \left(\text{per cent } Na_3PO_4 \times \frac{Na_2CO_3}{Na_3PO_4}\right) = \text{per cent sodium carbonate}$$

The sodium carbonate may be determined by evolution (page 224). As a primary method this is more accurate than titration or the preceding difference method, which accumulates the errors of two other determinations. If time and the importance of the determination permit, it is well to determine by two methods as a cross-check, bearing in mind that the evolution method takes considerable time.

Judgment must be exercised if the determined amount of carbonate is small, for each of the other substances determined will normally contain more or less carbonate as impurity, depending on the degree of previous exposure. In general, unless more than 1 part of carbonate to 20 of the sum of the other substances is present, it is probably due to reaction with carbon dioxide in the air.

Sodium Tetraborate. This material, also known as borax, gives a lower pH than the majority of the salts that have previously been considered in this class. Nevertheless, it is a common ingredient of some washing compounds, self-polishing waxes, etc., as well as many medicinals. Although its 0.1 per cent solution gives a pH of 9.2, it is an acid salt. The usual formula $Na_2B_4O_7.10H_2O$ can be written $Na_2HBO_3.3H_3BO_3.5H_2O$ to illustrate this. The essentially acid nature of the salt is important in the second method of titration.

DIRECT ACID TITRATION. The alkalinity of borax may be approximately titrated to the methyl red or methyl orange end point. Because the total alkalinity is small with respect to the size of the molecule, titration of a larger sample than usual is desirable. Therefore, dilute an aliquot containing about 5 grams of sample to about 150 ml., and titrate with standard acid. The end point is sharper when titrated with hydrochloric or nitric acid than when sulfuric acid is used. For this titration the borax, $Na_2B_4O_7$, is being titrated to metaboric acid as though it were $2NaBO_2$ and $2HBO_2$. $Na_2B_4O_7 + 2HCl + H_2O = 4HBO_2 + 2NaCl$.

CALCULATION.

$$\text{Ml. of acid} \times \text{normality of acid} \times 0.1006 \times \text{aliquot} \times \frac{100}{\text{weight of sample}} = \text{per cent borax}$$

For calculation directly as the decahydrate, use the factor 0.1907.

TITRATION IN THE PRESENCE OF POLYHYDRIC ALCOHOL. This is a method for determination of boric acid applied here on the assumption that whatever boric acid is present is there as borax. It follows that the original solution must be sufficiently alkaline so that the presence of borax and not boric acid in the sample is indicated. For the procedure the alkaline borate is then neutralized to liberate boric acid as in the previous method. While boric acid is so weak an acid that titration is unsatisfactory because of hydrolysis of the sodium borate produced, it forms esters with polyhydric alcohols, which are more strongly acid and can therefore be titrated with suitable accuracy. Thus the reaction of glycerol and boric acid appears to form a monobasic ester

$$H_3BO_3 + C_3H_5(OH)_3 = HBO_3C_3H_5OH + 2H_2O$$

This titrates normally with a strong alkali

$$HBO_3C_3H_5OH + NaOH = NaBO_3C_3H_5OH + H_2O$$

Suitable alcohols are mannitol, glycerol, or glucose. Practically, it is most convenient to use solid mannitol, sometimes called mannite.

PROCEDURE. Take an aliquot containing about 5 grams of sample. If sodium tetraborate has been titrated, that sample is suitable. It takes only a few minutes first to titrate the sample to be used by this method, and thus checks by two methods are available. This method is much more accurate. If carbonates are present, make just acid, and boil gently, covering with a watch glass or, preferably, under a reflux condenser. Boric acid will be lost by volatilization from acid solution without this precautionary measure. If ammonia is present, make alkaline with sodium hydroxide, boil until all the ammonia is volatilized, and then neutralize. The sample should not be covered for this.

Phosphates, iron, aluminum, and sulfates will react with the alkali in titration. If found present by qualitative test, they must be removed. To the slightly acid solution from which the carbon dioxide has been boiled off add sufficient 10 per cent barium chloride solution to precipitate sulfates and about 10 ml. in excess. Now add powdered barium hydroxide in sufficient amounts to render the solution alkaline, but avoid any substantial excess. Boil for about 5 minutes to coagulate the barium sulfate. Filter, and render just acid with 1:10 hydrochloric acid added dropwise. Cool, and wash down the cover glass and the sides of the beaker.

Adjust to exactly the neutral point, using methyl red or methyl orange as indicator, and add 25 to 50 ml. of neutral glycerin or about 1 or 2 grams of solid neutral mannitol. The mannitol offered on the market for analytical purposes is normally neutral. Add phenolphthalein indicator, and titrate to the end point with standard sodium hydroxide solution. When the end point is reached, add a few more milliliters of the glycerin or a gram of the mannitol. If the pink color disappears, continue the titration. Repeat until the color is permanent. If maximum accuracy is essential, run a blank titration on the amount of glycerin or mannitol used and subtract the value from the total titration. Commercially this is not usually necessary.

CALCULATION. In this calculation, while one molecule of boric acid is titrated by one equivalent of alkali, four molecules of boric acid are yielded by each molecule of sodium tetraborate. This determines the factor for calculation as $Na_2B_4O_7/4{,}000 = 201.27/4{,}000 = 0.05032$.

$$\text{Alkali titration in ml.} \times \text{normality of alkali} \times 0.05032 \times \text{aliquot} \times \frac{100}{\text{weight of sample}} = \text{per cent borax}$$

The corresponding factor for the decahydrate is 0.0954.

Sodium Carbonate and Sodium Tetraborate. Determine the total titration to a methyl red or methyl orange end point, as described for mixtures of sodium hydroxide and sodium carbonate (page 262). An attempt to titrate this directly for boric acid would be in error owing to the dissolved carbon dioxide, which would titrate as an acid. Render the solution distinctly acid, cover with a watch glass or place under a reflux condenser, and boil gently as for the usual titration in the presence of polyhydric alcohol. Then proceed as usual for that method, and calculate the results as sodium tetraborate.

In calculation of sodium carbonate, consideration must be given to the fact that sodium tetraborate has an equivalence of 2 when its alkalinity is titrated but of 4 when it is titrated as boric acid.

CALCULATION.

$$\left[(\text{Total alkali titration in ml.} \times \text{normality of acid}) - \left(\text{titration of boric acid in } \frac{\text{ml.}}{2} \times \text{normality of alkali}\right)\right] \times 0.053 \times \text{aliquot} \times \frac{100}{\text{weight of sample}} = \text{per cent sodium carbonate}$$

Sodium Carbonate, Trisodium Phosphate, and Sodium Tetraborate. Obtain the total titration and boric acid titration, as in the preceding method. Because of the presence of the polyhydric alcohol, take another aliquot, and determine phosphate in it. Any of the three methods may be used; but in the sample calculation use of the simplest, phosphomolybdate titration has been assumed.

CALCULATION.

$$\left[(\text{Total alkali titration in ml.} \times \text{normality of acid}) - \left(\frac{\text{titration of boric acid in ml.}}{2} \times \text{normality of alkali}\right) - \left(\text{titration of phosphomolybdate in ml.} \times \frac{2}{23} \times \frac{\text{aliquot for alkali titration}}{\text{aliquot for phosphate titration}}\right)\right] \times 0.053 \times \text{aliquot} \times \frac{100}{\text{weight of sample}} = \text{per cent sodium carbonate}$$

If phosphorus is determined gravimetrically as magnesium pyrophosphate, the data within the third set of parentheses inside the brackets would be replaced by the following corresponding calculation:

$$\text{Weight of } Mg_2P_2O_7 \times 1.4731 \times \frac{1}{0.082} \times \frac{\text{aliquot for alkali titration}}{\text{aliquot for phosphate determination}}$$

For a high degree of accuracy the sodium carbonate should be determined by evolution, particularly if the amount is small; otherwise, the error can easily approach 100 per cent owing to accumulation of errors from two difference methods applied in the calculation.

The foregoing are by no means all the possible combinations of alkaline salts, as indicated at the beginning of the chapter. Only a few tertiary combinations and no quaternary ones have been attempted. The methods cover the most common combinations. What is more significant is that the determination of alkaline salts may apply to only part of the material present in a commercial sample. As a single example, consider the crystalline compound composed of trisodium phosphate and sodium hypochlorite. The bleaching agent will destroy the indicator, and so acid titration is not feasible. Many more complex examples are encountered.

CHAPTER 13

CEMENT AND MORTAR

Portland cement is the product obtained by finely pulverizing clinker produced by calcining to incipient fusion an intimate and properly proportioned mixture of argillaceous and calcareous materials, with no additions subsequent to calcination, except water and calcined or uncalcined gypsum. In the main, Portland cement consists of calcium silicate. It is soluble in hydrochloric acid.

American Society for Testing Materials (ASTM) specifications give the following chemical limits with regard to its composition:

Type	*I*	*II*	*III*	*IV*	*V*
Loss on ignition, max. per cent	3.0	3.0	3.0	2.3	3.0
Insoluble residue, max. per cent	0.75	0.75	0.75	0.75	0.75
Magnesia (MgO), max. per cent	5.0	5.0	5.0	5.0	4.0
Sulfur trioxide (SO_3), max. per cent					
when $3CaO{\cdot}Al_2O_3$ is 8 per cent or less	2.5	2.5	3.0	2.3	2.3
when $3CaO{\cdot}Al_2O_3$ is more than 8%	3.0	...	4.0	...	...

A typical composition of a Portland cement is as follows:

	Per Cent
SiO_2	21.16
Al_2O_3	5.65
Fe_2O_3	2.23
CaO	63.36
MgO	4.20
SO_3	1.61
Loss on ignition	0.97
Insoluble residue	0.09

In general, the methods follow those of the ASTM, given in more detail in Specification C114-61.

Cement. Loss on Ignition. Heat 1 gram of the cement in a platinum crucible, placed in a hole in an asbestos board, to a full red heat, 900 to 1000°, for 15 minutes. Repeat the heating until a constant weight is attained. Alternatively, use a muffle furnace at the same temperature.

Insoluble Residue. To 1 gram of the cement add 30 ml. of 1:5 hydrochloric acid in the cold. Heat the solution, and grind up the sample well with a rod. Add 20 ml. of water, and digest on a steam bath for 15 minutes. Filter and wash well with hot water. Reserve this filtrate for the determination of sulfur trioxide. Digest the residue and the filter paper with 30 ml. of a 5 per cent sodium carbonate solution, at just below the boiling point for 15 minutes. Filter, and wash twice with hot water, then with a few drops of 1:9 hydrochloric acid, and finally eight to ten times with hot water. Ignite the filter paper and residue, weigh, and calculate the per cent so found.

Sulfur Trioxide. Dilute the filtrate and washings from the insoluble residue to 250 ml., and precipitate sulfate as described on page 160, starting at "Bring to a boil, and add with constant stirring sufficient boiling 10 per cent barium chloride solution . . . ". Calculate the result to sulfur trioxide.

Soluble Silica. Weigh 0.5 gram of the cement in a platinum dish. Moisten with 10 ml. of cold water to prevent lumping, add 8 ml. of concentrated hydrochloric acid, and digest on the steam bath until fully dissolved. Determine silica on this by the hydrochloric acid method, as described, starting on page 189, and applying the method in full.

Iron and Aluminum Oxides. Add 20 ml. of concentrated hydrochloric acid to the filtrate from the silica determination, and dilute to 250 ml. Use the method on page 192 as though phosphorus were absent. Start at "Just neutralize the solution to methyl orange or methyl red indicator with 1:4 ammonium hydroxide, and add a slight excess . . . ". Save the filtrate for determination of calcium and magnesium. Defer the calculations until phosphorus has been determined. Subtract the phosphorus calculated as the pentoxide present in the weight of sample used for iron and aluminum oxides from that found, and assume that the balance is iron and aluminum oxides.

Iron Oxide. Determine this on the precipitate of iron and aluminum oxides. Add 2 to 3 grams of C.P. potassium bisulfate or pyrosulfate to the platinum crucible containing the mixed oxides, mix well, and fuse. Continue as on page 194, starting at "The use of platinum ensures the high temperature . . . ". Use the permanganate method. Subtract the iron oxide so determined from the corrected mixed oxides to get the aluminum oxide.

Calcium. Determine by titration as though phosphorus had been absent in the sample, using the solution from which iron and aluminum hydroxides have been precipitated. Ordinarily, the entire filtrate is suitable. Proceed as on page 216, starting at "Heat the sample solution to boiling, and add 20 ml. . . . ". Always reprecipitate.

Magnesium. Determine as the pyrophosphate on the combined filtrates from calcium oxalate precipitation, (page 218). Follow the entire procedure, including reprecipitation.

ALKALIES. Mix 5 grams of the finely powdered cement with 1 gram of ammonium chloride in an agate mortar. Add 4 grams of calcium carbonate, and mix again. Place a 0.5-cm. layer of calcium carbonate in the bottom of a large platinum fusion crucible, and add the above mixture. Rinse out the mortar with 1 gram of calcium carbonate. Finally, cover with 0.5 cm. of calcium carbonate. This is essentially the start of the method for alkalies in insoluble silicates (page 220) except for the shape of the crucible and size of the sample. Proceed as there indicated, starting with "Cover the crucible, incline, and heat gently for about 15 minutes . . . ".

PHOSPHORUS. To 1 to 3 grams of the cement, the latter if the phosphorus pentoxide content is less than 0.1 per cent, add 60 ml. of water and then 12 ml. of concentrated nitric acid. If appreciable manganese is present, shown by the presence of a red or brown residue, add 5 ml. of a 5 per cent potassium nitrate solution to aid in solution. Boil to get rid of nitrous fumes, but do not concentrate so that silica precipitates.

Filter the solution into a 500-ml. conical flask, and wash to give a total volume of about 150 ml. Add 10 grams of ammonium nitrate, and dissolve. Continue as for the phosphomolybdate-pyrophosphate method (page 202), starting at "Heat to about 90°, and add about 70 ml. of ammonium molybdate solution . . . ". The volumetric method cannot be applied, for silica has not been removed from the solution.

FREE LIME. Uncombined lime is dissolved in a hot alcohol and glycerol solution and titrated with an alcoholic solution of ammonium acetate, $CaO + 2CH_3COONH_4 = (CH_3COO)_2Ca + H_2O + 2NH_3$.

This method does not differentiate between calcium oxide and calcium hydroxide and is therefore less valuable for aged cement since calcium hydroxide, formed from the reaction of calcium oxide and calcium silicates and aluminates with atmospheric moisture, has no deleterious effect on the cement.

Prepare a solution consisting of 1 volume of U.S.P. glycerol and 5 volumes of absolute alcohol, SD3A or SD2B will do. Add 2 ml. of a 1 per cent solution of phenolphthalein in absolute alcohol to each liter of this mixture. Now add a slight excess of a dilute absolute alcohol solution of sodium or potassium hydroxide. Just remove the pink color by the standard alcoholic solution of ammonium acetate made as follows:

As the reagent for titration, dissolve 16 grams of crystalline ammonium acetate in 1 liter of absolute alcohol, SD3A or SD2B will do. This gives approximately a $0.2N$ solution. Standardize this solution as follows against pure calcium oxide prepared by heating pure calcite or calcium oxalate in a platinum crucible at 900 to 1000° to constant weight. Commercial C.P. calcium oxide is not sufficiently pure.

Accurately weigh 0.1 gram of the calcium oxide immediately after attaining constant weight, and transfer to a 250-ml. conical flask containing 60 ml. of the neutralized glycerol-alcohol solution. Shake well to disperse.

Reflux the mixture for 20 minutes. Titrate at near the boiling point with the ammonium acetate solution.

Transfer 1 gram of the finely powdered cement into a 250-ml. conical flask containing 60 ml. of the neutral glycerol-alcohol solution. Proceed as in standardization in the preceding paragraph, starting at "Shake well to disperse . . .". Toward the end, if any difficulty is experienced in obtaining the end point, add a drop of indicator slowly to the solution, and at once observe the color as it strikes the solution. No color appears at the end point.

CALCULATION.

$$\text{Ml. titration} \times \text{standardization of ammonium acetate} \times \frac{100}{\text{weight of sample}} = \text{per cent free lime}$$

Mortar and Concrete. Mortar consists of cement, sand, and water. The usual ratio specified for cement and sand is 1:3 by volume. Since cement is more expensive than sand, there is always a tendency to use more than that amount of sand. While the amount of moisture used has an effect on strength, samples to be examined having usually lost moisture are not suitable for determination of the cement-water ratio.

Concrete consists of cement, sand. and gravel, the latter often including fair-sized pebbles. The ratio usually specified is 1:3:6, but the same tendency to use a low cement content applies here as in mortar. Sand and gravel are separated by sieving. Physical examination of the sand for mesh analysis (page 70) and for desired sharpness of grain as contrasted with a worn, rounded condition is often a part of the report.

In either mortar or concrete the building laws of many districts permit replacement of a substantial portion of the cement by what is designated as "lime putty," a well-aged slaked lime in wet condition, or by dry hydrated lime.

The following scheme is adapted to the chemical analysis of any of these mixtures:

PROCEDURE. *Stone, Gravel, Sand.* Since cement sets through a process of hydration, thoroughly blast the well-disintegrated sample, until no further loss in weight occurs. This process also decomposes any carbonates formed by reaction with the air.

Weigh out a representative 10-gram sample at once after cooling, and boil with 20 ml. of 1:10 hydrochloric acid. Do not use stronger acid, for soluble silica from the cement may then be thrown out of solution. Decant and filter the hot solution. Heat the residue with 100 ml. more of 1:10

hydrochloric acid, and decant through the same filter. Repeat if necessary. Dilute the filtrate and washings to 500 ml. in a calibrated flask, and save. Ignite the filter paper and residue, cool, and weigh. This gives the stone, gravel, and sand. If any pieces are larger than will go through a sieve with ¼-inch openings, perform the sieving operation. All the aggregate remaining on the sieve is considered as stone or gravel, also called aggregate; that passing through is considered as sand. When lime is absent, the difference between the ignited sample and the aggregate may be considered as cement. Otherwise, calculate from the soluble silica.

Soluble Silica. Determine this on an aliquot, depending in size on the expected amount of cement, exactly as described for cement analysis (page 276). This soluble silica is due to the cement present and must be determined accurately, for the calculations of the original composition depend largely upon the figure obtained.

Iron and Aluminum. Dilute the filtrate from soluble silica to 250 ml. Take 100 ml. as sample, and reserve the balance for other determinations. Add 8 ml. of concentrated hydrochloric acid to the aliquot. Add 1:4 ammonium hydroxide until neutral to methyl orange or methyl red, and then a slight excess. Boil for 2 to 3 minutes, and filter at once through a free-flowing paper. Wash and save the filtrate, but discard the precipitate, for the iron and aluminum are not needed quantitatively.

Calcium. Use the titrimetric method on page 216 in full, with the entire filtrate as sample. Reprecipitation is not necessary. Calculate as the oxide.

Magnesium. Use the filtrate from calcium as sample. Precipitate as the phosphate, using the entire procedure on page 218. Calculate as the oxide.

Sulfur Trioxide. An undue amount of sulfur trioxide, appreciably more than that normally found in the cement present, tends to retard setting and also to weaken the mortar. The filtrate from soluble silica has previously been diluted to 250 ml. Take a 100-ml. aliquot of this as sample. Add 1 ml. of concentrated hydrochloric acid, and determine sulfate gravimetrically as on page 160. Start at "Bring to a boil, and add with constant stirring sufficient boiling 10 per cent barium chloride solution . . . ". Calculate the result as sulfur trioxide.

Calculation of Original Composition. In building practice, cement, sand, and lime mixes are made up on a volume basis, usually as so many cubic yards. Hence, it is necessary to know the specific gravities of the original constituents. The data given below are commonly accepted as a basis for calculation. In making mortars, a lime putty containing 40 per cent of calcium hydroxide, $Ca(OH)_2$, and 60 per cent of water is frequently used and is employed in making these calculations. When both calcium and magnesium are present, use the total figure, and subtract the sum of calcium and magnesium normally present in the amount of cement found. If only calcium is found, subtract only the calcium normally present in the amount of cement found. For simplicity, only calcium is considered as being present in the calculations given.

Normally, high-magnesium limes do not hydrate well; therefore, the calcium may be completely hydrated, but the magnesium incompletely, producing a weak mortar. This is a factor which cannot be expressed quantitatively but to which judgment must be applied in the interpretation.

Example. Analysis of Cement, Aggregate, and Lime Mixture on Ignited Basis.

	Per Cent
Aggregate	71.06
Soluble silica	2.25
Calcium oxide	13.78

Portland Cement. This contains 21.16 per cent silica and 63.36 per cent calcium oxide. Hence,

$$\frac{2.25}{0.2116} = 10.64 \text{ per cent Portland cement}$$

Lime Putty.

$$10.64 \times 0.6336 = 6.74 \text{ per cent calcium oxide due to cement}$$

$$\text{Hence } 13.78 - 6.74 = 7.04 \text{ per cent calcium oxide due to lime putty}$$

$$= 7.04 \times \frac{74.1}{56.1} = 9.30 \text{ equivalent per cent calcium hydrate}$$

$$= \frac{9.30}{0.4} = 23.25 \text{ equivalent per cent 40 per cent lime putty}$$

Therefore, the original ratio was

Aggregate	71.06, or 67.71 per cent
Lime putty, 40 per cent	23.25, or 22.15 per cent
Portland cement	10.64, or 10.14 per cent

Specific Gravities of Commercial Materials.

Sand	1.55
Lime putty, 40 per cent	1.33
Portland cement	1.50

Introducing these values and calculating for sand as aggregate,

Sand........................ $\frac{67.71}{1.55}$ = 43.69, or 6.47 (6.5) parts by volume

Lime putty, 40 per cent....... $\frac{22.15}{1.33}$ = 16.65, or 2.46 (2.5) parts by volume

Portland cement............. $\frac{10.14}{1.50}$ = 6.76, or 1.00 part by volume

In terms of ratios, a volume of 40 per cent lime putty is accepted as equivalent to a volume of dry cement. Therefore, the ratio is 3.5:6.5, or about 1:2. One volume of dry hydrated lime contains so nearly the same amount of lime as a volume of putty that they may for practical purposes be considered identical.

CHAPTER 14

FERROUS METALS

The analysis of steel, cast iron, open-hearth steel, wrought iron, and ferrous alloys is dealt with in full in the American Society for Testing Materials (ASTM) "Methods of Chemical Analysis of Metals," revised every 3 years. This chapter must necessarily be limited to some of the more important methods, many ferrous alloys and alternative methods being omitted.

Sampling. Always obtain drillings of the sample from as many representative parts as possible. Take drillings either completely or exactly halfway through the sample, with borings at right angles to each other. Materials that give some fine powder in the drillings should be wholly reduced to as small a size as possible and very carefully mixed and sampled, for the fine substance is usually richest in carbon, the coarse material next, and the intermediate the poorest.

Total Carbon. PROCEDURE. *General.* Use a combustion train similar to that described for coal analysis (page 598). A temperature of 1000 to 1100° must be attainable, and up to 1370° is often required for refractory alloys.

Boats may be of porcelain, quartz, Alundum, clay, platinum, or nickel containing not more than 0.1 per cent carbon. To prevent injury from spattering, a platinum or nickel cover open at both ends and allowing free access of oxygen is desirable. Preheat new boats and covers in an atmosphere of oxygen before use. Line boats with specially prepared R R Alundum of 60 mesh. Either ignited, low-silica chrome ore or zirkite, a mixed natural oxide and silicate of zirconium, properly sized and ignited, may be used.

Excluding dust, use as fine a sample as possible except for rapid-burning alloys. Drillings or millings, between 14 and 60 mesh, are satisfactory. In general, pack the sample in a small groove, or furrow, made in the bedding material in the boat. A preheating period of 0.5 to 1.5 minutes at the temperature of the furnace usually causes the sample to burst into a bright flame and to burn completely upon the admission of oxygen. Sufficient oxygen should be run in to ensure a current of gas through the absorber at all stages of the combustion

The purifiers that follow the combustion tube must remove finely divided oxides of iron, sulfur, and selenium, dry the gases before they enter the weighed absorber, and protect the absorber from outside effects. Iron oxide is screened out during the passage of the gases through the liquids or columns of solids that precede the weighed absorber. The small amount of sulfur dioxide that is given off from low-sulfur steels can be satisfactorily removed by sulfuric acid saturated with chromic acid. High-sulfur steels need a 50 per cent aqueous chromic acid solution, a 5 per cent potassium permanganate solution or heated platinized silica gel to convert the dioxide to trioxide. The latter is condensed and absorbed during its passage through the liquids or columns of solids in the train.

The most desirable absorbent for carbon dioxide is 20- to 30-mesh soda-asbestos, Ascarite, followed by magnesium perchlorate, Anhydrone. The latter is needed to absorb the water that is formed during the reaction and not held by the unused Ascarite. The usual precautions of making preliminary blank runs on the apparatus are essential.

Carbon Steel, Sulfur Less Than 0.1 *Per Cent.* Having properly set up and tested the apparatus, spread 1 to 3 grams of the sample on the bed material in the boat so that the particles are in intimate contact. Cover the sample, and introduce the boat into the preheated combustion tube. Close the tube, and allow the sample to heat for 1 to 2 minutes, depending on the size of the particles. Then admit the oxygen, at a rate of 300 to 400 ml. per minute while combustion is going on, using a furnace temperature of 1100° or above. After 1.5 to 2 minutes, when combustion is complete, slow the current so that the gases leave the absorbents at the rate of about 150 ml. per minute for 6 to 8 minutes, in order to sweep out the carbon dioxide. Withdraw the absorption tube filled with oxygen, place it in a balance case for 10 minutes, open momentarily, and weigh against a similar tube used as a counterpoise. The increase in weight is due to carbon dioxide. If the drillings are not thoroughly fused in a solid pig, reject the determination. Run a blank in special cases or in analyses of very low carbon material.

CALCULATION.

$$\text{Weight of carbon dioxide} \times 0.2729 \times \frac{100}{\text{weight of sample}} = \text{per cent total carbon}$$

Other Ferrous Materials. For high-sulfur steels the train must be modified by insertion of a sulfur dioxide oxidant. For pig iron, cast iron, and malleable iron, use about 1 gram of sample. For wrought iron and open-hearth steel use 3 to 4 grams of sample. Otherwise, proceed as for carbon steel.

If an alloy steel burns with difficulty, either place a small ⅛-inch pellet of tin, as an accelerator, on each end of the sample, or spread 1 gram of pure tin drillings over the sample. Alternatively, mix with 1 to 2 grams of fine millings of open-hearth iron, and proceed as usual. High-chromium high-nickel steels require an accelerator, but more certain combustion is obtained if a temperature of 1250° or higher is employed. With selenium steels a tower packed with asbestos fibers placed in the train between the exit end of the combustion tube and the absorption tube or a 5 per cent solution of potassium permanganate should be used to absorb the volatile selenium oxides.

Graphite in Cast Iron. Dissolve 1 to 3 grams of the prepared sample in 50 ml. of nitric acid, specific gravity 1.20, made by mixing 380 ml. of concentrated acid and 620 ml. of water. Heat gently on a steam bath, with occasional stirring, until action ceases. If silicic acid separates, add 1 to 2 ml. of 48 per cent hydrofluoric acid to facilitate filtration. Collect the residue in a fused-silica Gooch crucible containing a mat of asbestos and previously gently ignited, cooled, and weighed. Wash the residue well with hot water, then with a hot solution of potassium hydroxide, containing 120 grams in 1,000 ml. of water, then with hot water, then with 1:20 hydrochloric acid, and finally with hot water. Dry at a temperature not exceeding 150°, and determine the graphite by direct combustion, starting at 500° and gradually raising the temperature to approximately 900° in the apparatus used for the determination of total carbon, taking care to close the combustion tube immediately after inserting the sample.

Manganese in the Absence of Cobalt. BISMUTHATE METHOD. This method depends on the quantitative conversion of the manganese to permanganate.

$$2Mn(NO_3)_2 + 5NaBiO_3 + 16HNO_3 = 5NaNO_3 + 5Bi(NO_3)_3 + 7H_2O + 2HMnO_4$$

Carbon Steel. Dissolve 1 gram of the sample in 50 ml. of 1:3 nitric acid. Boil to expel oxides of nitrogen, and cool somewhat. Add about 0.5 gram of sodium bismuthate, and boil for 2 to 3 minutes. High-carbon steels require about 1 gram of bismuthate, or an amount sufficient to effect either a pink color or a precipitate of manganese dioxide. Clear the solution of color or precipitate by adding a few drops of a fresh saturated solution of sulfur dioxide or other suitable manganese- and chloride-free reducing agent, and boil to expel oxides of nitrogen and sulfur. Cool to 15° or lower, add 0.5 gram of sodium bismuthate, or an amount equal to twenty-six times the weight of manganese present but not less than 0.5 gram, and agitate for 1 minute.

Prepare nitric acid free from nitrous acid as follows: Boil the concentrated acid for 1 minute, cool somewhat, and pass in a current of purified air for 5 minutes. Mix 30 ml. of this acid with 970 ml. of water, add 1 gram of sodium bismuthate, and let stand to settle. Add 50 ml. of this acid to the sample solution. Now filter the sample through prepared asbestos or through glass frit or Alundum crucibles of the correct porosity. Wash with the prepared diluted nitric acid until the washings are colorless.

The filtrate must be clear and free from bismuthate particles. Titrate by either of the following methods.

Ferrous Sulfate–Permanganate Titration. Treat the filtrate with 2 ml. of sirupy phosphoric acid, which gives a more definite end point by decolorizing the iron present. Then from a buret add enough ferrous ammonium sulfate solution, prepared by dissolving 12 grams of $Fe(NH_4)_2(SO_4)_2.6H_2O$ in a cold mixture of 950 ml. of water and 50 ml. of sulfuric acid, to discharge completely the color of the permanganic acid and 1 to 2 ml. in excess. Titrate with standardized approximately $0.03N$ potassium permanganate solution to the appearance of a faint pink color. In exactly the same manner, carry through a blank determination using the same amounts of acid and bismuthate as with the sample. Finally, add the exact volume of ferrous ammonium sulfate solution that was used, and titrate with the standard permanganate solution. The difference between the volumes of permanganate solution required in the two titrations represents the manganese in the sample.

CALCULATION.

$$(\text{Titration of blank} - \text{titration of sample}) \times \text{normality of permanganate} \times 0.01099 \times \frac{100}{\text{weight of sample}} = \text{per cent manganese}$$

Arsenite Titration. To prepare the standard sodium arsenite solution, approximately $0.035N$, dissolve 2 grams of sodium arsenite in water, filter if not clear, and dilute to 1 liter. The solution is preferably standardized against a standard steel of approximately the same manganese content as the sample. As an alternative, titrate 20-ml. portions of the standard $0.03N$ permanganate solution in a solution containing 1 gram of iron as ferric nitrate, under the same conditions of acidity and dilution as in the above method. The theoretical titer of the arsenite solution cannot be used because manganese is not quantitatively reduced to the bivalent stage. After filtering off the bismuthate, immediately titrate with the arsenite solution to a clear greenish yellow color that does not change upon the addition of another drop of arsenite.

CALCULATION.

$$\text{Ml. titration} \times \text{normality of sodium arsenite against permanganate} \times 0.01099 \times \frac{100}{\text{weight of sample}} = \text{per cent manganese}$$

Chromium-vanadium, Stainless and Similar Steels. Dissolve 1 gram of the sample in a 300-ml. covered conical flask in 20 ml. of 1:9 sulfuric acid. After dissolution, dilute to 100 ml. with boiling water. In order to remove chromium and vanadium, from a buret add an 8 per cent sodium bicarbonate solution until a permanent precipitate is formed, requiring approximately 36 ml., and then 4 ml. more. For 18 per cent chromium steel use a 6-ml. excess. Cover the flask, boil for 1 minute, and let the precipitate settle. Filter rapidly, a cone and paper containing a little paper pulp being a convenient means, wash the flask, and precipitate five times with hot water. If the precipitation has been properly performed, there will be no more precipitate than can be conveniently handled on an 11-cm. paper. The filtrate will become cloudy in the funnel stem and in the receiving vessel owing to hydrolysis and oxidation, which does no harm. Discard the precipitate. Heat the filtrate to boiling, and oxidize with small portions of concentrated nitric acid, adding a total of 12 ml. Evaporate to about 50 ml. Add about 0.5 gram of sodium bismuthate, and boil for 2 to 3 minutes. Clear the solution by adding a suitable reducing agent, boil, and complete the determination as for carbon steel (page 284), starting at "Cool to 15° or lower, add 0.5 gram of sodium bismuthate . . . ".

Other Alloy Steels. Steels containing nickel, molybdenum, and less than 1 per cent of chromium may be treated like carbon steels, although it is desirable in the most accurate work to remove chromium if it is present to the extent of over 0.5 per cent. Do this by the bicarbonate separation described in the immediately preceding method. An alternative procedure for high-chromium alloys, which will not be given in detail, consists in dissolving the alloy in 1:1 hydrochloric–nitric acid, adding 20 ml. of perchloric acid, fuming strongly for 10 minutes after the appearance of a red color to oxidize chromium, diluting, precipitating iron and manganese by ammonium hydroxide and ammonium persulfate, filtering, and washing. Dissolve the washed precipitate in 50 ml. of 1:3 nitric acid, and treat as for carbon steels.

Cast Iron. Transfer 1 gram of the sample to a small conical flask, add 25 ml. of 1:3 nitric acid, heat, and boil 2 to 3 minutes after action ceases. Cool, filter, and wash the flask and paper with 25 ml. of 1:3 nitric acid. Wash the paper a few more times with hot water, and then complete the determination as for carbon steel (page 284), starting at "Cool to 15° or lower, add about 0.5 gram of sodium bismuthate . . . ".

Open-hearth Iron and Wrought Iron. Proceed as for carbon steel (page 284), using 2 to 3 grams of the sample. It is advisable, before filtering the sample, to treat the asbestos pad with a weak permanganate solution and wash it free from the latter with 1:30 nitric acid. Very low manganese irons must not be filtered through pads used previously for higher manganese alloys.

Colorimetric determinations by the periodate method (page 200) on solutions of 0.5- to 1-gram samples of open-hearth iron are applicable for routine work.

Manganese in the Presence of Cobalt. PERSULFATE METHOD. *Carbon Steel.* Prepare an acid mixture as follows: Slowly add 100 ml. of concentrated sulfuric acid to 525 ml. of water, with stirring, cool, and add 125 ml. of 85 per cent phosphoric acid and 250 ml. of concentrated nitric acid. To 1 gram of the sample for a steel containing 1 per cent of manganese or less or 0.5 gram for a steel of higher manganese content, in a 500-ml. conical flask, add 30 ml. of the acid mixture. Heat until solution is complete, boil until oxides of nitrogen have been expelled, and add 100 ml. of water. Add 10 ml. of an 0.8 per cent silver nitrate solution as a catalyst and 10 ml. of a fresh ammonium persulfate solution. Prepare the latter by dissolving 25 grams of the salt, $(NH_4)_2S_2O_8$, in 85 ml. of water. This solution must be kept no longer than 12 hours.

Boil briskly for 1 to 1.5 minutes. Add 75 ml. of water, cool to 30° or lower, and titrate rapidly with standard sodium arsenite solution to a clear yellow end point that does not change upon the addition of more arsenite. If the solution is not titrated rapidly, part of the manganese may be reoxidized by the persulfate during titration and thus yield high results.

To standardize the sodium arsenite, dissolve 1 gram of electrolytic or open-hearth iron of a low and known manganese content, such as National Bureau of Standards standard sample 55a of ingot iron, in 30 ml. of the acid mixture. Dilute to 100 ml. with warm water, add exactly 20 ml. of standard 0.03*N* potassium permanganate solution, and then add just enough sulfurous acid to reduce the permanganate. Boil for a few minutes to remove all oxides of nitrogen. Add 10 ml. of 0.8 per cent silver nitrate solution and 10 ml. of the ammonium persulfate solution. Boil, cool, and titrate as in the procedure already described. To obtain the manganese equivalent, divide the amount of manganese present in the added permanganate and open-hearth iron by the milliliters of arsenite required.

Open-hearth Iron. Use the method for carbon steel with a 2 to 3-gram sample.

Wrought Iron. Use the method for carbon steel with a 1-gram sample.

Nickel, Chromium-nickel, Chromium-vanadium, and Chromium-vanadium-molybdenum Steels. Chromium steels with over 4 per cent chromium must have the greater part of the chromium separated by the bicarbonate separation as described under the Bismuthate Method. Start at the beginning of the procedure on page 285, and continue until the precipitate has been washed five times with hot water. To the filtrate add 30 ml. of the sulfuric–phosphoric–nitric acid mixture described for carbon steel, on page 286. Evaporate to 125 ml., and proceed as for carbon steel (above), starting at "Add 10 ml. of 0.8 per cent silver nitrate solution and . . .". With carbon over 0.8 per cent and chromium over 4 per cent, test the bicarbonate precipitate for manganese before discarding.

Chromium-tungsten Steel. Treat 0.5 gram of the sample with 50 ml. of 1:9 sulfuric acid and 3 ml. of 85 per cent phosphoric acid. Heat until all action ceases. Add 40 ml. of water and 5 ml. of concentrated nitric acid. Boil until the carbides have dissolved.

If tungsten steels of over 0.8 per cent carbon and over 4 per cent chromium content are not completely decomposed by this treatment, dissolve the sample with 10 ml. of a mixture of equal parts of concentrated hydrochloric and nitric acids, and 4 to 5 drops of 48 per cent hydrofluoric acid to dissolve silica. When action ceases, add 5 ml. of 60 per cent perchloric acid, and evaporate just to dryness. Cool, add 5 ml. of concentrated sulfuric acid, and fume until the perchloric acid is expelled. Cool, add 50 ml. of water and 5 ml. of 85 per cent phosphoric acid. Dilute to 100 ml., and heat until salts dissolve. If manganese dioxide appears, dissolve it with a few drops of saturated sulfurous acid.

Proceed as for carbon steel (page 287), starting at "Add 10 ml. of 0.8 per cent silver nitrate solution and . . . ". The arsenite solution is preferably standardized against a steel of similar type of known manganese content.

Cast Iron. Dissolve as for carbon steel (page 286). When 100 ml. of water is added, filter through a rapid paper. Wash with hot water, and dilute to 125 ml. Complete as for carbon steel, starting with "Add 10 ml. of 0.8 per cent silver nitrate solution and . . . ".

Phosphorus. TITRATION METHOD. *Carbon Steel and Open-hearth Iron.* Transfer 2 to 3 grams of the sample to a 300-ml. conical flask, and dissolve in 65 ml. of 1:3 nitric acid. With high-silicon steels, add 3 to 4 drops of 48 per cent hydrofluoric acid during dissolution. Oxidize organic matter by adding 10 ml. of 2.5 per cent potassium permanganate solution, and boil for 2 to 3 minutes. If no precipitate forms, add more permanganate and boil again. Dissolve the precipitate by the dropwise addition of a saturated solution of sodium sulfite or sulfurous acid, and boil for a few minutes to expel oxides of nitrogen.

Adjust the temperature to 45° and proceed with the titrimetric phosphate method (page 203), starting at "To 100 ml. of the usual ammonium molybdate reagent . . . ".

Nickel, Chromium-nickel, Stainless, and Similar Alloy Steels. If these contain neither tungsten nor vanadium, treat as carbon steel. With high-chromium-nickel or other steels insoluble in 1:3 nitric acid, treat 2 grams of the sample with 50 ml. of a mixture of equal parts of concentrated nitric and hydrochloric acids, add 4 to 5 drops of 48 per cent hydrofluoric acid, heat until action ceases, and then add 20 ml. of 60 per cent perchloric acid. Evaporate to fumes and heat for 5 to 10 minutes. Filter, and wash the flask, paper, and silica if present, with 55 ml. of 1:1 nitric acid. Add a few drops of 2 per cent potassium permanganate solution to the filtrate and boil 3 to 5 minutes. Add saturated sulfurous acid to dissolve oxides of manganese and to reduce all the chromium.

Adjust the temperature to 45°, and proceed with the titrimetric phosphate method (page 203), starting at "To 100 ml. of the usual ammonium molybdate reagent . . . ".

Austenitic Steel Containing More Than 10 *Per Cent of Manganese.* Transfer 2 grams of the sample to a 400-ml. beaker, add 35 ml. of 1:3 nitric acid, and digest until action ceases. Add 20 ml. of 60 per cent perchloric acid, and evaporate to fumes. Add 48 per cent hydrofluoric acid dropwise until all the silica is dissolved, and then add 5 drops in excess. Fume so that the perchloric acid refluxes on the side of the beaker for 20 to 25 minutes. Cool, and add 50 ml. of water and 10 ml. of concentrated nitric acid. Add a few drops of 2 per cent potassium permanganate solution, and boil until oxides of manganese are precipitated. Dissolve the precipitated manganese with saturated sulfurous acid, and boil to expel oxides of nitrogen.

Adjust the temperature to 45°, and proceed with the titrimetric phosphate method (page 203), starting at "To 100 ml. of the usual ammonium molybdate reagent . . . ".

Chromium-vanadium Steel or Other Steels Containing Vanadium but No Tungsten. Proceed as for carbon steel until the solution is ready for the addition of molybdate reagent. At this point, cool to 10°, and add 5 ml. of a ferrous sulfate solution containing 100 grams of $FeSO_4.7H_2O$ in a liter of 1:19 sulfuric acid, and 2 to 3 drops of saturated sulfurous acid. Mix, adjust the volume to 60 ml., but do not heat. Proceed with the titrimetric phosphate method (page 203), starting at "To 100 ml. of the usual ammonium molybdate reagent . . . ".

High-speed Steels or Other Steels Containing Tungsten and Vanadium. Decompose 2 grams of the sample in 125 ml. of 1:3 nitric acid, add 30 ml. of concentrated hydrochloric acid, and evaporate to dryness. Take up the residue in 20 ml. of 1:1 hydrochloric acid, dilute to 100 ml. with hot water, and filter off the tungstic acid. Evaporate the filtrate twice with 20 ml. portions of nitric acid to expel the hydrochloric acid, taking the second evaporation just to a sirup. Add 65 ml. of 1:3 nitric acid, and filter the solution if it is not entirely clear. Cool to 10°, and add 50 ml. of a solution of ferrous sulfate containing 100 grams in a liter of 1:19 sulfuric acid, and 2 to 3 drops of saturated sulfurous acid. Adjust the volume of the sample solution to 60 ml. and the temperature to 45° and proceed with the titrimetric phosphate method (page 203), starting at "To 100 ml. of the usual ammonium molybdate reagent . . . ".

Cast Iron. Dissolve 0.5 to 2 grams of the sample in 65 ml. of 1:3 nitric acid. Filter through a paper of loose texture, and catch the filtrate in a 300-ml. conical flask. Wash the paper a few times with 1:50 nitric acid and then with about 50 ml. of hot water. Add 12.5 ml. of 2 per cent potassium permanganate solution to the filtrate, and boil for 3 to 5 minutes. Dissolve the precipitated oxides by the dropwise addition of a saturated solution of sodium sulfite or sulfurous acid, and boil for a few minutes to expel oxides of nitrogen. Cool to 45°, and proceed with the titrimetric phosphate method (page 203), starting at "To 100 ml. of the usual ammonium molybdate reagent . . . ".

Wrought Iron. Proceed as for carbon steel, using a 1 gram sample. This gives the total phosphorus present. In wrought iron, phosphorus is also present as phosphide and as phosphate in slag inclusions. If a differentiation is desired, determine the phosphate remaining after driving off phosphine with hydrochloric acid from another sample. The phosphorus occurring as phosphide is then obtained by difference. The accuracy of the results obtained is somewhat doubtful, but the method is the only one available.

Sulfur. GRAVIMETRIC METHOD. *Carbon Steel, Open-hearth Steel, Cast Iron, and Wrought Iron.* Dissolve 5 grams of sample in 75 ml. of concentrated nitric acid in a covered beaker. Addition of 2 to 5 ml. of bromine as oxidizing agent may effect more complete recovery of sulfur. If solution is slow, add concentrated hydrochloric acid dropwise at intervals. Steels that dissolve too rapidly must be added in small portions to the cooled acid. When solution is complete, add 0.5 gram of sodium carbonate, and carefully evaporate to about 10 ml. in a sulfur-free atmosphere. Cool, add 30 ml. of concentrated hydrochloric acid, and evaporate to a sirupy consistency. Add 10 ml. of concentrated hydrochloric acid and 25 ml. of water. Then add 5 grams of 20 to 30-mesh sulfur-free zinc, and warm on the steam bath until the iron is reduced and the evolution of hydrogen has nearly ceased. Filter off silica and undissolved zinc through a tight filter paper, and wash with 75 ml. of 1:99 hydrochloric acid. Proceed to determine sulfate gravimetrically (page 160), starting at "Bring to a boil, and add with constant stirring sufficient boiling 10 per cent barium chloride solution . . . ". The result so obtained will ordinarily be low by 0.001 to 0.003 per cent of sulfur. If it is necessary to have an accuracy that includes this, evaporate the filtrate and washings to dryness. Dissolve the slight residue in 1 ml. of 1:1 hydrochloric acid, dilute to 25 ml. with hot water, add 1 ml. of 10 per cent barium chloride solution, and digest at 70 to 80° for several hours, avoiding evaporation. Filter on a small filter paper, and wash free from chlorides with hot water. Combine this paper with that carrying the previous precipitate. Take the precaution of removing silica with hydrofluoric acid and carry out blanks.

Alloy Steel. Alloy steels can often be dissolved in nitric acid or nitric acid plus small portions of hydrochloric acid. When dissolved proceed as for carbon steel. Dissolve high chromium-nickel steels in 125 ml. of a 1:1 mixture of concentrated nitric and hydrochloric acids, in a 400-ml. beaker. Mix the acids first; and when these have reacted to a red color, add them to the sample. Chill the beaker if the reaction is too vigorous. Then start as above at "When solution is complete, add 0.5 gram of sodium carbonate . . . ".

Tungsten Steel. Transfer 5 grams of the sample to a 600-ml. beaker or conical flask, and add 75 ml. of concentrated nitric acid. Cool or heat as necessary. If the reaction is very slow, add 5 ml. of concentrated hydrochloric acid or a drop or two of 48 per cent hydrofluoric acid occasionally as the solution is heated on the steam bath. When nitric acid alone is used, the particles of steel tend to become covered with a film of tungstic acid. Stirring or rubbing the coated granules speeds up the decomposition. When the reaction is complete, digest until the residue is pure yellow and contains no dark material, adding fresh portions of acid if necessary. Evaporate to a sirupy consistency. Cool, add 30 ml. of concentrated hydrochloric acid, and again evaporate to a sirupy consistency. Add 100 ml. of hot water, and boil until soluble salts are in solution. Filter, and wash the residue with 1:10 hydrochloric acid, catching the filtrate and washings in a 600-ml. casserole. Evaporate to dryness, add 30 ml. of concentrated hydrochloric acid, and again evaporate to dryness. Cool, add 60 ml. of 1:1 hydrochloric acid, warm until salts are in solution, and add 50 ml. of boiling water. Add 10 ml. of a solution of 125 grams of cinchonine in 1 liter of 1:1 hydrochloric acid, and let stand overnight. This prevents formation of acid tungstates not easily decomposable, and of colloidal solutions. Filter, and wash with a solution prepared by diluting 30 ml. of cinchonine solution to 1 liter. Evaporate the filtrate and washings until a slight film begins to form, and proceed as for carbon steel (page 290), starting at "Then add 5 grams of 20 to 30-mesh sulfur-free zinc . . . ".

In this determination it is particularly important for corrections to be made as determined by blanks carried through all steps of the method. As cinchonine sometimes contains sulfate, this solution should be measured so that the same amount is used in the blank as in the determination. If the cinchonine contains excessive quantities of sulfate, wash the crystals with water on a Büchner funnel until the washings no longer give a precipitate with barium chloride.

Open-hearth, Cast Iron, and Wrought Iron. Proceed as for carbon steel.

EVOLUTION METHOD. Most samples of carbon steel, open-hearth iron, and wrought iron give good results with this method (pages 163 to 165). Some steels dissolve too slowly, and some do not yield all their sulfur. In general, the method is not satisfactory for alloy steels and cast iron.

Silicon. SULFURIC ACID METHOD. *Carbon Steel.* Transfer 5 grams of sample or 2.5 grams if the silicon is over 0.5 per cent to a 300-ml. covered casserole. Dissolve the sample in 40 ml. of 1:1 hydrochloric acid, and evaporate to a small volume. Add 100 ml. of 1:4 sulfuric acid, and evaporate to copious fumes, taking care to avoid spattering. Replace the cover, and continue the fuming for 2 to 3 minutes but no longer, since insoluble sulfates may be formed. Cool somewhat, and add at one time 100 ml. of water at 45°. Stir until salts are in solution, heating gently if necessary, but never boiling. Filter immediately, using a rapid, ashless filter paper for high-silicon steels and a tighter paper for low-silicon steels. Transfer all the residue to the paper, scrubbing the casserole thoroughly with a policeman. Wash the paper and residue alternately with hot 1:19 hydrochloric acid and hot water until salts of iron have been removed. This usually requires six washings with 5-ml. portions of each solution. Transfer the paper and residue to a platinum crucible, heat carefully until the carbon is gone, and then cover and ignite for 10 to 15 minutes, at 1100 to 1200°. Cool, and weigh. Add 2 to 3 drops of 1:1 sulfuric acid and 3 to 5 ml. of 48 per cent hydrofluoric acid. Evaporate to dryness, and ignite at 1000°. Cool, and weigh. The difference in weight is due to silica.

Some silica passes into solution, the amount depending chiefly on the silicon content of the alloy. This is about 0.005 per cent silicon for a 0.1 per cent silicon steel, approximately 0.02 per cent for a 0.4 per cent silicon steel, and about 0.05 per cent or more for a 4.5 per cent silicon steel. Hence, in umpire work it is necessary to evaporate the filtrate and washings to fumes of sulfuric acid, dilute, filter, and ignite the washed paper and contents with the first portion. Except for umpire work this is unnecessary.

Calculation.

$$\text{Weight of silica} \times 0.4672 \times \frac{100}{\text{weight of sample}} = \text{per cent silicon}$$

Alloy Steel. Proceed as for carbon steel. If insoluble in the usual acid, employ 80 ml. of an acid mixture consisting of 600 ml. of water, 200 ml. of concentrated sulfuric acid, 100 ml. of concentrated nitric acid, and 100 ml. of concentrated hydrochloric acid. In this method, add 5 ml. of concentrated hydrochloric acid to the partly cooled residue before dilution with warm water.

Tungsten Steels. Transfer 2.5 grams of the sample to a beaker or casserole provided with a cover glass. Add a mixture of 20 ml. of concentrated hydrochloric acid and 20 ml. of concentrated nitric acid, and warm until all the steel has dissolved. Add 60 ml. of warm 1:1 sulfuric acid, and evaporate to copious fumes. Cool somewhat, cautiously add 5 ml. of concentrated hydrochloric acid, and swirl gently until well mixed. Add 125 ml. of water at 45°, and proceed as for carbon steel (page 291), starting at "Stir until salts are in solution . . . ". The precipitate is a mixture of tungstic acid and silica. Ignite together at a temperature of about 1000°, and weigh. Treat with a few drops of concentrated sulfuric and 48 per cent hydrofluoric acids, ignite at 800°, and reweigh.

Cast Iron. Proceed as for carbon steel, using a 2.5-gram sample. With gray or mottled pig iron it is desirable to use a finely divided sample and to boil vigorously with slightly more acid at the start, as the particles tend to become coated with gelatinous silica.

Open-hearth Iron. Proceed as for carbon steel, using a 10- to 20-gram sample and corresponding amounts of acids. Fume gently for 12 to 15 minutes instead of 1 to 2 minutes. Umpire analyses of open-hearth iron must be made in platinum dishes.

Wrought Iron. Dissolve 5 grams of the sample in 80 ml. of the acid mixture provided for alloy steel. Evaporate, and fume for 1 to 2 minutes. Cool somewhat, and cautiously add to the warm solution 5 ml. of concentrated hydrochloric acid as the solution is swirled gently. When well mixed, add 125 ml. of warm water, and complete as for carbon steel (page 291), starting at "Stir until salts are in solution . . . ".

Copper. THIOSULFATE-IODIDE METHOD. *Copper Steel.* Transfer 5 grams of the sample, which must contain not more than 0.25 per cent of molybdenum, to a 600-ml. beaker, add 100 ml. of 1:9 sulfuric acid, and heat gently until action ceases. Dilute to 250 ml., and heat to boiling. Add 10 ml. of a 50 per cent sodium thiosulfate, $Na_2S_2O_3.5H_2O$, solution, and continue boiling for 5 to 10 minutes or until the precipitate settles rapidly. Filter immediately, transfer all the precipitate to the paper, and wash the paper and precipitate well with 1:99 sulfuric acid saturated with hydrogen sulfide. Place the paper and precipitate in a porcelain or silica crucible. Dry, and ignite at 520 to 550°, a muffle furnace being a convenient means, until all carbon is destroyed. Cool, and transfer the contents to a 250-ml. beaker. Add 5 to 6 ml. of 3:5 nitric acid to the crucible, warm gently, and pour upon the residue in the beaker. Rinse the crucible with a little water, and warm the beaker and contents until the copper oxide has dissolved. Carefully evaporate the solution to a volume of 2 to 3 ml. in order to expel most of the acid.

Alternatively, dissolve the copper sulfide precipitate from the filter paper in hot 1:2 nitric acid, catch the solution in the original beaker, and add 5 ml. of 1:1 sulfuric acid. Evaporate to copious fumes.

Cool, and add 30 ml. of water. Add either 5 ml. of a 20 per cent solution of ammonium bifluoride or 1 gram of sodium fluoride to prevent interference by ferric iron. Then add 1:1 ammonium hydroxide until the solution just reacts alkaline to litmus, and cool the solution to room temperature. Acidify with 85 per cent acetic acid, and add 1 ml. in excess.

Prepare a special starch solution as follows: To 500 ml. of boiling water, add a cold suspension of 5 grams of soluble or arrowroot starch in 25 ml. of water. Cool, and add a cool solution of 5 grams of sodium hydroxide in 50 ml. of water. Then add 15 grams of potassium iodide and mix thoroughly.

To the sample solution add 3 to 4 grams of potassium iodide dissolved in a little water. One equivalent of iodine is liberated for each copper ion. Stir well, and immediately titrate with 0.1N thiosulfate solution. When the brown tints have nearly disappeared, add 5 ml. of starch solution, and continue to titrate until one drop changes the color from blue to yellowish white and remains permanent for 15 to 20 seconds.

For very accurate work, standardize the thiosulfate against copper. Dissolve 0.05 gram of pure metallic copper in a 150-ml. covered beaker with 4 to 5 ml. of 1:1 nitric acid. Boil gently to expel oxides of nitrogen. Cool, and add 1:1 ammonium hydroxide until just alkaline. Proceed as with the sample, starting as above at "Acidify with 85 per cent acetic acid, and . . . ".

Calculation.

$$\text{Ml. titration} \times \text{normality of sodium thiosulfate} \times 0.06357 \times \frac{100}{\text{weight of sample}} = \text{per cent copper}$$

Carbon Steel, Wrought Iron, Open-hearth Iron, and Alloy Steel. Use 10 grams of sample, and proceed as for copper steel. Most alloy steels may be analyzed as described for copper steel. When vanadium is present, reprecipitate the copper sulfide with hydrogen sulfide.

Tungsten Steel. Heat 5 grams of the sample with 75 ml. of 1:1 hydrochloric acid until dissolved. Carefully add 15 ml. of 1:1 nitric acid, and boil gently until the tungstic acid becomes bright yellow. Dilute the solution to 100 ml. with boiling water, and digest for several minutes. Filter, and wash with 1:9 hydrochloric acid.

To recover any copper retained in the tungstic acid, treat the residue with excess of 1:1 ammonium hydroxide, and then add 5 grams of tartaric acid and sufficient 1:1 sulfuric acid to have an excess of 1 ml. per 10 ml. of solution. Treat with hydrogen sulfide, and filter. Dissolve the precipitate in a little 1:1 nitric acid, and add to the main filtrate.

Add 10 ml. of concentrated sulfuric acid to the cool filtrate, and evaporate to fumes. Cool, dilute to 200 ml., and heat to boiling. Add 20 ml. of 50 per cent thiosulfate solution, and continue boiling for 5 to 10 minutes or until the precipitate settles readily. Complete as for copper steel (page 293), starting at "Filter immediately, transfer all the precipitate . . . ".

Cast Iron. Dissolve 5 grams of the sample in 100 ml. of 1:4 sulfuric acid. When solution is complete, evaporate to copious fumes. Cool somewhat, and dilute to 100 ml. with warm water. Heat until salts have dissolved, filter, and wash the residue with hot water. Dilute the filtrate to 250 ml., heat to boiling, and add 20 ml. of 50 per cent thiosulfate solution. Complete the determination as described for copper steel (page 293), starting at "continue boiling for 5 to 10 minutes or . . . ".

Nickel. Dimethylglyoxime Method. *Nickel Steel.* If the steel contains less than 1 per cent of nickel, use 2 to 3 grams of sample with correspondingly larger amounts of tartaric acid. For steels containing more than 3.5 per cent of nickel, use either a sample equivalent to 0.035 gram of nickel or a suitable aliquot of a large sample. Otherwise, transfer 1 gram of the sample to a 400-ml. beaker, cover, and treat with 60 ml. of 1:1 hydrochloric acid. If chromium present is less than 0.5 per cent, dissolve in 50 ml. of hot 1:3 nitric acid instead. When dissolved, cautiously add 10 ml. of 1:1 nitric acid, and boil until iron and carbides are oxidized and oxides of nitrogen are expelled. Dilute to 200 ml. with hot water. Add 20 ml. of a solution of tartaric acid made by dissolving 250 grams of the acid in 600 ml. of water, filtering, adding 10 ml. of nitric acid, and diluting to 1 liter. This addition prevents subsequent precipitation of iron. Neutralize with 1:1 ammonium hydroxide, add 2 ml. in excess and filter if any insoluble silica is present. Wash with hot water containing a little ammonium hydroxide and ammonium chloride.

Add hydrochloric acid until the sample solution is slightly acid. Warm to 70°, and add 20 ml. of dimethylglyoxime solution (page 718). Add 1:1 ammonium hydroxide until slightly alkaline, and digest solution and precipitate for 30 minutes at about 60°. Filter the precipitated nickel glyoxime through a weighed Gooch crucible under light suction, but do not allow the mat to run dry. Wash the precipitate thoroughly with hot water. Add 5 ml. of the oxime reagent to the filtrate and washings, stir, and allow to stand to be sure that precipitation is complete. If not, start with a fresh sample.

Dry the precipitate for 1 hour at 110 to 120°, and weigh as the salt, $NiC_8H_{14}O_4N_4$, which contains 20.32 per cent of nickel. Porous or glass frit crucibles may be used for filtration. The precipitate may also be caught on filter paper and ignited at about 800° to the oxide, NiO. In this case, take care to prevent sublimation of the glyoxime before it is decomposed by enclosing the wet paper and precipitate in a couple of wet ashless papers and igniting so that the papers char before taking fire.

If the amount of nickel is under 0.2 per cent, or if much cobalt is present, allow the precipitated glyoxime and solution to stand overnight and filter cold. In routine analysis involving more than 2 mg. of nickel and no appreciable amount of cobalt, the solution may be filtered within 5 minutes after adding the oxime and stirring.

When over 1 per cent of cobalt or over 4 per cent of copper is present, add sufficient oxime to take care of them as well as the nickel, and reprecipitate the nickel as follows: After the precipitate has been washed, discontinue the suction, place the original beaker under the funnel, and add 25 ml. of a hot mixture of 20 ml. of 1:1 hydrochloric acid and 5 ml. of concentrated nitric acid. After 1 minute, apply suction until dry. Repeat the treatment with 25 ml. of acid mixture, drain, and wash thoroughly with 50 ml. of hot water containing 1 gram of tartaric acid. Nearly neutralize the absolutely clear solution with concentrated ammonium hydroxide, and precipitate with the oxime and ammonium hydroxide as before. A glyoxime precipitate contaminated by cobalt is a darker red than pure nickel dimethylglyoxime.

Calculation.

$$\text{Weight of nickel glyoxime} \times 0.2032 \times \frac{100}{\text{weight of sample}} = \text{per cent nickel}$$

The factor as nickel oxide is 0.2586.

High-chromium High-nickel Steel. Transfer 0.35 to 0.5 gram of the sample to a 400-ml. beaker, and treat with 20 ml. of 1:1 hydrochloric acid and 20 ml. of 1:1 nitric acid. Heat until solution ensues, add 15 ml.

of 1:1 sulfuric acid, and evaporate just to light fumes. For low-carbon low-silicon alloys, the latter step may be omitted. Cool somewhat, and add 100 ml. of water. Warm until salts dissolve, filter, and proceed as directed for nickel steel (page 294), starting with "Add 20 ml. of a solution of tartaric acid . . . ". Usually 20 to 40 ml. of 1 per cent oxime solution will be required for complete precipitation.

Carbon Steels and Other Steels with Less Than 0.05 *Per Cent Nickel.* Transfer 5 grams of sample to a 400-ml. beaker, and add 40 ml. of 1:1 hydrochloric acid. When dissolved, carefully add 15 ml. of 1:1 nitric acid, evaporate to about 15 ml., and add 50 ml. of 1:1 hydrochloric acid. Transfer to a 200-ml. separatory funnel, rinse the beaker with several 15-ml. portions of the same acid, and cool to 10°. Add 120 ml. of alcohol-free ethyl ether, and carefully shake for 1 to 2 minutes in a stream of cold water. The ether will dissolve most of the ferric chloride present. It is necessary to have this correct acidity. Let settle for several minutes, and then draw off the lower clear solution into the original beaker. Gently heat the solution to expel ether, avoiding a free flame. Add 0.3 gram of potassium chlorate, and boil until this is decomposed. Dilute to 100 ml., and add 3 grams of tartaric acid. Make the solution ammoniacal, and filter off any silica. Acidify with 1:10 hydrochloric acid, and complete the determination as directed for nickel steel (page 295), starting with "Warm to 70°, and add 20 ml. of . . . ".

Cast Iron. Dissolve 5 grams of the iron in 40 ml. of 1:1 hydrochloric acid, carefully add about 15 ml. of 1:1 nitric acid to oxidize the iron, and evaporate completely to dryness. Drench the hot dried mass with 10 ml. of concentrated hydrochloric acid, and then dilute with 75 ml. of hot water. When dissolved so far as is possible, filter, wash with 1:1 hydrochloric acid, and evaporate the filtrate to a sirupy consistency. Add 50 ml. of 1:1 hydrochloric acid, transfer to a 200-ml. separatory funnel, and rinse the beaker with several small portions of the same acid. Complete the determination as for carbon steel above, starting with "Add 120 ml. of alcohol-free ethyl ether, and carefully shake . . . ".

High-nickel Chromium-alloy Cast Irons. Transfer a 2.5-gram sample to a 400-ml. beaker, and treat with a mixture of 25 ml. each of concentrated hydrochloric and nitric acids. When dissolved, add 35 ml. of 60 per cent perchloric acid and 5 to 10 drops of 48 per cent hydrofluoric acid. Heat to fumes for 10 to 15 minutes after the chromium has been oxidized. Cool somewhat, add 100 ml. of water, and heat to boiling. Filter, and wash well with 1:19 hydrochloric acid, catching the filtrate and washings in a 250-ml. volumetric flask. The residue may contain appreciable nickel. If it is necessary to recover this for greater accuracy, ignite the residue. Then treat with an excess of 48 per cent hydrofluoric acid to volatilize silica. Fuse the ignited residue with the minimum feasible amount of potassium pyrosulfate, and dissolve. Filter, if necessary, and add to the main solution in the volumetric flask.

Mix the contents of the flask, and cool to room temperature. Adjust to the mark, and mix thoroughly. Pipet out 50-ml. aliquots, and proceed as directed for nickel steel (page 295), starting at "Warm to 70°, and add 20 ml. . . . ". Use the reprecipitation technic described in the last paragraph of that method.

Open-hearth Iron and Wrought Iron. Proceed as for nickel steel; but dissolve the first oxime precipitate, and reprecipitate as directed in the last paragraph of that method.

Chromium. PERSULFATE OXIDATION METHOD. *Chromium Steel.* With steels containing less than 2 per cent of chromium, transfer 2 grams of the sample to a 600-ml. beaker, and add 60 ml. of an acid mixture made as follows: Mix 320 ml. of 1:1 sulfuric acid, 80 ml. of 85 per cent phosphoric acid, and 600 ml. of water. Heat until action ceases, cautiously add 10 ml. of 1:1 nitric acid, and boil until all carbides are dissolved and oxides of nitrogen are expelled. With steels containing 2 to 5 per cent of chromium, dissolve a 1-gram sample, and evaporate until salts separate. Dilute with 50 ml. of warm water, add 5 to 10 ml. of concentrated nitric acid, and again carefully evaporate until salts separate. If carbides still persist, filter, and wash the paper with warm water. Ignite the paper and residue, fuse with a small amount of sodium carbonate, dissolve in 1:1 nitric acid, and add the solution of the melt to the main solution. After the sample is completely dissolved, dilute to 300 ml. with hot water. Add 5 ml. of a 1 per cent silver nitrate solution and 20 ml. of a fresh solution of 15 grams of ammonium persulfate dissolved in 100 ml. of water. Boil the solution for 8 to 10 minutes. If the color of permanganate does not develop, add more silver nitrate and ammonium persulfate, and again boil for 10 minutes. Add 5 ml. of 1:3 hydrochloric acid, and continue the boiling for 5 minutes after the pink color has just disappeared. If the color is not destroyed after 10 minutes' boiling or if a precipitate of manganese dioxide remains, add 2 to 3 ml. more of 1:3 hydrochloric acid, and boil as before. The total period of boiling after the addition of the persulfate must be not less than 15 minutes; in fact, 30 minutes will do no harm. Cool the solution, and dilute to 400 ml.

Prepare a ferrous ammonium sulfate solution as follows: Dissolve 32 grams of ferrous ammonium sulfate, $FeSO_4(NH_4)_2SO_4.6H_2O$, in a liter of cooled 1:19 sulfuric acid. To obtain the ratio of this solution to the standard permanganate solution, dilute 25 ml. of it to 350 ml. with 1:19 sulfuric acid, add 2 ml. of 85 per cent phosphoric acid, and titrate with approximately $0.05N$ potassium permanganate solution (page 728) to a faint permanent pink tint. Determine the blank on the same volume of water and acids, deduct, and calculate the volume of permanganate solution that is equivalent to 1 ml. of the ferrous ammonium sulfate solution. The ratio of one solution to the other must be determined daily unless the ferrous solution is kept under hydrogen. Stronger solutions of each may be prepared for high-chromium steels.

To the solution of sample add a measured volume of the standard ferrous ammonium sulfate solution, 25 ml. for each 1.5 per cent chromium present. Stir, and titrate with the potassium permanganate solution until an end point is obtained that is permanent upon continued stirring for 1 minute. If vanadium is present, the end point will fade at first owing to the slow oxidation from the quadrivalent to the quinquevalent stage in a cold solution. Reserve this solution for determination of vanadium.

The titration must be corrected for dilution effect and color interference in one of three ways.

1. Empirically. For this method of correction, ignore the dilution effect, and take the volume of permanganate used in overcoming the green color as equivalent to 0.6 per cent of the chromium present and the corrected titration as 99.4 per cent of that determined. The correction is usually applied to the titer of the solution, as, for example, by using the titer 0.01744 of chromium instead of 0.01734 gram per ml. of N solution.

2. By titrating the same volume of ferrous ammonium sulfate solution in a solution of like volume and acidity containing the same amounts of the coloring elements in their final valences.

3. By a second titration of the final solution. This last is the most convenient and satisfactory in occasional analyses. It may be performed by boiling for 10 minutes the solution that has just been titrated in order to destroy the slight excess of permanganate, cooling to room temperature, and then titrating with permanganate to the color that was originally taken as the end point. This solution is still suitable to reserve for later determination of vanadium.

CALCULATION.

[(Ml. ferrous solution added × volume of permanganate solution equivalent to 1 ml. of ferrous solution) − (ml. permanganate solution added − correction for the end point)] × normality of permanganate solution × 0.01734

$$\times \frac{100}{\text{weight of sample}} = \text{per cent chromium}$$

High-chromium High-nickel Steels. Transfer 0.5 gram to a 600-ml. beaker, and treat with 75 ml. of 1:4 sulfuric acid. When solution is complete, evaporate to salts, and cool. Dilute to 70 ml., dissolve the iron salts, and oxidize them by the cautious addition of concentrated nitric acid. Add 3 to 4 drops of 48 per cent hydrofluoric acid, and again evaporate to salts. Cool, dilute to 300 ml. with hot water, and add 10 ml. of 0.1 per cent silver nitrate solution. Add 15 ml. of an ammonium persulfate solution containing 15 grams in 100 ml. of water, and boil the solution for 10 to 12 minutes. Complete the determination as for chromium steel (page 297), starting with "Add 5 ml. of 1:3 hydrochloric acid, and continue the boiling . . . ". Use stronger ferrous and permanganate solutions than specified for the previous method.

The visual end point in high-chromium steels can be detected more easily when oxidation-reduction indicators are used. An example is the ortho-phenanthroline ferrous complex for the ferrous permanganate titration or diphenylamine sulfonic acid for the ferrous dichromate titration. With 1 to 2 drops of a 0.025*M* solution of ortho-phenanthroline the end point is indicated by the change in color from pink to clear green, permanent for 1 minute. With diphenylamine sulfonic acid, the chromic acid is titrated with the ferrous solution to a clear green end point. The ferrous solution in this case is standardized against either a standard chromium steel or pure potassium bichromate. Vanadium, if present, will also be titrated.

Steels and Irons Containing Less Than 0.15 *Per Cent Chromium.* Transfer 10 grams, more or less, of the sample to a 500-ml. conical flask, and add 11 ml. of 1:9 sulfuric acid for each gram of sample. Boil until reaction is complete, and then dilute with 100 ml. of boiling water. Add an 8 per cent sodium bicarbonate solution from a buret until a permanent precipitate appears, approximately 36 ml. for carbon steels, and then 4 ml. in excess. Boil for 1 minute, let settle, and filter on a rapid filter. Quickly wash the flask and precipitate two or three times with hot water. If the precipitation has been properly performed, there will be no more precipitate than can be conveniently handled on an 11-cm. paper. The filtrate will become cloudy in the funnel stem and in the receiving vessel because of hydrolysis and oxidation. Ignite the residue in a nickel or iron crucible free from chromium, and fuse with ten to twelve times its volume of sodium peroxide. Dissolve the cooled melt by immersing it in 100 ml. of cold water. Remove the crucible, add 1 gram of sodium peroxide, and boil for 5 to 10 minutes or allow to stand on the steam bath for 30 minutes. Filter through an asbestos pad or a filter paper that is first well washed with 5 per cent sodium hydroxide solution to remove soluble organic matter, preferably on a small Büchner funnel. Wash with a cold 2 per cent sodium hydroxide solution containing 1 per cent of sodium sulfate. Dilute to a measured volume, and compare the color (page 61) with a standard solution of potassium bichromate solution containing approximately the same concentration of chromium and alkali.

To prepare the standard solution, dissolve 0.283 gram of potassium bichromate in water, and dilute to exactly 1 liter. Each milliliter will correspond to 0.1 mg. of chromium. Make this solution alkaline to match that of the sample solution. Solutions containing 2 to 10 per cent of sodium hydroxide and 1 mg. of chromium per 100 ml. are suitable for colorimetric comparisons.

If the color of the unknown solution is too deep for convenient colorimetric comparison, boil the solution thoroughly to decompose all the peroxide, acidify, and apply one of the titration procedures.

Cast Iron, Over 0.15 *Per Cent Chromium.* Transfer 2 grams of the sample to a 600-ml. beaker, and add 60 ml. of a mixture of 320 ml. of 1:1

sulfuric acid, 80 ml. of 85 per cent phosphoric acid, and 600 ml. of water. Heat until action ceases, add 15 ml. of 1:1 nitric acid, and boil until the oxides of nitrogen are expelled. Evaporate until salts separate, dilute with 50 ml. of warm water, and digest until salts are dissolved. Filter, and wash with warm water. If the chromium is less than 0.75 per cent, the residue will be practically free from chromium. If more than that amount of chromium is present and its recovery is necessary for sufficient accuracy, ignite the paper and insoluble matter, and treat the residue with a few drops of 48 per cent hydrofluoric acid and concentrated sulfuric acid. Fuse with a small amount of sodium carbonate, and add the solution of the melt to the main solution. Dilute to 300 ml. with hot water, and proceed as for chromium steel (page 297), starting at "Add 5 ml. of a 1 per cent silver nitrate solution . . . ".

High-nickel Chromium-alloy Cast Iron. Transfer 2 grams of the sample to a 500-ml. conical flask, and add 20 ml. of a mixture of equal parts of concentrated hydrochloric and nitric acids. Heat until action ceases. Then add 20 ml. of 60 per cent perchloric acid and 5 drops of 48 per cent hydrofluoric acid, and evaporate to fumes on a hot plate. Fume 1 to 2 minutes over an open flame and then 10 minutes more on a hot plate. Cool somewhat, add 50 ml. of water, and transfer to a 600-ml. beaker. Add 20 ml. of concentrated sulfuric acid, and dilute to 300 ml. Complete the determination by the method for chromium steel (page 297), starting at "Add 5 ml. of a 1 per cent silver nitrate solution . . . ".

Vanadium. REDUCTION METHOD. *Chromium-vanadium Steel.* After the determination of chromium by the persulfate oxidation method, add 5 ml. of 85 per cent phosphoric acid unless already present. Then add 15 ml. of an approximately 0.03*N* ferrous ammonium sulfate solution (page 726), 1 ml. of which will reduce 1.5 mg. of vanadium. If the vanadium is over 0.8 per cent, add proportionately more. When enough has been added, a drop of the solution will immediately give a blue color with a drop of fresh potassium ferricyanide solution. Stir the solution thoroughly, add 8 ml. of a freshly prepared 15 per cent solution of ammonium persulfate, and stir for 1 minute. Titrate with standard 0.05*N* potassium permanganate solution (page 728) to a definite pink tint that does not fade upon continued stirring for 1 minute. This is a titration of tetravalent to pentavalent vanadium. Subtract the same blank as was determined for chromium in chromium steels.

If a correction for the end point has not been determined in the determination of chromium or if the chromium value is unknown and an empirical correction is therefore impossible, obtain the blank by method 3 (page 298).

The vanadium may again be reduced by an excess of ferrous solution and the solution treated with persulfate, etc., as a check on the first titration.

CALCULATION.

$$(\text{Ml. permanganate used} - \text{blank}) \times \text{normality of permanganate} \times 0.05095 \times \frac{100}{\text{weight of sample}} = \text{per cent vanadium}$$

Chromium-tungsten-vanadium Steel. Transfer 2 grams of the sample to a 600-ml. beaker, and add 100 ml. of 1:9 sulfuric acid. Heat until action ceases, add 10 ml. of 1:1 nitric acid and 4 ml. of 1:1 hydrochloric acid, and boil the solution gently for about 30 minutes or until the tungsten has been oxidized to yellow tungstic acid. Stir frequently to break up the film adhering to the sides and bottom of the beaker. Dilute to 100 ml., filter through a tight filter paper, and wash with 1:200 sulfuric acid.

Dilute the filtrate to 300 ml., and add 5 ml. of 85 per cent phosphoric acid and then permanganate solution until the solution is pink. From this point proceed as described for chromium-vanadium steel (page 300), starting at "Then add 15 ml. of an approximately 0.03*N* ferrous ammonium sulfate solution . . .".

Carbon Steel, Cast Iron, Open-hearth Iron and Wrought Iron Containing Less than 0.05 *Per Cent Vanadium.* Transfer a 10-gram sample to a 500-ml. conical flask, and add 11 ml. of 1:9 sulfuric acid for each gram of sample. Proceed as for chromium in steels and irons with less than 0.15 per cent chromium (page 299), starting with "Boil until reaction is complete and . . . ". Carry this on through "Wash with a cold 2 per cent sodium hydroxide solution containing 1 per cent of sodium sulfate . . . ".

Acidify the solution with 1:1 sulfuric acid, and add 5 ml. of 85 per cent phosphoric acid. Proceed as for vanadium in chromium-vanadium steel (page 300), starting at "Then add 15 ml. of an approximately 0.03*N* ferrous ammonium sulfate solution . . . ".

High-chromium High-nickel Steel. Treat 10 grams of the sample containing less than 0.05 per cent vanadium, in a 400-ml. beaker with 250 ml. of 8:92 sulfuric acid, and heat gently until action ceases. For vanadium steels use a 2-gram sample and 150 ml. of 1:19 sulfuric acid. Cool the solution to 15°, and add a considerable amount of paper pulp. Then add dropwise with constant stirring a cold, freshly prepared 6 per cent solution of cupferron, ammonium nitrosophenylhydroxylamine, $C_6H_5NNOONH_4$, until the precipitate just assumes a reddish brown color. Filter through an 11-cm. paper containing some paper pulp, and wash ten to twelve times with cold 1:99 sulfuric acid. Transfer the paper and contents to the original beaker, and add 20 ml. of concentrated nitric acid and 10 ml. of concentrated sulfuric acid. Evaporate to fumes of sulfuric acid. Cool, dilute to 300 ml., and add 3 ml. of 85 per cent phosphoric acid. Then add permanganate solution until the solution is pink. From this point proceed as described for chromium-vanadium steel (page

300), starting with "Then add 15 ml. of an approximately 0.03N ferrous ammonium sulfate solution . . . ".

Molybdenum. PRECIPITATION AS SULFIDE, WEIGHING AS OXIDE. *Molybdenum Steel, Carbon Steel, Open-hearth Iron, and Wrought Iron in the Absence of Tungsten.* Transfer 2 to 10 grams of the sample, containing approximately 0.03 gram of molybdenum, to a 600-ml. beaker, treat with 100 ml. of 1:5 sulfuric acid, and warm until action ceases. Add 20 ml. of a 25 per cent solution of ammonium persulfate, and boil the solution for 8 to 10 minutes to oxidize the molybdenum and part of the iron. Cool somewhat, add 5 grams of tartaric acid, and approximately neutralize with 1:1 ammonium hydroxide. Add 1:1 sulfuric acid until acid and then 10 ml. in excess for each 100 ml. of solution. Heat to boiling, and pass in a rapid stream of hydrogen sulfide for 10 minutes. Dilute with an equal volume of hot water, and pass in the gas for 5 minutes longer. Digest at 50 to 60° for 1 hour. Filter and wash the sulfur and sulfides with 1:99 sulfuric acid saturated with hydrogen sulfide.

In umpire analyses, boil the filtrate to expel hydrogen sulfide gas and to reduce the volume to about 450 ml. Add 20 ml. of 25 per cent ammonium persulfate solution, and boil for 8 to 10 minutes. Then pass in a rapid stream of hydrogen sulfide for 10 to 15 minutes. Digest for 1 hour, filter, wash, and combine with the main precipitate.

Place the paper and precipitate or precipitates in the original beaker. Add 5 ml. of concentrated sulfuric acid and 20 ml. of concentrated nitric acid. Cover and heat to fumes of sulfuric acid. Cool somewhat, add 10 ml. of concentrated nitric acid, and again evaporate to fumes. If the solution is not clear and of a light color, repeat the treatment with nitric acid. Cool, dilute to 100 ml., and add a slight excess, 10 to 12 drops, of a 20 per cent sodium hydroxide solution. Heat to boiling, and set aside for 5 minutes. Filter, and wash the paper and residue with hot water. Heat the filtrate to boiling, and treat with hydrogen sulfide for 10 minutes. Add 1:1 sulfuric acid until acidified and then a 4-ml. excess per 100 ml. of solution. Again treat with hydrogen sulfide for 5 minutes, and digest at 50 to 60° for 1 hour. Filter through a tight 9-cm. filter paper, and wash thoroughly with 1:99 sulfuric acid saturated with hydrogen sulfide. Test the filtrate for traces of molybdenum by boiling to expel hydrogen sulfide, oxidizing with bromine water, boiling to expel excess bromine, and again passing in hydrogen sulfide.

Transfer the precipitate and paper to a small porcelain crucible. Heat carefully until carbon is destroyed and then at 500 to 525° until constant weight is obtained. Test the ignited oxide for impurities by dissolving in concentrated ammonium hydroxide. If copper is indicated, determine the amount colorimetrically (page 206). Calculate to copper oxide, CuO, and deduct. If a residue remains insoluble in ammonium hydroxide, filter, wash with water, ignite, weigh, and deduct from the weight of the molybdenum oxide, MoO_3.

In routine analyses the impure sulfide precipitate may be ignited to constant weight at 500 to 525° and the weight so obtained corrected as follows: To the crucible, add 1:1 hydrochloric acid, and heat until the oxide is dissolved. Transfer the solution to a small beaker, dilute to about 100 ml., and add a slight excess, 10 drops, of a 20 per cent sodium hydroxide solution. Boil for 1 minute, allow the precipitate to settle, and filter. Dissolve the precipitate in 1:1 hydrochloric acid, and precipitate with sodium hydroxide as before. Wash the paper and precipitate thoroughly with water to remove alkali salt; ignite the paper and precipitate in the original crucible. The difference between the two weights represents molybdenum oxide, MoO_3.

CALCULATION.

$$\text{Weight of molybdenum oxide, corrected} \times 0.6666 \times \frac{100}{\text{weight of sample}} = \text{per cent molybdenum}$$

Tungsten Steel. Dissolve 2 to 10 grams of the sample, containing approximately 0.03 gram of molybdenum, in 100 ml. of 1:1 hydrochloric acid, and cautiously add 20 ml. of 1:1 nitric acid. Boil gently until the tungstic acid becomes bright yellow. Dilute to 150 ml., and heat to boiling. Filter, and wash the residue with 1:9 hydrochloric acid. Reserve the precipitate. Add 15 ml. of concentrated sulfuric acid to the filtrate, evaporate to fumes of sulfuric acid, cool, and add 10 ml. of water. Digest until soluble salts are in solution. If any tungstic acid separates, filter through a small filter, wash with a little 1:99 sulfuric acid, and combine with the reserved tungstic acid precipitate. Add 5 grams of tartaric acid to the clear filtrate, and neutralize with 1:1 ammonium hydroxide. Add concentrated sulfuric acid until acidified, then 5 ml. per 100 ml. in excess, and pass in hydrogen sulfide as in the absence of tungsten.

Some molybdenum is always carried down by the tungstic acid and must be recovered. Dissolve the combined tungstic acid residues in a hot solution of 5 per cent sodium hydroxide, and wash the papers with a few milliliters of water and then with a few milliliters of hot 1:99 sulfuric acid. Add 5 grams of tartaric acid, then concentrated sulfuric acid until the solution contains 5 ml. per 100 ml. Precipitate with hydrogen sulfide as in the absence of tungsten. Filter, wash, and combine with the main sulfide precipitate. By precipitating only this small amount of molybdenum sulfide in the presence of tungstic acid, contamination is minimized.

Complete as in the absence of tungsten (page 302), starting at "Place the paper and precipitate or precipitates in the original beaker."

Cast Iron. Transfer 2 to 5 grams of the sample to a 600-ml. beaker, treat with 100 ml. of 1:4 sulfuric acid, and warm. When action ceases, add 1:1 nitric acid dropwise until rapid effervescence ceases, usually 5 to 10 ml., and then add 3 to 5 drops in excess. Evaporate the solution to sulfuric acid fumes. Cool somewhat, add 100 ml. of warm water, stir, and heat until salts are dissolved. Filter through a rapid filter paper, catching the filtrate in a 600-ml. beaker. Wash the paper well with hot water. Complete as for molybdenum steel (page 302), starting at "Cool somewhat, add 5 grams of tartaric acid, and . . . ".

COLORIMETRICALLY BY THIOCYANATE. If the molybdenum content is less than 0.25 per cent, a colorimetric method is applicable. For iron and steel containing 0.02 to 0.25 per cent of molybdenum, use 0.2 gram of sample. For alloys containing less than 0.02 per cent, use proportionate amounts of sample and reagents.

To prepare a standard molybdenum solution dissolve 0.5 gram of pure sodium molybdate, $Na_2MoO_4.2H_2O$, in water containing 5 ml. of concentrated sulfuric acid, and dilute to a liter. This solution should contain 0.0002 gram of molybdenum per ml. and if the salt is of good quality need not be further standardized. It standardization is necessary, transfer 150 ml. of the solution to a 600-ml. beaker, and add 50 ml. of 1:5 sulfuric acid. Analyze as the sulfide, (page 302), starting at "Add 20 ml. of a 25 per cent solution of ammonium persulfate . . . ".

Measure out a volume of the standard solution containing approximately the amount of molybdenum expected in the sample, and add to a weight of molybdenum-free iron or steel equal in weight to the sample. Alternatively, weigh out an amount of iron or steel of known molybdenum content like that of the sample. To sample and standard add 10 ml. of 1:1 perchloric acid per 0.2 gram, and warm until they have dissolved. Heat to boiling, cover, and continue to heat until all carbonaceous matter has been destroyed. If less than 0.02 per cent molybdenum is present, dissolve 0.5 to 1 gram in 20 ml. of 1:3 nitric acid, add 10 ml. of 60 per cent perchloric acid, and evaporate to fumes.

Cool somewhat, add 25 ml. of water to sample and standard, and boil to expel free chlorine. Add 2 grams of tartaric acid and a slight excess of 20 per cent sodium hydroxide solution to each of the cooled solutions. Heat at about 80° for a few minutes. Neutralize with 1:1 sulfuric acid, and then for each 8 ml. of solution add an excess of 2 ml. of the same acid. This will give solutions containing about 10 per cent of sulfuric acid by volume. Cool to room temperature, and transfer each solution to a 250-ml. separatory funnel, rinsing out the flasks with 10 per cent sulfuric acid. Add 10 ml. of a 5 per cent sodium thiocyanate solution, and shake vigorously for 30 seconds.

Prepare a solution of stannous chloride in hydrochloric acid as follows: To 350 grams of stannous chloride, $SnCl_2.2H_2O$, in a 500-ml. conical flask add 200 ml. of 1:1 hydrochloric acid, and heat at 65° until solution is practically complete. Transfer the cooled solution to a dark, liter bottle, and dilute to volume with freshly boiled water. Add a few pieces of metallic tin, and stopper. Add 10 ml. of the stannous chloride solution to sample and standard, and shake well for 1 minute. Use 20 ml. for a 1-gram sample.

Saturate butyl acetate with sodium thiocyanate and the acid stannous chloride solution. Alternatively, saturate ethyl ether in the same way. Cool the solutions to room temperature, and add 20 ml. of the saturated butyl acetate or ether to each. For satisfactory color comparisons the concentration of molybdenum in the extract being prepared should not exceed 0.05 mg. per ml. Stopper, and shake well. Allow the extracts to separate and draw off, and discard the lower layers. Add 10 ml. of sodium thiocyanate solution and 5 ml. of stannous chloride solution to each and shake well. Cool the funnel and contents, and repeat the shaking. Allow the layers to separate. Draw off and discard the lower layers.

If either extract is turbid, filter through a small pad of glass wool. Transfer the sample extract into a Nessler tube or a colorimeter, and compare with the standard (page 61).

Tungsten. ACID DIGESTION–CINCHONINE METHOD. *Tungsten Steel.* Transfer 2 grams of the usual sample, 5 grams of steel containing less than 5 per cent of tungsten, to a smooth 400-ml. beaker. Tungstic acid tends to stick to etched glass. Cover, add 50 ml. of concentrated hydrochloric acid, and warm gently. When decomposition is complete, cease heating and use a policeman to detach carbides and tungstic acid. Gradually add 10 ml. of 1:1 nitric acid. Digest at 100°, with occasional stirring, until the tungstic acid is bright yellow and free from black particles. Dilute to 150 ml., and add 5 ml. of a 12.5 per cent solution of cinchonine in 1:1 hydrochloric acid. This prevents formation of acid tungstates not easily decomposable and of colloidal solutions. Add a small amount of paper pulp and digest at 90 to 95° for 30 minutes or longer, with occasional stirring. Let the solution and precipitate stand 18 to 24 hours if less than 2 per cent of tungsten is present. Decant the clear solution through an ashless paper containing a little ashless paper pulp. Wash by decantation with two or three 30- to 40-ml. portions of hot dilute cinchonine solution consisting of 30 ml. of the reagent solution diluted to 1 liter. Transfer the residue to the paper, and wash the paper and residue thoroughly with hot dilute cinchonine solution. Dissolve any tungstic acid that still adheres to the beaker with a few drops of 1:1 ammonium hydroxide, and evaporate just to dryness. Add 2 to 3 ml. of 1:9 hydrochloric acid and 0.5 ml. of the reagent solution. Heat to boiling, digest a few minutes, and pour onto the filter.

Transfer the paper and precipitate to a weighed platinum crucible, and ignite at as low a temperature as possible until carbon is gone. Treat with 1 or 2 drops of 1:1 sulfuric acid and 1 to 3 ml. of 48 per cent hydrofluoric acid. Evaporate to dryness, ignite at 750 to 850°, cool, and weigh as impure tungstic acid. Add 4 grams of sodium carbonate, and heat gradually until fusion is complete. Cool, take up the melt in 100 ml. of water, and filter. Thoroughly wash the crucible and residue, and reserve the filtrate. Transfer the residue to the crucible, ignite, and fuse with 1 gram of sodium carbonate. Cool, and take up with 25 ml. of water. Filter, and thoroughly wash the crucible, paper, and residue with hot water. Combine the filtrate with the reserved filtrate. Again transfer the residue to the crucible, ignite, cool, and weigh. Correct this weight for the water-insoluble residue obtained from 5 grams of sodium carbonate, which may exceed 1 mg., and subtract the corrected weight from the weight of impure tungstic acid.

If the combined filtrates show a yellow color, evaporate to less than 100 ml., and dilute to that volume. Determine chromium colorimetrically (page 299). Calculate to chromic oxide, Cr_2O_3. On an aliquot, determine molybdenum oxide colorimetrically (page 304). On another aliquot determine vanadium pentoxide titrametrically (page 300), adding the amount of chromate to the comparison solution that was found by colorimetric analysis in the sample. Separate tin, tantalum, columbium, etc., from a fourth aliquot by acidifying, adding 1:1 ammonium hydroxide in moderate excess, and boiling. Filter, wash thoroughly, ignite, and weigh. Calculate the weight of the oxides found in the aliquot portions to the amount in the entire sample. Add these weights of oxide, and subtract from the weight of the impure tungstic acid.

Calculation.

$$\text{Corrected weight of tungstic acid} \times 0.7930 \times \frac{100}{\text{weight of sample}} = \text{per cent tungsten}$$

Steel Containing Less Than 0.2 *Per Cent of Tungsten.* Transfer 5 grams of the sample to a 600-ml. beaker, add 75 ml. of 1:6 sulfuric acid, and warm until action ceases. Carefully add just enough 1:1 nitric acid to decompose the carbides and to oxidize the iron. Add about 5 mg. of molybdenum, to aid the precipitation of tungsten; a convenient means is to add 25 ml. of a solution containing 0.5 gram of sodium molybdate, $Na_2MoO_4.2H_2O$, in 1 liter of water containing 5 ml. of concentrated sulfuric acid. Dilute to 150 ml., cool to 5°, and precipitate by slowly adding 10 ml. of α-benzoinoxime solution prepared by dissolving 10 grams of the reagent in 500 ml. of alcohol. Determine tungsten subsequently in the mixed oxides as described for tungsten steels (page 305), starting at "Add a small amount of paper pulp and digest at 90 to 95° for 30 minutes or longer . . . ".

Cobalt. Zinc Oxide–α-Nitroso-β-Naphthol Method. *Cobalt Steel.* Transfer 1 gram of sample to a 400-ml. beaker; add 25 ml. of 1:1 hydrochloric acid; heat; and when decomposition is complete, add 5 ml. of 1:1 nitric acid to oxidize the iron. If tungsten is present, continue the digestion until all the tungsten has been converted to yellow tungstic acid. Evaporate until salts begin to separate. Add 100 ml. of hot water, and digest on the steam bath for about 5 minutes. Dilute to about 200 ml., and add a freshly prepared suspension of 50 grams of zinc oxide in 300 ml. of water, in portions of about 5 ml. until the iron is precipitated and a slight excess of zinc oxide is present. Shake thoroughly after each addition of the precipitant, and avoid a large excess. When sufficient zinc oxide has been added, further addition causes the brown precipitate to appear lighter in color upon thorough shaking. A sufficient excess is also indicated by a slightly milky supernatant liquid upon standing. Allow the precipitate to settle for a few minutes, and filter the solution through a 12.5-cm. rapid filter paper. A little zinc oxide coming through is not objectionable. Wash the beaker and the precipitate three times with cold water. Reserve the filtrate and washings. When the filter has drained, transfer the paper and precipitate to the beaker in which the precipitation was carried out. Add 12 ml. of concentrated hydrochloric acid, and stir the paper to a pulp. The iron should now be in solution; if it is not, add more concentrated hydrochloric acid, but avoid a large excess. Dilute to 200 ml., and repeat the precipitation with zinc oxide. Filter on a 15-cm. paper, and wash four or five times with cold water.

For routine work, a single precipitation will often suffice. In this case take a 2-gram sample, dilute the solution to exactly 500 ml. after the addition of zinc oxide, mix thoroughly, and filter through a dry filter into a 250-ml. calibrated flask. With one precipitation by zinc oxide, the percentages obtained for cobalt in high-speed steel will be 0.1 to 0.3 per cent too low as the result of retention of cobalt in the bulky precipitate.

To the combined filtrate and washings from the zinc oxide separations, add 10 ml. of concentrated hydrochloric acid, and adjust the volume to 400 ml. Heat the solution to boiling, and add 8 ml. of α-nitroso-β-naphthol solution made by dissolving 1 gram of the reagent in 15 ml. of acetic acid and filtering. Further add 3 ml. for every 0.01 gram of cobalt present. Allow the solution to cool for at least 30 minutes, and filter through a rapid paper. Transfer all the precipitate to the filter, and wash with hot 1:3 hydrochloric acid and then thoroughly with hot water.

Transfer the wet paper and precipitate to a weighed porcelain crucible. Heat gently at first, preferably in a muffle furnace, and finally ignite to constant weight at 750 to 850°. Heating above 900° tends to convert Co_3O_4 to CoO. Cool and weigh as Co_3O_4. In very accurate work in which more than 0.01 gram of cobalt is involved, reduce the oxide in hydrogen, cool in an atmosphere of hydrogen, and weigh the cobalt as metal.

Nickel accompanies cobalt almost completely in the zinc oxide separation; hence, in very accurate work when much nickel is present, dissolve the ignited cobalt oxide in concentrated hydrochloric acid, and reprecipitate the cobalt as before.

Run a blank through all steps of the determination. A 1-gram sample of the National Bureau of Standards standard sample 10d of Bessemer steel or sample 50a of chromium-tungsten-vanadium steel is satisfactory for this purpose. It is especially important for the same quality of

α-nitroso-β-naphthol to be used in the blank run as in the determination. A little macerated paper added to the blank after the reagent facilitates filtration and washing.

CALCULATION.

$$\text{Weight of cobalt oxide, } Co_3O_4 \times 0.7343 \times \frac{100}{\text{weight of sample}} = \text{per cent cobalt}$$

Steels and Irons Containing Less Than 0.10 *Per Cent Cobalt.* Dissolve 10 grams of the sample in concentrated hydrochloric acid, and cautiously oxidize with just enough concentrated nitric acid. Cool, adjust acidity to 1:1 hydrochloric acid, and extract successively with 50, 25, and 25 ml. of alcohol-free ethyl ether, reserving the acid solution. Wash the ether extract with 25 ml. of 1:1 hydrochloric acid, and add to the acid solution. Warm the combined acid solutions to drive off residual ether. Add about 0.2 gram of potassium chlorate, and heat to boiling until excess oxides of chlorine have been driven off. Precipitate with zinc oxide, and continue as directed for cobalt steel (page 306), starting at "Dilute to about 200 ml., and add a freshly prepared suspension . . . ". In material containing very little cobalt, it is advantageous to combine the extracted acid solution obtained in ether separations of a number of separate 10-gram samples.

Titanium. COLORIMETRICALLY AFTER CUPFERRON SEPARATION. *Titanium Steel.* Transfer 0.5 to 1 gram of the sample to a 400-ml. beaker, add 100 ml. of 1:4 hydrochloric acid, cover, and heat gently until action ceases. Cool the solution to 15 to 20°, and add dropwise a cold, freshly prepared solution made by dissolving 2 grams of cupferron in 35 ml. of cold water. Stir constantly until the precipitate just assumes a reddish brown color. Additional cupferron only causes more iron to be precipitated. Add a considerable amount of ashless paper pulp, and filter through an 11-cm. rapid filter paper. Wash twelve to fifteen times with cold 1:9 hydrochloric acid. Transfer the paper and residue to a 50-ml. platinum crucible, dry, and ignite at a temperature below 500° just sufficiently to destroy the carbon of the filter paper. Fuse the contents of the crucible with 1 gram of potassium pyrosulfate, and dissolve the cooled melt in 25 ml. of 1:9 sulfuric acid. To this solution add sirupy phosphoric acid dropwise until the color due to iron is eliminated or reduced to a minimum. Avoid excess.

Transfer to a 50-ml. Nessler tube, and determine titanium colorimetrically as on page 198, starting at "To a similar tube add 20 ml. of water and salts . . . ".

Copper Steel. If the steel contains an appreciable amount of copper, prepare the sample solution as for titanium steel, but filter the original solution dissolved in 1:1 sulfuric acid in place of hydrochloric acid, through a small filter paper containing some paper pulp. Wash well with hot 1:9 sulfuric acid, and reserve the filtrate.

Transfer the paper containing the acid-insoluble material to a 250-ml. beaker. Add 25 ml. of 3:7 nitric acid, and heat until the residual copper and titanium have dissolved. Add 50 ml. of hot water and a slight excess of 1:1 ammonium hydroxide. Heat to boiling, filter, and wash with hot water. Reserve this precipitate on the paper.

Cool the reserved filtrate to 15 to 20°, and precipitate with cupferron as for titanium steel (page 308), starting at "Cool the solution to 15 to 20° . . . ". When ready to burn off the paper, add the precipitate with ammonium hydroxide previously reserved.

Vanadium Steel. Proceed as for titanium steel until the cupferron precipitate has been ignited. Then transfer the ignited cupferron precipitate to a 100-ml. platinum dish, and add 5 ml. of 48 per cent hydrofluoric acid and 12 ml. of 60 per cent perchloric acid. Evaporate to a volume of 5 ml. or less. Cool somewhat, and dilute to 50 ml. Neutralize with 10 per cent sodium hydroxide solution, and add 5 ml. excess. Boil for several minutes, let settle, and filter on a tight 9-cm. filter paper. Wash the paper and precipitate with hot water. Ignite and fuse with 1 gram of potassium pyrosulfate. Dissolve the cooled melt in 25 ml. of 1:9 sulfuric acid, and add sirupy phosphoric acid dropwise until any color due to iron is eliminated or reduced to a minimum. Avoid excess.

Transfer to a 50-ml. Nessler tube, and determine titanium colorimetrically as on page 198, starting at "To a similar tube add 20 ml. of water and salts . . . ".

Carbon Steel, Open-hearth Iron, and Wrought Iron Containing Less Than 0.05 *Per Cent of Titanium.* Transfer 5 grams of the sample to a 400-ml. beaker, add 150 ml. of 1:4 hydrochloric acid, cover, and heat gently until action ceases. Proceed as described for titanium steel (page 308), starting at "Cool the solution to 15 to 20° . . . ".

Cast Iron for Total Titanium. Treat 5 grams of the sample with 100 ml. of 1:2 hydrochloric acid, cover, and warm. When all action has ceased, cool to 10°, and add 2 ml. of cupferron reagent prepared by dissolving 2 grams in 35 ml. of water. Filter on a tight paper, and wash with water. Transfer the paper and residue to a platinum crucible, dry, and ignite under good oxidizing conditions in the uncovered crucible until all carbon is gone. Treat with 1 to 2 ml. of 48 per cent hydrofluoric acid and 1 ml. of 1:5 sulfuric acid, and evaporate to dryness. Fuse the residue with 1 to 2 grams of sodium carbonate. Dissolve the melt in about 50 ml. of water, and digest for 15 minutes at 90 to 95°. Filter, and wash with water. Ignite the residue in platinum, and fuse with 1 to 3 grams of potassium pyrosulfate. Cool, dissolve the melt with 25 ml. of 1:9 sulfuric acid, and add sirupy phosphoric acid dropwise until color due to iron is eliminated or reduced to a minimum. Avoid excess.

Transfer to a 50-ml. Nessler tube, and determine titanium colorimetrically as on page 198, starting at "To a similar tube add 20 ml. of water and salts . . . ".

Zirconium. CUPFERRON-PHOSPHATE METHOD. *Zirconium Steel.* Transfer 2 to 3 grams of the sample to a 250-ml. beaker, and add 100 ml. of 1:4 hydrochloric acid. Cover, and heat until action ceases. Dilute to 150 ml., cool to 15 to 20°, and precipitate with a solution of 2 grams of cupferron in 35 ml. of water. Stir constantly until the precipitate shows a reddish brown color. Excess cupferron precipitates more iron. Add ashless paper pulp, and filter through an 11-cm. rapid filter paper. Wash twelve to fifteen times with cold 1:9 hydrochloric acid, and transfer the paper and precipitate to a 50-ml. platinum crucible. Dry, and ignite below 500°, at a temperature just sufficient to destroy the carbon of the filter paper. Fuse the residue with 1 gram of potassium pyrosulfate. Dissolve the cooled melt in 100 ml. of 1:9 sulfuric acid, and filter through a small paper to separate any siliceous matter. Wash with 1:9 sulfuric acid. To the filtrate add 2 ml. of 30 per cent hydrogen peroxide solution and a solution of 3 grams of diammonium phosphate, $(NH_4)_2HPO_4$, in 25 ml. of water. Stir vigorously, and let stand for 1 to 2 hours at a temperature of 60 to 65°. An excess of peroxide must be present at all times. With amounts of zirconium under 0.01 per cent the solution should be allowed to stand at room temperature overnight. Filter through a 9-cm. paper containing ashless pulp, and wash thoroughly with a cold 5 per cent ammonium nitrate solution. Transfer the paper and residue to a platinum crucible, and ignite very carefully so that the paper does not flame. When the paper has charred, gradually increase the temperature until all carbon is gone, and then heat at about 1050° for 15 minutes. Cool in a desiccator and weigh as zirconium pyrophosphate, ZrP_2O_7.

In very accurate analysis determine titanium in the pyrophosphate. Fuse with 4 to 5 grams of potassium pyrosulfate, and dissolve the melt in 25 ml. of 1:9 sulfuric acid. Transfer to a 25-ml. Nessler tube and proceed as on page 198, starting at "To a similar tube add 20 ml. of water and salts . . . ". Calculate the titanium to $Ti_2P_2O_9$, and deduct from the weight of zirconium pyrophosphate.

CALCULATION.

$$\text{Corrected weight of zirconium pyrophosphate} \times 0.3440 \times \frac{100}{\text{weight of sample}} = \text{per cent zirconium}$$

Carbon Steel, Open-hearth Iron, Wrought Iron, and Cast Iron. Transfer 5 grams of the sample to a 400-ml. beaker, and add 150 ml. of 1:4 hydrochloric acid. Proceed as for zirconium steel above, starting at "Cover, and heat until action ceases."

CHAPTER 15

ALUMINUM AND LIGHT ALUMINUM ALLOYS

The analysis of aluminum and its alloys is dealt with in full in the American Society for Testing Materials (ASTM) "Methods of Chemical Analysis of Metals." The methods here are selected ones, alternatives having been eliminated when the ASTM gives more than one method. The great and growing use of aluminum alloys makes this subject an important one in the field of nonferrous metallurgy. Unless the approximate composition is known, it is necessary to do a complete qualitative analysis according to the technics in Chap. 5.

Sampling. Always obtain drillings of the sample from as many representative parts as possible. Make drillings either completely or exactly halfway through the sample and also at right angles to each other. Remove any iron from the drillings with a magnet and any oil or grease with petroleum ether.

Silicon. MIXED ACID METHOD. Use a sample that will contain not less than 0.5 mg. of silicon. Mix with 350 ml. of water, in order, 250 ml. of 1:1 sulfuric acid, 200 ml. of concentrated nitric acid, and 200 ml. of concentrated hydrochloric acid. Employ the following ratio of sample weight to milliliters of this acid mixture: 1 gram:35 ml., 2 grams:70 ml., 3 grams: 100 ml., 4 grams:120 ml., 5 grams:150 ml. Add the acid mixture in small portions, particularly if the sample is finely divided, and digest until the metal is dissolved. Evaporate until heavy fumes of sulfuric acid have been evolved for 15 minutes; then cool somewhat. Add 10 ml. of 1:3 sulfuric acid, followed by 60 ml. of water, and boil until the salts are dissolved. Add paper pulp, filter, and use a policeman on the reaction vessel. Wash thoroughly with 1:99 sulfuric acid. Reserve the combined filtrates if lead has been found qualitatively. Wash the residue with a hot solution of ammonium acetate prepared by dissolving 25 grams of the salt in 100 ml. of water containing 1 ml. of glacial acetic acid. This wash solution must be excluded from the silicon filtrate if it is to be used for the determination of other elements.

Place the washed paper and residue in a platinum crucible, dry, and ignite at 500° until free from carbon. Finish the ignition to constant weight at 1100°, cool, and weigh. Moisten the residue with a few drops of 1:4 sulfuric acid and a few ml. of 48 per cent hydrofluoric acid. Evaporate to dryness, ignite at 1000°, cool, and weigh. Repeat this acid treatment to constant weight. The loss in weight represents silica. Subtract the weight of silica determined in a blank covering all steps of the procedure.

There is a slight loss of silica in the filtrates. For work of the highest accuracy, this must be recovered by repeated evaporations, excluding the acetate washings if any. If samples give a dark silica residue, they are preferably analyzed by an alkaline attack on the sample. For this, refer to the ASTM methods.

Calculation.

$$(\text{Weight of silica} - \text{blank}) \times 0.4672 \times \frac{100}{\text{weight of sample}} = \text{per cent silicon}$$

Iron. Sulfide Separation. If there is any appreciable residue after the determination of silica, fuse it with a small amount of potassium pyrosulfate and dissolve in 1:20 sulfuric acid. Add it to the combined filtrates from the silica determination, and dilute to a known volume. Take an aliquot of this equivalent to about 1 gram of original sample.

Stir in 25 ml. of a 25 per cent aqueous solution of tartaric acid. This forms complexes with aluminum, iron, vanadium, chromium, titanium, etc., preventing their precipitation as hydroxides. Pass in hydrogen sulfide, and filter off the precipitate of Group 2 metals, washing with 1:99 sulfuric acid saturated with hydrogen sulfide.

Add concentrated ammonium hydroxide to the filtrate until it is slightly alkaline. Treat with hydrogen sulfide for 1 minute, and warm to coagulate the precipitated sulfides consisting chiefly of iron, titanium, and zinc. Cool, filter, and wash with a solution containing 1 per cent each of ammonium sulfide and ammonium sulfate. Dissolve the sulfide from the filter with 40 ml. of warm 1:6 sulfuric acid, washing the paper thoroughly with hot water. Boil until all hydrogen sulfide is expelled, and adjust the volume to 150 ml. To the hot solution add a strong solution of potassium permanganate drop by drop until a small excess is shown by a persistent pink color. Pass a rapid stream of hydrogen sulfide through the solution for 5 minutes or add 50 ml. of cold water saturated with the gas. Filter any precipitate of residual Group 2 metals, and wash with 1:99 sulfuric acid saturated with hydrogen sulfide. To prevent bumping add 4 or 5 solid glass beads, and boil the filtrate and washings vigorously for at least 30 minutes to remove hydrogen sulfide. The removal of all hydrogen sulfide is essential. Cool the flask, and test for completeness of reduction by removing a drop of the solution and testing with potassium thiocyanate solution. No pink color should result. Titrate the iron with 0.02*N* potassium permanganate solution (page 728). Reserve the titrated solution for determination of titanium.

If the iron is found too low for satisfactory titration with permanganate, the titrated solution is a sample for its colorimetric estimation. Usually the solution is diluted to 250 ml. in a calibrated flask and 25 ml. used as sample. This permits saving an aliquot for titanium or nickel. Occasionally an even smaller aliquot is needed. To the aliquot in a 50-ml. Nessler tube add a drop or two of 0.1*N* potassium permanganate to ensure complete oxidation. Proceed as on page 196, starting at "Now add 5 ml. of a 10 per cent ammonium or potassium thiocyanate solution . . . ".

CALCULATION.

$$\text{Ml. titration} \times \text{normality of permanganate} \times 0.05585 \times \frac{100}{\text{weight of sample}} = \text{per cent iron}$$

Titanium. COLORIMETRIC METHOD. Use as sample the solution in which reduced iron has been titrated with potassium permanganate. This may have been diluted to 250 ml. for colorimetric determination of iron, and in this case use a 25-ml. aliquot. Otherwise, dilute or concentrate the sample solution to 50, 100, or 250 ml. according to the titanium content, and use 25 ml. Transfer this to a 50-ml. Nessler tube, and add 20 ml. of 1:9 sulfuric acid. To a similar tube transfer 45 ml. of 1:9 sulfuric acid. The sample may have a faint yellow color due to iron. If so, make the standard match by adding a sufficient amount of a solution of 1.8 grams of ferric sulfate in 100 ml. of 1:99 sulfuric acid.

If enough colored salts are present, other than iron, to interfere with the matching of colors, make a cupferron separation of the titanium and iron. In that case dilute the aliquot to 200 ml., and add sufficient sulfuric acid to bring the acidity to about 10 per cent by volume. Cool with ice, and precipitate with a slight excess of 6 per cent cupferron solution. Add paper pulp, and filter. Wash with 1:10 hydrochloric acid, dissolve the precipitate with mixed concentrated sulfuric and nitric acids, and fume completely to remove organic matter. Take up with 15 ml. of 1:3 sulfuric acid and sufficient water to hold the titanium in solution, transfer to a 50-ml. Nessler tube, and dilute to 45 ml.

Complete as on page 198, starting at "Add 5 ml. of 3 per cent commercial hydrogen peroxide solution . . . ".

Copper, Lead, Bismuth, and Zinc. Use a sample of 1 to 10 grams, depending on the copper content. Dissolve the sample in 25 ml. of 1:1 hydrochloric acid per gram of sample. Add the acid cautiously; and, near the end of the reaction, add 1 ml. of concentrated nitric acid. Boil to complete the solution and to expel the oxides of nitrogen. Dilute with water, filter, and wash with hot water. Ignite the residue in a platinum crucible at 500°. Add a few drops of 1:1 sulfuric acid and 1:1 nitric acid, then a few ml. of 48 per cent hydrofluoric acid. Volatilize the silica present. Bring the nonvolatile residue into solution with a few drops of concentrated nitric acid, and add it to the filtrate. To the combined filtrates add 25 ml. of a 25 per cent aqueous solution of tartaric acid per gram of sample. Dilute with hot water, and neutralize with 1:1 ammonium hydroxide, using methyl red as indicator. Add 25 ml. of an ammonium formate–formic acid mixture prepared by diluting 200 ml. of formic acid, specific gravity 1.20, to 970 ml. with water and adding 30 ml. of concentrated ammonium hydroxide. Heat nearly to boiling, and pass in a rapid stream of hydrogen sulfide for 15 minutes. Stir in a little paper pulp, and heat to aid coagulation of the precipitate. Filter, and wash eight times with a solution made by diluting 25 ml. of the ammonium formate–formic acid mixture to 1 liter and saturating with hydrogen sulfide.

Ignite the precipitate of sulfides in a porcelain crucible at 500°. Transfer the ignited residue to the original beaker with the aid of a little 1:1 nitric acid. Add 10 ml. of 1:1 nitric acid, and evaporate to a small volume. If tin or antimony is present, filter, and wash with 1:99 nitric acid. If bismuth is present, separate, and determine as follows:

Bismuth. OXYCHLORIDE METHOD. Add concentrated ammonium hydroxide to the filtrate until it becomes slightly cloudy. Add 1 ml. of 1:1 hydrochloric acid, dilute to 300 ml. with water, and allow to stand for 1 hour at a temperature just below boiling. Filter through a weighed Gooch crucible, and wash with hot water. Dry the bismuth oxychloride at 110° for 1 hour, and weigh.

CALCULATION.

$$\text{Weight of bismuth oxychloride} \times 0.8024 \times \frac{100}{\text{weight of sample}} = \text{per cent bismuth}$$

Copper and Lead. ELECTROLYTIC METHOD. After the bismuth separation, add 5 ml. of a 25 per cent aqueous solution of tartaric acid to the filtrate. Neutralize with concentrated ammonium hydroxide, using methyl orange as indicator. Add 5 ml. of the ammonium formate–formic acid solution prepared for an earlier step. Pass in hydrogen sulfide. Stir in a little paper pulp, and heat to aid coagulation of the precipitate. Filter, and wash with the ammonium formate–formic acid wash solution. Ignite the precipitate in a porcelain crucible at 500°. Transfer the ignited residue, with the aid of a few drops of concentrated nitric acid, to the electrolytic vessel, and dissolve with a few milliliters of concentrated nitric acid. Evaporate the solution to about 2 ml., dilute to 150 ml., and electrolyze, using weighed electrodes and a current of from 3 to 5 amperes. For general details of this step, see page 323. When the solution is colorless and 1 ml. of the solution gives no color when tested with hydrogen sulfide solution on a spot plate, remove the solution from the electrodes quickly without interrupting the current. Save this electrolyte for the determination of zinc. Quickly rinse the electrodes twice in water, and then dip them in two successive baths of alcohol. Shake off excess alcohol, and dry the cathode a few minutes at 100°. Cool, and weigh as metallic copper. Dry the anode at 220°, cool, weigh, and calculate lead dioxide to lead by use of the empirical factor 0.866.

Zinc. Ammonium Mercuric Thiocyanate Method. Add 5 ml. of concentrated sulfuric acid to the electrolyte from which the lead and copper have been removed. Evaporate to fumes, cool, take up with 50 ml. of water, and add 25 ml. of water saturated with hydrogen sulfide. Filter, and wash the filter thoroughly with 1:99 sulfuric acid saturated with hydrogen sulfide. Boil the filtrate until hydrogen sulfide has been completely removed. Dilute to 150 ml., and add potassium permanganate solution until a pink tinge persists for a minute.

Prepare ammonium mercuric thiocyanate solution by dissolving 32 grams of ammonium thiocyanate and 27 grams of mercuric chloride in 500 ml. of water. Filter to remove any residue. Add 25 ml. of this solution, with vigorous stirring, to the clear oxidized filtrate. Allow the precipitate to stand overnight, collect on a weighed Gooch crucible, and wash with a solution containing 10 ml. of the precipitating agent per liter. Dry at 105°, cool, and weigh. Deduct a determined blank. As the precipitate is not of exact theoretical composition, use the empirical factor 0.1289 to convert the weight of zinc mercuric thiocyanate to zinc.

Calculation.

$$\text{Weight of zinc mercuric thiocyanate} \times 0.1289 \times \frac{100}{\text{weight of sample}} = \text{per cent zinc}$$

Zinc Oxide Method. If the electrolyte from which copper and lead have been removed is not available, the following steps must precede the ammonium mercuric thiocyanate method or, if the amount of zinc is small, the oxide method. Select a sample of 1 to 10 grams according to zinc content. Dissolve in a covered 400-ml. beaker in 1:1 hydrochloric acid, added cautiously, followed by 1 ml. of concentrated nitric acid toward the end of the reaction. For 1 gram, use 25 ml. of acid; for 2 grams, 40 ml. Boil to complete solution and to expel oxides of nitrogen. Dilute to 125 ml., and add 10 ml. of 1:1 hydrochloric acid. Pass hydrogen sulfide through the solution for 3 minutes. Add a little paper pulp, filter, and wash with 1:999 sulfuric acid saturated with hydrogen sulfide. Boil the filtrate for 10 minutes to expel hydrogen sulfide. Cool somewhat, add 25 ml. of a 12.5 per cent aqueous solution of tartaric acid per gram of sample, and dilute to 250 ml. Neutralize with concentrated ammonium hydroxide, using methyl red indicator; then add 25 ml. of a mixture of 200 ml. of formic acid, specific gravity 1.20, 770 ml. of water, and 30 ml. of concentrated ammonium hydroxide. Heat nearly to boiling, and pass in a rapid stream of hydrogen sulfide for 15 minutes. Stir in a little paper pulp. Prepare an ammonium formate–formic acid wash solution by diluting 25 ml. of the formic acid mixture to 1 liter and saturating with hydrogen sulfide. Filter the sulfides on a moderately close paper, containing a little paper pulp, and wash with the formic acid wash solution. Dissolve the zinc sulfide from the filter into the original beaker with hot 1:3 hydrochloric acid, and thoroughly wash the paper.

Add 5 ml. of concentrated sulfuric acid, and evaporate to fumes. Take up with 50 ml. of water, bring to a boil, and add 25 ml. of hydrogen sulfide solution. Filter, and wash. Boil 5 to 10 minutes to remove the hydrogen sulfide.

If the zinc is over 0.1 per cent, it may be determined by the ammonium mercuric thiocyanate method (page 315), starting at "Dilute to 150 ml., and add potassium permanganate . . . ".

If the zinc is under 0.1 per cent, dilute the solution to 100 ml. Add 5 ml. of 25 per cent tartaric acid solution, and make the solution just alkaline to methyl red with ammonium hydroxide. Heat, and pass hydrogen sulfide through the solution for 3 minutes. Add 10 ml. of the formic acid mixture, and continue to pass in hydrogen sulfide for 5 minutes more. Coagulate and filter as before, and wash five times with the formic acid wash solution. Ignite the precipitate in a weighed porcelain crucible at 700°. Cool, and weigh as zinc oxide. Deduct a determined blank.

Calculation.

$$\text{Weight of zinc oxide} \times 0.8034 \times \frac{100}{\text{weight of sample}} = \text{per cent zinc}$$

Manganese. Pyrophosphate Method. Weigh a sample of 1 to 2 grams into a 250-ml. conical flask. Dissolve by the use of 4.5 to 6 grams of sodium hydroxide per gram of sample. A solution of 200 grams of sodium hydroxide per liter or sodium hydroxide pellets and water added in small quantities may be used. Make additions cautiously as the reaction ceases. When reaction is complete, add about 2 ml. of 6 per cent hydrogen peroxide solution dropwise to complete the oxidation of silicon, zinc, and chromium, and digest a few minutes. Dilute to 150 ml. with hot water, filter on close paper, and wash with hot water. To the filtrate add a little ammonium persulfate to oxidize and precipitate residual manganese, and heat for a short time. Filter, and wash. Return the combined residues to the original container, and dissolve in 20 ml. of 1:1 hydrochloric acid to which a few drops of concentrated nitric acid have been added. Filter, and wash. Heat the filtrate to boiling, add methyl red indicator, and neutralize with concentrated ammonium hydroxide. Add 2 ml. of concentrated hydrochloric acid, and pass in hydrogen sulfide to precipitate the metal sulfides. Filter, and wash with 1:999 sulfuric acid saturated with hydrogen sulfide. Boil the filtrate until the hydrogen sulfide is completely removed. Make the filtrate ammoniacal, and add 5 ml. of concentrated ammonium hydroxide in excess. Add 2 grams of ammonium persulfate, and boil for a short time, being sure that the solution remains alkaline. Filter, and wash with hot water. Return the precipitate to the beaker, and dissolve in 30 ml. of 1:1 hydrochloric acid and a few drops of concentrated nitric acid. Dilute to 150 ml., and boil to remove the liberated chlorine. With the solution about 150 ml. in volume and heated nearly to boiling, introduce methyl red indicator, and add concentrated ammonium hydroxide carefully until the color just changes to a distinct yellow. Filter, and wash with hot 1 per cent ammonium chloride wash solution made just yellow to methyl red with ammonium hydroxide. Return the paper and precipitate to the beaker, and dissolve the precipitate in 15 ml. of 1:1 hydrochloric acid. Dilute to 50 ml., precipitate as before, filter, and wash with the 1 per cent ammonium chloride solution. Combine the filtrates, and add 5 ml. of concentrated ammonium hydroxide together with 2 grams of ammonium persulfate. Make a manganese dioxide separation by boiling for a short time and filtering. Dissolve the precipitate from the paper into the beaker, using 1:1 hydrochloric acid and a few drops of concentrated nitric acid, and wash with hot water. Dilute to 100 ml., and boil to remove chlorine. Repeat the ammonium hydroxide precipitation exactly as before. Filter, and wash thoroughly with the ammonium chloride solution. All these steps are necessary to separate the manganese from residual iron, aluminum, and titanium.

Make the filtrate slightly acid with hydrochloric acid, and add 10 ml. of a 10 per cent diammonium phosphate solution. Add ammonium hydroxide slowly, with vigorous stirring, until the precipitate becomes crystalline and its silky appearance remains unchanged. Add 15 ml. of concentrated ammonium hydroxide, and allow to stand overnight. Filter, and wash with 1:19 ammonium hydroxide. Dissolve the precipitate from the paper into the original vessel with a little 1:3 hydrochloric acid. Dilute to 100 ml., and add 1 ml. of the diammonium phosphate solution. Repeat the precipitation by adding 1:1 ammonium hydroxide very slowly, with constant stirring, until a crystalline precipitate of silky appearance forms. Add 10 ml. of 1:1 ammonium hydroxide in excess per 100 ml. of volume, and allow to stand for 3 hours. Filter, and wash with cold 1:19 ammonium hydroxide. Dry, and ignite at 1100° to constant weight in a weighed porcelain or platinum crucible. Weigh as manganese pyrophosphate, $Mn_2P_2O_7$. Deduct a determined blank.

CALCULATION.

$$\text{Weight of manganese pyrophosphate} \times 0.3870 \times \frac{100}{\text{weight of sample}} = \text{per cent manganese}$$

Calcium and Magnesium. PERMANGANATE AND PYROPHOSPHATE METHODS. In aluminum alloys the calcium will usually be small and the amount only approximated. The magnesium content controls the size of sample. Use 5 grams for a magnesium content of 0.001 to 0.01 per cent; 2 grams for 0.01 to 0.2 per cent; 1 gram for 0.2 to 2.0 per cent; 0.5 gram or less above 2 per cent. Preferably, use a sample containing not more than 50 mg. of magnesium. The method of dissolving is varied according to the silicon content.

Samples Not High in Silicon. Place the sample in a 250-ml. wide-mouthed conical flask; and add 25 to 30 ml. of 20 per cent sodium hydroxide solution. Make the additions in small portions, allowing time between for the reaction to subside. After violent reaction subsides, heat the flask. When visible hydrogen evolution ceases, cautiously add a few drops of 30 per cent hydrogen peroxide to the hot concentrated solution to complete the oxidation of silicon and chromium. If chromium constituents are not completely decomposed, an insoluble residue containing some magnesium will remain after treatment with hydrochloric acid. Dilute to 150 ml., filter, and wash with hot 0.5 per cent sodium carbonate. Dissolve the magnesium in the residue into the original flask, using 40 ml. of 1:1 hydrochloric acid, and wash the filter with hot water.

In dissolving alloys high in silicon in sodium hydroxide solution as described, a precipitate, presumably of aluminum silicate, settles out.

Samples High in Silicon. Place the sample in a 250-ml. widemouthed conical flask, and add 125 ml. of cold water. Add a total of 25 ml. of 20 per cent sodium hydroxide solution in small portions so that the reaction proceeds slowly. Let stand until hydrogen evolution ceases. Filter through close-textured paper, and wash twice with hot 0.5 per cent sodium carbonate solution. Wash the precipitate from the paper back into the original flask with a minimum amount of the sodium carbonate solution, add 5 grams of solid sodium hydroxide, and digest, adding a few drops of hydrogen peroxide from time to time to the hot concentrated solution, until the silicon is dissolved. Dilute to 150 ml., filter through the original paper, and wash with the sodium carbonate solution. Dissolve the magnesium from the filter into the original flask with 40 ml. of 1:1 hydrochloric acid, and wash the filter with hot water.

Further Treatment of Samples. If the sample contains no manganese or nickel and not more than 3 mg. of copper, add a few drops of methyl red indicator and concentrated ammonium hydroxide until the solution is just yellow in color. Allow the precipitate to coagulate. Filter, and wash with a hot 1 per cent ammonium chloride solution made just yellow to methyl red by addition of 1:1 ammonium hydroxide. If the magnesium content is over 20 mg., place the paper with the precipitate in the original flask, and dissolve with 15 ml. of 1:1 hydrochloric acid. Dilute to 50 ml. with hot water. Reprecipitate as above, filter, and wash with the ammonium chloride solution. Make the combined filtrate just acid to methyl red, and proceed to precipitation of calcium and magnesium.

If manganese or nickel is present, or more than 3 mg. of copper, proceed as follows: Neutralize the hydrochloric acid solution containing the magnesium with concentrated ammonium hydroxide, and add a few drops in excess. Prepare ammonium sulfide solution by saturating a mixture of 100 ml. of ammonium hydroxide and 900 ml. of water with hydrogen sulfide. Add 5 ml. of this ammonium sulfide solution to the sample. Filter into a 250-ml. beaker, and wash with a solution of 10 grams of ammonium chloride and 10 ml. of ammonium sulfide solution diluted to 1 liter. If more than 20 mg. of magnesium is present in the sample, some magnesium may be carried down with the sulfide. In this case, dissolve the precipitate with 15 ml. of 1:1 hydrochloric acid, precipitate as before, filter, and wash the sulfides. Combine the filtrates.

Boil the filtrate vigorously to drive off the ammonium sulfide. If a noticeable metal sulfide precipitate forms as the solution boils and becomes acid, usually indicating nickel or tin, filter. Wash the precipitate, and boil the filtrate. In either case, oxidize the remaining sulfur by adding an excess of bromine water, and continue boiling until the color is gone. Adjust the volume of solution to 75 ml. Remove from the hot plate, and add 10 ml. of bromine water. While stirring, make the solution alkaline with concentrated ammonium hydroxide, and add 15 drops in excess. Digest until any manganese present flocks out as manganese oxide. Stirring just as the solution begins to boil will aid coagulation. Maintain excess of ammonium hydroxide in the solution. Filter, and wash with hot water. Make the filtrate acid to methyl red with 1:1 hydrochloric acid. If the solution does not hold the color of this indicator, boil until all the bromine is gone.

Calcium. If calcium is present in the sample, it is determined as the oxalate. Usually, in aluminum it can be neglected. Make the solution just alkaline with 1:1 ammonium hydroxide, and proceed as on page 216, starting at "Heat the sample solution to boiling, and add . . .".

Magnesium. Use the filtrate from precipitation of calcium oxalate or the original solution if calcium is negligible. Proceed as on page 218, starting at "Add 10 ml. of 10 per cent disodium hydrogen phosphate solution . . .".

Chromium. Ferrous Ammonium Sulfate Method. Place 1 gram of sample in a 400-ml. beaker. Add 30 ml. of an acid mixture prepared by slowly stirring 400 ml. of concentrated sulfuric acid into water, cooling, adding 400 ml. of concentrated nitric acid, and diluting to 2 liters. Then add 20 ml. of 0.3 per cent silver nitrate solution, and heat gently until the sample is dissolved. If free silicon is present, add a few drops of 48 per cent hydrofluoric acid. Boil to expel oxides of nitrogen.

The sample may also be decomposed in a Pyrex beaker with sodium hydroxide and a little hydrogen peroxide as in the determination of calcium and magnesium (page 318). Then cautiously add 20 ml. of the above-mentioned acid mixture.

Dilute the solution of sample to 300 ml. with hot water, add 2 grams of ammonium persulfate, and heat to boiling. After the chromium is oxidized, boil for 10 minutes, add 0.5 ml. of 1:1 hydrochloric acid, and continue boiling until any manganese present is reduced. If necessary, add further 0.5-ml. portions of 1:1 hydrochloric acid, and boil for 20 minutes after the final addition. Cool, and run in a measured volume of 0.1N ferrous ammonium sulfate solution (page 726), until the color changes from yellow to green; 7 ml. will reduce about 10 mg. of chromium. Titrate back with 0.1N potassium permanganate solution (page 728) until a faint pink appears in the chrome-green solution, that will persist for 1 minute.

CALCULATION.

$$(\text{Ml. } 0.1N \text{ ferrous ammonium sulfate} - \text{ml. } 0.1N \text{ potassium permanganate}) \times 0.001734 \times \frac{100}{\text{weight of sample}} = \text{per cent chromium}$$

Nickel. DIMETHYLGLYOXIME METHOD. Samples from various sources may be used for the nickel determination.

Original Alloy. Use 1 gram of sample containing up to 5 per cent of nickel or a proportionately smaller weight if larger nickel contents are expected. Dissolve in 15 ml. of 20 per cent sodium hydroxide solution. When the reaction is complete, dilute to 150 ml. with hot water. Allow the residue to settle for a short time, filter, and wash with hot water. Discard the filtrate and washings. Dissolve the residue with a minimum amount of hot 1:1 hydrochloric acid, together with a few drops of concentrated nitric acid. Wash the paper thoroughly with hot water. Add 10 ml. of 1:1 sulfuric acid, evaporate to fumes of sulfur trioxide, and dilute to 100 ml. with hot water. Saturate with hydrogen sulfide. Filter, and wash with 1:999 sulfuric acid saturated with the gas. Boil the filtrate until the hydrogen sulfide is completely removed. Add 5 ml. of concentrated nitric acid, and boil to oxidize the iron. Dilute to 200 ml., add 10 ml. of 25 per cent tartaric acid solution, and neutralize with concentrated ammonium hydroxide.

Solution from Iron Determination. The silicon filtrate from which copper, bismuth, and lead have been removed and in which iron was titrated is often used for titanium determination. In the absence of titanium it or an aliquot is a suitable sample for nickel determination. Add 10 ml. of 25 per cent tartaric acid solution, and dilute to 300 ml. Neutralize with concentrated ammonium hydroxide.

PROCEDURE. Heat nearly to boiling, and stir in a dimethylglyoxime solution made by dissolving 1 gram of the oxime in 100 ml. of concentrated ammonium hydroxide and filtering. Use 20 ml. for up to 0.02 gram of nickel and 5 ml. for each additional 0.01 gram. If the nickel content is over 0.25 per cent, filter through a weighed Gooch crucible 30 minutes after precipitating; if less than 0.25 per cent, allow to stand 1 hour before filtering. Wash with hot water, dry for 2 hours at 110 to 120°, cool, and weigh.

Modifications of this procedure should be made in special cases. If much cobalt is present, precipitate nickel as described, make just acid with glacial acetic acid, and add a further 3 ml. of the oxime reagent. Allow to stand for 1 hour, filter on a hardened paper, and wash with hot water. Dissolve the precipitate in cold 1:3 hydrochloric acid, and neutralize with 1:1 ammonium hydroxide. Add 10 ml. of the oxime reagent, and finish the determination as described.

If the original sample contains less than 0.01 per cent of nickel, special handling will be required. Weigh out sufficient sample to give at least 1 mg. of nickel. Follow the main procedure until the oxime reagent is added. Make the solution acid to methyl red with 1:1 hydrochloric acid and then just alkaline with 1:1 ammonium hydroxide, and allow to stand overnight. Filter on a paper, and wash with hot water. Dissolve the nickel glyoxime into the original container with 1:3 hydrochloric acid, add 5 ml. of 1:1 sulfuric acid, and evaporate to fumes. Add 2 ml. of concentrated nitric acid, and take up with water. Boil to dissolve soluble salts, filter, and wash the residue with hot water. To the filtrate add 5 ml. of 25 per cent tartaric acid solution, and neutralize with 1:1 ammonium hydroxide. Adjust the volume to 100 ml., heat nearly to boiling, and stir in 10 ml. of the oxime reagent. Allow to stand for 1 hour. Filter, wash with hot water, and dry for 2 hours at 110 to 120°. Cool, and weigh.

CALCULATION.

$$\text{Weight of nickel glyoxime} \times 0.2032 \times \frac{100}{\text{weight of sample}} = \text{per cent nickel}$$

Tin. BROMATE METHOD. Place 1 to 3 grams of sample in a 500-ml. flask. Add 0.25 gram of pure powdered antimony and 150 ml. of 1:2 hydrochloric acid. After the reaction slackens, heat the flask, finally boiling for about 10 minutes to ensure complete solution of the tin. If any appreciable insoluble residue remains, filter, and wash with 1:4 hydrochloric acid. Discard the residue, and return the filtrate to the original flask. Add 0.25 gram of pure powdered antimony and 0.5 gram of pure aluminum.

Fit a three-hole rubber stopper to the flask. In one hole place a stopper. Through another hole pass a glass tube that reaches the bottom of the flask. Attach this tube to a cylinder of carbon dioxide. Leave the third hole open. Start a slow current of the gas through the flask. Heat gently until the aluminum is completely dissolved, and boil for about 10 minutes. Increase the flow of gas enough to prevent back pressure, and cool below 25°, preferably below 10°. As soon as cold, with the flow of gas still passing, open the stoppered hole, and from a pipet add 5 ml. of fresh starch solution containing 2 grams of potassium iodide. To prepare the starch solution, suspend 1 gram of soluble starch in a little cold water, dilute to 150 ml. with hot water, boil for 2 minutes, and add 5 mg. of mercuric iodide as preservative. Now insert the tip of a buret containing standard 0.1N or weaker potassium bromate solution (page 727). Titrate to a blue color. Deduct a determined blank. The tin is changed from bivalent to tetravalent and the bromate reduced to bromide.

Standardize the bromate solution against pure tin in the same manner as for the analysis.

If the sample contains sufficient silicon to obscure the end point, dissolve a fresh sample in 50 ml. of 1:1 hydrochloric acid. Boil, and filter into a 500-ml. flask, washing with hot water. Proceed as above, starting at "Add 0.25 gram of pure powdered antimony and 0.5 gram of pure aluminum . . . ".

CALCULATION.

$$(\text{Ml. potassium bromate solution} - \text{blank}) \times \text{normality} \times 0.05935 \times \frac{100}{\text{weight of sample}} = \text{per cent tin}$$

CHAPTER 16

COPPER-BASE ALLOYS

This group of methods is designed for use with the commercial alloys known as Muntz metal, yellow brass, red brass, commercial bronze, free turning brass, naval brass, and admiralty metal, and with similar compositions. They are selected from the American Society for Testing Materials (ASTM) "Methods of Chemical Analysis of Metals," with some modifications.

These methods apply to alloys of the following composition and ranges:

Alloy	Per cent	Alloy	Per cent
Copper	55.0 to 95.0	Antimony	0.0 to 0.1
Tin	0.0 to 1.5	Zinc	Balance
Lead	0.0 to 4.0	Iron	Impurity—less than 0.25
Phosphorus	0.0 to 0.1	Nickel	Impurity—less than 0.25
Arsenic	0.0 to 0.1		

In brasses, manganese bronzes, and other alloys containing 5.0 per cent or more of zinc, it is usual to report this element by difference. A method for its determination is included, however. In many methods variations are made according to the elements present and their relative quantity. Therefore, unless a sample is furnished with an approximate composition, it is essential to do a complete qualitative analysis first, according to the technics in Chap. 5.

Sampling. Discard any drillings containing oxide. Make drillings completely or exactly halfway through the samples and also at right angles to each other, from as many representative parts as possible. Sift all drillings in a screen with 250 meshes per square centimeter or 40 meshes per linear inch in order to remove material that is ground between the drill and sides of the hole. Remove any iron from the drillings with a magnet and any oil or grease with ether.

Copper. Electrolytic Method. *Lead Less than* 0.1 *Per Cent.* In a covered 250-ml. electrolysis beaker dissolve 2 grams of the sample in 60 ml.

of an acid mixture consisting of 500 ml. of concentrated sulfuric acid, 300 ml. of concentrated nitric acid, and 1,700 ml. of water. Allow to remain on the steam bath until solution is complete; then boil gently until the oxides of nitrogen are removed. Wash down the cover glass and sides of the beaker, and dilute to about 150 ml. with water. If arsenic or antimony is present, stir in 1 to 2 ml. of 9 per cent hydrogen peroxide at the start and another 1 to 2 ml. toward the end of the electrolysis.

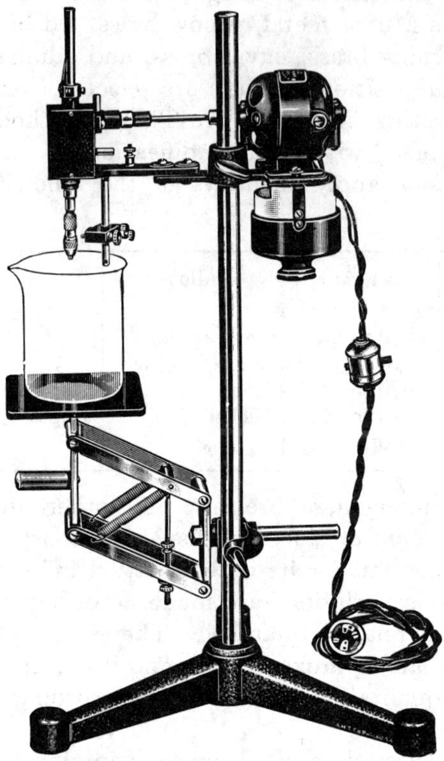

FIG. 94.—Electrolytic unit for single determination.

Insert sandblasted platinum electrodes, cover with a pair of split watch glasses, and electrolyze with a unit such as those shown in Figs. 94 and 95. Use a current density overnight of 0.5 amp. per sq. dm., 4.5 amp. per sq. ft. Alternatively use gauze electrodes and a current density of 4 amp. per sq. dm., 36 amp. per square foot, for 2.5 hours. When the solution is colorless, wash down the cover glasses, electrodes, and sides of the beaker, raising the level of the liquid slightly, and continue the electrolysis for about 15 minutes, noting whether or not copper is deposited on the newly exposed surface of the platinum. If copper appears, raise the level slightly, and continue the electrolysis until none appears on the stem. The electrolyte is ordinarily to be used for subsequent determinations. If not, remove a few drops, place on a spot plate, and treat with a saturated solution of hydrogen sulfide. Electrolysis is complete when no copper sulfide is observed.

When all the copper has been plated from the solution, remove the cathode quickly while washing with water from a wash bottle, rinse it in water, and then dip it in two successive baths of alcohol. Dry in an oven at 110° for 3 to 5 minutes, cool, and weigh as metallic copper.

Lead More Than 0.1 *Per Cent.* Dissolve 2 grams of sample in a covered 250-ml. beaker in 60 ml. of a mixture of 500 ml. of concentrated sulfuric

Fig. 95.—Electrolytic unit for multiple determinations.

acid, 300 ml. of concentrated nitric acid, and 1,700 ml. of water. Allow to remain on the steam bath until solution is complete, and then boil gently until oxides of nitrogen are removed. Let the sample stand at room temperature 2 hours or longer, and then filter off lead sulfate through a Gooch crucible. Wash the residue in the crucible thoroughly with 3:97 sulfuric acid to remove all the copper. Transfer to a 200-ml. electrolysis beaker, and dilute to about 150 ml. with water. If arsenic or antimony is present, stir in 1 to 2 ml. of 9 per cent hydrogen peroxide before starting the electrolysis and another 1 to 2 ml. toward the end of the electrolysis.

Proceed as for lead under 0.1 per cent (page 324), starting at "Insert sandblasted platinum electrodes . . . ".

Copper and Lead Simultaneously. ELECTROLYTIC METHOD. *Absence of Tin. Lead* 0.1 *to* 4.0 *Per Cent.* In a covered 250-ml. electrolysis beaker, dissolve 2 grams of the sample in 25 ml. of 1:1 nitric acid, and boil gently to remove oxides of nitrogen. After allowing to cool somewhat, wash down the cover glass and sides of the beaker. Add 1 drop of approximately 0.1*N* hydrochloric acid, and dilute to 150 ml. Proceed as for copper by the electrolytic method (page 324), starting at "Insert sandblasted platinum electrodes . . . ". Continue until the copper is all deposited. At that time it can be assumed that deposition of lead is complete.

To recover the electrolyte for subsequent determinations, with the current still running, slowly lower the beaker, at the same time washing both the anode and the cathode with water. Remove the anode, rinse it thoroughly with water, and dry it at 210° for 15 to 20 minutes. Weigh and calculate the lead present by using the empirical factor 0.866. The deposit of lead peroxide is fragile and must be handled carefully. Remove the cathode, and rinse it with water and in two successive baths of alcohol. Dry at 110° for 3 to 5 minutes, cool, and weigh as metallic copper.

If the electrolyte is not to be saved, it is more convenient to siphon it off with the current still running, at the same time adding water to the beaker.

Presence of Tin. Lead 0.1 *to* 4.0 *Per Cent.* In a covered 250-ml. beaker, dissolve 2 grams of the sample in 25 ml. of 1:1 nitric acid, and boil gently to remove oxides of nitrogen. Add 50 ml. of hot water, and allow to stand on a steam bath for 1 hour. Add paper pulp, and filter off the metastannic acid through a close-textured paper into a 250-ml. beaker. Wash a few times with hot 1:99 nitric acid. Reserve the filtrate.

Transfer the filter paper and contents to the original beaker, and add 12 to 15 ml. of concentrated sulfuric acid and 25 ml. of concentrated nitric acid. Heat until all organic matter is decomposed. Transfer the digested and cooled solution to a 400-ml. beaker, and dilute to 250 ml. Then add a 25 per cent solution of sodium hydroxide until the tin hydroxide first precipitated has dissolved. The solution will now be strongly alkaline. Add 20 ml. of a 25 per cent solution of sodium sulfide, stir thoroughly, and digest on the steam bath for several hours, or until the supernatant liquid is clear. Cool to room temperature, filter through a paper of close texture, and wash the precipitate with a 2 per cent solution of sodium sulfide. Discard this filtrate. Dissolve the final sulfides in a few milliliters of 1:1 nitric acid, and add the recovered copper and lead to the reserved nitric acid filtrate containing the large amount of copper and lead. Add 1 drop of 0.1*N* hydrochloric acid, and adjust the volume to 150 ml. Proceed as in the usual electrolytic copper determination (page 324), starting at "Insert sandblasted platinum electrodes . . . ". Remove the electrodes by the technic described for copper and lead in the absence of tin (page 326), starting at "To recover the electrolyte for subsequent determinations . . .".

Lead. ELECTROLYTIC METHOD. *Absence of Tin.* With a lead content of less than 1 per cent, use a 5-gram sample, and dissolve in 50 ml. of 1:1 nitric acid in a 250-ml. beaker. With a lead content of 1 to 4 per cent, use a 1-gram sample, and dissolve in 20 ml. of 1:1 nitric acid in a 250-ml. beaker. After solution is complete, boil gently to remove oxides of nitrogen, wash down the cover glass and sides of the beaker, and dilute to 150 ml.

Insert the sandblasted platinum electrodes with reverse connections; that is, use the larger electrode, which would be the cathode for plating of copper, as the anode for deposition of lead peroxide. Use a current density of 1.25 to 1.50 amp. per sq. dm., 11.25 to 13.5 amp. per square foot, for 1 hour. Wash down the cover glasses, electrode stems, and sides of the beaker, raising the level of the beaker slightly, and continue the electrolysis for 15 minutes. If no darkening of the newly exposed surface can be detected, the lead has been entirely deposited.

Continue the electrolysis until no darkening of the platinum can be detected when the current has been passed for 15 minutes after the level of the liquid has been raised. With the current still running, siphon off the electrolyte, at the same time filling the beaker with water. Remove the anode quickly, rinse it thoroughly in water, and dry it at 210° for 15 to 20 minutes. Weigh and calculate to lead by using the empirical factor 0.866.

Presence of Tin. With a lead content less than 1 per cent, use a 5-gram sample and dissolve in 50 ml. of 1:1 nitric acid in a 250-ml. beaker. With a lead content of 1 to 4 per cent, use a 1-gram sample, and dissolve in 20 ml. of 1:1 nitric acid in a 250-ml. beaker. When solution is complete, boil gently to remove oxides of nitrogen. Proceed as for copper and lead simultaneously in the presence of tin (page 326), starting at "Add 50 ml. of hot water, and allow to stand on a steam bath for 1 hour." Follow this method until the solution is ready for electrolysis. Then turn to the method for lead in the absence of tin above, starting at "Insert the sandblasted platinum electrodes with reverse connections . . .".

Zinc. PRECIPITATION AS THE SULFIDE. Dilute the electrolyte, from which copper or copper and lead have been exhausted, to a known volume. Usually it is convenient to evaporate to somewhat under 200 ml., cool, and dilute to 200 ml. in a volumetric flask. Normally this will be from a 2-gram sample. If the zinc content of the sample is under 20 per cent, remove 100 ml. of the solution with a dry pipet for this determination, rinsing the pipet with water back into the flask. This leaves the aliquot from a 1-gram sample in the flask for the later determination of tin. If the zinc in the sample is over 20 per cent, pipet out only 50 ml. with a dry pipet and in this event do not rinse the pipet into the flask.

If the aliquot is from the determination of copper, add 3 ml. of 1:1 sulfuric acid; if from simultaneous determination of copper and lead, add 15 ml. In either case, evaporate to strong fumes of sulfuric acid.

Cool the solution, and add 15 to 25 ml. of 40 per cent hydrobromic acid solution. To volatilize tin, arsenic, and antimony, evaporate the solution cautiously to fumes of sulfuric acid and then heavily until the residue becomes light-colored. Cool, wash down the cover and beaker walls, and fume again. Cool, dilute with water, and digest until sulfates are in solution. Then allow the solution to cool. Neutralize with concentrated ammonium hydroxide, and make acid with 1:1 sulfuric acid, adding an excess of 3 to 4 ml. per 100 ml. of solution. Cool to room temperature, and pass in a rapid stream of hydrogen sulfide for 15 minutes. This treatment will remove any last traces of copper and lead, if present, and traces of tin, arsenic, and antimony. Allow to stand for 2 hours, and filter through a close-textured paper into a 600-ml. beaker. Wash the sulfide precipitate thoroughly with 1:99 sulfuric acid saturated with hydrogen sulfide. If there are appreciable amounts of copper and lead sulfides in the precipitate, the earlier erroneous determinations can sometimes be corrected by dissolving this sulfide precipitate in 1:3 nitric acid and electrolyzing for copper and lead. It is not good technic to do so, however, and the precipitate should ordinarily be discarded.

Boil the filtrate and washings for 30 minutes to remove hydrogen sulfide. Cool the solution, and carefully neutralize with 1:2 ammonium hydroxide, using methyl orange as indicator. Adjust the volume to approximately 400 ml. Now make the solution just $0.01N$ acid by adding 1 ml. of approximately N sulfuric acid for each 100 ml. of solution. This N sulfuric acid can be prepared with sufficient accuracy by dilution of 28 ml. of concentrated sulfuric acid to 1 liter. Cool to room temperature, and pass in a rapid stream of hydrogen sulfide for at least 30 minutes. The gas should pass through the solution at the rate of at least 6 bubbles per second. Allow the white precipitate of zinc sulfide to settle in the cold, and filter through an ashless paper of close texture, washing the precipitate thoroughly with cold water. Use a policeman on the beaker.

Place the filter and precipitate in a weighed porcelain crucible, dry, and ignite carefully over a Bunsen flame, making sure that the sulfur fumes do not burn. Finally, place the crucible and contents in an electric muffle at 950° for 15 minutes. Cool and weigh as zinc oxide, repeating the ignition until constant weight is obtained.

The purified zinc sulfide precipitate may also be redissolved and determined by titration with potassium ferrocyanide (page 504). Not less than 100 mg. of zinc must be present, or else the end point will not be clear. The upper limit is generally regarded as 600 mg. The size of the aliquot used must accordingly be determined by these limits. Carry one or two standards along with each lot of zinc determinations, for end points vary somewhat according to the analyst.

CALCULATION.

$$\text{Weight of zinc oxide} \times 0.8034 \times \text{aliquot} \times \frac{100}{\text{weight of sample}} = \text{per cent zinc}$$

Tin. VOLUMETRIC METHOD. Use the aliquot of exhausted electrolyte equivalent to a 1-gram sample from the determination of copper or copper and lead. Evaporate to strong fumes. Allow to cool somewhat, wash down the sides of the beaker with water, and fume again in order to drive off all traces of nitric acid. Cool and transfer to a 300-ml. conical flask. Add 80 ml. of concentrated hydrochloric acid and sufficient water to make a total volume of 100 ml.

If it is necessary to start with a fresh sample of alloy for the determination of tin, weigh 1 to 5 grams into a 250-ml. beaker. Add 60 ml. of a mixture of equal volumes of concentrated hydrochloric and nitric acids. When the solution is complete, boil until oxides of nitrogen are driven off, and dilute to 125 ml. with water. Add 2 to 3 ml. of a 10 per cent solution of ferric chloride, and neutralize with concentrated ammonium hydroxide, adding a slight excess. Bring the solution to an incipient boil, and filter it through a loose-textured paper, washing the original beaker and paper several times with a hot 1 per cent solution of ammonium sulfate. Discard this filtrate and the washings. Place the original beaker beneath the funnel, and dissolve the precipitate from the paper with hot 1:1 hydrochloric acid. Wash the paper several times with hot water, once with 1:1 hydrochloric acid, and then several times more with water. Reprecipitate with concentrated ammonium hydroxide, and filter and wash as previously. Discard this filtrate and the washings. Now place a 300-ml. conical flask beneath the funnel, and dissolve the precipitate remaining on the beaker and in the paper. For this purpose use 160 ml. of hot 1:1 hydrochloric acid, and wash with about 20 ml. of water. By this means, the copper, soluble in ammonium hydroxide, is separated from the tin, which is coprecipitated with the iron and gives an alternative solution to that obtained by electrolysis for copper and lead.

Add 2 drops of a solution of antimony trichloride prepared by dissolving 2 grams of the crystals in 50 ml. of concentrated hydrochloric acid and diluting to 100 ml. with water. Reduce the tin by boiling with test lead, iron, or nickel for 20 to 30 minutes. As an alternative method of reduction use a gram of metallic aluminum in the form of sheet or rolled foil as reductant. After adding the 2 drops of antimony trichloride solution, add the aluminum to the solution in the flask while at room temperature.

Provide a three-hole rubber stopper for the flask. One carries an inlet tube leading to the bottom of the flask to carry carbon dioxide. One is open for escape of the carbon dioxide. The third is stoppered and will later take a buret for titration. If test lead, iron, or nickel is used for reduction, boil until it has dissolved. If aluminum is used, let stand until nearly dissolved. Insert the rubber stopper, and boil gently for about 10 minutes with a slow stream of gas passing through the flask. Then proceed as on page 321, starting at "Increase the flow of gas enough to prevent back pressure, and cool . . .".

The titration may also be made with standard iodine solution. An atmosphere of carbon dioxide must always be present. Subtract the volume of iodine or iodate solution required in a blank run. Bivalent tin is oxidized to tetravalent.

CALCULATION.

$$(\text{Ml. iodine solution} - \text{blank}) \times \text{normality} \times 0.05935 \times \text{aliquot} \times \frac{100}{\text{weight of sample}} = \text{per cent tin}$$

Iron. VOLUMETRIC METHOD. *Absence of Tin, Arsenic, and Antimony.* Weigh a 5-gram sample into a 400-ml. beaker, and dissolve in 50 ml. of 1:1 nitric acid. Heat until all fumes have been driven off, and then dilute to 150 ml. with water. Add concentrated ammonium hydroxide until distinctly alkaline, and boil gently for a few minutes. Remove the source of heat, and allow the precipitate of ferric hydroxide to settle for a few minutes. Filter while hot through a paper of close texture. Wash the beaker and precipitate alternately with a 2 per cent ammonium chloride solution and hot water, to remove the bulk of soluble salt. Finally, wash a few times with hot water. Discard the filtrate and washings. Redissolve the precipitate on the paper with hot 1:1 hydrochloric acid, finally washing the paper very thoroughly with hot water to remove the last traces of iron, and catch the solution and washings in the original beaker. Dilute to 150 ml., and reprecipitate with concentrated ammonium hydroxide as before. Filter and wash as before. Dissolve the precipitate from the paper in a small amount of hot 1:1 hydrochloric acid, and wash the paper well with hot water, catching the solution and washings in the original beaker.

Evaporate to a volume of 15 to 20 ml., cool somewhat, and wash down the sides of the beaker. Prepare a solution of 12.5 grams of stannous chloride, $SnCl_2.2H_2O$, in 25 ml. of concentrated hydrochloric acid, and dilute to 250 ml. with water. Add this drop by drop while swirling the beaker over a white background until the yellow color of the ferric chloride disappears. Add 2 drops of stannous chloride solution in excess, but not more. Dilute the reduced solution to 50 ml., and cool to room temperature. Add 10 ml. of a saturated solution of mercuric chloride while stirring. The resultant precipitate should be pure white in color and in amount should preferably be not more than enough to give a pearly opalescence. If the precipitate is gray or dark, this shows that too large an excess of stannous chloride was used in reducing the iron; in this case it is necessary to start again with a fresh sample.

To the properly reduced solution, add 10 ml. of a solution containing 250 ml. of 85 per cent phosphoric acid diluted to 1 liter with water and 3 drops of a 1 per cent solution of diphenylamine in sulfuric acid. This indicator retains its properties even after becoming discolored. The volume of the solution should not exceed 100 ml. at this point. Titrate with standard $0.01N$ potassium dichromate solution until a purple or black color persists, and add 5 to 10 drops in excess. The end point may not be sharp, but the change is unmistakable. Do not carry over the end point by more than 1 ml., for the indicator may be destroyed. Back-titrate with a $0.01N$ ferrous ammonium sulfate solution recently standardized against the dichromate solution in the presence of phosphoric acid and diphenylamine indicator. Add the ferrous solution, drop by drop, with stirring. The first few drops should deepen the violet color, but at the end point the color will sharply fade to a clear or faintly turbid green.

CALCULATION.

$$(\text{Ml. } 0.01N \text{ dichromate solution} - \text{ml. } 0.01N \text{ ferrous solution}) \times 0.0005585 \times \frac{100}{\text{weight of sample}} = \text{per cent iron}$$

Presence of Tin, with or without Arsenic or Antimony. It is desirable to remove tin, for this facilitates later handling. Arsenic and antimony must be removed because they interfere in the titration. In a 250-ml. beaker, dissolve 5 grams of the sample in 50 ml. of 1:1 nitric acid, and boil to remove oxides of nitrogen. Dilute to 100 ml. with hot water, and digest the metastannic acid on the steam bath for 1 hour. Filter through double paper of close texture into a 400-ml. beaker, and wash with hot 1:99 nitric acid. Reserve the filtrate.

Transfer the precipitate and paper to the original beaker, and treat first with 15 ml. of concentrated nitric acid and then with 10 ml. of 60 per cent perchloric acid. Heat gently until fumes of perchloric acid appear, and cool. Wash down the sides of the beaker and add 10 ml. of 40 per cent hydrobromic acid. Boil gently without a cover glass until fumes of perchloric acid again appear. Repeat the hydrobromic acid treatment to expel the last traces of tin, arsenic, and antimony. Filter off silica, if present, and add this solution to the original filtrate. There is also a satisfactory alkaline sulfide method for separation of the tin, arsenic, and antimony, but details will not be given here.

The solution from which interfering elements have been removed is now suitable for volumetric determination of iron. Dilute to 150 ml., and proceed as on page 330, starting at "Add concentrated ammonium hydroxide until distinctly alkaline, and boil gently for a few minutes."

Phosphorus. ALKALIMETRIC METHOD. *Absence of Tin and Arsenic.* Weigh 5 grams of the sample into a 500-ml. conical flask, and dissolve in 25 ml. of concentrated nitric acid. Heat gently until oxides of nitrogen are driven off; then add 100 ml. of water. Heat the solution to boiling, and add a saturated potassium permanganate solution, drop by drop, until a deep purple color persists. Boil for 5 minutes; then destroy the excess permanganate and precipitated manganese dioxide by adding, drop by drop, a 10 per cent solution of ammonium nitrite until the solution is a clear blue color. Boil again for 5 minutes to destroy excess nitrite.

Cool the sample to 45°, and add 5 grams of ammonium nitrate. Complete as on page 203, starting at "To 100 ml. of the usual ammonium molybdate reagent . . . ".

Presence of Tin. Absence of Arsenic. Weigh a 5-gram sample into a 500-ml. conical flask, and dissolve in 5 ml. of concentrated hydrochloric acid and 20 ml. of concentrated nitric acid. When solution is complete, add 20 ml. of water and heat gently until oxides of nitrogen are driven off. Then add 80 ml. of water. Take care not to boil off enough hydrochloric acid to allow the tin to precipitate. Cool the sample to 45°, and add 5 grams of ammonium nitrate. Complete as on page 203, starting at "To 100 ml. of the usual ammonium molybdate reagent . . . ".

Presence of Arsenic or Tin and Arsenic. Weigh a 5-gram sample into a 400-ml. beaker, and dissolve in 5 ml. of concentrated hydrochloric acid and 20 ml. of concentrated nitric acid. When solution is complete, add 20 ml. of water and boil gently until oxides of nitrogen are removed. Prepare a ferric chloride solution by dissolving 25 grams of ferric chloride, $FeCl_3.6H_2O$, in 400 ml. of water. Filter if not clear. Always run a blank on each new lot of this reagent, which usually contains small amounts of phosphorus even if labeled "special phosphorus-free." Add 10 ml. of this reagent solution to the sample.

Add 5 grams of ammonium chloride, dilute the solution to 200 ml., make just alkaline with concentrated ammonium hydroxide, and bring to a boil. Let the precipitate settle on the steam bath for 5 minutes; then filter on a loose-textured paper. Wash the beaker and precipitate with hot 1 per cent ammonium chloride solution until most of the blue copper salts have been washed out. Transfer the paper and precipitate to the original beaker, add 30 ml. of 1:2 hydrochloric acid, and digest on a steam bath until the precipitate is entirely dissolved. Bring the solution to a boil, filter off the paper pulp on a loose-textured paper, and wash ten times with hot 1:49 hydrochloric acid. Dilute the filtrate to 300 ml., and pass in hydrogen sulfide for 30 minutes. Let the precipitated sulfides coagulate for 2 to 3 hours, and filter on a close-textured paper. Wash the precipitate five times with hydrogen sulfide solution containing a few drops of hydrochloric acid, and discard it. Evaporate the filtrate to approximately 10 ml., and transfer to a 500-ml. conical flask. Add 15 ml. of concentrated nitric acid, and boil gently for 10 minutes. Dilute to 100 ml. In this procedure the oxidized phosphorus is precipitated as iron phosphate and then retained in solution after the removal of arsenic and tin. Cool the sample to 45°, and add 5 grams of ammonium nitrate.

Complete as on page 203, starting at "To 100 ml. of the usual ammonium molybdate reagent . . . ".

Nickel. DIMETHYLGLYOXIME METHOD. If an aliquot of the electrolyte from which copper and lead were removed is available, use this. Otherwise, start with a fresh 2-gram sample, and carry through the electrolytic separation of copper and lead.

Transfer the copper-free solution to a 400-ml. beaker, and add 10 ml. of a 25 per cent tartaric acid solution. Neutralize the solution with concentrated ammonium hydroxide, and add 1 ml. excess. Heat to 70°, and stir in 0.4 ml. of alcoholic dimethylglyoxime solution (page 718) for each milligram of nickel present, plus 3 to 5 ml. in excess. Add the reagent directly to the sample solution, not down the wall of the beaker. Allow to cool to room temperature, stirring occasionally. If the nickel is very low, allow the solution to stand overnight. If only a very small amount of nickel is present, the precipitate will probably be contaminated with excess of oxime crystals. In that event, filter, and ash. Dissolve the ash in a small volume of 1:1 nitric acid, dilute with water, and boil off oxides of nitrogen. Dilute to about 50 ml., and add 10 ml. of 25 per cent tartaric acid solution. Reprecipitate as above, starting at "Neutralize the solution with concentrated ammonium hydroxide . . . ", but add only 1 to 2 ml. of reagent.

Filter the nickel dimethylglyoxime through a weighed Gooch crucible or medium-porosity fritted-glass crucible, and wash well with cold water. Dry at 120 to 140° to constant weight. This usually requires about 1 hour. Cool and weigh.

CALCULATION.

$$\text{Weight of nickel glyoxime} \times 0.2032 \times \frac{100}{\text{weight of sample}} = \text{per cent nickel}$$

Arsenic and Antimony. DISTILLATION-VOLUMETRIC METHOD. This method requires special apparatus, as shown in Fig. 96. Approximate results are obtainable by assembling a similar unit with stoppers and tubing connections.

The flask is of 200 ml. capacity and has a neck 2.5 cm. in diameter. The neck is provided with a side tube, a thermometer well, and a delivery tube for conveying the acids and carbon dioxide to the flask. The side tube connects the apparatus to the condenser, the thermometer well is sealed to the top of the neck of the flask, and the delivery tube passes through the wall in the upper part of the neck of the flask and is connected to the acid bulb, which has a capacity of 100 ml. Both the thermometer well and delivery tube reach within 3 mm. of the bottom of the flask. The distance between the bottom of the flask and the exit tube leading to the condenser is 17 cm. An outlet on the side of the delivery tube is provided for the carbon dioxide. The apparatus is clamped to a ring stand. The flask rests on an asbestos board, about 18 by 18 cm., having a round hole 4.5 cm. in diameter.

The thermometer should be about 12 inches long and should have a range up to 200 or 250°. A rubber band can be placed about the thermometer at a position that will suspend it from the top of the well and allow the bulb to be near the bottom. This permits protection against

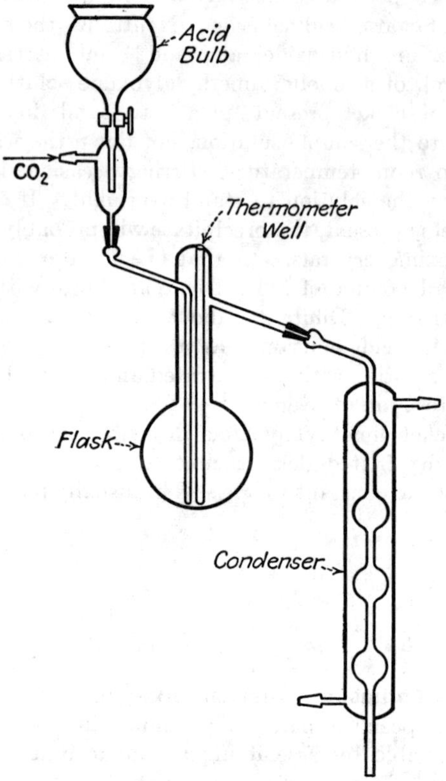

FIG. 96.—Distillation apparatus.

breakage of the bottom of the well. A small plug of asbestos fiber carefully pushed to the bottom of the well is also a good protection.

Arsenic. The method depends on quantitative distillation of arsenic as arsenic trichloride. In a 400-ml. beaker dissolve a 5-gram sample with 25 ml. of concentrated nitric acid and 10 ml. of concentrated hydrochloric acid. Add 10 ml. of a ferric nitrate solution made by dissolving 200 grams of the salt in 100 ml. of water. Boil for 1 to 2 minutes, and dilute to 200 ml. Neutralize with concentrated ammonium hydroxide, and add a distinct excess. Boil for 5 minutes, allow the precipitate to settle, and filter through an 11-cm. loose-textured paper. Wash several times with hot

1:49 ammonium hydroxide. It is desirable to leave some copper with the precipitate, for this aids in the distillation.

Dissolve the precipitate into the original beaker, using 80 ml. of 1:3 sulfuric acid, and wash the paper well with hot water. Transfer to the flask of the still, and evaporate to fumes of sulfuric acid, the carbon dioxide gas passing through at the rate of 6 to 8 bubbles per second. If any particles of paper pulp have contaminated the sample, oxidize these by the addition of concentrated nitric acid before fuming has developed.

Allow to cool, and place a 400-ml. beaker containing 50 ml. of water under the condenser, submerging the tip for about 7 mm. Then through the funnel carefully add 20 ml. of water and then 35 ml. of concentrated hydrochloric acid containing 1 ml. of 50 per cent hypophosphorous acid. Place 75 ml. of concentrated hydrochloric acid in the acid bulb, and bring the temperature of the solution in the flask to 112°. Drop acid from the bulb at a rate that will maintain the temperature at 115°. Continue the distillation until the acid in the bulb is exhausted. If the amount of arsenic is unexpectedly high and too much hypophosphorous acid has been inadvertently added, a black precipitate of arsenic will form in the flask. If this happens, remove the beaker from beneath the condenser and allow the flask to cool. Add 5 to 10 ml. of concentrated nitric acid to the flask, and evaporate to sulfuric acid fumes. Replace the 400-ml. beaker, and continue as before.

If antimony is present, retain the solution in the still. Remove the distillate, cool to below 30°, and neutralize to methyl red with 20 per cent sodium hydroxide solution. Then make just acid with 1:1 hydrochloric acid. Cool and add 10 to 12 grams of sodium bicarbonate, 3 ml. of a 0.25 per cent potato or arrowroot starch solution, and 3 drops of a 10 per cent potassium iodide solution. Titrate to a permanent blue tint with 0.02*N* standard iodine solution. Two equivalents of iodine are required to convert the trivalent arsenic to pentavalent. Run a blank on the same amounts of reagents as are used with the sample.

CALCULATION.

$$(\text{Ml. iodine solution} - \text{blank}) \times \text{normality} \times 0.03746 \times \frac{100}{\text{weight of sample}} = \text{per cent arsenic}$$

Antimony. This method provides quantitative distillation of antimony trichloride at a higher temperature than that required for arsenic trichloride. After the 400-ml. beaker has been removed from beneath the condenser in the arsenic determination, replace it with another containing 50 ml. of water, submerging the tip of the condenser about 7 mm. With the carbon dioxide gas flowing, evaporate the solution to a temperature of 160°. Place 75 ml. of concentrated hydrochloric acid in the acid bulb, and proceed with the distillation while dropping in the acid at a sufficient rate to maintain the temperature at $160 \pm 5°$. Continue the distillation until the acid in the bulb is exhausted.

If no arsenic is present, distillation for antimony can proceed directly after the ammoniacal precipitation, dissolving, and fuming; but 35 ml. of concentrated hydrochloric acid containing 1 ml. of hypophosphorous acid must then be added to the flask prior to the distillation of antimony.

To the distillate add 1 gram of sodium sulfite, and boil the solution down to a volume of 40 to 50 ml. Wash down the cover glass and sides of the beaker with water, and further boil down to a volume of 20 ml. The volume must not be appreciably lower, or antimony may be volatilized. All sulfur dioxide must have been expelled at this point.

Dilute the sample to about 100 ml., and add 10 ml. of 10 per cent tartaric acid solution. Make alkaline with 20 per cent sodium hydroxide solution, using methyl red indicator, and then just acid with 1:9 hydrochloric acid. Cool and add 10 to 12 grams of sodium bicarbonate, 3 ml. of 0.25 per cent starch solution and 3 drops of 10 per cent potassium iodide solution. Titrate to a permanent blue tint as in the arsenic determination. The trivalent antimony is converted to pentavalent. A blank must be determined on the same amounts of reagents as are used with the sample.

CALCULATION.

$$(\text{Ml. iodine solution-blank}) \times \text{normality} \times 0.06088 \times \frac{100}{\text{weight of sample}} = \text{per cent antimony}$$

CHAPTER 17

SAPONIFIABLE FATS AND OILS

The fatty glycerides of animal or vegetable origin constitute a large class of substances with varied applications in foods, protective coatings, the manufacture of soap, medicinal uses, etc. While in lay terms those which are solid at room temperature are fats and those which are liquid are oils, it is convenient to refer to both as saponifiable oils. This avoids the complication of designating a substance coconut oil on a warm day and coconut fat in prolonged cold weather. Fatty fractions that prove to be saponifiable and therefore fall into this class are frequently separated from commercial products, by ether extraction, for example.

A preliminary degree of identification can often be obtained from the odor, either cold or on warming; the general appearance; etc. For true identification, definite chemical or physical constants are required. Of these, acid value and refractive index are readily obtained. Saponification value and iodine number usually supplement these. For special cases many other determinations are applied, such as acetyl value and thiocyanogen number. Saponification gives the soap and glycerin; therefore, the fatty acid is often separated and constants obtained on it. This may be because by manipulation the sample has been saponified when it is impossible to separate the true fat or oil as such. An example is identification of the stock from which a soap has been produced.

Tables of constants are given in all the usual chemical handbooks and may serve in place of Table 15, given at the end of this chapter. The constants of mixtures of fats and oils are a weighted average of the oils mixed. Thus, a 1:1 mixture of an oil with an iodine value of 86 and of an oil with an iodine value of 136 would be 111; a mixture of 90 per cent of the first and 10 per cent of the second would have a value of 91. However, since these are natural products, at best only approximations are obtained. A range to allow for natural variations is applied to the constants. This necessarily introduces the factor of judgment in the interpretation of data.

While the majority of fats and oils are composed of glycerides of fatty acids of the C_{16} and C_{18} series, there are some, such as butter

TABLE 14.—FATTY-ACID DISTRIBUTION IN SOME COMMON FATS AND OILS. APPROXIMATE PERCENTAGE OF INDIVIDUAL FATTY ACIDS IN TOTAL*

Fat or oil	Approximate iodine value	Palmitic	Stearic	Oleic	Linoleic	Linolenic	Myristic	Lauric	Capric	Caprylic	Palmitolic	Arachidic	Total accounted for	
1. Beef tallow	35–42	25	24	42	2	..	3	..	..	..	2	..	98	1
2. Mutton tallow	48–61	27	31	32	4	..	2	..	..	..	3	..	99	2
3. Horse fat	75–86	27	5	50	13	5	..	..	..	..	..	..	100	3
4. Sow, outer back	....	20	8	54	13	..	4	..	..	..	..	..	97	4
5. Sow, inner back	....	26	11	44	14	..	4	..	..	..	..	..	99	5
6. Leaf lard	....	28	17	36	14	..	4	..	..	..	..	..	99	6
7. Palm oil	53–58	40	4	44	10	..	2	..	..	..	..	..	100	7
8. Olive oil	79–86	10	2	81	7	..	..	..	..	..	..	..	100	8
9. Corn oil	110–130	8	4	46	42	..	..	..	..	..	..	..	100	9
10. Cottonseed oil	104–113	21	2	30	46	..	1	..	..	..	..	..	100	10
11. Soybean oil	125–140	8	5	28	54	4	..	..	..	..	..	..	99	11
12. Linseed oil	170–185	5	4	10	43	38	..	..	..	..	..	..	100	12
13. Peanut oil	85–95	8	4	61	21	..	..	..	..	..	..	6	100	13
14. Coconut oil	8–10	10	2	7	2	..	18	46	7	8	..	..	100	14
15. Palm-kernel oil	15–20	9	1	18	1	..	14	47	7	3	..	..	100	15

* *Food Investigation Special Rept.* 46, Dept. *Sci. Ind. Research* (1938), His Majesty's Stationery Office.

or coconut oil, with lower fatty acids present. This results in their having a high saponification value.

The saponifiable oils may be roughly divided into three classes, nondrying oils, semidrying oils, and drying oils. Nondrying oils are chiefly glycerides of straight-chain fatty acids containing at most only one unsaturated linkage. Typical examples of these fatty acids are as follows:

Oleic acid........ $CH_3(CH_2)_7CH:CH(CH_2)_7COOH$
Stearic acid....... $CH_3(CH_2)_{16}COOH$
Palmitic acid...... $CH_3(CH_2)_{14}COOH$

Drying oils consist largely of glycerides of straight-chain fatty acids containing two or more unsaturated linkages, in particular linoleic and linolenic acids.

Linoleic acid.... $CH_3(CH_2)_4CH:CHCH_2CH:CH(CH_2)_7COOH$
Linolenic acid.... $CH_3CH_2CH:CHCH_2CH:CHCH_2CH:CH(CH_2)_7COOH$

Semidrying oils, as the name implies, are intermediate and contain varying proportions of the preceding type of fatty-acid glycerides. Cottonseed oil, for example, contains about 21 per cent of palmitic acid, 30 per cent of oleic acid, and 40 per cent of linoleic acid as glycerides.

Castor oil is in a class by itself as the only oil composed mainly of glycerides of a hydroxy fatty acid,

Ricinoleic acid, $CH_3(CH_2)_5CHOHCH_2CH:CH(CH_2)_7COOH$

Table 14 shows the relative distribution of different fatty acids in many of the common fats and oils.

Many of the methods that follow are those given by the Association of Official Agricultural Chemists, the American Oil Chemists Society, and the American Society for Testing Materials (ASTM). All the methods should seldom be applied to any one sample; selections must be made. Often specific gravity, refractive index, iodine value, acid value, and saponification value are determined initially. Then, if there are indications of a high unsaponifiable content, the unsaponifiable matter may be determined but this is a laborious determination to be avoided if possible.

Sample. More often than not, the sample will be an extract from a commercial product, such as an extract from a meal, a cosmetic, or a food. Soxhlet extraction (page 67) is typical. The previous treatment will have to be considered. Moisture on a sample of such derivation will mean nothing, but on a commercial sample of oil it may be important.

Filter the fat or oil through a dry paper to free it from suspended matter if present, but be sure to work at a sufficiently high temperature so that the lower melting material will not be separated from the higher melting. Practically, 50 to 60° is sufficient. This is not done if moisture is to be determined. Store such samples in a reasonably cool place to prevent development of rancidity.

Volatile Matter. This usually means moisture; but, with as general a consideration of samples as is involved here, the presence of other volatile matter is often a possibility.

Oven Method. The amount of sample required depends on the expected water content. Use a few grams when this is high and 20 to 100 grams when low. Greater care is necessary to avoid spattering than in the case of water-soluble substances. The accuracy of the method is low, for drying and semidrying oils will absorb oxygen unless heated in inert gas. Thus successive weighings after partial drying can result in an increase in weight, rather than a decrease. Proceed as on page 40, starting at "Weigh a suitable sample into a tared dish." It is often convenient to use the vacuum-oven method (page 41). Report the percentage loss in weight as moisture and volatile matter.

Solvent Distillation. This is a very convenient method where other procedures are not official. The upper limit of weight of sample is 100 grams. Use the method given on page 41 with either toluene or xylene.

Tube Distillation. This method has no official standing but is very convenient for determination of volatile matter in oils that will stand 130° without decomposition.

Weigh 5 grams into a 160- by 18-mm. Pyrex test tube supported on the balance beam by a wire passing around its neck and forming a loop. Heat carefully over a low flame, using a test-tube holder, until the water present distills out of the oil and condenses on the upper sides of the test tube. Now flame the upper sides with just sufficient heat to drive off the water. Repeat the heating of the oil and subsequent evaporation of the water until no more moisture is evolved. Cool, and reweigh.

The more intense heating for a relatively brief time replaces heating in the oven and largely avoids both oxidation of unsaturated glycerides and slow volatilization of the lower molecular weight glycerides or small amounts of free fatty acids present. Therefore, under properly selected conditions, it may be more accurate than the official methods.

Specific Gravity. Determine the specific gravity with the pycnometer (page 43); or if frequent determinations are made, use a Westphal balance (page 48). For oils liquid at that temperature, use 25° or as near that as feasible. The methods of the Association of Official Agricultural Chemists call for this at 25°/25° and designate it the apparent specific gravity. If determined at other than the standard temperature, correct to 25°/25° by the formula

$$G = G' + 0.0007(T - 25°)$$

in which G = specific gravity at 25°, G' = specific gravity at T/25°, T = temperature at which the specific gravity was determined, and 0.0007 = mean correction for 1°.

For oils that are solid at 25°, take the specific gravity just above the melting point as compared with water at the same temperature. This will usually be 40°/40°.

Refractive Index. This is one of the simplest constants to determine on a saponifiable oil and as Table 15 shows is in many cases quite distinctive. The determination has been fully described on pages 53 to 57.

Acid Value. The acid value, sometimes called the acid number, is obtained by direct titration and shows the amount of free fatty acid present. If any titratable acid other than a fatty acid is present, it will be in error; and if the molecular weight of the fatty acid is not known, the value may be misleading.

Procedure. Weigh 20 grams of oil or fat into a flask, add 50 ml. of 95 per cent alcohol that has been neutralized to the phenolphthalein end point with 0.1N alkali solution, and heat to boiling. Shake or swirl the flask thoroughly in order to dissolve the free fatty acids as completely as possible. Titrate with 0.1N alkali solution, shaking or swirling thoroughly after an end point is first reached, until the pink color persists after vigorous shaking or swirling. Oils containing much fatty acid when titrated with alkali give soaps that precipitate on cooling and occlude some of the solution. Heating dissolves the soap so that titration may be completed. For consistency always keep the solution hot, even when no precipitation occurs, completing the determination near the boiling point.

On occasion, fugitive end points are obtained. This may be due to gradual extraction of the fatty acid or to saponification of sensitive oils or fats. Judgment must be exercised in determining which is the cause. As a general rule, consider the end point to have been reached when the pink color of the alkaline phenolphthalein persists for 30 seconds in the hot solution.

While applicable in general, this routine procedure is subject to numerous modifications. When the available quantity of oil is limited, a 5-gram sample or less is adequate to obtain results, necessarily with a somewhat lower order of accuracy. If the sample is largely fatty acid, reduce the sample to not over 5 grams; satisfactory results are obtainable with 1 to 2 grams.

Except when the sample consists solely of fatty acid or castor oil, which are soluble in alcohol, a mixed solvent is advantageous. Mix equal parts of 95 per cent alcohol and benzene, heat to boiling, and titrate to neutrality to phenolphthalein with 0.1N alkali. Use as in the routine method in place of neutralized alcohol. Fats and oils not completely soluble in alcohol will dissolve in this solvent and thus give a clearer end point.

With dark oils there is sometimes difficulty in detecting the end point. To minimize this, use as large an amount of solvent as feasible, in a conical flask. Add 1 ml. of the phenolphthalein solution, and proceed to add the alkali solution dropwise from a buret, with one hand. At the same time swirl the contents of the flask with the other hand sufficiently vigorously so that a thin layer of solution constantly appears on the side of the flask above the main body of the liquid. This is more easily carried out with a widemouthed flask, for the sides slant less. With experience one can easily detect a change in the color of this thin layer that would not be detectable in a thicker layer. While a spot plate may be used for tests at intervals, such a procedure is more cumbersome and experience indicates that it is less accurate.

Calculation. There are three forms of expression of results applicable under varying conditions. The acid value, or acid number, is the number of milligrams of alkali calculated as potassium hydroxide required to neutralize the free fatty acids in 1 gram of oil or fat.

$$\text{Ml. of } 0.1N \text{ alkali} \times \frac{5.61}{\text{weight of sample}} = \text{acid value}$$

If, in titration of fatty acid samples or samples that are high in fatty acids, 0.5N alkali is used, the factor is 28.05.

The amount of fatty acid present may be calculated in percentage, in which case it is usually calculated as oleic acid.

$$\text{Ml. of alkali} \times \text{normality} \times 0.2825 \times \frac{100}{\text{weight of sample}} = \text{per cent fatty acids as oleic acid}$$

As stearic acid, the factor would be 0.2845. Actually, calculation as oleic acid is within the accuracy of the method unless lower molecular weight fatty acids such as myristic or lauric acids are present. Expression in terms of some other fatty acid is occasionally resorted to.

Degree of acidity is an arbitrary value expressed as the milliliters of normal alkali required to neutralize the fatty acids in 100 grams of fat or oil.

Ml. of alkali × [illegible] = degree of acidity in ml.

Saponification Value. This is a measure of the amount of saponifiable matter present. It is also called the saponification number or the Koettstorfer number. It is expressed in terms of milligrams of potassium hydroxide consumed per gram of sample. It follows that, the higher the molecular weight of the material being saponified, the lower the saponification value. Therefore the value serves a dual purpose. If one knows that the sample is completely saponifiable, it gives an indication of the average molecular size of the material. Conversely, since the majority of saponifiable oils have saponification values of the order of 190 to 195, it is useful to estimate the approximate amount of oil in a fraction isolated from a sample. In making such interpretations it must be remembered that castor oil has a saponification value of around 180 and coconut and palm-kernel oils around 260. It follows that the molecular weight of the fatty glyceride of castor oil is higher than normal and that of coconut oil lower than normal. Since fatty acids form soaps by neutralization and fatty glycerides saponify to form soaps and glycerin, it follows that the saponification value must always equal or exceed the acid value since the saponification value includes the neutralization of fatty acid with the saponification of glyceride. When the acid value and saponification value are substantially identical, it follows that fatty glycerides are absent. When present in a sample, volatile esters such as ethyl, butyl, and amyl acetates, being of relatively lower molecular weight, will greatly increase the saponification value. Synthetic resins, which in general are polymers of low molecular weight unsaturated compounds, also have high saponification values. Therefore, the method has many applications other than to saponifiable oils.

Procedure. Weigh an accurate sample of about 5 grams into a 250-ml. conical flask. Pipet 50 ml. of approximately $0.5N$ alcoholic potassium hydroxide solution into the flask and an equal volume into a similar flask containing no sample. It is a convenient expedient to suck up the solution with a filter pump, since it is highly caustic. Drain the pipet for the same length of time into each flask. Connect the flasks to air condensers, and boil for not less than 30 minutes, longer if necessary to saponify the sample completely. Cool the sample and blank, and titrate with 0.5*N* hydrochloric acid. Note that the alcoholic potassium hydroxide need not be of exact normality since the titration is obtained by difference from the blank. Since all glassware, even of the Pyrex brand, is attacked to some degree by alcoholic alkali, it is desirable to heat the blank in much the same way as the sample, for a little less time if there is any deviation.

There are several alternatives to fit special cases encountered in general work. If soap separates on cooling, conduct the titration of sample and blank while as near boiling as feasible. This is contrary to recommendations of the Association of Official Agricultural Chemists. However, the reasons for this hot titration in special cases are the same as those discussed under Acid Value. When mixtures of saponifiable and unsaponifiable material are present in the sample, dissolve it in 25 ml. of benzene. It is desirable to use *N* instead of 0.5*N* alcoholic potassium hydroxide under those conditions to avoid too great dilution. Longer boiling before titration is also desirable with samples high in unsaponifiable matter. Should the sample prove to be insoluble in benzene, try higher boiling aromatic solvents such as xylene, high-flash naphtha, or cumene in place of the benzene.

Some oils, particularly those which have been polymerized by heating, as in varnish making, and most waxes, are very difficult to saponify. Higher boiling solvents such as diethylene glycol or diethylene glycol monoethyl ether, known commercially as Carbitol, may advantageously be used to replace alcohol. Reaction proceeds more rapidly at elevated temperatures. Prepare a special potassium hydroxide solution as outlined (page 725) but replace the alcohol with the higher boiling solvent. If necessary, increase the strength to *N* or 2*N*. If in doubt as to whether saponification is complete, saponify three samples, boiling for 1, 3, and 5 hours, respectively. When two samples that have boiled for different periods of time check, saponification is established as substantially complete.

CALCULATION.

$$\text{(Titration of blank in ml.} - \text{titration of sample in ml.)} \times \frac{28.05}{\text{weight of sample}} = \text{saponification value}$$

Ester Value. The ester value, which is rarely used, is a relative measure of the amount of esters present. It is expressed in the same terms as the saponification value and the acid value and is obtained by subtracting the latter from the former. The saponification value includes both saponification of the esters and neutralization of the fatty acids.

Iodine Value. The iodine value, or iodine absorption number, is a relative measure of the unsaturated bonds present in the oil or fat. It is the percentage of halogen absorbed, calculated in terms of iodine. The reaction takes place by addition of halogen to unsaturated bonds; and as there is no assurance that reaction is complete, it is necessary to standardize the method rigidly. Therefore, since there are two recognized methods in use, it is desirable for results to specify not only the iodine value obtained but the method used. Results by the two methods are similar, and both are official with the Association of Official Agricultural Chemists.

It is important to consider the history of a sample of fat or oil before obtaining the iodine value, for misleading conclusions are thus avoided. As a means of identification of a fat or oil, the iodine value must be applied to a sample that has not been unduly exposed to the air. Drying oils oxidize in contact with the air, and the iodine value is accordingly reduced. Semidrying oils oxidize to some extent on long exposure. Oil or fat fractions isolated during an analysis must be dried *in vacuo* or in an inert atmosphere to rid them of solvent if a high iodine value is expected. Otherwise, the absorption of oxygen during evaporation of the solvent may proceed to such an extent that the iodine value is not only valueless but misleading.

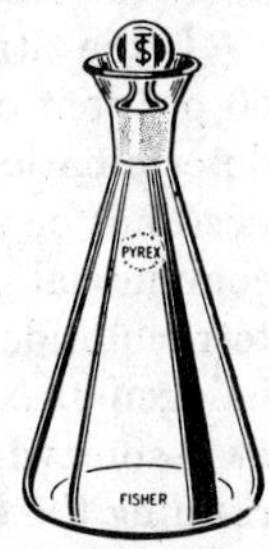

FIG. 97. Stoppered flask for determination of iodine number.

PROCEDURE. *Hanus Method.* The weight of sample to be taken depends on the degree of unsaturation of the oil. In the absence of preliminary information one must select a weight of sample and if more than 40 per cent of the halogen is consumed repeat with a smaller sample. The larger the sample within the prescribed limits of the method the more accurate the results.

Weigh accurately 0.1 to 0.5 gram of oil or fat, the higher the less the iodine value expected, into a 250- to 300-ml. glass-stoppered flask such as that shown in Fig. 97. Tiny beakers, with a capacity of 0.5 to 1.0 ml., are available under the name of "timesavers." They are convenient for weighing such samples as those for iodine values. The sample is weighed into them, and they are dropped into the reaction flask. They are recovered after the determination and may be used repeatedly. Dissolve the sample in 10 ml. of chloroform or carbon tetrachloride, warming slightly if necessary. If warmed, cool completely before adding the Hanus solution. While temperature has some effect on the iodine value, the range of normal room temperature will not introduce serious error. Transfer the same volume of chloroform or carbon tetrachloride to a similar flask containing no sample. Pipet 25 ml. of Hanus solution (page 719) into the flask containing the sample and an equal volume into the blank. Shake each flask vigorously, and let the sample stand for exactly 30 minutes, with occasional shaking. Add 50 to 100 ml. of water and 10 ml. of 15 per cent potassium iodide solution to the blank and the sample, rinsing the stopper and sides of the flask with the water.

Titrate the solutions with $0.1N$ sodium thiosulfate solution (page 730). Add the solution slowly until the yellow color nearly disappears. Shake thoroughly to extract iodine from the solvent and again titrate until the yellow color almost disappears. At this point add 1 ml. of starch solution (page 717), and titrate to disappearance of the blue color, shaking vigorously after each addition when near the end point.

If the titration of unreacted Hanus solution in the sample is not at least 60 per cent of the titration of the blank, discard the results and repeat the determination with a new sample. If this is done immediately, the blank need not be repeated. Some dark oils, particularly fractions isolated from commercial products, may impart a dark color to the chloroform or carbon tetrachloride. In such cases there may appear to be a greater content of halogen unextracted than is actually present. If so, add the starch solution at a somewhat earlier stage than usual, using precautions to avoid overrunning the end point owing to sorption of iodine by the starch.

Calculation.

$$(\text{Ml. of thiosulfate required by blank} - \text{ml. of thiosulfate required by sample}) \times \text{normality of thiosulfate} \times \frac{12.692}{\text{weight of sample}} = \text{iodine value}$$

Wijs Method. This method uses iodine monochloride in glacial acetic acid. While less used than the Hanus method, it is included because it is accepted as standard by both the Association of Official Agricultural Chemists and the American Oil Chemists Society.

Weigh out a sample as in the Hanus method, and proceed as in that method, except that the oil solution must stand in a dark place after addition of the Wijs reagent. Observe the same precautions for dark-colored oils as have already been outlined for the Hanus procedure.

Bromine Value. In the iodine value the total halogen that disappears is determined. This may be either by addition at a double bond or by substitution. The bromine value, a little-used determination, corrects for any substitution and is therefore a more accurate measure of the degree of unsaturation.

Procedure. As reagent, dissolve 26.6 grams of bromine in carbon tetrachloride, and dilute to 1 liter. This is approximately $0.33N$ and keeps indefinitely. For determination, weigh out 0.2 to 1 gram of fat or oil, and transfer to a 250- to 300-ml. iodine-value flask. Add 10 ml. of carbon tetrachloride, and warm if necessary to dissolve the sample, cooling when completely dissolved.

Accurately pipet 20 ml. of the reagent into the flask containing the sample and an equal volume into another flask as blank. Put the stoppers in place, and pour a few milliliters of 25 per cent potassium iodide solution into the space between the stopper and the flared neck. None of this should be drawn into the flask; for if it is, the glass stopper does not fit properly. At the end of 2 minutes, immerse the body of the flask containing the sample in ice water, and loosen the stopper so that the potassium iodide solution can be sucked in. Restopper, and shake to absorb gases; then remove the stopper, and add more 25 per cent potassium iodide solution to give a total of 20 to 30 ml. Titrate the liberated iodine in the same way as for the iodine value. If emulsification of the solvent obscures the end point, add 75 ml. of 10 per cent sodium chloride solution during this titration.

When the end point has been reached and recorded, add 5 ml. of a potassium iodate solution containing 21.4 grams per liter. Hydrobromic acid from substitution reactions is converted into the equivalent as free iodine. Mix well, and repeat the titration. The blank may be titrated after the sample has been completed, care being taken to use the same amount of reagents as with the sample as it is also a blank on the reagents. It gives two values, just as the main determination does, although the second one should be small.

In the addition of bromine there is no liberation of hydrobromic acid. When bromine substitutes in the compound, for every unit substituted another is converted into hydrobromic acid. Therefore twice the amount found as hydrobromic acid must be deducted from the total bromine absorbed to get the amount that is added to the sample.

CALCULATIONS.

$$(\text{Ml. of thiosulfate equivalent to liberated hydrobromic acid} - \text{blank}) \times \text{normality of thiosulfate} \times 0.0799 \times \frac{100}{\text{weight of sample}} = \text{bromine substitution value}$$

$$(\text{Blank} - \text{ml. of thiosulfate equivalent to residual free bromine}) \times \text{normality of thiosulfate} \times 0.0799 \times \frac{100}{\text{weight of sample}} = \text{per cent total bromine absorbed}$$

$$\text{Per cent total bromine absorbed} - (2 \times \text{bromine substitution value}) = \text{bromine addition value}$$

Unsaponifiable Matter. When the glycerides of a fat or oil have been saponified to soap and glycerol, the unsaponifiable fraction remains. Fats and oils normally contain little unsaponifiable matter. A few oils such as sperm oil and wool fat, or lanolin, contain large amounts, separation of unsaponifiable matter thus serving as a means of identification. Higher alcohols such as cholesterol or phytosterol are a minor unsaponifiable ingredient often of importance in determining whether an oil is of animal or of vegetable origin.

Fractions isolated from commercial products are often mixtures of saponifiable and unsaponifiable matter. Determination of the latter serves a dual purpose, to permit calculation of the corrected saponification value of the saponifiable part of the fraction and to provide a new fraction consisting of the unsaponifiable matter for further identification. The latter applies only if the amount of unsaponifiable matter is substantial, of course.

PROCEDURE. *Cylinder Method.* Weigh 5 grams of the sample into a 250-ml. conical flask, and saponify by refluxing for 1 hour or more with 30 ml. of alcohol and 5 ml. of 50 per cent potassium hydroxide solution. Alternatively, if the sample is limited, as is often the case in dealing with fractions isolated from commercial products, use the solution that has been titrated for determination of saponification value, and add about 2 ml. of 50 per cent potassium hydroxide solution to this. The alternative method also saves time and is quite generally employed. A substantial excess of alkali should be present during saponification. Only a slight but definite excess should be present to suppress hydrolysis of the soap during extraction. A large excess tends to make the washing of the extract difficult.

Transfer the saponified solution to a glass-stoppered cylinder graduated at 40, 80, and 130 ml. Wash the flask used for saponification with successive small portions of alcohol, adding these to the cylinder until the volume is 40 ml. Complete the transfer by washing the flask with a small volume of hot water and then cold water until the total volume is 80 ml. Wash out the flask with 50 ml. of petroleum ether boiling below 75°, and add to the cooled cylinder and contents. Shake the cylinder vigorously for 1 minute, and let stand until both layers are clear. Siphon off the petroleum ether layer as closely as possible into a 500-ml. separatory funnel, using a slender glass siphon. Repeat this extraction with six further portions of 50 ml. each of petroleum ether. Wash the combined petroleum ether extracts with three 25-ml. portions of 10 per cent alcohol, shaking well. Drain off the water layers thoroughly, and transfer the ether solution to a flask. Evaporate to dryness, preferably recovering the ether in a Soxhlet tube (page 69). Finally heat to constant weight in an oven at 100 to 110° or under a vacuum. Displace any ether vapors in the flask with air, cool, and weigh. Deduct any blank from this weight. The use of petroleum ether leaving no residue on evaporation will make a blank unnecessary.

If the unsaponifiable matter is substantial and is to be used for further determinations, dissolve the weighed residue in 50 ml. of petroleum ether at room temperature, filter, and evaporate as before.

Separatory Funnel Method. Use the saponified sample obtained by either of the alternatives in the preceding method.

Evaporate off most of the alcohol, and benzene if present, by immersing the conical flask in the water bath. A convenient device is to grip the neck of the flask with a large test-tube holder, which may rest on the cover of the bath when the flask is nearly full and may be weighted down to hold the flask submerged when the latter is nearly empty. Such a method of evaporation is far more rapid than placing the flask on the bath because it substantially avoids refluxing from the sides. Since the flask is heated on all sides, boiling usually proceeds smoothly. It is advisable to agitate occasionally before boiling starts to make sure this is not too violent initially. The soaps, being potassium soaps rather than sodium soaps, show less tendency to gel.

When nearly dry, dilute with hot water. If carbonation of the alkali has occurred because of long delays in the analysis, it may be desirable to add additional alkali to suppress hydrolysis of the soap. Wash the contents of the flask into a 500-ml. separatory funnel with warm water, dilute to about 300 ml., and cool. Add 75 ml. of ethyl ether. Petroleum ether boiling below 75° may be substituted unless the unsaponifiable fraction is somewhat volatile, a high-boiling naphtha, for example. In such cases, ethyl ether must be used.

Shake thoroughly. When large amounts of saponifiable matter are present, too vigorous shaking may produce stubborn emulsions. When such emulsions are obtained, add a few milliliters of alcohol from a wash bottle, just around the sides of the funnel and then on top of the emulsion. Usually a clear layer forms at once and gradually increases until complete separation occurs. Carefully drain off the water layer into another separatory funnel. If ethyl ether was used, the solubility in water will have decreased the volume to about 50 ml.

Wash the ether layer with 10 to 15 ml. of water, and add the washings to the water layer. Repeat the extraction of the water portion at least twice more with 50-ml. portions of solvent, washing each extract as before. Combine the ether or petroleum ether extracts, and wash with three 10-ml. portions of water, discarding these washings. Shake the solvent extract with 50 ml. of 1:10 hydrochloric acid to decompose any traces of soaps. Withdraw this, and wash with 10-ml. portions of water until the aqueous extract is no longer acid to methyl orange.

Drain well; and if the ether is at all turbid, filter through a dry paper. Transfer the ether to a flask, and evaporate. A few drops of water that had dissolved in the ether may be present in the flask with the extract. If put in the oven at that stage, spattering is apt to occur. When the unsaponifiable fraction is substantially nonvolatile, to eliminate this water add a few milliliters of alcohol, preferably absolute, and evaporate on a water bath. Then heat to constant weight in an oven at 110°, cool, and weigh.

When the unsaponifiable matter is appreciably volatile, substantially lower accuracy must be accepted and a special technic used. In such cases evaporate the ethyl ether on a water bath until the odor disappears. Water usually separates. Take up the unsaponifiable matter in the minimum volume of anhydrous ether, and decant from separated water. Wash the water with a few milliliters of ether. Occasionally it is necessary to filter through a dry paper. Evaporate this second ether solution on a water bath, dry the flask, and weigh. When substantial amounts of volatile unsaponifiable matter are present, it may be advisable to make a steam distillation from a solution before extraction.

Even though the extraction takes place from alkaline solution, correction should be made for fatty acids. These are in part extracted as such and in part liberated from soap by the acid wash. The amount may be large relative to a small amount of unsaponifiable matter present, as in a high-grade fat or oil. To correct, take up the weighed residue in neutral alcohol, and titrate with $0.1N$ alkali to a phenolphthalein end point. Calculate the titration to oleic acid, and deduct from the total weight of unsaponifiable matter found.

Dry Extraction. This method is not suitable for quantitative determination of very small amounts of unsaponifiable matter. It prevents the formation of troublesome emulsions in wet extractions.

Dry extraction is primarily designed for isolation of unsaponifiable matter when present as from a few per cent to a substantial part of the total. It finds real application in dealing with large quantities of material that cannot be handled in solution on a laboratory scale. Large enough quantities of unsaponifiable matter for investigation can thus be isolated, usually by combining the result of several qualitative extractions after the quantitative one.

Evaporate the saponified sample to dryness on a water bath, adding inert material such as diatomaceous earth or fine sand if needed. Grind the residue to as fine a powder as feasible. Transfer the sample or a weighed aliquot to a Soxhlet thimble. Extract for several hours with petroleum ether. The time required for extraction depends on the fineness of the ground sample and the amount of unsaponifiable matter present. Those materials with little extractable matter frequently require a long time for the solvent to penetrate the sample thoroughly and extract all the unsaponifiable material. If a high degree of accuracy is essential, wash the extract with water to remove traces of soap extracted and wash with acid to decompose soaps, following this with water washes. This step may often be eliminated.

Complete as in the case of extraction with a separatory funnel, taking precautions to prevent loss of volatile matter if present. The correction for fatty acids extracted is applied.

CALCULATIONS.

Uncorrected:

$$\text{Weight of unsaponifiable residue} \times \frac{100}{\text{weight of sample}} = \text{per cent unsaponifiable matter}$$

Corrected:

$$\text{Weight of unsaponifiable matter} - (\text{titration of fatty acid in ml.} \times \text{normality of alkali used} \times 0.2825) \times \frac{100}{\text{weight of sample}} = \text{per cent unsaponifiable matter}$$

Acetyl Value. The acetyl value is an indication of the hydroxyl groups present in an oil or fat. Of the common oils and fats, castor oil, the glyceride of ricinoleic acid, is the only one that has a high acetyl value. Acetylation takes place at the hydroxyl group and is measured by the milligrams of potassium hydroxide required to neutralize the acetic acid obtained by saponification of 1 gram of the acetylated oil or fat.

PROCEDURE. Reflux 50 ml. of freshly distilled acetic anhydride with 50 ml. of sample for 2 hours. Loss of acetic anhydride is not significant within reasonable limits, and so an air condenser is adequate. If the available sample is limited, relatively smaller amounts may be used. There is a large excess of acetic anhydride provided; greater excess will do no harm.

Pour the acetylated oil into a beaker containing 500 ml. of water, and boil for 15 minutes. To prevent bumping, bubble air or carbon dioxide through the solution or use an ebullator (page 11). Siphon off the water as completely as possible. Repeat this operation twice, which will complete hydrolysis of the acetic anhydride and substantially remove the resulting acetic acid. These operations do not attack the acetylated hydroxyl group. Cool, transfer to a separatory funnel, and drain off any residual water. Wash with two 200-ml. portions of warm water to remove the last traces of free acid. Drain thoroughly, add a few grams of anhydrous sodium sulfate, and let stand at least 1 hour, shaking occasionally. Filter through a dry paper, or decant the clear oil. If at all cloudy, dry in the oven at 110° until crystal clear.

Obtain saponification values on both the original and the acetylated oils. Any hydroxyl groups are measured. Those in free alcohols such as glycerol from partial hydrolysis of the fatty glycerides are one source of error. Such hydrolysis is usually associated with rancidity although not a cause of it. Blown or oxidized oils or those sulfated at a double bond contain hydroxyl groups and may give acetyl values approaching that of castor oil. In such cases differentiation from castor oil by the saponification and acid values is simple.

CALCULATION.

$$\frac{\text{Saponification value of the acetylated oil} - \text{saponification value of the oil}}{1 - (0.00075 \times \text{saponification value of the oil})} = \text{acetyl value}$$

This formula is derived by consideration of the increase in molecular weight of the oil after acetylation.[1]

ALTERNATIVE PROCEDURE. The acetylated oil may be saponified and the liberated acetic acid titrated. Acetylate the oil and saponify a weighed portion as in the previous method. Evaporate off the alcohol, and take up the soaps in water. Add standard sulfuric acid in an amount exactly equivalent to the alkali used to saponify the acetylated oil. Warm just enough to melt the liberated fatty acids but not enough to drive off the liberated acetic acid. Filter through a wet paper and wash with warm water until the washings are neutral. Titrate the acetic acid in the filtrate and washings to a phenolphthalein end point with standard alkali.

The same errors occur as in the preceding method. Water-soluble acids from the fat or oil will introduce additional error. After acidification in this method the acetic acid may be distilled with suitable precautions to prevent loss. This is tedious; and since the acids that are water-soluble are usually also volatile, little improvement in accuracy is obtainable. If water-soluble or volatile acids are separately determined, a correction should be introduced.

CALCULATION.

$$\text{Ml. of alkali} \times \text{normality of alkali} \times \frac{56.1}{\text{weight of acetylated oil}} = \text{acetyl value}$$

Titer Value. This term is applied to determination of the solidifying point of the mixed fatty acids derived from an oil or a fat. The titer value requires a substantial sample and is therefore not determinable on small fractions isolated from commercial samples.

PROCEDURE. Saponify approximately 75 grams of sample in a metal or porcelain dish or in a beaker. Use 60 ml. of 30 per cent sodium hydroxide solution with 75 ml. of alcohol or 120 ml. of water in place of the alcohol. The use of alcohol hastens saponification, but precautions must be taken to remove it completely. After saponifying, evaporate to dryness, with constant stirring, over a very slow flame or on a steam bath. Dissolve the dry soap in 1 liter of boiling water. If alcohol has been used to saponify, boil for 40 minutes to complete its removal. Add 100 ml. of 1:3 sulfuric acid, and boil until the liberated fatty acids form a clear transparent layer.

[1] For a more complete explanation see Leon W. Cook, *J. Am. Chem. Soc.*, **44**, 392–394 (1922).

In this operation bumping must be carefully guarded against, by stirring or by the use of an ebullator.

An alternative method of saponification is to mix 50 ml. of sample and 75 ml. of a solution containing 25 grams of potassium hydroxide to 100 ml. of U.S.P. glycerol, in an 800-ml. beaker. Heat to not much over 150° for 15 minutes, for saponification. After the solution becomes homogeneous, indicating complete saponification, mix with 500 ml. of water that is nearly at the boiling point. Add 50 ml. of 1:3 sulfuric acid, and heat with frequent stirring until the fatty acids separate as a clear layer, exercising the same precautions as in the other method.

Wash the fatty acids by decantation of the water layer and repeated washings with hot water. Filter the fatty acids through a dry paper into a dry beaker. Dry at 100° for 20 minutes. If there is doubt about saponification being complete, place 3 ml. of fatty acids in a test tube and add 15 ml. of alcohol. Boil, and add 15 ml. of 1:2 aqua ammonia. A clear solution indicates complete saponification of the sample.

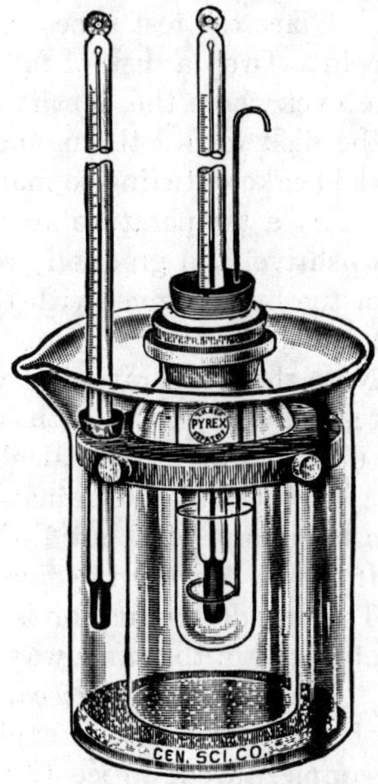

Fig. 98.—Apparatus for titer value.

Cool the fatty acids to 15 to 20° above the expected titer, and transfer to a tube 25 by 100 mm. Place this tube in a 16-ounce widemouthed bottle 70 by 150 mm. and fitted with a cork to keep the titer tube steady. This air jacket serves to slow down the rate of cooling. The apparatus is shown in Fig. 98.

The standard thermometer required reads from −2 to +66° in 0.2° units and must meet a detailed set of specifications of the Association of Official Agricultural Chemists. Suspend this standard thermometer in the fatty acids, and stir the mass slowly until the mercury remains stationary for 30 seconds. The fatty acids will then have started to solidify and liberate the latent heat accompanying that change of state. Cease stirring, and observe within 0.1° the highest point to which the mercury rises. This is the titer value of the fatty acids and therefore, by definition, of the fat from which they were derived.

Melting Point. Although the titer value is the solidifying point of the fatty acids and can be in error only by being too low, there is often occasion to obtain a melting point on the fatty acids. This is applicable to small samples.

Procedure. *Wiley Method.* This uses one physical property in determining another, the point at which the fat assumes a spherical form being taken as its melting point. Prepare four disks 1 to 1.5 cm. in diameter and weighing about 200 mg. by allowing the melted and filtered fat to fall a distance of 15 to 20 cm. upon a piece of ice or a cold mercury surface. Remove when solid, and allow to stand 2 to 3 hours before testing.

Boil 95 per cent alcohol and water separately to drive out dissolved gases. While it is still hot, nearly half fill 3.5- by 30-cm. test tubes with the water. Pour the alcohol down the sides of the tubes so as to avoid much mixing. The lighter liquid, alcohol, floats on the surface of the water with more or less mixing at the interface to give tubes of solutions of graded specific gravity.

Place the test tubes in a beaker containing ice and water, and leave till cold. Drop a disk of fat into one of the prepared tubes. It will sink to a level where the density is equivalent to its own. Stir gently just above the disk with a thermometer reading to 0.1°, and slowly heat the water in the beaker, stirring to maintain an even temperature.

At a temperature about 6° below its melting point, the fat will begin to shrivel and gradually roll up. Lower the thermometer until the center of the bulb is level with the fat mass, and rotate it slowly. Regulate the heat so that 10 minutes is required for the last 2° rise in temperature. Note the temperature at which the mass becomes spherical. Remove the test tube from the bath, and cool the bath by placing in it a second test tube containing the alcohol and water mixture and a fresh disk of fat. After the first determination, regulate the temperature of the bath so that a maximum of about 1.5° above the melting point of the fat is obtained. If a disk touches the side of the test tube, the determination is valueless. The first determination is only an approximation, but two additional values obtained in the same way should agree within a small fraction of a degree.

Capillary-tube Method. By careful suction draw a column of 1 to 2 cm. of melted fat into a capillary tube. Seal one end without burning the sample. Cool on ice 12 to 15 hours. Attach the tube by a small rubber band to a thermometer so that the bulb is level with the fat in the tube. Suspend in a beaker or test tube of water, and heat very slowly. Take as the melting point the temperature at which the fat becomes transparent. This is a simple modification of standard capillary-tube technic (page 50).

Peroxide Value. Traces of reactive peroxides in oils and fats appear to be proportional to the degree of rancidity of the fat.[1]

PROCEDURE. Weigh 1 gram of the oil or fat into a 150- by 25-mm. test tube, and take another tube as blank. Add 1 gram of powdered potassium iodide, followed by 20 ml. of a 2:1 by volume mixture of glacial acetic acid and chloroform or carbon tetrachloride. Add 1 or 2 grams of sodium

[1] LEA, C. H., Rancidity in Edible Fats, *Food Investigation Special Report* 46, Department of Science and Industrial Research, His Majesty's Stationery Office, p. 107, London, 1938.

bicarbonate, or pass in carbon dioxide or nitrogen gas, using a rubber stopper as closure. This precaution to remove the air present, which may oxidize the sample, may be dispensed with if the next step is performed promptly. Boil the tubes vigorously, shaking well to break up any undissolved sample. As soon as the chloroform vapors reach the tops of the tubes, insert a plug of glass wool and plunge into a boiling water bath. Shake for a few seconds, remove, and cool under running water. Pour the contents into conical flasks containing 30 ml. of water. Rinse with a few milliliters of water, and titrate the liberated iodine, using 0.02*N* sodium thiosulfate solution (page 730) with starch solution added near the end as indicator.

The peroxide value is the millimoles of peroxide present in the sample per kilogram of oil, which is the same as the titration of the sample in milliliters of 0.002*N* thiosulfate divided by the weight of the sample in grams.

Experimentally, this technic is very satisfactory for comparing the relative susceptibility of oils to development of rancidity. For this determination, absorb a suitable weight of sample on a fat-free filter paper, and suspend this in an oven at 70 or 100°. The time of exposure to give substantial reproducible values must be determined experimentally; the less the unsaturation, the longer the time. After exposure of the samples for the time established experimentally, transfer the filter paper to the test tube, and proceed according to the technic outlined.

Soluble and Insoluble Acids. The fats containing substantial amounts of low-molecular-weight fatty acids such as butterfat are most readily identified by their relative content of water-soluble and water-insoluble acids.

PROCEDURE. Place the flask in which the saponification value has been determined in a water bath, as described for unsaponifiable matter, and evaporate the alcohol. Add an amount of 0.5*N* hydrochloric acid equivalent to the corrected titration of the saponification value plus 1 ml. more. This converts the soaps produced in determining the saponification value into their free fatty acids and provides a known excess of free acid. Heat until the fatty acids form a clear layer on the surface of the liquid. Fill the flask with hot water to float the fatty acids on the top; then cool in ice water until they are solid. Break the fatty-acid layer, and decant the water through a filter paper into a large conical flask. Add more hot water to the original flask to melt the fatty acids, and again chill and decant the water layer through the filter paper. Repeat the whole procedure three more times. Titrate the combined filtered washings with 0.1*N* standard alkali, using phenolphthalein as indicator. When in the calculation 5 is subtracted, it is equivalent to the excess of 1 ml. of 0.5*N* acid in terms of 0.1*N* alkali.

Calculation.

$$(\text{Ml. } 0.1N \text{ alkali} - 5) \times 0.0088 \times \frac{100}{\text{weight of sample}}$$
$$= \text{per cent soluble acids calculated as butyric}$$

Insoluble Acids. The insoluble acids remaining from the previous determination can be isolated and determined. This is the Hehner number. Because the soluble acids are arbitrarily calculated as butyric, the sum of soluble and insoluble acids will not necessarily be 100 per cent.

Procedure. Drain the flask containing the bulk of the insoluble fatty acids in the above determination and the filter paper through which the soluble fatty acids were filtered, and allow to dry for 12 hours or overnight. Transfer as much of the solid acids as possible to a weighed wide-mouthed conical flask. Wash the original flask with hot absolute alcohol, and pour through the filter paper, catching the washing in the weighed flask. Thoroughly wash the filter paper with the alcohol. Evaporate the alcohol, dry for 2 hours at 100°, cool in a desiccator, and weigh. Heat again for 2 hours, cool, and weigh. Repeat until there is little decrease in weight.

Calculation.

$$\text{Weight of insoluble acids} \times \frac{100}{\text{weight of sample used for saponification value}}$$
$$= \text{per cent insoluble acids}$$

Water-soluble Volatile Acids. As a further refinement on the differentiation between soluble and insoluble acids the volatile acids may be separated and their soluble portion reported. As outlined below, this is the Reichert-Meissl value. The result depends strictly on the conditions of determination, and therefore each detail must be scrupulously followed.

Procedure. Accurately weigh 5 grams of the sample into a clean, dry 250- to 300-ml. flask. Add 2 ml. of 1:1 sodium hydroxide solution and 18 ml. of glycerin. Heat until complete saponification occurs, when the mixture becomes clear, shaking as necessary to prevent foaming. Add 135 ml. of boiled water, a little at a time at first to prevent foaming, and then 6 ml. of 1:4 sulfuric acid together with a few pieces of pumice stone previously well ignited and then plunged into water to prevent air from reentering the pores.

The distillation is conducted with special equipment illustrated in Fig. 99. The receiver is a flask calibrated at 100 ml. and 110 ml. Rest the flask to be used for distillation on an asbestos board having a hole 5 cm. in diameter in its center. Connect to the distillation setup before heating.

Distil slowly, regulating the flame so as to collect 110 ml. of distillate in about 30 minutes, the distillate being no hotter than 18 to 20° when entering the receiving flask. Turn out the flame, remove the calibrated flask, and immediately replace with a 25-ml. cylinder to catch any further drops. Mix the 110 ml. of distillate without too violent shaking, chill in water at 15° for 15 minutes, and filter through a dry 9-cm. filter paper. Titrate 100 ml. with 0.1N sodium hydroxide solution using phenolphthalein indicator. Run a blank determination using all the chemicals employed. This is especially necessary if the alkali used is not quite free from carbonate, for carbon dioxide will distill over and be titrated by the standard alkali.

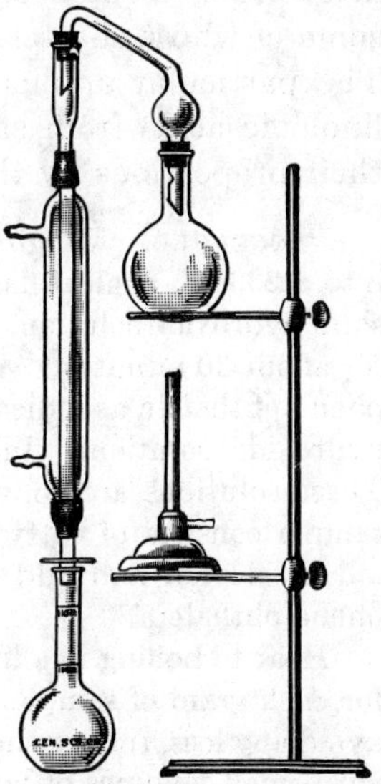

FIG. 99.—Apparatus for determination of water-soluble volatile acids.

CALCULATION.

(Ml. of alkali solution used by sample
 − ml. of alkali solution used by blank)
× normality × 10 × 1.1 = Reichert-Meissl value.

Water-insoluble Volatile Acids. The volatile acids that were not soluble in water in the determination of the Reichert-Meissl value may be titrated as the Polenske number. Details as given in the following must be observed exactly:

PROCEDURE. Wash the condenser, 25-ml. graduate, receiving flask, and filter paper with three successive 15-ml. portions of water, and discard these rinsings. Wash the condenser, graduate, flask, and filter paper with 15-ml. portions of neutralized alcohol until the fatty-acid residues have been completely dissolved. Titrate these combined alcoholic washings with 0.1N sodium hydroxide solution, using phenolphthalein as indicator. This titration expressed in milliliters of 0.1N alkali is the water-insoluble volatile acids, or Polenske number.

CALCULATION.

Ml. of alkali used × normality × 10 = Polenske number

Ratio of Saturated and Unsaturated Fatty Acids. The iodine value of a fat or oil gives a measure of unsaturation; but since an acid with two double bonds will be expected to absorb twice as much halogen as one with a single double bond, the iodine value does not permit accurate estimation of the ratios of the two types of acids. Even though this value is finally used for a correction, much more accurate results can be obtained if saturated and unsaturated acids are first separated.

Lead salts of many unsaturated fatty acids are soluble in ethyl ether, while lead salts of many saturated fatty acids are not. The method is not applicable in the presence of erucic, elaeostearic, chaulmoogric, hydnocarpic and related acids; to hydrogenated products containing notable amounts of iso-oleic acid; or to the fatty acids from coconut and palm-kernel oils the lead salts of some of whose shorter chain saturated fatty acids are ether-soluble. The particular applicability is to separation of oleic, linoleic, and linolenic acids from saturated acids and subsequent estimation of their proportions by the thiocyanogen number.

PROCEDURE. Weigh accurately 10 or 20 grams of oil or fat sample into a 200-ml. conical flask. Add 30 ml. of alcohol and 8 ml. of 1:1 potassium hydroxide solution. Mix thoroughly, and saponify on the steam bath for about 30 minutes. Add a slight excess of 10 per cent acetic acid, using phenolphthalein as indicator, and then add sufficient 15 per cent potassium hydroxide solution, while rotating the flask, to produce a pink color. These solutions are conveniently added from a buret or a pipet. If the sample consists of fatty acids, dissolve the weighed sample in about 30 ml. of alcohol and add potassium hydroxide solution until just pink to phenolphthalein.

Heat to boiling in a liter flask 12 ml. of 10 per cent lead acetate solution for each gram of sample. Add the neutralized soap solution cautiously to avoid any loss, rinsing the saponification flask with 5 ml. of alcohol and then with small volumes of hot water.

Boil the mixture gently for about 5 minutes, shake thoroughly, and cool under running water, rotating the flask to cause all the precipitated lead soaps to adhere to the sides and bottom of the flask. When the mixture is cold, pour off the aqueous solution into a large beaker in order to examine the solution for particles of lead soap. Usually the liquid is slightly turbid, owing to some basic lead acetate, and no particles or globules of lead soap are seen. Should any be found, they must be quantitatively transferred to the flask and caused to adhere to the sides. Wash the flask and lead soap twice with cold water, and invert the flask carefully and allow it to drain for 10 minutes. Remove the last drops of water by means of a thin roll of filter paper held by forceps, being careful to press the paper only lightly against the precipitate. Add about 120 ml. of ethyl ether, and shake by rotating the flask about 5 minutes.

Boil the contents gently under a reflux condenser until the lead soap is completely disintegrated or dissolved. Remove the flask, and rinse down the sides with sufficient ether to make a final volume of about 150 ml. Invert a close-fitting beaker over the neck of the flask and place in an icebox for at least 15 hours. Place a 7-cm. qualitative filter paper in a 7.5-cm.-diameter Büchner funnel, turn on full suction, and fit an 8-cm. hardened filter paper as snugly as possible to the sides of the funnel. Decant the ether solution from the separated lead soaps, using only sufficient suction to draw the ether through the filter. Too much suction causes ether to evaporate so rapidly that the filter may become clogged with separated unsaturated acids, lead soaps, or ice.

Transfer the precipitate to the filter by rinsing the flask with small portions of ether. Keep the funnel covered as much of the time as possible, to prevent evaporation of the ether. If at any time filtration proceeds so fast as to cause the mass of lead soap to crack, close the cracks by pressing with a spatula or spoon; otherwise, the precipitate cannot be properly washed. Rinse the spoon, and wash the precipitate eight or ten times with ether, finally allowing suction to continue until the precipitate cracks into numerous pieces. Without delay, separate as much of the precipitate as possible with a spatula or spoon, and transfer it without loss to a 500-ml. separatory funnel containing about 50 ml. of ether, washing off with ether any precipitate adhering to the spoon and to the neck of the separatory funnel. Transfer the filter paper to the liter flask. Shake the contents of the separatory funnel thoroughly to disintegrate lumps of lead soap, and allow to stand for about 20 minutes. Add 30 ml. of 2:1 hydrochloric acid solution, and shake thoroughly for 2 minutes to decompose all the lead soap. Add 5 to 10 ml. of 2:1 hydrochloric acid solution to the liter flask containing the filter paper, and shake thoroughly to decompose any precipitate adhering to the flask and paper. Wash into the separatory funnel with small alternate portions of ether and water until all the fatty acids and lead chloride are removed from the flask. Again shake the contents of the separatory funnel with a rotary motion, and allow to stand for 10 minutes. Withdraw the aqueous lower layer slowly, taking precautions not to remove any emulsion or undecomposed lead soap. When undecomposed lead soap is present, shown in the form of lumps floating on top of the aqueous solution, add 10 ml. of 2:1 hydrochloric acid solution, and shake again. Add about 20 ml. of water, shake, and allow to stand until the layers have separated. Withdraw the aqueous solution, and wash the ether layer with successive 25-ml. portions of water until the washings are free from hydrochloric acid. Dehydrate the ether solution with about 2 grams of anhydrous sodium sulfate, and transfer the solution to a weighed 300-ml. conical flask. Rinse the separatory funnel and the sodium sulfate with several small portions of ether to remove all the fatty acids, taking care not to allow any of the sodium sulfate to fall into the weighed flask. Distill the ether, and heat in an oven at 110° to constant weight. Obtain the weight of these saturated acids, and save them for later investigation.

Transfer the ether solution of the soluble lead soaps to a 500-ml. or liter separatory funnel, rinsing the Büchner funnel and the filter flask with small quantities of ether. Add a mixture of 30 ml. of hydrochloric acid and 95 ml. of water, and shake with a rotary motion, ensuring thorough contact of the ether and the acid solution for 2 minutes. Allow to stand for 10 minutes, and then slowly withdraw the aqueous layer into a beaker. If drops of the ether solution are entrapped by the lead chloride precipitate and removed with it, decant the solution from the precipitated lead chloride that has settled into the separatory funnel. Rinse the beaker and precipitate with small quantities of ether, adding the washings to the separatory funnel. Rotate the contents of the funnel, and allow to stand for 10 minutes. Withdraw the aqueous layer, and wash the ether layer with successive 50-ml. portions of water until it is free from hydrochloric acid. Transfer the ether solution to a 300-ml. weighed conical flask. Distill off the ether, and place the flask in an oven at 110° for about 1 hour while passing a stream of carbon dioxide gas into the flask to prevent oxidation of the unsaturated acids. Alternatively, place the flask in a vacuum desiccator in the oven, and maintain a pressure of about 20 to 30 mm. Liability to oxidation is greater by this method than by using carbon dioxide gas, but good results may be obtained if the fatty acids are not at all volatile and the flask is heated for a minimum time. With gentle suction, considerably less time than 1 hour is normally necessary to remove all the ether. Cool in an atmosphere of carbon dioxide or in a vacuum. In the latter case before weighing very gradually let in air to avoid spattering of liquid fatty acids. Flush out the carbon dioxide gas from the flask, and weigh. Obtain the weight of the unsaturated acids.

Determine in duplicate the iodine values of both the saturated and the unsaturated fatty-acid fractions. Owing to incomplete separation, the saturated acid fraction will contain a small quantity of unsaturated acid and will give a small iodine value.

To correct for these unsaturated acids present in the saturated acid fraction, use the following formula:

$$\frac{\text{Iodine value of saturated acid fraction}}{\text{Iodine value of unsaturated acid fraction}} = \text{fraction of unsaturated acids in saturated acid fraction}$$

Subtract this value multiplied by the percentage of the impure saturated acids, as found by analysis, from the percentage of impure saturated acids, and add this same amount to the percentage of unsaturated acids actually determined.

In applying this method the facts must be taken into account that results are duplicable to barely 1 per cent and that fats and oils vary in composition. Therefore it is applicable only to approximate determination of percentages of oils differing widely in composition according to Table 14.

Thiocyanogen Number. Thiocyanogen is absorbed somewhat like halogen, but with the difference that only one of the two double bonds in linoleic and two of the three double bonds in linolenic acid react. The thiocyanogen number in conjunction with the Hanus iodine value therefore permits approximation of the percentages of oleic, linoleic, and linolenic acids. The variations in composition of commercial fats and oils must be considered.

Procedure. Weigh 0.1 to 0.2 gram of oil or fat into a glass-stoppered flask such as is used for determination of iodine value. The weight of sample should be such that the titration of excess is 50 to 60 per cent of the total added. Set aside two similar flasks for blanks. To the sample and each blank add 25 ml. of the thiocyanogen reagent (page 720). Rotate the flask containing the sample until homogeneous, and allow to stand in the dark for 20 to 24 hours. Add 10 ml. of 10 per cent potassium iodide solution quickly and at one time, and shake the bottle to avoid hydrolysis of the thiocyanogen solution. Add 100 ml. of water, and titrate the liberated iodine with $0.1N$ sodium thiosulfate solution, using starch indicator. Similarly, titrate the blanks, which should require 24 to 26 ml. of $0.1N$ sodium thiosulfate solution. After determination of the difference in titrations, it is expressed in terms of iodine.

Calculations.

$$(\text{Titration of blank in ml.} - \text{titration of sample in ml.}) \times \text{normality of thiosulfate} \times 0.1269 \times \frac{100}{\text{weight of sample}} = \text{thiocyanogen number}$$

Using the cyanogen number together with the Hanus iodine value, calculate the composition of oils or fats composed of glycerides of oleic, linoleic, and saturated acids by the following formulas:

$$\begin{aligned} x + y + z &= 100 - \text{per cent unsaponifiable matter} \\ 173.3x + 86.1y &= 100 \times \text{Hanus iodine value} \\ 86.7x + 86.1y &= 100 \times \text{thiocyanogen number} \end{aligned}$$

where x = per cent linoleic acid glycerides, y = per cent oleic acid glycerides, and z = per cent saturated acid glycerides.

If the percentage of unsaturated acids present in an oil has been determined, calculate the percentages of oleic, linoleic, and linolenic acids present by the following formulas:

$$\begin{aligned} a + b + c &= \text{per cent unsaturated acids} \\ 273.7a + 181.2b + 89.9c &= 100 \times \text{Hanus iodine value} \\ 167.1a + 96.7b + 89.3c &= 100 \times \text{thiocyanogen number} \end{aligned}$$

where a = per cent linolenic acid, b = per cent linoleic acid, and c = per cent oleic acid.

It will be noticed that the iodine and thiocyanogen values for the fatty acids are somewhat higher than the values for the corresponding glycerides.

Cholesterol and Phytosterol in Mixtures of Animal and Vegetable Fats. Cholesterol and phytosterol are found in animal and vegetable fats, respectively. It is frequently important to distinguish animal from vegetable fats. This is done by determining the presence of one or the other of these sterols or, in mixtures of animal and vegetable fats, of both. They are both unsaponifiable and being alcohols may be acetylated. Cholesteryl acetate melts at 114°, and phytosteryl acetate melts at 125 to 137°. In the method given the sterols are extracted with alcohol. Any saponifiable matter is saponified and removed. In another method, not given, the sterols are precipitated with a solution of digitonin in alcohol and identified as the acetates.

PROCEDURE. *Alcohol Extraction Method.* Introduce 200 to 300 grams of the oil or melted fat into a flatbottomed liter flask. If only a restricted sample is available, reduce the amount of reagents used accordingly. As the amount of unsaponifiable matter is usually small, always use as large a sample as is conveniently possible. Insert in the neck a three-hole rubber stopper. Connect (1) a reflux condenser; (2) a right-angled glass tube one arm of which reaches to a point 6 mm. above the surface of the melted fat, the other arm being closed a short distance from the flask by means of a short piece of rubber tubing and a pinchcock; and (3) a glass tube bent so that one arm reaches to the bottom of the flask and the other serves as a delivery tube to a 700-ml. round-bottomed flask containing 500 ml. of alcohol.

Place the flasks on a steam bath, and heat so that alcohol vapor passes through the melted fat in the liter flask and is condensed in the reflux condenser, finally collecting in a layer over the melted fat. After all the alcohol has passed in this manner into the flask containing the fat, disconnect the flask from which the alcohol layer was distilled and connect it to the tube ending just above the fatty layer. Siphon the alcohol back into this flask. Reconnect as at first, and again distill the alcohol through the fat. When all the alcohol has been distilled, siphon it again into the alcohol flask and extract in the same manner a third time.

Discard the fat, and retain the alcohol, which now contains practically all the cholesterol and phytosterol originally present in the fat. Concentrate the alcoholic solution to about 250 ml. To the boiling liquid, add 20 ml. of 1:1 potassium hydroxide solution, and boil for a minimum of 10 minutes to ensure complete saponification of any fat that was dissolved by the alcohol during the extraction. Cool to room temperature, and pour into a large separatory funnel containing 500 ml. of warm ether. Shake to ensure thorough mixing. Add 500 ml. of water, and mix the whole thoroughly, but with gentle shaking to prevent the formation of troublesome emulsions. A clear sharp separation should take place. Draw off the lower soap layer, and wash the ether layer with 300 ml. of water, avoiding shaking, which may induce emulsification. Repeat the washings with small amounts of water until the washings show no pink coloration with phenolphthalein, indicating that all the soap is removed from the ether layer. Transfer the ether solution to a flask, and distill off the ether to a volume of about 25 ml. Transfer this residue to a 50-ml. beaker, and evaporate to complete dryness. If desired, the unsaponifiable matter so obtained may now be weighed.

Add 3 to 5 ml. of acetic anhydride to this unsaponifiable matter, cover the beaker with a watch glass, and boil for a few seconds over a free flame. Remember that acetic anhydride is volatile and very inflammable. Cool, add 35 ml. of 60 per cent by volume alcohol, and mix thoroughly. Filter the alcoholic solution, and wash the precipitate with 60 per cent alcohol in which the sterol acetates are insoluble. Dissolve the precipitate on the filter with a stream of hot 80 per cent by volume alcohol, and wash any insoluble residue with the 80 per cent alcohol. Acetates of cholesterol and phytosterol are dissolved in the 80 per cent alcohol, leaving on the filter the greater portion of any impurities, including paraffin and paraffin oil, should they be present. Cool the combined 80 per cent alcohol filtrate and washings to 10 to 12°, and allow to stand at that temperature for 2 to 3 hours. Collect the crystallized acetates of cholesterol and phytosterol upon a filter, wash with cold 80 per cent by volume alcohol, and then dissolve in a minimum quantity of hot absolute alcohol. Collect the alcoholic solution of the acetates in a small glass evaporating dish, add 2 or 3 drops of water, and heat if not perfectly clear. Allow the alcohol to evaporate spontaneously, stirring contents of dish occasionally to mix the deposit of crystals that form upon the edges with the main body of liquid. As soon as a good deposit of crystals has formed, collect them upon a hardened filter; wash twice with cold 90 per cent alcohol and dry by suction. Dry finally at 100° for 30 minutes. Determine the melting point by the capillary-tube method (page 50) using glycerin as the heating medium.

The melting point of the first crop of crystals usually gives definite information as to the presence or absence of phytosterol, but the conclusion indicated should be confirmed by recrystallizing from absolute alcohol and again determining the melting point. If the crystals are pure cholesteryl acetate, the melting point of the second crop should agree closely that of the first, about 114°. If phytosterol acetate is present, however, a higher melting point will be noted, for phytosteryl acetate is less soluble in alcohol than cholesteryl acetate, approaching 125°.

If the unsaponifiable matter is known to be wholly cholesterol and/or phytosterol, crystals may be obtained directly from an alcoholic solution. Cholesterol crystallizes from either alcohol or ether in crystals that contain one molecule of water of hydration, as leaflets or rhomboidal tables. Phytosterol crystallizes from alcohol in crystals that contain one molecule of water of hydration, as needles forming stars or bundles.

Maumené Number. This determination is a measure of the heat evolved on mixing sulfuric acid with the oil being examined. As, under this treatment, unsaturated fatty oils evolve more heat than saturated ones, this test is sometimes used in the comparison of samples, as a simple substitute for the iodine value, for it is more rapid and convenient to perform.

The strength of sulfuric acid may be varied in comparative tests. This is especially desirable in dealing with the highly reactive drying oils, which may give a relatively violent reaction. Alternatively, half the sample of oil may be replaced by mineral oil.

Results are also sometimes expressed as the specific-temperature reaction, which is the rise with the oil divided by that with water and multiplied by 100.

Procedure. Weigh 50 grams of sample into a 200- to 250-ml. beaker, and note the temperature. Add 10 ml. of concentrated sulfuric acid or sulfuric acid of selected lower concentration, measured at the same temperature, from a pipet, and allow it to drain for 5 seconds. Meanwhile, stir well with a thermometer, and note the highest temperature at which the thermometer remains constant for any appreciable time.

Calculations.

$$\text{Final temperature} - \text{initial temperature} = \text{Maumené number}$$

$$\frac{\text{Temperature rise with oil}}{\text{Temperature rise with water}} = \text{specific-temperature reaction}$$

In comparative tests when the test oil has been diluted with an equal volume of mineral oil, the following calculation is used:

$$\left(\text{Temperature rise with mixture} - \frac{\text{temperature rise with mineral oil alone}}{2}\right) \times 2 = \text{Maumené number under these specific conditions}$$

Identification of Hydrogenated Oils. Oils are converted to more or less hardened fats with different constants according to the degree to which they are hydrogenated. A nickel catalyst is used for this purpose, and traces of catalyst are detectable in the product.

Procedure. Reflux 20 grams of fat with 25 ml. of 1:1 hydrochloric acid for several hours. Filter through a wet paper to separate the acid extract. Evaporate the filtrate to a small volume, and destroy dissolved organic matter by adding a few milliliters of concentrated nitric acid. Evaporate almost to dryness, take up with water, and test qualitatively for nickel with dimethylglyoxime (page 99).

TABLE 15.—PHYSICAL AND CHEMICAL CONSTANTS OF FATS AND OILS

	Solidifying point, °C. glycerides	Titer of fatty acids	Specific gravity		Refractive index		Saponification value	Acid value		Iodine value		Acetyl value	Reichert Meissl number	Maumené number	Unsaponifiable matter, %	
			Value	Temp., °C.	Value	Temp., °C.		Glycerides	Fatty acids	Glycerides	Fatty acids					
1. Almond oil	−15 to −20	...	0.914–0.921	15	1.4592–1.4646	40	183–207	0.5–3.5	196–207	93–103	93–96.5	9.6	0.5	51–54	0.75	1
2. Apricot-kernel oil	−17	...	0.915–0.926	15	1.4636–1.4705	25	191–198	3.5	197	100–109	99–108	12.2	0.2	42.5	...	2
3. Babassu oil	...	...	0.869	100	...	...	246–250	...	...	12–16	...	...	...	...	0.5	3
4. Bayberry wax	25	...	0.880	100	1.4783	25	198–199	26.3	...	68–80	82	...	1.6	116	...	4
5. Beechnut	−17	17	0.922	15	1.4696	25	191–196	...	...	97–111	114	...	...	64	...	5
6. Beef tallow	27 to 38	38–40	0.943–0.952	15	1.4582–1.4587	40	193–200	0.25	197–200	35–46	41.3	2.7–8.6	0.25	...	...	6
7. Bone fat	15 to 17	39–43	0.914–0.916	15	...	...	185–198	...	...	46–56	...	11.3	0.2–1.7	...	0.5–1.8	7
8. Butter fat	20 to 23	33–39	0.935–0.940	15.5	1.459–1.462	25	216–233	0.5–35.4	210–233	26–38	28–52	1.9–8.6	17–34.5	...	0.33–0.56	8
9. Castor oil	−10 to −18	3	0.950–0.968	15.5	1.4799–1.4803	15	177–186	0.1–0.8	192	83–88.5	87–93	146–150.5	1.4	46–47	0.6	9
10. Chaulmoogra oil	20 to 25	39.6	0.943–0.954	15	1.4777–1.4779	25	196–213	0.8–21.5	215	97.6–110.4	103	...	...	...	2.4–2.6	10
11. Chinese vegetable tallow	24 to 34	45–47	0.913–0.918	15	1.4510	...	198–206	2.4	200–208	28–50	30–55	...	0.2–0.9	...	...	11
12. Coca butter	21 to 23	20–23	0.950–0.976	15	1.456–1.458	40	192–202	1.1–1.9	190–198	32–41	33–39	2	0.3–1	...	...	12
13. Coconut oil	14 to 22	21–25	0.925–0.935	15	1.4481–1.4497	40	246–268	2.5–10	258–272	8–10	8.4–9	2.3–6.9	6.6–7.5	21	0.2	13
14. Cod-liver oil	−10 to −0	17–24	0.920–0.925	15.5	1.4800–1.4852	15	182–189	5.6	204–207	135–168	140–170	1.15	0.2	102–115	0.5–9.9	14
15. Cod oil, Newfoundland	...	...	0.925–0.935	15.5	...	...	About 190	...		About 129	...	...			...	15
16. Corn oil	−10 to −20	18–20	0.920–0.926	15.5	1.4760–1.4768	15.5	188–193	1.4–2.0	198.4	113–129	113–130	7.5–11.5	4.6	74–86	1.35–2.86	16
17. Cottonseed oil	−1 to 0	17–19	0.922–0.930	15	1.4737–1.4757	15.5	191–198	0.6–0.9	202–208	104–115	111–115	21–25	0.95	75–90	0.73–1.64	17
18. Cottonseed stearin	16 to 22	32–38	0.919–0.923	15	1.4700–1.4725	25	195	0.10	...	89–94	94	...	0.22	...	...	18
19. Croton oil	−8 to −18	17–19	0.937–0.943	15	1.4757–1.4768	27	210–215	27–31	201	102–107	112	19.8–38.6	12–13.6	...	0.55	19
20. Dolphin oil	−3 to 5	...	0.908–0.930	15	1.4708	15	197–203	2–12	...	100–125	...	...	...	...	2.0	20
21. Halibut-liver oil	...	...	0.928	...	1.47–1.48	...	...	...	...	114–161	...	...	...	...	6.34–17.6	21
22. Hempseed oil	−15 to −28	15.6–16.6	0.925–0.928	15	1.4740–1.4745	40	192–195	0.5	...	148–160	141	...	...	97	1.08	22
23. Herring oil	...	23–27	0.920–0.939	15.5	1.4665–1.4729	25	171–194	1.8–44	179	123–142	...	...	...	...	0.99–10.7	23
24. Horse fat	20 to 45	34–45	0.916–0.922	15.5	1.4603–1.4717	40	195–200	0–24	203	71–86	72–87	...	1.6–2.1	46–54	...	24
25. Japan wax	40 to 46	54–59	0.970–0.989	15.5	1.4560–1.4591	40	210–237	11–12	214	5–12	...	17.3–26.5	...	...	1.1–1.6	25
26. Lanolin	38 to 40	...	0.932–0.945	15	1.4781–1.4822	40	84–127	60	...	15–29	17	23	5–8	...	39–44	26
27. Lard	27 to 30	36–43	0.935–0.940	15.5	1.4584–1.460	40	193–200	...	...	50–64	...	...	...	...	0.23	27
28. Lard oil	−2 to 4	27–33	0.913–0.919	15.5	1.4702–1.4720	15.5	195–198	0.1–2.5	...	65–80	...	2.6	0–0.2	40–47	0.6	28
29. Linseed oil, boiled	...	...	0.935–1.186	...	...	...	186	...	...	...	...	...	...	...	2.5	39
30. Linseed oil, raw	−17 to −27	19–21	0.931–0.938	15	1.4820–1.4852	15	190–195	1–3.5	196–199	171–201	179–210	...	0.95	103–126	0.4–1.9	30

TABLE 15.—PHYSICAL AND CHEMICAL CONSTANTS OF FATS AND OILS.—(*Continued*)

	Solidifying point, °C. glycerides	Titer of fatty acids	Specific gravity		Refractive index		Saponification value	Acid value		Iodine value		Acetyl value	Reichert Meissl number	Maumené number	Unsaponifiable matter, %	
			Value	Temp., °C.	Value	Temp., °C.		Glycerides	Fatty acids	Glycerides	Fatty acids					
31. Menhaden oil	−5	31–33	0.925–0.935	15.5	1.478	25	189–193	3–11		139–173			1.2	123–128	1.6–6.7	31
32. Mutton tallow	32 to 41	37–46	0.937–0.952	15	1.4531	60	192–197	1.7–14	198–210	35–46	35					32
33. Neat's-foot oil	−2 to 10	16–26.5	0.913–0.916	15.5	1.4695–1.4708	15	192–200	0.1–0.6	201–206	66–73	62–77	7.7–9.3	0.9–1.2	43–58	0.1–0.7	33
34. Oleo oil	2 to 8		0.914–0.919	15			193–198			56–80						34
35. Olive oil	−6 to 2	18–25	0.916–0.920	15.5	1.4698–1.4716	15	185–196	0.3–1.0	193–198	77–94	86–90	10.5	0.6–1.5	35–52	0.46–1.0	35
36. Oiticia oil			0.969	15			188–189			180						36
37. Palm oil	35 to 42	38–47	0.920–0.945	15.5	1.4510	60	196–205	10	204–207	51–57	53	15.7	0.9–1.9			37
38. Palm-kernel oil	21 to 28	20–23	0.925–0.935	15.5	1.4431	60	242–250	5–22	251–265	10–23	12	7.6	5–6.8			38
39. Peach-kernel oil	−20	13–13.5	0.918–0.922	15	1.4697–1.4705	25	189–193	1–1.5	201–210	93–109	94–102	6.5		42.5		39
40. Peanut oil	−3 to 3	28–32	0.911–0.922	15.5	1.4707–1.4731	15.5	186–197	0.8	202	85–103	95–103	3.5	0.4	44–67	0.54–0.94	40
41. Perilla oil		12–17	0.930–0.987	15	1.4825	15	189–197			185–206	200–211			124	0.43	41
42. Poppyseed oil	−16 to −18	17–19	0.924–0.927	15.5	1.4766–1.4774	15.5	189–197	2.5	199	133–157.5	139		0.6	71–88	0.43	42
43. Porpoise-jaw oil			0.9258	15			254–272			22–50					16–17	43
44. Rapeseed oil	−2 to −10	11.7–13.6	0.913–0.916	15.5	1.4720–1.4757	15	167–169	0.4–1.0	185	96–104	99–106	14–75	0–0.8	50–67	0.58–10	44
45. Salmon oil			0.9258	15.5			182–188			161.4						45
46. Sardine oil	20 to 22	28–34	0.927–0.933	15	1.4802–1.4808	20	189–196	4–25	177–185	170–193		21–22	0.5–1.0		0.5–1.4	46
47. Seal oil	−2 to −3	13–17	0.924–0.926	15	1.474	25	189–196	1.9–40	190–198	127–159	186–202	33–34	0.2		0.38–1.4	47
48. Sesame oil	−4 to −6	23–32	0.921–0.925	15.5	1.4748–1.4762	15.5	188–193	9.8	196–202	103–114	109–116		1.1–1.2	61–68	0.95–1.32	48
49. Shark-liver oil			0.910–0.919	15			161–164			114–144					5.5–10.2	49
50. Shea butter			0.917	15			179–192			54–62					5–9	50
51. Soybean oil	−10 to −16	20–21	0.920–0.925	15.5	1.476–1.4775	15.5	190–200	0.3–1.8	198	121–148	115–140	4.9	0.5–2.8	59–87	1.3–1.5	51
52. Stearin, oleo	54 to 56		0.914–0.919	15			193–199			56–60						52
53. Sunflower oil	−17 to −19	17–20	0.924–0.926	15	1.4659–1.4721	25	188–194	11.2	202	119–135	124–134		0.5	60–75	0.31	53
54. Sweet-almond oil	−15 to −20		0.916–0.920	15.5	1.4685–1.4693	25	189–195			93–104						54
55. Tea-seed oil			0.920	15			190–194			91						55
56. Tung oil	17	37–38	0.936–0.943	15	1.5179	20	190–197	2	189–198	150–170	144–159		0.35–1.1		0.44	56
57. Turtle oil					1.4677	30	209–211									57
58. Walnut oil	−15 to −27		0.925–0.926	15	1.469–1.471	40	188–197	2.5	200	142–152	150		0.92	96–110	0.5–1.0	58
59. Whale oil	−2 to 0	22–24	0.920–0.925	15.5	1.4723–1.4749	25	188–194	1.9		110–136	131	11–23	14	85–92	0.92–3.7	59
60. Wheat-germ oil			0.9268	25	1.4762	20	186.5			125.6					4.7	60

CHAPTER 18

SULFATED OIL AND RELATED PRODUCTS

The task of the analyst given a bottle of amber liquid with the instructions "Analyze this sulfonated oil" is not difficult. If he is given a commercial product and finds by methods here indicated that it is sulfated or sulfonated, his task is much more complex.

Processing of castor oil, the triglyceride of ricinoleic acid, by mixing the oil with sulfuric acid of a carefully controlled strength at a correspondingly controlled temperature replaces the hydroxyl group to form an organic sulfuric acid ester group. Sulfonyl chloride reacts similarly. Less definite reactions give products designated as sulfonated neat's-foot oil, sulfonated olive oil, sulfonated tallow, sulfonated oleic acid, etc. The amount of combined organic sulfur trioxide varies widely as between different grades. Excess sulfuric acid is in part washed out with saturated sodium sulfate solution and in part neutralized with alkali. In this processing the sulfuric acid ester is converted to the sodium salt. In relatively rare cases the ammonium salt is present, and in even rarer cases the potassium salt. The finished commercial product, therefore, normally contains sodium sulfate, the sodium salt of the sulfated oil, the sodium salt of the fatty acid corresponding to the oil, soap, and unsulfated fatty matter. They are usually transparent liquids; a few are pasty. Occasionally, owing to lack of proper temperature control during sulfation, true sulfonates are present, and the methods provide for their estimation in the presence of organic sulfates.

Another class of compounds, of related structure but by no means so important tonnagewise, are the sulfated alcohols. These are produced by treating the fatty alcohol by a procedure analogous to the formation of sulfated oil. They contain up to 50 per cent of sodium sulfate in the commercial product and are usually white or brown solids. Their use as detergents, textile wetting agents, and emulsifying agents causes them to be widely distributed in compounded products.

When petroleum fractions are treated with sulfuric acid, addition at the double bond of the olefins present to form organic sulfates leads to a class of sulfated products of less definite composition. These are isolated as by-products and commercially are usually associated with substantial and often predominating amounts of unsulfated mineral oil.

A type of products that are structurally sulfonamides are similar in properties to those already cited. While less important than the preceding, if nitrogen is present as well as organic sulfate and is not otherwise accounted for these should be suspected. There is an analogous class of organic esters. Typical products of this class have the structures $CH_3(CH_2)_7CH:CH(CH_2)_7COOCH_2CH_2SO_2ONa$ and $CH_3(CH_2)_7CH:CH(CH_2)_7CONHCH_2CH_2SO_2ONa$ and are known as Igepon A and Igepon T, respectively.

In the discussion of this class of products the term sulfuric ester or organic sulfate will be used for materials having this structure, but it must be remembered that they are always spoken of commercially as sulfonated products. Thus the term sulfated oil, which is correct, refers to what is known industrially as sulfonated oil. This leaves the term sulfonate for those products which are true sulfonates.

True sulfonates are also important commercially, many water-soluble dyestuffs possessing this structure. Besides these, many sulfonates are used in compounded products as wetting and foaming agents. A typical commercial product of this class has a cyclic structure, keryl benzene sulfonate. In this designation, keryl refers to a side chain radical of an average molecular weight comparable with the hydrocarbons in kerosene. This type of product must be watched for if for no other reason than that they are lower in cost than sulfated alcohols, although not less expensive than sulfated oils. This also applies to true sulfonates that are synthesized from petroleum fractions. These possess the characteristics of sulfonates as distinct from the sulfated petroleum oils.

Chemically, the distinction in structure is that organic sulfates are $ROSO_2OH$, while sulfonates are RSO_2OH. The difference in stability between the carbon-oxygen-sulfur linkage and the carbon-sulfur linkage serves for qualitative and quantitative differentiation. A third form of sulfur, always present to a major or minor extent and requiring consideration, is inorganic sulfate.

QUALITATIVE

Organic Sulfates. These are distinguished by the presence of their characteristic sulfate group. If the amount present is appreciable, the product, diluted with water if originally concentrated, will usually foam on shaking. The foam is not so stable as that from a soap, although the distinction is difficult to make without considerable experience. Acidification with hydrochloric acid will at once stop foaming due to soap or sulfated oils and will usually cause fatty sulfuric esters of fatty glycerides to separate from solution, even in the cold. The fatty acid from soap usually will not separate as a definite layer without heating. Sulfated fatty glycerides and sulfated alcohols can be distinguished by this property, for the sulfuric acid esters of the alcohols liberated on acidifying are sufficiently soluble in water to foam on shaking. Thus even small amounts of sulfated alcohols foam considerably on shaking before or after acidification in the cold.

If sulfated oil or sulfated alcohol is suspected, acidify a portion with hydrochloric acid and divide into two parts. Filter one part, and test with barium chloride. Either product is normally associated with substantial amounts of inorganic sulfates. To confirm the presence of organic sulfate, boil the other half of the acidified solution vigorously for several minutes or, better, a half hour. If a considerable amount of an organic sulfate is present, a much greater precipitate is now obtained from this filtrate. This is a qualitative application of a well-recognized quantitative method, but it is difficult to apply in the presence of only a fraction of a per cent of organic sulfate.

To establish the presence of organically combined sulfur and, presumably, if not an organic sulfate, then a sulfonate, ash a portion of the original sample with Eschka mixture (page 162), extract with water, oxidize with bromine water, and test the extract for sulfate. Aside from sulfuric acid esters or true sulfonates, the presence of sulfur compounds in this test could be accounted for by various other sources, inorganic sulfate, organic sulfides, or ring compounds, for example. Sources other than sulfates or sulfonates must be rigorously excluded by other methods of testing to permit proper interpretation of this test.

The time taken for acid to hydrolyze the compound is a valuable indication of what is present. Sulfated oils in general separate out as an oil in the cold on acidification. Sulfated alcohols usually require boiling with acid for at least 1 and usually 2 to 3 hours before foaming ceases. The solution gradually becomes opaque and finally clears up, with the fatty alcohol floating on the top. On the other hand, many true sulfonates, as, for example, the keryl benzene sulfonates and sulfonated petroleum products, will not break down under normal conditions of hydrolysis with acid but continue to foam abundantly even after days of boiling. There is consequently no separation of fatty substance.

A qualitative test for ammonium compounds should be run, for some of the methods require modification to be suitable if these are present.

QUANTITATIVE

General. While the sulfated or sulfonated product may be the sole ingredient present in an unknown sample, it is also common to find them in complex proprietary mixtures for wetting out, in emulsions, in detergents, in compounds going to the tanning industry, etc. Therefore, the methods that follow must logically be modified to fit the special case. With minor differences they are simplified forms of standard methods established by the American Association of Textile Chemists and Colorists, American Oil Chemists Society, and American Leather Chemists' Association as well as those privately published by the Sulfonated Oil Manufacturers Association.

Moisture. DISTILLATION METHOD. The moisture content of sulfated or sulfonated products can be reliably determined by the usual xylene distillation method (page 41) modified as indicated below. To be directly applicable, mineral acids, free sulfonic acids, free sulfuric acid esters, alcohol, glycerol, diethylene glycol, acetone, and other water-miscible volatile compounds must be absent. Errors due to the presence of free acidity from the first three sources can be avoided by addition of a known small excess of standard alkali to the sample before distillation. The results obtained must then be corrected for the moisture so introduced. Volatile organic solvents such as alcohol and acetone will be present in the distillate with the water and can be corrected for, as discussed later (page 371).

PROCEDURE. Transfer to the distillation flask sufficient sample to yield about 4 ml. of water. This will necessarily vary widely according to the nature of the sample. Add 100 ml. of xylene and glass beads to prevent bumping. To this mixture add about 2.5 times the weight of the sample of best-quality oleic acid that has been dried by heating to 130 to 135° for 5 to 10 minutes with a free flame. This prevents both foaming and solidification of the contents of the flask.

Proceed as on page 42, starting at "Heat the flask in an oil bath . . . ". Duplicate determinations can be expected to agree within 0.05 ml. of water.

Organic solvents in the aqueous distillate are determined by specific gravity, a very small pycnometer being used (see page 43). The volume of water distilled is then correspondingly corrected. Such solvents are rarely present in commercial sulfated oils, but their presence is far from a rarity in sulfated or sulfonated fractions isolated from commercial products.

Direct Drying. This method is subject to the same limitations with respect to sample as the xylene distillation. Except for volatile solvents, the limitations can be avoided by the following procedure:

Procedure. Transfer about 5 grams of good-quality oleic acid to a 150-ml. low-form beaker. Heat this gradually over a well-spread source of heat, stirring with a small thermometer until 130° is reached. The thermometer should be a special one, about 3 inches long and calibrated for 90 to 150°. Remove from the heat, place in an oven at 105 to 110° for 15 minutes, cool in a desiccator, and weigh. Repeat until checks within 15 mg. are obtained. The heating medium so prepared will mix with the sample without gelling. The preparation removes moisture and fatty acids volatile at the temperatures used.

Weigh 5 grams of sample into the prepared beaker of oleic acid, and repeat the procedure until constant weight within the defined limit is obtained. The loss in weight is assumed to be moisture. Accuracy to about 0.4 per cent on a 5-gram sample is to be expected.

Inorganic Salts. This procedure is suitable only for accurate application to a sample of sulfated oil. A fraction isolated from a commercial product can normally be expected to have gained or lost inorganic contaminants to such a degree that the determination would have little meaning.

Procedure. Weigh a sample of 3 to 5 grams into a 250-ml. beaker. Add an equal amount of oleic acid of good quality, and heat in an oil bath at 105 to 110°, stirring with a thermometer. When the greater part of the water has been driven off, raise the temperature to about 118 to 120° and hold at that temperature for about 5 minutes. On cooling, the sample should remain liquid; if it does not, add more oleic acid and reheat until clear and liquid.

Dissolve the dehydrated sample in 100 ml. of carbon tetrachloride by warming to 50 to 55°. Inorganic salts are insoluble under these conditions. Filter through a weighed filter paper or a Gooch crucible. To meet the specifications of standard methods a Gooch crucible should have a 1-mm. layer of acid- and alkali-washed asbestos and should be dried, ignited, washed with carbon tetrachloride, and ignited again, the washings and ignitions being repeated to constant weight. In most practical work, however, this is unnecessary. Wash the residue in the beaker with three

15-ml. portions of 2 per cent solution of oleic acid in carbon tetrachloride, then with six 15-ml. portions of carbon tetrachloride at about 50°. Use these wash solutions to transfer the residue from the beaker to the filter. Finally wash the beaker and filter with two 15-ml. portions of ether; and if any evidence of oil remains, continue this until both are oil-free. Pay particular attention to the top of the crucible to ensure freedom from oil. If any residue remains in the beaker, allow the solvent to evaporate and it will normally be free-flowing.

Dry the residue, after the ether has evaporated, at 125 to 130° for 45 minutes, cool, and weigh. Ignite to constant weight, 15 minutes being usually sufficient, and weigh.

If preliminary qualitative examination showed ammonium salts to be absent, there should be no greater than 0.25 per cent difference in weight between the residue as dried at 130° and that ignited. In that case, calculate the weight of inorganic salts from the ignited weight. If ammonium salts are present, calculate the total inorganic salts from the weight on drying at 130°. The loss on ignition may be used as a rough measure of the amount of ammonium salts present in the residue, always with the reservation that interactions may occur in ignition so that this is not very accurate. Such loss is a valuable check against the amount of ammonium salt as calculated from quantitative distillation of ammonia from a diluted sample, as outlined on page 149.

Total Alkalinity. The total alkalinity of a sample of sulfated oil is a measure of the soap, free alkali, and alkaline salts present. This will normally be applicable to a sulfated fraction isolated from more complex compounds and will indicate roughly the amount of soap present. The method of isolation must be considered in interpreting it.

PROCEDURE. Dissolve 10 grams of sample in 100 ml. of water in a 250-ml. conical flask, by warming if necessary. Add 30 grams of pure sodium choride, 25 ml. of ether, and 5 drops of 0.1 per cent methyl orange solution. Add 0.5*N* sulfuric acid from a buret, with swirling, until a slight excess is present. Swirl vigorously. Back-titrate with 0.5*N* alkali added several drops at a time until the solution is just definitely alkaline. Now complete the titration to an acid end point drop by drop.

The definite excess of acid is added to ensure destruction of alkaline materials present. Alkaline salts will react, and carbonates will be decomposed. Soaps will be liberated as free fatty acid that will pass into the ether layer and permit titration of the aqueous layer without interference. Titration from the alkaline side to the acid side gives a more sensitive end point.

Express the difference in alkaline and acid titrations as the equivalent in milligrams of potassium hydroxide per gram of sample. If alkaline salts are present, they may be separately determined by known methods and checked against this total alkalinity, in some complex cases giving a determination of soap by difference.

CALCULATION.

$$(\text{Acid titration in ml.} \times N) - (\text{alkali titration in ml.} \times N) \times \frac{56.1}{\text{weight of sample}}$$
$$= \text{alkalinity in mg. of potassium hydroxide per gram of sample}$$

Free Fatty Acid. Occasionally, a sample of sulfated oil is acid. More frequently, sulfated fractions isolated from mixtures contain free fatty acid. So long as this is not partly neutralized with ammonia, it is titratable.

PROCEDURE. Dissolve a weighed sample of about 10 grams in a neutralized mixture of 50 ml. of alcohol and 25 ml. of ether. Add a few drops of phenolphthalein-indicator solution, and titrate with standard sodium hydroxide solution selected as to normality according to the probable amount of free fatty acid. Calculate the results to milligrams of potassium hydroxide per gram of sample; thus there is a correspondence in form of expression with the acid value of saponifiable oils (page 341).

CALCULATION.

$$\text{Titration in ml.} \times \text{normality of alkali} \times \frac{56.1}{\text{weight of sample}}$$
$$= \text{free fatty acid in mg. of potassium hydroxide per gram of sample}$$

On occasion it is more convenient for use in interpreting results to calculate the free fatty acid as percentage of oleic acid as follows:

$$\text{Titration in ml.} \times \text{normality of alkali} \times 0.2825 \times \frac{100}{\text{weight of sample}}$$
$$= \text{per cent free fatty acid as oleic acid}$$

Organic Ester Sulfur Trioxide. BY HYDROLYSIS. The first step provides for extraction from sodium acetate or other materials that hinder accurate acid titration with a methyl orange indicator. For sulfated fractions isolated from commercial products such extraction is a wise precaution. If these interferences are absent, sulfated oils may be analyzed without the preliminary extraction.

The sulfated material is boiled with a known amount of free sulfuric acid. This splits off sulfate groups as sulfuric acid. The acid so produced is determined by titration and calculated as sulfur trioxide. This method does not determine the sulfur present as sulfonates, only that present as organic sulfuric acid esters.

PROCEDURE. Transfer 5 to 10 grams, preferably the upper limit, of weighed sample to a 250-ml. separatory funnel containing 50 ml. of saturated sodium chloride solution, a little excess sodium chloride, and 50 ml. of ether. Shake the mixture well, and add standard sulfuric acid until about 0.2 ml. excess is present. Let this stand for 5 minutes, and draw off the lower layer into another separatory funnel. Wash the ether layer with successive 25-ml. portions of saturated sodium chloride solution until one drop of 0.5*N* sodium hydroxide solution will render the washings very definitely alkaline to methyl orange. Allow each extraction to stand for 5 minutes for separation to be complete.

Combine the salt solutions used for extraction, and extract them with two 25-ml. portions of ether. Combine the ether extracts, and wash with saturated salt solution until free from acid by the standard previously given. Combine all the ether layers in one flask, and evaporate the ether.

This procedure has isolated the sulfated oil in the ether solution, while sodium acetate, alkaline salts, and related materials have been separated and discarded in the water layer. Soap has been decomposed, but the fatty acid is with the sulfated oil because it may be a sulfated fatty acid. Since the determination is fundamentally titrimetric, precautions to avoid the presence of sulfuric acid in the ether layers are very essential.

The extract is continued as the sample in the next step. Sulfated oils not so extracted are weighed out as sample at this point. To such a sample in a conical flask, add the amount of *N* sulfuric acid shown necessary by separate determination to neutralize the free alkalinity and make it comparable with an extracted sample.

Fit the flask containing the sample with an air condenser about 36 inches long. While it is desirable for the air condenser to be joined to the flask by a ground-glass joint and this is specified in the standard methods, in practice a clean, dense cork stopper serves satisfactorily. Also, the usual water condenser is convenient, and the sample is less apt to go to dryness.

Add 25 ml. of *N* sulfuric acid to the sample, and transfer a duplicate amount to a similar flask as a blank. Boil each so that vigorous ebullition takes place, with only a little evaporation. At the end of 90 minutes' boiling, rinse the condensers with small volumes of water, and let the flasks and contents cool well. Add 30 grams of sodium chloride, 25 ml. of ether, 50 ml. of water, and 5 ml. of 0.1 per cent methyl orange solution to each. Titrate with *N* alkali, mixing well, and if necessary stopper and shake the flask at intervals. Reserve the titrated sample for determination of total fatty matter.

The difference in the two titrations in terms of sulfur trioxide is then calculated to percentage of sulfur trioxide in the original sample. This gives a basis of evaluation and relative comparison of commercial sulfated oils. As applied to fractions from industrial compounds, it is a basis for wise guessing, to be confirmed in other ways. The organic sulfur trioxide in commercial sulfated oils varies from a fraction of a per cent to around

16 per cent. Only by assuming a specific organic sulfur trioxide content of the sulfated oil entering into a compound can this be then calculated to percentage of sulfated oil. Results may also require interpretation in terms of sulfated alcohols, depending on qualitative indications.

The accuracy of the method permits duplicate determinations to agree to about 0.04 per cent of combined sulfur trioxide, but in other than umpire analyses agreement within about 0.15 per cent is a reasonable tolerance.

CALCULATION. In explanation it should be said that the reaction to be calculated is the following very simple one:

$$ROSO_2ONa + H_2O = ROH + NaHSO_4$$

Thus, for every molecule of combined sulfur trioxide there is liberated one hydrogen equivalent to mineral acid in the form of sodium bisulfate.

$$(\text{Sample in ml.} - \text{blank in ml.}) \times N \times 0.080 \times \frac{100}{\text{weight of sample}}$$
$$= \text{per cent sulfur trioxide present as organic esters}$$

GRAVIMETRIC METHOD. This procedure differs from those which precede it in the determination of the total organic sulfur trioxide. The sample is dissolved in a solvent, and acidified and interfering substances, notably inorganic sulfates, are removed by washing with saturated sodium chloride solution. The purified extract so isolated is ashed and the sulfur trioxide determined gravimetrically as sodium sulfate or by precipitation as barium sulfate.

PROCEDURE. If the sample contains ammonium compounds, dissolve a 5- to 8-gram sample in about 80 ml. of water in a beaker. Add 10 ml. of *N* sodium hydroxide solution, boil gently until ammonia is no longer given off as indicated by the reaction to wet litmus paper, and let cool. This step is omitted if ammonium compounds are known to be absent.

Proceed with the sample from which ammonium compounds have been removed, or with an original sample free from ammonium compounds, according to the procedure for extraction and hydrolysis until the purified sample is present in solution in ether.

Combine the ether layers in a separatory funnel, and let any water separate. Add 5 grams of anhydrous sodium sulfate, and shake for 5 minutes. Filter through paper into a beaker, and evaporate the ether. Wash the separatory funnel and paper with ether until all oil has been transferred. To avoid loss by creeping, the beaker used for evaporation should at no time be more than one-third full of ether solution. Evaporate the ether solution to about 20 ml.

Transfer the ether solution of purified sample in a weighed 50-ml. high-form crucible. Heat this in a 100-ml. beaker of water to drive off the balance of the ether. Rinse the beaker successively with at least two 10-ml. and three 5-ml. portions of ether, more if necessary, to transfer all the purified oil to the crucible. Evaporate each ether rinse before adding the next one to the crucible. It is advisable to use a small stirring rod in the crucible during evaporation. If this is done, wipe the rod with quantitative paper and add the paper to the crucible before ashing.

After all the ether is evaporated, dry the water from the outside of the crucible with a towel and ignite the contents to ash at a low temperature. Let the crucible cool, and moisten the ash with 30 per cent hydrogen peroxide. Dry and ignite again. This helps to remove traces of carbon and oxidizes sodium sulfide, which may have been produced from the sulfate by carbon reduction, to sodium sulfate.

The weight of residual sodium sulfate may be calculated to sulfur trioxide in the sample as representing the total combined sulfur trioxide. Note that this results from ignition of the bisulfate to sulfate, that only half the original sulfur remains as sulfate.

An alternative step is to ash in a crucible that has not been weighed, dissolve the ash in water, and precipitate as barium sulfate. For this, dilute the solution to about 250 ml., and add 2 ml. of concentrated hydrochloric acid. Proceed as on page 160, starting with "Bring to a boil, and add with constant stirring sufficient boiling 10 per cent barium chloride solution . . . ". This offers some advantages in accuracy and is a particularly desirable procedure in dealing with fractions isolated from commercial products.

Manipulative losses in this procedure are such that duplicate samples usually agree within 0.1 to 0.2 per cent.

CALCULATION.

From sodium sulfate:

$$\text{Ash} \times 1.1267 \times \frac{100}{\text{weight of sample}} = \text{organic sulfur trioxide}$$

From barium sulfate:

$$\text{Barium sulfate} \times 0.6860 \times \frac{100}{\text{weight of sample}} = \text{organic sulfur trioxide}$$

Inorganic Sulfur Trioxide. DIRECT DETERMINATION. If the sample is known to contain no inorganic sulfate other than that introduced with the sulfated oil, a simple method of direct determination is applicable.

PROCEDURE. Weigh out a sample selected according to the probable inorganic sulfate content, usually 1 to 5 grams. Dilute with water, and titrate with hydrochloric acid to a methyl orange end point If soap is present, heat the solution of sample during this titration. Discard this, and weigh out another sample of the same amount. Dilute, and add the determined amount of acid in the cold. Filter the separated fatty sulfuric acid ester of the glyceride on a wet filter paper. The solubility of fatty alcohol sulfuric esters in water prevents the accurate application of this method, but the alternative method by difference can be used. Wash the residue on the filter paper with cold saturated sodium chloride solution. Alternatively, this separation and washing may be carried out in a separatory funnel. These precautions are to prevent hydrolysis of the organic sulfate.

Dilute the filtrate to about 250 ml., and add 2 ml. of concentrated hydrochloric acid. Proceed as on page 160, starting with "Bring to a boil, and add with constant stirring sufficient boiling 10 per cent barium chloride solution . . . ".

BY DIFFERENCE. Determine total sulfate present as follows: Weigh out a sample as for organic ester sulfur trioxide, without extraction add 25 ml. of *N* hydrochloric acid, and boil gently for 90 minutes. At the end of this time, cool and extract the separated fatty matter with ether to prevent interference by it. Dilute the aqueous layer to about 250 ml., and add 1 ml. of concentrated hydrochloric acid. Proceed as on page 160, starting with "Bring to a boil, and add with constant stirring sufficient boiling 10 per cent barium chloride solution . . . ". The value for total sulfate so determined less that determined as organic ester sulfur trioxide gives the amount present as inorganic sulfate.

BENZIDINE METHOD. This method for organic and inorganic sulfates is found on pages 167–8. It is very satisfactory. Both organic and inorganic sulfate are precipitated. They can be separated and each determined.

CALCULATION.

$$BaSO_4 \times 0.6086 \times \frac{100}{\text{weight of sample}} = \text{per cent sodium sulfate}$$

The corresponding factor for ammonium sulfate is 0.5661.

Total Fatty Matter. By hydrolyzing off the sulfate groups, the fatty matter is isolated by normal extraction procedures and determined. If aliphatic sulfonates are present along with the sulfates, this method is not applicable, nor indeed is any other, with satisfactory accuracy.

PROCEDURE. To the residue from titration of total organic sulfur trioxide or another sample similarly boiled with sulfuric acid and neutralized, add about 1 ml. of *N* sulfuric acid. Transfer this solution to a 250-ml. separatory funnel, and extract with 50 ml. of ether. Draw off the water layer into another separatory funnel, and extract with two 25-ml. portions of ether. Combine the ether layers, and wash with 15 ml. portions of water until the wash water does not react acid to methyl orange. Transfer the ether layer to a weighed beaker, and evaporate the ether on a water bath. Dry in an oven at 105 to 110° to constant weight, usually for about 30 minutes. Calculate this weight to percentage of the original sample used for determination of organic sulfur trioxide.

Evaporation of ether leaves some residual water, which has an appreciable solubility in ether at room temperature. The drying operation at 105 to 110° is to remove this. It is convenient to add a few milliliters of absolute alcohol and evaporate off the diluted alcohol on a steam bath before drying in the oven. This prevents spattering.

Theoretically, the fatty matter so isolated is the same structurally as the oil originally sulfated. Practically, this is only a rough approximation. Some hydrolysis of the fatty glyceride to free fatty acid occurs in sulfation. In washing the sulfated oil, the sulfated glycerides of the lower fatty acids are more soluble, and some are selectively extracted in the sodium sulfate solution washes. With these limitations of the sample in mind, the usual determinations in Chap. 17 may be applied for identification of the fatty matter isolated. If the sulfated product is a sulfated alcohol, the fatty matter will be a higher alcohol and unsaponifiable. In the case of a sulfated petroleum derivative, the saponification value will be low.

The method is normally accurate to 0.1 to 0.2 per cent difference between duplicate samples.

Total Sulfated Fatty Matter. The results of this determination should equal the sum of total fatty matter and organically combined sulfur trioxide calculated as the sodium salt of the sulfuric acid ester. The factor 1.487 is readily obtained by considering the effect of hydrolysis. The sulfate group, $—OSO_2ONa$, is split off and, in the determination, calculated as SO_3. The former group has a molecular weight of 119.06 and the latter of 80.06. So to convert SO_3 to OSO_2ONa it is necessary to multiply the percentage of SO_3 by 119.07 divided by 80.06, that is, 1.487. Interfering substances are extracted before this determination by a saturated sodium chloride solution.

Procedure. Weigh a sample equivalent to about 5 grams of anhydrous fat into a separatory funnel containing 50 ml. of saturated sodium chloride solution, a little solid sodium chloride, 50 ml. of ether, and 5 drops of 0.1 per cent methyl orange solution. Shake and add *N* sulfuric acid from a buret until about 0.2 ml. excess is present. Use of too much acid causes excessive washing later, which decreases the accuracy of the method. Let settle for 5 minutes, and draw off the lower layer into another separatory funnel. For the case in which three layers are present, see a later paragraph near the end of this method.

Wash the ether layer with successive 25-ml. portions of saturated sodium chloride solution until one drop of 0.5*N* sodium hydroxide removes the pink color of the wash water with methyl orange. Let the layers stand for 5 minutes after shaking with each wash water before drawing off. Combine the wash waters, and extract with two 25-ml. portions of ether. Combine the ether extracts, and wash with saturated sodium chloride solution until free of acidity by the test previously described.

Combine the ether layers in a flask, being sure that no traces of water layers remain. Shake vigorously for 5 minutes with 5 grams of anhydrous sodium sulfate to dehydrate. Filter through paper into a weighed beaker. Wash the flask, and filter with ether until free from oil. Since the oil in ether has a tendency to creep, the beaker should at no time be more than one-third full. Evaporate the ether solution to about 20 ml. Add exactly 2 ml. of $0.5N$ standardized alcoholic potassium hydroxide solution. Mix well, and evaporate the remaining ether. Dry to constant weight at 108 to 112°, usually for 1.5 hours. Weigh, and correct the result for the amount of potassium hydroxide in the alcoholic potash solution added.

Some highly sulfated oils give three layers instead of two when shaken with ether and salt solution. Wash the two layers as though they were a single layer. Extract the combined washings with 10 ml. of alcohol and 20 ml. of ether. Wash this extract with saturated sodium chloride solution as usual. If a third layer forms during this washing, add alcohol in 5-ml. portions until the upper layers merge. Wash the flask and filter paper after filtration with a mixture of 2 parts of ether and 1 of alcohol. In evaporation, remove the major part of the solvent, and add 1 ml. of $0.5N$ alcoholic potassium hydroxide solution. Heat on a water bath until practically free of solvent, preferably with stirring; a stirring rod, weighed with the beaker, should be used. Then add 5 ml. more of $0.5N$ alcoholic potassium hydroxide solution, and heat to constant weight.

Accuracy to 0.1 to 0.2 per cent is a normal expectation.

Calculation.

$$[\text{Weight of sulfated fatty matter} - (\text{alcoholic potassium hydroxide} \times N \times 0.0391)] \times \frac{100}{\text{weight of sample}} = \text{per cent sulfated fatty matter}$$

Unsaponifiable Matter. To the weighed total fatty matter as determined add 50 ml. of approximately $0.5N$ alcoholic potassium hydroxide solution. Boil under a reflux condenser for 1 hour. Let cool, and make up to the original volume with alcohol. Transfer to a separatory funnel, and wash out the flask with four 10-ml. portions of water, adding these to the separatory funnel. Extract the contents with three 50-ml. portions of petroleum ether boiling at 40 to 75°. Combine the petroleum ether extracts, and wash with 50 ml. of 50 per cent alcohol containing 1 ml. of $0.5N$ alcoholic potassium hydroxide solution. Wash twice with 50-ml. portions of water. Filter the petroleum ether extract through paper if necessary, and evaporate in a weighed beaker. Dry to constant weight at 105 to 110°, usually for about 30 minutes.

If mineral oil was present with a sulfated oil, it will show here as high unsaponifiable matter. Thus it would appear as an indicator of petroleum sulfates having been present in a compound. This would also be checked by a low saponification value on the total fatty matter.

Possible variations in the amount of unsaponifiable matter are so wide that comment on the expected accuracy is almost impossible. Only careful attention to every stage of the manipulation can give reasonable accuracy.

CHAPTER 19

SOAP AND SOAP PRODUCTS

Soaps are the salts of fatty acids. They are produced by saponification of fats or oils; by neutralization of fatty acids; or, in the case of insoluble soaps, from metathetical precipitation by a soluble soap reacting with another salt. Conventionally, the water-soluble soaps of sodium are usually called soaps, and others are prefixed by the name of the alkali used, as potassium soaps, aluminum soaps, or morpholine soaps. Salts of abietic acid, the acid of rosin, are considered as soaps.

If made by saponification of fats and oils, the usual process, the glycerol produced at the same time is generally recovered. In sodium soaps made by the cold or half-boiled processes and in potassium soaps the glycerol remains in the soap as sold and used. Soap is made in the production of many commercial items by separately adding fatty acid and a neutralizing agent, ammonia, for example. Soaps of ammonia and the amines are relatively unstable and in analysis may often be recovered as the free fatty acids.

The functions of different forms vary: as toilet and laundry detergents; as emulsifying agents in such products as cosmetics; as lubricants in wire drawing; as scouring agents for textiles; etc.

Confusion with sulfated products is possible unless these are eliminated by tests already described (see Chap. 18). There are many other types of surface-active agents that could be confused with soaps. Hence, qualitative identification is desirable.

QUALITATIVE

Soaps of the alkali metals in usual concentration have a pH value of about 10.2. Dissolve a few milligrams of solid sample or a few drops of liquid sample in a couple of milliliters of water, and add a few drops of phenolphthalein solution. Upon radical dilution, absence of an alkaline reaction eliminates soap. A positive reaction does not prove its presence, for other alkaline substances may be present in the sample or fraction being tested.

Evaporate a few drops of liquid sample or take a few milligrams of solid, and dissolve in neutralized alcohol. Soap in alcohol does not hydrolyze and give a color with phenolphthalein. If the aqueous solution of the sample gave such a reaction and the solution in alcohol was complete and did not so react, soap is present. A pink color in the alcohol solution shows the presence in the sample of free alkali or a highly alkaline salt, such as sodium orthosilicate.

Take about 10 ml. of liquid or concentrated aqueous solution of the solid sample. Acidify with 1:1 hydrochloric acid, and boil vigorously for about 1 minute, being careful to prevent spattering. If soap is present, a layer of fatty acid will rise to the surface. Other indications of the presence of soap are the formation of a precipitate on addition of dilute solutions of calcium, magnesium, aluminum, or iron salts.

On heating a sample suspected to be sodium or potassium soap in a porcelain crucible or on a spatula, a residue of the corresponding carbonate is left, and a characteristic odor from decomposition of the fatty acids is obtained.

ANALYSIS OF SOAPS OF FIXED ALKALIES

The methods that follow are largely those of the American Oil Chemists Society, the American Society for Testing Materials (ASTM), and the American Chemical Society (ACS). As such, they apply to analysis of samples containing mainly soap. Modifications permit examination of fractions isolated from commercial products.

Matter Volatile at 105°. OVEN METHOD. Weigh about 5 grams of the soap into a porcelain or glass dish about 6 to 8 cm. in diameter and about 2 to 4 cm. high, and dry to constant weight in an air oven at 105°. Constant weight is taken as that at which successive heating for 1-hour periods shows a loss of not more than 0.1 per cent, or even a gain.

There are serious limitations to the application of this method. With soaps containing salts with water of hydration, sodium silicate, for example, dehydration is incomplete. There is no remedy for this. Soaps of fatty acids from readily oxidizable oils, linseed or soybean, for example, absorb oxygen during the drying operation so that an unduly low value is obtained. This can be corrected by drying in an inert atmosphere at 105° or at 80 to 100° *in vacuo*. If substantial amounts of glycerol are present as in cold-process and half-boiled soaps, it will partly volatilize. This cannot be remedied.

DISTILLATION METHOD. For soaps containing 5 to 20 per cent of moisture and volatile matter, use 20 grams of the sample. For soaps containing more than 20 per cent of moisture and volatile matter, use 10 grams of the sample. Transfer carefully to a 500-ml. short-necked round-bottomed flask, and add 10 grams of anhydrous fused sodium acetate to prevent violent frothing. Proceed as on page 41, starting at "Add 100 to 150 ml. of toluene or xylene . . . ". If the soap contains glycerol, toluene is preferred, for results will be about 1 per cent high with xylene owing to volatilization of glycerol at the higher temperature.

This method is generally more satisfactory than the oven method. However, when salts containing water of hydration are present, there is no method of determining that all this water has been driven off, even at the temperature of boiling xylene, which is about 140°. In such cases there is no method available by which a direct determination can be made. The only alternative is to use intricate methods of calculation after analyzing for the salts present.

Matter Insoluble in Alcohol. STANDARD METHOD. Soap is soluble in alcohol. Salt and some adventitious contaminants are insoluble. Added builders such as carbonates, borates, silicates, and phosphates are substantially insoluble although they are preferably determined on a separate sample of soap. To minimize such solubility it is desirable to use absolute alcohol, specially denatured grades being suitable. Some decomposition of the more alkaline silicates, particularly orthosilicate, occurs so that results tend to be low and the free alkali determined on the filtrate correspondingly high.

PROCEDURE. Digest 2 to 10 grams of the sample with 200 ml. of freshly boiled neutral alcohol, 94 per cent or higher, in a covered beaker on a steam bath until the soap is dissolved. Filter through a counterpoised filter paper neutralized to phenolphthalein by washing with very dilute alkali in alcohol followed by neutral alcohol until the washings are neutral. Alternatively, use a weighed Gooch crucible with suction. In either case, during the operation protect the solution from carbon dioxide and other acid fumes by covering it with a watch glass. Wash the residue on the paper or in the crucible with hot neutral alcohol until it is free from soap. It is not easy to remove the last traces of soap, and this washing, therefore, must be thorough. In filtering through paper, the use of a hot-water funnel (Fig. 100) is advantageous. Always keep the alcohol solution hot. or the pores of the filter paper may be closed with gelled soap solution, Reserve the filtrate and washings for determination of free alkali. Dry the filter paper or crucible with the residue at 100 to 105° for 3 hours, cool, and weigh the total matter insoluble in alcohol.

DIRECT ALCOHOLIC EXTRACTION. In the case of commercial samples, in which the soap content may be small and the alcohol-insoluble residue large, a modified procedure is convenient.

PROCEDURE. Disperse 5 grams of sample in hot neutral alcohol, preferably absolute, in a tared bottle, and centrifuge at once. When the insoluble matter has been thrown down, decant the clear alcohol layer. This may be used for determination of free alkali or free acid. Add a second portion of hot neutral alcohol, and a third if necessary. Use as little alcohol as possible to avoid errors due to the slight solubility of some ingredients such as highly alkaline silicates.

Dry the centrifuge bottle in the oven at 100 to 110°, and weigh. Alternatively, the alcohol-insoluble matter may be brushed out and weighed. Such a residue from a commercial sample is usually employed as a fraction for further analytical determination of its composition.

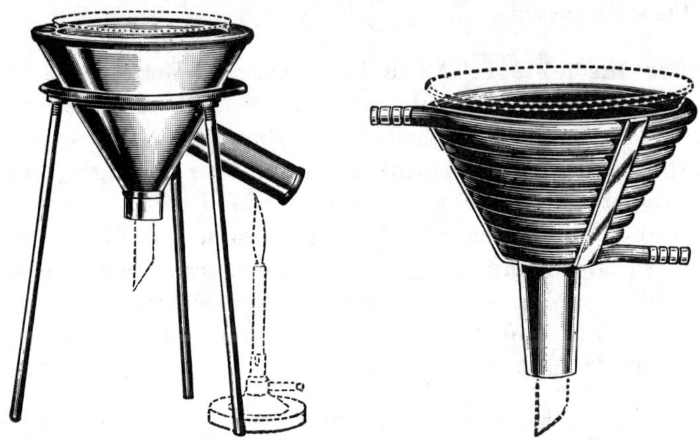

FIG. 100.—Two types of hot-water funnel. That on the right may also be heated with steam.

Free Alkali or Free Acid. The first procedure is carried out on the filtrate from matter insoluble in alcohol and is subject to substantial errors already discussed if it contains highly alkaline builders. In dealing with toilet soaps containing only a very small amount of free alkali, usually well under 0.05 per cent, another source of error enters. Filter paper sorbs considerable amounts of alkali, and it is difficult to prepare a paper known to be neutral to phenolphthalein. Shielding the solution with such small contents of free alkali from carbon dioxide fumes is difficult. The second, or centrifuge, procedure largely avoids both these sources of error.

PROCEDURE. *Standard Method.* Heat the filtrate reserved from the determination of total matter insoluble in alcohol to incipient boiling, and add 0.5 ml. of a 1 per cent solution of phenolphthalein. If the solution is pink, the usual condition, titrate to the end point with $0.05N$ or $0.1N$ acid. If no color is shown by the indicator, titrate similarly with alkali.

Alternative Method. Dissolve 5 grams of the sample in hot neutral alcohol, and at once centrifuge until the alcohol solution is clear. Decant the alcohol without disturbing the precipitate, heat to incipient boiling, and titrate with standard acid or alkali according to the reaction. If the amount of alcohol insoluble is small, it is not necessary to wash it. When a large precipitate is thrown down in the centrifuge, washings with hot neutral alcohol are desirable.

CALCULATIONS.

$$\text{Titration in ml.} \times \text{normality of acid} \times 0.040 \times \frac{100}{\text{weight of sample}} = \text{free alkali as per cent sodium hydroxide}$$

This is sometimes expressed as sodium oxide, with the factor 0.031, or in the titration of potassium soaps as potassium hydroxide with the factor 0.0561.

$$\text{Titration in ml.} \times \text{normality of alkali} \times 0.2825 \times \frac{100}{\text{weight of sample}} = \text{free acid as per cent oleic}$$

Matter Insoluble in Water. Some matter soluble in alcohol is insoluble in water, unsaponifiable and unsaponified matter in small amounts, for example. More strictly, therefore, this matter insoluble in water is the part of the alcohol-insoluble matter that is not water-soluble. Since drying the alcohol-insoluble matter may produce changes in it, a new sample is required.

PROCEDURE. *Standard Method.* Proceed as in the determination of matter insoluble in alcohol, using a fresh sample of soap. After filtering and thoroughly washing the residue with alcohol, change the receivers, extract the residue with water at 60°, and wash the filter thoroughly. When the matter insoluble in alcohol is all inorganic, use boiling water for the extraction and washing. Reserve the water solution for analysis for inorganic salts. Dry the filter and residue at 100 to 105° for 3 hours, cool, and weigh. Calculate the result as matter insoluble in water. After weighing, examine the material on the filter paper for possible contaminants that may have been introduced during the procedure. If the original sample contained pigment, as in some white toilet soaps, or abrasive as in mechanics' hand soaps, it will be present in this fraction, ready for analysis as a separate step.

Direct Aqueous Extraction. Soap samples for analysis may vary from 100 per cent soap at the one extreme to nearly 0 per cent at the other. The procedure that follows is designed for the determination of matter insoluble in water when this fraction is high and the soap content is low, for analysis of an abrasive hand soap, for example.

Disperse 5 grams of sample in hot neutral alcohol, and centrifuge at once until the insoluble matter has all been thrown down. The residue from determination of free alkali or free acid by the analogous procedure can be taken at this point as sample. Decant the alcohol, and repeat the extraction. If necessary, extract further until free from soap. Add water at 60°, shake, and centrifuge. If the insoluble matter is entirely inorganic, operations may be somewhat accelerated at this stage by the use of boiling water. Decant, and repeat the extractions until only water-insoluble matter is left. Usually the aqueous extracts are saved and diluted to standard volume as a fraction for analysis of the water-soluble matter. When extraction of the water-soluble matter is complete, dry the centrifuge bottle in the oven at 100 to 110°. Brush out the residue, and weigh; or if a tared bottle has been used, weigh by difference. This water-insoluble matter is a fraction that is usually further analyzed. If it is an abrasive, for example, a routine mineral analysis would be run on it after microscopic inspection for possible identification by that means.

Inorganic Salts. The aqueous solution from water extraction of the material insoluble in alcohol will contain the alkaline salts and, in addition, any neutral salts such as the small amount of sodium chloride present in kettle soap.

It therefore of itself constitutes a sample for analysis according to the procedures described in Chap. 12, Fixed Alkalies and Alkaline Salts. Usually additional extraction of comparatively large portions of the original sample will be needed to isolate a large enough fraction. Conventional soap analysis methods provide for titration of such solutions, and this is an integral part of the analysis if the solution is to be completely analyzed.

Alkalinity of Water-soluble Matter. PROCEDURE. Titrate the sample of water-soluble matter in the filtrate from determination of water-insoluble matter. First titrate in the cold to the phenolphthalein end point with standard acid; then add methyl orange or methyl red and also titrate cold to that end point. The total titration is used for calculation of alkalinity; the relative amounts of the two and the sharpness of the end points are a clue as to amounts and kinds of alkaline salts present. For example, sodium carbonate and phosphate titrate half their value to phenolphthalein; sodium bicarbonate and sodium metaphosphate, none of their value; sodium silicate, all its value; and tetrasodium pyrophosphate and sodium borate, indeterminate amounts of their value.

CALCULATION.

$$\text{Titration in ml.} \times \text{normality of acid} \times 0.031 \times \frac{100}{\text{weight of sample}} = \text{per cent of alkalinity as sodium oxide}$$

The factor for potassium oxide is 0.0471.

Combined Alkali and Total Anhydrous Soap. Since soap frequently contains at least minor amounts of several other materials, the most accurate method of determining the amount of true soap is to isolate the fatty acids and convert these to anhydrous soap. With a fresh sample, the method is not satisfactory, if large amounts of insoluble material or of soluble silicates are present. In that case, however, the filtrate from the determination of total matter insoluble in alcohol is substantially free from the silicates and insoluble matter. To use it, neutralize any free acid present, evaporate to dryness to free from alcohol, and take up in water. A very approximate result is also obtainable by exactly neutralizing this alcohol solution and evaporating to dryness, subject to the errors discussed under Matter Insoluble in Alcohol (page 41). Titration methods follow the standard method, one applicable to the fatty acids and the other to an aqueous soap solution.

PROCEDURE. *Standard Method.* Dissolve sufficient sample in 100 ml. of water to give about 5 grams of anhydrous soap. Add 1:1 sulfuric acid in slight excess, insert a small funnel in the neck of the flask, and heat to a temperature not greater than 60° until the fatty acids separate as a clear layer. Cool, and transfer to a separatory funnel, washing out the flask with ethyl ether and adding this to the contents of the funnel. Draw off the acid layer into another separatory funnel, and extract it with two 20-ml. portions of ethyl ether. Dissolve the fatty acids in the first separatory funnel in this ether, and wash with 10-ml. portions of water until they are no longer acid to methyl orange. Unite the water portions used for washing, and shake with 20 ml. of ether. Wash this ether until the wash water is neutral to methyl orange. Reserve the acid water and washings for the determination of chlorides.

Unite the ether solutions, and, if necessary, filter through a dry paper, washing the paper with ether. Add 100 ml. of neutral, freshly boiled, and cooled alcohol; add phenolphthalein; and titrate to exact neutrality with standard $0.5N$ alkali solution. Usually this will be sodium hydroxide for sodium soap. Occasionally the equivalent analysis is performed on potassium soaps, and in this case potassium hydroxide is used for titration. Evaporate the alcohol, and dry to constant weight as in the determination of matter volatile at 105°.

This value is in error by an amount equal to any unsaponifiable matter and unsaponified fat present. The correction for a high-grade soap will be practically negligible. If lower grade soap is being analyzed, particularly a soap fraction containing mineral oil, subtract the percentages of unsaponifiable matter and of unsaponified fat present in the weight of sample taken from the total weight of soap to give the corrected anhydrous soap.

Various additional corrections can be applied to this. Titration may be with potassium hydroxide when results as sodium soap are desired, or vice versa. Correction must then be made to calculate to the form to be reported. If the original soap contained free fatty acid, this will appear in the results as the alkali salt, as soap. If the free fatty acid was substantial, which would not occur with the usual commercial soaps, correction must be made by calculation of the titration to the equivalent as soap and deducting from the total soap determined.

For very accurate work a blank for neutral salts may be made on the alkali used. For this, titrate the equivalent of the amount of alkali used with standard acid to methyl orange neutrality, and evaporate to dryness. Weigh, and subtract the calculated amount of salt that would theoretically be formed, the difference being contamination. With usual grades of reagents, this is less than the limit of accuracy in determining the blank.

CALCULATIONS.

$$\text{Total weight of anhydrous soap} \times \frac{100}{\text{weight of sample}} = \text{per cent anhydrous soap present}$$

$$\text{Titration in ml.} \times \text{normality of alkali} \times 0.031 \times \frac{100}{\text{weight of sample}} = \text{per cent sodium oxide}$$

In the case of potassium soaps, substitute the factor 0.0471, and the result is potassium oxide.

Per cent anhydrous soap − per cent alkali oxide = per cent anhydride of fatty acid, usually reported as oleic anhydride

Titration of Fatty Acids. When extreme accuracy is not required, the fatty acids may be isolated and titrated. Incidentally, this gives an acid value (page 341) on the fatty acids, a constant which is of value in identifying the kind of fat from which the soap was made.

Dissolve the sample in water, cool without causing gelation, and transfer with water to a 500-ml. separatory funnel. Add sufficient hydrochloric acid so that the water solution is definitely acid to methyl orange. Do not add too much excess, or subsequent washing will be more difficult. Extract at least three times with ethyl ether, washing each extract with 15 ml. of water and returning the wash water each time to the separatory funnel containing the acid solution. If fatty matter such as lanolin is present with the fatty acids, several more extractions with ether will be required.

Combine the ether extracts, wash with 15-ml. portions of water until the washings are neutral to methyl orange, and drain well. Draw off the drained water; and if the ether extract is not clear, filter through a dry paper, finally washing the paper free of fat with ether. Evaporate off the ether in a weighed fat-extraction flask. Should any globules of water be present, add a few milliliters of absolute alcohol and drive off the alcohol on a steam bath. Dry in the oven at 105° until all the alcohol is gone, cool, and weigh. If the original sample selected was of the proper size, dissolve the entire fatty-acid fraction in hot neutral alcohol; otherwise, take a weighed portion. Titrate to a phenolphthalein end point as is usual in the determination of the acid value of a fat (page 341). To calculate the results to soap, supplementary evidence must be available as to the predominant type of fatty acids. The difference between stearic and oleic acids is not important, for the difference in molecular weight is within the error of the method.

The foregoing is subject to numerous variations. If the nature of the fatty acids is well established from other data, the first weighing of total fatty acids may be omitted. In that case, add neutral alcohol to the ether extracts, which have been washed free of acid, and titrate for the final result.

The fatty acids in ether solution contain any unsaponified and unsaponifiable matter just as in the previous method. Subsequent separation of unsaponified and unsaponifiable matter gives a soap solution that may be acidified for isolation and titration of the liberated fatty acid.

The acid value calculated from the titration of fatty acids often gives significant information. If the value approximates that of the fatty acid known to predominate, this suggests the presence of only minor amounts of unsaponified and unsaponifiable matter. Much more frequently, particularly in dealing with fractions isolated from commercial products, a very low value indicates that something other than fatty acid is present. In that event no time has been wasted as the titrated sample may be used for determination of unsaponified and unsaponifiable matter.

CALCULATION.

$$\text{Titration in ml.} \times \text{normality of alkali} \times \frac{\text{total weight of fatty acids}}{\text{weight of fatty acids used}} \times 0.3045 \times \frac{100}{\text{weight of original sample}} = \text{per cent sodium oleate}$$

Substitute 0.3065 for the factor 0.3045 to convert to sodium stearate, and 0.2337 to convert to sodium coconut-oil soap. The corresponding factors for potassium soaps are 0.3205, 0.3225, and 0.2498.

Titration of Soap. This method is useful as a method of check analysis when the highest degree of accuracy is not required. It depends on the fact that soap may be completely titrated with acid in water solution, before a methyl orange end point is reached. Sodium and potassium soaps cannot be titrated in alcohol solution, but ammonia soap is sufficiently unstable so that it can. However, the latter titration is not reliable.

Dissolve a sample of about 5 grams in 150 ml. of hot water. Add 1 ml. of methyl orange or methyl red indicator solution, and titrate with 0.5N acid while hot. A good end point is usually obtained. It is essential to keep the solution near to boiling in order to prevent occlusion of the soap solution by precipitated fatty acids. An exception is made when ammonium soaps are present. The titration should then be made at as low a temperature as possible; finally heating should be done before adding the last small amount of acid.

If alkaline salts are present, the methyl orange titration obtained will include titration of alkaline salts to that end point. Since a titration of the alkalinity of the water-soluble matter has been obtained, this need only be subtracted to give the value for titration of the soap. If the amount of alkaline salts is large, titration of the fatty acids is preferable. The method is particularly applicable when soap and no other material titratable by acid is present.

Calculations are subject to the same limitation as in the titration of fatty acids, that a factor for the type of fatty acid present is used and therefore the fatty acids must have been identified.

CALCULATION.

$$\text{Titration in ml.} \times \text{normality of acid} \times 0.3045 \times \frac{100}{\text{weight of sample}} = \text{per cent sodium oleate}$$

Other factors are sodium stearate 0.3065, sodium coconut-oil soap 0.2337. Equivalent potassium-soap factors are 0.3205, 0.3225, and 0.2489.

Soap from the Ashing of Soap and Soap Mixtures. When sodium soap is ashed at a low temperature, it gives an equivalent amount of sodium carbonate. The same is true of a potassium soap. The applicability of this in the analysis of commercial products is evidenced almost daily. Thus if the alkaline salts in a product are known, by analysis for phosphates or silicates, for example, then titration of a low-temperature ash gives a quick determination of the soap present and a check against that found by more complex methods.

Conversely, when the soap is known and the alkaline salt is not, this is equally applicable. Assume, for example, that a grit soap paste being analyzed consists mainly of pumice with minor amounts of soap and sodium carbonate in a vehicle of water. A sample has been dried and the soap extracted with alcohol. If, now, a direct titration is made of the ash and this titration is corrected for the amount of soap present, it will indicate the content of sodium carbonate.

This method is not a primary method of analysis; that is, it is seldom relied on as a determination. Rather, it is a secondary method used to cross-check the accuracy of results obtained by the primary method.

Several precautions must be observed. In ashing a mixture of soap and sulfated oil the sodium carbonate from the soap will be unreliable owing to reaction with acid derived from the decomposition of the sulfated oil, even if it has been neutralized. One may visualize the sulfated oil as first giving an ash of sodium bisulfate that reacts with sodium carbonate to form sodium sulfate. This is not necessarily the mechanism of the reaction, but the result is the same. The interaction is not quantitative. In ashing mixtures containing soap and borax the formation of carbonate from soap is not reliable: the borax reacts as an acid salt in the ash. There are various other examples of similar interferences.

Total Fatty Matter. The fatty matter must be separated for use in several determinations. Although the residue is not weighed quantitatively, the separations must be nearly quantitative; otherwise, the loss at some stage may not be representative of the total sample and thus may modify the characteristics of the final sample.

PROCEDURE. Weigh out about 50 grams of sample. If the soap to be used contains alcohol, remove it completely by evaporation. Dissolve the sample in 500 ml. of hot water, and add about 100 ml. of 1:3 sulfuric acid. Stir well and heat gently until the free fatty acids separate as a clear layer. Siphon off the aqueous acid layer, and add 300 ml. of hot water. Boil gently for a few minutes, and siphon off the aqueous acid layer. Repeat this washing twice more. During these operations keep the beaker covered to minimize oxidation of the acids and carry through the operations as rapidly as possible.

When water has been removed from the fatty acids as completely as is feasible, filter the fatty acids through a dry paper and dry at 105° for 45 to 60 minutes. If the acids are highly unsaturated, it will be necessary to do this *in vacuo*, at a lower temperature if lower molecular weight fatty acids, lauric, for example, are present. If the titer value is known, the filtration should be at not more than 15° above the titer point.

Reserve this total fatty matter in a suitable sealed container, preferably with the air displaced with nitrogen. Before taking a sample of part of the batch, warm the acids until they are clear, and mix well until homogeneous. Sometimes more than one such batch must be separated if a comprehensive analysis is being done, for the total amount of fatty matter so separated is used as the sample in some determinations.

Unsaponified and Unsaponifiable Matter. The unsaponified matter determined in a soap is residual fat or oil that could be converted but was not. The unsaponifiable matter is organic but of a hydrocarbon, alcoholic, or other nature such as not to react to form a soap. The unsaponified matter is obtained by difference when the unsaponifiable matter is separated.

PROCEDURE. *Cylinder Method.* Weigh 5 grams of the sample into a 250-ml. conical flask containing approximately 0.1 gram of sodium bicarbonate. Dissolve in 100 ml. of 50 per cent alcohol. Warm to dissolve, but not above 60°. Filter off any undissolved residue in a Gooch crucible with an asbestos pad or in a funnel using an asbestos pad deposited on a perforated porcelain disk. Wash three times with hot 50 per cent alcohol. Wash with a small amount of petroleum ether to remove any traces of unsaponified and unsaponifiable matter. In the method of the ASTM this is required to be nearly pure normal pentane. Ordinarily petroleum ether of the 30 to 70° boiling range will do.

Transfer the entire alcohol, water, and ether filtrate to a 250-ml. graduated glass-stoppered cylinder, and make up to the 160-ml. mark with 50 per cent alcohol. Add 50 ml. of petroleum ether, shake vigorously for 1 minute, and allow to settle until both layers are clear. The volume of the upper layer should be about 40 ml. Draw off the petroleum ether layer as closely as possible, by means of a slender glass siphon, into a 500-ml. separatory funnel. Repeat the extraction at least six times, using 50 ml. of petroleum ether each time. After the first extraction has saturated the solution, the volume of the upper layer should be about the same as the volume of petroleum ether added. Thorough and vigorous shaking is necessary in order to secure accurate results. Soaps that contain substances of the nature of lanolin will require a greater number of extractions and even then will often give low results.

Wash the combined ether extracts, first with a mixture of 15 ml. of 0.1N sodium hydroxide solution and 15 ml. of alcohol and then three times with 25-ml. portions of 10 per cent alcohol, shaking vigorously each time. This treatment removes most of the fatty acids that may have been extracted from the soap solution owing to hydrolysis of the soap. The use of a 50 per cent alcohol solution and the sodium bicarbonate added with the sample tend to suppress hydrolysis of the soap.

The next operation is evaporation of the petroleum ether. This may be in a weighed beaker on a steam bath under a well-ventilated hood to carry away fumes; the process may be assisted by a current of air.

Usually some small amount of soap will have dissolved in the petroleum ether owing to its being saturated with alcohol and water. Therefore, add 50 ml. of cold petroleum ether to the residue to dissolve it. Unless this is absolutely clear, which rarely will be the case, filter on a dry paper and wash the beaker or flask and the paper with several small portions of cold petroleum ether, making sure that all residues are transferred from the tared container. Evaporate as before; when free from odors of petroleum ether, transfer to an oven regulated at 100 to 105°; and dry for 30 minutes. Weigh, and return to the oven. Continue to reweigh at 15-minute intervals until constant weight is obtained. Even with the precautions cited, some fatty acid will be present in this residue and a correction must be applied for this. On the assumption that the fatty acid is oleic acid, dissolve the weighed residue in 50 ml. of warm, neutralized alcohol, and at once titrate with dilute standard sodium hydroxide solution, 0.01N to 0.05N is suitable, to the same end point as was given by the neutralized alcohol. Obtain the correction as shown under Calculations.

For correction in very accurate work, heat 0.25 gram of stearin or other hard fat to constant weight in a 110° oven. Then dissolve in 250 ml. of the petroleum ether being used, and evaporate to dryness in the same oven. The increase in weight is the blank. Deduct such a blank from the total weight of residue left in the determination. Practically, if a good grade of petroleum ether is used, this blank is negligible.

CALCULATIONS.

Ml. of alkali × normality of alkali × 0.2825 = correction in grams of oleic acid

$$(\text{Total weight of residue} - \text{correction for free acid blank if any}) \times \frac{100}{\text{weight of sample}} = \text{per cent unsaponified and unsaponifiable matter}$$

Wet Extraction. Use the total fatty matter from 5 grams of soap as sample. Transfer to a 500-ml. separatory funnel, using either ethyl ether or petroleum ether for the determination. Add 250 ml. of water and a few drops of 1 per cent phenolphthalein solution. Add sufficient 50 per cent potassium hydroxide solution to render the whole definitely alkaline, but avoid a large excess. Addition should be dropwise. Sodium soaps such as sodium stearate are difficultly soluble in cold water, and in all cases potassium soaps are more soluble than those of sodium. For these reasons, potassium hydroxide should be used. Any large excess of free alkali may saponify some unsaponified fat and is therefore undesirable.

Extract the unsaponified and unsaponifiable matter by the technic outlined for extraction of unsaponifiable matter from fats. Start on page 349 at "Add 75 ml. of ethyl ether." The same corrections could be applied as in the preceding cylinder method, but usually when this method is applied such a high degree of accuracy is neither expected nor necessary. Calculate as for the cylinder method.

Dry Extraction. When no volatile organic matter is present, dry the sample, powder it, and proceed as for the corresponding method of determination of unsaponifiable matter in oils (page 350), starting at "Transfer the sample or a weighed aliquot to a Soxhlet thimble."

Unsaponifiable Matter. The final fraction separated in the determination of unsaponified and unsaponifiable matter was in all cases extracted from soap. It follows that, if the saponifiable matter in the original sample were first saponified to convert it to soap, the same procedures would then give a determination of unsaponifiable matter. Alternatively, particularly in dealing with commercial products, the unsaponified and unsaponifiable fraction may be similarly treated as sample. Some tall oil is used in the lower grades of industrial soaps. Since this contains a substantial amount of unsaponifiable resinous material, its presence will be detected by a high unsaponifiable value. This can be confirmed by a rosinlike odor on heating the unsaponifiable fraction after it has been weighed.

PROCEDURE. *Cylinder Method.* Weigh 5 grams of the original sample into a 250-ml. conical flask. Alternatively, transfer the unsaponified and unsaponifiable matter from a similar sample to such a flask. Add 30 ml. of alcohol and 5 ml. of 50 per cent aqueous potassium hydroxide solution, and boil for 1 hour under a reflux condenser. Transfer to a 250-ml. graduated glass-stoppered cylinder, and wash to the 40-ml. mark with alcohol. Complete the transfer, first with warm and then with cold water, until the final volume is 80 ml. and finally with a small quantity of petroleum ether. Cool the cylinder to room temperature, and proceed with the extraction as outlined for the determination of unsaponified and unsaponifiable matter (page 392), starting at "Add 50 ml. of petroleum ether, shake vigorously for 1 minute, and let stand until both layers are clear." Apply the correction for fatty acids in the residue as there outlined.

Wet Extraction. Use the total fatty matter from 5 grams of soap (page 391) or the unsaponified and unsaponifiable matter separated from such a sample by wet extraction (page 393). Saponify as in the cylinder method described in the immediately preceding section. Evaporate off the bulk of the alcohol; this can be done most safely and expeditiously by immersing in a steam bath (page 349).

Take up the residue with warm water, and transfer to a 500-ml. separatory funnel. Add sufficient 1:1 hydrochloric acid dropwise to just acidify, and then dilute potassium hydroxide solution to make just alkaline. Cool, and proceed as outlined for the determination of unsaponifiable matter by the separatory-funnel method (page 349), starting at "Add 75 ml. of ethyl ether." The remarks on the accuracy of that method also apply to this determination.

Dry Extraction. Saponify an original sample or the unsaponified and unsaponifiable fraction extracted by dry extraction by boiling with an excess of 0.5*N* alcoholic potassium hydroxide. When reaction is complete, titrate qualitatively to neutrality with 1:10 hydrochloric acid; then add just a few drops of the alcoholic potassium hydroxide to maintain alkalinity. Proceed as on page 350, starting at "Evaporate the saponified sample to dryness on a water bath . . . ".

Unsaponified Matter. This value is obtained by deducting the amount of unsaponifiable matter from the total of unsaponified and unsaponifiable matter, both determinations having been conducted by the same method so that any errors will be similar in type.

Rosin. In addition to the fatty acids, abietic acid from rosin is often present in soaps. A very small amount can be present without causing serious discoloration, but soaps that contain substantial amounts are yellow to brown in color. Methods are provided both for qualitative detection and for quantitative estimation when rosin is present.

PROCEDURE. *Liebermann-Storch Test.* Heat 1 to 2 grams of the total fatty matter, which should be completely dry, with 5 to 10 ml. of C.P. acetic anhydride. As the vapors of this reagent are corrosive, and highly inflammable, keep it from contact with the skin and free flames. Cool and filter if necessary. Transfer a few drops to a white spot plate. Add a drop or two of 1:1 sulfuric acid at room temperature. If rosin is present, a violet coloration is immediately produced, ranging from a very deep violet when much rosin is present to a fugitive violet changing to a brownish tinge when little is present. If a test suspected to be positive is obtained, add known amounts of soap containing rosin to rosin-free soap and treat similarly to confirm. This will serve also to give a rough approximation of the amount of rosin present.

In dealing with commercial samples it must be remembered that some other resins not used in soaps give a positive reaction. Examples are Manila gums and gum pontianak. Ester gum is a synthetic glyceride of abietic acid from rosin and gives a positive reaction. Tall oil reacts like rosin.

McNicoll Method. The two available quantitative methods for rosin are applied to the free acids from the soap. Both depend on the fact that under the conditions specified rosin acids are not esterified and fatty acids are. Both are subject to inaccuracies. The method given is the simpler of the two.

Weigh out about 2 grams of the total fatty matter (page 391); or, to simplify later calculations, obtain the total fatty matter from an equivalent amount of soap. Transfer this sample to a 50-ml. conical flask, and add 25 ml. of a solution containing 40 grams of pure naphthalene-β-sulfonic acid in 1 liter of C.P. absolute methanol. At the same time transfer 25 ml. of the reagent to a similar flask as a blank. Add a few glass beads to each to ensure smooth boiling, and reflux for 30 minutes. This esterifies the fatty acids but not the abietic acid. The reagent is only a catalyst to promote esterification and does not alter its concentration.

Cool the sample and blank, and add 0.5 ml. of a 0.5 per cent solution of phenolphthalein in alcohol. Titrate both immediately with 0.2*N* alcoholic potassium hydroxide solution. The difference between the two titrations is due to the rosin acids. The method consistently gives high results approximating 1 per cent, and the ASTM and the American Oil Chemists Society therefore recommend that 1 per cent shall be deducted from the result as representing a closer approximation to the correct value than is otherwise obtainable. If less than 5 per cent of rosin acids is found present, qualitative confirmation is essential and the results are of doubtful accuracy.

Calculations. Using a weighed sample of total fatty matter as sample,

$$\left[(\text{Ml. titration of sample} - \text{ml. titration of blank}) \times \text{normality of alcoholic potassium hydroxide} \times 0.346 \times \frac{100}{\text{weight of fatty acids}}\right] - 1$$
$$= \text{per cent rosin in total fatty acids}$$

$$\text{Per cent rosin in total fatty acids} \times \frac{\text{per cent total fatty acids in soap}}{100}$$
$$= \text{per cent rosin on basis of original sample}$$

$$\left(\text{Per cent rosin in total fatty acids} \times \frac{368}{346}\right) \times \frac{\text{per cent total anhydrous soap}}{100}$$
$$= \text{per cent sodium rosin soap on basis of original sample}$$

Using the total fatty matter from a weighed sample,

$$\left[(\text{Ml. titration of sample} - \text{ml. titration of blank}) \times \text{normality of alcoholic potassium hydroxide} \times 0.346 \times \frac{100}{\text{weight of sample}}\right] - 1$$
$$= \text{per cent rosin in sample}$$

By substitution of the factor 0.368 the value will be reported as sodium rosin soap; with 0.384, as potassium rosin soap.

Titer Value. Obtain the total fatty matter (page 391), and proceed as on page 353, starting with "Cool the fatty acids to 15 to 20° above the expected titer, and transfer . . . ".

Acid Value of Fatty Acids. Weigh 1 to 2 grams of total fatty matter into a flask, add 50 ml. of 95 per cent alcohol that has been neutralized to the phenolphthalein end point with $0.1N$ alkali solution, and heat to boiling. Proceed as on page 341, starting at "Shake or swirl the flask thoroughly . . . ". Since only free fatty acids are present, the titration will be much smoother than with a fat or oil.

Iodine Value of Fatty Acids. Using a sample of total fatty matter (page 391), follow the procedure for the Hanus method (page 345) or the Wijs method (page 346).

Glycerol, Sugars, and Starch. Since these are all carbohydrates and are therefore structurally related, the possible presence of each must be considered in determining the others and methods modified to fit the circumstances. Glycerol is a normal ingredient of soap, low in full-boiled soap and approaching or even exceeding 10 per cent of the soap content in half-boiled or cold-process soap. Sugars are rarely present, and if they occur it is usually in transparent soaps. Starch is rarely present, for there is no specific reason for its being added.

Glycerol by Bichromate Method. Weigh out a sample of soap equivalent to not more than 3 grams of glycerol. If starch is present, but no sugar, separate the alcohol-insoluble matter, which will include the starch. If the amount of alcohol-insoluble matter is low, the glycerol will be in the filtrate with sufficient completeness for the determination. If it is high, sorption of glycerol will be substantial. In that event, obtain the water-soluble matter, and combine that filtrate with alcohol-soluble matter. Use cold water for extraction, or the starch will be swelled and colloidally dispersed. Then proceed as though starch had not been present. If sugar is present, it is determined with the glycerol and corrected by a separate determination of sugar.

Dissolve the sample in 200 ml. of hot water in a 600-ml. beaker. Add 25 ml. of 1:4 sulfuric acid to decompose the soap and provide excess acid. If any residual alcohol is present, boil for 20 to 30 minutes to complete its removal, for alcohol if present will be determined as glycerol. If sugar is present, boil for this 20- to 30-minute period to hydrolyze it completely to invert sugar. Cool, and, if the titer value is high enough, fatty acids will solidify to a cake. Remove the acid-glycerol liquid to a 500-ml. calibrated flask, and rinse the fatty acids with water, adding the washings.

Dilute the contents of the flask to about 400 ml., and add about 0.25 gram of silver sulfate. This precipitates traces of soluble fatty acids and chlorides. A convenient device is to add about 30 ml. of a saturated solution and, after the precipitate settles, a few more milliliters to be sure that precipitation is complete. All chlorides and fatty acids must be removed. Dilute to volume, mix well, and allow the precipitate, which is negligible in volume, to settle out.

Filter a portion of the contents of the calibrated flask, discarding the first 10 to 15 ml. and protecting from evaporation. Pipet 50 ml. of the filtrate into a 250-ml. beaker. Measure 50 ml. of water into a similar beaker as a blank. To the sample add 75 ml. of a solution containing 74.553 grams of potassium bichromate per liter, and to the blank add 25 ml. of the same solution. This is a factor solution of which 1 ml. is equivalent to 0.0100 gram of glycerol. Add 25 ml. of concentrated sulfuric acid to each, stir thoroughly, and cover with watch glasses. To oxidize the glycerol and invert sugar if present, immerse the beaker in a water bath, and gradually raise the temperature to 90 to 100°, holding it there for not less than 2 of the 3 hours of heating. Similarly, heat the blank in another water bath. Oxidation of the glycerol may take place rapidly, and attention must therefore be given to this during the heating to prevent spattering and frothing.

Let the beakers cool, and transfer the contents of each to a 1-liter calibrated flask. Dilute each to volume, and mix well. From each flask pipet out 50 ml. of solution, and add about 50 ml. of water and 20 ml. of 10 per cent potassium iodide solution. Titrate each with $0.1N$ sodium thiosulfate solution, using starch indicator when near the end point. The final color is green owing to chromate.

CALCULATIONS. The calculation of results in terms of glycerol is complex, even after balancing out factors that recur if all the data are set up. The following detailed calculation will illustrate:

50 ml. aliquot of blank solution = a ml. of $Na_2S_2O_3$ solution.

$$1{,}000 \text{ ml. of blank solution} = a \times \frac{1{,}000}{50} = 20a \text{ ml. of } Na_2S_2O_3 \text{ solution.}$$

This is equivalent to 25 ml. of the original $K_2Cr_2O_7$ solution. Hence the 75 ml. of the $K_2Cr_2O_7$ solution added to the glycerol solution aliquot $= 3 \times 20a = 60a$ ml. of $Na_2S_2O_3$ solution.

50 ml. aliquot of oxidized glycerol solution = b ml. of $Na_2S_2O_3$ solution.

$$1{,}000 \text{ ml. of oxidized glycerol solution} = b \times \frac{1{,}000}{50} = 20b \text{ ml. of } Na_2S_2O_3$$

solution.

Hence, $60a - 20b$ = equivalent $K_2Cr_2O_7$ solution used, in terms of $Na_2S_2O_3$ solution.

Since 75 ml. of $K_2Cr_2O_7$ solution = $60a$ ml. of $Na_2S_2O_3$ solution, $(60a - 20b)$ ml. $Na_2S_2O_3$ solution $= [(60a - 20b)/60a] \times 75$ ml. of $K_2Cr_2O_7$ solution.

As 1 ml. of $K_2Cr_2O_7$ solution = 0.0100 gram glycerol,

$$75\left(1 - \frac{b}{3a}\right) \times 0.01 \times \frac{500}{50} \times \frac{100}{\text{weight of sample}} = \text{per cent of glycerol}$$

The exact concentration of the sodium thiosulfate is immaterial provided that the same concentration is used for sample and blank.

SUGAR. Dissolve 10 grams of the sample in 200 ml. of hot water in a 600-ml. beaker. If starch is present, remove as described for the determination of glycerol, and remove the alcohol completely. Decompose with 25 ml. of 1:4 sulfuric acid, and boil gently for 20 minutes to invert the sucrose completely. Use an ebullator (page 11), to ensure even boiling and to break the surface film of fatty acids. Cool, rinse, and remove the fatty acids. If the fatty matter is not solid, filter through a paper wet with water and wash with warm water. Extract the acid solution with 25 ml. of ether, and discard the extract. Remove the dissolved ether from the acid solution by heating on a steam bath, and neutralize with 10 per cent sodium hydroxide solution. Transfer to a 500-ml. calibrated flask, dilute to the mark, and thoroughly mix. Determine the invert sugar in 50 ml. of this solution by the Munson and Walker method (page 659). To calculate the sugar present in terms of sucrose, multiply the amount of invert sugar found by 0.95.

GLYCEROL IN THE PRESENCE OF SUGAR. The determination of glycerol as outlined will be in error by the amount of sugar present in the same solution. To correct, subtract from the glycerol as determined a factor for sugar found by the Munson and Walker method (page 659).

CALCULATION.

$$1 \text{ ml. of } K_2Cr_2O_7 \text{ solution} = 0.0100 \text{ gram of glycerol}$$
$$1 \text{ ml. of } K_2Cr_2O_7 \text{ solution} = 0.01142 \text{ gram of invert sugar}$$
$$\text{Determination in terms of glycerol} - \left(\text{per cent sugar} \times \frac{0.0100}{0.01142}\right) = \text{corrected glycerol}$$

STARCH. Separate the matter insoluble in alcohol, using a sample that will give not more than 3 grams of starch. Transfer the insoluble matter, without drying, to a 500-ml. conical flask with 200 ml. of water. Add 20 ml. of a mixture of 2 parts of hydrochloric acid with 1 part of water, and reflux for 2.5 hours. Cool the solution, and nearly neutralize with 10 per cent sodium hydroxide solution. If there is much turbidity, add a few milliliters of neutral aluminum hydroxide clarifier (page 718) and dilute to 250 ml. Filter, discarding the first 10 ml. Determine the dextrose present on 50 ml. of the filtrate, using the Munson and Walker method (page 659). This value multiplied by 0.90 equals the amount of starch.

Matter Volatile with Steam. A standard method of the ASTM is designed primarily for the estimation of naphtha in soaps. For this specific determination, if the highest accuracy is essential the standard methods should be consulted. The method that follows is more generally applicable:

The apparatus is shown in Fig. 101 and is largely self-explanatory. Steam is generated in a can or flask and is led into a 500-ml. Kjeldahl flask, selected because of the reflux action from condensation on the long neck. A long straight condenser is adequate. It delivers the condensate by means of an adapter into a graduated buret without stopcock and suspended by a slit stopper in a 500-ml. separatory funnel. Fill this funnel with water to just above the opening in the tip of the buret.

PROCEDURE. Place a suitable amount of sample, such as 100 grams of soap, in the flask with 100 ml. of distilled water. Acidify the sample with 1:1 sulfuric acid or strong phosphoric acid, and put in place. Pass steam through the flask containing the sample. Heat the flask containing the sample by a small Bunsen flame regulated so that the volume of liquid remains constant.

If the water-insoluble distillate is lighter than water, as is the case with naphthas, draw off water from the buret from time to time so that the upper layer remains between the graduations. Continue to pass steam until the volume of distillate no longer increases.

In the few cases in which this is applied to distillates heavier than water, the distillate accumulates in the bottom of the separatory funnel. Excess condensate may be allowed to overflow the separatory funnel if necessary. When distillation is complete, withdraw the insoluble distillate from the bottom of the separatory funnel, with some small amount of water, into a graduated cylinder of suitable size.

The distillate so separated may contain some volatile fatty acids. By taking a specific gravity (page 43) on the distillate the weight is deter-

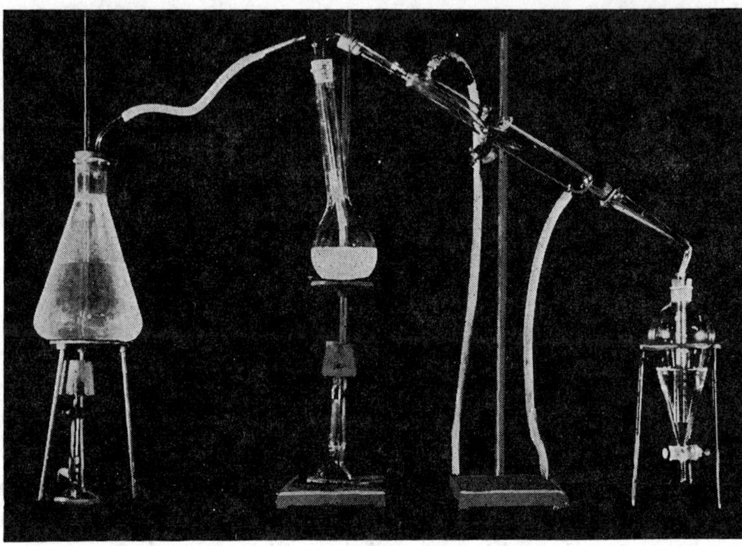

FIG. 101.—Setup for steam distillation.

mined; and by titration of the acid value (page 341), assuming that the nature of the acid is known from other parts of the determination, the weight present is corrected for the fatty acid.

Chlorides. Chlorides, as sodium chloride, are a normal ingredient of kettle soap to the extent of 0.1 to 0.5 per cent and may be present in half-boiled and cold-process soaps as contaminants. The amount present is often a convenient indicator of the process by which a soap was made. Two alternative methods are available.

PROCEDURE. *After Combined Alkali.* Carefully neutralize the acid filtrate that was left from determination of combined alkali and total anhydrous soap. The alkali used must be free from chlorides. Titrate the resulting solution with 0.1N silver nitrate to the end point with potassium chromate indicator (page 717). In the titration, chloride is first precipitated, and when this is complete red silver chromate is formed with excess and shows the end point. This is a difficult end point to identify without previous experience.

Direct Determination. Dissolve 5 grams of sample in about 300 ml. of hot water, boiling if necessary to complete solution. Add an excess of 20 per cent magnesium nitrate solution, which must be chloride-free. Usually about 25 ml. will be required. This precipitates the soap as magnesium soap. Without cooling or filtering off the precipitate add the potassium chromate indicator solution, and titrate with 0.1N silver nitrate solution.

CALCULATION.

$$\text{Titration in ml.} \times \text{normality of silver nitrate} \times 0.05845 \times \frac{100}{\text{weight of sample}} = \text{per cent sodium chloride}$$

To calculate as potassium chloride, use the factor 0.07455.

Sulfates. Sulfates will be present in soaps only by accidental contamination or introduction with the salt used to salt out the soap. Such salt reused from the glycerol department contains appreciable sulfate.

PROCEDURE. Digest the material insoluble in alcohol with 100 ml. of water, and cover the beaker with a watch glass. Carefully neutralize with concentrated hydrochloric acid, and add 2 ml. excess. Filter, wash the residue thoroughly, and dilute to about 250 ml. Complete as on page 160, starting at "Bring to a boil, and add, with constant stirring, sufficient boiling 10 per cent barium chloride solution . . . ". Calculate to sodium sulfate or potassium sulfate according to the type of soap.

Alkaline Builders. While Chap. 12, Fixed Alkalies and Alkaline Salts, has covered this subject in considerable detail, there are specific methods for determining some of these builders designed especially for application to soaps. They are therefore included here.

PROCEDURE. *Borax.* Weigh 10 grams of the sample, or 5 grams if more than 5 per cent of borax is present, into a platinum dish. Add 12 to 15 grams of a fusion mixture consisting of 200 grams of sodium carbonate and 15 grams of finely ground silica. This mixture gives a free-flowing readily soluble melt. With high percentages of borax relative to the carbonate present from the soap, glassy difficultly soluble fusions are obtained without the use of this fusion mixture. Add 15 ml. of alcohol to assist in making the mixture homogeneous. Mix with a glass rod, wash the rod with a little alcohol, and evaporate the mass to dryness on a water bath. Ignite until the combustible matter is destroyed.

Completely disintegrate the fusion by boiling with water, and transfer the solution to a 250-ml. round-bottomed flask. Acidify with 20 ml. of 1:1 hydrochloric acid, adding this carefully to prevent spattering. Heat nearly to boiling, and add a moderate excess of dry precipitated calcium carbonate. This precipitates the silica present and at the same time neutralizes the excess of hydrochloric acid. Connect the flask with a reflux condenser, for boric acid is appreciably volatile, and boil vigorously for 10 minutes to drive off all the carbon dioxide present. Filter off the precipitate on a folded filter, washing several times with hot water, but keeping the total volume of the liquid below 100 ml.

Return the filtrate to the flask, add a pinch of calcium carbonate to ensure complete neutralization, and again boil under a reflux condenser. Remove the flame, and connect the top of the condenser with a water pump. Apply suction until the boiling has nearly ceased. This treatment removes any carbon dioxide remaining. Cool to room temperature, add 50 ml. of neutral C.P. glycerol or 1 to 2 grams of C.P. mannitol, and titrate the solution with 0.1N sodium hydroxide solution, using phenolphthalein as indicator. After the end point is reached, add 10 ml. more of glycerol or 0.5 gram of mannitol and again titrate. Repeat this process until no further titration is obtained.

CALCULATION.

$$\text{Titration in ml.} \times \text{normality of alkali} \times 0.0955 \times \frac{100}{\text{weight of sample}}$$
$$= \text{per cent borax as } Na_2B_4O_7.10H_2O$$

Silica Present as Alkaline Silicates. When the material contains no mineral matter that is insoluble in water, ignite a sample of the soap, containing not to exceed 0.2 gram of silica, in a platinum dish at a low temperature. When this is charred, extract the soluble salts with water, return the paper and charred residue to the dish, and complete the ignition. Unite the residue in the dish and the water extract. Carefully acidify with hydrochloric acid, and finally add 5 to 10 ml. of concentrated hydrochloric acid in excess. Cover the container with a watch glass while adding the acid to prevent loss by spattering.

When the material contains mineral matter insoluble in water or when a determination of highest accuracy is not necessary, take a portion or all of the solution after titrating the matter insoluble in alcohol but soluble in water. This should be of a size to contain not more than 0.2 gram of silica. Add 5 to 10 ml. of concentrated hydrochloric acid to this for use as sample.

Proceed with the determination as described on page 189, using the entire procedure.

Calculation. In soaps it is desirable to calculate silica to the actual type present. This involves a knowledge of the $Na_2O:SiO_2$ ratio. The formula for calculation is

$$\text{Weight of silica} \times \frac{x\text{Na}_2\text{O}.y\text{SiO}_2}{y\text{SiO}_2} \times \frac{100}{\text{weight of sample}} = \text{per cent sodium silicate}$$

The most common ratio in silicated bar soaps is $1Na_2O:3.25SiO_2$; in this case $x\text{NaO}.y\text{SiO}_2/y\text{SiO}_2$ equals 1.308. In the absence of specific information it is well to make a trial calculation on this basis. For the small amounts used as stabilizers in toilet soaps one may use this ratio with reasonable safety, the difference from other ratios that might have been used being inconsiderable. In granules, beads, etc., the form is usually sodium metasilicate, a 1:1 ratio. Here, too, in the absence of more specific information, the 1:1 ratio should be assumed. For sodium metasilicate in its anhydrous and hydrated forms, Na_2SiO_3 and $Na_2SiO_3.5H_2O$, $x\text{Na}_2\text{O}.y\text{SiO}_2/y\text{SiO}_2$ is 2.032.

Phosphates. Use the solution of matter insoluble in alcohol but soluble in water for qualitative detection of the presence of ortho-, meta-, pyro-, and other complex phosphates.

Orthophosphate. Determine this on the filtrate after separation of silicates; phosphate cannot be determined until after the silicate is removed. Select an aliquot containing 0.04 to 0.1 gram of phosphorus pentoxide, and follow the procedure on page 203 starting at the beginning of the procedure.

Pyrophosphate. Boil a sample containing an equivalent of 1 gram of tetrasodium pyrophosphate with absolute alcohol and filter through a hot-water funnel. Wash well with hot absolute alcohol. Dissolve the insoluble salts in cold water. The filter paper may be added to the water in the beaker. Proceed exactly as on page 267.

Carbonates. These are insoluble in alcohol and are left on the filter when the alcohol-soluble matter is filtered (page 383). If carbonates are the only alkaline salts present, they can be titrated in the filtrate from determination of water-insoluble material (page 385). Otherwise, separate the alcohol-insoluble matter, and proceed as described in Chap. 12, taking into consideration the other salts found present.

Analysis of Ammonia and Amine Soaps. Soaps of this type are rarely, if ever, met with *per se.* They are employed as emulsifying agents in polishes, waxes, and cosmetics and sometimes in dry cleaners' naphtha-base soaps. As such, the amount present is often minor in quantity but important as an ingredient.

Ammonia, triethanolamine, and morpholine soaps are those most commonly used. Mono- and diethanolamine soaps and some tertiary amines are less frequently found, but the possibility of their presence must not be overlooked.

The ammonia and amine soaps are not commonly made by combination of the theoretically equivalent amounts of alkali and fatty acid; usually the alkali is in excess. Since these alkalies do not give a high pH, such an excess is seldom injurious. More often it is helpful.

The presence of ammonia soap or amine soap is detected qualitatively by tests described on pages 542 and 543, respectively. Such soaps are commonly found in wax emulsions, and suggestions for their determination are given on page 558.

CHAPTER 20

HEAVY-METAL SOAPS

The term heavy-metal soap is applied to soaps of a radical other than sodium or potassium. These metal radicals may cover a large range but only those of the metals shown in Table 16 are in common use. This table should not be given too great weight, for the proportion of heavy metal and fatty acid in commercial heavy-metal soaps is often not stoichiometric. The fact that they are made *in situ* in many products tends to cause even greater deviation in proportions.

Heavy-metal soaps differ from sodium, potassium, and amine soaps in that they are insoluble in water. The acid radicals usually present are naphthenic, oleic, palmitic, stearic, or resin acids. Because of the indefinite nature of the resin acids used, data such as are shown in Table 16 are not available for resinates.

Metallic soaps are used for a variety of purposes. They act as driers in paints, varnishes, lacquers, and inks; cobalt, lead, and manganese soaps being most commonly used for this purpose. As flatting agents to remove gloss from paints and inks, aluminum stearate is used when transparency is desired and calcium, zinc, and titanium soaps when transparency is not important. Copper and lead soaps are commonly employed in antifouling paints.

Aluminum, zinc, and calcium stearates and palmitates find use as thickening or suspension agents in nonaqueous media. A specialized field is the thickening of petroleum oils to form greases; this is dealt with in a separate chapter. Calcium and zinc stearates have been used in synthetic plastic molding powders to improve internal lubrication and to exert a plasticizing effect. They are also employed as mold lubricants. Aluminum stearate finds considerable use as a waterproofing agent in the textile industry.

Zinc stearate and, to a more limited extent, magnesium stearate are found in various cosmetic creams and in face powders to lend adhesiveness. To prevent adhesion of uncured rubber stock, articles are sometimes dusted with zinc and other stearates. As water-proofing agents for porous stone and cement, the metallic soaps, particularly those of calcium, find application. They also act as aids in grinding, plasticizing, and working, as wood-screw lubricants, in ceramics to produce different shades of color and luster, in wire drawing as lubricants, and in colored crayons to increase hardness and writing smoothness.

METHODS OF ANALYSIS

To analyze commercial metallic soaps as such is a relatively simple matter. When, however, the metallic soap is incorporated in a complex product in a minor amount of the order of 1 per cent or less, as is often the case, the analysis becomes entirely different. Normally, under these circumstances, the acid radical of the soap cannot be determined. It is even difficult to detect qualitatively except in the case of rosin, for which the Liebermann-Storch reaction may be used (page 395). The metal itself will commonly be found in the ash and is determined by the usual methods. The following methods are for metallic soaps themselves or for the fraction isolated from a more complex analysis:

Moisture. Weigh 5 grams of the sample in a dish, and dry for 1 hour at 105°. Cool, weigh, and calculate the percentage loss in weight. Moisture is ordinarily less than 1 per cent in commercial heavy-metal soaps. Higher amounts cause cloudiness and prevent gelling if used for grease manufacture.

Total Ash. Weigh 1 gram of the sample into a porcelain crucible. Ignite very carefully, preferably over a low flame, and finally more strongly until all carbon is gone. Placing a crucible lid on top of a nichrome triangle on the crucible will hasten the ashing process. Cool, weigh, and calculate the percentage of ash.

This method is not satisfactory for lead soaps; for some lead will be volatilized, and the rest will be present as metallic lead and a mixture of oxides. In that case determine lead by the following method:

Lead Soaps. Dissolve 1 gram of the sample as far as possible in toluene, and reflux with 1:5 sulfuric acid until the soap is completely broken down. Dilute with an equal volume of water, and filter off the precipitated lead sulfate through a weighed Gooch crucible. Wash with water and then with alcohol, dry in the oven, cool, and weigh.

CALCULATION.

$$\text{Weight of lead sulfate} \times 0.683 \times \frac{100}{\text{weight of sample}} = \text{per cent lead}$$

Insoluble Ash. Place the crucible containing the total ash in a beaker, cover with water, break up the ash with a glass rod, and boil for 20 minutes.

Filter, wash, dry, and ignite the insoluble ash in the same crucible. Cool, weigh, and calculate the percentage of insoluble ash. This is usually the oxide of the metal forming the soap.

Soluble Ash. This is the difference between the total ash and insoluble ash. In the case of aluminum stearate, the ash analysis affords much information as to its properties. A low insoluble ash indicates a high percentage of free stearic acid and poor gelling power in greases. A high insoluble ash points to low free stearic acid and good gelling power. The soluble ash should be as low as possible. It may consist of sodium sulfate, carbonate, or hydroxide, depending on the care used in manufacture. Large quantities cause cloudiness and break down the gel structure of greases.

Analysis of Ash. Ash another sample, burning out the carbon as completely as possible. Analyze as a mineral sample (see Chap. 10).

Free Stearic Acid in Stearate Soaps. This is present in smaller or larger quantities in nearly all types of aluminum stearate and in many other stearates. It acts as a peptizing agent and induces gelling. Low percentages give firm gels and good suspension. Higher amounts tend to thin gels. Excess free stearic acid in varnishes and lacquers will separate out in cold storage as gritty matter or will weaken the protective film.

To 2 grams of sample in a 250-ml. beaker add 50 ml. of neutral alcohol, SD3A will do. Allow to soak at room temperature for 30 minutes, stirring occasionally. The free stearic acid dissolves, the aluminum soap remaining insoluble. Filter through a paper, and wash well with neutral alcohol. Titrate the filtrate and washings with $0.1N$ sodium hydroxide, using phenolphthalein as indicator. Do not heat the sample in the alcohol. Any slight solubility of the metallic stearate in the alcohol will often titrate as free stearic acid, for the heavy metals used are frequently weak bases.

CALCULATION. Calculation is frequently carried out on the basis of the theoretical value for stearic acid. Practically, in commercial stearic acid the actual stearic of molecular weight 284.5 is usually less than half the total, the balance being palmitic acid of molecular weight 256.5. Therefore, as a practical matter the most satisfactory factor to use is about 270.

$$\text{Ml.} \times N \times 0.270 \times \frac{100}{\text{weight of sample}} = \text{per cent commercial stearic acid}$$

Bulkiness. Place 10 grams of the sample in a 100-ml. graduate, and note the volume. Tap sixty times on a rubber surface, and again note the volume. This gives a rough indication of particle size and shape.

Fineness. Place 5 grams of the sample in a screen, using 325 mesh for zinc soaps and 200 mesh for others. See page 70 for details of the technic. Wash with denatured alcohol until the washings are no longer milky. Dry, weigh, and calculate the percentage residue.

CHAPTER 21

LUBRICATING GREASES

In general, greases are lubricating oils bodied by soaps. For stiff greases, sodium stearate soap is commonly used, sometimes modified by the addition of calcium stearate. Aluminum, lead, potassium, and zinc and, less frequently, barium, lithium, iron, and strontium are also employed, giving particular characteristics to the greases. Often a grease contains two types of soap. Aluminum-base greases are usually transparent, and less of the soap is needed to give the same body and consistency than in a calcium- or sodium-base grease. They are water-insoluble and consequently anhydrous, whereas calcium-base greases required 0.5 to 1.5 per cent of water to act as an emulsifier. Lead soaps give greases with little body and are used for added film strength.

Occasionally rosin oil, beeswax, montan wax, degras, candle pitch, spermaceti, and other materials such as asphaltic material are added with the lubricating oil. Fillers, such as asbestos, chalk, gypsum, mica, talc, and wood flour, are sometimes used in low-grade greases. Graphite is occasionally present in moderate amounts in expensive types.

METHODS OF ANALYSIS

The methods of analysis that follow involve much work, the technics being far from easy. Even when the data are all obtained, it is often not possible to formulate a grease exactly from the analysis, especially if more than one type of soap is present. In such cases the final adjustments are made on the basis of experimental laboratory production to match the sample.

Ash. The quantitative determination of ash is in itself usually not significant. Interaction between the alkali of the soap and the filler or even the porcelain crucible may occur. Carbon is difficult to burn off, and calcium may be present in the ash as carbonate or oxide, or both. It is usually helpful to ash a weighed amount of the grease in porcelain if lead and zinc soaps are present or in platinum if they are absent. Heat at as low a temperature as possible. Avoid ignition of the sample. After complete charring,

TABLE 16.—CONSTANTS OF SOME COMMERCIAL METALLIC SOAPS

Soap	Appearance	Theoretical formula	Molecular weight	Specific gravity	Melting point, °C.	Insoluble ash, %	Soluble, ash %	Solubility
				Naphthenates				
Cobalt	Purple	...	....	1.268	90	12.0 (Co)	...	Benzene, petroleum naphtha, methanol
Lead	Viscous brown liquid	...	653	.....	50	30.0 (Pb)	...	Benzene, petroleum naphtha, methanol
Manganese	Brown	...	....	1.189	90	11.0 (Mn)	...	Benzene, petroleum naphtha
				Oleates				
Aluminum	Light brown	...	....	1.012	180	6.5 (Al_2O_3)	...	Gels in petroleum naphtha at 150°
Copper	Hard waxy, blue	...	....	.....	Indefinite			
Lead	Viscous brown liquid	...	....	1.378	60	23.0 (Pb)	...	Benzene, petroleum naphtha
Nickel	Green-blue oil	...	....	.....	20	....	...	Partly soluble in water
				Palmitates				
Aluminum	White powder	$Al(OH)_2.C_{16}H_{31}O_2$	316	1.072	200	17.0 (Al_2O_3)	1.0	Gels in benzene at 18°, in petroleum naphtha at 50°
Copper	Voluminous, crumbly	...	....	.....	110	....	...	Slightly soluble in water
Lead	Viscous brown	...	....	1.378	60	23.0 (Pb)	...	Benzene, petroleum naphtha
Nickel	White, crystalline, opaque	...	....	.....	45			
Zinc	Voluminous, white	$Zn(C_{16}H_{31}O_2)_2$	575	.....	103	18.0 (ZnO)	1.0	Partly soluble in benzene
				Stearates				
Aluminum (mono-)	White powder	$Al(OH)_2.C_{18}H_{35}O_2$	344	.....	158–163	10.0	1.0	Gels in benzene, petroleum naphtha
Aluminum (di-)	White powder	$Al(OH)(C_{18}H_{35}O_2)_2$	610	1.009	145	8.0	0.6	Gels in benzene, petroleum naphtha
Aluminum (tri-)	White powder	$Al(C_{18}H_{35}O_2)_3$	876	.....	91–99	6.0	0.6	Gels in benzene, petroleum naphtha
Barium	White powder	$Ba(C_{18}H_{35}O_2)_2$	703	.....	140–150	26.8	1.7	Insoluble
Calcium	White powder	$Ca(C_{18}H_{35}O_2)_2$	606	1.035	150	....	...	Insoluble
Cerium	White powder	$Ce(C_{18}H_{35}O_2)_3$	989	.....	100–110	18.5	0.2	Partly soluble in benzene, petroleum naphtha
Chromium	Green solid	$Cr(C_{18}H_{35}O_2)_3$	901	.....	95–100	8.4	2.7	Partly soluble in benzene, petroleum naphtha
Cobalt	Purple powder	$Co(C_{18}H_{35}O_2)_2$	625	.....	115–130	12.6	1.7	Gels in benzene, petroleum naphtha
Copper	Green powder	$Cu(C_{18}H_{35}O_2)_2$	629	.....	100–110	12.8	0.6	Partly soluble in benzene, petroleum naphtha
Lead	White powder	$Pb(C_{18}H_{35}O_2)_2$	774	.....	100	23.0 (Pb)	...	Insoluble
Magnesium	White powder	$Mg(C_{18}H_{35}O_2)_2$	590	1.028	140	8.0	0.6	Insoluble
Manganese	Pink powder	$Mn(C_{18}H_{35}O_2)_2$	621	.....	100–110	13.2	0.2	Gels in benzene, petroleum naphtha
Nickel	Green powder	$Ni(C_{18}H_{35}O_2)_2$	625	.....	150–160	11.4	1.6	Benzene, gels in petroleum naphtha
Strontium	White powder	$Sr(C_{18}H_{35}O_2)_2$	654	.....	130–140	21.4	0.5	Partly soluble in benzene, petroleum naphtha
Tin	White powder	$Sn(C_{18}H_{35}O_2)_2$	684	.....	45–50	22.7	0.1	Benzene, petroleum naphtha
Titanium	White powder	$Ti(C_{18}H_{35}O_2)_4$	1180	.....	55–70	17.7	0.2	Petroleum naphtha
Zinc	White powder	$Zn(C_{18}H_{35}O_2)_2$	631	1.095	120	15.5	0.5	Insoluble

place a clean Nichrome triangle on the dish, and put a porcelain cover on it to reflect the heat downward. Continue heating until the ash becomes almost white. It is not necessary and is often not possible to burn off the last traces of carbon. If only sodium or potassium soaps and no fillers are present, a higher temperature may be employed. Cool and weigh as an approximation, bearing in mind that carbon is often present and that ash from the soaps and filler may have reacted. Cross-checking calculations with later data are often useful.

PROCEDURE. *Alkaline Carbonates.* Dissolve the soluble part of the ash in hot water, filter, and wash. Cool and titrate the filtrate and washings to phenolphthalein and methyl orange end points. A simple calculation (page 260) will give the amount of sodium or potassium carbonate if only one of these is present, provided, of course, that no interaction has occurred with the filler. If lime soaps are present, a small amount of calcium oxide will be dissolved in the water and will be titrated also.

In extreme cases, determine the alkali in the ash and then the total soap fatty acids by a method to be described later (page 413). From the fatty acids calculate the corresponding amount of alkali required, and compare with the figures obtained by analysis of the ash. A reasonably accurate estimation can then usually be made.

Other Metals. Now heat the insoluble residue, if any, with 1:10 hydrochloric acid, and again filter and wash with hot water. Add the filtrate and washings to the titrated solution. Reserve this solution for mineral analysis.

As not more than two soaps are usually present together in a grease, the analysis of this ash solution is not difficult unless reaction has taken place between alkaline carbonates and filler. Graphite, on ashing and treating with acid, may give tests for small amounts of iron and calcium. Talc and asbestos will give small amounts of aluminum, which will be significant, for the amount of aluminum in an aluminum soap is small. Gypsum will dissolve in acid.

Under some difficult circumstances more accurate results can be obtained on the acid layer after breaking down the grease as described later (page 411). This treatment prevents the interaction of the alkaline carbonate with the filler that may occur on ashing. In other cases, treatment with 1:10 hydrochloric acid will effect partial or complete decomposition of the filler.

Dilute the solution obtained by one of the preceding methods to 250 ml. in a volumetric flask, and perform the necessary qualitative tests to determine what metals are present. Use 50-ml. aliquots as samples. Silica, if present, is due to reaction with a filler and indicates that a different method of preparation of the sample should be elected.

Barium. If barium is present, determine as the sulfate by the procedure on page 189.

Aluminum and Iron. If aluminum is present, phosphorus should be absent and the procedure on page 192 is applicable. Start at "To 50 ml. of the sample solution or other aliquot . . . ". If iron is also present, which is rare in greases, determine it in the precipitate by one of the methods that immediately follow.

Calcium. When calcium is present, it will not be complicated by the presence of phosphorus. Aluminum, if present, will already have been determined on the same aliquot. Therefore it can be titrated as the oxalate by the procedure given on page 216. Start at "Heat the sample solution to boiling, and add 20 ml. of boiling ammonium oxalate solution . . . ".

Magnesium. If calcium and/or aluminum were present, they will have been precipitated. Use the same aliquot and the procedure on page 218. Start at "Add 10 ml. of 10 per cent disodium hydrogen phosphate solution . . . ".

Lead. If lead is present, determine it on an aliquot. If calcium was absent, use the method on page 501, starting at "Add concentrated ammonium hydroxide until just alkaline . . . ". If calcium is present, use the method on page 512, starting at "Add 20 ml. of 1:1 sulfuric acid, and evaporate to copious fumes of sulfur trioxide . . . ".

Zinc. If zinc is present, determine it volumetrically on an aliquot according to the procedure on page 504, checking for possible interfering substances.

GENERAL PROCEDURE

Light-colored Greases. This procedure should be employed for cup, fiber, and sponge greases, with or without graphite, and all comparatively light-colored greases, including axle greases.

SEPARATION. Use a well-mixed 10 to 20-gram sample for No. 3 cup grease. For thin transmission and other greases use 20 to 30 grams. Weigh to within 0.1 gram in a 250-ml. conical flask. If thick, warm slightly in an oven. Add 50 ml. of 1:3 hydrochloric acid, and warm on the steam bath. Stir thoroughly to break down lumps, or shake carefully after inserting a rubber stopper in the flask. Heat until the upper layer is clear. Filter both layers successively through a weighed Gooch crucible or through one crucible for the aqueous layer and another for the petroleum ether layer. Wash the flask and crucible or crucibles successively with water and with petroleum ether until all the oil is washed through. Reserve this extract for the analysis of the grease. Finally, wash the crucible or crucibles with alcohol, collecting these washings separately and discarding them. Dry at 120°, cool, and weigh. The insoluble matter may be asbestos, talc, mica, wood pulp, etc. Microscopic examination will often give valuable clues to the composition. Gypsum will have been largely dissolved and alkaline earth carbonates completely and talc attacked to a considerable extent. Analysis of this insoluble matter is sometimes desirable as a supplement to analysis of the ash.

Modified Separation. The filtration of two immiscible liquids is often a very difficult operation. To avoid this weigh out 5 grams of the grease, add 75 ml. of trichlorobenzene, and heat to about 165°. Centrifuge while still hot. Decant off the bulk of the solvent. Heat the residue with 50 ml. of fresh trichlorobenzene, and repeat the operation. Shake the residue with 75 ml. of a hot 1:1 alcohol-benzene mixture, filter through a tared filter paper, and wash with the same mixture. Dry the residue in the oven, and weigh.

The trichlorobenzene solution can be used for the grease analysis after carefully fuming off the solvent from a hot plate, the last traces being disposed of by repeated additions of small amounts of benzene. Normally, however, it is better to start with the original sample and proceed as later described for dark-colored greases. Difficult filtrations are thus avoided.

Dark-colored Greases. This method should be used for dark greases containing residuum, asphaltic oils, asphalt, tar, etc., because of their tendency to form stubborn emulsions when an attempt is made to shake them out with acid and petroleum ether.

Separation. Weigh 8 to 30 grams of the sample, depending on its consistency, into a 3-inch porcelain dish, and add 10 grams of technical granulated potassium acid sulfate and 10 grams of ignited sand. Heat on the steam bath, with frequent stirring, until all water is driven off and the grease is broken down, about 2 hours usually being sufficient. Cool, break up any lumps, and transfer to an extraction thimble, using petroleum ether to wash out the dish into the thimble, which should already be in the Soxhlet tube. Extract the contents of the thimble thoroughly with petroleum ether, until the extract contains no more color. Reserve this extract.

Extract the thimble a second time with carbon disulfide until the extract is colorless. Evaporate the solvent and heat the extract at 120° for 1 hour. Cool and weigh the asphaltic and tarry matter. Perform qualitative tests to identify the extract. Use the residue in the thimble for examination for fillers as described for light-colored greases (page 411).

Gypsum. Because of its solubility in acid, gypsum must be determined separately. Weigh 5 grams of grease in a 150-ml. beaker, and decompose with 50 ml. of petroleum ether and 25 ml. of concentrated hydrochloric acid, warming on the steam bath until all the gypsum is dissolved. Cool, transfer to a separatory funnel, and rinse the beaker with water and a little petroleum ether. Allow the two layers to separate, and draw off the acid layer into the beaker. Wash the petroleum ether layer with two 20-ml. portions of 1:2 hydrochloric acid, and add the washings to the strong acid in the beaker. Evaporate the acid solution on the steam bath to near dryness; then dilute with 150 ml. of water. Complete as on page 160, starting at "Bring to a boil, and add with constant stirring sufficient boiling 10 per cent barium chloride solution . . . ". Calculate the gypsum, $CaSO_4.2H_2O$.

Soap. With dark greases use the petroleum ether extract obtained after decomposition with potassium acid sulfate.

If fillers were found in light greases, use the combined washings from that determination. If none were found, it is preferable to start with a new grease sample of the amount specified for the determination of fillers. In the latter case, transfer this sample to a 250-ml. conical flask. Warm slightly in an oven, if necessary to thin, and carefully add 75 ml. of petroleum ether. When possible, break up any lumps with a rod; otherwise, insert a rubber stopper in the flask, and shake well. Add 50 ml. of 1:2 hydrochloric acid, and again shake carefully until complete decomposition results. Occasionally, stubborn greases require refluxing with the petroleum ether and acid in order to completely break them down.

Titration of Fatty Acids. The fatty acids may be titrated, although more commonly they are recovered gravimetrically. For titration, transfer the cold petroleum ether and acid layers to a separatory funnel. Wash the flask with petroleum ether and with water. Allow the layers to separate completely, and draw off the lower acid layer into another separatory funnel. Wash the petroleum ether layer with three 25-ml. portions of water, and add the washings to the acid layer. Wash the acid layer with two 20-ml. portions of petroleum ether, and reserve the washed acid layer for later glycerin determination, if desired. Now wash this 40 ml. of petroleum ether with 15 ml. of water. Discard the wash water, and add the petroleum ether to the main bulk of petroleum ether solution. Let stand for complete separation of the last trace of aqueous layer. Titrate the petroleum ether solution with $0.5N$ alcoholic potassium hydroxide, using phenolphthalein indicator. Add a little water if the petroleum ether and alcohol layers do not readily separate, but not enough to dilute the alcohol to less than 50 per cent. By so doing, a clearer end point in the alcohol-water layer will be obtained. Shake thoroughly after each addition of alcoholic potassium hydroxide, and allow to settle.

Calculation.

$$\text{Ml. of alcoholic potassium hydroxide used} \times \text{normality} \times 0.2825 \times \frac{100}{\text{weight of sample}} = \text{per cent free and soap fatty acids as oleic}$$

Gravimetric Separation of Fatty Acids. Normally, it is desired to recover and weigh the fatty acids. They then furnish a sample for identification of the fat or fatty acid originally used. For this purpose titrate the petroleum ether layer roughly as in the preceding method, add a slight excess of alkali, and draw off the lower alcoholic layer after complete separation. Wash the petroleum ether layer thoroughly with 30-, 25-, and 20-ml. portions of 50 per cent alcohol, and add these washings to the first alcoholic layer. Wash the combined alcoholic solution with 25 ml. of petroleum ether, and draw off into a beaker, draining carefully. Add the petroleum ether wash to the main solution. Evaporate off the alcohol in the beaker, and take up the soap in water. Transfer to a separatory funnel, and cool. Just acidify with 1:2 hydrochloric acid, and extract with 50- and 25-ml. portions of ethyl ether. Wash the combined ethyl ether extracts in a separatory funnel with two 20-ml. portions of water. Discard these washings. Drain the water from the ethyl ether, and evaporate the ether on a steam bath. Finally finish for a few minutes in the oven until a residual trace of ether is gone. If necessary, while on the steam bath, add a few milliliters of absolute alcohol to drive off any water present. Cool, and weigh the fatty acids.

Identification of Fatty Acids. The methods described in Chap. 17, Saponifiable Fats and Oils, are applicable at this point. Normally, an acid value (page 341) will indicate the extent to which acids other than oleic, stearic, and palmitic are present. An iodine value (page 345) will give a very approximate indication of the relative amounts of saturated and unsaturated acids originally present, approximate because the previous history of the sample, as well as laboratory manipulation, will have affected this value. To the experienced worker, appearance and odor are valuable guides.

Calculation. From the total fatty acid present, subtract the fatty acid found present as free acid by a method described later in the analysis (page 417). The remaining fatty acid is that originally present as soap. The nature of the soap will be that indicated by the ash analysis.

$$\text{Weight of fatty acid} \times \text{factor} \times \frac{100}{\text{weight of sample}} = \text{weight of soap}$$

Since much palmitate is often present with the stearate, it is an open question which factor to use, that based on molecular weight 284.5 for stearic acid or that based on 270. Because the soaps are often made *in*

Fatty acid	Factor stearic acid = 284.5	Factor stearic acid = 270
To aluminum stearate	1.028	1.030
To calcium stearate	1.067	1.071
To lead stearate	1.361	1.380
To zinc stearate	1.111	1.112
To barium stearate	1.238	1.251
To strontium stearate	1.150	1.159
To lithium stearate	1.021	1.022

situ from commercial fats containing considerable oleate, it is most common to use the higher factor. If the soap is shown by the analysis to have been made from commercial stearic acid, use the lower factor.

Glycerin. The presence of glycerin in the grease indicates that it was made from whole fats and not from fatty acids. Evaporate the acid solution obtained by the decomposition of the soap in the grease to a small volume. To prevent loss of glycerin do not evaporate all the water. Add sufficient dry sodium carbonate to neutralize the residual acid and to precipitate calcium and other metals. Evaporate to dryness at as low a temperature as possible. Extract the dried residue several times with small portions of absolute alcohol. This may be done by washing on a filter paper. Evaporate the alcohol; and, if desired, weigh the extracted glycerin after heating a few minutes in the oven. A little sodium chloride may be present owing to moisture retained in the dried salts by the deliquescent glycerin. A second absolute alcohol extraction of the first extract will separate most of the sodium chloride. The initial evaporation of the acid solution before neutralization minimizes the amount of acid to be converted to sodium chloride.

This method is far from exact, owing to losses of glycerin in evaporation and elsewhere. Carefully applied, the glycerin obtained should roughly check the amount of metallic soap found if the soap was made from fats rather than fatty acids. Confirm the identity of the residue separated as glycerin, after weighing, by the acrolein test (page 466).

Neutral Fat. The petroleum ether solution from the soap determination contains any free neutral fat that may have been present in the grease, as well as the petroleum oil and any other unsaponifiable matter. Concentrate the solution to about 125 ml. in a 250-ml. conical flask, preferably by evaporation on a steam bath. Add 10 ml. of 0.5*N* alcoholic potassium hydroxide and 50 ml. of neutral alcohol. Reflux under a condenser for 1 to 2 hours. Titrate the excess alkali with 0.5*N* hydrochloric acid, and calculate the alkali used to saponify the fat to the corresponding amount of fatty acid in exactly the same way as in the soap fatty-acid determination.

Add a slight excess of alkali, and draw off the lower alcoholic layer after complete separation. Complete as for gravimetric separation of fatty acids from soap (page 414), starting at "Wash the petroleum ether layer thoroughly with . . ."

Multiplying the weight of fatty acid by 1.045 gives a very close approximation of the weight of neutral fat from which it is derived. The factor varies little with the molecular weight of the fat.

Petroleum Oil. Evaporate the petroleum ether solution from the neutral fat determination. Place on a water bath for a few minutes and finally in an oven at 110° until constant weight is obtained. Should the oil be partly volatile, as is sometimes the case, dry until all odor of petroleum ether has gone, to obtain an approximate value. The final weight will be that of the petroleum oil and any other unsaponifiable matter that may be present. Normally, it is not possible to separate such mixtures.

When constants on the oil such as viscosity at various temperatures are required, as is frequently the case, break down a much larger sample of 150 to 200 grams of the grease, omitting all quantitative operations, and following the method on page 412. As it is difficult to handle large volumes of acid and petroleum ether solution and break down such large amounts properly in this manner, the method for dark greases is advisable for such large quantities.

It is not feasible to obtain an accurate flash point (page 669) on the petroleum oil so obtained owing to the treatment it has undergone. This constant is often obtained on the grease itself. A moderate degree of accuracy may be expected with a rather viscous oil of high flash point. It is usually necessary to keep large amounts of such separated oils in the air oven at 110° at least overnight to remove the last traces of petroleum naphtha, before getting the flash point by the Cleveland open-cup method.

Free Alkali and Free Acid. FREE ALKALI, INCLUDING CARBONATES. Weigh 10 to 30 grams of the grease into a 250-ml. conical flask, add 75 ml. of petroleum ether, insert a clean rubber stopper, and shake vigorously to disintegrate the grease. Remove the stopper, and wash down with a little of the solvent. Add 50 ml. of neutralized alcohol containing phenolphthalein, and again shake thoroughly. If, after settling a few minutes, the lower alcoholic layer is pink, the grease contains either free caustic alkali or an alkaline carbonate. A qualitative test on the grease will show which is present.

Add 10 ml. of 0.5N hydrochloric acid, or more if necessary. Boil very carefully on a steam bath or hot plate to expel any carbon dioxide, taking care not to ignite the volatile petroleum ether. Then titrate the excess acid with 0.5N alcoholic potassium hydroxide. Calculate the difference between the volume of 0.5N acid used and this titration, in terms of the hydroxide or carbonate present. If both are present, an approximation must be taken; there is no method of calculation to carbonate in the presence of hydroxide, or vice versa.

CALCULATION.

$$\text{Volume of acid used} \times 0.5 \times 0.040 \times \frac{100}{\text{weight of sample}} = \text{per cent sodium hydroxide}$$

Other factors:

Potassium hydroxide	0.0561
Calcium hydroxide	0.03705
Sodium carbonate	0.0530
Potassium carbonate	0.06902
Calcium carbonate	0.05005

FREE ACID. Should the original alcoholic layer not be pink, it is neutral or contains free fatty acid. If soaps of iron, aluminum, zinc, or other weak bases are present, they will also titrate. For the determination titrate the warm solution carefully with $0.5N$ alcoholic potassium hydroxide, shaking well after each addition. Calculate this determination to free stearic acid exactly as illustrated for free alkali, using the factor 0.2845 for stearic acid and the alkaline titration in place of the acid titration called for. In the presence of soaps of weak bases calculate the amount present, as shown by analysis of the ash, to stearic acid, and subtract this correction from the free acid as titrated. For example, suppose that 2.5 per cent of sodium carbonate and 0.5 per cent of calcium carbonate based on the original sample have been found in the ash and that the total fatty acids found are 18.0 per cent. For simplicity, the more common bases are used. The amount of fatty acid, which may be assumed for purposes of illustration to be stearic, that is equivalent to the sodium carbonate present will be

$$\frac{\text{Molecular weight of stearic acid}}{\text{Half molecular weight of sodium carbonate}} \times 2.5$$

or

$$\frac{284.45}{53} \times 2.5 = 13.42 \text{ per cent}$$

then substituting the values for calcium carbonate, the equivalent fatty acids are

$$\frac{284.45}{50.05} \times 0.5 = 2.84 \text{ per cent}$$

The total soap fatty acids are $13.42 + 2.84 = 16.26$ per cent. The free fatty acids are therefore $18.0 - 16.26 = 1.74$ per cent. Owing to the determinations by difference such a value is only an approximation.

Water. SOLVENT DISTILLATION. Weigh a sample of grease, up to 50 grams depending on the moisture expected, into a 500-ml. round-bottomed shortnecked flask. Complete as on page 41, starting at "Add 100 to 150 ml. of toluene or xylene, and . . . ". To prevent frothing it is well to add to the flask 10 to 20 ml. of oleic acid that has been freed from water by heating for 5 to 10 minutes over a free flame at a temperature of about 130°.

Alternatively, use a 500-ml. conical flask connected to a vertical condenser, the lower end of which dips into a 100-ml. American Society for Testing Materials (ASTM) centrifuge tube (page 691). Use well-fitting corks, covered with metal foil, or ground glass joints for connecting the glass tubing to the flask and the condenser.

Heat the flask so that the distillation will be carried on as rapidly as possible without carry-over of the xylene to the condenser. When water is no longer observed to condense on the walls of the connecting tube, drain the cooling water from the condenser and continue the distillation until approximately 100 ml. of liquid has been collected in the receiver. The time of distillation should not exceed 15 minutes.

Centrifuge the receiver as described under water and sediment in petroleum products by the centrifuge (page 690), until the water and xylene layers separate clearly. Read the volume of water, estimating to the nearest one-half of the scale division of the receiver.

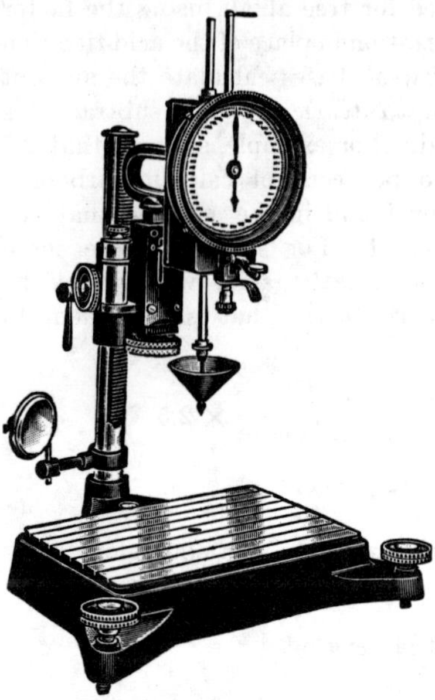

FIG. 102.—ASTM penetrometer for greases.

This method has the advantage over the use of the Dean-Stark tube that any water adhering to the side of the receiver is centrifuged to the bottom. This is especially helpful because greases frequently contain very little water.

When a grease contains free alkali, correct for the amount of water formed by reaction of the fatty acid and any free alkali present. For example, multiply the percentage of free alkali as sodium hydroxide by 18/40 = 0.45 or as calcium oxide by 18/56.1 = 0.32, and subtract this figure from the percentage of water found.

Consistency of Lubricating Greases and Petrolatum. This test is a measure by an ASTM method of the worked or unworked consistency of lubricating greases having a worked consistency of less than 400 and of the original consistency of petrolatum.

The original, or unworked, consistency of lubricating greases depends on a number of difficultly controllable factors, such as the soap content, the kind of fats used, the method of manufacture, the final water content, the rate of cooling, and the temperature of filling containers. This test must be made on the sample in its original container, for any working or remelting will change the consistency. A 1-pound can, approximately 8 cm. in diameter, is suitable. The consistency changes also on standing, most rapidly during the first 48 hours after manufacture. Therefore, the test should not be made on fresh grease. An exception to the limitation on remelting applies in the case of petrolatum. It may be melted, poured into a suitable container, and cooled to the temperature for testing.

The worked consistency, which is the value of practical importance for most lubricating greases, bears no definite relationship to the unworked consistency. However, for very hard greases, such as railroad greases, slabs of which may be tested, the unworked consistency is generally the property of practical importance.

The apparatus used is a penetrometer with a special cone (Fig. 102), which allows the cone to drop without measurable friction and accurately measures the depth of penetration. The scale of the penetrometer is calibrated in tenths of a millimeter.

The cone, of stainless steel or brass, with a detachable hardened-steel or stainless-steel tip, is shown in Fig. 103. Cones of the dimensions required are made and sold specifically for this purpose. The total moving weight of the cone and attachments is 150 grams.

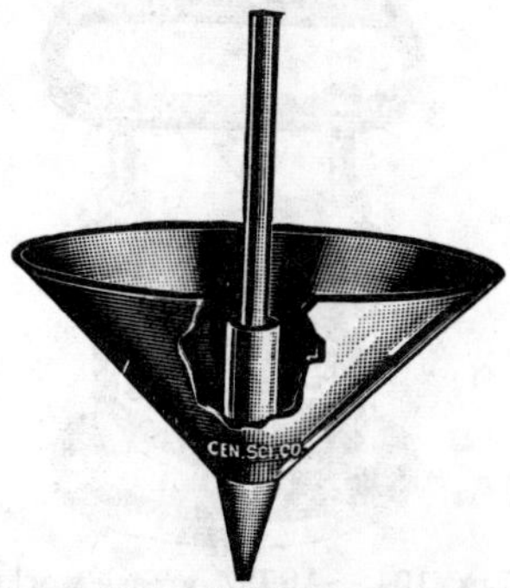

FIG. 103.—ASTM penetrometer cone.

PROCEDURE. *Unworked Consistency of Lubricating Grease and Petrolatum.* Bring the temperature of the sample to 25 ± 0.5°. If the sample is initially within 1.5 to 2° of this temperature, place it in a water bath maintained at 25° for 30 to 40 minutes to ensure that the sample is of the same temperature throughout. When the initial temperature is outside this range, place in a constant-temperature bath for at least 90 minutes. If the room temperature is more than 1.5 to 2° from 25°, place a lid on the can, seal with grease or petrolatum to prevent entrance of water, and immerse the can in the bath for the required temperature. This ensures the surface of the sample being at the correct temperature.

Cut the sample to a level and very smooth surface with a sharp knife, being careful to avoid working the surface. Place the can on the penetrometer table, and lower the cone until the tip just touches the surface. This is done accurately by noting when the tip and its shadow just meet. It is particularly important with soft greases with unworked consistency of 310 or more for the cone tip to be placed as nearly as possible in the center of the sample. With older penetrometers it is necessary to set the scale at zero. With newer models the scale and cone move as a unit, and no setting is needed. Now quickly release the cone for 5 seconds. Read the total penetration in tenths of a millimeter and report as the unworked consistency.

The total surface area disturbed by the test will have a diameter about equal to the measured depth of the penetration. In order to prevent one test from being affected by the disturbed area of a previous test or by the sides of the can, never place the tip nearer the sides of the can or the edge of the previous hole than the penetration distance of that particular lubricating grease or petrolatum. The sample must not be smoothed over for further tests. Therefore, several samples will be needed for soft lubricating greases or petrolatum.

Make 5 tests and report the average as the unworked consistency if the mean deviation of the readings does not exceed 3 per cent. Should this value be exceeded, take the average of 10 tests.

FIG. 104.—ASTM grease worker.

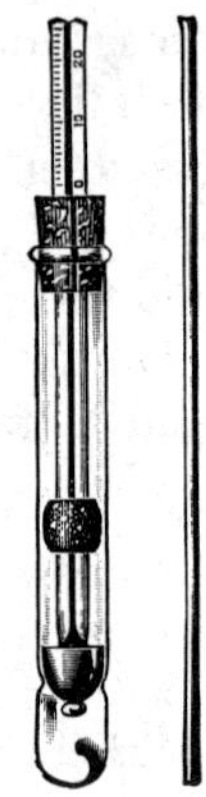

FIG. 105.—ASTM apparatus for dropping point of lubricating greases D566-42T.

Worked Consistency of Lubricating Greases. For this test a grease worker such as is shown in Fig. 104 should be employed. Routine tests can be made without it but do not strictly conform to ASTM standards. When no grease worker is available, work the sample thoroughly for 5 minutes on a smooth surface such as a marble slab, using a large spatula. Avoid occlusion of air.

To conform to ASTM standards fill the grease worker with a minimum inclusion of air, bring to 24 ± 1°, and work with 60 full double strokes of the plunger. For stiff greases a wall bracket to hold the grease worker is helpful. Remove the top and the plunger, and, if necessary, bounce the grease worker a few times to remove occluded air from soft greases. Adjust the temperature to 25 ± 0.5°. Smooth the surface, and proceed with the test exactly as described for unworked consistency. When one test is completed, smooth over the surface, avoiding air pockets, and make a further test. The number of tests and the mean deviation are the same as for unworked consistency.

Owing to the rapid change in consistency of cold-set lubricating greases after working, it is not satisfactory to attempt to determine their worked consistency.

Dropping Point of Lubricating Grease. The ASTM dropping point is the temperature at which the grease passes from a semisolid to a liquid state under the conditions of the test.

For this test use a special chromium-plated brass cup as shown in Fig. 105. A special Pyrex test tube with rim, 4 inches in length by $^{15}/_{32}$ inch inside diameter and provided with three indentations equally spaced, about $^{3}/_{4}$ inch from the bottom, is also needed. This assembly is shown in Fig. 105. Two ASTM partial-immersion thermometers reading -5 to $+300°$ are required.

PROCEDURE. Adjust the thermometer in the Pyrex tube so that the tip of the bulb is about 3 mm. above the bottom of the grease cup. Suspend the assembled apparatus by means of a clamp in an oil bath consisting of a 400-ml. beaker and suitable oil. Suspend the second thermometer in the oil bath so that its bulb is approximately at the same level as the bulb of the thermometer in the test tube.

Remove the grease cup, and fill by pressing the larger opening into the grease to be tested until the cup is filled, taking care to avoid working of the grease as much as possible. Remove excess grease with a spatula. Hold the cup in a vertical position with the smaller opening at the bottom, and gently press it down over a polished metal rod $^{3}/_{64}$ to $^{1}/_{16}$ inch in diameter and 6 inches in length until the latter protrudes about 1 inch. Press the rod against the cup in such a manner that the rod makes contact at both the upper and lower peripheries of the cup. Maintain this contact, and rotate the cup on the rod along the index finger so as to give it a spiral-like motion down the rod to remove a conical section of the grease that adheres along the rod. When the cup is slipped finally over the end of the rod, a smooth film of reproducible thickness will be left inside the cup.

Reassemble the apparatus, and suspend in the oil bath with the oil level within $^{1}/_{4}$ inch of the rim. If the cork holding the thermometer in the test tube is properly chosen, the 3-inch immersion mark on the thermometer will coincide with the lower edge of the cork and the assembly should be immersed to this point.

Stir the oil bath, and heat at the rate of 4.5 to 7° per minute until the bath reaches a temperature approximately 15° below the expected dropping point of the grease. Now reduce the rate of heating so that the temperature in the test tube will be within 2° or less of the temperature in the oil bath before the oil-bath temperature increases an additional 3°. Continue heating at such a rate that the difference between the temperatures in the test tube and in the oil bath is maintained at 1 to 2°. This condition is established when the oil bath is heated at a rate of about 1 to 2° per minute. As the temperature increases, grease will gradually protrude through the orifice of the grease cup. When a drop of this grease falls, note the temperatures on the two thermometers and record their average as the dropping point of the grease.

Certain greases, such as aluminum-base greases, form upon melting a drop with a tailing thread, which may break off or which may hold until the drop reaches the bottom of the test tube. In any case, take the dropping point as that at which the drop reaches the bottom of the tube.

The dropping points of some greases, particularly those containing aluminum soaps, are known to decrease upon aging, the change being much greater than the deviation permitted in results obtained by different laboratories. Therefore, comparative tests between laboratories must be made within a period of 6 days. Two determinations may be made simultaneously in the same bath.

The ASTM requirements are that a sufficient number of determinations shall be made for the average deviation from the mean to be 3° or less. The average results so obtained by different operators with different apparatus shall agree within 3°.

CHAPTER 22

WAXY SUBSTANCES

Superficially the true waxes are similar to hard fats. They are in general noncrystalline esters, in some cases only a little harder than the fats. True waxes are esters of monohydric alcohols with higher fatty acids, the alcohols often being of large molecular weight. Table 17 contains some of the typical acids and alcohols present in esterified form in the true waxes, with such constants as are available. In contrast, the fats are esters of a trihydric alcohol, glycerol, with fatty acids of not over 18 carbon atoms, in the majority with stearic, oleic, or palmitic acids.

Some synthetic waxlike bodies are intermediate, being esters of fatty acids with dihydric alcohols such as ethylene glycol and its derivatives. Many others are complex mixtures, frequently high in hydrocarbons.

Waxy substances, roughly in the order of their commercial importance, fall into the following classes. The term wax is applied generally to waxlike substances regardless of whether they are waxes structurally.

Mineral waxes are hydrocarbons, the most familiar example being paraffin. *Vegetable waxes* such as carnauba and *insect waxes* such as beeswax are true waxes. *Mineralized vegetable waxes* are intermediate between mineral and vegetable waxes; an example is montan wax. The only solid *animal wax* is spermaceti. *Synthetic waxes*, classified according to their properties, may fall in any of the classes.

General Properties of Waxes. Waxes and mixtures of waxes often cannot be as completely identified by their physical and chemical constants as can fats and oils. This is particularly because such waxes as are being examined in the laboratory are commonly extractives that have been isolated from commercial products in which a part of the wax may have been separated from another part owing to difference in either chemical properties or solubilities. For example, in isolating montan wax one might necessarily extract the hydrocarbons as one fraction, the acids as another fraction.

Characteristics peculiar to these waxes such as odor on burning and general physical appearance are useful as identifying properties. Therefore, brief descriptions of the commercial waxes and their uses are given. Table 18 summarizes their physical constants.

TABLE 17.—TYPICAL ACIDS AND ALCOHOLS OCCURRING IN WAXES

	Formula	Specific gravity	Approx. melting point, °C.	Soluble, usually slightly, in	Occurs in
Palmitic acid	$C_{15}H_{31}COOH$	0.846(70°/4°)	64	Alcohol, ether	Spermaceti as cetyl palmitate; beeswax as myricyl palmitate
Cerotic acid	$C_{25}H_{51}COOH$	0.836(70°/4°)	82	Hot alcohol, hot ether	Carnauba wax as myricyl cerotate; Chinese insect wax as ceryl cerotate; as free acid in beeswax
Montanic acid	$C_{27}H_{55}COOH$		83	Methyl alcohol	Montan wax as free acid; I.G. wax as montanyl montanate
Melissic acid	$C_{30}H_{61}COOH$		91	Hot alcohol, ether	
Cetyl alcohol	$C_{16}H_{33}OH$	0.818(50°/4°)	50	Alcohol, ether, benzene	Spermaceti as cetyl palmitate
Octadecyl or stearyl alcohol	$C_{18}H_{37}OH$	0.812(60°/4°)	59	Alcohol, ether	Free alcohol in spermaceti; made synthetically
Ceryl alcohol	$C_{26}H_{53}OH$		80	Alcohol	Chinese insect wax as ceryl cerotate
Myricyl alcohol or melissyl alcohol	$C_{31}H_{63}OH$	0.777(95°)	88	Alcohol, ether, benzene	Beeswax as myricyl palmitate; carnauba wax as myricyl cerotate

Bayberry wax is a green waxy fat composed of the glycerides of palmitic, myristic, and lauric acids. It is added to candles because of the odor it imparts when burned. Analytically it will be found with the fats and oils.

Beeswax is a relatively hard, brittle insect wax obtained from the honeycomb. It consists of myricyl palmitate, melissic and cerotic acid and homologues, and myricyl and ceryl alcohols. Grades vary from crude yellow through refined yellow to bleached white. The bleached grade is sometimes known as white wax and is used in ointments, cosmetics, candles, and molded wax products. The yellow grades find use in leather dressings, candles, sizings, polishes, lithographing inks, etc.

TABLE 18.—CONSTANTS ON WAXES

	Melting point, °C.	Specific gravity, 15°C.	Refractive index	Acid value	Saponification value	Iodine value	Acetyl value	Unsaponifiable matter	
1. Beeswax	61–63	0.958–0.970	1.430–1.445 (75°)	16.8–20.6	90–102	8.8–11	15.2	52–56	1
2. Candelilla	65–70	0.950–0.993	1.455 (75°)	16–19	46–65	14–37	....	74	2
3. Carnauba	80–87	0.990–0.999	1.463 (60°)	4–8	78–88	13.5	....	55	3
4. Ceresin	54–77	0.88 –0.92	1.441–1.446	0	< 2	0	7–9	100	4
5. Chinese insect	65–80	0.926–0.970	....	13	80–93	1.4–2	....	49.5	5
6. Montan	76–92 crude 72–77 dist'd 77–84 ref'd	1.0 crude	....	25 crude 73–85 dist'd 15–20 ref'd	58 crude 75–89 dist'd 70–80 ref'd	16 crude 10–15 dist'd	....	.....	6
7. Ouricuri, American refined	83.5	....	....	17	58.3	....	....	.....	7
8. Ozocerite	58–100	0.85 –0.95	1.4415–1.446	0	0	7.8	....	100	8
9. Spermaceti	41–49	0.905–0.960	1.440 (60°)	0.5–2.8	121–135	3–9	2.6	.....	9

Candelilla wax is harder than beeswax but not so hard or brittle as carnauba wax. This vegetable wax is brown in the crude form but may be refined to light yellow. On burning it gives an odor slightly resembling that of beeswax. Because of good dielectric properties it is used in electrical insulating compositions. It also modifies or replaces carnauba wax and beeswax in many of their uses.

Carnauba wax is the most expensive commercial wax used on a large scale and is obtained from the leaves of a Brazilian palm. It is very hard and brittle, consisting mainly of myricyl cerotate with free myricyl alcohol, free cerotic acid, and hydrocarbons. The crudes are deep yellow to brown and are refined to light yellow, entering the market in numerous grades. The melting point of 83 to 91° is the highest of all natural waxes except some grades of ozocerite. The specific gravity is very slightly less than 1. Warming reveals a characteristic aromatic odor similar to that of new-mown hay.

Carnauba wax is difficult to saponify and requires the use of a solvent with a higher boiling point than alcohol. A major use of this wax is in floor wax, with which it gives a hard lustrous surface. It is also used to a lesser extent in electrical insulating compositions, polishing materials, carbon paper, and lithographic inks.

Ceresin is an amorphous mineral wax. It is relatively hard and consists of saturated hydrocarbons; therefore, it is unsaponifiable. It is often described as being refined from ozocerite but sells at a much lower price than crude or bleached ozocerite. Ceresin is sold in yellow to white grades with a melting point range of 130 to 170°F. In contrast to the paraffins, it produces mixtures with oils that cannot be separated by filtration or sweating. The odor on burning is similar to, but somewhat sweeter than, that given by paraffin. Ceresin is used in candles, polishes, cosmetics, lubricants, and insulating materials.

Chinese insect wax is a relatively hard wax composed largely of ceryl cerotate. This yellow to white wax is available only in limited amounts and finds use in sizings, polishes, candles, etc.

Japan wax is a hard, brittle, pale yellow to white waxlike solid. Sometimes brownish or greenish colors are present. The material resembles tallow in appearance and odor. It is glyceryl palmitate with up to 12 per cent of free palmitic acid. Therefore Japan wax is a hard fat rather than a wax and will be found analytically with the fats and oils. Uses are similar to those for beeswax.

Montan wax is a hard wax melting at 80 to 90° and extracted from lignite. It is only partly mineralized, is brittle, and shows a conchoidal fracture. It contains montanic acid and its esters, along with hydrocarbons. The crude dark brown grade can be refined to light colors. Montan wax is used as a hard wax in many applications, in insulating compounds, for example, as a replacant for carnauba wax.

Ouricuri wax is a vegetable wax from a part of Brazil other than that yielding carnauba wax. Beyond this little is known about it, for substantial amounts appeared on the American market only in the late 1930's. The crude is gray in color but is refined to a light yellow. It sells somewhat lower than carnauba wax and replaces it in whole or in part in a few uses.

Ozocerite is a greenish brown to black wax composed of unsaponifiable hydrocarbons, probably deposited from evaporation of petroleum. This hard wax is available in a range of colors varying from green to white, with melting points of 66 to 78°. The lighter grades are sold as bleached ozocerite. Crude and refined grades are used to replace beeswax, carnauba wax, or ceresin in polishes, printing inks, electrical insulating compounds, sizing preparations, and cosmetics.

Paraffins are mineral waxes derived from petroleum. Like the ceresins they are saturated hydrocarbons and will not saponify. As distinct from the ceresins they possess a crystalline structure. Commercial grades are normally white, semitranslucent, and of various melting point ranges from 105 to over 170°F. The price increases rapidly with the melting point. On burning, a petroleum oil odor is in evidence. Paraffins are the lowest priced waxes available and are used in the manufacture of candles with added harder waxes, for waterproof coatings on many surfaces, in various compounds, in short, for a majority of the uses that call for wax. A low-melting grade is used to increase the inflammability of matches.

Amorphous paraffins are also sold, not as paraffins but under trade names. One of the most widely sold is Cerese wax. These are noncrystalline or microcrystalline and have negligible saponification and acid values. They are considerably more expensive than paraffin and are used when the crystalline structure of paraffin would be objectionable.

Spermaceti is an animal wax obtained from the head cavities of the sperm whale. The crude is light yellow and consists mainly of ceryl palmitate. Refined spermaceti occurs as lustrous, white, translucent masses, very brittle and readily powdered. It has a broad, leafy, crystalline structure and is practically odorless. The main uses of spermaceti are in candles and cosmetics.

Synthetic waxes of a number of varying types are shown in Table 19, with such constants as are available. They vary in hardness and other properties. One type such as Opal wax consists of hydrogenated fats. Another such as Lanette wax appears to consist of mixed high-molecular-weight alcohols. Esterification of montanic acid gives a series of I.G. waxes, some of which contain montanyl montanate. An exception is I.G. wax V, which is unsaponifiable. Low-melting-point waxlike solids are obtained by partial esterification of polyhydric alcohols. These are in composition and properties somewhat intermediate between fats and waxes. Examples are glyceryl monostearate and diethylene glycol stearate, both of which are dispersible in hot water. Another class of synthetic waxes consists of naphthalenes chlorinated to various degrees and known as Halowaxes.

TABLE 19.—SOME SYNTHETIC WAXES

Name	Specific gravity	Melting point, °C.	Acid value	Saponification value	Acetyl value
Opal wax	0.98–1.0(20°)	77–81	2	180	
Rilan wax	0.98(15°)	84–85	25 approximately	155 approximately	
Lanette wax	0.81 liquid	about 50	0	0	180–190
I.G. wax E	1.01–1.02	80–83	15–30	135–180	
I.G. wax KP	1.02–1.03	79–82	25–35	130–145	
I.G. wax O	1.03–1.04	102–108	20–35	100–145	
I.G. wax OP	....	104–110	20–30		
I.G. wax S	....	83–85	140		
Glyceryl monostearate	0.97	55–57	5–14	165–183	
Diethylene glycol stearate	0.96	51–54	96–104	175–185	
Halowax No. 1012 ...	1.53–1.59	91–96	0.1 maximum		
Halowax No. 1013 ...	1.65–1.71	115–120.6	0.1 maximum		
Halowax No. 1014 ...	1.75–1.81	132–137.8	0.1 maximum		

METHODS OF ANALYSIS

In general, the methods of analysis applied to fatty glycerides are modified to a greater or lesser degree for application to true waxes. Some of those applicable to liquid petroleum hydrocarbons apply to the waxy hydrocarbons. Additional determinations depending on the solubilities of waxes in solvents are useful in identifying mixtures.

Specific Gravity. Determine this by the general method (page 48), using a specific-gravity bottle with pieces of wax and a liquid which is of lighter gravity than the sample and in which it is not soluble. Alcohol is usually a suitable liquid, for most waxes are insoluble in it at room temperatures. Make the weighing as quickly as possible after addition of the alcohol to avoid possible slight solubility of the wax or contaminants present in the wax fraction isolated. Accuracy to the third decimal place is readily obtainable.

Acid Value. Proceed as for fats and oils (page 341) if feasible. Usually the preparation of the sample for titration must be modified as follows: Weigh the sample of wax into a conical flask. Heat until the wax is melted, exercising care to prevent decomposition. Separately heat to boiling a neutralized mixture of equal parts by volume of alcohol and benzene. Remove both to a place well away from flames because of the inflammable nature of the solvent, and add the solvent at the boiling point to the melted wax. Reflux until the wax is all dissolved if possible—if not, until it is well disintegrated. Titrate as usual for the acid value while as hot as possible, reheating if further precipitation of wax occurs during titration. Even with these precautions, it is difficult to obtain a reproducible acid value for high-melting waxes. The accuracy therefore varies with the nature of the sample.

Saponification Value. Proceed as for fats and oils (pages 343 and 344), using the modifications for difficultly saponifiable fats with the solution of potassium hydroxide in ethylene glycol or other suitable high-boiling solvent. Reflux with the alkali for at least 1 hour; a longer period is preferable. The final solution may contain suspended matter. Titrate hot, and reheat if further precipitation begins during titration.

For some waxes, particularly those having a high melting point and low saponification value, this method is unsatisfactory, and the following modification should be used:

PROCEDURE. Weigh into 250-ml. conical flasks two equal amounts of a saponifiable oil that contains little unsaponifiable matter. Castor oil is suitable. The weights, about 2 to 3 grams, must be exact. The time used to ensure that the amounts are the same will be justified, for it will eliminate the need for troublesome calculations later. To one of the flasks add an accurately weighed sample of 3 to 5 grams of the wax. To each flask add 50 ml. of a 0.5*N* or stronger solution of potassium hydroxide in alcohol. Reflux the contents of the two flasks containing oil for 1 hour or more. Titrate the contents of the refluxed flasks with standard acid while hot. The soap formed by saponification of the oil helps to emulsify the wax, and to promote its saponification.

The amount of alkali reacting with the oil is the same in both flasks, provided that identical amounts were added. Therefore the amount of acid to titrate the contents of the flask which contained the wax is subtracted from that for the flask to which no wax was added. The saponification value can then be calculated directly from this difference. The saponification value of a wax is rarely reproducible within better than 2 to 3 units.

Melting Point. Employ the capillary tube open at both ends (page 51). Water is the medium to use for most waxes. A few synthetics are water-dispersible, and in such cases a solvent in which they are insoluble must be selected. In rare cases waxy solids melt above 100°, and a liquid of higher boiling point than water must then be selected.

Mixtures of waxes tend to give erratic results. The lower melting point wax sometimes softens sufficiently to cause the mixture to rise in the tube long before the melting point of the other constituent is reached. Visual observation of the transparency of the wax when rising in the tube will give valuable information as to heterogeneity.

The method in which a sliver of the wax is placed on the surface of mercury in a crucible and the temperature observed at which it collapses may also be used (page 53).

Unsaponifiable Matter. Saponify the wax completely. If alcohol serves as a suitable solvent, evaporate in the water bath (see page 349) nearly to dryness, take up in hot water, and transfer to a 500-ml. separatory funnel. When cool, dilute to about 300 ml., wash out the flask with 75 ml. of ethyl ether or petroleum ether, and transfer the ether to the separatory funnel. Proceed as for fats on page 349, starting at "Shake thoroughly."

If ethylene glycol was used for the saponification, dilute to 300 ml. with water and use 75 ml. of petroleum ether, starting at the same place as for the previous technic. This avoids three-phase systems such as are obtained with ethylene glycol, ethyl ether, and water.

The unsaponifiable matter from waxes of high melting point will often itself melt so high that the preceding method cannot be used. In such cases the suspended unsaponifiable solid matter promotes emulsification or forms a layer at the interface. If this is so, saponify the wax with alcoholic potash, using a saponifiable oil to promote the reaction as described under Saponification Value. Evaporate the alcohol, and dry the mixed soaps. Break up the dried mixture, and transfer to a cellulose thimble. Extract this with either a Soxhlet extractor and petroleum ether or, if necessary to use a higher extraction temperature, with a rubber extractor (page 67). The accuracy obtainable in this determination is somewhat variable, depending on the nature of the sample.

Iodine Value. Use the method for fats and oils (page 345) with either Hanus or Wijs solution. Warm the flask containing the wax, to melt it before adding the solvent. Be sure that it is all in solution and cooled to room temperature before adding the reagent. The majority of waxes are more soluble in carbon tetrachloride than in chloroform. In some cases they are replaced in both sample and blank by trichloroethylene. This should be duplicable within 1 to 2 units.

Aniline Point. This is the temperature at which under specific conditions turbidity due to separation of phases occurs in a mixture of wax, aniline, and petroleum naphtha. The method is most useful in checking a synthetic mixture against the commercial sample it is believed to match. Table 20 gives some data on commercial waxes and mixtures. For specific problems such data will have to be determined for the range suspected.

PROCEDURE. To determine, heat one volume of wax, conveniently weighed after ascertaining the specific gravity, one volume of naphtha of specific gravity 0.750 at 60°F., and two volumes of redistilled aniline in a test tube until a clear homogeneous liquid is obtained. Insert the test tube in an air bath, and cool gradually. Stir with the thermometer to maintain even temperature throughout the mixture. The aniline point is the temperature at which complete opacity of the liquid occurs. This point is normally sharp and definite. With dark waxes some difficulty in observing the point may occur.

TABLE 20.—ANILINE POINTS OF SOME WAXES AND MIXTURES

	Degrees Centigrade
Beeswax	58.5
Carnauba wax	64
Candelilla wax	66.5
Paraffin m.p. 135°F.	93.5
44 per cent carnauba wax 56 per cent paraffin m.p. 135°F.	79
67 per cent carnauba wax 33 per cent paraffin m.p. 135°F.	69.5
45 per cent carnauba wax 15 per cent beeswax 40 per cent paraffin m.p. 125°F.	72
60 per cent paraffin m.p. 118°F. 40 per cent stearic acid	61
60 per cent paraffin m.p. 140°F. 40 per cent stearic acid	58.8
50 per cent carnauba wax 50 per cent cerese wax	84.5
60 per cent carnauba wax 40 per cent cerese wax	79

Precipitation from Normal Butyl Alcohol or Normal Heptane. Characteristic precipitation temperatures for waxes have been reported from these solvents under standardized conditions.[1] Values are shown in Table 21.

PROCEDURE. To determine, dissolve by warming 0.100 gram of wax in 10 ml. of the solvent in a test tube. Air-jacket this with a larger test tube, insert a thermometer, and immerse the assembly to well above the level of the solvent in a water bath at a temperature higher than the precipitation temperature. Allow the bath to cool at a rate not greater than 1° per minute, and record the temperature at which formation of crystals first occurs. In general, this point is fairly sharp.

The precipitation temperatures from normal butyl alcohol of carnauba wax and ozocerite are much higher than those of other waxes. To differentiate between them repeat with normal heptane in which the order of solubility is reversed. Beeswax does not precipitate from normal heptane even at room temperature. Candelilla wax forms cloudy solutions in boiling normal heptane.

TABLE 21.—PRECIPITATION TEMPERATURES FOR 0.1 GRAM OF WAX IN 10 ML. OF SOLVENT SPECIFIED

Wax	Melting point, °C.	Precipitation temperature, °C.	
		n-Butyl alcohol	*n*-Heptane
Beeswax	60–62	42.5	
Montan, bleached	68	46.5	36
Candelilla	70.5	48	
Carnauba, refined	81.5	58	42.5
Carnauba No. 1 yellow	83	64	49.5
Ozocerite	65–67	66	24
Ozocerite	76–78	76	36

Quantitative Solubility in Cold Alcohol. The quantitative solubility of waxes in cold 95 per cent alcohol is a useful indicator for their identification.[2] Treat a quantity of wax with approximately ten times its weight of boiling 95 per cent alcohol. Decant from the insoluble melted wax. Cool the alcohol extracts to 25°, and filter either through a coarse quantitative paper or through a Büchner funnel. Wash twice with cold alcohol, and evaporate the filtered alcohol solution and washings. Dry in the oven for as short a period as possible, and weigh. Obtain the melting point,

[1] ZWEIG, S., and TAUB, A., *Ind. Eng. Chem., Anal. Ed.*, **12**, 9–14 (1940).
[2] *Loc. cit.*

refractive index at 90°, and acid value of the alcohol-soluble fraction. Typical data are shown in Table 22.

Qualitative Solubility of Waxes. Methods of determination already outlined become complicated in dealing with commercial mixtures of waxes. Some qualitative solubilities in different solvents are helpful in identifying or confirming the composition. Usually parallel confirmatory tests are best conducted on a synthetic mixture made up according to a prior interpretation of the composition of the sample.

TABLE 22.—TYPICAL DATA ON PORTIONS OF WAXES SOLUBLE IN 95 PER CENT ALCOHOL AT 25°

Wax	Percentage of wax	Melting point, °C.	Acid value	Refractive index n_D 90°C.
Beeswax	5	60	88.5	1.4391
Candelilla	19	....	24	1.5115
Carnauba	3	....	19	1.4916
Montan, bleached	3	52	29.5	1.4386
Spermaceti	5	32.5–33	10	1.4295

Acetone. In the cold all waxes are insoluble. Beeswax and spermaceti are soluble in the hot solvent. Carnauba wax is partly soluble and Japan wax is insoluble in hot acetone.

Alcohol. In the cold all waxes are insoluble in 95 per cent alcohol. Beeswax, candelilla wax, carnauba wax, Japan wax, montan wax, and spermaceti are soluble in the hot solvent. Chinese wax is slightly soluble and ceresin and ozocerite are insoluble in hot alcohol.

Benzene, Toluene, and Xylene. Ceresin, Chinese wax, Japan wax, montan wax, and ozocerite are soluble.

Ethyl Ether. Beeswax, Japan wax, ozocerite, and spermaceti are soluble. Carnauba wax is insoluble in the cold but soluble in the hot solvent on repeated extraction.

Petroleum Naphtha. Ceresin, Japan wax, montan wax, and spermaceti are soluble. Beeswax is insoluble. Carnauba wax is soluble only in the hot solvent.

Trichloroethylene. All waxes are soluble in trichloroethylene.

CHAPTER 23

SOLVENTS AND THINNERS

By far the most common and the least costly solvent is water. But there are numerous other types that may be conveniently classified as *esters; ketones; alcohols,* whether monohydric or polyhydric; *ethers; halogenated hydrocarbons; petroleum hydrocarbons; coal-tar hydrocarbons;* and *turpentine* or related terpenes. Individual solvents may belong to more than one class, such as the methyl ether of ethylene glycol, which is both an ether and an alcohol.

While the words have different meanings, one can hardly make a sharp distinction between solvents and thinners. In general, a solvent is used to dissolve a material. A thinner serves to make the solution less expensive; otherwise, the process of thinning would be one of adding more solvent. These classes merge with plasticizers, which are solvents with boiling points sufficiently high so that they will not evaporate at ordinary temperatures. Since the operations discussed in this chapter generally involve distillation, it follows that the plasticizer, insofar as a dividing line can be drawn, will remain as undistilled residue. Solvents and thinners, as classes, merge with each other; the commercial products in which they appear are usually mixtures and may contain suspended as well as dissolved matter.

For the purposes of this chapter, it is assumed that the solvent has been freed from suspended matter, as by filtration or centrifuging, and from dissolved matter, as by distillation. In dealing with commercial products generally, the methods are for analysis of fractions previously isolated. A typical example is the separation of lacquer thinners (page 447). Because alcohols have other applicability than as solvents and thinners, they are discussed only in part in this chapter. Methods of analysis for individual alcohols are left for the next chapter.

In general, a solvent with a flash point below 100°F. is designated as inflammable. If it flashes above that temperature, it is called combustible. Table 26 (page 454) lists many solvents frequently met in commerce, with their physical properties. As a matter of completeness, it includes many compounds not specifically considered in this chapter. On the other hand, such a table cannot be complete since there are many solvents of lesser importance and new ones are constantly being supplied commercially by the synthetic organic chemical industry.

One must often evaluate each of the fractions resulting from a fractional distillation. Therefore comparatively rapid methods are desirable, if not absolutely essential, for qualitative examination. Experience with the odors is of great assistance in rapid identification.

Generally, when dealing with solvents of these classes, one identifies the class qualitatively and then proceeds to quantitative determination. Therefore, qualitative and quantitative methods are given for each class. In all cases the specific gravity needs to be determined for accurate identification.

Qualitative and quantitative examination of relatively pure fractions will first be discussed, then methods for identification and separation of mixtures.

RELATIVELY PURE COMPOUNDS

Esters. This class of compounds results from interaction of an alcohol and an organic acid with loss of water. Those commonly used are most frequently acetates, but propionates, butyrates, and lactates are not unusual. Similarly, ethyl alcohol is most commonly combined with the acid, but methyl, propyl, butyl, and other alcohols are used. Because the boiling point goes up rather rapidly, the number of low-boiling esters is reasonably limited.

QUALITATIVE. Esters have fruity odors in general, ethyl acetate being somewhat apple-like, amyl acetate resembling bananas. They saponify readily on warming with aqueous alkali to give the alcohol and a salt of the acid. The odor on heating the saponification mixture is often identifiable as that of the alcohol. Similarly, on partial evaporation to drive off the alcohol, they may often be acidified and a clue obtained as to the acid by the odor. They burn with blue to yellow flames.

QUANTITATIVE. If it is established that only one type of ester is present and is the only substance present, its identity can be confirmed by density and boiling point. More frequently, the type of ester present is established, but it is blended with unsaponifiable material not readily separable by distillation. In that case the amount of ester is determined by saponification.

PROCEDURE. Being careful to avoid loss by evaporation, pipet 5 ml. of the solvent of known density into a 250-ml. conical flask containing 25 ml. of approximately $2N$ or 50 ml. of approximately N alcoholic potassium hydroxide solution. Boil under a reflux condenser for 45 minutes to 1 hour. With low-boiling esters such as methyl and ethyl acetates, very efficient condensers are necessary, and the mixture should only just boil. Similarly boil a blank containing the same amount of alcoholic potassium hydroxide for the same time. Cool, dilute sample and blank with 100 ml. of water, and titrate to a phenolphthalein end point with $0.5N$ or N acid.

CALCULATION.

$$(\text{Ml. blank titration} - \text{ml. sample titration}) \times \text{normality of acid} \times \text{factor} \times \frac{100}{5 \times \text{specific gravity of sample}} = \text{per cent ester by weight}$$

The factor to use is found in Table 23 and consists of the milliequivalent of the ester, which, for monobasic derivatives such as these, is the molecular weight divided by 1,000.

For example, assume that analysis showed the fraction to contain approximately 90 per cent of butyl acetate, with the balance largely butyl alcohol. This can then be confirmed by the density of the mixture. Thus $(0.883 \times 0.9) + (0.810 \times 0.1) = 0.8757$. If the density of the sample checks this with reasonable accuracy, the analysis is confirmed.

To convert per cent by weight of an ester to per cent by volume, as the first step divide the per cent by weight by the density of the ester. This gives the volume in 100 grams of sample. Calculate the volume of 100 grams of sample by dividing 100 by the density. Then the volume of the ester in 100 grams of sample multiplied by 100 and divided by the volume of 100 grams of sample is the per cent by volume of ester in the sample.

Ketones. While many ketones are possible, those used industrially are limited mainly to three. Acetone and methylethyl ketone are structurally the expected chemical individuals in relatively high purity. Methyl acetone is a mixture, usually 48 per cent acetone, 28 per cent methyl acetate, and 24 per cent methyl alcohol, Thus the latter contains a mixture of ketone, ester, and alcohol. All ketones are readily soluble in water, alcohol, and esters. Ketones are unsaponifiable. If free from contaminants with intense odors, they are identifiable by their odors.

QUALITATIVE. *Iodoform Test.* All ketones containing a methyl group respond to the iodoform test. While this reaction is similar to the characteristic test for ethyl alcohol, ketones give it at room temperature while ethanol must be heated to 60° to give the reaction. Isopropyl alcohol reacts like acetone. The reactions with acetone are

$$3I_2 + 6NaOH = 3NaOI + 3NaI + 3H_2O$$

and

$$CH_3COCH_3 + 3NaOI = CH_3COONa + CHI_3 + 2NaOH$$

These can be combined as

$$CH_3COCH_3 + 3I_2 + 4NaOH = CH_3COONa + 3NaI + CHI_3 + 3H_2O.$$

For the test, mix about 1 ml. of the solution and 1 ml. of *N* sodium hydroxide solution. Then add sufficient dilute aqueous iodine solution to cause a brown color due to excess of unabsorbed iodine. If a light yellow silky-looking precipitate forms, a ketone containing a methyl group is present. This has a characteristic odor.

TABLE 23.—FACTORS AND SPECIFIC GRAVITIES FOR CALCULATION OF PERCENTAGES OF ESTER-TYPE SOLVENTS

Ester	Factor	Specific gravity ester 20°/20°
Methyl acetate	0.07408	0.935
Ethyl acetate	0.0881	0.902
Isopropyl acetate	0.1021	0.874
Butyl acetate	0.1162	0.883
Amyl acetate	0.1302	0.866
Methyl cellosolve acetate	0.1181	1.0067
Cellosolve acetate	0.1321	0.975
Methyl carbitol acetate	0.1622	1.0396
Carbitol acetate	0.1762	1.0114
Butyl cellosolve acetate	0.1601	0.943
Ethyl propionate	0.1021	0.896
Butyl propionate	0.1301	0.870 (90–92 per cent)
Ethyl lactate	0.1181	1.031
Butyl lactate	0.1461	0.974–0.984 (95 per cent)

Nitroprusside Test. Mix 1 ml. of a solution containing about 1 per cent of the suspected ketone with 5 drops of a 5 per cent solution of sodium nitroprusside and 2 ml. of a normal solution of sodium hydroxide. Add a slight excess of acetic acid. A deep red coloration is produced with ketones. This develops a violet tint when diluted with several volumes of water.

QUANTITATIVE. The iodoform reaction is applied quantitatively for determination of ketones containing a methyl group. Weigh about 1.75 ml. of ketone from a 2-ml. pipet graduated in 0.1 ml. into a small, conical, glass-stoppered weighing bottle of 12 to 15 ml. capacity. Invert the weighing bottle, and hold the mouth under the surface of freshly boiled distilled water in a liter beaker. Remove the stopper under water by means of a glass rod with a hook on the end. Thoroughly wash out the bottle, transfer the water and ketone to a liter flask, and dilute to the mark.

Pipet three 25-ml. aliquots from the liter flask prepared as above, into separate 750-ml. conical flasks containing 50 ml. each of N sodium hydroxide solution. Into each flask, while shaking, constantly run 50 ml. of $0.1N$ iodine solution. Stopper the flasks, and allow to stand at about 20° for at least 20 minutes. Then pour 26 ml. of $2N$ sulfuric acid solution into the first flask. This slight excess of acid must be added just before the titration with thiosulfate. The amounts stipulated give the correct excess; too great an excess will give low results owing to decomposition of the thiosulfate solution. Titrate immediately with $0.05N$ sodium thiosulfate solution. When the iodine color is nearly gone, add starch solution and titrate until the blue color disappears and the bright yellow color of the iodoform suspension is reached. Similarly, titrate the other two aliquots. Check the accuracy of the procedure by using a water blank in place of the ketone solution. With these amounts, the 35 to 40 per cent excess of iodine found to be optimum for completion of the iodoform reaction will be present.

Because of the use of aliquots from such a small sample, it is desirable to run another separately weighed sample as a check; even then, accuracy to better than 1 per cent is obtained only with difficulty.

CALCULATION

$$(\text{Blank} - \text{sample}) \times N \times \text{factor} \times \frac{1{,}000}{25} \times \frac{100}{\text{weight of ketone taken for dilution}} = \text{per cent ketone by weight}$$

For acetone the factor is 0.00968; for methylethyl ketone, 0.01201.

Monohydric Alcohols. QUALITATIVE. When pure, the alcohols are largely distinguishable by their odors, but the differences are clear only to the experienced worker. While pure methanol has relatively little odor, that obtained from wood distillation has a characteristic odor of pyroligneous acid. Ethanol is best described as "winy"; propanol and isopropanol are somewhat similar but less pungent. The butyl and amyl alcohols appear more intense and give a characteristic effect at the back of the throat when smelled.

The alcohols burn with a clear colorless to blue flame. The iodoform reaction described for acetone identifies isopropanol in the cold or ethanol on warming to 60°. They do not saponify but form acetates of distinguishable odor when mixed with slightly less than an equivalent of glacial acetic acid and a few drops of concentrated sulfuric acid. Additional information on their properties is contained in Chap. 24, Alcohols.

QUANTITATIVE. The quantitative estimation of individual alcohols is dealt with in the next chapter. A convenient and useful method for obtaining the amount of alcohol in a mixed solvent is that in which the amount of hydroxyl groups present is measured. This method is valuable when the substance to be tested is soluble in pyridine. It is quicker than the determination of acetyl value (page 351).

PROCEDURE. *Hydroxyl Value.* Prepare pyridine–acetic anhydride reagent as follows: Add exactly 3.5 ml. of water to 1 liter of pyridine that has previously been dried over solid potassium hydroxide for at least 72 hours. Mix and add 140 ml. of acetic anhydride. It is important that both the amount of water and the order of addition shall be strictly adhered to; otherwise, low results are likely to be obtained. The presence of the water inhibits the formation of dark-colored resins in the blank owing to reaction between dry pyridine and acetic anhydride during the refluxing period. The addition of more than 0.4 per cent of water gives low results; therefore, the pyridine must first be dried.

Dry the substance to be tested, using anhydrous sodium sulfate, potassium carbonate, or ignited calcium oxide. The latter are applicable only if acids are absent. Weigh a sample estimated to be sufficient to give approximately half the titration value of the blank, or 15 ml. of 2*N* sodium hydroxide solution, into a 250-ml. flask having a standard ground-glass joint. Add exactly 25 ml. of the pyridine–acetic anhydride reagent from a pipet. Grease the ground-glass joints to prevent creeping of reagent, and attach the condenser to the flask. At the same time, transfer 25 ml. of reagent to another flask as a blank. Water must not be absorbed through the condenser, and therefore it is desirable to fit the tops of the condensers with calcium chloride or soda-lime tubes. The experienced worker can get satisfactory results by limiting diffusion of air by placing a wad of oven-dried absorbent cotton in the tops of the condensers, which in that case are desirably of the coil type. Reflux the solution for 30 minutes, preferably from a hot plate to avoid unduly high humidity in the air. With condenser attached, remove the flasks from the heat. Flush the condensers with 30 to 50 ml. of water. Cool the flasks to room temperature. Finally, remove the condensers, rinse the glass joints with a few milliliters of water, and titrate. Use phenolphthalein indicator and 2*N* sodium hydroxide.

CALCULATION.

$$\text{(Ml. of alkali used for blank} - \text{ml. used for sample)} \times \text{normality} \times 0.017 \times \frac{100}{\text{weight of sample}} = \text{per cent of hydroxyl}$$

To get the result in terms of ethanol substitute the factor 0.04607. Equivalent factors can be used for other alcohols.

To obtain the hydroxyl value of the sample in conventional terms, that is, in milligrams of potassium hydroxide per gram of substance, calculate as follows:

$$\frac{(\text{Ml. of alkali used for blank} - \text{ml. used for sample}) \times \text{normality} \times 56.1}{\text{weight of sample}} = \text{hydroxyl value}$$

Glycols and Ether Alcohols. While there are other polyhydric alcohols, glycerol, for example, only ethylene glycol and propylene glycol are sufficiently volatile for use as solvents of the type being considered. They are products of the synthetic organic chemical industry. When one of the alcohol groups of a glycol is converted to an ether, the boiling point is lowered. Several of these ether alcohols are important such as the methyl, ethyl, and propyl ethers of ethylene glycol, sold under trade names. Similarly, ethers of diethylene glycol are used.

QUALITATIVE. The members of this class are very soluble in water. Many have characteristic odors which, with their boiling points and densities, serve for identification when one takes into consideration the product being analyzed.

QUANTITATIVE. All alcohols acetylate with acetic anhydride. This reaction is variously used for their estimation, as in the acetin method for glycerol. Determine the hydroxyl value as described for monohydric alcohols. Calculate as there given (page 439). The factor to be used is the milliequivalent, which is the molecular weight divided by one thousand times the number of hydroxyl groups present. Values for a series are shown in Table 24.

Chlorinated Hydrocarbons. The outstanding property of this class of synthetic compounds is their high specific gravity, always greater than 1, so that they sink when poured into water. The density increases with higher degrees of chlorination. The introduction of chlorine decreases inflammability; therefore, those containing three or more chlorines may be expected to be noninflammable. Industrially, carbon tetrachloride is the most important, trichloroethylene a poor second, and the others of much less importance. All are toxic, and sustained exposure to their vapors is therefore to be avoided.

QUALITATIVE. The somewhat sweetish odors serve to identify the members when taken in conjunction with their densities. Positive identification of chlorinated hydrocarbon depends on a green color of the flame when associated with copper. As one method of testing, heat a clean stout copper wire in a Bunsen flame until no green color, due to the volatilization of copper salts, is seen. Cool, and immerse in the liquid to be tested. Burn as slowly as possible if other inflammable liquids are present; other-

TABLE 24.—FACTORS AND SPECIFIC GRAVITIES FOR CALCULATION OF PERCENTAGES OF GLYCOLS AND ETHER ALCOHOLS

Glycol or ether alcohol	Commercial name	Factor	Specific gravity of pure compound
Ethylene glycol	Ethylene glycol	0.0310	1.1155
Diethylene glycol	Diethylene glycol	0.0531	1.1184
Propylene glycol	Propylene glycol	0.0380	1.0381
Ethylene glycol monoethyl ether	Cellosolve	0.0901	0.9311
Ethylene glycol monomethyl ether	Methyl cellosolve	0.0761	0.9664
Ethylene glycol monobutyl ether	Butyl cellosolve	0.1182	0.9019
Diethylene glycol monoethyl ether	Carbitol	0.1342	1.9898
Diethylene glycol monomethyl ether	Methyl carbitol	0.1202	1.0211
Diethylene glycol monobutyl ether	Butyl carbitol	0.1622	0.9536

wise, volatilize as slowly as possible in the Bunsen flame. Then heat strongly. The presence of a green flame indicates the presence of chlorinated compounds, and the intensity of color is a rough indication of the amount of chlorine in the sample. As another modification, pass illuminating gas through a wash bottle of the sample cut into the line of tubing to a gas burner. Burn the gas as usual with a Bunsen burner. Traces of volatile chlorinated compounds will impart an intense green color to the flame where it impinges on a copper gauze. This gauze must previously be freed from volatile copper compounds by prolonged heating. Often sufficient chlorinated hydrocarbon can be introduced into the gas from holding a piece of cotton saturated with solvent close to the air hole of a Bunsen burner.

QUANTITATIVE. The chlorinated hydrocarbons are usually estimated by application of rather general methods. A pure sample can be identified solely by density. Fractional distillation may isolate the pure chlorinated hydrocarbon quantitatively. If this is not feasible and the other ingredient that cannot be separated is known, the density of each fraction is taken and the amount of chlorinated hydrocarbon thus calculated with a rather low degree of accuracy. In special cases, for example, in legal work, it may be necessary to determine the amount of combined chlorine. Then, provided that the chlorinated hydrocarbon has been identified, the amount present may be calculated.

PROCEDURE. *Bomb Method.* Weigh a known sample, avoiding evaporation, and add an equal amount of chlorine-free carbonaceous material such as sugar. The apparatus to be used is described on page 161. A bomb lining that will not be attacked by hydrochloric acid fumes is essential. Follow the procedure on page 161 to the statement "Preferably the volume is 200 to 250 ml." But in this case dilute to about 100 ml. Determine the chloride gravimetrically, starting with the procedure on page 129.

Carius Method. This is described in full, starting on page 131. The results have a high degree of accuracy; but only small samples can be used, and selective volatilization during weighing is a factor.

Petroleum Solvents. There is a general impression that all petroleum solvents consist solely of paraffinic hydrocarbons. Actually, the qualities vary with the source of the crude oil, the degree and method of refining, and the temperature at which cuts are taken. Some may contain a good many per cent of naphthenes that are aromatic rather than paraffinic in properties.

The grades range from light naphthas or petroleum ethers with specific gravities of 0.627 to 0.72 to petroleum solvents of gravities up to 0.82. This range of densities includes many different fractions, particularly when considered in terms of trade names, but the great bulk is in terms of the four members, petroleum naphtha; VM and P naphtha; safety solvent, sometimes known as Stoddard solvent; and kerosene. By the standard of flash point cited earlier in this chapter the first two are inflammable and the last two combustible. Note the absence of gasoline, which because of its wide boiling range is seldom used. Above specific gravity 0.82 are light mineral oils and heavier oils, which are not usually spoken of as solvents. Such fractions, considered in Chap. 37, Petroleum Products, are occasionally used as plasticizers of special types.

QUALITATIVE. The various petroleum solvents have related types of odors. When they are poorly refined, the odor tends to be intense. Highly refined fractions of relatively high density may have relatively low odor intensity. The odor is greatly intensified by heating. All petroleum products when heated will burn with a more or less yellow flame, sometimes giving a little smoke. Low specific gravity of a fraction is a suggestion that it may be of petroleum origin. The distillation range will usually identify the fraction.

Unsaturated paraffins when suitably processed give cyclicized hydrocarbons, which have a modified odor more like that of the aromatic hydrocarbons. These are aromatics of petroleum origin rather than coal-tar origin. As would be expected, they show the same sulfonation characteristics and relatively high density as coal-tar hydrocarbons. Another class of petroleum hydrocarbons of lesser importance is produced by hydrogenation and is distinguishable only by odor and boiling range.

Petroleum solvents are miscible with absolute alcohol but become insoluble on dilution with water. They are miscible with aromatic hydrocarbons and with ethyl ether. They are not saponifiable.

QUANTITATIVE. Methods for quantitative determination of petroleum hydrocarbons are largely of a somewhat negative nature. Obtain the distillation range (page 665) of the solvent. If the sample consists of a series of fractions from distillation, the range of each will be known. Compare this and the densities of the fractions with those recorded for commercial petroleum solvents in Table 26. While this sounds indefinite, it is quite simple when one has acquired a reasonable familiarity with the petroleum solvents available.

Aromatic Hydrocarbons. These are coal-tar distillates of which the most commonly encountered are benzene, toluene, xylene, and high-flash naphtha. The latter is a fraction distilling above xylene and is composed of a mixture of related compounds. Crude coal-tar distillates of gravity up to nearly 1 are occasionally found. While benzene and toluene are chemical individuals, the commercial products are mixtures with a specific gravity about that of the pure compound and having a boiling range including it.

QUALITATIVE. Aromatic hydrocarbons burn with a very smoky flame, an outstanding characteristic. Because of its toxicity, benzene is rarely used in products sold to the public. These solvents are soluble in alcohol and in ether. They are not saponifiable.

Nitration. Benzene reacts when mixed with a well-cooled mixture of 30 per cent by volume of concentrated nitric acid, 55 per cent by volume of concentrated sulfuric acid, and 15 per cent by volume of water to give nitrobenzene. The mixture should be kept well cooled during the mixing and then carefully warmed. On dilution the odor is characteristic. Toluene and xylene give solid compounds under these conditions.

QUANTITATIVE. *By Sulfonation.* By far the most useful method of qualitatively establishing that a sample is all aromatic hydrocarbon or determining quantitatively how much of it is aromatic hydrocarbon is by sulfonation. Turpentine also reacts. While many other materials sulfonate, they are eliminated from consideration by their qualitative reactions.

PROCEDURE. When only small samples are available, pipet 5 or 10 ml. into a Babcock bottle graduated in the neck in 0.1 ml. Cool in water, and very carefully and slowly add 20 per cent fuming sulfuric acid with constant shaking. It is essential to follow these precautions, for uneven heating is apt to crack the bottle, which will not only ruin the sample but is dangerous. When carefully performed, this operation is not difficult. Add the fuming acid until no further evolution of heat occurs, and let stand for about 30 minutes at about 40 to 50°. Dilute with concentrated rather than fuming sulfuric acid until the neck of the bottle is nearly full. Pack the bottle well in sand, and centrifuge. The sulfonated aromatic compound is soluble in the acid, and unsulfonated hydrocarbons are insoluble. If there is no insoluble layer, the sample was all aromatic hydrocarbon; if such a layer is present, it is a measure of the amount of paraffin hydrocarbon. Except in precise work no attention should be paid to an unsulfonated residue much under 5 per cent of the total volume. Similarly, with larger residues, the results can hardly be expected to be accurate to better than 2 to 3 per cent. If such a substantial residue is obtained, isolate it, wash it free from acid, and proceed with its identification as a petroleum solvent fraction.

If desired, the sulfonation may be carried out on a larger scale. The purpose is usually to obtain a large enough unsulfonated fraction for analysis. Take one volume of the solvent in a Kjeldahl flask cooled in ice water. Very slowly add, with continuous shaking, twice its volume of 20 per cent fuming sulfuric acid. Oleums of other strength are permissible as long as the sulfonation reaction is complete. Do not allow to become hotter than is comfortable to the hand, particularly if highly volatile material is present. In the latter case, it may be desirable to attach an air reflux to minimize loss. On completion of the sulfonation, determined by the fact that no heat is evolved on further addition of acid, adjust to room temperature, and transfer to a measuring cylinder or a buret. Let stand for the unsulfonated residue to separate. The total volume is too large to centrifuge with safety.

If the entire sample was sulfonatable, the specific gravity and distillation range will be sufficient for quantitative estimation of the composition, it being remembered that the commercial hydrocarbons will distill over a fairly short range rather than have the boiling point of a pure compound.

Turpentine and Related Terpenes. While there are many chemical compounds or mixtures of this class that might be present, in practice wood turpentine and gum turpentine constitute the great majority of cases. If the sample is a fraction separated from a product containing soap, the possibility must be considered that it constitutes the 5 to 20 per cent of terpenes associated with tall oil and that the soap was derived from such an oil. Odor is a distinctive qualitative characteristic, and specific gravity another.

QUALITATIVE. When very small amounts of turpentine are suspected, test as follows: Dissolve 2 drops of the solvent in 2 to 3 ml. of chloroform in a dry test tube, and add 1 drop of antimony pentachloride. In the absence of turpentine a slow or slight change in color will take place. A rapid change in color to a dark red or purple usually indicates turpentine.

QUANTITATIVE. In the absence of aromatic hydrocarbons and other solvents whose identification has been provided for, determine turpentine by sulfonation as described for aromatic hydrocarbons. Sulfonation is not complete, and the method is accurate only within about 10 per cent. If over 90 per cent is sulfonatable, consider the fraction as containing little else than turpentine. If more than 10 per cent residue is left, divide the loss in volume by 0.9 to give an approximation of the turpentine content.

MIXTURES

The analysis of single solvents by odor, specific gravity, boiling point or boiling range, and other properties has been discussed. Mixtures are usually far more difficult to evaluate. The very fact that they are mixtures suggests the need of separation, which in many cases is exceedingly difficult. Odor cannot always be relied upon, for one solvent may modify the odor of another to such an extent that only one or neither can be recognized.

Boiling range may be useful but unless the facts are known may also be very deceptive. Many pairs of solvents form isotropic mixtures which have a boiling point below that of either. In general, the definition of boiling point shows that only with very efficient fractionation can one hope to isolate a comparatively pure ingredient from a mixture unless the boiling points are 50 to 60° apart.

Considering these difficulties when only two solvents are present, it follows that they are multiplied when three or more are mixed. Separations may be made by solubilities, by taking advantage of chemical properties, or, failing these, by fractional distillation with all its faults as a laboratory procedure. Examination will first be discussed qualitatively, then quantitative methods applied.

Qualitative. ODOR. Many solvents have characteristic odors that may at least identify the presence of that solvent in the mixture. It is advisable, if possible, to obtain the opinion of other competent observers. Chemists often have idiosyncrasies in regard to detection of odors, one readily recognizing an odor which another cannot but missing an odor to which the other is sensitive. Much more important is the background of experience, many chemists being able to identify a trace of underlying odor because of previous familiarity with it.

SPECIFIC GRAVITY. This simple determination is of considerable value in itself. A gravity less than 0.780 would normally indicate the presence of ethyl ether or a petroleum solvent, while a gravity of greater than 1 will almost surely show that chlorohydrocarbons are present.

Consideration together of odor and specific gravity in the light of experience will usually give a general indication of the classes of materials to be looked for.

BOILING POINT OR RANGE. Clamp a test tube containing 5 ml. of sample in a buret clamp, and suspend a thermometer just above the liquid surface. Gently boil the mixture, being careful not to ignite any volatile vapors. Obtain the initial boiling point and, observing the rate of rise in temperature, the highest temperature attained. These data when carefully interpreted may give much information. A similar technic can be applied with lesser accuracy by using a test-tube clamp and holding a thermometer in the vapors.

SOLUBILITY IN WATER. To 5 ml. of sample in a 25-ml. graduated cylinder, add 20 ml. of water. Shake well, separate, and read the volume of solvent remaining. Any loss in volume indicates the possible presence of alcohols, ketones, glycols, and various other solvents, some of which are only partly water-soluble. No appreciable loss usually means that only hydrocarbons and/or higher alcohols are present, classes readily distinguishable by odor.

Quantitative. The specific gravity determined for qualitative purposes will be valuable in quantitative estimation. Be sure that the temperature at which this is taken is recorded.

FRACTIONAL DISTILLATION. The scale on which this determination can be conducted will depend on the size of sample available and the accuracy of the results needed. Generally it is inadvisable to use a larger amount than 500 ml., 250 ml. is more convenient, and 100 ml. will often serve. By giving careful attention to details of technic and using small units, values accurate to within 5 per cent can often be obtained with a 10-ml. sample. Use of a sample under 100 ml. should be avoided in the interests of accuracy. The larger samples are mainly used when it is necessary to isolate a solvent present to only a minor extent. In setting up the distillation unit use an efficient fractionating column (page 72), and follow the general details given there.

PROCEDURE. Distill in such a manner that the distillate comes over in drops and not in a continuous stream. Record the initial boiling point as that at which the first drop issues from the condenser. No set rule can be given as to when fractions are to be cut. One may wish to cut at every 10 per cent of the total volume. In other cases the temperature may hold relatively constant while a fraction is being distilled and the material coming over at a constant boiling point may be taken as one fraction, even though it constitutes 60, 70, or 80 per cent of the whole. Generally speaking, the first technic is more commonly applied by an inexperienced person, the latter by an experienced one who knows how to save time in arriving at results. Often there will be a substantial undistillable fraction to be recorded as residue not distilled at the highest temperature attained.

Obtain specific gravities on all the fractions, if necessary, but at any rate on the first, the middle, and the last and also on the residue in the flask after the fractionating column drains. Note the odors of the various fractions for further qualitative identification or confirmation. By utilizing the boiling points, specific gravities, and qualitative data many mixtures can be calculated at this point, always provided that no isotropic mixture distills or, if it does, that it is recognized. Frequently, fractionation of a fraction thus obtained may be necessary.

Type Mixtures. The detailed analysis of all possible combinations of solvents cannot be considered here. Practically, the great majority of solvent mixtures fall into a limited number of classes. Of these classes, the three most frequently encountered are lacquer solvents and thinners, mixtures of hydrocarbon solvents with chlorinated hydrocarbons, and mixtures of hydrocarbon solvents with turpentine.

Lacquer Solvents and Thinners. The most common solvents and thinners found in such mixtures are any of the lower alcohols up to amyl; the methyl, ethyl, propyl, and butyl esters of acetic acid and, less commonly, esters of lactic and propionic acids;

various petroleum naphthas; and toluene and occasionally benzene, xylene, acetone, and some of the higher ketones.

FRACTIONAL DISTILLATION. Using a good fractionating column, distill 200 ml. of the mixture slowly. Stop the distillation when the vapor temperature reaches 93°.[1] At this temperature all ethyl compounds will have distilled over. If iso- or secondary butyl compounds are present, small amounts of these compounds may come over with ethyl acetate below 93°.

Record the volume of distillate, and obtain that of the residue by difference. The two fractions so separated are to be further analyzed by the procedures that follow:

SOLUBILITY IN 20 PER CENT SODIUM CHLORIDE SOLUTION. Each of the two fractions from distillation is to be extracted with a 20 per cent solution of commercial-grade salt. Shake 50 ml. of the fraction in a stoppered 100-ml. graduated cylinder with 50 ml. of 20 per cent sodium chloride solution. Allow to settle, and read. For the fraction below 93°, the loss in volume will be due to acetone, methyl alcohol, or ethyl alcohol. Isopropyl alcohol and methyl and ethyl acetates are largely insoluble in this salt solution. For the fraction above 93° the loss in volume will be due to ethylene glycol and diethylene glycol or to methyl or ethyl ethers of ethylene glycol. Butyl alcohol, isopropyl acetate, ethylene glycol monobutyl ether, and the acetate of ethylene glycol monoethyl ether are largely insoluble.

This extraction can be conducted on a smaller scale by using 20 ml. of each in a buret or even 5 ml. of each in a 10-ml. buret. Necessarily, accuracy is sacrificed in dealing with small volumes. The salt solution is used in place of water to lessen the solubility of slightly water-soluble ingredients. Even as outlined, the separations are only relative rather than absolute. Both fractions of each extraction should be saved so that the original sample now comprises four fractions from this procedure.

SOLUBILITY IN 80 PER CENT SULFURIC ACID. For this determination, carefully add 80 ml. of concentrated sulfuric acid to 20 ml. of water, cooling during the addition. Deviation by more than 3 per cent from the specified concentration will introduce errors. Use this acid for extraction of each of the two fractions obtained from the original by distillation.

Thoroughly shake 50 ml. of the fraction in a stoppered 100-ml. graduated cylinder with 50 ml. of the 80 per cent sulfuric acid. Allow to settle overnight if necessary, and read. In both fractions the insoluble liquid will consist of coal tar and petroleum hydrocarbons. The fraction below 93° may contain benzene and the lower fractions of petroleum naphtha. The fraction above 93° may contain toluene, xylene, and the higher fractions of petroleum naphtha. As described for extraction with salt solution, if necessary these operations may be carried out on a smaller scale with lower accuracy.

[1] WATTS, C. E., *Ind. Eng. Chem., Anal. Ed.*, **6**, 262 (1934).

Obtain the specific gravities of the insoluble fractions. If below 0.86, which is the lowest specific gravity of any coal-tar hydrocarbon, the presence of petroleum naphtha is indicated. As before, reserve each fraction of each extraction for further analytical examination.

Saponification Value. This gives a determination of the esters present, since they are the only solvents that react in this determination. Carry out this procedure on the fraction distilling below 93° and that remaining from the distillation. Follow the procedure starting on page 343. Calculate the saponification value of the fraction distilling below 93° to ethyl acetate of 88 per cent purity, the balance being mainly ethyl alcohol but with 1 per cent of denaturant required by law. Thus the calculated percentage by volume of pure ethyl acetate divided by 0.88 will give the result in terms of the commercial form regardless of the nature of the other 12 per cent. An exception to this calculation is when there is clear evidence of a preponderance of methyl acetate or isopropyl acetate. In that event, calculate to those esters, assuming approximate purity in the absence of evidence as to the presence of the commercial grades. Calculate the saponification value of the fraction not distilling below 93° to butyl acetate of 92 per cent purity unless there is clear evidence of the preponderance of any one of the higher acetates.

Approximate Composition of Thinner. The data already available from two forms of extraction and a saponification value permit a calculation of approximate composition that will often be sufficient for industrial purposes. Even if the analysis is to be carried further, the approximate composition is a valuable clue for use.

In Table 25, data have been assumed and calculations made from them.[1] Unless the lacquer to which this thinner is applied is a cellulose acetate, it is probable that the second item should read ethyl alcohol since acetone is rarely used except with cellulose acetate.

Further Fractionation of Salt-solution Soluble Fraction. Rather than using the extracts reserved from the two 50-ml. extractions it is desirable to extract on a larger scale. To obtain larger fractions, the two salt extractions of 50-ml. samples are combined with the larger extract described in the following. If adequate sample is available, transfer 200 ml. to a 500-ml. separatory funnel and shake with an equal volume of 20 per cent salt solution. Separate and perform a fractional distillation on the salt solution, taking one fraction up to 75°, another at 75 to 83°, and a third at 83 to 100°. Usually after fractionating it is desirable to refractionate in a smaller unit, adding the whole of each successive fraction slowly through a separatory funnel when the preceding one has about 75 per cent distilled. The rate of addition should be so controlled that boiling does not entirely cease. The interpretation of the fractions assumes an excellent fractionation, and the quality of results will deteriorate rapidly with decline

[1] *Loc. cit.*

in the quality of that step, one that requires experience in both manipulation and interpretation of the point at which to cut.

TABLE 25.—SAMPLE CALCULATION OF MIXED THINNERS

	Volume, %	Hydro-carbons, %	Esters, %	Alcohols and ketones
As received	100	55.5	28.5	16.0
First fractionation				
Boiling below 93°	32	47.75	21.32	42
Residue at 93°	68	58.75	29.64	12
Calculated to percentage of original thinner				
Boiling below 93°	100	15.28	6.82	13.44
Residue at 93°	100	39.95	20.16	8.16
		55.23	26.98	21.60

The total at this stage is 103.81 per cent, and the error is in the alcohols and ketones. Taking them by difference gives the following corrected values:

Boiling below 93°	15.28	6.82	9.90
Residue at 93°	39.95	20.16	7.89
	55.23	26.98	17.79

This leads to the following approximation of the composition as obtained from these data.

	Per Cent
Hydrocarbons	55.0
Ethyl alcohol and acetone	10.0
Ethyl acetate	7.5
High-boiling alcohols	7.5
High-boiling esters	20.0
	100.0

The fraction boiling below 75° will normally contain acetone, methyl alcohol, and methyl acetate. It is quantitatively analyzed for acetone by the iodoform method (page 437) and for methyl acetate by saponification (page 449). The fraction boiling at 75 to 80° will contain ethyl alcohol, methylethyl ketone, and ethyl acetate. The ketone can be determined in the same way as acetone, the ester by saponification, and the alcohol by hydroxyl value (page 439). Between 83 and 100° the distillate contains isopropyl acetate, butyl acetate, and possibly higher ketones. The total esters can be determined by saponification, qualitative tests (page 435) being depended on to identify whether the ester is isopropyl or butyl acetate or both. Usually a qualitative test will eliminate ketones. Ethylene glycol, diethylene glycol, and their derivatives cannot be satisfactorily fractionated from the salt solution. If they are present, other means must be devised for their determination, as by fractionation of an original sample.

FURTHER FRACTIONATION OF HYDROCARBON FRACTION. The extracts from 50-ml. samples with 80 per cent sulfuric acid are often small for fractionation, even when combined. If adequate sample is available, extract 200 ml. with an equal volume of 80 per cent sulfuric acid in a 500-ml. separatory funnel. Draw off the acid layer, and reserve for later use. If desired, combine the hydrocarbon layer with those derived from the two 50-ml. fractions. Wash this hydrocarbon layer with water until the extract is not acid to methyl orange or methyl red. Drain well, and either filter through a dry filter paper or dry over potassium carbonate or anhydrous sodium sulfate. Obtain the specific gravity of the dried mixed solvents.

Carefully fractionate the dried solvent. In this case no instructions can be given as to the levels at which cuts are to be taken. This must depend on the ranges in which the rate of rise of temperature in the fractionating column falls off. Usually a refractionation is desirable as described in considering the fractions distilled from the salt-solution extract. The boiling range, odor, and specific gravity of each fraction will normally be sufficient for reasonably close identification.

Supplementing this, take a separate sample of the dried hydrocarbon fraction, and sulfonate (page 444) to determine the amounts of petroleum and aromatic hydrocarbons. If sample is not available for this, parts of the fractions distilled may be recombined in the proper ratio with part of the undistillable residue, if any. Avoid overheating during the sulfonation, for petroleum naphtha would thus be lost. Usually the odor and density of such unsulfonated residue taken in conjunction with the prior fractional distillation of hydrocarbons will be adequate. If necessary, enough can be collected by a series of sulfonations, washed with water until no longer acid to methyl orange or methyl red and fractionally distilled by the American Society for Testing Materials (ASTM) method (page 665). The density of the original before sulfonation and that of the unsulfonatable fraction can be used to calculate approximately the density of the sulfonated material. An unduly low value for the sulfonated material may indicate cyclicized petroleum naphtha or a petroleum naphtha high in naphthenes.

If a sample contains both aromatic and petroleum naphtha of similar boiling ranges, while the above gives a separation of the petroleum naphtha, there is no way of isolating the aromatic fraction unchanged.

Further Fractionation of Material Soluble in 80 Per Cent Sulfuric Acid. Use the portion soluble in 80 per cent sulfuric acid as set aside during the previous determination, adding, if desired, the two earlier extracts of 50-ml. portions. Dilute slowly with at least three times its volume of water, taking care to prevent overheating, which would cause volatilization of the dissolved solvents.

Distill this mixture without fractionation, stopping when a temperature of 100° indicates that only water is being distilled. The distillate contains water-soluble solvents such as alcohols, ketones, and esters. Add sufficient commercial salt to nearly saturate the distillate. The esters will separate as an insoluble layer. If large quantities of water-soluble matter are present but only small amounts of esters, wash this layer twice with its own volume of 20 per cent sodium chloride solution. Take the specific gravity of the separated ester layer.

This ester layer could be fractionated, but in general the esters do not give such clear-cut fractions as the corresponding alcohols. To saponify take a volume of diethylene glycol equal to the volume of the esters. To this add an amount of potassium hydroxide or sodium hydroxide approximating one-third the weight of esters. When dissolved and cool, add the ester layer. Reflux gently for at least 1 hour to saponify. Diethylene glycol is used as solvent for the alkali because its high boiling point will permit sharp separation by distillation of the alcohols produced. Connect to a condenser through a fractionating column, and distill the alcohols. No rule can be set as to the cuts to be taken; judgment must be applied according to the results obtained. Refractionate as described earlier to obtain as sharp cuts as possible, repeating if necessary. Identify the alcohol fractions by qualitative tests and density.

Hydrocarbons and Chlorinated Hydrocarbons. Commercially, the degree of fire hazard is closely correlated with flash point. Therefore chlorohydrocarbons are often added to hydrocarbon solvents to raise the flash point so that an otherwise inflammable solvent receives the rating of a combustible solvent. Since the chlorinated hydrocarbon costs several times as much as the hydrocarbon per gallon, only the minimum amount permissible is used. The experienced worker can often identify the hydrocarbon and predict from it, with a high degree of accuracy, how much chlorohydrocarbon will be present. Regardless of this, however, separations and identifications are required.

Fractionation. Carefully fractionate a sample of suitable size. This will depend somewhat on the amount available; 250 ml. is usually sufficient, and results of reasonable accuracy can be obtained with 100 ml. Commercial chlorohydrocarbons boil within a narrow temperature range; but when mixed with solvents having a wide distillation range, such as petroleum naphtha, they cannot be sharply separated even with extreme care and repeated fractionation. Take as many fractions as the distillation characteristics justify. Obtain the specific gravity of the various fractions as well as of the original.

The chlorohydrocarbon will have been qualitatively identified; this is no great problem. There is normally no other reason for adding chlorohydrocarbon to a product than to raise the flash point. The problem is then to identify the hydrocarbon. Sulfonation of one or two of the later fractions will readily indicate whether it is of petroleum or aromatic origin.

If of petroleum origin it may be partly sulfonatable but will not sulfonate completely. Usually the two are not mixed in this type of product. When the type, whether of aromatic or petroleum origin, has been identified, the specific gravity of the naphtha in commercial mixtures may often be obtained with a fair degree of accuracy by observing the initial and final boiling points of the mixture and comparing these with known types of naphtha. Chlorohydrocarbons frequently have boiling points intermediate between the initial and final boiling points of naphthas. If the distillation temperature of the chlorohydrocarbon is below the initial point of the hydrocarbon, little of the latter will distill with the chlorohydrocarbon and fractionation will be reasonably effective. Therefore from these data an approximation of the density of the hydrocarbon is arrived at and used for calculation of proportions from the density of the mixture and that of the chlorohydrocarbon. One must assume that no volume change occurs in mixing, an assumption that is approximately correct. This is most easily explained by an example.

Given a naphtha of specific gravity 0.765 and carbon tetrachloride, specific gravity 1.595, a mixture of these has a specific gravity of 1.014.

Let x = volume per cent of carbon tetrachloride in this mixture. Then $1.595x + 0.765(100 - x) = 1.014 \times 100$ or $x = 30$.

The composition by volume is therefore 30 per cent carbon tetrachloride and 70 per cent naphtha.

Because of the high density of the chlorohydrocarbon and the low value for naphtha, the relation by weight will be radically different. To obtain per cent by weight of carbon tetrachloride, divide its density times per cent by volume by the specific gravity of the mixture or

$$\frac{(1.595 \times 30)}{1.014} = 47.2 \text{ per cent.}$$

The composition by weight is therefore 47.2 per cent carbon tetrachloride and 52.8 per cent naphtha.

A convenient check is obtained by noting the varying densities of the successive distillates, but this requires experience. As a mathematical method of checking, perform similar calculations on the various fractions. This has an element of inaccuracy, for the earlier fractions contain hydrocarbon of lower specific gravity than the later fractions. However, error at one end of the distillation will be counterbalanced by an error at the other end. Add up all the various percentages of each ingredient so found, and the result should be an approximate check on the calculation from the density of the original mixture.

Hydrocarbon Solvents and Turpentine. Mixtures of petroleum hydrocarbons and turpentine are used for such diverse purposes as paint thinners, vehicles for metal polishes, and solvents for liquid and paste waxes. While mixtures of aromatic hydro-

TABLE 26.—CONSTANTS ON COMMERCIAL SOLVENTS

	Formula	Molecular weight pure	Boiling point, °C.		Specific gravity, 20°/20°		Solubility in water, per cent weight pure compound	Flash point, °F.	
			Pure	Commercial	Pure	Commercial			
1. Petroleum naphtha	.....	.....	.....	96–129	.....	0.735	Insoluble	Below 70	1
2. V.M. and P. naphtha	.....	.....	.....	110–155	.....	0.750	Insoluble	Below 80	2
3. Stoddard solvent	.....	.....	.....	149–210	.....	0.780	Insoluble	Above 100	3
4. Kerosene	.....	.....	.....	175–275	.....	0.820	Insoluble	.....	4
5. Benzene, industrial pure	C_6H_6	78.06	80.4	78–82	0.879	0.875–0.886 (15.5°/15.5°)	0.08	10–15	5
6. Toluene, industrial pure	$C_6H_5CH_3$	92.13	111	108–112	0.866	0.864–0.874 (15.5°/15.5°)	Insoluble	35–40	6
7. Xylene, industrial (solvent naphtha)	$C_6H_4(CH_3)_2$	106.16	138–142	130–155	0.866–0.881	0.860–0.870 (15.5°/15.5°)	Insoluble	63	7
8. High-flash naphtha 185 dry point	.....	.....	.....	150–185	.....	0.855–0.890 (15.5°/15.5°)	Insoluble	Above 100	8
9. Methyl acetate, 97 per cent	CH_3COOCH_3	74.08	57.1	55–58	0.935	0.930 –0.940	24.2	25	9
10. Ethyl acetate, 88 per cent	$CH_3COOC_2H_5$	88.10	77.1	71–85	0.902	0.884 –0.887	8.7	40	10
11. Ethyl acetate, 99 per cent	$CH_3COOC_2H_5$	88.10	77.1	75–85	0.902	0.899 –0.902	8.7	40	11
12. Isopropyl acetate	$CH_3COOCH(CH_3)_2$	102.13	88.4	84.5–90	0.874	0.866 –0.871	2.9	60	12
13. Butyl acetate, 90 per cent	$CH_3COOC_4H_9$	116.16	126.3	118–128	0.883	0.873 –0.876	0.68	100	13
14. Amyl acetate, primary	$CH_3COOC_5H_{11}$	130.19	.....	.....	.....	.....	.....	.....	14
15. Amyl acetate, secondary, iso	$(CH_3)_2CHCHOHCH_3$	130.19	.....	121–144	.....	0.862	0.19	111	15
16. Amyl acetate, mixed isomers	.....	130.19	.....	127–155	.....	0.862	0.17	111	16
17. Ethyl propionate, 96 per cent	$CH_3CH_2COOC_2H_5$	102.08	.....	90–118	.....	0.891	1.92	70	17
18. Isopropyl butyl propionate	$CH_3CH_2COOC_4H_9$	130.11	.....	124–171	.....	0.874	0.15	105	18
19. Ethyl lactate	$CH_3CH(OH)COOC_2H_5$	118.08	.....	119–179	.....	1.034	Complete	140	19
20. Isopropyl lactate	$CH_3CH(OH)COOCH(CH_3)_2$	132.09	.....	149–167	.....	0.988	Complete	130	20
21. Butyl lactate	$CH_3CH(OH)COOC_4H_9$	146.11	.....	145–230	.....	0.980	4.0	176	21
22. Ethylene glycol monomethyl ether acetate*	$CH_3COOCH_2CH_2OCH_3$	118.13	144.5	132–152	1.0067	1.003 –1.008	Complete	140	22
23. Ethylene glycol monoethyl ether acetate*	$CH_3COOCH_2CH_2OC_2H_5$	132.16	156.3	145–165	0.975	0.971 –0.976	23	140	23
24. Diethylene glycol monomethyl ether acetate*	$CH_3COOC_2H_4OC_2H_4OCH_3$	162.18	209.1	.....	1.0396	.....	Complete	180	24
25. Diethylene glycol monoethyl ether acetate*	$CH_3COOC_2H_4OC_2H_4OC_2H_5$	176.21	217.7	208–223	1.0114	1.008 –1.013	Complete	230	25
26. Diethylene glycol monobutyl ether acetate*	$CH_3CO(OC_2H_4)_2OC_4H_9$	204.26	246.4	235–250	0.981	0.975 –0.985	6.5	240	26
27. Methanol	CH_3OHOH	32.04	64.5	.....	0.7924	0.7929	Complete	60	27
28. Ethanol 95 per cent	C_2H_5OH	46.07	78.3	77–80	0.8126	0.8089–0.8128	Complete	75	28
29. Ethanol 100 per cent	C_2H_5OH	46.07	78.3	77–80	0.7905	0.7905–0.7914	Complete	70	29
30. Isopropanol 91 per cent	$(CH_3)_2CHOH$	60.09	80.4	1° range	0.8179	0.8175–0.8185	Complete	75	30
31. Isopropanol 99.4 per cent	$(CH_3)_2CHOH$	60.09	82.4	1.5° range	0.7363	0.7863–0.7893	Complete	70	31
32. Butanol, normal	$CH_3C_2H_4CH_2OH$	74.12	117.7	116–119	0.8109	0.8100–0.8120	7.7	115	32
33. Butanol, iso-	$(CH_3)_2CHCH_2OH$	74.12	.....	107–111	.....	0.803	10.0	111	33
34. Butanol, secondary	$CH_3CH_2CHOHCH_3$	74.12	99.8	96–103	0.819	0.808	12.5	88	34

35. Amyl alcohol, normal	$CH_3(CH_2)_3CH_2OH$	88.15	137.8	134–139	0.817	0.820	Slightly	136	35
36. Amyl alcohol, mixed isomers (pentasol)	....	88.15	.....	110–140	....	0.815	16 per cent	113	36
37. Amyl alcohol, tertiary	$(CH_3)_3CCH_2OH$	88.15	102.5	99–104	0.814	0.814	Slightly	75	37
38. Hexanol, normal	$CH_3(CH_2)_4CH_2OH$	102.17	157.2	153–160	0.8208	0.819 –0.823	0.58	165	38
39. Octyl alcohol (2-ethyl-hexanol)	$C_4H_9CH(C_2H_5)CH_2OH$	130.23	183.5	180–190	0.834	0.832 –0.837	0.10	185	39
40. Acetone	CH_3COCH_3	58.08	56.1	Within 1°	0.7911	0.7915–0.7935	Complete	15	40
41. Methyl acetone	48 per cent acetone 28 per cent methyl acetate 24 per cent methyl alcohol	.....	.....	50–67	....	0.830 –0.836	Complete	23	41
42. Methyl ethyl ketone	$CH_3COC_2H_5$	72.06	.....	77–82	....	0.809	25.6	34	42
43. Diacetone alcohol	$(CH_3)_2COHCH_2COCH_3$	116.16	167.9	135–170	0.9406	0.937 –0.942	Complete	170	43
44. Diacetone alcohol, technical	....	.....	.....	60–170	....	0.915 –0.920	Complete	...	44
45. Mesityl oxide	$(CH_3)_2C{:}CHCOCH_3$	98.14	128	....	0.8569	0.852 –0.856	2.8	90	45
46. Ethylene glycol	$HOCH_2CH_2OH$	62.07	197.2	190–210	1.1155	1.1151–1.1156	Complete	240	46
47. Diethylene glycol	$HOCH_2CH_2OCH_2CH_2OH$	106.12	244.8	230–270	1.1184	1.1170–1.120	Complete	290	47
48. Propylene glycol	$CH_3CHOHCH_2OH$	76.09	188.2	180–210	1.0381	1.037 –1.040	Complete	225	48
49. Monomethyl ether of ethylene glycol*	$CH_3OCH_2CH_2OH$	76.09	125	122–126	0.9664	0.963 –0.967	Complete	115	49
50. Monoethyl ether of ethylene glycol*	$C_2H_5OCH_2CH_2OH$	90.12	135.1	132–137	0.9311	0.928 –0.933	Complete	130	50
51. Monobutyl ether of ethylene glycol*	$C_4H_9OCH_2CH_2OH$	118.17	171.2	166–173	0.9019	0.900 –0.905	Complete	165	51
52. Monomethyl ether of diethylene glycol*	$CH_3OCH_2CH_2OCH_2CH_2OH$	120.15	194.1	187–197	1.0218	1.030 –1.040	Complete	200	52
53. Monoethyl ether of diethylene glycol*	$C_2H_5OCH_2CH_2OCH_2CH_2OH$	134.17	201.9	185–205	0.9898	1.024 –1.030	Complete	210	53
54. Monobutyl ether of diethylene glycol*	$C_4H_9OCH_2CH_2OCH_2CH_2OH$	162.23	230.7	230–235	0.9555	0.954 –0.960	Complete	240	54
55. Ethyl ether	$C_2H_5OC_2H_5$	74.12	34.5	About 35	0.7146	0.818 –0.721	6.9	40	55
56. Isopropyl ether	$(CH_3)_2CHOCH(CH_3)_2$	102.17	68.4	63–69	0.7252	0.722 –0.726	1.2	15	56
57. Diethyl ether of ethylene glycol*	$C_2H_5OCH_2CH_2OC_2H_5$	118.17	121.4	115–140	0.8424	0.8405–0.8455	21.0	95	57
58. Diethyl ether of diethylene glycol*	$(C_2H_5OC_2H_4)_2O$	162.23	187.9	180–190	0.9094	0.906 –0.911	Complete	180	58
59. Dioxane	$O{:}(CH_2CH_2)_2{:}O$	88.10	101.1	95–103	1.0343	1.030 –1.038	Complete	65	59
60. Methylene chloride	CH_2Cl_2	84.93	29.8	29.2–40.0	1.336	1.333–1.336 (15°/4°)	0.027	...	60
61. Chloroform	$CHCl_3$	119.37	61.2	60–61.5	1.489	1.478–1.482 (25°/4°)	0.021	...	61
62. Carbon tetrachloride	CCl_4	153.83	76.7	76.2–77	1.595	1.595–1.597 (20°/155°)	.....	...	62
63. Trichloroethylene	$CHCl{:}CCl_2$	131.38	86.7	86–87.5	1.462	1.472–1.476 (15°/4°)	0.002	...	63
64. Dichloroethylene, mixed isomers	$CHCl{:}CHCl + CH_2{:}CCl_2$	96.93	48.4–60.3	51–61	1.272	1.279–1.285 (15°/4°)	0.004	...	64
65. Perchloroethylene	$CCl_2{:}CCl_2$	165.83	120.8	120–122	1.623	1.628–1.632 (15°/4°)	0.008	...	65
66. Ethylene dichloride	CH_2ClCH_2Cl	98.95	83.7	82.1–84.6	1.257	1.252	0.092	...	66
67. Propylene dichloride	$CH_3CHClCH_2Cl$	112.96	95.9	94.4–97	1.155	1.161	0.038	...	67
68. Monochlorobenzene	C_6H_5Cl	112.5	131.6	Within 1°	1.105	....	0.049	81	68
69. Dichlorobenzene, ortho-	$C_6H_4Cl_2$	147.00	180.2	Within 4°	1.304	1.303 (25°/25°)	0.008	154	69
70. Trichlorobenzene, 1,2,3 and 1,2,4	$C_6H_3Cl_3$	181.45	214.8	Within 5°	1.460	1.465–1.475	Insoluble	212	70

* The monoethyl ether of ethylene glycol is Cellosolve and the monoethyl ether of diethylene glycol is Carbitol commercially. See Table 24 for indication of other commercial names.

carbons and turpentine have no specific application, such a mixture might be encountered. The possibility, however, may be reasonably assumed absent unless substantially the entire fraction is sulfonatable.

PROCEDURE. As the initial step with such a mixture, sulfonate according to the method on page 444 at as low a temperature as possible. Since the turpentine will leave about 10 per cent of unsulfonatable residue, the loss in volume after sulfonation divided by 0.9 will give the volume of turpentine in the sample. Upon eliminating aromatic hydrocarbons, the residue will consist of petroleum naphtha on the assumption that the sulfonation loss is small and that the unsulfonatable residue from turpentine is unimportant. Ordinarily this petroleum naphtha fraction can be washed with water until the washings no longer react acid to methyl orange or methyl red and then identified with sufficient accuracy by specific gravity and distillation range. The possible presence of cyclicized petroleum naphthas must not be overlooked.

Another method that is useful when the most accurate results obtainable are not required is to shake 25 ml. of the sample with 25 ml. of 90 per cent alcohol in a 50-ml. graduated cylinder. Allow to separate. Should this not readily occur, use half volumes of each in a 25-ml. graduated cylinder, shake thoroughly, and centrifuge. Turpentine is soluble in alcohol but petroleum solvents are not readily so in 90 per cent alcohol. Some partition effect will usually occur, but in many instances sufficient information may be obtained by this method. Obtain a gravity on the insoluble petroleum solvent for identification.

Still another method of approximating the turpentine in a sample is by determination of the iodine value by the Wijs method (page 346). Turpentine has an iodine value of about 340. An iodine value of 20 or more will indicate the presence of turpentine, and an approximate calculation of the amount present may be made. Such a calculation can be only approximate, for some absorption of iodine occurs with any petroleum solvent, even if only a few units.

CHAPTER 24

ALCOHOLS

Alcohols are characterized by containing at least one hydroxyl group. All are colorless liquids, and the lower members have characteristic odors. Increase in molecular weight in the series causes a change from complete miscibility with water to limited solubility and then to insolubility.

As the water solubility increases, the solubility of oil in the alcohol increases. The higher alcohols, some of which contain multiple hydroxyl groups, are relatively nonvolatile and odorless liquids or even solids. Some properties of alcohols used as thinners are given in Table 26. Since the hydroxyl group forms no derivatives that are readily isolated, methods of determination are largely by specific gravity of solutions.

Ethanol. Ethanol is the most common alcohol. It is also known as Cologne spirits and, more often, as just alcohol. It is made by the fermentation of grain or the sugar in molasses and synthetically. Although experienced workers in the industry claim to be able to differentiate organoleptically between alcohol from grain and that from molasses, analytically they are indistinguishable.

Ethanol is completely miscible with water and with most organic solvents as well as with castor and croton oils. Many gums, natural and synthetic, shellac, rosin, cellulose ethers, soaps, and some forms of nitrocellulose are soluble in absolute alcohol and less so in the usual commercial grade containing 95 per cent of alcohol by volume.

Completely denatured alcohol is alcohol made unfit for beverage purposes. The Bureau of Internal Revenue has prescribed denaturants not removable by distillation. Since these are modified from time to time, the chemist should have available a booklet of completely and specially denatured alcohol formulas, obtainable from any of the large alcohol distillers without charge. Such completely denatured, or C.D., alcohol is sold without permit or restriction.

Specially denatured alcohol also is nonpotable. It contains specific additions for specific industrial purposes as shown in the booklet of denatured alcohols mentioned. Such specially denatured, or S.D., alcohol is not taxed; but a permit is necessary before purchase, and purchases are limited to specific uses. Pure alcohol is highly taxed, except for university laboratories. It may be purchased without Federal permit, but some states require a license.

Ethanol is the most common solvent, other than water, used in industry. It serves for extraction, purification, and crystallization and also as an ingredient in antifreeze mixtures, emulsions, lacquers, enamels, soaps, varnishes, cements, plastics, and many other compounds. It has effective germicidal properties.

QUALITATIVE. The odor of ethanol is characteristic. In more dilute solutions, test as for its presence in thinners and solvents (page 438).

QUANTITATIVE. Since alcohol is rarely available for analysis without contaminating substances, it is usually separated by distillation, an operation that would otherwise be unnecessary. The methods that follow are often applicable to fractions distilled from complex samples. Other volatile solvents must be absent.

PROCEDURE. Accurately pipet or measure 25 ml. of the sample containing about 50 per cent of alcohol, or up to 100 ml. if a less amount is present, into a flask. Alternatively, a weighed sample may be taken. A Kjeldahl flask is convenient because of the fractionating effect of the long neck. If the sample is acid, it may contain volatile acids that would distill over and vitiate the results. Prevent this by adding sufficient sodium hydroxide solution to render it slightly alkaline. If aldehydes are present, remove them by addition of a few milliliters of dilute silver nitrate solution and sufficient sodium hydroxide solution to precipitate silver oxide. If the sample contains oily or fatty matter, first transfer to a separatory funnel. Saturate this solution with sodium chloride, and shake with 25 ml. of petroleum ether. This extracts the oil or fat without extracting the alcohol. Draw off the aqueous salt solution into the flask. Wash the petroleum ether with two 10-ml. portions of saturated salt solution, adding each to the flask.

Add sufficient water to dilute the sample to about 125 ml. Connect with an efficient water-cooled condenser, and slowly distill into an accurately calibrated 100-ml. flask. When organic matter is present, do not heat the flask above the level of the liquid or decomposition will occur and the results will be vitiated. The end of the condenser must project well into the neck of the flask. Stop the distillation a few milliliters below the calibration mark, and dilute nearly to volume with water. Finally adjust to exactly 100 ml. at the same temperature as that at which the original solution was measured. This gives a solution containing the same amount of alcohol per milliliter as the original sample if 100 ml. was taken as sample or a definite fractional amount if a volume less than 100 ml. was diluted to 125 ml. for distillation.

Avoiding evaporation, accurately determine the specific gravity to five decimal places, by the pycnometer method, preferably at the same temperature at which the sample was taken. From tables found in most handbooks, but much more fully in the Methods of Analysis of the Association of Official Agricultural Chemists, obtain the corresponding percentage of alcohol by volume. For accurate work it is essential to measure volumes and obtain specific gravities at the same temperature. If the distillate is much larger than the sample, a large multiplying factor is used and the error is thus increased. It has been shown that even the tables of the Association of Official Agricultural Chemists are strictly exact only at 60°F., at which the percentages of alcohol by volume are given. In summer, because of condensation at 60°F., a higher temperature must be used.

CALCULATION.

$$\text{Alcohol in distillate by volume at } t^\circ \times \frac{100}{\text{volume of sample at } t^\circ} = \text{per cent by volume of alcohol at } t^\circ$$

If tables are available, giving alcohol in grams per 100 ml.,

$$\text{Alcohol in distillate in grams} \times \frac{100}{\text{weight of sample}} = \text{per cent by weight of alcohol}$$

ACCURATE METHOD INVOLVING WEIGHING ONLY. The preceding method involves accurately measuring certain volumes. Hence the pipet and flask must be highly accurate. A method employing weight only eliminates this necessity and gives a higher degree of accuracy.

PROCEDURE. Carefully weigh the sample in a covered beaker, transfer without evaporation to the Kjeldahl flask, and dilute as in the previous method. Distill into an accurately weighed, clean, dry 100-ml. flask. After the distillation weigh the flask and distillate, and so get the weight of the latter. Mix thoroughly without diluting to exactly 100 ml., and obtain the specific gravity of the distillate and of the original sample, preferably at 60°F. For reasons already stated, if dilutions are involved or this is at a different temperature, the accuracy will be lowered.

CALCULATION.

$$\frac{\text{Weight of sample}}{\text{Specific gravity of sample at } t^\circ} = \text{volume of sample at } t^\circ$$

$$\frac{\text{Weight of distillate}}{\text{Specific gravity of distillate at } t^\circ} = \text{volume of distillate at } t^\circ$$

$$\frac{\text{Volume of distillate at } t^\circ}{\text{Volume of sample at } t^\circ} \times \text{per cent alcohol content by volume of distillate at } t^\circ = \text{per cent alcohol by volume at } t^\circ$$

If tables are available, giving alcohol by weight,

$$\frac{\text{Weight of distillate at } t^\circ}{\text{Weight of sample at } t^\circ} \times \text{per cent alcohol content by weight of distillate at } t^\circ = \text{per cent alcohol by weight}$$

HYDROMETER METHOD. If the alcohol content is high enough and interfering substances are absent, the use of a set of specially calibrated hydrometers furnishes another convenient method. In case of doubt as to accuracy, check the hydrometer used at the correct temperature against alcohol solutions of known gravity obtained by the more accurate pycnometer method. This is the method used in routine control of alcohol content of spirits. Therefore complete conversion tables are published by the Bureau of Internal Revenue to give the percentage of alcohol present when the hydrometer is read at varied room temperatures.

IMMERSION REFRACTOMETER METHOD. Obtain the reading of the refractive index at 17.5° with the Zeiss immersion refractometer (page 57). Refer to tables such as those of the Association of Official Agricultural Chemists or in a chemical handbook to translate into per cent of alcohol by volume or by weight.

For routine work on samples of which the alcohol and solids content is approximately known, as with beers and wines, the refractometric reading may be taken directly, subject to correction. The correction for the given product is determined by comparing the alcohol content of various samples differing in alcohol and solids content, such determinations being by distillation and refractometric readings as contrasted with direct reading. These are then plotted as a curve or tabulated. The correction so developed, which is valid only for the product on which it was determined, is known as the obscuration factor.

PROOF SPIRITS. By definition American proof spirits is held to be that alcoholic liquor which contains one-half of its volume of alcohol of a specific gravity of 0.7939 at 60°F. referred to water at 60°F. as unity. Proof spirit has at 60°F. a specific gravity of 0.93426, 100 parts by volume of the same consisting of 50 parts of absolute alcohol and 53.73 parts of water. The difference of the sum of the parts of alcohol and water and the resulting 100 parts of proof spirits is due to contraction, which takes place when alcohol and water are mixed.

Methanol. Methanol is also called wood alcohol or wood spirits because it was formerly obtained entirely from the distillation of hardwood. A much purer grade is now made synthetically by catalytically combining carbon monoxide and hydrogen under pressure.

Methanol will dissolve low-viscosity nitrocellulose, ethyl cellulose, soaps, many resins, shellac, and castor and croton oils. It is miscible not only with water but also with alcohols, esters, ketones, and halogenated hydrocarbons. Methanol is largely used as an antifreeze, as a denaturant for ethyl alcohol, and as solvent in many compositions, such as celluloid cements.

QUALITATIVE. When present in appreciable quantities, the characteristic odors will not only suffice for identification of methanol but also distinguish between the pure grade and that obtained by wood distillation. When present in small amounts, usually in the presence of large amounts of ethanol as in denatured alcohol, convert the methanol to an aldehyde and test for it with Schiff's reagent.

PROCEDURE. Dilute the solution under test to about 5 per cent total alcohol by volume. To 5 ml. of this, add 2 ml. of a solution prepared by dissolving 3 grams of potassium permanganate in 15 ml. of 85 per cent phosphoric acid, adding water to make 100 ml. Let this stand for 10 minutes, and then remove the excess permanganate by adding 2 ml. of a solution of 5 grams of oxalic acid in 100 ml. of 1:1 sulfuric acid solution. When the solution is decolorized, add 5 ml. of Schiff's reagent (page 720). Mix well, and allow to stand for 10 minutes. If methanol is present, a characteristic reddish purple color is produced, varying in intensity according to the amount. Run samples of ethanol and of ethanol containing 0.5 per cent of methanol, for comparison.

QUANTITATIVE. The procedure for separation from impurities is identical with that cited for ethanol. Then determine the specific gravity or the refractive index, and refer to tables that translate these values in terms of methanol rather than ethanol.

Methanol in the Presence of Ethanol. Use the preceding qualitative technic to ascertain if more than 0.5 per cent of methanol is present. If so, dilute with approximately 50 per cent ethyl alcohol to not over 0.5 per cent methanol. Fractionally distill a 25-ml. aliquot with an efficient column, collecting about 8.5 ml. in seven fractions of 1.2 ml. each. Collect the fractions together in the same receiver; but keep the column under total reflux 30 minutes before taking off the first fraction, and reflux 15 minutes between fractions. This is done to produce a distillate containing about 94 per cent of total alcohols. Take a specific gravity on the distillate; and, assuming that it is entirely ethyl alcohol, calculate the dilution of the distillate necessary to give 22 per cent of ethyl alcohol. Dilute to that volume, and then dilute to any convenient volume with 22 per cent ethyl alcohol.

Use the same reagents as are used in the qualitative test for methanol in ethyl alcohol. Transfer accurately 0.25 ml. of the diluted alcoholic solution to a 6-inch Nessler tube containing 4.75 ml. of water. A 6-inch Pyrex test tube is suitable. Prepare a series of standard 22 per cent alcohol solutions containing 0.02, 0.04, 0.06, 0.08, 0.10, and 0.12 per cent of methanol. Transfer 0.25 ml. of each of these to similar tubes. To each tube add 2 ml. of the potassium permanganate solution, and allow to stand for 10 minutes, with occasional shaking, without inverting the tube. Then add 2 ml. of the oxalic–sulfuric acid solution to each. Add 5 ml. of Schiff's reagent, mix thoroughly, by inverting each tube three times, stopper, and allow to stand for 1 hour. Compare the depth of color developed in the sample with that in the standards developed at the same time. Intermediate percentages may be estimated by eye.

Ethanol in the Presence of Methanol. BY DIFFERENCE. If the methanol content has been determined and is relatively low as compared with ethanol, which is the usual case, a fair approximation can be obtained by a difference method.

Specific gravities for corresponding strengths of the alcohols are quite similar but not identical. Obtain the ethanol content from the specific gravity on the assumption that only that alcohol is present. Then subtract from this total alcohol strength the amount of methanol found colorimetrically with Schiff's reagent. The difference is the ethanol present.

BY DENSITY AND REFRACTIVE INDEX. Using the Zeiss immersion refractometer, obtain the reading of the solution at 15.56°. Table 27 gives corresponding specific gravities of ethanol and methanol solutions. Reference to specific-gravity tables will give the corresponding methanol content if that only is present. When ethanol is also present, the corresponding scale reading is higher. The addition of methanol to ethanol decreases the refractive index in direct proportion to the quantity added. Hence a quantitative calculation may be made by interpolating in the table figures for pure ethanol and methanol at the same specific gravity as the sample.

For example, the mixture has a specific gravity at 15.56° of 0.9625 and a refractometer reading at 17.5° of 43.1. Interpolation gives the readings for ethanol and methanol of this gravity as 65.2 and 31.7, respectively. The difference is 33.5.

$$\text{Since} \quad 65.2 - 43.1 = 22.1 \qquad \text{hence} \qquad (22.1 \div 33.5)\,100 = 66.0$$

which is the percentage of methanol present in the total alcohol.

Propanol. Isopropanol, or secondary propyl alcohol, is a water-white liquid with a characteristic odor somewhat similar to that of ethanol. Commercially, it is supplied either as an 80.8° constant-boiling mixture, containing approximately 91 per cent of isopropanol and 9 per cent of water by volume, or as a grade containing at least 99.4 per cent by volume.

Isopropanol is frequently substituted for ethanol. It has similar chemical properties and the advantage that no permit is necessary for its use. Like ethanol, it has some germicidal properties and is used in rubbing compounds, skin lotions, cosmetics, hair tonics, lacquers, perfumes, etc. Owing to its toxic nature, it is not used in food products.

QUALITATIVE. *Iodoform Test.* Isopropanol gives the iodoform test at room temperatures (page 436).

Mercuric Sulfate Test. Prepare the reagent solution by mixing 5 grams of yellow mercuric oxide, HgO, with 40 ml. of water, adding, with stirring, 20 ml. of concentrated sulfuric acid. Now add 40 ml. of water, and stir until completely dissolved.

To 2 ml. of a dilute solution of the sample add 3 ml. of water and 10 ml. of the reagent solution. Heat in a steam bath for 3 minutes. A white or yellow precipitate forming within 3 minutes indicates the presence of isopropanol, acetone, ketones, or tertiary butyl alcohol.

QUANTITATIVE. *Iodoform Method.* Use the method as described for ketones (page 437). The factor for calculation as isopropanol is 0.0075.

Butanol. Butanol, or normal butyl alcohol, is a water-white combustible liquid. Other butyl alcohols are commercially produced, but the normal alcohol is most commonly used and is of special interest since it serves as the basis for making esters. It is largely made by selective fermentation processes. Butanol is miscible with most organic solvents and is used as a blending

TABLE 27.—ZEISS IMMERSION REFRACTOMETER READINGS AT 15.56° FOR METHANOL AND ETHANOL

Sp. gr. 15.56°/15.56°	Scale readings*		Differences	Sp. gr. 15.56°/15.56°	Scale readings*		Differences
	Ethyl alcohol	Methyl alcohol			Ethyl alcohol	Methyl alcohol	
1.0000	15.0	15.0	0.0	0.9720	51.5	27.0	24.5
0.9990	15.8	15.3	0.5	0.9710	53.0	27.5	25.5
0.9980	16.6	15.6	1.0	0.9700	54.6	28.1	26.5
0.9970	17.5	15.9	1.6	0.9690	56.1	28.7	27.4
0.9960	18.5	16.2	2.3	0.9680	57.6	29.2	28.4
0.9950	19.4	16.5	2.9	0.9670	59.1	29.6	29.5
0.9940	20.4	16.9	3.5	0.9660	60.6	30.1	30.5
0.9930	21.4	17.2	4.2	0.9650	62.0	30.6	31.4
0.9920	22.5	17.5	5.0	0.9640	63.3	31.0	32.3
0.9910	23.6	17.9	5.7	0.9630	64.6	31.5	33.1
0.9900	24.7	18.2	6.5	0.9620	65.8	31.9	33.9
0.9890	25.9	18.6	7.3	0.9610	67.0	32.4	34.6
0.9880	27.1	19.0	8.1	0.9600	68.1	32.8	35.3
0.9870	28.4	19.5	8.9	0.9590	69.2	33.3	35.9
0.9860	29.6	19.9	9.7	0.9580	70.2	33.7	36.5
0.9850	31.0	20.4	10.6	0.9570	71.2	34.1	37.1
0.9840	32.4	20.8	11.6	0.9560	72.1	34.5	37.6
0.9830	33.8	21.3	12.5	0.9550	73.0	34.9	38.1
0.9820	35.2	21.8	13.4	0.9540	73.8	35.3	38.5
0.9810	36.7	22.3	14.4	0.9530	74.6	35.6	39.0
0.9800	38.3	22.8	15.5	0.9520	75.4	35.9	39.5
0.9790	39.9	23.4	16.5	0.9510	76.2	36.2	40.0
0.9780	41.5	24.0	17.5	0.9500	76.9	36.5	40.4
0.9770	43.1	24.5	18.6	0.9490	77.6	36.8	40.8
0.9760	44.8	25.0	19.8	0.9480	78.3	37.0	41.3
0.9750	46.5	25.5	21.0	0.9470	79.0	37.3	41.7
0.9740	48.2	26.0	22.2	0.9460	79.7	37.6	42.1
0.9730	49.8	26.5	23.3				

* The scale readings are applicable only to instruments calibrated in the arbitrary scale units proposed by Pulfrich, *Z. angew. Chem.*, 1899, p. 1168. According to this scale, 14.5 = 1,33300, 50.0 = 1.34650, and 100.0 = 1.36464.

medium. Many natural and synthetic resins, camphor, oils, waxes, and ethyl celluloses are soluble in butanol. It has latent solvent properties for nitrocellulose and is used as a "medium boiler" in pyroxylin lacquers, to which it imparts flow, gloss, blush prevention, and lower viscosity. In solutions of synthetic resins and oleoresins, it imparts similar properties. Its uses are as a defoamer, penetrant, and denaturant and in printing inks, cleaners, polishes, varnish removers, motor fuel, perfumes, etc.

Butanol dissolves up to 20.1 per cent of its weight of water at 25° and forms with it a 92° constant-boiling mixture containing 63 per cent of butanol. This property makes it a useful dehydrating agent and accounts for its use as a blush preventive.

QUALITATIVE. The characteristic odor is the best method of detecting butanol. When it is present with ethanol, extract the sample several times with 20 per cent sodium chloride solution, in which butanol is insoluble. Isopropyl, isobutyl, amyl, and other higher alcohols are also separated from ethanol by this method.

QUANTITATIVE. No simple method is available. Extraction with salt solution may be made quantitative by measuring or weighing the insoluble liquid, provided that no other insoluble matter is present. The class determination, by hydroxyl value (page 439), is applicable.

Amyl Alcohols. These include eight isomeric forms. Fusel oil, obtained as a by-product in the manufacture of ethanol, is a mixture of a number of these alcohols with smaller amounts of lower alcohols, esters, furfural, pyridine, etc., which account for its very disagreeable odor. Refined amyl alcohol contains approximately 85 per cent of isoamyl alcohol. Synthetic amyl alcohols are made from chlorination products derived from the pentane fraction of gasoline and consist of a mixture of isomeric alcohols.

These alcohols are excellent solvents for resins, possess unusual blending properties, and impart flow and gloss to lacquers, in which they are largely used. They also impart polishing properties and blush resistance.

QUALITATIVE. As with butanol, the characteristic odor and physical and chemical constants are a means of identification.

Pettenhoffer Furfural Reaction. Mix 0.5 ml. of the sample under test with 5 ml. of water and sufficient ethanol to dissolve the alcohols in the water. Place the water-soluble portion in a test tube, and add a drop of a 0.1 per cent aqueous furfural solution. Mix thoroughly, and carefully add 3 ml. of concentrated sulfuric acid down the side of the tube so as to form a layer below the aqueous alcohol layer. A purplish ring at the junction of the two liquids indicates the presence of amyl alcohol. A brownish decomposition ring must not be confused with the purple-red color. Many substances give this color reaction, but none that might be confused with amyl alcohol are lacquer constituents. This test will distinguish amyl alcohol in the presence of other higher alcohols.

QUANTITATIVE. There is no recognized method for quantitative determination. The comments under Butanol are applicable.

Polyhydric Alcohols. This class of compounds is theoretically numerous but practically consists of only a few glycols and glycol-ethers, glycerol and sugars, or sugar alcohols. Not all members can be considered here.

TABLE 28.—COLORS DEVELOPED WITH CATECHOL AND SULFURIC ACID

Alcohol	Color
Ethyl alcohol	Water white
Ethylene glycol	Water white
Diethylene glycol	Water white
Propylene glycol	Faint pink
Diethylene glycol	Faint pink
Trimethylene glycol	Dark brown
Glycerol	Blood orange
Pentaerythritol	Dark purple-red
Erythritol	Faint pink
Mannitol	Red-orange
Sorbitol, alone	Blue
Sorbitol plus sulfuric acid	Faint pink

QUALITATIVE. *Reaction with Catechol and Sulfuric Acid.* This test serves to detect several members of the class, as indicated in Table 28. Aldehydes interfere.

Add 3 ml. of the sample solution to a test tube, and to this add 3 ml. of a freshly prepared 10 per cent aqueous solution of catechol. Shake thoroughly, and then add 6 ml. of concentrated sulfuric acid. Heat the tube gently for about 30 seconds. Compare the color developed with those in the table. Reheat, boiling down the solution, and compare again. The test will detect minor amounts of glycerol.

Acrolein Test for Glycerol. Concentrate the solution of sample to a very small volume in a porcelain dish. Add 3 ml. of concentrated sulfuric acid and a few crystals of potassium acid sulfate. These absorb the water given off on heating to dehydrate the glycerol to form acrolein, CH_2:CHCHO, boiling point 52.5°. Heat slowly and cautiously. The solution blackens; and if glycerol is present, the sharp, very unpleasant odor of acrolein, to which the eyes are particularly sensitive, is noticed. Do not confuse this odor with the less acrid but irritating odor of sulfur dioxide, which is often evolved at the same time. This test is not specific, for other glycols act similarly, giving off other aldehydes, which are also acrid.

QUANTITATIVE. *Glycerol.* The bichromate method, applicabie in the absence of other oxidizable substances, is shown on page 397 as applied to soaps. This is the method always used for dilute glycerol solutions. Concentrated glycerol is analyzed by acetylation, the acetin method, which will not be given here.

Other Alcohols. This chapter like any system of analysis is incomplete as regards substances having some of the properties here discussed. *Ethylene glycol* and *diethylene glycol* are a dihydric alcohol and an alcohol-ether, respectively, considered in Chap. 23, Solvents and Thinners, and often present as humectants and plasticizers. *Benzyl alcohol* is a high-boiling solvent and plasticizer not considered because of its high price and therefore relative rarity. *Phenol* is an aromatic hydroxy compound with substantial solubility, considered on page 484. *Hexahydrophenol* and *methyl cyclohexanol* are quite similar in properties, as are also *furfuryl alcohol* and *tetrahydrofurfuryl alcohol.*

CHAPTER 25

PAINTS, VARNISHES, LACQUERS, AND RESINS

Paint, in general, is pigment in a drying type of vehicle. The vehicle is, rarely, raw oil to which drier has been added; more often, it is a processed oil such as a boiled or blown linseed. The action of the drier is apparently catalytic, the resultant hardened oxidized film containing less oxygen than if normal air oxidation had taken place. As the terminology has developed, a pigment in a varnish vehicle is often designated as a paint; but, particularly if it is to be heat- rather than air-dried, it may also be known as *enamel.*

Broadly speaking, a *varnish* is the product of heating together at an elevated temperature a fatty oil and a natural or synthetic resin with or without rosin or its glyceryl ester, ester gum, and subsequently diluting with petroleum naphtha, sometimes containing some turpentine. When called a varnish it is not pigmented. The heating process is known as bodying or kettling and increases the viscosity of the mixture owing to complex interreaction. The oils and resins are oxidized or polymerized or both.

An alcohol solution of shellac or other alcohol-soluble natural resin is sometimes called a *spirit varnish.* The term varnish as used in this chapter does not designate this type of product.

Lacquers may be clear or pigmented. They differ from paints and varnishes in drying entirely by evaporation of a volatile vehicle rather than in part by air oxidation of an oil or varnish. The base of the lacquer is usually nitrocellulose modified or plasticized by high-boiling organic solvents, the whole being dissolved in a mixture of solvents and nonsolvent thinners.

Thus the group has in common only its use as protective coatings, the compositions being diverse. Natural resins are included in the chapter because protective-coating formulation is one of the major uses.

PAINTS

Water. Weigh 100 grams of the well-mixed paint into a 250-ml. round-bottomed flask, and proceed by the solvent-distillation method described on page 41, starting at "Add 100 to 150 ml. of toluene or xylene, and . . .".

Volatile Thinner. Weigh 3 to 5 grams of the sample into a tared flat-bottomed dish about 8 cm. in diameter, spreading it over the bottom. Heat for 1 hour at 105 to 110°, cool, and weigh. Subtract the equivalent weight of moisture found from this loss. The difference is volatile thinner. Since some oxidation of the paint vehicle will take place during this drying period, the method is a compromise, an attempt to drive off all the volatile thinner without unduly increasing the weight of the residual oil due to oxidation.

Identification of Thinner. Weigh 100 to 200 grams of sample into a Kjeldahl flask. Connect with a source of steam, a condenser, and a receiver. It is desirable to heat the flask in an oil bath at 130°, but direct heating may be used with care. Steam-distill, taking care not to let water accumulate in the sample flask, until about 300 ml. of distillate has been collected or until no more thinner distills over. Collect and measure the upper layer of thinner that separates. Either add anhydrous sodium sulfate to dry the separated thinner, or filter through a dry paper.

Details for identification of such thinners are given in Chap. 23. If the specific gravity is below 0.76, the thinner is largely or entirely petroleum naphtha. The odor is extremely helpful in identification.

Separation of Pigment. Strain the well-mixed paint through a No. 80 sieve to remove any skins, and accurately weigh about 15 grams of the sieved paint into a weighed 100-ml. test tube. Add 25 ml. of a mixture of solvents consisting of 10 parts of ethyl ether, 10 parts of benzene, 4 parts of methyl alcohol, and 1 part of acetone by volume. When pigments such as Prussian blue do not readily separate with this mixture, increase the benzene content or use benzene alone. Mix well with a rod, wash the rod with a stream of the same mixed solvents, and dilute the contents of the tube to about 60 ml. Centrifuge at as slow a speed as is feasible to settle the pigment.

Decant the clear supernatant liquid, and reserve for use in the determination of nonvolatile matter. Repeat the extraction with 40 ml. of the extraction mixture at least twice more, or until the color of the centrifuged liquid shows no yellow tinge due to the presence of oil or varnish. Wash the pigment with ethyl ether, and centrifuge. Decant the ether, and dry the pigment in the tube, carefully at first to prevent spattering due to volatilization of the ether, and finally in the oven at 105 to 110° for 2 hours. Cool, and weigh as total pigment. Grind the pigment sample, being sure not to leave any large particles, and transfer to a weighing bottle, which should in turn be stored in a desiccator. If extraction of the vehicle has been complete, this grinding should proceed easily. The analysis of the pigment is described in Chaps. 27 to 29.

Nonvolatile Vehicle. To determine this from the washings of the pigment, evaporate the more volatile solvents, but do not evaporate to dryness. Dilute the remaining solution to 100 ml. or 250 ml. with benzene. Pipet an aliquot of suitable size into a crystallizing dish, and reserve the balance for use in its identification. Addition of absolute alcohol to this aliquot will aid evaporation and prevent creeping of the oil. Evaporate to dryness on a water bath, and heat for just sufficient time in an oven at 105 to 110° to drive off all remaining solvent. Slight oxidation may occur, but usually not enough to vitiate the results. Weigh the residue; and, allowing for the aliquot, calculate it as the nonvolatile vehicle.

With determinations of volatile thinner, pigment, and nonvolatile vehicle, the sum should be close to 100 per cent. In less exact work, having determinations of two, the third, usually the thinner, which, being volatile, is less accurately determined, may be taken by difference.

Ash and Driers. Transfer a suitable portion of the solution of nonvolatile vehicle to give 5 to 10 grams of sample to a porcelain dish. Alternatively, platinum may be used if only a low flame is employed. Heat until well charred. Then place a Nichrome triangle on top of the dish and on this a porcelain lid. This serves to reflect the heat downward, so that a luminous flame which does not touch the dish is sufficient to burn off all the carbon without fusing the ash. Cool and weigh the ash.

Dissolve the weighed ash in 1:4 nitric acid, filter if necessary, and dilute to a standard volume. Use aliquots for qualitative testing for the kinds of driers used. Usually lead, manganese, or cobalt will be found by tests in Chap. 5. For quantitative determination of lead use an aliquot, and proceed to determine lead as on page 511, starting with "Saturate this with hydrogen sulfide . . . ". Manganese is best done colorimetrically on an aliquot according to pages 198 to 201. For cobalt, use the method that follows. If the amounts of these account for only a relatively small fraction of the ash, determine calcium as on page 216. If present in substantial amounts, the use of limed rosin in a varnish vehicle is indicated.

COBALT BY α-NITROSO-β-NAPHTHOL. Add 2 ml. of concentrated hydrochloric acid to a suitable aliquot, and evaporate to dryness on a water bath. Take up the residue in 2 ml. of concentrated hydrochloric acid, and again evaporate to dryness. Take up in about 25 ml. of 1:10 hydrochloric acid.

If copper is present, precipitate it as the sulfide. If nickel is present, add 10 ml. of 50 per cent ammonium citrate solution, add a slight excess of alcoholic dimethyl glyoxime (page 718), and render the solution alkaline with 1:1 ammonium hydroxide. In that case evaporate the filtrate to dryness, take up with 1:4 nitric acid as in the previous treatment of the ash, and volatilize the nitric acid with hydrochloric acid. Manganese will not interfere unless the amount exceeds the cobalt content. If it has to be removed, add an excess of sodium bismuthate and digest on a steam bath until the pink color disappears. Filter, and take the filtrate to dryness. Take up with 2 ml. of concentrated hydrochloric acid, and again evaporate to dryness. Then take up with 25 ml. of 1:10 hydrochloric acid. The precautions in this paragraph seldom have to be applied.

PROCEDURE. Transfer the cobalt solution to a 50-ml. Nessler tube, and dilute to about 45 ml. In a similar tube take 25 ml. of 1:10 hydrochloric acid diluted to about 40 ml. Add 5 ml. of reagent solution to each, the solution containing 0.05 gram of α-nitroso-β-naphthol dissolved in 5 ml. of $0.1N$ sodium hydroxide and dilute to 100 ml. Add standard solution containing 0.4936 gram of cobalt nitrate, $Co(NO_3)_2.6H_2O$, per liter to the blank. Each milliliter of the standard is equivalent to 0.1 mg. of cobalt. Complete by duplication, as on page 62.

Rosin. Apply the Liebermann-Storch reaction to a portion of the nonvolatile vehicle substituted for the total fatty matter in the technic on page 395. This determines qualitatively whether rosin is present. Its presence indicates a varnish type of vehicle although its absence does not indicate that the vehicle may not be a varnish.

IDENTIFICATION OF NONVOLATILE VEHICLE

In the absence of positive evidence of a varnish vehicle, such as calcium from limed rosin or a positive Liebermann-Storch reaction, assume an oil vehicle, and proceed by the technic that follows. Otherwise, analyze by the technic for a natural-resin or synthetic-resin varnish.

Oil-type Vehicle. SAMPLE. There are three possible sources of the sample for this examination. One is the nonvolatile vehicle already available, which is particularly satisfactory if it is a raw oil but is unsatisfactory if bodied.

The second is to convert the nonvolatile vehicle already available to free fatty acids. For this purpose weigh out a sample of the solution of nonvolatile vehicle that will contain 25 grams of vehicle. Evaporate nearly but not quite to dryness, thus preventing oxidation. Then saponify and separate the fatty acids by the methods provided for a fat or oil (page 352), depending on the evaporation of the alcohol in the first method or heating to 150° in glycerol in the second method to volatilize most of the residual volatile solvent. Carry through "Filter the fatty acids through a dry paper into a dry beaker . . . ". Finally heat the recovered fatty acids in an uncovered casserole for a minimum period to limit oxidation, to drive off residual volatile thinner.

With some pigments and vehicles, there is difficulty in separation by the usual methods for fats and oils, and the third method must be used. Transfer a weighed sample of about 50 grams of the paint to a porcelain casserole. Add 30 ml. of 30 per cent sodium hydroxide solution and 125 ml. of alcohol. Mix and heat on the steam bath till dry. Transfer to a 400-ml. beaker, and boil with 200 ml. of water. Add sufficient 1:3 sulfuric acid to decompose the soap, boil, stir, and filter through a large, wet, coarse filter paper. Drain, and either scrape the fatty acids from the paper into a centrifuge bottle or add the whole paper containing the fatty acids. Shake thoroughly with ethyl ether, centrifuge, and decant the ether. Repeat with successive portions of ether until the extraction is complete. Wash the ether solution free of mineral acid with small portions of water, and drain off the water. Add anhydrous sodium sulfate, and let stand until the ether solution is clear. Decant and distill off the ether, finally heating the residue at not above 75° until all the ether is gone. Transfer the fatty acid to a weighing bottle, and use as sample.

The fatty acids as isolated will contain any unsaponifiable matter that was present. In the solution of the nonvolatile fraction those acids are present as fatty glycerides, and similarly unsaponifiable matter is present in a greater or lesser amount.

In the description that follows it will be assumed that aliquots of the benzene solution of the nonvolatile vehicle are being used. If fatty acid is used, except as noted, substitute an equivalent weight of the separated fatty acids.

Iodine Value. Pipet an aliquot of the benzene solution containing about 0.3 gram of nonvolatile vehicle into an iodine flask. Warm on a water bath until the volume is reduced to 10 to 20 ml. Determine the iodine value as on page 345, starting at "Dissolve the sample in 10 ml. of chloroform or carbon tetrachloride, warming slightly if necessary." As the blank in this case, use 10 ml. of benzene and 10 ml. of chloroform. Although not standard, this method has the advantages of simplicity and prevention of oxidation of the drying oils.

Acid Value. While this value frequently has little importance, its determination takes but a moment and it may assist in the interpretation, if only in a negative sense. The sample so titrated can then be used for the saponification value. It is of no value on a sample of vehicle that has been converted to free fatty acid.

Pipet a sample of the benzene solution of nonvolatile vehicle that will contain about 5 grams of vehicle into a flask. Add an equal volume of neutral alcohol and then phenolphthalein indicator, and heat to boiling. Proceed as on page 341, starting with "Shake or swirl the flask thoroughly . . . ".

Saponification Value. This determination is also valueless on the isolated fatty acid. Add sufficient standard alcoholic potassium hydroxide to the sample titrated for acid value to furnish an excess, and reflux for at least 1 hour to saponify completely. As the blank to be similarly treated, take the same volume of alcoholic potassium hydroxide plus the amount of $0.1N$ alkali used for titration to obtain the acid value. Complete as on page 344, starting at "Cool the sample and blank, and titrate with $0.5N$ hydrochloric acid . . . ". If saponification is unsatisfactory, use the alternative on page 344. If the value found is much below 180, either considerable amounts of unsaponifiable matter are present or the oil vehicle is not raw oil.

UNSAPONIFIABLE MATTER. Follow the technic for the separatory-funnel method on page 349. If fatty acids of the nonvolatile fraction are being used, it will be necessary to weigh out a sample and neutralize it at this point. If considerable unsaponifiable matter is present, it is apt to be of a petroleum oil type and can usually be identified by specific gravity and appearance (Table 26, page 454).

INTERPRETATION. If a paint vehicle were only a raw drying oil with added drier, constants to indicate this could be obtained. In practice, in the simplest cases boiled or blown oils or varnishes are commonly used.

The iodine value will usually be much below that for a saponifiable oil or the fatty acids isolated from such an oil. This may be due to the presence of unsaponifiable matter; if so, it will be confirmed by the method shown. It may be due to prior partial oxidation of a boiled or blown oil. It may be due to the vehicle's being a varnish; in this case, further details of the examination of varnishes are applicable, although they hardly give conclusive results. An acid value on the nonvolatile residue of the order of 10 or higher is likely to correlate with a strongly positive Liebermann-Storch reaction and indicates the use of rosin. A lower acid value but strongly positive Liebermann-Storch reaction will usually indicate a varnish containing large amounts of ester gum.

The odor of the nonvolatile residue, particularly when warmed, is frequently distinctive. The presence of linseed oil or tung oil is usually detectable without great difficulty by the experienced worker. But while a paint pigment may be quantitatively analyzed, as indicated in Chaps. 27–29, the analysis of the vehicle usually ends up as a more or less wise guess, according to the degree of experience of the guesser. The composition of the pigment and the type of paint are often a useful lead in estimating the type of vehicle likely to be present.

NATURAL RESIN VARNISHES

For application of this procedure it is assumed that the sample has been examined qualitatively for synthetic resins (page 492), although this procedure must often be applied with a low degree of accuracy when they are present. Some of the steps are applied to a natural-resin vehicle from a paint. Water is not commonly present.

Volatile Thinner. Use the procedure for paints (page 469).

Identification of Thinner. As a preliminary indication of the grade of thinner, determine the flash point of the varnish by the open cup (page 671). Then use the procedure for paints (page 469). The varnish solids must not be decomposed; therefore, an oil bath at not over 130° must be used, and a vacuum of 150 to 300 mm. is helpful.

Acid Value. To 10 to 20 grams of the varnish in a 150-ml. conical flask add 50 ml. of neutral alcohol, and reflux for about 15 minutes or until thoroughly dispersed. Titrate to a phenolphthalein end point with 0.1N alkali solution, and calculate the acid value (page 341). When the oils and resins are determined later, it is often helpful to calculate the acid value on the basis of their weight, assuming the value for the volatile vehicle to be zero, an approximately correct assumption.

Ash and Driers. Use 5 to 10 grams of the varnish as described for the residue from paint (page 470), starting at "Heat until well charred."

Oils and Resins. The total amount of oils and resins is synonymous with the usual nonvolatile vehicle and may be determined by drying to minimum weight at 105 to 110°. Hourly weighings should be made; after the minimum is reached, the weight will increase. Because some oxidation occurs before evaporation is complete, this value tends to be high.

The sum of the percentages of oil and resin, and volatile thinner should approximate 100 per cent; usually is high rather than low.

Separation of Oils and Resins.

Separation of Oils and Resins. Because the oils and resins will have reacted and polymerized to a considerable extent in kettling, the separation is only approximate. Two methods are given.

Rough Separation. Mix 15 to 20 grams of the varnish with 250 ml. of petroleum ether, and chill in a refrigerator. The resins, in the main, are insoluble in this solvent and will settle out. Decant off the petroleum ether solution, add more at room temperature, stir well, and again chill. Decant and wash the residue well with chilled petroleum ether. Unless the oils have been high-oxidized or polymerized, they and any rosin present will be present in the petroleum ether. Evaporate the petroleum ether solution, and weigh the oils. Having the total oils and resins and this determination of oils, get the resins by difference.

If the varnish shows a negative Liebermann-Storch reaction and the oils are not too greatly altered, examine as described for the nonvolatile vehicle of paint and roughly interpret as there described (page 473). If rosin is present follow a method for its determination that is given later (page 477) and that is applied to an original sample.

Boughton Method. A more involved method for the better separation of the oils and resins has been worked out. It depends on the fact that oxidized and polymerized oils are saponifiable with alcoholic potassium hydroxide, whereas heat-treated resins, other than rosin, are not. Differential esterification will then separate rosin acids from fatty acids.

Weigh about 5 grams of the varnish into a 150-ml. conical flask, add 25 ml. of water, and boil off about 15 ml. of water, thus eliminating most of the volatile matter. Add 25 ml. of 0.5*N* alcoholic potassium hydroxide and 25 ml. of benzene. Reflux for at least 0.5 hour. Be sure an excess of alkali is present. Evaporate to about 15 ml., and add 10 ml. of alcohol. Transfer to a separatory funnel, using a policeman if necessary to com-

TABLE 29.—CONSTANTS ON SOME NATURAL RESINS AND THEIR GLYCERIDES

Resin	Specific gravity 20°/20°	Refractive index 20°	Direct acid value	Saponification value	Iodine value	Soluble in	
1. Congo, No. 1 pale	1.06	1.540	97	120	116	Insoluble	1
2. Congo, hard amber	1.06–1.07	1.545	95–102	123–132	123	Insoluble	2
3. Congo, hard amber run, 25 per cent loss	1.04	1.545	60–70	72	85	Most solvents	3
4. Congo ester	1.076	1.506	......	......	......	....	4
5. Damar, No. 1 Singapore	1.04–1.06	1.52 –1.54	17	36–39	103–123	Petroleum, coal tar, naphthas, terpenes	5
6. Kauri, pale No. 2	1.04	1.546–1.565	67	94	120	Alcohol, ketones	6
7. Kauri, pale No. 2 run, 20–25 per cent loss	1.06	....	20–30	36	87	Most solvents	7
8. Manila, hard amber	1.07–1.08	1.539	122–141	175	110–133	Alcohols, ketones	8
9. Manila DBB, run, 25 per cent loss	1.05	....	50–60	67	96	Most solvents	9
10. Manila ester	1.09	1.506	......	......	......	....	10
11. Mastic	1.04–1.06	....	50–70	62–90	64–142	Alcohols, coal tar, naphthas	11
12. Pontianak, nubs	1.07–1.08	1.54	112–118	151–180	106	Alcohols, ketones	12
13. Pontianak, bold, run, 20–25 per cent loss	1.06	....	75–85	86	99	Most solvents	13
14. Rosin	1.07–1.09	1.525	150–180	150–200	112–117	Most solvents; partly in petroleum naphtha	14
15. Rosin ester	1.095	1.496	......	......	......	....	15
16. Sandarac	1.05–1.09	1.545	117–155	145–157	112–141	Alcohols	16
17. Shellac	1.08–1.15	1.516–1.534	48–64	194–213	8	Alcohol, glacial acetic acid; partly in most solvents	17

plete the transfer, and wash the flask with water. Dilute with water to 100 ml., add 100 ml. of ether, and shake. If separation is not sharp, add small amounts of alcohol from a wash bottle around the sides and over the surface of the liquid. Draw off the aqueous layer containing the soaps of the oils and any resin acids. This will include all the rosin as soaps and some emulsified unsaponifiable matter. Wash the ether layer containing the unsaponifiable resinous matter three times with 25-ml. portions of water, and combine these washings with the previous aqueous extract of soaps. Reserve this ether solution of unsaponifiable resinous matter.

Acidify the aqueous extract with a slight excess of 1:1 hydrochloric acid, and extract successively with 50, 25, and 25 ml. of ether. Discard the water layer, and transfer the ether extract, after twice washing with 10-ml. portions of water, to a conical flask. Distill off the ether, and thoroughly dry the residue.

To this dry mixture of fatty and resin acids, add 20 ml. of absolute alcohol. Dissolve, and add 20 ml. of a mixture of 1 volume of concentrated sulfuric acid and 4 volumes of absolute alcohol. Boil for 2 minutes under a reflux condenser. The fatty acids are esterified, but the resin acids are not.

Completely transfer the contents of the flask to a separatory funnel, washing the flask with 10 ml. of water and 10 ml. of ether. Add 10 ml. of ether, and shake. Add 100 ml. of a 10 per cent sodium chloride solution, and again shake. Draw off the aqueous layer containing the sulfuric acid, and extract it with 50 ml. of ether. Combine the ether solutions, and wash with water until free of acid. The ether solution now contains both the resin acids and the ethyl esters of the fatty acids.

To separate these, add 50 ml. of $0.2N$ potassium hydroxide and 10 ml. of alcohol, shake, and draw off the aqueous layer into a second separatory funnel. Wash the ether layer with 50 ml. of water containing 5 ml. of $0.2N$ potassium hydroxide and 5 ml. of alcohol. Extract the combined aqueous solutions containing the potassium soaps of the resin acids with two 50-ml. portions of ether, and finally wash the combined ether solutions, which contain the ethyl ester of the fatty acids and small amounts of unsaponifiable matter, with water.

For direct determination of the resin acids, acidify the combined aqueous soap solution and ether washings with 1:1 hydrochloric acid. Extract with three successive 25-ml. portions of ether. Evaporate the ether, and weigh the residue as resin acids present in the varnish as combined acids or as rosin.

Evaporate the ether solution of the ethyl esters of the fatty acids, and saponify the residue with 25 ml. of $0.5N$ alcoholic potassium hydroxide for 30 minutes under a reflux condenser. Evaporate off most of the alcohol on a steam bath, and take up the residue in 100 ml. of water. Transfer completely to a separatory funnel, and extract with four successive 25-ml. portions of ether. Wash the combined ether extracts with water four times, and add the washings to the solution of the fatty-acid soaps.

Add the ether solution to the ether solution of unsaponifiable resinous matter obtained after the first saponification in this process. Evaporate this ether solution to obtain the unsaponifiable resinous matter.

To the soap solution, add a slight excess of 1:3 hydrochloric acid, and extract with 50-, 25-, and 25-ml. portions of ether. Wash the combined ether extracts with three 10-ml. portions of water. Evaporate off the ether. dry, and weigh the fatty acids.

By this method, the varnish solids are separated into fatty acids, resin acids, and unsaponifiable resinous matter in a fairly quantitative manner. Using portions of the resin acids as sample, determine the acid value (page 341). This is but little different for rosin acids and rosin, 150 to 180. Similarly, determine the iodine value of a portion of these resin acids (page 345). The usual range is 112 to 117.

To identify the fatty acids, determine the iodine value and other usual constants for fats and oils and their fatty acids (Chap. 17). The unsaponifiable resinous matter may be identifiable by comparison with the properties of natural resins shown in Table 29.

For some purposes, it is sufficient to determine the fatty acids only and calculate the total resins by difference. Since separations have to be made in this method, it is usually well worth while to complete the segregation of the resin acids and unsaponifiable matter also, a cross-check on the accuracy of separation being thus obtained by the total of the three determinations.

Rosin. Heat-treated oils and resins are to a large extent insoluble in cold alcohol. Therefore, if rosin is indicated as present, it may be roughly but rapidly estimated by this method. Thoroughly shake a 20-gram sample of the varnish in a separatory funnel with several 100-ml. portions of alcohol. Evaporate the alcoholic extracts, and weigh.

For a more accurate determination of the rosin, use the method described for soap. For this, weigh 5 grams of sample into a flask, add an excess of aqueous alkali, and reflux until saponification is complete. Let cool, and transfer the aqueous solution quantitatively to a 500-ml. separatory funnel. Ignore any gummy material on the walls of the flask. Extract the aqueous alkaline solution with 25-, 15-, and 10-ml. portions of ether, and discard these ether extracts. Transfer the contents of the funnel quantitatively to a 250-ml. beaker, and add a slight excess of 1:1 sulfuric acid. Heat until the fatty acids collect in a clear layer. Continue by the McNicoll method on page 395 using this fatty-acid layer as sample. Start at "Transfer this sample to a 50-ml. conical flask . . . ".

SYNTHETIC RESIN VARNISHES

In recent years the use of synthetic resins in varnish formulation has become widespread. The synthetic resins as such form the subject of a separate chapter. Those used in varnishes are largely the glycerophthalates and the phenol-formaldehyde types. Even these are not well enough known analytically to the chemist to permit of reliable quantitative analysis. The only advice that can be given is to follow the procedure for natural-resin varnishes. Depending on the heat-treatment they have received, the synthetic-resin varnishes will be more or less modified. The amount of resin present will be the minor ingredient in the vehicle, even in a short-oil varnish.

LACQUERS

Since separation of solvents or mixtures of esters and oils is not readily carried out a little prior qualitative examination is effective. Odor gives a general clue to the solvents and thinners to be expected, particularly if they are observed at several stages during drying. Burning the residue after evaporation of solvents will give a clue to some resins, oils, and plasticizers. Some qualitative methods for identification of plasticizers, as given later (page 482), may be applied to the solids, but the only ones worth while are those for phthalates, phenol, and phosphates.

Total Solids. Stir the sample thoroughly, and from a weighing bottle weigh about 5 grams into a crystallizing dish containing a stirring rod. Now add about 50 ml. of acetone, and mix well. From a pipet slowly add 10 ml. of water, with stirring. This throws the nitrocellulose and resins out of solution and prevents the formation of a smooth, solid crust, which would prevent efficient evaporation. Evaporate to dryness on a steam bath. If the residue is hard and not well broken up, redissolve in acetone, reprecipitate with water, and evaporate again. Place in the oven at 110° until all traces of water, acetone, and volatile solvent are removed, usually for about 1 hour. Avoid overheating or too prolonged heating or plasticizers will gradually evaporate, owing to an appreciable vapor pressure at 110°. Also, while commercial nitrocellulose used for lacquers is supposed to be stable well above 110°, there is always the possibility of deflagration or explosion of such a precipitate. This dictates not running such determinations in large multiples in the same oven, particularly on samples of unknown history. Cool, and weigh. If pigments are absent, this weight represents the resins, nitrocellulose, and plasticizers.

Ash. This determination is hardly necessary for a clear lacquer, but it takes little time and will confirm the fact that no inorganic substances are present. Weigh about 5 grams of sample into a porcelain dish. If no injurious pigment is present, a platinum dish may be used and will expedite ignition. Carefully evaporate off solvents, and heat over a low flame until fully charred. Cover with a porcelain dish resting on a Nichrome triangle, and continue heating until all the carbon is burned off. Cool, and weigh as ash. If the amount is substantial, grind the ash and use for qualitative examination for the pigments present. Also, by calculation, the ash, pigment analysis, and total pigment serve as a cross-check on each other. Driers are not present in lacquers.

Pigments. Weigh 50 or 100 grams of the sample into a 250-ml. centrifuge bottle. Normally a sensitive trip balance reading to 0.1 gram will suffice for this purpose. Nearly fill the bottle with a mixture of 40 per cent acetone, 50 per cent benzene, and 10 per cent methyl alcohol. Shake thoroughly, and centrifuge at as high a speed as convenient until the supernatant liquid is clear or nearly so. Colloidal pigment particles may remain suspended, but nothing can be done about this source of error. Decant, and save the solution. Add more of the solvent mixture, and shake thoroughly, making sure that the pigment is loosened from the bottom of the bottle and thoroughly disintegrated. Centrifuge, and decant as before. Repeat this process at least once more, and always until the solvent mixture layer is clear and uncolored. Save the extracts and washings as a sample of the lacquer vehicle. Dry the precipitated pigment, first on the steam bath and finally in the oven to constant weight. The percentage present may not check with the ash owing to decomposable pigments.

Some Prussian blue pigments found in blue and green lacquers are soluble in the solvent mixture specified; in this case use 10 per cent acetone, 40 per cent benzene, and 50 per cent methyl alcohol. Should nitrocellulose precipitate with this mixture, add just sufficient acetone to prevent precipitation. Red toners are extracted from their lake pigments with either of these mixtures. However, the amount of red dyestuff is negligible as compared with the lake pigment and true pigment.

Carry out the analysis of the separated pigment as described in Chaps. 27 to 29.

Solvents and Thinners. The amount and kind of solvents and thinners in a lacquer are of particular importance because they are often indicative of the types of plasticizers, oils, and gums present with the cellulose ester. They may be vacuum-distilled with reasonable safety provided that precautions are taken not to overheat the contents of the flask. This method tends to lose some low-boiling fractions, and the distillation of high-boiling fractions tends to be incomplete. Solvents and thinners may be steam-distilled from lacquers with less tendency to retain solvent in the residue. At the same time, some hydrolysis of esters is known to occur. These are both routine technics that need receive no detailed description.

By far the most satisfactory method is to dilute the sample with high boiling nonsolvent. For this transfer a 100-gram sample to a 500-ml flask. Add a light mineral oil such as a light spindle oil, a petroleum fraction giving no significant distillate below 200°. The addition of about 200 ml. is usual, with vigorous stirring during slow addition so that as the cellulose ester is precipitated it may be in granular form. Let this precipitate settle, and decant the clear upper layer into a 750-ml. distilling flask through a filter protected from evaporation of the solvent. Wash the residue in the original flask with two additional 50-ml. portions of the oil. No more than a negligible amount of cellulose ester is now in the distilling flask if the technic has been properly carried out; if the cellulose ester has not precipitated, discard the sample and use another method.

Make a fractional distillation from an oil bath, utilizing the technics of Chap. 23, and identify the esters and thinners present.

Plasticizers and Oils. Evaporate the solvent from the extract obtained in the pigment separation or from 50 or 100 grams of the original clear lacquer. These solvents are very inflammable; therefore, exercise every precaution during the evaporation. Finally heat on, or preferably in, a steam bath. By weighing at this point the percentage of extract plus the pigment should check the total solids. Dissolve the residue in the minimum amount of a mixture of equal parts of acetone and benzene. If not completely soluble, add as small a quantity of methyl alcohol as will clear the solution.

With vigorous stirring, gradually add petroleum ether until no further precipitation takes place. At this point the precipitated nitrocellulose and resins coalesce, and the mixed solvents become clear. If the stirring is not vigorous, the precipitate will ball up and occlude some of the acetone-benzene solution, resulting in incomplete precipitation. Pour off the clear liquid, which contains the plasticizers and oils. Drain and wash the precipitate 3 times with 10-ml. portions of petroleum ether. Add this to the decanted liquid. Dissolve the precipitate in the same amount and kind of mixed solvents in which it was first dissolved, and reprecipitate in the same manner. Repeat this cycle at least once more, and oftener if necessary, to obtain a clear, colorless liquid after precipitation.

Reserve the precipitated nitrocellulose and resins for later separation. Evaporate off all the solvent in the combined petroleum ether–acetone-benzene mixture with the same care as in the removal of solvent from the original lacquer. Dissolve the residual plasticizers and oils in the same mixture of solvents first used, and precipitate any nitrocellulose and resins remaining, by the addition of petroleum ether. Decant, wash with petroleum ether, redissolve the precipitate, and again precipitate with petroleum ether. Decant, wash finally with petroleum ether, and add the precipitate to the first precipitate.

Combine the mixed solvents and washings, and evaporate as before. Dry on the steam bath and finally in the oven at 110° until all the added solvents have been removed. Cool and weigh as total plasticizers and oils. Reserve the sample for other determinations. Some of the common plasticizers are shown in Table 30.

Acid Value of the Plasticizers and Oils. Determine the acid value in the usual way as outlined on page 341, the plasticizers and oils being used in place of the oil sample specified there. An appreciable acid value suggests the presence of rosin or dammar resin, which are partly soluble in petroleum ether. If these are suspected, test for rosin by the Liebermann-Storch reaction (page 395).

Iodine Value. Determine the iodine value in the usual way as described on page 345, substituting 0.3 gram of the oils and plasticizers for the sample specified. Unsaturated vegetable oils give a substantial iodine value, of the order of 80 to 130. Many synthetic plasticizing resins react with the halogen solution under the conditions of the test so that an iodine value of that magnitude cannot be accepted as more than an indication of the presence of unsaturated oils.

TABLE 30.—SOME SYNTHETIC PLASTICIZERS FOR LACQUERS

	Formula	Molecular weight	Crystallizing point, °C.	Specific gravity	°C.	Boiling point, °C.	Refractive index	°C.	Solubility in 100 parts of water	°C.	Saponification value	
1. Tricresyl phosphate	$(CH_3C_6H_4)_3PO_4$	368.4	Noncrystalline solid at −35	1.165	20	295/13 mm.	1.556	20	Less than 0.002	85	151	1
2. Triphenyl phosphate	$(C_6H_5)_3PO_4$	326.3	48.5*	1.268	60	245/11 mm.			0.001	34	171	2
3. Dibutyl phthalate	$C_6H_4(COOC_4H_9)_2$	278.3	−35	1.047	20	206/20 mm.	1.490	25	0.001	30	202	3
4. Diethyl phthalate	$C_6H_4(COOC_2H_5)_2$	222.2	− 0.3	1.123	20	295	1.503	20	0.058	30	250	4
5. Dimethyl phthalate	$C_6H_4(COOCH_3)_2$	194.2	0	1.196	15.6	282	1.515	20	0.4	32	290	5
6. Diphenyl phthalate	$C_6H_4(COOC_6H_5)_2$	318.3	69	1.28	25	....	1.572	74	Insoluble	..	176	6

* Melting point.

Saponification Value. Determine the saponification value as outlined on page 343, starting at the beginning of the procedure. Vegetable oils, resins, and high-boiling esters such as dibutyl phthalate and tricresyl phosphate give saponification values varying inversely with their molecular weights for unpolymerized substances. Synthetic resins frequently give high values. Typical values of some synthetic plasticizers are shown in Table 30.

Unsaponifiable Matter. Evaporate the alcohol from the sample used for saponification value, or, if necessary, saponify a larger sample. An odor of butyl alcohol at the end of this evaporation suggests that dibutyl phthalate was originally present. Extract the unsaponifiable matter by the separatory-funnel method, starting on page 349 at "When nearly dry, dilute with hot water." The unsaponifiable matter will usually be mineral oil and unsaponified resin. Appreciable unsaponifiable matter and negligible iodine value are an indication that the unsaponifiable matter was mineral oil.

Fatty Acids. After extraction of unsaponifiable matter, render the solution distinctly acid with 1:1 hydrochloric acid. Extract the fatty acids with 25, 15, and 10 ml. of ethyl ether. Wash the combined ether extracts twice with 10-ml. portions of water, and discard the aqueous extracts. Evaporate the ether. The odor and iodine value (page 345) will serve to identify the resulting fatty acids. If their acid value (page 341) is around 200, this indicates that only fatty acids are present.

Modified fatty-acid resins will have been extracted with the unsaponifiable fraction. Phthalic acid will be present with this fatty-acid fraction but has an acid value of about 336. Cresols may be present but are readily detectable by odor.

Organic Phosphates. To obtain a qualitative test for phosphate it is necessary to decompose the sample by acid digestion or in a bomb. As little more time is taken in accurately weighing the sample, it is almost always worth while to make the determination quantitative up to the steps after decomposition is complete and to test an aliquot qualitatively at that point.

Acid Digestion. To 0.5 gram of sample add 5 ml. of concentrated nitric acid, 5 ml. of concentrated sulfuric acid, and 1 gram of sodium sulfate. Heat until fumes of sulfur trioxide are evolved, cool somewhat, add 5 ml. more of concentrated nitric acid, and heat. Repeat the operation until no further charring occurs when the mixture is evaporated to fumes. Fume off the sulfuric acid to near dryness, cool, and dissolve in 25 ml. of water and 5 ml. of concentrated nitric acid. Boil for a few minutes, and dilute to 250 ml. Test a portion qualitatively for phosphate (page 258). If this is found, use an aliquot for quantitative determination by the volumetric molybdate method (page 203), starting at "Transfer an aliquot of sample equivalent to about 0.04 to 0.1 gram of phosphorus pentoxide to a 250-ml. beaker."

Assuming the phosphorus to be calculated as phosphorus pentoxide, multiply this figure by 5.190 to obtain the percentage of tricresyl phosphate in the plasticizers and oils or by 4.592 to convert to triphenyl phosphate.

BOMB METHOD. If the equipment is available and set up, this is the most convenient method. Use the apparatus described on page 161. Mix 0.5 gram of sample with about 1.5 grams of sugar to ensure complete combustion. Carry this along until the washings of the bomb and cup are transferred to a beaker. At this point dilute accurately to 250 ml. Test qualitatively, and determine quantitatively if necessary as outlined for a sample from acid digestion.

Organic Phthalates. QUALITATIVE TEST WITH RESORCINOL. Heat a little of the sample in a test tube, with two to three times its weight of pure resorcinol, at the boiling point of resorcinol, 205–210°, for a few minutes. Cool, dissolve the melt in water, and filter if necessary. Make the solution slightly alkaline with sodium hydroxide solution. A greenish yellow fluorescence will result if phthalates are present. With very small samples or in the presence of very little phthalate the fluorescence produced may be readily visible under ultraviolet light although invisible to the eye in daylight.

This method is not entirely reliable, for fluorescent substances are often formed by substances other than phthalates. As a confirmatory test, when in doubt, heat another portion of sample with twice its weight of thymol and a drop of concentrated sulfuric acid at 150° for 10 minutes, and proceed as when resorcinol was used. A blue fluorescence of thymolphthalein indicates a positive test. This test is more specific than the previous one. Phenol under similar conditions gives the pink of phenolphthalein.

QUANTITATIVE. *Kappelmeier Method.* Weigh 0.5 to 3.0 grams of sample, depending on the phthalate content, into a 250-ml. flask. Dissolve in 5 ml. of benzene, and add 50 ml. of $0.5N$ absolute alcoholic potassium hydroxide from a graduate. Mix well, attach to a condenser, preferably using glass-jointed apparatus, and place in a bath of water held at 60° for 1 hour. Partial reaction takes place, and some potassium phthalate monoalcoholate crystallizes out. Remove the water bath, heat the flask directly, and gently reflux for 3 hours. Without removing the flask from the reflux condenser, cool in ice water. Wash down the condenser with a 3:1 mixture of absolute alcohol and absolute ethyl ether. Remove the flask, and add 15 to 20 ml. of absolute ethyl ether to complete the precipitation of the potassium phthalate monoalcoholate.

Filter rapidly through a Gooch crucible prepared with an asbestos mat, dried at 105°, and weighed. Loosen any precipitate clinging to the sides of the flask. Wash out the flask thoroughly with about 30 ml. of the 3:1 alcohol-ether mixture, quantitatively transferring all the precipitate to the crucible. Finally wash with 15 to 20 ml. of the absolute alcohol-ether mixture. Do not suck air through the crystals, for they are strongly hygroscopic.

Dry the Gooch crucible at 60° for 10 minutes only, then in an evacuated desiccator over sulfuric acid for 2 to 3 hours until it shows constant weight.

For very accurate work the purity of the potassium phthalate monoalcoholate should be checked. Dissolve the precipitate from the Gooch crucible with water, using suction. Wash thoroughly, and dry as before. The final weight subtracted from the first weight of the crucible containing the precipitate gives the weight of the pure potassium phthalate monoalcoholate. Multiply this weight by 0.5139 to obtain the equivalent phthalic anhydride in the sample.

As the glyceryl phthalate, the ratio is somewhat indefinite. The factor is usually given as 0.665. Factors for conversion of the potassium phthalate monoalcoholate to other phthalate plasticizers are as follows: dibutylphthalate, 0.9649; diethylphthalate, 0.7705; dimethylphthalate, 0.6732; diphenylphthalate, 1.104.

Phenols. QUALITATIVE. Sodium phenolates are formed from plasticizers when the saponification value is determined on oils and plasticizers. Phenols and cresols are detected by odor on acidification. Frequently, destructive heating of the plasticizers and oil on a spatula or spoon will produce a definite phenolic odor in the presence of such substances.

QUANTITATIVE. *Indophenol Method.* This method is very satisfactory for determination of small amounts of phenols. For larger amounts, the errors of a large dilution enter in. It is used because no better method is available. It is generally applicable to phenolic bodies, including many that are more complex than phenol. It is most important for the same type of phenolic compound to be used as comparison standard as that in the sample being tested. The shade of color produced varies with the type of compound. But what phenol is present is often a matter of guesswork or of judgment from factors too indefinite to be presented here.

PROCEDURE. Prepare a buffer solution by dissolving 15 grams of anhydrous sodium tetraborate or 28.4 grams of C.P. borax in 900 ml. of boiled and cooled water. Add 60 ml. of 10 per cent sodium hydroxide solution, and dilute to 1,000 ml. Store in a Pyrex bottle.

Make a 1 per cent solution of the phenol believed present by dissolving it with an equivalent amount of sodium hydroxide solution. If necessary, this may be found by titration. Keep in a Pyrex bottle. From this prepare a solution containing 10 ppm., using in every case freshly boiled and distilled water, first by diluting 10 parts to 1,000 and then making a further dilution of 10 parts to 100.

Dissolve 0.08 gram of 2,6-dibromoquinone chloroimide in 50 ml. of 95 per cent ethyl alcohol as reagent, and filter through paper. Keep in a brown bottle in a cool, dark place. Make up fresh at least every other day.

Saponify a suitable sample with alcoholic alkali, and evaporate the alcohol. Dissolve the residue in water and dilute to such a volume that not more than 10 ppm. of the phenol will be present. Filter, if necessary, to remove suspended solids.

Prepare a set of standards in 10-ml. tubes of uniform diameter containing 0, 1, 2, 3, 4, 5, 6, 7, 8, 9, and 10 ml., respectively, of the 10-ppm. standard phenol solution. In each case dilute with water to 10 ml., and add 0.2 ml. of the borate buffer solution. To a similar tube transfer 10 ml. of the sample solution, and add 0.2 ml. of the borate buffer solution. Mix well, and add 0.2 ml. of the 2,6-dibromoquinone chloroimide solution. Mix well and allow the sample and standards to stand for 10 minutes. Compare the sample with the standards. With the more highly colored solutions, it may be preferable to compare by looking across the tubes instead of down them. The lower limit that can be detected is 0.1 ppm.

Separation of Resins from Nitrocellulose. Dissolve the precipitated nitrocellulose and resins separated from the plasticizer and oils in the least quantity of a mixture of equal parts of acetone and toluene. Should any resins be insoluble, add just sufficient butyl alcohol to dissolve them. Place on a water bath to evaporate, being careful that the volatile acetone vapors do not ignite. As soon as nitrocellulose begins to precipitate, stir vigorously, meanwhile evaporating off more acetone. The solution of the resins is usually easily decanted from and washed out of this finely divided nitrocellulose.

Wash the precipitate with a little toluene, and add the washings to the decanted solution. Dissolve the precipitate as before in an acetone-toluene mixture with butanol if necessary, and reprecipitate. Repeat again if necessary, as is usually evidenced by the color of the solution. Put aside the precipitate, and evaporate the combined decantates and washings to about the minimum volume of solvents in which the resins will dissolve. Carefully add toluene in small amounts until it ceases to precipitate any trace of nitrocellulose that may have been carried over. Decant from this, redissolve, and reprecipitate the nitrocellulose. Unite the precipitate with the balance of the nitrocellulose, and combine the resin solutions. Evaporate the resin solutions until partly precipitated. Redissolve by adding methyl alcohol, and evaporate with vigorous stirring so as to reprecipitate in finely divided form. Complete the evaporation of solvent, dry at 110°, and weigh as resins.

IDENTIFICATION. Because of the large number of possibilities, no outline of methods can be given. It is advisable to get iodine value (page 345), acid value (page 341), saponification value (page 343), and solubility in several solvents. An approximation of the melting point by the ball-and-ring method (page 52) is often helpful. Polymerized oils are insoluble in alcohol and petroleum ether and will therefore be present with the resins. Rosin is detected by the Liebermann-Storch test (page 395). Ester gum, which is rosin–acid glyceride, gives the same test. Natural and synthetic resins may be identified to some extent by reference to Tables 29 and 32. See Chap. 26 if a synthetic resin is suspected. The process of separation will have changed the characteristics of the resins to some extent. The segregated resin portion may contain two or more resins.

NITROCELLULOSE. Add absolute alcohol, and dry the combined precipitates on the steam bath until constant weight is obtained. While usually safe it is inadvisable to dry this presumably pure nitrocellulose in the oven. Drying may be promoted by blowing a stream of warm, dry air across the dish. Nitrocellulose is hygroscopic, and so contact with moisture should be avoided. When dry, weigh.

Test the purity of the nitrocellulose by igniting a little on the end of a spatula. If it is pure, a quick yellow flash occurs and no residue is left. If any resins or plasticizers are present, the sample will burn more or less slowly and a gummy residue will remain.

Shellac. This alcohol-soluble resin is widely used. It should be suspected when the solvent is alcohol, although there are several other natural resins that are soluble in alcohol. The best means of identification is by iodine value.

In general, follow the technic starting on page 345. Weigh a 0.2-gram sample of the resin solids suspected to be shellac into an iodine flask. Since this resin is not soluble in chloroform or carbon tetrachloride, employ glacial acetic acid. Dissolve the sample with gentle heat in 20 ml. of glacial acetic acid. Then add 10 ml. of chloroform, and cool to room temperature. Add a similar amount of acetic acid to the blank. Continue either as on page 346, starting at "Pipet 25 ml. of Hanus solution into the flask containing the sample . . . ", or use Wijs solution by the same technic.

The iodine value of shellac is low, as indicated in Table 29.

CHAPTER 26

SYNTHETIC RESINS

Synthetic resins are used for molding, in varnish formulation, and for many miscellaneous purposes. The trade brands are numerous, but chemically the types are more limited although each is modified in many ways.

Representatives of the most common type will be mentioned and general methods, usually qualitative, of analysis indicated. A commercial synthetic resin may consist of a mixture of resins, of resin or resins compounded with one or more plasticizing agents, or of a solution in a solvent. To complicate this the resins and plasticizers are rarely used alone, such cases being practically limited to transparent articles. Usually the resin is a minor ingredient with a large amount of extender such as wood flour, paper pulp, inorganic filler, or drying oil.

Synthetic resins have the common property of being polymers. The degree of polymerization has a profound effect on the physical properties of a resin. Some are thermosetting; they irreversibly polymerize on heating. Others are thermoplastic, reversibly softening on warming; they are not further polymerized by this treatment.

A list of the more common types of polymer, omitting the more rubberlike, which are briefly discussed in Chap. 32, Miscellaneous Rubber Analyses and Synthetic Elastomers, is given in Table 31.

Preparation of Sample. The sample may be a plastic article, varnish solids, or a fraction from the analysis of a mixture suspected to contain resins. Grind it as finely as possible, and extract in a Soxhlet with petroleum ether to remove as much foreign material as possible. The great merit of many of the synthetic resins is their insolubility and infusibility, properties that correspondingly prevent their isolation by differential solubility. If a substantial petroleum ether extract is obtained, the method of analysis will depend on what qualitative examination shows to be present.

General Examination of the Petroleum Ether–Insoluble Matter. Note the relative hardness, color, and any solubility in 95 per cent alcohol, benzene, ethyl acetate, butyl acetate, Cellosolve, and

TABLE 31.—TYPICAL CLASSES OF SYNTHETIC RESINS*

	Derivation	Type	Typical commercial name
Linear polymers	$CH_2{=}C(H)COOC_2H_5$ Ethyl acrylate	$\ldots CH{-}CH(COOC_2H_5)\ldots$	Acryloid
	$CH_2{=}C(CH_3)COOCH_3$ Methyl methacrylate	$\ldots CH_2{-}C(CH_3)(COOCH_3)\ldots$	Crystalite Lucite†
	$C_6H_5{-}CH{=}CH_2$ Styrene	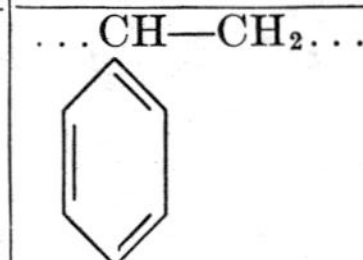	Styron
	$CH_2{=}CH{-}OC({=}O){-}CH_3$ Vinyl acetate	$\ldots CH_2{-}CH(O{-}C({=}O){-}CH_3)\ldots$	Gelva Vinylite A
	$\ldots CH_2{-}CH(O{-}C({=}O){-}CH_3)\ldots$ Polyvinyl acetate	$\ldots CH_2{-}CH(OH)\ldots$	PVA
	Coal-tar distillates	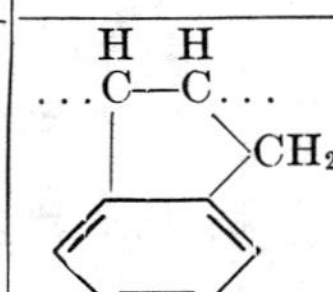	Cumar Nevindine Piccolyte

* POWERS, P. O., *Ind. Eng. Chem., News Ed.*, July 10, 1941. † Plexiglas is the analogous methyl acrylate resin.

TABLE 31.—TYPICAL CLASSES OF SYNTHETIC RESINS.*—(*Continued*)

	Derivation	Type	Typical commercial name
Condensation polymers	C_6H_5OH + HCHO Phenol, Formaldehyde	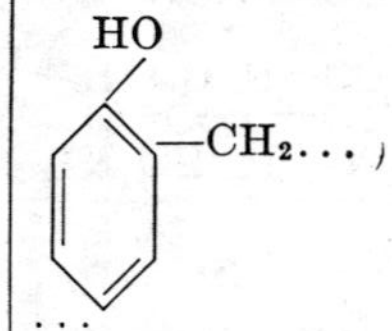	Bakelite Durez Resinox
	$CO(NH)_2$ + HCHO Urea, Formaldehyde	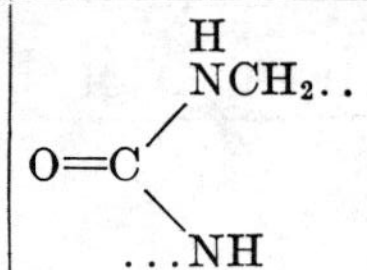	Beetle Plaskon
	$C_3N_3(NH_2)_3$ + HCHO Melamine, Formaldehyde	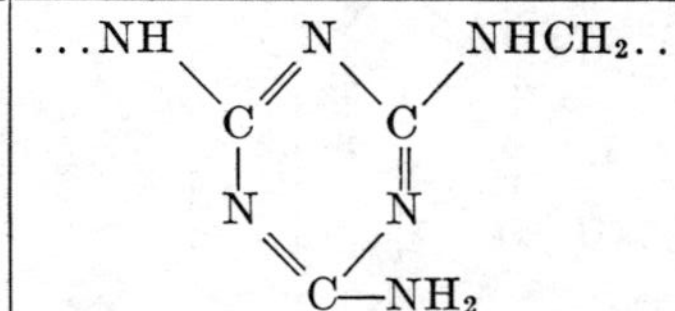	Melamine resin
	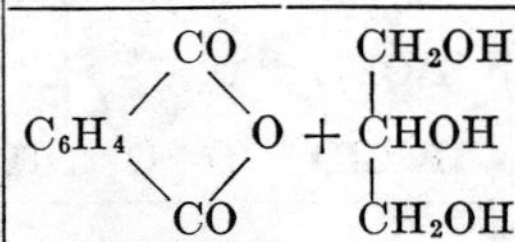 + $C_{17}H_{31}COOH$ Phthalic anhydride, Glycerol, Linoleic acid	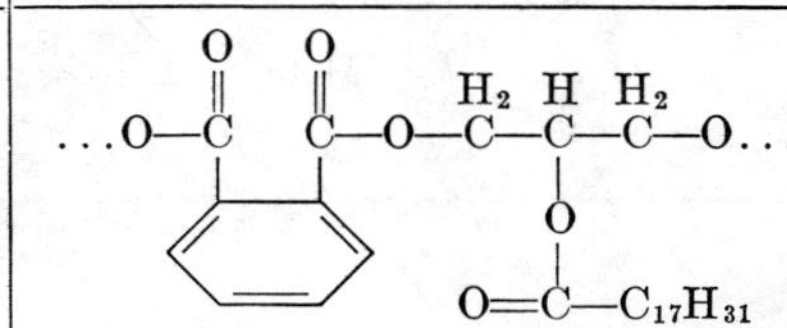	Beckosol Glyptal Duraplex

* POWERS, P. O., *Ind. Eng. Chem., News Ed.*, July 10, 1941.

TABLE 31.—TYPICAL CLASSES OF SYNTHETIC RESINS.*—(*Continued*)

	Derivation	Type	Typical commercial name
Condensation polymers	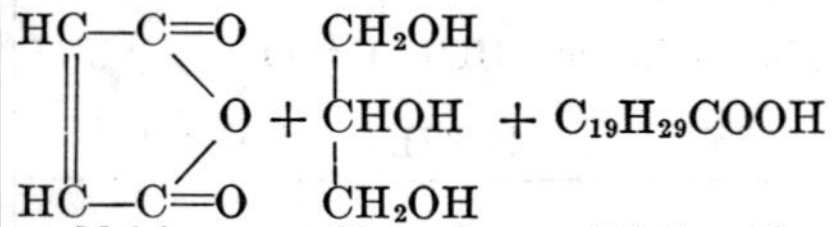 Maleic anhydride, Glycerol, Abietic acid	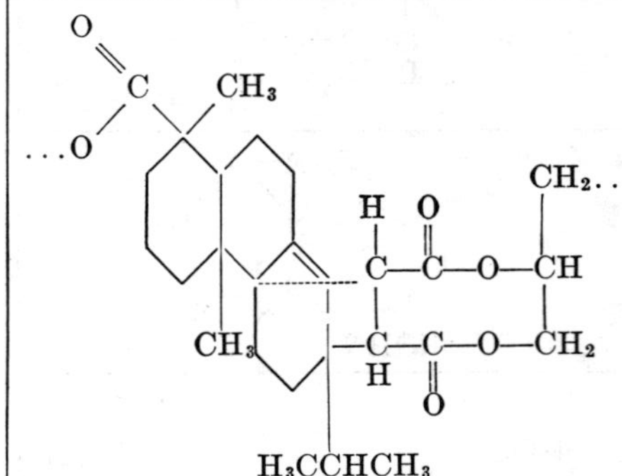	Amberol Beckacite Maleic resins
Copolymers	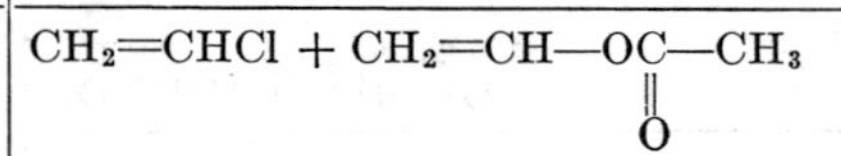 Vinyl chloride, Vinyl acetate	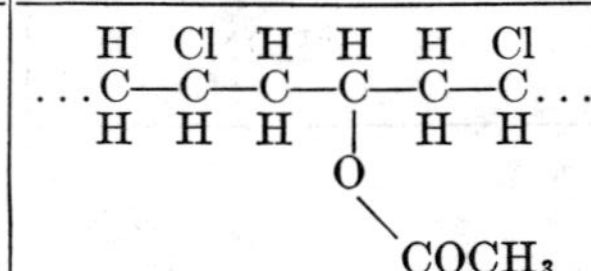	Vinylite V
	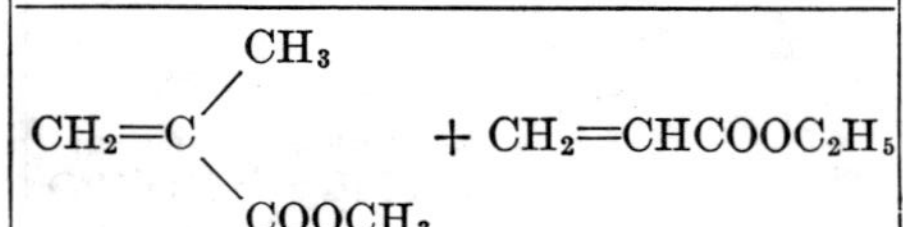 Methyl methacrylate, Ethyl acrylate	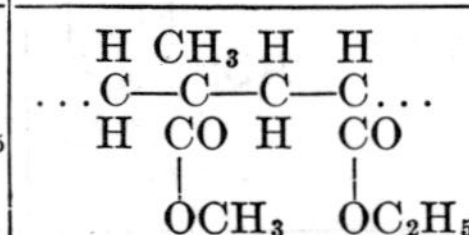	Acryloid

* POWERS, P. O., *Ind. Eng. Chem., News Ed.*, July 10, 1941.

turpentine. For resins that have an appreciable solubility in a solvent, the viscosity of solutions in the solvents may at times be useful when compared with solutions of known resins. This implies the availability of samples of many resins.

TABLE 32.—PHYSICAL CONSTANTS OF SOME SYNTHETIC RESINS

Resin type	Specific gravity, 20°C.	Refractive index, 20°C.
Chlorinated diphenyl	1.157–180	1.613–1.703
Coumarone-indene	1.05–1.15	1.617–1.645
Cresylic acid–formaldehyde	1.20	1.645
Ethylene glycol phthalate	1.31–1.36	1.56–1.57
Glycerol phthalate	1.33–1.42	1.56–1.58
Phenol-formaldehyde	1.21–1.27	1.62–1.66
Phenol-formaldehyde, modified	1.10–1.21	1.545–1.70
Phthalic anhydride, modified	1.21	1.573
Toluene–sulfonamide–formaldehyde	1.35	1.596
Urea-formaldehyde	1.44	1.54–1.90
Urea-thiourea-formaldehyde	1.477	1.66
Vinyl acetate polymer	1.15–1.19	1.473

SPECIFIC GRAVITY. Determine by any convenient method for solids (page 48). For rapid work, immersion of samples in solutions of known specific gravity has been found useful. Comparison of data on natural resins in Table 29 and on synthetic resins in Table 32 indicates that, if the specific gravity of a resin is above 1.15, the resin is of synthetic origin but may still contain oils or fatty-acid modifying agents. Shellac occasionally has a density of about this value. If the specific gravity is between 1.10 and 1.15, the resin may comprise a coumarone-indene polymer, shellac, or a modified synthetic resin. If less than 1.10, the resin is probably of natural origin but may be a soft grade of coumarone-indene resin or a highly modified synthetic resin.

REFRACTIVE INDEX. Obtain the refractive index with the microscope, using the Becke method (page 60). This constant is particularly useful in distinguishing synthetic resins from natural resins as indicated by Tables 29 and 32. Owing to the known variation in the quality of natural resins and to the numerous modifications and grades of the synthetic resins, the data in these tables should not be considered as absolute but as a general guide

Most synthetic resins possess higher refractive indices than do the natural resins. Certain vinyl polymers, chiefly acetates, fall below the natural resin range and some cyclohexanone-formaldehyde resins fall in the natural resin class so far as this constant will show. Modification of synthetic resins with oils or natural resins may bring the refractive index of the mixture within the natural resin range. It is therefore at times advisable to determine the refractive index of both the original and the unsaponifiable part of the resin.

Distillates and Odors on Heating. Strongly heat a small piece of the resin in a Pyrex test tube. White needle-like crystals of phthalic anhydride forming upon the upper part of the tube indicate a phthalic anhydride resin.

Carefully note any characteristic odors. The acrid, penetrating odor of acrolein suggests the presence of glycerin or glycerides. This does not necessarily indicate a glyceryl resin; the glycerin may be a concomitant plasticizer. If necessary, heat another portion with fused sodium bisulfate to assist in the dehydration. If glycerin or glycerides are present, a larger amount of acrolein will be evolved. Acetaldehyde, indicative of glycols or vinyl derivatives or of acetaldehyde resins, may be detected. A formaldehyde resin will give strong evidence of formaldehyde. Phenol may sometimes be detected but not nearly so readily as formaldehyde. Vinyl resins often char upon heating.

For more accurate identification, destructively heat a larger amount, absorb the vapors in water, and test the solution according to the resin suspected. Fuse a further portion with 50 per cent of its weight of solid sodium hydroxide. The strong permeating odor of capryl alcohol is indicative of castor oil, used as a plasticizer in some resin products. Phenolic and vinyl resins and chlorinated diphenyl will yield products that may be examined after absorbing the vapors from destructive distillation in water.

Most of the pure synthetic resins have characteristic odors on heating with which the analyst should become familiar. It should be borne in mind, however, that many resins are modified and that one odor may disguise a less penetrating one.

Test for Halogens, Nitrogen, and Sulfur. Strongly, but carefully, fuse a piece of resin with an equal volume of metallic sodium in a small Pyrex test tube. The whole should be no more than a couple of grams. Let cool, and add alcohol, dropwise, to destroy the excess of sodium. When the reaction is complete, add water, heat to boiling, and filter.

HALOGENS. Acidify a portion of the filtrate with 1:3 nitric acid, and add silver nitrate solution. A white precipitate indicates chloride. A yellow precipitate indicates bromide or iodide, which are not usually present, or a mixture of halogens. A positive chloride test may indicate a chlorinated diphenyl resin, especially if the resin is alcohol-soluble. Alternatively, a vinyl chloride polymer may be present or, on occasion, chlorophthalates or chlorophenols.

NITROGEN. To a portion of the filtrate add 2 drops of freshly prepared ferrous sulfate solution. Ferrous hydroxide will precipitate. Boil for 1 minute. If nitrogen is present, fusion with sodium produces a cyanide, which is converted to sodium ferrocyanide on boiling in alkaline solution with ferrous hydroxide.

Cool, add a drop of ferric chloride solution, and acidify with hydrochloric acid. A blue coloration or precipitate of ferric ferrocyanide indicates the presence of nitrogen. Sometimes a faint blue coloration occurs initially, which forms a precipitate on long standing. A positive test for nitrogen suggests the presence of a urea resin or a melamine resin.

SULFUR. To another portion of the filtrate add a freshly prepared solution of sodium nitroprusside. A deep violet coloration indicates the presence of a sulfide in the solution. All sulfur compounds yield sodium sulfide by the reducing action of sodium at high temperatures. A sulfonamide or thiourea resin is indicated by a positive test.

Acetic Anhydride–Sulfuric Acid Color Test. Test for rosins and resinates by the Liebermann-Storch method (page 395). With coumarone-indene polymers this test gives a bright and fairly permanent pink color, which may be masked by the violet-blue given by rosin. The fugitive color given by rosin may fade, leaving the more permanent pink color of this type of resin.

Cyclohexanone and cyclohexanone-formaldehyde resins give a less permanent but deeper color than that given by coumarone resins, a rose red. The rose may change to a dirty orange within a few minutes. More concentrated sulfuric acid than that used in the Liebermann-Storch reaction tends to increase the color in both the above cases.

Vinyl resins give no color when the more dilute sulfuric acid is used. The addition of 2 drops of concentrated sulfuric acid to a solution of a vinyl resin in acetic anhydride placed in a spot plate will give, on stirring, a bluish green color slowly changing to green.

Test for Phthalates. Heat with resorcinol, thymol, or phenol (page 483). Previous petroleum ether extraction will have removed phthalic acid esters used as plasticizers in protective coatings from the synthetic phthalic type resins.

Test for Phenols. The indophenol test (page 484) is most satisfactory. It can be applied to an aqueous solution containing products of destructive distillation but will show a test with all phenols.

Quantitative Examination. SAPONIFICATION VALUE. Follow the usual procedure (page 343). It is often necessary, however, to use stronger alkali than $0.5N$ and a higher boiling solvent (page 344)

A saponification value greater than 200 indicates that the substance is or contains a resinous ester of the polybasic acid type. While coconut oil, highly oxidized oils, and esters of comparatively low molecular weight may give high saponification values, they are usually not present. Polybasic acid resinous esters may be present and still yield saponification values of less than 200 when modified by phenolic resins or unsaponifiable matter.

When a value higher than 200 is obtained for the saponification value, saponify sufficient sample to identify the several components if possible. Evaporate off the alcohol, and dilute the saponified sample with water. Extract the unsaponifiable matter with 25, 15, and 10 ml. of ether, wash free from alkali, and evaporate the combined ether extracts to dryness. Identify the unsaponifiable matter by specific gravity, refractive index, melting point, and solubility.

After extraction, just acidify the remaining alkaline solution with 1:1 acetic acid. This is selected to retain phthalic acid in the aqueous phase as sodium acid phthalate. Extract the fatty and rosin acids with ether. Wash, evaporate, and examine as usual for acids from fats and oils (Chap. 17).

Neutralize the acid filtrate with 10 per cent sodium hydroxide solution. Precipitate the polybasic acids as a suitable metallic salt. Calcium chloride under these conditions precipitates a portion of the phthalic acid and some other polybasic acids such as citric, tartaric, and malic acids. Ferric chloride precipitates phthalic acid more effectively and also precipitates benzoic acid. Lead and silver give better quantitative precipitation. Filter, and wash the metallic salts. Decompose with mineral acid, and recover the organic acid, applying tests such as those for melting point and neutralization equivalent.

To recover polyhydric alcohols from the final filtrate, carefully evaporate just to dryness, and extract with 1:1 absolute ether and ethanol or other appropriate solvents.

Ash. This determination is of particular importance if the resin contains inorganic matter as filler. Weigh a suitable sample into a porcelain dish, and ignite carefully until the carbon is burned out. Quantitative mineral analysis of this ash is desirable with some samples.

Chlorine and Sulfur. Use the type of bomb described on page 161. Weigh out a 0.5-gram sample, and mix with 1.5 grams of sugar. Follow the procedure until the bomb washings are obtained, and dilute to 250 ml. in a volumetric flask.

To determine sulfur in a 50 or 100-ml. aliquot of the solution, add 1 ml. of concentrated hydrochloric acid. Complete as on page 160, starting at "Bring to a boil, and add with constant stirring sufficient boiling 10 per cent barium chloride solution . . . ".

To determine chloride in a 50- or 100-ml. aliquot of the solution follow the gravimetric procedure on page 129.

Nitrogen. Follow the Kjeldahl method on page 143. Drastic conditions for decomposition are sometimes required.

Interpretation. When the data are all assembled, interpretation still depends to a considerable degree on guesswork, except in the unusual case. In considering the data and trying to interpret them in terms of Table 29 or 32, the available synthetic elastomers and even natural rubber merit at least brief consideration. The products merge, as types, one into the other, in such a way that there are borderline cases. For example, hard rubber closely resembles the heat-polymerized resins.

CHAPTER 27

WHITE PIGMENTS, EXTENDERS, AND MINERAL FILLERS

There is no sharp dividing line between the various classes of inorganic substances covered in this and the two succeeding chapters, but they differ in a general way. *Pigments* are usually substances with refractive indices substantially different from those of paint oils and varnishes so that they can be used for pigmentation of paints as well as of rubber and many other products. The majority of pigments are white, but some are colored. *Extenders* are similar to pigments but have refractive indices nearer to those of paint vehicles. Often the particle size is not so fine as that of pigments. The term is applied largely to paints in this connotation. *Mineral fillers* may be used as added mineral matter in products as diverse as plastics, hard rubber, and pastes. Mineral colors are a class of naturally occurring materials, usually of relatively low tinctorial value, that are widely used. The term pigment is sometimes used more broadly as covering this entire group.

In general, if a product contains an inorganic precipitate that can be filtered or inorganic matter that can be isolated, these methods are applicable. It follows that there is a close parallelism to mineral analysis. The class is taken up in this chapter first as substantially pure white pigments, then in the next chapter as various mixtures of white pigments, and finally as colored pigments and mineral colors. Since the true colored pigments are expensive, a colored pigment often consists largely of white pigment with some colored pigment added. This is especially true as colors approach light pink or cream. Even whites that have been originally yellowish are often toned with blue. The possible combinations are so numerous that only some of the most common ones can be discussed. Other combinations of methods must be made to fit the circumstances.

Since a majority of the pigments isolated from commercial products have been in an oil vehicle, a sorbed layer of oil or surface coating of a product of reaction of the pigment with the oil is often present. The error so introduced is minor, but the problem of wetting the pigment in order to attack it with reagents is often introduced.

WHITE PIGMENTS

Qualitative. Wet the pigment with water, or if it repels water add a few drops of alcohol. Add dilute acetic acid until any effervescence ceases. If gas is given off, this may be carbon dioxide, hydrogen sulfide, or both. It is convenient to test for carbon dioxide with a stirring rod wet with barium hydroxide or calcium hydroxide solution. A positive test is given when the carbonate precipitates promptly on exposure. To test for small amounts pass the evolved gas into barium hydroxide solution. Sulfides will be distinguished by the odor or by the blackening of paper wet with lead acetate solution. When the evolution of gas is complete, add a large excess of ammonium acetate reagent (page 718). Heat to boiling to dissolve lead, zinc, calcium, and acid-soluble barium salts. Filter and test the filtrate for barium by addition of very dilute sulfuric acid. Test it for calcium by rendering a portion alkaline with 1:1 ammonium hydroxide and adding ammonium oxalate solution. Zinc and lead will be detected elsewhere.

Wash the matter insoluble in ammonium acetate with more of the solution and then with water. This contains any titanium oxide, barium sulfate, and siliceous matter. Add to a substantial amount of this insoluble residue or to some of the original pigment 20 ml. of concentrated sulfuric acid and 7 to 8 grams of ammonium sulfate. Heat to fumes, and continue heating for several minutes. Let cool, and add 100 ml. of water. Heat to boiling, filter, and wash with 1:20 sulfuric acid. Insoluble matter from the residue will be barium sulfate and siliceous matter. These will be identified in another part of the procedure. If the original pigment was used, lead sulfate may be present. Add 3 per cent hydrogen peroxide to a portion of the filtrate. A clear orange-yellow color indicates that titanium was present. Boil another portion of the filtrate with metallic zinc or tin. A pale blue to violet solution further confirms the presence of titanium.

Boil another portion of pigment with 20 ml. of 1:1 hydrochloric acid solution for a few minutes. Add 25 ml. of hot water, filter, and wash. This solution will be used for many tests. Neutralize a portion of the filtrate with 1:1 ammonium hydroxide, and then make just acid with 1:1 hydrochloric acid. Add barium chloride solution. A precipitate will indicate acid-soluble sulfate. Add a little dilute sulfuric acid to another portion of the strongly acid filtrate. A white precipitate indicates the presence of lead or acid-soluble barium such as barium carbonate, or both. Calcium sulfate may also precipitate but is readily differentiated by the experienced worker. Filter and wash the precipitate. A yellow or black color on the addition to the precipitate of potassium iodide or hydrogen sulfide solution, respectively, indicates that lead was present.

TABLE 33.—SPECIFIC GRAVITY AND BULKING VALUE OF PIGMENTS, EXTENDERS, AND FILLERS*

Pigment	Specific gravity	Pounds per solid gallon	Gallons bulked per pound
Alizarin lakes	1.74	14.49	0.0690
Aluminum flake	2.54	21.16	0.0473
Aluminum hydrate	2.45	20.41	0.0490
Antimony oxide	5.75	47.90	0.0209
Asbestine	2.85	23.74	0.0421
Barytes, blanc fixe	4.36	36.32	0.0275
Barytes, natural	4.45	37.07	0.0270
Bentonite	2.41	20.08	0.0498
Black, bone	2.29	19.08	0.0524
Black, carbon	1.79	14.91	0.0671
Black, lamp-	1.77	14.74	0.0678
Blanc fixe, barytes	4.36	36.32	0.0275
Bronze powder, aluminum	2.51	20.91	0.0478
Bronze powder, gold	2.50	20.83	0.0480
Cadmium red	4.30	35.82	0.0279
Cadmium yellow	4.25	35.40	0.0282
Calcium carbonate, chalk, whiting	2.70	22.49	0.0445
Calcium sulfate, gypsum	2.36	19.66	0.0509
Chalk, calcium carbonate, whiting	2.70	22.49	0.0445
China clay	2.60	21.66	0.0462
Chrome green, dark	3.30	27.49	0.0364
Chrome green, light	5.12	42.65	0.0234
Chrome green, medium	4.06	33.82	0.0296
Chrome orange, dark	6.85	57.06	0.0175
Chrome orange, light	6.69	55.73	0.0179
Chrome orange, medium	7.04	58.64	0.0170
Chrome oxide, green	5.10	42.48	0.0235
Chrome oxide, green, hydrated, Guignet's green	3.32	27.66	0.0361
Chrome yellow, light	5.83	48.56	0.0206
Chrome yellow, medium	5.96	49.65	0.0201
Cobalt blue	3.83	31.90	0.0313
Cuprous oxide	5.79	48.23	0.0207
Diatomaceous earth	2.31	19.24	0.0520
Grinder's green, 25 per cent	3.76	31.32	0.0319
Guignet's green, hydrated, chrome oxide green	3.32	27.66	0.0361
Gypsum, calcium sulfate	2.36	19.66	0.0509
Hansa yellow	1.48	12.33	0.0811
Indian red (see Iron oxide, red)			

* H. A. Gardner, "Physical and Chemical Examination of Paints, Varnishes, Lacquers and Colors," Institute of Paint and Varnish Research, Washington, D.C., 1939.

TABLE 33.—SPECIFIC GRAVITY AND BULKING VALUE OF PIGMENTS, EXTENDERS, AND FILLERS.*—(*Continued*)

Pigment	Specific gravity	Pounds per solid gallon	Gallons bulked per pound
Iron black, magnetic oxide of iron	4.83	40.23	0.0249
Iron oxide, red, 98 per cent	5.08	42.32	0.0236
Iron oxide, red, 80 per cent	4.35	36.24	0.0276
Iron oxide, red, 50 per cent	3.63	30.24	0.0331
Iron oxide, red, precipitated	4.48	37.32	0.0268
Iron oxide, yellow, precipitated	3.90	32.49	0.0308
Jobber's green, less than 20 per cent	3.75	31.24	0.0320
Keystone filler	2.75	22.91	0.0436
Lead dust	11.09	92.39	0.0108
Litharge	9.40	78.30	0.0108
Lithol red	1.59	13.24	0.0755
Lithopone, calcium base	3.10	25.82	0.0387
Lithopone, high-strength, 50 per cent zinc sulfide	4.20	34.99	0.0286
Lithopone, regular, 28 to 30 per cent zinc sulfide	4.30	35.82	0.0279
Lithopone, titanated, 15 per cent titanium dioxide	4.25	35.40	0.0282
Magnesium carbonate	1.81	15.08	0.0663
Magnetic oxide of iron, iron black	4.83	40.23	0.0249
Mercuric oxide	11.14	92.80	0.0108
Metronite	2.85	23.74	0.0421
Ocher, domestic	4.03	24.82	0.0402
Ocher, French	2.99	24.91	0.0401
Orange mineral	8.80	73.30	0.0136
Para red	1.49	12.41	0.0806
Para red, 10 per cent on barytes	4.10	34.15	0.0293
Paris green	3.27	27.24	0.0367
Prussian blue	1.83	15.24	0.0656
Red lead, 85 per cent	8.90	74.14	0.0135
Red lead, 95 per cent	8.73	72.72	0.0137
Sienna, burnt	3.56	29.65	0.0337
Sienna, raw	3.14	26.16	0.0382
Silica	2.65	22.07	0.0453
Slate flour	2.80	23.32	0.0429
Talc	2.77	23.07	0.0433
Titanium dioxide	3.90	32.49	0.0308
Titanium-barium pigment	4.30	35.82	0.0279
Titanium-calcium pigment	3.10	25.87	0.0387
Toluidine red	1.42	11.83	0.0854
Toluidine red, 10 per cent on barytes	4.10	34.15	0.0293

* H. A. Gardner, "Physical and Chemical Examination of Paints, Varnishes, Lacquers and Colors," Institute of Paint and Varnish Research, Washington, D.C., 1939.

TABLE 33.—SPECIFIC GRAVITY AND BULKING VALUE OF PIGMENTS, EXTENDERS, AND FILLERS.*—(*Continued*)

Pigment	Specific gravity	Pounds per solid gallon	Gallons bulked per pound
Tuscan red	3.63	30.24	0.0331
Ultramarine blue	2.34	19.49	0.0513
Umber, burnt	3.64	30.32	0.0330
Umber, raw	3.20	26.66	0.0375
Vandyke brown	1.61	13.41	0.0746
White lead, basic carbonate	6.70	55.81	0.0179
White lead, basic sulfate	6.40	53.31	0.0189
Whiting, calcium carbonate, chalk	2.70	22.49	0.0445
Zinc chromate	3.46	28.82	0.0347
Zinc dust	7.06	58.81	0.0170
Zinc oxide, lead-free	5.65	47.06	0.0212
Zinc oxide, 3 to 5 per cent leaded	5.70	47.48	0.0211
Zinc oxide, 8 to 12 per cent leaded	5.75	47.90	0.0209
Zinc oxide, 35 per cent leaded	5.95	49.56	0.0202
Zinc sulfide	4.00	33.32	0.0300
Zinc yellow	3.46	28.82	0.0347
Zirconium oxide	5.69	47.40	0.0211

* H. A. Gardner, "Physical and Chemical Examination of Paints, Varnishes, Lacquers and Colors," Institute of Paint and Varnish Research, Washington, D.C., 1939.

Neutralize another portion of the strongly acid solution with 1:1 ammonium hydroxide, and add acetic acid until just acid. A yellow precipitate on addition of potassium chromate solution indicates the presence of lead or acid-soluble barium, or both.

Partly neutralize another portion of the strongly acid solution with 1:1 ammonium hydroxide, and add a few drops of potassium ferrocyanide solution. A white precipitate, frequently rendered somewhat bluish by a trace of iron, indicates zinc.

Pass hydrogen sulfide gas into the strongly acid filtrate for 5 to 10 minutes, and dilute with an equal volume of water. Again pass the gas through for 5 minutes. Filter, wash with hydrogen sulfide water, and digest the precipitate with ammonium polysulfide solution. Filter, acidify the filtrate with hydrochloric acid, and warm. The presence of antimony is indicated by the separation of an orange-colored precipitate. The filtrate from precipitation with hydrogen sulfide will show a white precipitate with dilute sulfuric acid. To test for calcium, neutralize with 1:1 ammonium hydroxide, and test with ammonium oxalate. To test for magnesium, add diammonium phosphate after testing for calcium and if present precipitating it with excess of oxalate.

It is often convenient to heat a small amount of white pigment on a spatula. If reducing matter and lead are present, tiny globules of lead may form. Zinc oxide turns bright yellow while hot but on cooling is white. Titanium oxide reacts similarly, but the color is cream only while hot.

Quantitative. SPECIFIC GRAVITY. Whatever the pigment, obtain the specific gravity by the method given for powders (page 48). Check this against the value in Table 33 for the pigment suspected. To the experienced worker, this value is often a reliable indication of the pigment present and may render part of the wet-way analysis unnecessary.

Basic Lead Carbonate, or White Lead. LEAD AS LEAD SULFATE. Dissolve 1 gram of sample in 20 ml. of 1:1 nitric acid in a covered beaker. Boil until the evolution of gas ceases. Wash down the cover, dilute somewhat, filter if necessary, and add 20 ml. of 1:1 sulfuric acid. Evaporate to fumes of sulfur trioxide. Cool, and add 150 ml. of water and 150 ml. of alcohol. Let stand in cold water for an hour, and filter on a weighed Gooch crucible. Wash with alcohol, dry at 110°, and weigh as lead sulfate. This method is reasonably accurate unless calcium is present; results then are high owing to occlusion of calcium sulfate in the precipitate of lead sulfate.

CALCULATION.

$$\text{Weight of lead sulfate} \times 0.8527 \times \frac{100}{\text{weight of sample}} = \text{per cent white lead}$$

The factor as lead oxide is 0.7360.

LEAD AS LEAD CHROMATE. Boil 1 gram of sample with 20 ml. of glacial acetic acid. Dilute with water to about 200 ml., and filter if necessary. Add concentrated ammonium hydroxide until just alkaline and then a few drops of glacial acetic acid until just acid. Heat to boiling, and add 10 to 15 ml. of a boiling 10 per cent solution of potassium dichromate. Boil until the precipitate assumes an orange color. Add a few drops more of the reagent to the supernatant liquid to be sure that precipitation is complete. Let the precipitate settle, and decant the upper layer through a Gooch crucible. Wash the precipitate in the beaker by decantation with hot water until the washings are colorless; then transfer the precipitate to the crucible. Wash the crucible and contents with 95 per cent alcohol and then with ether. Dry at 110°, and weigh as lead chromate.

CALCULATION.

$$\text{Weight of lead chromate} \times 0.7999 \times \frac{100}{\text{weight of sample}} = \text{per cent white lead}$$

The factor as lead oxide is 0.6906.

LEAD CHROMATE VOLUMETRICALLY. Precipitate, separate, and wash the lead chromate as for gravimetric determination. Thoroughly rinse all soluble chromate from the filter flask, and replace the crucible in the holder. Add 6 to 10 ml. of 1:4 hydrochloric acid to the Gooch crucible containing the precipitate. Allow to dissolve, swirling gently if necessary. Apply the suction, and wash the crucible with a few more milliliters of 1.4 hydrochloric acid. Wash the crucible with hot water. Cool the filter flask with its contents, and add 20 ml. of 20 per cent potassium iodide solution.

The reaction is

$$2PbCrO_4 + 6KI + 16HCl = 2PbCl_2 + 2CrCl_3 + 6KCl + 8H_2O + 3I_2$$

Titrate the liberated iodine with standard sodium thiosulfate solution (page 730), adding starch solution (page 717) when the iodine color becomes faint. A green color due to reduced chromium will be present.

CALCULATION.

$$\text{Ml. titration} \times \text{normality of sodium thiosulfate} \times 0.08619 \times \frac{100}{\text{weight of sample}} = \text{per cent white lead}$$

The factor for lead oxide is 0.0744.

BASIC CARBONATE. Using 1 to 2 grams of pigment, determine the carbon dioxide by evolution with dilute hydrochloric acid (page 224). Commercial white lead may not correspond to the theoretical value assumed in the calculation of lead, as determined, to white lead. The carbon dioxide and the lead as lead oxide permit a calculation of the ratio of carbonate to hydroxide actually present.

CALCULATIONS. First calculate the carbon dioxide to per cent in the sample.

$$\text{Weight of carbon dioxide} \times \frac{100}{\text{weight of sample}} = \text{per cent carbon dioxide}$$

Next calculate this per cent carbon dioxide to per cent lead carbonate.

$$\text{Per cent carbon dioxide} \times 6.0718 = \text{per cent lead carbonate}$$

Also, calculate the equivalent of lead oxide to the carbon dioxide found.

$$\text{Per cent carbon dioxide} \times 5.0730 = \text{per cent lead oxide to form carbonate}$$

At this point the lead oxide calculated and the carbon dioxide together equal the calculated lead carbonate as a check on the mathematics.

Next subtract the lead oxide as carbonate from the total lead as oxide.

$$\text{Total lead oxide} - \text{carbonate as lead oxide} = \text{lead as lead oxide, present as hydroxide}$$

Calculate this lead oxide to hydroxide.

Lead oxide present as hydroxide $\times$ 1.0806 = per cent lead hydroxide

The sum of the lead carbonate and the lead hydroxide is the basic lead carbonate, or white lead. Comparison of this with lead calculated from the total lead oxide present to the theoretical white lead composition gives an indication of the deviation from the theoretical and also of any gross error in calculation.

The ratio of carbonate to hydroxide can also be calculated as follows, though this is only for confirmation of results and is not usually reported:

$$\frac{\text{Per cent lead carbonate}}{\text{Molecular weight of lead carbonate}} = \text{lead carbonate ratio figure}$$

$$\frac{\text{Per cent lead hydroxide}}{\text{Molecular weight of lead hydroxide}} = \text{lead hydroxide ratio figure}$$

$$\frac{\text{Lead carbonate ratio}}{\text{Lead hydroxide ratio}} = \text{moles lead carbonate per mole lead hydroxide}$$

Values here should be closely scrutinized unless they fall between 1.8 and 2.2. Quite commonly they fall between 1.9 and 2.0. The compound approached is $Pb_3(OH)_2(CO_3)_2$, also expressed as $Pb(OH)_2.2PbCO_3$.

Basic Lead Sulfate. This is a commercial pigment having an average composition of 78.5 per cent lead sulfate, 16 per cent lead oxide, and 5.5 per cent zinc oxide.

INSOLUBLE MATTER. Moisten 1 gram of dry pigment with a few drops of alcohol. Add 50 ml. of acid ammonium acetate reagent. Boil for 2 minutes and decant through a filter paper, leaving any residue in the beaker. Add 50 ml. more of the acid ammonium acetate solution, and boil again for 2 minutes. Filter through the same paper, transferring the precipitate to the paper. Wash well, and save the combined filtrates for determination of lead. Ignite and weigh the precipitate, reporting as insoluble matter. If substantial in amount, examine qualitatively after weighing.

TOTAL LEAD. Use the combined filtrates from the insoluble matter as sample for determination of lead as lead chromate (page 501), starting at "Heat to boiling, and add 10 to 15 ml. of a boiling 10 per cent solution of potassium dichromate." Results may be completed gravimetrically as there shown or volumetrically as in the method that follows the gravimetric one. Calculate the lead as lead oxide.

LEAD AS SULFATE. Weigh about 0.5 gram of dry pigment into a 400-ml. beaker. Wet with a few drops of alcohol, and dissolve by heating on a steam bath for 5 minutes with 10 ml. of bromine water, 10 ml. of 1:1 hydrochloric acid, and 3 grams of ammonium chloride. Dilute with 200 ml. of hot water, and boil for 5 minutes. All the basic lead sulfate should dissolve under this treatment. If there is any insoluble matter, filter and wash well. Cover the solution or filtrate, and add sufficient dry sodium carbonate to complete neutralization and about 1 gram excess. Boil for 10 to 15 minutes. Wash off the watch glass, and allow the precipitated lead carbonate to settle. Filter the solution, and wash the precipitate by decantation, saving the washings. Pour about 10 ml. of 1:1 hydrochloric acid through the filter paper into the beaker containing the precipitate, so that it is completely redissolved. Cover the beaker, and reprecipitate with sodium carbonate. Filter through the same filter paper into the same receiving beaker, and wash well with hot water. Acidify the combined filtrates with 1:1 hydrochloric acid, adding 1 ml. excess. Boil to expel the bromine, and to the boiling solution slowly add 15 ml. of boiling 10 per cent barium chloride solution. Let stand for at least 1 hour on a steam bath, and filter through a weighed Gooch crucible or a retentive filter paper. Wash well with hot water, dry, and ignite carefully. Cool and weigh the barium sulfate, which is equivalent to the lead sulfate in the sample.

Calculation.

$$\text{Weight of barium sulfate} \times 1.299 \times \frac{100}{\text{weight of sample}} = \text{per cent lead sulfate}$$

Calculate the lead sulfate to lead oxide, and subtract from the total lead as lead oxide to give the amount present as oxide.

$$\text{Per cent lead sulfate} \times 0.7360 = \text{per cent lead sulfate as lead oxide}$$
$$\text{Total lead oxide} - \text{lead sulfate as lead oxide} = \text{lead oxide in sample}$$

The sum of lead sulfate and lead oxide as such is the basic lead sulfate, subject to the usual amount of added zinc oxide.

Zinc as Zinc Oxide. Accurately weigh about 1 gram of pigment into a 400-ml. beaker, add 30 ml. of 1:2 hydrochloric acid solution, and boil for 2 to 3 minutes. Add 200 ml. of water, and make slightly alkaline with 1:1 ammonium hydroxide. Then make just acid with 1:2 hydrochloric acid. Add 3 ml. excess of the hydrochloric acid. Heat to boiling, and titrate with standard potassium ferrocyanide solution as described for the zinc oxide determination in the next method. Calculate as zinc oxide.

The sum of basic lead sulfate, zinc oxide, and insoluble matter should approximate 100 per cent. If the insoluble matter is substantial, it should be separately analyzed by the methods of Chap. 10 and reported by composition rather than as insoluble matter.

Zinc Oxide. Zinc Volumetrically by Ferrocyanide. The standard solution must first be prepared. Dissolve 22 grams of potassium ferrocyanide in water, and dilute to 1 liter. Each milliliter will be equivalent to about 0.005 gram of zinc, but the purity of the salt is such as to require standardization. Weigh about 0.2 gram of pure zinc or pure zinc oxide into a 400-ml. beaker. Dissolve in 30 ml. of 1:2 hydrochloric acid; then render just alkaline with 1:1 ammonium hydroxide. Make just acid with a few drops of concentrated hydrochloric acid, and add 3 ml. in excess. Dilute to 250 ml., and heat to boiling. Set aside about 50 ml. of the solution, and while stirring run the standard ferrocyanide solution slowly from a buret into the main volume. A complex zinc ferrocyanide,

$$K_2Zn_3[Fe(CN)_6]_2$$

will be precipitated. While this should be white, a trace of iron not suffi cient to interfere will usually render it bluish until near the end point, when the precipitate becomes white. This change in color is a guide but is not a reliable indicator. It is so useful that if the highest accuracy is not essential sometimes a drop of very dilute ferric chloride solution is introduced to provide this impurity.

Use a 5 per cent solution of uranyl nitrate or uranyl acetate, slightly acidified with acetic acid, on a spot plate as indicator. The end point is shown by a brown tinge appearing within 1 minute. When the end point is approximately located, add the balance of the solution and again heat to boiling. The end point can now be approached accurately and with reasonable promptness. A false end point can be obtained by titrating too rapidly. It is convenient to calculate the standardization in terms of zinc, zinc oxide, and zinc sulfide. If it is protected, this solution is stable for months.

Calculation of Standardization.

$$\frac{\text{Weight of zinc or zinc oxide}}{\text{Volume of ferrocyanide in ml.}} \times \frac{\text{equivalent weight of zinc compound desired}}{\text{equivalent weight of zinc or zinc oxide used}} = \text{factor for ferrocyanide per ml.}$$

Procedure. Weigh out 0.2 to 0.3 gram of sample, dissolve in 30 ml. of 1:1 hydrochloric acid, and cool. If manganese, iron, or copper is present, it must be removed as follows: Add 35 ml. of a solution containing 50 ml. of concentrated ammonium hydroxide, 20 grams of ammonium chloride, and 75 ml. of water. This precipitates ferric and manganous hydroxides. Boil very gently for 1 to 2 minutes, and add excess of bromine water. This oxidizes the manganous to manganic hydroxide. Filter and wash thoroughly with a solution containing 10 per cent of ammonium chloride and 5 per cent of concentrated ammonium hydroxide. The precipitate can be reserved for later determination of the iron titrimetrically (page 194), or colorimetrically (page 195), and the manganese colorimetrically (page 198). Neutralize the filtrate with concentrated hydrochloric acid, and add 3 ml. in excess. Dilute to about 200 ml. with hot water, and heat nearly to boiling. Add 50 ml. of water saturated with hydrogen sulfide to precipitate small amounts of copper, or pass in hydrogen sulfide gas to precipitate large amounts. If small amounts were precipitated with hydrogen sulfide solution, use without removal of copper sulfide although its presence obscures the end point somewhat. If large amounts of copper are present so that hydrogen sulfide was passed in, filter and wash. The copper sulfide may then be saved for determination by conventional methods such as electrodeposition (page 323). If hydrogen sulfide gas was used, heat to boiling to remove most of the excess, for a large amount interferes. This separation of iron, manganese, and copper is unnecessary in a commercial zinc oxide pigment unless toners have been added. It is introduced here to clarify methods described later and to provide for analysis of paint pigments that have been toned with traces of mineral color.

Make the solution of sample, from which interfering substances have been removed if necessary, just alkaline with 1:1 ammonium hydroxide and then just acid with concentrated hydrochloric acid, adding 3 ml. in excess. Proceed as described for standardization (page 505), starting at "Dilute to 250 ml., and heat to boiling." Calculate as zinc oxide.

If the zinc oxide did not require separation of iron, manganese, or copper, for very accurate work a blank is automatically taken care of by standardizing the ferrocyanide at the same time. If other reagents had to be added in purification, for the highest accuracy a blank must be run on them, but not on the reagents used in standardization, and subtracted from the results obtained. This precaution is rarely required.

CALCULATION.

$$\text{Ferrocyanide solution in ml.} \times \text{factor for zinc oxide} \times \frac{100}{\text{weight of sample}} = \text{per cent zinc oxide}$$

If the pigment is straight zinc oxide, this should approximate 100 per cent.

Lithopone. Lithopone is a white pigment consisting of barium sulfate and zinc sulfide obtained by reacting zinc sulfate with barium sulfide and subsequent calcination. Therefore, the proportions are roughly chemically equivalent, 70 per cent barium sulfate and 30 per cent zinc sulfide. Sometimes zinc oxide or titanium oxide is also present.

BARIUM SULFATE OR INSOLUBLE MATTER. To 1 gram of pigment in a 250-ml. beaker add 10 ml. of concentrated hydrochloric acid and then, in small quantities, about 1 gram of potassium chlorate. This oxidizes the sulfide. Evaporate to about half volume on a steam bath; then dilute with 100 ml. of water. Add 5 ml. of 1:10 sulfuric acid, and boil until the precipitate settles readily. Let stand a few minutes to settle, filter, wash, ignite, cool, and weigh. Save the filtrate for determination of zinc.

Qualitative examination of the original sample should have established whether titanium oxide is present and may have shown whether silica and alumina are present. If not, part of the precipitate can be tested for silica with hydrofluoric acid. In this treatment always add a few drops of sulfuric acid. Similarly test for alumina by fusion of a part with potassium acid sulfate. Dissolve this in water, filter, and add a small excess of ammonium hydroxide. If free from silica, titanium, and alumina, this insoluble matter is entirely barium sulfate.

TOTAL ZINC. Use the ferrocyanide method applied to the total filtrate and washings from barium sulfate. Render just alkaline by addition of 1:1 ammonium hydroxide, and proceed as described for standardization (page 504) starting at "Make just acid with a few drops of concentrated hydrochloric acid, and add 3 ml. in excess."

ZINC OXIDE. Zinc oxide is soluble in cold dilute acetic acid, but zinc sulfide is not. Weigh out about 1 gram of sample, and wet with a few drops of alcohol. Add about 100 ml. of cold 2 per cent acetic acid, and stir vigorously. Cover with a watch glass, and stir at 5-minute intervals for the first half hour. Then let stand undisturbed for 18 to 24 hours. Do not heat at any time, or some zinc sulfide will dissolve. Filter and wash on the filter, first with cold 2 per cent acetic acid, then with water. Dilute the filtrate to about 200 ml. with water, and add 30 ml. of 1:2 hydrochloric acid. Then render just alkaline with 1:1 ammonium hydroxide, and proceed as in the ferrocyanide method, starting at the point under standardization (page 504) "Make just acid with a few drops of concentrated hydrochloric acid, and add 3 ml. in excess."

CALCULATION.

$$\text{Ferrocyanide in ml} \times \text{factor as zinc oxide} \times \frac{100}{\text{weight of sample}}$$
$$= \text{per cent of zinc oxide in original sample}$$

ZINC SULFIDE. Normally, in dealing with lithopone, the total zinc as zinc oxide less the zinc oxide present as such gives a value that can be calculated to zinc sulfide.

If the barium sulfate, or insoluble matter, zinc oxide, and zinc sulfide total approximately 100 per cent, further determination is unnecessary. If it is necessary to determine the zinc sulfide as such, evolve the sulfide as hydrogen sulfide and absorb in a suitable medium. The determined substance may be lead sulfate gravimetrically (page 163) or cadmium sulfide gravimetrically (page 164) or titrimetrically (page 171). It may even be barium sulfate, page 164. Calculate the sulfide as determined to zinc sulfide and to zinc oxide. The zinc sulfide calculated as zinc oxide should check with the total zinc oxide if free zinc oxide was not present. Otherwise, the zinc sulfide calculated as oxide plus the determined free zinc oxide should check the total zinc oxide.

Titanium Dioxide. While titanium dioxide is sold as such and has the greatest obscuring power of any white pigment, it is also sold in conjunction with barium sulfate, calcium sulfate, and magnesium silicate, which with the exception of magnesium silicate are part of the original pigment in much the same sense as the barium sulfate in lithopone, although not in stoichiometric proportions.

BARIUM SULFATE. Transfer 0.5 gram of pigment to a 250-ml. beaker, and add 20 ml. of concentrated sulfuric acid and 7 to 8 grams of ammonium sulfate. Heat with occasional stirring until sulfur trioxide fumes are given off. Continue to heat until solution is complete, or nothing more is dissolving. No substantial loss of sulfuric acid as fumes should occur in heating. Any insoluble matter at this stage indicates the presence of silica or siliceous matter. Usually it will be absent. Cool, and carefully add 100 ml. of water. This causes the precipitation of barium sulfate if present. Heat to boiling, and digest on a steam bath until the precipitate settles readily. Let settle and filter on a retentive paper, washing well with small volumes of 1:20 sulfuric acid. Reserve the filtrate for determination of titanium. Ignite and weigh the barium sulfate. Unless siliceous matter was found, barium sulfate is all that is present in this precipitate. If silica as such was found, add 0.5 ml. of 48 per cent hydrofluoric acid and a few drops of concentrated sulfuric acid to the precipitate and heat at first gently, later igniting strongly. The loss in weight is silica. If clay or other siliceous mineral were present, the routine procedure for mineral analysis (Chap. 10) must be applied to the original sample.

Titanium Dioxide. The filtrate from barium sulfate contains the titanium. Calcium sulfate, if present in the sample, is also in solution but will not interfere. Dilute the filtrate to 200 ml., and reduce the acidity to about 5 per cent sulfuric acid by addition of 10 ml. of concentrated ammonium hydroxide. Determine the titanium titrimetrically in this solution, starting as on page 197 at "Wash out a Jones reductor with 1:20 sulfuric acid . . . ". The amount of titanium found is subject to a later correction for any iron present.

Iron. Dissolve 1 gram of sample as described for the determination of barium sulfate on page 307. Continue through "Cool and carefully add 100 ml. of water . . . ". Without filtering, transfer to a 250-ml. calibrated flask, and dilute to volume when cool. Pipet out a 25-ml. aliquot for the determination of iron. Use the thiocyanate method, starting on page 196 at "To the aliquot in a 50-ml. Nessler tube add . . . ".

Calculate the iron as ferric oxide, and multiply by 1.001 to get the equivalent as titanium dioxide. Subtract this value from the amount of titanium dioxide found by titration to correct for the iron present.

Calcium Sulfate. If calcium sulfate is present, weigh a sample of 1 gram and transfer to a 250-ml. beaker. Add 50 ml. of 1:4 hydrochloric acid, and stir vigorously for 5 minutes. Let this solution stand overnight, with occasional stirring. Filter, and wash on the paper with 25 ml. of 1:4 hydrochloric acid. Heat this filtrate and washings to boiling, and add 20 ml. of boiling ammonium oxalate solution, saturated at room temperature. To this solution slowly add 1:2 ammonium hydroxide, with vigorous stirring, until the solution is alkaline to litmus. Complete as on page 216, starting with "Boil for 1 minute, and allow to stand at least 20 minutes or until the precipitate has settled."

The total of titanium dioxide and barium sulfate or calcium sulfate and any other acid-insoluble matter should approximate 100 per cent. If not, some other pigment, magnesium silicate, for example, was present.

CHAPTER 28

MIXED WHITE PIGMENTS, EXTENDERS, AND MINERAL FILLERS

The methods available for separate analysis of the white pigments are adapted to mixed pigments and to separation of the various mixtures that occur commercially.

Loss on Ignition. Ignite a sample of 1 to 5 grams below red heat until constant weight is reached. This determination is meaningless when lead and sulfides are present. Avoid the use of platinum if lead is suspected.

Acid-insoluble Matter. Wet 1 gram of pigment with a few drops of alcohol, and add 40 ml. of 1:1 hydrochloric acid. Cover with a watch glass, and boil gently for 5 to 10 minutes. Wash down the cover glass, evaporate to dryness, and heat to about 150° in an oven for 30 to 60 minutes to dehydrate the residue. Moisten with 4 ml. of concentrated hydrochloric acid, let stand for a few minutes, and dilute with 100 ml. of hot water. Boil a few minutes, and filter through paper while still hot. Wash the paper with hot water until all the lead salts are washed through. Save the filtrate and combined washings for other determinations, diluting to a standard volume so that aliquots can be taken. Ignite the insoluble residue, cool, and weigh. This represents acid-insoluble matter present in the original pigment unless both barium carbonate and lead sulfate were present. In that case reaction will have occurred to precipitate some or all of the barium as sulfate. A trace of titanium dioxide may have dissolved.

Titanium. In case titanium dioxide is present, transfer the ignited residue of acid-insoluble material to a 250-ml. beaker. Add 20 ml. of concentrated sulfuric acid and 7 to 8 grams of ammonium sulfate. Proceed as for the separation of titanium and barium sulfate from titanium pigments described on page 507, starting at "Heat with occasional stirring until sulfur trioxide fumes are given off." Continue to "Reserve the filtrate for determination of titanium." This separation gives an insoluble residue free from titanium and a filtrate containing the titanium. Reserve both.

Dilute this filtrate to a known volume, and determine titanium in an aliquot. Some is to be reserved for later determination of other positive radicals present, although the amounts are usually small. For titanium, use the method on page 197, starting at "Wash out a Jones reductor with 1:20 sulfuric acid . . . ".

To correct this, determine colorimetrically the amount that has dissolved. For this use an aliquot of the filtrate obtained in the separation of the acid-insoluble matter, and follow the procedure on page 197. Add the amount so found, calculated back to the original sample, to the titanium titrated and calculated to the same basis.

Barium Sulfate. Mix the residue containing barium sulfate and possibly silica or silicates in a platinum crucible with ten times its weight of anhydrous sodium carbonate. Cover, and fuse. When it is cool, transfer the crucible and cover to a porcelain casserole. This is preferable to a beaker because of some extraction of silica from glass by strongly alkaline solutions. Boil with 100 ml. of water until the melt is completely disintegrated. If the silica content is high, this will require considerable time. To aid solution, break up any lumps with the flattened end of a rod.

Filter through a quantitative paper into a 300-ml. porcelain casserole. Wash the crucible and cover thoroughly with hot 1 per cent sodium carbonate solution. Similarly wash the casserole in which the melt was dissolved, and finally the residue on the paper. Put aside this filtrate, which may contain sulfate, sodium silicate, and some metals other than barium.

The residue on the paper not only contains all the barium as carbonate but may contain some silica. Place the casserole used for disintegrating the melt under the funnel, and pierce the tip of the paper with a pointed glass rod. Dissolve the barium carbonate, and wash through any insoluble matter with hot 1:1 hydrochloric acid. Finally wash the paper with hot water. Add 20 ml. of 1:1 hydrochloric acid, and evaporate to dryness. Bake at 150° for 30 to 60 minutes to dehydrate silica. Moisten the residue with 2 ml. of concentrated hydrochloric acid, and dilute with 100 ml. of hot water. Boil and filter on a quantitative paper while hot. Reserve this paper, which may contain some silica. Dilute the filtrate to about 250 ml., and heat to boiling. Slowly add 5 ml. of boiling 1:4 sulfuric acid to precipitate the barium as sulfate, stirring vigorously during the addition. Boil for a few minutes to increase the size of particles; then let it stand for at least an hour and preferably overnight. Filter, ignite, and weigh as barium sulfate.

Silica. The acid-insoluble matter less the titanium oxide and barium sulfate gives the total siliceous matter but not necessarily the silica. To determine silica, acidify the alkaline filtrate and washings reserved from fusion of the insoluble matter with sodium carbonate. Add 40 ml. of 1:1 hydrochloric acid in excess, and evaporate to dryness. Bake at 150° for 30 to 60 minutes to dehydrate the silica. Moisten the residue with about 2 ml. of concentrated hydrochloric acid, add about 100 ml. of water, and boil for 2 minutes. Filter through the paper reserved as possibly containing some silica isolated before precipitation of barium sulfate. For quick filtration use a fresh paper if the reserved paper contains much silica. Wash thoroughly with hot 1:20 hydrochloric acid, ignite the paper or papers, and weigh. The weighed residue should be entirely silica. Confirm this by adding a few milliliters of 48 per cent hydrofluoric acid and a few drops of concentrated sulfuric acid. When gently fumed off and then blasted, the silica is volatilized.

OTHER METALS IN ACID-INSOLUBLE MATTER. If the sum of titanium dioxide, barium sulfate, and silica does not account for the acid-insoluble matter, the silica must have been present as a silicate. The positive radical of this silicate is present in the filtrate from the determination of silica and the filtrate from the precipitation of barium sulfate. This may be aluminum from clay or pyrophyllite, magnesium from talc or asbestine, or some iron from a brown oxide or silicate. Some may also be in the titanium solution.

Combine the filtrates from silica and barium sulfate. Phosphorus will ordinarily be absent. Therefore proceed with this solution as sample according to the methods described for mineral analysis, starting at page 191 with iron and aluminum and continuing with the method for magnesium, if it is present.

Also, precipitate an aliquot of the filtrate containing titanium. Render it just alkaline with 1:1 ammonium hydroxide. Bring to a boil, and filter. Wash, ignite, and weigh. The precipitate is titanium oxide with any aluminum or iron oxides that had dissolved. Since the titanium dioxide is known from the titration, any small amount of aluminum or iron oxide is obtained by difference. If magnesium is present in the filtrate, precipitate it as the phosphate by the method on page 218, starting at "Add 10 ml. of 10 per cent disodium hydrogen phosphate solution . . . ". These amounts found in the titanium solution are to be added to the amounts found in the silica filtrates.

Lead in the Presence of Antimony. SULFIDE METHOD. The reserved filtrate and washings from the separation of acid-insoluble matter will total 150 to 200 ml. Saturate this with hydrogen sulfide, add an equal volume of water, and again saturate. These precautions ensure more complete precipitation. Filter the sulfides, and wash the precipitate on the filter with water saturated with hydrogen sulfide. Reserve this filtrate for determination of acid-soluble barium, aluminum, and total zinc.

Wash the sulfides into a porcelain dish or casserole, and add 25 ml. of ammonium polysulfide solution. Retain the filter paper undamaged. Cover, and warm to 50° for 10 to 15 minutes. Stir frequently to promote extraction of the antimony sulfide. Rinse off the cover, and filter through the paper that originally contained the sulfides. Wash the precipitate with ammonium sulfide solution diluted 1:40. Reserve the filtrate if antimony is to be determined.

Dissolve the lead sulfide from the paper with hot 1:3 nitric acid, and wash thoroughly with water. Add 10 to 20 ml. of 1:1 sulfuric acid to the filtrate, and evaporate down to copious fumes of sulfur trioxide. Cool, dilute to 100 ml., add about 150 ml. of 50 per cent alcohol, and stir well. Filter on a Gooch crucible that has previously been heated to incipient redness. Wash well with 50 per cent alcohol, and dry. Ignite to incipient redness, cool, and weigh, as lead sulfate.

To determine antimony on the same sample, render the filtrate just definitely acid to precipitate antimony sulfide with some free sulfur. Filter on a Gooch crucible, and heat in an atmosphere of carbon dioxide at 280 to 300°. This volatilizes sulfur and converts the antimony sulfide to the stable black variety. Cool in carbon dioxide, and weigh. This method of determining antimony is not ordinarily applied; determination by titration on an original sample of pigment is more usual.

ALTERNATIVE METHOD. Boil 1 gram of the original pigment in a covered beaker with 40 ml. of 1:1 hydrochloric acid for 5 to 10 minutes. Wash off the cover glass, and carefully evaporate to dryness. To the residue add 10 to 20 ml. of concentrated hydrochloric acid and 50 ml. of hot water. Boil for a few minutes to dissolve the lead sulfate. Filter through a paper, and wash until free from chlorides.

Add 20 ml. of 1:1 sulfuric acid, and evaporate to copious fumes of sulfur trioxide. Cool to 60°, and slowly add 50 ml. of water with stirring. Heat to boiling, and continue to boil for several minutes to dissolve the antimony sulfate completely. Allow the lead sulfate to settle to give a clear supernatant liquid, but do not allow the temperature to drop below 60°. If the liquid does not clear promptly on standing, reheat to boiling for a few minutes.

Decant the clear solution through a Gooch crucible. Add 10 ml. of concentrated sulfuric acid to the wet lead sulfate in the beaker, and heat to boiling for several minutes. Cool to 60°, and slowly add 30 ml. of water. Heat to boiling for a few minutes, cool to about 60°, and completely transfer the precipitate to the Gooch crucible.

Prepare "lead acid." Mix 300 ml. of concentrated sulfuric with 1,800 ml. of water, and add to the hot mixture 1 gram of C.P. lead acetate dissolved in 300 ml. of hot water. Let this stand for not less than 24 hours, and siphon the clear upper layer through a thick asbestos filter. The result is a diluted sulfuric acid for washing purposes that has been saturated with lead sulfate.

Wash the precipitated lead sulfate with this lead acid to free from acid-soluble sulfates. Then wash with 50 per cent alcohol until free from sulfates, and dry at 105 to 110°. If the original sample contained acid-soluble barium compounds, the lead sulfate will contain the barium as sulfate, quantitatively. Silica may also be present. Calcium, if present, originally, is incompletely present in the precipitate. Dissolve the lead sulfate from the filter with ammonium acetate reagent (page 718), and determine lead as the chromate as on page 501, starting at "Add concentrated ammonium hydroxide until just alkaline, then a few drops of glacial acetic acid until just acid."

Alternatively, if calcium is absent, lead may be calculated from the difference in weight of the Gooch crucible before and after the ammonium acetate treatment.

If calcium is not present in the pigment, the residue on the Gooch crucible after extraction of lead will be practically quantitatively barium sulfate and siliceous matter, a valuable cross-check on the determination by other means.

Lead. Acid-soluble Antimony, Barium, and Calcium Absent. Add 25 ml. of 1:1 nitric acid to 1 gram of pigment in a covered beaker. After reaction appears to be complete, boil gently for a few minutes. Rinse down the cover glass, and evaporate the solution to dryness on a steam bath. Moisten the residue with about 5 ml. of 1:1 nitric acid, add 50 ml. of hot water, and heat to dissolve. Filter, and wash the filter with hot water until free from lead. Add 10 to 20 ml. of 1:1 sulfuric acid to the filtrate, and evaporate down to copious fumes of sulfur trioxide. Cool, and dilute to 100 ml. Add about 150 ml. of 50 per cent alcohol, and stir well. Filter on a Gooch crucible that has previously been heated to incipient redness. Wash well with 50 per cent alcohol, and dry. Ignite to incipient red heat, cool, and weigh as lead sulfate.

Lead. Acid-soluble Antimony, Iron, Aluminum, and Barium Absent. Prepare the solution of pigment as in the previous method through "Filter, and wash the filter with hot water until free from lead." Dilute this clear filtrate to about 200 ml., and add 1:1 ammonium hydroxide until approximately neutral. Add 5 ml. of glacial acetic acid, and heat to boiling. Slowly add 10 to 15 ml. of hot 10 per cent potassium bichromate solution, with vigorous stirring. Continue to heat until the originally yellow precipitate becomes definitely orange and settles readily. Transfer the supernatant liquid to a Gooch crucible that has been dried at 110°. Wash the precipitate by decantation until the washings are colorless. Transfer the precipitate quantitatively to the crucible, and wash first with alcohol, then with ether. Dry at 100°, and weigh as lead chromate.

Antimony Oxide. Iron Absent. Antimonious and antimonic oxide can be determined on either a mixed pigment or a pure one by the same technic, provided that iron is absent.

ANTIMONIOUS OXIDE. To 0.3 gram of straight oxide pigment or 0.5 gram of mixed pigment in a 500-ml. conical flask, add 15 ml. of water and 25 ml. of concentrated hydrochloric acid. Cover and warm on a steam bath for 10 to 15 minutes to dissolve the antimonious oxide. Wash off the cover, and add 250 ml. of water and 15 ml. of concentrated sulfuric acid. Boil for 2 minutes, cool to 10 to 15°, and titrate to a faint pink with $0.1N$ potassium permanganate solution. The reaction is

$$5Sb_2(SO_4)_3 + 4KMnO_4 + 24H_2O = 10H_3SbO_4 + 2K_2SO_4 + 4MnSO_4 + 9H_2SO_4$$

CALCULATION.

$$\text{Ml. titration} \times \text{normality of permanganate} \times 0.07288 \times \frac{100}{\text{weight of sample}} = \text{per cent antimonious oxide}$$

ANTIMONIOUS AND ANTIMONIC OXIDES. Transfer 0.3 gram of the antimony oxide pigment or 0.5 gram of mixed pigment to a 500-ml. conical flask. Add 15 ml. of concentrated sulfuric acid, 10 grams of potassium sulfate, and a 9-cm. qualitative filter paper to furnish carbon to act as reducing agent. Place a funnel in the neck of the flask, and heat until all the filter paper is oxidized and the solution becomes colorless. Cool, wash the funnel, and dilute to 250 ml. with water. Add 20 ml. of concentrated hydrochloric acid, and boil for 2 minutes. Cool to 10 to 15°, and titrate as in the preceding method.

If the digestion is continued after the solution becomes colorless, which is when all reducing matter is gone, some of the antimony may be oxidized from the *-ous* to the *-ic* condition. In such cases, cool, wash the funnel, dilute to 100 ml. with water, add 1 to 2 grams of sodium sulfite, and boil until all the sulfur dioxide is expelled. This is shown by no blue appearing with starch-iodate paper. The volume will be reduced to about one-half. Dilute to 250 ml., add 20 ml. of concentrated hydrochloric acid, and boil for 2 minutes. Cool and titrate as in the preceding.

In this procedure all the antimony has been determined in antimonious form. Calculate as usual for antimonious oxide. Subtract the antimonious oxide found as such by the preceding method. The difference is the amount of antimonic oxide present, expressed in terms of antimonious oxide. Multiply by 1.110 to convert to terms of antimonic oxide.

Total Antimony. Iron Present. Transfer 0.5 to 1.0 gram of pigment to a 500-ml. conical flask. Add 5 ml. of water and 25 ml. of concentrated hydrochloric acid, and cover the flask. Dissolve the antimony by heating on a steam bath for about 15 minutes, and cool. Rinse down the cover, and add 3 grams of tartaric acid to the solution. This will keep the iron in solution as the complex during separation of the antimony. Add 100 ml. of hot water, and digest for a few minutes. If an insoluble residue remains, filter, and wash the residue with hot water. Dilute the filtrate and washings to 300 ml. with hot water, and pass in hydrogen sulfide until precipitation is complete. Filter the antimony sulfide, and wash with water saturated with hydrogen sulfide until the washings are no longer acid. Transfer the paper and precipitate to the original flask, and add 15 ml. of concentrated sulfuric acid and 10 grams of potassium sulfate. Continue as in the preceding method, starting with "Place a funnel in the neck of the flask, and . . . ".

Acid-soluble Barium. CALCIUM ABSENT. The conventional form in which this would occur is as barium carbonate. Use the combined filtrate and washings from the precipitation of total lead and antimony as sulfides as sample. Dilute to volume, and take an aliquot. Boil until all the hydrogen sulfide has been driven off. Then add a moderate excess of hot 1:4 sulfuric acid to the boiling solution to precipitate barium sulfate. Continue to boil until the precipitate settles readily, and digest on the steam bath for 1 hour. Filter, wash, ignite, and weigh as usual. Reserve the filtrate for determination of acid-soluble aluminum. The factor to convert barium sulfate to barium carbonate is 0.8456; to barium oxide, 0.6570.

CALCIUM PRESENT. If calcium is present, calcium sulfate may contaminate the barium sulfate. Therefore, in the presence of calcium, precipitate barium sulfate as in the preceding method. Let the precipitated barium sulfate settle, and decant the clear upper layer through the filter. Add 50 ml. of 1:20 hydrochloric acid to the precipitate, and boil. Filter while hot, and wash with hot 1:20 hydrochloric acid, then with hot water until acid-free. Filter, wash, ignite, and weigh as usual. Reserve the combined filtrates for acid-soluble aluminum and calcium.

Acid-soluble Aluminum. If acid-soluble barium was present, use the filtrate from that determination. Otherwise, use the filtrate from which lead and antimony were removed as sulfides. In either case, heat to boiling, and if hydrogen sulfide is present boil it off. Add a few drops of concentrated nitric acid to the boiling solution to oxidize any iron present. Add 5 grams of ammonium chloride and 8 ml. of concentrated hydrochloric acid per 200 ml. of solution. Add a few drops of 0.2 per cent alcoholic solution of methyl red, and if necessary heat again just to boiling. Add 1:2 ammonium hydroxide dropwise until the color changes from red to a distinct yellow. Boil for only 1 to 2 minutes, and filter immediately. Reserve this filtrate for determination of zinc. This technic inhibits the formation of the gelatinous precipitate so characteristic of aluminum. For very accurate results, redissolve the precipitate from the filter with 40 ml. of 1:4 hydrochloric acid, add 5 grams of ammonium chloride, and dilute to 200 ml. Add methyl red, and precipitate as before. Wash the precipitate on the filter thoroughly with 2 per cent ammonium chloride solution, ignite, cool, and weigh. This precipitate of aluminum oxide would be contaminated with iron oxide, titanium dioxide, and phosphate if they were present.

If necessary, fuse the ignited precipitate with potassium acid sulfate or pyrosulfate, and determine contaminants. Dissolve the melt in water, dilute to a known volume, and use aliquots. If considerable iron is present, determine titrimetrically with permanganate (page 194). For smaller amounts use the colorimetric thiocyanate method described on page 196, starting at "To the aliquot in a 50-ml. Nessler tube add a few drops . . . ". If titanium is present, determine colorimetrically by the procedure on page 198. If phosphate is present, determine volumetrically as on page 203, starting at "Transfer an aliquot of sample equivalent to about 0.04 to 0.1 gram of phosphorus pentoxide . . . ".

Total Zinc. IN FILTRATES. The filtrate from acid-soluble aluminum can be used without pretreatment. If the filtrate from acid-soluble barium is to be used, neutralize with 1:1 ammonium hydroxide and add 5 grams of ammonium chloride per 100 ml. of solution. If the filtrate from total lead and antimony as sulfide is used, boil off the hydrogen sulfide, neutralize with 1:1 ammonium hydroxide, and add 5 grams of ammonium chloride per 100 ml.

To the neutral solution, add 1 gram of ammonium acetate, make just acid with glacial acetic acid, and precipitate zinc sulfide by saturating with hydrogen sulfide. Let the precipitate settle, and filter through paper. Wash with 2 per cent acetic acid saturated with hydrogen sulfide. Reserve this filtrate for determination of acid-soluble calcium. Transfer the paper containing the precipitate to the original beaker used for precipitation. Add 30 ml. of water and 10 ml. of concentrated hydrochloric acid, and heat until solution of the zinc sulfide is complete. Dilute with 200 ml. of water, and neutralize with 1:1 ammonium hydroxide. Add 3 ml. of concentrated hydrochloric acid, and proceed to determine zinc by the ferrocyanide titration method described on page 505, starting at "Dilute to 250 ml., and heat to boiling."

ORIGINAL SAMPLE, IRON ABSENT. Transfer 1 gram of pigment, or more if necessary to give an adequate titration, to a 400-ml. beaker. Add 30 ml. of 1:3 hydrochloric acid, and boil for a few minutes to complete solution. In the absence of lead, antimony, and cadmium, filter, wash the filter with hot water, and neutralize the filtrate with 1:1 ammonium hydroxide. In that case, use the preceding method, starting with "To the neutral solution, add 1 gram of ammonium acetate . . . ".

If antimony, lead, or cadmium is present, dilute to 150 to 200 ml. and saturate with hydrogen sulfide. Add an equal volume of water, and again saturate. This ensures more complete precipitation. Filter the sulfides, and wash with water saturated with hydrogen sulfide. Discard the precipitate, and boil off the hydrogen sulfide. Neutralize the solution with 1:1 ammonium hydroxide, and add 5 grams of ammonium chloride per 100 ml. Then follow the preceding method, starting with "To the neutral solution, add 1 gram of ammonium acetate . . . ".

ORIGINAL SAMPLE, IRON PRESENT. Prepare the sample as for the preceding method until free of lead, antimony, and cadmium and available in acid solution. Dilute to about 150 ml. in a 400-ml. beaker. Add 2 grams of ammonium chloride, heat to boiling, and add 1:1 ammonium hydroxide dropwise until alkaline to methyl red. Filter while hot into a 400-ml. beaker, and wash the precipitated hydroxides once with hot water. Set aside the filtrate, and place a 250-ml. beaker under the funnel. Redissolve the precipitated hydroxides with about 10 ml. of 1:2 hydrochloric acid, and wash the paper with about 50 ml. of hot water. Add 1 gram of ammonium chloride, and reprecipitate with 1:1 ammonium hydroxide. Filter through the same paper into the filtrate reserved in a 400-ml. beaker.

Be sure that the combined filtrates are approximately neutral, and add 3 ml. of concentrated hydrochloric acid. Proceed by the ferrocyanide titration method for zinc, described on page 505, starting with "Dilute to 250 ml., and heat to boiling."

ZINC OXIDE IN THE PRESENCE OF ZINC SULFIDE. Follow the procedure for zinc oxide in lithopone (page 507).

Acid-soluble Calcium. This will normally be calcium carbonate, calcium sulfate, or both. Use the filtrate from the precipitation of zinc sulfide if available. Otherwise, substantially the same steps must be taken for removal of possible interfering elements although the precipitates need not be ignited and weighed. Concentrate, if the volume is excessive, and add 1 ml. of concentrated ammonium hydroxide. Complete as on page 216, starting with "Heat the sample solution to boiling, and add 20 ml. of boiling ammonium oxalate solution . . . ". If strontium is present, which is rare, it will be titrated with calcium.

Acid-soluble Magnesium. Magnesium carbonate is readily acid-soluble. Talc and asbestine show an appreciable solubility on heating with hydrochloric acid. The sample to be used is the filtrate from acid-soluble calcium. Follow the method on page 218.

Carbon Dioxide. If the pigment contains carbonates but no sulfide, determine carbon dioxide by the evolution method (page 224).

If both carbonates and sulfides are present, mix 1 to 2 grams of potassium dichromate intimately with the pigment to oxidize the sulfides when acidified. To ensure complete removal, it is advisable to pass the gas through a small wash bottle containing copper sulfate solution prior to drying and absorption.

Total Acid- and Water-soluble Sulfur. This determination includes sulfides, sulfites, and sulfates other than barium sulfate. If barium carbonate and a soluble sulfate are present, they will react in acid solution to produce barium sulfate and the acid- and water-soluble sulfate determination will be in error. In that case, correction must be made in the calculations.

To 1 gram of pigment in a 400-ml. beaker add 10 ml. of concentrated hydrochloric acid saturated with bromine and 5 grams of ammonium chloride. Cover, and digest on a steam bath for 5 minutes. Dilute to 200 ml., and boil for 5 minutes. Filter off any insoluble matter, and thoroughly wash the residue on the paper with hot water. Add 10 per cent sodium hydroxide solution carefully under the cover glass until nearly neutral. Complete the neutralization with dry sodium carbonate, and add 2 grams in excess. Boil for 10 to 15 minutes, and wash the cover glass. Let settle, and filter off the insoluble carbonates. Wash them with hot water. Redissolve the precipitate in hot 1:1 hydrochloric acid, and reprecipitate with dry sodium carbonate as before. Filter, and wash thoroughly. Acidify the combined filtrates with hydrochloric acid, adding 2 ml. in excess. Boil to expel the bromine, at the same time concentrating or diluting to about 100 ml. All the sulfur is now present as sulfate. Determine gravimetrically as on page 160, starting at "Bring to a boil, and add with constant stirring . . .".

Acid-soluble Sulfate. To 1 gram of pigment add 10 ml. of concentrated hydrochloric acid and 5 grams of ammonium chloride. Boil until the hydrogen sulfide is gone, adding more acid if necessary. Dilute with hot water to 200 ml., and boil for 5 minutes. Filter off any insoluble matter, and wash thoroughly with hot water. Complete as in the preceding method, starting at "Add 10 per cent sodium hydroxide solution carefully under the cover glass until nearly neutral." There is no bromine to boil off when acidified in a later stage of the process.

Sulfide Sulfur. Determine this by one of the evolution methods given on pages 163 to 165, and 171.

Water-soluble Matter. To 2.5 grams of pigment add 100 ml. of water, and boil. Transfer to a 200-ml. calibrated flask, cool, and dilute to volume. Allow to settle, and pour the clear liquid through a dry filter paper, discarding the first 20 ml. of filtrate. Transfer 100 ml. of the filtrate to a weighed dish, and evaporate to dryness on a steam bath. Dry for 1 hour at 105 to 110°, cool, and weigh. Determine the nature of the solids. Calcium sulfate is soluble to the extent of 0.20 gram in 100 ml. of water at 20° and if present in the pigment will be extracted here.

Calculation of Results on Mixed White Pigments. Calculation of analytical data to the commercial pigments present is an involved process at best.

Commercially, one rarely finds more than one white pigment and one or two extenders. But the provision for calculation of mixtures must take into account possible more complex cases. With complex mixtures, it is often necessary to make several trial calculations to determine on what basis the results balance. All determinations are subject to some errors. In a series of calculations a series of errors may be accumulated in the same figure, both analytical errors and necessary approximations in calculation of such commercial mixtures as white lead and basic lead sulfate. As in so many other cases, judgment and experience are the best guide.

Commercial silicas usually contain relatively small amounts of iron and aluminum. Clay is aluminum silicate, but the ratio of aluminum to silica varies over a considerable range in natural clays and cannot always be calculated to a stoichiometric ratio. Add any small amounts of aluminum and iron found in the acid-soluble portion to the acid-insoluble siliceous matter. Alternatively, if silica was determined as such, add any small amounts of aluminum and iron to it provided that they amount to less than 5 per cent of the silica. Report this as *silica*. In dealing with the silica fraction, microscopic examination may serve to identify the presence of diatoms; in this case, part or all of the silica may be *diatomaceous earth*.

If the amount of iron and aluminum, of which the iron is only a minor part, is over 20 per cent and not over 60 per cent of the silica, add the two values and report the total as *clay*, or *natural aluminum silicate*. If the aluminum exceeds 60 per cent of the silica, or, more often, if considerable aluminum is present but no substantial amount of silica, then a part or all of the aluminum should be reported as *alumina*, or *aluminum oxide*.

A small amount of acid-soluble magnesium will usually correlate with a large amount of acid-insoluble magnesium as magnesium silicate. The total of magnesium from all sources and of silica may then agree in ratio fairly closely to $H_2Mg_3(SiO_3)_4$, which indicates the presence as *talc*. With somewhat less magnesium the form is probably as *asbestine*.

A large amount of acid-soluble magnesium will usually correlate with carbonates and show *magnesium carbonate*. The magnesium is likely to be slightly deficient from the stoichiometric proportion owing to a small amount of calcium present as impurity in the carbonate. Similarly, if acid-soluble calcium and carbon dioxide are present, but no acid-soluble sulfate, probably the values will correlate for *calcium carbonate*, or *whiting*, which may be contaminated with magnesium carbonate. This may go quite high as in dolomitic limestone, which contains about 4 parts of magnesium carbonate to 3 of calcium carbonate. A small amount of calcium may be present in siliceous matter and extracted from it with acid.

Calculate the sulfide to zinc sulfide and the equivalent as zinc oxide. Subtract that amount of zinc oxide from the total present to indicate the amount left for later distribution. Zinc may be present without sulfide but not sulfide without zinc in white pigments.

If barium sulfate is present but no zinc sulfide or titanium oxide, report the amount as *barium sulfate*, or *barytes*. When zinc sulfide is present, multiply the value by 2.33 to estimate the equivalent amount of barium sulfate to form lithopone. If the zinc sulfide and barium sulfate approximate this ratio within 10 per cent, report the sum of the two as *lithopone*. When the zinc sulfide is found to be in substantial excess to this ratio, while it is possible that zinc sulfide as such has been added as well as lithopone, it is more probable that an error has occurred in the analysis. When the barium sulfate is in substantial excess over that equivalent to the zinc sulfide, report the excess over the stoichiometric proportion as barium sulfate. Added barium sulfate is not so improbable as added zinc sulfide.

When titanium oxide is present without barium sulfate, calcium sulfate, or magnesium silicate, report it as *pure titanium oxide*. If barium sulfate is present, multiply the titanium oxide by 2.33 to determine the equivalent amount of barium sulfate. If it agrees within 10 per cent, report the sum of titanium oxide and barium sulfate as *titanium oxide–barium sulfate* pigment. Similarly, if the calcium sulfate corresponds approximately to 2.33 times the titanium oxide, the probable composition of the sum of the two is *titanium oxide–calcium sulfate* pigment. Likewise, if the percentage of magnesium silicate is 2.33 times that of titanium oxide, then *titanium oxide–magnesium silicate* pigment should be reported.

If carbon dioxide is present and lead is absent, the calcium carbonate, barium carbonate, and magnesium carbonate will usually account for it. Conversely, with lead and carbon dioxide present and the carbonates just mentioned absent, calculate each to basic lead carbonate of the composition $(PbCO_3)Pb(OH)_2$ to see if they agree within 10 per cent. If they do, accept the higher of the two or the one that for reasons connected with the analysis appears the more accurate and report it as *basic lead carbonate*, or *white lead*. Commercially this often deviates considerably from that formula so that the best result obtainable must be an approximation.

Basic lead sulfate and white lead are rarely present in the same mixture. If lead and sulfate are both present, the lead will usually be in excess. Calculate the amount of sulfate to lead sulfate and the balance of lead as lead oxide. They will usually approximate a ratio of 78:16. If they agree with this within 10 per cent, use the sum of lead sulfate and lead oxide for further calculation. Multiply this by 0.058 to obtain the amount of zinc oxide normally present in basic lead sulfate. Then the sum of lead sulfate, lead oxide, and the zinc oxide so calculated is *basic lead sulfate*. This pigment averages 78.5 per cent lead sulfate, 16.0 per cent lead oxide, and 5.5 per cent zinc oxide. Excess of lead oxide over the ratio for basic sulfate and absence of carbon dioxide would indicate an error in the analysis. Lead oxide as such is never reported in a white pigment other than one in which all the ingredients are reported as oxides rather than probable compounds. There may be acid-soluble sulfate remaining at this stage.

Deduct from the total the zinc oxide in the basic lead sulfate found, and the balance is *zinc oxide* present as such. Lithopone and lead pigments are not used together because contact with any acid would darken the mixture, lead sulfide being formed by reaction.

As in water analysis, there is a standard arbitrary method of calculation, but it is far less dependable than that applied to minerals in water. Calculate the acid-soluble sulfate to calcium sulfate. If the calcium sulfate does not use all the acid-soluble sulfate, calculate the remainder to basic lead sulfate. Conversely if the calcium sulfate does not use all the calcium, calculate the balance to calcium carbonate. Then, in turn, balance excess carbon dioxide against available lead as white lead. If excess carbon dioxide remains, calculate to magnesium carbonate or barium carbonate according to the acid-soluble positive radical available.

The cases considered in these calculations cannot take in all the possibilities and are largely confined to mixtures of paint pigments and extenders. Adaptations must often be made to fit the multitude of possible mixtures for various purposes. A trace of acid-soluble barium can be barium carbonate in calcium carbonate or a trace of solubility of barium sulfate in the relatively high acid concentrations used in some of the determinations.

The total carbon dioxide should have been accounted for in white lead, barium carbonate, magnesium carbonate, and calcium carbonate. The total lead should have been accounted for in white lead or basic lead sulfate. If the sample contains barium, calcium, or magnesium carbonate and basic lead sulfate, or calcium sulfate and basic lead carbonate, it is possible that they occur in such proportions that positive and accurate identification of the mixture is not possible from the analysis.

If acid-soluble sulfate is present but no carbonate or if the carbon dioxide has all been used elsewhere, calculate sulfate not used as basic lead sulfate to calcium sulfate, and check it against the amount of calcium available. If such calculations agree within 10 per cent, report as *calcium sulfate,* which will normally be the unhydrated material, $CaSO_4$. Since calcium sulfate is appreciably water-soluble, this determination serves as a cross-check on its presence. While the determination of water-soluble matter may not account for all the calcium sulfate if the content is large, the method can then be modified by use of a smaller sample to obtain a cross-check.

If antimony is present, calculate it as antimonious oxide, Sb_2O_3. Report the loss at 110° as such, and consider it in the total of pigments accounted for. This total should be within 2 per cent of 100 per cent. Practically, unless for some special reason, the total accounted for is usually much better and is almost as likely to run over 100 per cent as under it. Report water-soluble matter, but do not consider it in the total. Report loss on ignition and consider it in terms of a rough cross-check on the compounds reported, but do not consider it in the total accounted for. Sulfites are so rarely present that they have not been considered.

CHAPTER 29

COLORED PIGMENTS AND MINERAL COLORS

Colored pigments or mineral colors in themselves present no serious problem. The difficulty arises in connection with the commercial products, which are likely to involve all the problems of a mixture of white paint pigment and extender plus a colored pigment. Often the colored pigment is present in only a minor amount.

The colored pigments are usually readily identified by the experienced worker. Therefore, they will be taken up roughly in the sequence of the spectrum and either qualitative tests for each given, prior to the quantitative method, or an indication as to where the analysis will fit into the scheme for mixed white pigments.

Red Lead. Red lead varies in color from orange to red and has the composition Pb_3O_4, present as Pb_2PbO_4. It is expensive and is thus rarely used as a color in minor admixtures. Red lead as a straight pigment or slightly toned in order to differentiate successive coats is used as red lead in oil for protecting structural steel against corrosion. The toning agent is often a small amount of carbon black. Qualitatively it is detected by extraction of the pigment with 1:1 nitric acid, neutralization, and precipitation as the sulfate or chromate. This may be confirmed by some of the pigment in 5 per cent acetic acid liberating iodine on addition of potassium iodide solution.

Total Lead. Boil 1 gram of sample with 15 ml. of 1:1 nitric acid. This will dissolve the bivalent lead of plumbous plumbate, Pb_2PbO_4, but will leave lead dioxide. Add 3 per cent hydrogen peroxide slowly and carefully as a reducing agent until the brown color of the lead dioxide disappears. Filter from the insoluble material, which may be analyzed as directed for acid-insoluble material with reference to mixed white pigments (page 509). If carbon black was present as toner, this will be found in the insoluble matter. Analyze the filtrate for lead as either the sulfate or the chromate. For determination as the sulfate, add 20 ml. of 1:1 sulfuric acid, and proceed as on page 501, starting at "Evaporate to fumes of sulfur trioxide." For determination as the chromate dilute the filtrate to 100 ml., neutralize with 1:1 ammonium hydroxide, and then add a few drops of glacial acetic acid. Proceed as on page 501, starting with "Heat to boiling, and add 10 to 15 ml. of a boiling 10 per cent solution of potassium dichromate." The process may be completed either gravimetrically or volumetrically.

LEAD PEROXIDE. Dissolve the lead from 0.5 gram of the pigment, more if this is low in red lead, in 5 ml. of 5 per cent acetic acid containing 1.2 per cent of potassium iodide and 10 grams of sodium acetate. Warm as little as feasible, and filter from any insoluble matter. Wash with small amounts of warm water, and dilute the filtrate to 25 ml. Do not use more than the specified amount of potassium iodide, or lead iodide may precipitate. Titrate at once with standard sodium thiosulfate solution (page 730). Tetravalent lead is reduced to the bivalent form, liberating iodine which is titrated.

CALCULATION.

$$\text{Ml. titration} \times \text{normality of sodium thiosulfate} \times 0.1196 \times \frac{100}{\text{weight of sample}} = \text{per cent lead peroxide}$$

Calculate the amount of lead oxide equivalent to the lead peroxide in red lead by the ratio $2PbO:PbO_2 = 1.866:1$

$$\text{Per cent lead peroxide} \times 1.866 = \text{per cent lead oxide present as red lead}$$

Calculate the lead peroxide to its equivalent as lead oxide.

$$\text{Per cent lead peroxide} \times 0.9331 = \text{per cent lead oxide equivalent to lead peroxide}$$

If the sum of the lead peroxide calculated as lead oxide and the lead oxide to combine with it as red lead agree within 10 per cent with the total lead, assume that all the lead present is present as red lead. To obtain the total red lead, take the sum of the lead peroxide and the total lead less that present as lead peroxide but calculated to lead oxide.

If the lead present as lead oxide and as lead peroxide is far from the proper ratio of red lead, basic lead carbonate or lead chromate will probably be found present. Each has a comparatively simple characteristic test.

Vermilion. Vermilion is mercuric sulfide, a brilliant red but expensive pigment. Adulteration is fairly common; thus, the amount of a commercial vermilion used may differ from the amount of mercuric sulfide found by analysis. The pigment is very resistant to attack by acids or alkalies. For qualitative detection, the pigment should be heated in a test tube with sodium carbonate. Globules of mercury will condense on the upper part of the test tube and can be dissolved in 1:1 nitric acid and qualitatively identified by the usual reactions of mercury. As negative indications, alcohol and water should not extract organic dyestuffs, and carbon disulfide should not extract free sulfur.

MERCURIC SULFIDE. If other pigments volatilized or oxidized at dull red heat are absent, heat a 1-gram sample in a well-ventilated muffle. The loss in weight less the loss at 110° is the mercuric sulfide, to a low order of accuracy.

MERCURY AS THE SULFIDE. Dissolve a 1-gram sample by heating with 6 ml. of concentrated hydrochloric acid and 2 ml. of concentrated nitric acid. Dilute this solution to about 25 ml., and filter from any insoluble matter. A substantial amount of insoluble matter is probably barium sulfate. Dilute the filtrate to 100 ml. in a calibrated flask.

Since the solution has been obtained by dissolving the mercury sulfide in *aqua regia*, it is not possible to determine the mercury by precipitation with hydrogen sulfide. Boiling off the nitric acid in the presence of hydrochloric acid would induce appreciable loss of mercury by volatilization as mercuric chloride.

Nearly neutralize an aliquot, containing about 0.1 gram of mercury, with sodium carbonate, and treat with a slight excess of fresh ammonium sulfide solution. Rotate the solution and add 20 per cent sodium hydroxide solution until the dark liquid begins to lighten. Heat the solution to boiling and add more sodium hydroxide solution until the liquid is clear. This procedure forms sulfo salt, $Hg(SNa)_2$. For each 100 ml. of solution add 5 grams of ammonium nitrate, and boil until the ammonia is almost entirely expelled. Allow the precipitate to settle. The following reaction takes place:

$$Hg(SNa)_2 + 2NH_4NO_3 = 2NaNO_3 + (NH_4)_2S + HgS$$

Decant the liquid through a weighed Gooch crucible, and wash the precipitate by decantation until the washings give no reaction with silver nitrate solution. Transfer the precipitate to the crucible. As free sulfur is likely to be present, without drying the crucible wash it three times with alcohol. Then place it on a glass tripod in a beaker containing some carbon bisulfide. Cover the beaker with a round-bottomed flask containing cold water, and place the whole assembly on a hot water bath. Allow the carbon bisulfide to reflux and condense on the flask, dropping onto and through the crucible. After 1 hour the sulfur is extracted. Remove and wash the crucible once each with alcohol and ether. Finally dry the precipitate at 110°, cool, and weigh. The weight is directly that of mercuric sulfide. To convert to mercury, the factor is 0.8622.

Red Lakes. Many of the brilliant red colors are dye lakes precipitated on an inorganic base. This will usually be barium

TABLE 34.—REACTIONS OF COMMON RED DYES USED AS LAKES

Dye	Concentrated sulfuric acid	Concentrated hydrochloric acid	Concentrated sodium hydroxide	Alcohol	Concentrated sodium hydroxide and alcohol
Alizarin	Dark brown with reddish undertone becoming light yellow on diluting	Muddy color	Dark reddish brown; little change on diluting	Insoluble	Purplish; dark brown on diluting
Eosin	Orange to yellow; fluorescent solution with excess sodium hydroxide	Changes to yellow; reddish fluorescence with excess sodium hydroxide	Little change; fluorescent on diluting	Reddish fluorescent solution	
Paranitraniline	Purple; color returning on diluting	Color slightly darkened; lighter on diluting	Color changed to brownish red; light red on diluting	Slight yellowish orange solution	Purple; color returning on diluting
Orthoanisidine	Purple; red on diluting	Dark purplish red; lighter on diluting	Little change	....	Little change
Scarlet 2R	Darkened; lighter on diluting	Darkened; lighter on diluting	Reddish solution on diluting	Slight reddish solution	Color darker; reddish solution on diluting

sulfate, calcium carbonate, or clay but is occasionally red lead or zinc oxide.

Table 34 gives the reactions of a few of the more common reds. The amount of dye present is quite small. Only an estimate by difference is usually necessary, the base and dye being identified and a commercial grade of dye deposited on that base being then depended on to give a color match.

Iron Pigments. By far the largest variety of the less brilliant colors is obtained with iron earth pigments and iron oxides. Browns, yellows, and reds are often due to ochers, siennas, umbers, and natural or synthetic iron oxide pigments. Ordinarily iron will dissolve as a whole or at least in part on boiling with 1:1 hydrochloric acid, but occasionally one encounters a form such as a dead-burned sienna that dissolves only on fusing with potassium acid sulfate.

Ochers and siennas are mixtures of oxides and silicates, sometimes with organic matter. Umbers tend more toward being entirely iron oxide, but even natural iron oxides contain some silicate. The synthetic iron oxides are produced by calcining of ferrous sulfate, may contain more or less undecomposed iron sulfates, and in some grades are associated with a stoichiometric amount of calcium sulfate.

Of these iron pigments, the ochers are yellow to light brown, the siennas and umbers brown to dark brown, the natural iron pigments red to black, and the synthetic iron pigments yellow, pink, red, light and dark brown, and even black.

Analytically, small amounts are determinable with aluminum oxide and aluminum silicate as in the scheme for mixed white pigments. The ratios of iron oxide to silicates are so variable that no specific ratios can be predicted. Variation will be from practically 100 per cent iron oxide to less than 40 per cent. Since many of the iron pigments are relatively inexpensive, they may be present with little or no other pigment, or extender. Therefore, methods of analysis are provided for straight iron pigments. Manganese pigments resemble the iron browns, and greater or lesser amounts of manganese are usually present. Therefore, the methods include determination of manganese. The procedures are generally similar to those applied to iron ores.

Loss on Ignition. Ignite a sample, dried at 110°, at well below red heat to avoid conversion to a form difficultly soluble in acid.

IRON. *Permanganate Method.* Add 25 ml. of concentrated hydrochloric acid to the ignited sample, and cover with a watch glass. Heat for about 2 hours on a steam bath. Add 10 ml. of 1:1 sulfuric acid, washing down the watch glass. Evaporate first on the steam bath and then on a hot plate, until fumes of sulfur trioxide are evolved and all the hydrochloric acid is removed. Cool, dilute with 80 ml. of water, and boil to dissolve the iron sulfate. Filter through a retentive paper into a 250-ml. calibrated flask, and wash thoroughly with hot water.

If the residue is not colorless, fuse with sodium carbonate in a platinum crucible and dissolve in water. Acidify with sulfuric acid, and evaporate to sulfur trioxide fumes. Cool, dilute with water, and filter, adding the filtrate to the calibrated flask containing the previous filtrate. Reserve the insoluble residue.

Dilute the iron solution to volume, and take a 50-ml. aliquot as sample. Follow the usual procedure for iron by either the permanganate or the bichromate method, starting on page 194 at "Wash out the reductor with 1:20 sulfuric acid . . . ".

ANALYSIS OF INSOLUBLE RESIDUE. If the residue is white and contains barium sulfate, ignite, weigh, and analyze it as insoluble matter by the procedure for mixed white pigments, starting on page 510 with "Mix the residue containing barium sulfate and possibly silicates in a platinum crucible . . . ".

If the residue is free from barium sulfate, ignite and weigh. Then moisten with a few milliliters of 48 per cent hydrofluoric acid and a few drops of concentrated sulfuric acid. The loss on evaporating and igniting will be silica. Any substantial residue will either be barium sulfate, indicating that the alkaline fusion should have been used, or aluminum and possibly iron oxides, indicating that the acid extraction of the sample was incomplete. In the latter case, the residue must be fused with potassium acid sulfate or pyrosulfate, dissolved in 1:10 sulfuric acid and added to the iron solution. The quantitative figure for silica will be of value only if the residue after treatment with hydrofluoric acid is small.

MANGANESE. *Volhard Method.* Transfer a 100-ml. aliquot of sample, prepared for the iron determination, to a conical flask. Add 10 per cent sodium hydroxide solution dropwise until nearly neutralized. Complete the neutralization by the addition of an excess of a concentrated suspension of zinc oxide in water. Agitate vigorously until the iron is completely precipitated. Heat to boiling, and titrate to a faint pink color with 0.1N potassium permanganate. During the titration, zinc manganite is precipitated and contains all the manganese in quadrivalent form. The reaction is

$$4KMnO_4 + 5ZnSO_4 + 6MnSO_4 + 14H_2O = 4KHSO_4 + 7H_2SO_4 + 5[Zn(OH)_2.2MnO_2]$$

Some experience is necessary to obtain the end point in this determination.

When freshly agitated before titration, the contents of the flask are white and opaque. As manganese is titrated, they become brown. The pink end point is seen, after mixing the contents of the flask well, in the shallow layer at the surface from which the precipitate has settled.

CALCULATION.

$$\text{Ml. titration of sample} \times \text{normality of permanganate} \times 0.02608 \times \text{aliquot} \times \frac{100}{\text{weight of sample}} = \text{per cent manganese dioxide}$$

The factor for manganese is 0.01648.

Williams Method. Evaporate 100 ml. of the sample solution prepared for the iron determination until salts start to crystallize. Add 5 ml. of concentrated nitric acid, and evaporate nearly to dryness, or until hydrochloric acid has been completely volatilized. Add 75 ml. of concentrated nitric acid and 5 grams of solid potassium chlorate. Boil for 15 minutes, and add 50 ml. more of concentrated nitric acid and 2 grams of potassium chlorate. Boil until yellow fumes cease to come off; then cool in ice water. The manganese is precipitated as manganese dioxide by the reaction $Mn(NO_3)_2 + 2KClO_3 + H_2O = MnO_2,H_2O + 2KNO_3 + 2ClO_2$. Filter on a Gooch crucible, and wash with colorless concentrated nitric acid. Suck dry, and wash with water until free from nitric acid.

Prepare a ferrous sulfate solution containing 10 grams of the crystallized salt in 900 ml. of water and 100 ml. of concentrated sulfuric acid. Transfer the precipitate with the asbestos to a beaker. Add a known excess of the ferrous sulfate solution to the precipitate, an amount gauged according to the size of the precipitate, and an identical amount of the ferrous sulfate solution to another beaker. Stir the solution with the sample until the brown color of manganese dioxide has disappeared by the reaction

$$MnO_2 + 2FeSO_4 + 2H_2SO_4 = MnSO_4 + Fe_2(SO_4)_3 [Fe_2(SO_4)_3] + [2H_2O]\, H_2O$$

Titrate both amounts of ferrous sulfate with $0.1N$ potassium permanganate.

CALCULATION.

$$(\text{Blank} - \text{sample}) \times N \times 0.04347 \times \text{aliquot} \times \frac{100}{\text{weight of sample}} = \text{per cent } MnO_2$$

The factor for manganese is 0.02747.

Colorimetric Method. In some cases, for example, in rubber pigments, the presence of amounts of manganese of the order of 10 ppm. is of major importance. In the determination of such small amounts, use any one of the three colorimetric methods described on pages 198 to 201 applied to aliquots of the solution prepared for the determination of iron.

COPPER. *Colorimetric.* Traces of copper are often important. In pigments to be used in rubber it has about one-fifth the deleterious effect of manganese. Use an aliquot of the solution prepared for the iron determination by the procedure described on page 206.

Calculation of Iron Pigments. In general, the percentage of iron oxide plus that of silica is taken to be the iron pigment. With only a very small amount of iron and a large amount of silica the latter may in part be present as free silica. Another type of pigment is that containing calcium sulfate. If the iron content is low and calcium sulfate is suspected, the latter is easily extracted from a fresh sample with 1:1 hydrochloric acid by the technic described on page 509 for the separation of acid-soluble and acid-insoluble matter. Then dilute the filtrate to a known volume, and determine sulfate as follows: Dilute the solution to about 250 ml., and bring to a boil. Add sufficient boiling 10 per cent barium chloride to precipitate all the sulfate and a small excess. The solution may be too acid for quantitative precipitation. Therefore, nearly neutralize the solution with 1:1 ammonium hydroxide, and then add 2 ml. of concentrated hydrochloric acid. Continue as on page 160, starting at "The barium sulfate as originally precipitated is in too finely divided form for filtration."

Successful interpretation of the nature of the pigment depends largely on experience. Manganese would normally be shown as manganese dioxide present as impurity.

Chrome Yellow. Lead chromate, a pigment of brilliant yellow tone, is often known as chrome yellow. A deep orange tone is obtained in qualities containing basic chromate and is often called chrome orange. Lead chromate has high covering power and is therefore usually present with white pigments. The most commonly associated pigment is lead sulfate; but barium sulfate or calcium sulfate is entirely satisfactory, and even barium phosphate is occasionally encountered. For qualitative detection, lead chromate can be extracted from a pigment with concentrated hydrochloric acid. On neutralizing the solution and adding silver nitrate, it first reacts to precipitate silver chloride, and then an excess shows a brick red precipitate of silver chromate.

Acid-insoluble Matter. Heat 1 gram of the pigment in a covered beaker with 20 ml. of concentrated hydrochloric acid, for 15 minutes on a steam bath. Add 100 ml. of water, and boil. All the pigment should dissolve unless barium sulfate or other insoluble matter such as silica is present. Filter, and wash well with hot water, reserving the filtrate and washings for determination of lead and chromium. Ignite and weigh the insoluble matter. Analyze as for insoluble residue on page 527.

Lead. Calcium Sulfate Absent. Evaporate the filtrate from the insoluble matter to about 2 ml., but do not take to dryness. If this occurs, take up in more 1:1 hydrochloric acid and reevaporate. Add 100 ml. of 95 per cent alcohol, stir well, and let stand for 1 hour in ice water. Filter on a Gooch crucible, and wash with small amounts of 95 per cent alcohol. Reserve this filtrate and washings for determination of chromium. It will be subsequently referred to as the lead chloride filtrate. Dry the lead chloride at 110° for 1 hour, cool, and weigh. The lead chloride is soluble to the extent of 0.0085 gram per 100 ml. of 95 per cent alcohol.

CALCULATION.

$$(\text{Weight of lead chloride} + 0.0085) \times 0.8026 \times \frac{100}{\text{weight of sample}} = \text{per cent lead as lead oxide}$$

CHROMIUM. CALCIUM SULFATE ABSENT. Evaporate the filtrate from the separation of lead chloride to dryness on a steam bath to drive off all the alcohol. Dissolve the residue in 50 ml. of 1:10 hydrochloric acid by warming. Cool, transfer to a 250-ml. calibrated flask, and dilute to volume. This solution is also to be used for other determinations.

Transfer a 100-ml. aliquot to a conical flask, and add sodium peroxide in small amounts until the solution is just alkaline. This will oxidize the chromium back to chromate. Evaporate to about 20 ml., which will provide for decomposition of the excess of sodium peroxide. Cool, and add 1:1 sulfuric acid until neutral and 2.5 ml. in excess. Add an excess of a 1 per cent ferrous sulfate solution in 10 per cent sulfuric acid to the sample, and transfer an identical amount of the same ferrous sulfate solution to another flask. The reaction with bichromate is

$$Na_2Cr_2O_7 + 6FeSO_4 + 7H_2SO_4 = 3Fe_2(SO_4)_3 + Na_2SO_4 + Cr_2(SO_4)_3 + 7H_2O$$

Titrate each with $0.1N$ potassium permanganate solution. The green color of reduced chromium in the sample makes the end point somewhat difficult to identify. Alternatively, the titration may be with potassium dichromate.

CALCULATION.

$$(\text{Sample} - \text{blank}) \times N \times 0.03334 \times \frac{250}{100} \times \frac{100}{\text{weight of sample}} = \text{per cent } CrO_3$$

The factor for calculation as lead chromate is 0.10774.

Calculate the chromate to chromic acid and to lead chromate. The difference between the two is the combined lead oxide in the lead chromate. This should approximately check the determination of total lead if lead chromate is the only lead compound present. Otherwise, subtract the amount of lead oxide as lead chromate from the total lead oxide to give that present as other lead compounds, probably as the sulfate.

ZINC. CALCIUM SULFATE ABSENT. Pipet a 50-ml. aliquot of the solution, prepared from the lead chloride filtrate for the determination of chromium, into a 400-ml. conical flask. Neutralize with 1:1 ammonium hydroxide, make just acid with 1:1 hydrochloric acid, and add 6 ml. excess. Proceed as for ferrocyanide titration (page 505), starting with "Dilute to 250 ml., and heat to boiling."

SULFATE. CALCIUM ABSENT. Use 50 ml. of the solution prepared for the determination of chromium. Dilute to 100 ml., and proceed as on page 160, starting at "Bring to a boil, and add with constant stirring sufficient boiling 10 per cent barium chloride solution . . . ". The factor for barium sulfate to lead sulfate is 1.2992; to lead oxide, 0.9563. Subtract the lead oxide as lead sulfate from that remaining from the lead chromate. The balance should check the excess lead oxide in basic lead sulfate within 10 per cent (page 503).

LEAD. CALCIUM SULFATE PRESENT. The method for determination of lead as chloride cannot be applied in the presence of calcium sulfate, which would precipitate lead sulfate under the conditions prevailing. Dissolve the pigment, and filter from insoluble matter. Dilute the filtrate to about 400 ml., and cool. Pass in hydrogen sulfide rapidly until the precipitation of lead as lead sulfide is complete. Slow addition of hydrogen sulfide may cause some precipitation of chromium. Filter and wash the lead sulfide with hydrogen sulfide water. Reserve this filtrate for the determination of zinc, chromium, and calcium.

Dissolve the lead sulfide from the filter with 10 ml. of hot 1:4 nitric acid, and wash it well with the hot acid. Wash the filter with hot water until free of acid. Add 10 ml. of 1:1 sulfuric acid, and complete as for the analysis of white lead as the sulfate on page 501, starting with "Evaporate to fumes of sulfur trioxide."

CALCIUM. CALCIUM SULFATE PRESENT. Transfer the filtrate from the precipitation of lead sulfide to a 500-ml. calibrated flask. Add 1:1 ammonium hydroxide until slightly alkaline, and pass in hydrogen sulfide until precipitation is complete. Dilute to volume, mix well, and let settle. The zinc and chromium are in the precipitate. Pipet off 100 ml. of the clear supernatant liquid for the determination of calcium, and set the flask and contents aside for the determination of zinc and chromium.

Render this sample distinctly acid with 1:1 hydrochloric acid, and boil until the hydrogen sulfide is removed. Then proceed as on page 216, starting with "Heat the sample solution to boiling, and add 20 ml. of boiling ammonium oxalate solution saturated at room temperature."

MAGNESIUM. CALCIUM SULFATE PRESENT. Determine magnesium on the filtrate from calcium as described on page 218, starting with "Add 10 ml. of 10 per cent disodium hydrogen phosphate solution . . . ".

ZINC. CALCIUM SULFATE PRESENT. When no further aliquots are needed for calcium, add 1:1 hydrochloric acid to the flask containing the zinc and chromium precipitate until neutral and 12 ml. in excess. Dilute to volume, and use a 100-ml. aliquot for the titration of zinc as the ferrocyanide as on page 505, starting with "Dilute to 250 ml., and heat to boiling."

CHROMIUM. CALCIUM SULFATE PRESENT. Transfer another 100-ml. aliquot of the solution in which zinc and chromium are present to a conical flask. Neutralize with 1:1 hydrochloric acid, and add 2 ml. in excess. Boil gently to drive off hydrogen sulfide. Add sodium peroxide in small amounts until the solution is just alkaline. Complete as when calcium sulfate is absent, as on page 530, starting with "Evaporate to about 20 ml., which will provide for decomposition of the excess of sodium peroxide."

SULFATE. CALCIUM SULFATE PRESENT. Dissolve 0.5 gram of the pigment in 10 ml. of 1:1 hydrochloric acid, dilute to about 50 ml. with water, and filter off insoluble matter. Wash the paper very thoroughly with hot water, and dilute the filtrate and washings to about 250 ml. Complete as on page 160, starting with "Bring to a boil, and add with constant stirring sufficient boiling 10 per cent barium chloride solution . . . ". Calculate the result as sulfur trioxide. This should check the amounts of sulfur trioxide present as lead sulfate and calcium sulfate.

Zinc Chromate. Zinc chromate, or zinc yellow, is a yellow pigment less commonly used than chrome yellow. The double salt approximates the composition $(ZnCrO_4)_3.K_2Cr_2O_7$. It is soluble in strong ammonium hydroxide, and qualitative tests on such an extract may be expected to establish its presence. Because it is expensive, white pigments and extenders also may be expected to be present.

INSOLUBLE MATTER. Weigh out 1 gram of oven-dried pigment, or more if the content of zinc chromate is expected to be small, and stir well with 25 ml. of concentrated ammonium hydroxide. Add 100 ml. of water, and stir frequently for about 30 minutes. Do not heat. Filter through a tared, ashless filter paper, and wash thoroughly with 1:4 ammonium hydroxide and then with water. Reserve the filtrate for the determination of potassium, zinc, and chromium. Dry the paper with the residue at 110°, and weigh.

This residue may be analyzed for mixed white pigments and extenders according to the methods of Chap. 28.

SODIUM AND POTASSIUM. Heat the filtrate from insoluble matter to boiling, and transfer to a 500-ml. calibrated flask. Heat in that flask to boiling, and slowly pass in hydrogen sulfide until precipitation of zinc as the sulfide and reduction of chromium to the chromic form and precipitation as the hydroxide is complete. Cool, dilute to volume, and mix well. Let settle, and pipet off 100 ml. of the clear supernatant liquid. Reserve the flask with the precipitate and residual solution for determination of zinc and chromium.

Add 1 ml. of concentrated sulfuric acid to the clear aliquot, and complete as on page 219, starting with "Evaporate on a water bath and then in an oven . . . ". The method is invalidated if other substances, soluble in concentrated ammonium hydroxide and not precipitated by ammoniacal hydrogen sulfide, are present. The result may otherwise be assumed to be the potassium present in zinc chromate. The factor as potassium oxide is 0.5406; as chromium trioxide, 1.1474.

Chromium. When the alkali metals have been determined, add 1:1 hydrochloric acid to the flask containing chromium and zinc until neutral and 5 ml. in excess. Heat to boiling until hydrogen sulfide has been driven off, cool, and dilute to volume.

Proceed with an aliquot as described for the analysis of lead chromate on page 530, starting at "Transfer a 100-ml. aliquot to a conical flask, and add sodium peroxide in small amounts until the solution is just alkaline." In addition to calculation as chromium trioxide as there provided, the resulting value multiplied by 1.6768 will give the percentage of zinc chromate, the commercial double salt.

Zinc. Use a 50-ml. aliquot of the solution prepared for the determination of chromium. Neutralize with 1:1 ammonium hydroxide, render just acid with 1:1 hydrochloric acid, and add 6 ml. in excess. Proceed as for ferrocyanide titration of zinc on page 505, starting with "Dilute to 250 ml., and heat to boiling."

The per cent as zinc oxide multiplied by 2.229 gives the per cent as zinc chromate. The percentages of potassium, zinc, and chromium trioxide so obtained should balance with the commercial salt within a 10 per cent variation; in this case the sum of potassium oxide, chromium trioxide, and zinc oxide may be accepted as a sufficiently accurate measure of the zinc chromate present.

Cadmium Sulfide. Cadmium sulfide is itself a yellow pigment, sometimes used as such. A yellow to orange cadmium lithopone is sold in which cadmium sulfide replaces zinc sulfide. A complex cadmium sulfide and selenide is red. The presence of these pigments can be established by extracting with 1:4 hydrochloric acid, which destroys the color, nearly neutralizing the solution and precipitating the characteristic cadmium sulfide. Because of their limited use, methods for cadmium pigments are not included here.

Prussian Blue. This blue pigment is theoretically ferric ferrocyanide, $Fe_4[Fe(CN)_6]_3$, produced by precipitation of a ferric salt with a ferrocyanide. In practice, the commercial products are mixtures more or less approximating the formula of *Williamson's blue*, $KFe[Fe(CN)_6]$, with sometimes other iron and aluminum cyanide complexes. Therefore, alkali metal is commonly present. Numerous desired shades result from differences in the method of manufacture. A so-called *soluble Prussian blue* results from adding oxalic acid or ammonium oxalate to the pulp color before drying and is water-dispersible. *Potash blue* is another form resulting from mixing a ferrocyanide with ferrous sulfate, followed by oxidation of the resulting ferrous ferrocyanide. It is likely to be higher in alkalies and is found mainly in carbon paper. The corresponding ferrous ferricyanide, $Fe_3[Fe(CN)_6]_2$, *Turnbull's blue*, is rarely found.

Qualitatively, all these blues are decomposed by acidifying strongly. On heating the acid solution the characteristic odor of hydrocyanic acid can often be detected. The presence of iron in the acid solution from their decomposition is easily demonstrated after neutralizing. A red shade may be due to added red lake, detectable by alcohol extraction of the original pigment. Analysis of the product is only one step in grading; often, tinting strength tests are more important.

Loss on Ignition. Weigh out 1 gram of the oven-dried pigment, and ignite at the lowest temperature that will suffice to remove the last traces of blue. Too high a temperature will often render the resulting iron oxide very difficultly soluble. Cool, and weigh.

Acid-insoluble Matter. Treat the residue from ignition with 15 ml. of concentrated hydrochloric acid, cover with a watch glass, and digest for 1 hour on a steam bath. Remove the watch glass, and evaporate to a sirupy consistency. Add 75 ml. of water, and heat to boiling. Filter, and wash well, reserving the filtrate for the determination of iron. Ignite and weigh the insoluble matter. This will usually be barium sulfate but may be any inert filler. It should be white unless an acid-insoluble pigment was mixed with the Prussian blue, carbon black for example. Usually it may be analyzed by the procedure for acid-insoluble matter in mixed white pigments described on page 510, starting with "Mix the residue containing barium sulfate and possibly silica or silicates in a platinum crucible with ten times its weight of anhydrous sodium carbonate."

Iron. Add 10 ml. of 1:1 sulfuric acid to the filtrate from the insoluble matter, and evaporate, first on the steam bath and then on a hot plate until fumes of sulfur trioxide are evolved and all the hydrochloric acid is removed. When cool, add about 50 ml. of water, and boil to ensure complete solution of the iron salts. There should be no appreciable insoluble matter. Dilute to 250 ml., take a 50-ml. aliquot as sample, and determine the iron by the permanganate method as described on page 194, starting with "Wash out the reductor with 1:20 sulfuric acid . . . ".

The amount of iron present in the dry pigment should be not less than 30 per cent if it is unreduced Prussian blue. The total iron times 3.03 approximates the commercial Prussian blue present.

ALUMINUM. Usually this is not determined. If necessary, precipitate the combined iron and aluminum oxides from a 50-ml. aliquot of the solution prepared for the determination of iron, as described on page 192, starting with "To 50 ml. of the sample solution, or other aliquot . . . ". Calculate the iron, determined previously, as ferric oxide, and subtract from the combined iron and aluminum oxides to give the amount of aluminum oxide. Such a difference method does not possess a high degree of accuracy.

NITROGEN. Weigh out 1 gram, and determine nitrogen by the Kjeldahl-Gunning method, as described on page 143, starting with "Introduce the sample into a 500-ml. Kjeldahl flask." At least 3 hours' digestion is normally required. The percentage of nitrogen so found multiplied by 4.4 approximates the amount of commercial Prussian blue present. It should cross-check the amount found from the total iron within a tolerance of less than 10 per cent.

SULFATE. This analysis is not usually carried out. If necessary, ignite and treat another sample, as described for the separation of insoluble matter. Precipitate iron and aluminum oxides as described on page 192, starting with "To 50 ml. of the sample solution, or other aliquot . . . ". Discard the precipitate, and dilute the filtrate in a calibrated flask to 250 ml. To a 100-ml. aliquot add 1:1 hydrochloric acid until acid and 4 ml. in excess. Complete as described on page 160, starting with "Bring to a boil, and add with constant stirring sufficient boiling 10 per cent barium chloride solution . . . ".

SODIUM AND POTASSIUM. This analysis is not usually necessary. For the determination, take 100 ml. of the solution prepared for sulfate. Add an excess of 1:1 sulfuric acid, usually about 5 ml. Complete as on page 219, starting with "Evaporate on a water bath and then in an oven . . . ".

After weighing the sulfate, determine qualitatively whether the alkali is sodium or potassium, and calculate the amount equivalent to the alkali sulfate found. This is the alkali metal present in the double alkali-iron-ferrocyanide compounds.

Ultramarine. This is a blue pigment of indefinite composition produced by roasting such mixtures as kaolin, soda ash, sulfur, charcoal, and silica. Sodium sulfate is sometimes added. The finished product is a complex silicate and sulfide of aluminum and sodium. It dissolves in hydrochloric acid with evolution of hydrogen sulfide, which readily distinguishes it from Prussian blue. Owing to the presence of sulfides, it is not found admixed with lead compounds. It breaks down on long heating with alkaline solutions. While chemical analysis is of little significance, relatively speaking, procedures are available. A typical range of compositions is as follows:

	Per Cent
Silica	39–43
Alumina	24–30
Sodium oxide	17–21
Sulfur	10–16

SILICA. Heat 1 gram of pigment with 20 ml. of 1:1 hydrochloric acid in a covered casserole until decomposition is complete. Evaporate to dryness and proceed as on page 189, starting with "If it is necessary to expedite the operation, evaporate . . .".

ALUMINUM. Dilute the filtrate from the separation of silica to about 100 ml., and proceed as on page 192, starting with "To 50 ml. of the sample solution, or other aliquot . . .".

CALCIUM. If calcium is present, precipitate it in the filtrate from the determination of aluminum as on page 216, starting with "Heat the sample solution to boiling, and add 20 ml. of boiling ammonium oxalate solution saturated at room temperature."

SODIUM. Render the filtrate from the determination of aluminum or of calcium distinctly acid with 1:1 sulfuric acid, and proceed as on page 219, starting with "Evaporate on a water bath and then in an oven . . .".

TOTAL SULFUR. Mix 1 gram of pigment with 4 grams of sodium carbonate and 4 grams of sodium peroxide in a nickel crucible. A platinum crucible would be attacked. Cover the fusion mixture with about 1 gram of sodium carbonate, and place in a hole in an asbestos shield to protect the crucible from combustion gases. Fuse well; and when all reaction ceases, let cool. Dissolve the residue in hot water, and dilute to about 100 ml. Neutralize with 1:1 hydrochloric acid, and add 2 ml. excess. Filter to remove insoluble matter if present, and wash the filter well. Dilute to about 250 ml., and continue as on page 160, starting with "Bring to a boil, and add with constant stirring sufficient boiling 10 per cent barium chloride solution . . .".

SULFUR AS SULFATE. Boil 1 gram of pigment with 10 ml. of 1:1 hydrochloric acid until sulfide fumes are no longer evolved. Dilute with hot water to 50 ml., and filter from insoluble matter. Wash well on the filter, and dilute the filtrate and washings to about 250 ml. Complete as on page 160, starting with "Bring to a boil, and add with constant stirring sufficient boiling 10 per cent barium chloride solution . . .".

SULFUR AS SULFIDE. Subtract the sulfate calculated as sulfur from the total sulfur to give sulfide.

Cobalt Oxide. Pure cobalt oxide is occasionally used as a blue pigment. Because of its rarity, methods of analysis are not given.

Smalt is a blue cobalt glass formed by fusion of cobalt oxide, silica, and potash. Ultramarine has largely displaced it.

Copper Phthalocyanine. A characteristic bright blue pigment is copper phthalocyanine, a complex organic material containing copper. It is very stable to light and heat. Aside from a rough approximation from the copper determination, methods of analysis are not readily available.

Chrome Green. The usual green pigment is called chrome green and is a mixture of Prussian blue and lead chromate, identifiable by tests already given. Other mixtures of blue and yellow pigments are possible, but the only one likely to be encountered is a mixture of ultramarine and zinc chromate.

A pure chrome green should contain only Prussian blue, lead chromate, and lead sulfate, the varying amounts determining the shade of green. Greens made by precipitating the ingredients together are superior to those made by mixing the blue and yellow separately. To distinguish the two types, put the pigment in oil, and allow to stand. A coprecipitated green will show no blue, but a green made by mixing will show yellow, blue, and green. Microscopically, a well-made green will show almost all green, with some blue particles, but no yellow. A badly made green will show yellow, blue, and green particles.

As different grades of lead chromate vary considerably in shade and as this is true to a somewhat limited extent with Prussian blue, actual determinations of the three major ingredients are not so important as analysis for extenders such as barium sulfate, silica, and clays.

Insoluble Matter. Weigh 1 gram of sample into a small casserole, and add 10 ml. of water. Add 1 gram of sodium peroxide in several small portions, stirring well with each addition. This decomposes the Prussian blue. Add 25 ml. of water, and heat to boiling for a few minutes to decompose excess of sodium peroxide. Add 1:1 hydrochloric acid until distinctly acid, and bring to boiling. Filter and wash the residue, saving the filtrate for the determination of sulfate.

Transfer the precipitate back to the casserole with 25 ml. of water, and add 1 gram of ammonium chloride and a drop or two of 1:1 ammonium hydroxide. This dissolves the last traces of lead salts. Filter and wash, combining this filtrate and washings with those previously obtained.

Ignite and weigh the insoluble matter. The residue will usually be barium sulfate and silica but may contain any inert filler. Usually it can be analyzed by the procedure for acid-insoluble matter in mixed white pigments, as described on page 510, starting with "Mix the residue containing barium sulfate and possibly silica or silicates in a platinum crucible with 10 times its weight of anhydrous sodium carbonate."

SULFATE. Dilute the filtrate from acid-insoluble matter to about 250 ml., and proceed as on page 160, starting with "Bring to a boil, and add with constant stirring sufficient boiling 10 per cent barium chloride solution . . . ".

LOSS ON IGNITION. Weigh 1 gram of sample in a porcelain crucible, and ignite at as low a temperature as will decompose all the blue color. Cool, and weigh.

LEAD. Add 15 ml. of concentrated hydrochloric acid to the residue from ignition, and digest on a steam bath for 1 hour. Cool, dilute to about 300 ml., and without filtering to remove acid-insoluble matter pass in a rapid stream of hydrogen sulfide to precipitate lead. Filter, and wash well with water saturated with hydrogen sulfide, saving the filtrate for the determination of iron, aluminum, calcium, and magnesium.

Determine lead in the precipitate containing lead sulfide as described on page 531, starting with "Dissolve the lead sulfide from the paper with 10 ml. of hot 1:4 nitric acid, and wash it well with the hot acid." Calculate as lead oxide.

IRON. Boil the filtrate from precipitation of lead sulfide until hydrogen sulfide is no longer given off. Proceed as on page 192, starting with "To 50 ml. of the sample solution, or other aliquot . . . ". The addition of excess of bromine water is required in this case to oxidize the iron to the ferric state. Use the double precipitation, and save the filtrates and washings for determination of calcium and magnesium.

Dissolve the precipitated but unignited iron and aluminum hydroxide in 100 ml. of 1:10 sulfuric acid, and dilute to 250 ml. Take a 50-ml. aliquot for the determination of iron by the permanganate method described on page 194, starting with "Wash out the reductor with 1:20 sulfuric acid and then with water . . . ". Save the balance of the solution for the determination of aluminum and chromium.

CHROMIUM. Use a 100-ml. aliquot of the foregoing solution for the determination of chromium as in that for lead chromate described on page 530, starting with "Transfer a 100-ml. aliquot to a conical flask, and add sodium peroxide in small amounts until the solution is just alkaline." For lead chromate the value is desired as chromic acid. For later correction of the mixed hydroxide precipitate to determine aluminum, multiply this by 0.76 to get the percentage as chromous oxide.

ALUMINUM. Precipitate the mixed hydroxides of iron, aluminum, and chromium from a 50-ml. aliquot of the solution according to the technic on page 192, starting with "To 50 ml. of the sample solution, or other aliquot . . . ". Double precipitation is not required, and the filtrate should be discarded. From the total weight of oxides so obtained subtract the iron oxide and chromium oxide as determined separately. The difference is a rather inaccurate estimate of the aluminum oxide present.

CALCIUM. If needed, determine on the filtrate from precipitation of iron and aluminum, as on page 216, starting with "Heat the sample solution to boiling, and add 20 ml. of boiling ammonium oxalate solution saturated at room temperature."

MAGNESIUM. If needed, determine on the filtrate from calcium determination by the method described on page 218.

NITROGEN. Determine on a fresh sample as described for Prussian blue on page 535, using the factor given there to estimate the Prussian blue content.

Chromic Oxide. This is a very stable, relatively expensive green pigment. Extenders are likely to be present in it as used. Heat 6 grams of potassium bisulfate in a platinum crucible until all frothing ceases. Carefully add 0.5 gram of the pigment, and gradually increase the heat until a clear melt is obtained or reaction ceases. Cool, dissolve in water, and dilute to 250 ml. Use a 100-ml. aliquot for the determination of chromium, as for that of lead chromate described on page 530, starting with "Transfer a 100-ml. aliquot to a conical flask, and add sodium peroxide in small amounts until the solution is just alkaline." Be sure sufficient peroxide is added to oxidize the chromium completely. The factor for calculation of the titration to chromic oxide, Cr_2O_3, is 0.02534.

CHAPTER 30

OIL AND WAX EMULSIONS

The possible ingredients in emulsions are widely varied—all types of saponifiable and unsaponifiable oils; essential oils; abrasives, in particular the finer ones; pigments both white and colored; soaps and alkaline salts; alcohols and solvents. Necessarily, therefore, the methods in this chapter in many cases refer in a general way to other chapters rather than give specific details. This is particularly true with respect to qualitative examination.

OIL EMULSIONS

From a theoretical standpoint the number of possible different oil emulsions is endless. Sometimes this possibility seems approached as a reality, when one considers the number of commercial preparations that contain some material classifiable as an oil in greater or lesser amount as a dispersed phase or as a dispersion medium.

All emulsions are fundamentally classified according to two types, oil in water and water in oil. For practical reasons, in operation they are more conveniently further classified in respect to the content of dispersed inorganic solids. For convenience of discussion in this chapter, they are separately considered according to the following types:

Oil in water.

Oil in water with added inorganic solids.

Water in oil.

Water in oil with added inorganic solids.

If sufficient wax is present so that a product is predominantly a wax dispersion, it is excluded from this classification and considered later.

Many emulsions are of such complexity as to test the ingenuity, knowledge, and technic of the worker. The methods of this chapter are necessarily not exact because of the large possible variations in percentage and proportion of ingredients. The analyst must often in a sense adopt a research technic, that is, base each step on the behavior of and results in the previous step. He must feel his way *about* and then *through* and be always ready to change his course when necessary.

QUALITATIVE EXAMINATION

Type. Simple qualitative tests indicate the type of emulsion. All oil-in-water emulsions are miscible with water because water is the external phase. If, on mixing, the emulsion is broken, the amount of emulsifying agent is either small or inefficient. If water is not the external phase, oil is. The sample is then miscible with the same type of oil as is present as the external phase. This is usually but not always a petroleum fraction.

Detection of dispersed inorganic solids is not usually difficult. Plain oil-in-water emulsions on moderate dilution usually produce a translucent emulsion. With dispersed solids present they are normally more opaque. Differentiation is often possible by means of microscopic examination or by the use of a centrifuge. The same generalizations apply to dispersions in oil when diluted with oil.

Solvent. If distillable solvents other than water are present, certain steps in quantitative examination usually precede qualitative examination. That is, some ingredients, such as alcohol, can ordinarily be detected and confirmed only by quantitative means. For this purpose, distill a portion of sample quantitatively, using the technics explained in greater detail on page 555. The distillate may or may not separate into two layers. For completeness, it will be assumed that two layers are obtained. Whether the aqueous layer is on top or at the bottom will indicate whether its density is greater or less than that of water. Withdraw portions of each layer after determining the volume of each. Determine the density of each, and apply suitable qualitative tests. Usually the water-insoluble layer will float and will be found to be a petroleum type. The water layer may be solely water as confirmed by its density, or it may contain ethyl or isopropyl alcohol or acetone as shown qualitatively and by density. Special cases of solvents partitioning between the layers must be watched for as in the case of ether and butyl alcohol. Fortunately, these are readily detected by odor. If the amount of water is relatively small and the insoluble layer has a high boiling point, it is to be expected that this will be only partly distilled, a partial steam distillation. Also, the density of the fraction distilled will not be the true density of the commercial material used.

Emulsifying Agents. The available emulsifying agents are constantly increasing in number. Some of the most common types follow.

1. Soaps. For oil-in-water emulsions, these may be sodium, potassium, ammonium, triethanolamine, or morpholine soaps. The negative radicals may be the fatty acids of stearate, oleate, or linoleate types derived from tallow, corn, cottonseed, or other common liquid oils or linseed or perilla oils. Ricinoleate soaps derived from castor oil are rarely used. For water-in-oil emulsions, calcium or occasionally other alkaline-earth or heavy-metal soaps of these same fatty acids may have been used. In general, water-soluble soaps produce oil-in-water emulsions, water-insoluble soaps produce water-in-oil emulsions. Oil-in-water emulsions stabilized by soap are broken down by rendering them slightly acid with mineral acid and heating.

Lack of a pink color on addition of phenolphthalein solution to an emulsion, either as received or on dilution, is not necessarily evidence of the absence of soap. Triethanolamine soaps do not give a pink color even on dilution. Therefore, in the absence of alkalinity but with other evidence of the presence of soap, a hot titration as described in connection with quantitative methods (page 389) is indicated. Possible presence of salts that are not alkaline to phenolphthalein but that hydrolyze in the range of pH 4 to 8 must be considered as an alternative; unadjusted sodium metaphosphate or disodium phosphate are examples. Fortunately, these are rarely met with in emulsions of this type.

Sodium, potassium, or ammonium soaps may also be present as acid soaps and may not react to phenolphthalein. This can result from the addition of soap and a later addition of fatty acid to reduce the pH below 8.0. These emulsions will usually be relatively unstable. Such a failure to react to phenolphthalein usually occurs when a relatively small amount of soap is present.

The presence of sodium and potassium soaps can be detected by the experienced worker by means of the flame test. Substantial amounts of ammonia soaps give off some ammonia on simple heating, usually with evidence of the instability of the emulsion. When heated with excess of sodium hydroxide solution, they liberate ammonia freely. This can be detected by odor, red litmus paper, or a glass rod dipped in 1:1 hydrochloric acid. Triethanolamine soaps under similar conditions yield small but detectable amounts of ammonia odor on considerable heating. Morpholine soaps yield less ammonia odor than ammonium soaps but more than triethanolamine soaps under these conditions. A detailed test for ammonia is given on page 86.

The mere presence of sodium, potassium, or volatile alkalinity is not evidence of a soap. Free alkalies or salts of these alkalies may be present. There may also be an excess of these over the amount to form a soap. In general, in conjunction with a prompt breaking down of the emulsion on rendering it slightly but definitely acid and heating, these tests do show that some form of soap is present.

2. Sulfated Oils, Synthetic Sulfated and Sulfonated Substances. Commonly miscalled sulfonated oils, sulfated oils are detectable by the fact that, on boiling with hydrochloric acid and filtering, the filtrate gives a test for sulfate. Usually, inorganic sulfate is present and may be a clue to the presence of sulfated oils, if no other source of inorganic sulfate is indicated. The detection of these compounds is more fully detailed in Chap. 18.

Other sulfated or sulfonated organic compounds may occur. Strictly speaking, they have no great emulsifying ability but are present as surface-tension depressants in order that the sample may readily wet the substance on which it is to be used. Sulfonated compounds do not hydrolyze on heating with acid, but these commercial products also usually contain sodium sulfate. Not more than about 1 per cent of such a product is commonly present. Such products can be recognized only as belonging to a type—sulfated or sulfonated compounds. Chapter 18 gives further information on the synthetic sulfated or sulfonated organic compounds, as distinct from sulfated oils.

3. Water-dispersible Gums. While the number of water-dispersible gums is large, those which are used commercially are limited. Gum karaya is by far the most common because of its low cost. Gum tragacanth is similar and not readily distinguishable at this stage of the work. Occasionally gum arabic is used, the term being applied here to various types similar in properties, having a high degree of water solubility, and chemically indistinguishable. Pectin also falls into this classification. Alginic acid and its sodium salt are used.

Ordinarily indication of the presence of gums is obtained either by failure to find other emulsifying agents or sufficient emulsifying agent to account for the stability of the emulsion. When the alcohol content is greater than about 70 per cent, the water-soluble gums precipitate as a flocculent mass, or gel, a device for detecting them.

The specific identification of water-dispersible gums is best left until they are separated in the course of the quantitative procedure. They should then be tested according to the following scheme: These reactions are given by water solutions prepared from the gum as it is sold commercially and often depend on the concentration. Thus 5 ml. of 0.5 per cent solution of locust kernel gum is jellied by a few drops of 4 per cent borax solution, but 5 ml. of 0.2 per cent solution is not jellied by this reagent. Table 35 gives the reaction of solutions of the gums that vary in concentration from 0.5 to 1 per cent. To make proper use of the table, care must be taken to notice the distinguishing characteristics of the precipitate formed, whether it is voluminous, flocculent, stringy, powdery, curdy, filamentous, etc. The word "gel" in the table indicates an actual jellied condition. The word "gelatinizes" implies that the mixture has been thickened. These are clear distinctions and are easily noticeable in making the tests. The special reagents used are prepared as indicated in Chap. 39.

The procedure in general is to add 2 to 3 drops of the reagent to 5 ml. of the test solution, note the result, and then add an excess of the reagent. In the case of some of the reagents, namely, potassium hydroxide, phosphoric acid, hydrochloric acid, and ferric chloride, a few drops of the reagent are added to the test solution and the mixture boiled. Then an excess of the reagent is added, and the mixture is boiled again. For noting the type of alcohol precipitate, the technic of Weinberger and Jacobs[1] is recommended. To 20 ml. of the test solution add 70 ml. of 95 per cent alcohol drop by drop, with constant stirring until precipitation is complete. Note the texture, quality, and the characteristics of the precipitate, as well as the point at which definite precipitation begins.

4. Starches and dextrines do not give very effective emulsification and are not often used for this purpose. Identification is by addition of a dilute solution of iodine in potassium iodide. Starches give a deep blue coloration. Dextrin produces a purplish red coloration. They are precipitated by alcohol in the way that

[1] Weinberger, Walter, and Jacobs, Morris B., *J. Am. Pharm. Assoc.*, **18**, 34 (1929).

TABLE 35.—QUALITATIVE REACTIONS OF COMMERCIAL GUMS

Gum	Stokes's acid mercuric nitrate reagent	Neutral lead acetate 20% soln.	Basic lead acetate	Potassium hydroxide 10% soln.	Neutral ferric chloride 5% soln.	Alcohol precipitate	Borax, 4% soln.	Millon's reagent	2 vols. acetic acid
Arabic	White, fine, opaque ppt., soluble in excess of reagent	No ppt.	White curdy ppt., insoluble in excess	Faint yellow tinge	Ppt., soluble in excess	Very fine, flocculent, nonadherent, 40-ml. point of definite pptn.	Negative	Yellow fine ppt.	Negative
Tragacanth	Voluminous, flocculent, translucent ppt.	Voluminous flocculent ppt., gels	Voluminous ppt., gels	Bright yellow stringy ppt.	Gelatinizes	Coagulated, long and stringy, adherent, 10 ml.	Negative	Voluminous flocculent ppt.	Negative
Agar-agar	Gelatinizes	Flocculent ppt. gels	Voluminous ppt.	Clarifies soln.	Gelatinizes, heat + excess ppt.	Heavy flocculent adherent to beaker, 20 ml.	Negative	Fine ppt.	Ppt.
Karaya	White curdy ppt., settles rapidly	Negative	Stringy ppt., settles rapidly	Negative	Ppt. coagulates on heating	Fine filamentous particles, nonadherent, 15–20ml.	Negative	Fine ppt.	Negative
Irish moss	Gelatinizes	Flocculent ppt., gels	Voluminous flocculent ppt., gels	Gels	Voluminous stringy ppt., gels	Coagulated, translucent, stringy, adherent, 20 ml.	Negative	Fine ppt.	Negative
Quince seed	Voluminous, flocculent, yellowish ppt.	Yellowish flocculent ppt., gels	Yellowish, voluminous, flocculent ppt., gels	Stringy ppt.	Stringy ppt.	Coagulated short, stringy, nonadherent, 25 ml.	Negative	Flocculent ppt.	
Locust kernel	Gelatinizes	Voluminous ppt., gels	Gels	Slight flocculent ppt.	Stringy ppt.	Stringy, clotty, opaque, nonadherent, start, 2 ml., complete, 15 ml.	Gels	Fine ppt.	Ppt.
Ghatti	No ppt.	No ppt.	Translucent flocculent ppt.	Negative	Negative	Fine flocculent ppt., nonadherent	Negative	Fine ppt.	Negative
Pectin	Flocculent ppt., gelatinizes	Flocculent ppt., gelatinizes	Flocculent ppt., gelatinizes	Ppt.	Ppt.	Flocculent ppt., translucent		Flocculent ppt.	Ppt.

From Chemical Analysis of Food and Food Products by Morris Jacobs.

gums are. They often contain a little nitrogen, only that due to the gluten present if a crude starch or a flour is used. The active type of starch present, such as wheat, potato, corn, rice, etc., is identified by microscopic observation. In the case of solubilized starches, the granules are likely to be so broken up that identification will be very difficult.

Oils. UNSAPONIFIABLE. Petroleum naphthas, recognizable by odor and ready inflammability on heating, are present as wetting, cleaning, and penetrating agents. Light mineral oils are used to produce a polish and may be present alone or in conjunction with a petroleum naphtha. If, after boiling off the naphtha, acidification shows an oil floating on top of the liquid, mineral oil is indicated provided that this layer is not fatty acid liberated from soap. In polishes, mineral oils of specific gravity around 0.82 are common.

Chlorinated hydrocarbons, discussed in Chap. 23, may be present and are recognizable by odor. They do not often occur in emulsions because of the difficulty of dispersing high-specific-gravity liquids. Turpentine has a distinguishing odor and is volatile with steam.

SAPONIFIABLE. In polishes only raw, blown, and sulfonated castor oils need ordinarily be considered. They give excellent luster, whereas other saponifiable oils do not. They are extracted and identified during the quantitative work. In other types of emulsion, almost any of the saponifiable oils may be present. In general, however, saponifiable oils in emulsions are relatively rare because they are comparatively expensive.

Pigments. These may be present to body the emulsion or to color, whiten, or blacken it. Among those which are used are zinc oxide, zinc sulfide, titanium oxide, iron oxide, carbon black, clay, talc, and chalk. Clearly, the heavier pigments are not suitable for use in emulsions, for they will not suspend easily. Identify by the usual chemical means, using a qualitatively ashed sample or the pigment obtained during the analysis. Ashing may alter the composition. See also Chaps. 27 to 29 and 5.

Abrasives. The most commonly used abrasives are silica, whether crystalline, semicrystalline, or amorphous, and diatomaceous earth. Feldspar and pumice are not uncommon. As the purpose of the abrasive is to clean without scratching, a very fine mesh material is usually necessary. A suitable product for cleaning furniture and automobiles will all pass through a 200-mesh sieve and leave at the most 1 or 2 per cent on a 325-mesh sieve. Identification is made microscopically (page 177), usually on the abrasive obtained during the quantitative work or, less frequently, on a separate qualitatively ashed sample.

Cover Odors. A good sense of smell is the best asset in detecting cover odors. They are present in too small a proportion to separate quantitatively. Nevertheless, they may cover the characteristic odor of another ingredient. Amyl acetate, oil of citronella, and lemon-grass oil are most frequently added. In a few older products nitrobenzene, known as oil of mirbane, gives a characteristic, somewhat almondlike odor. Blended perfumes are occasionally used in more expensive types of emulsion.

Water-soluble Liquids. The various alcohols soluble in water are tested according to the methods described in Chap. 24. Glycerin, which is often present in polishes to prevent too rapid evaporation, remains after boiling off the water. The water-soluble liquids also are best detected during the course of the quantitative examination.

Various Ingredients. Many unimportant substances, usually representing manufacturers' idiosyncrasies, may be present in minor amounts. In general, they do not add to the utility of the product and may be neglected in an analysis. However, always watch each step in the work carefully for unexpected results that may give a clue to unsuspected substances. Always read all the statements on the container, but do not always believe them.

QUANTITATIVE

Qualitative examination will have classified the type of the emulsion. In general, the methods that follow are based on extractions on the sample without obtaining total solids, the loss due to spattering and volatilization and the alteration of any ingredients due to oxidation being thus obviated. A quantitative ash is rarely needed. If it would be useful, this is indicated in the method.

Oil-in-water Emulsion Free from Suspended Solids. Usually white emulsions, these commonly contain a light mineral oil with perhaps some petroleum naphtha, sulfated castor oil as emulsifier, and sometimes castor oil. One or 2 per cent of ammonia or triethanolamine soap may replace the sulfated oil, and, less probably, sodium or potassium soaps. Water-soluble gums, alone or with soaps, occasionally occur in such emulsions used for industrial purposes. When the emulsion is to be administered internally, the emulsifying agent is usually a water-dispersible gum.

Separation. To 25 ml. of the sample in a 25-ml.-capacity narrow-necked graduated cylinder, add 0.5 ml. of concentrated hydrochloric acid. Shake thoroughly. Usually the emulsion is broken and on standing will separate into 2 or more layers. The upper layer normally consists of the mineral oil, together with any petroleum naphtha; the lower layer consists of the water, hydrochloric acid, and gums if present. If no clear separation occurs, centrifuge the cylinder, taking care to pack it sufficiently. Any soap present is decomposed by the acid, and the resulting fatty acid is in the oil layer.

If sulfated oil or blown castor oil is present, there may be three layers, the sulfated or blown oil forming a layer between the other oil and the water. With chlorohydrocarbon present, the sulfated or castor-oil layer may be on the bottom and in any event will be easily recognized by its different color and consistency.

An alternative procedure for samples giving indefinite separations is to place the cylinder in a can of cold water and gradually heat to near, but not to, boiling. Cool the cylinder and contents before reading. By any of these methods, sulfated oil is not decomposed, for no heating occurs until the acid is practically separated from the oil. Read the volumes, and calculate percentages by volume of the various oil layers after these have been allowed to stand overnight. Remember that the sulfated oil layer will be the acid sulfated oil and not its normal sodium salt and also that commercial sulfated oils usually contain 50 to 80 per cent of active ingredients, the rest being water and inorganic salts.

For further investigation, decompose a larger sample in a separatory funnel, using the method of heating in water after the main separation has occurred. When well separated, slowly run off the hot aqueous acid layer. Save this for determination of water-soluble matter. The viscous saponifiable oils flow more readily when hot. Allow to stand overnight, for saponifiable oils sometimes dissolve in hot mineral oil but separate on cooling and standing. Drain off any further aqueous layer. Do not attempt to wash the saponifiable oil layer if sulfated, or emulsions will form. Determine constants on this layer according to Chaps. 17 and 18 and depending on what qualitative tests have shown to be present. Some soap fatty acids may be present, and these cannot be separated from acid sulfated oils. When separations do not occur readily and when both sulfated and saponifiable oils are present, use the modified method described later (page 549).

A sample of the mineral oil, free from most of the saponifiable oil together with any petroleum naphtha, is obtained by pipetting from the upper layer. Steam-distill if naphtha is present. Extract the residue with ether to separate the mineral oil, and obtain such constants as are necessary according to Chap. 23. An acid value (page 341) near zero indicates no fatty acid from soap. Calculate any substantial acid value to fatty acid and to the corresponding soap.

When fixed alkali soaps are present, determine the alkalinity on a carefully ashed sample according to page 390, and calculate to the corresponding soap. Check by hot titration of an original sample to a methyl orange end point (page 389). If sulfated compounds are present, the first method is not applicable, and the second method in error by the amount of soap present in the commercial sulfated oil. Any alkalinity due to salts must be taken into account. Such alkaline salts will be included both in the ash and in the titration of the original sample.

Separation by Extraction. Weigh the sample into a separatory funnel, and dilute with water. Make just alkaline with 50 per cent potassium hydroxide solution to reduce hydrolysis of soap, and proceed as described with reference to unsaponified matter in oils on page 349, starting at "Add 75 ml. of ethyl ether." The extract will contain any petroleum naphtha, sulfated oil, and saponifiable oil. Separate, or estimate by constants as described in Chaps. 17, 18, or 23.

Acidify the aqueous layer after the ether extraction, and extract the fatty acids present, as such and as soap, with ether. Examine this extract as described in Chap. 17. If desired, obtain an acid value on an original sample, using a large excess of alcohol to dissolve or disperse it. Calculate the titration to free fatty acid. Deduct from the combined fatty acids, and calculate the remainder to the appropriate soap. The acid layer contains the water-soluble matter.

Modified Separation Method. To 16 ml. of the sample in a 25-ml. narrow-necked graduated cylinder, add 8 ml. of 95 per cent alcohol. Make distinctly alkaline with 50 per cent potassium hydroxide solution, shake thoroughly, and centrifuge. Unsaponifiable oils separate on top, and the volume is read. Transfer to a separatory funnel, and return the alcohol-saponifiable oil layer to the cylinder. Wash the mineral oil with 5 ml. of 50 per cent alcohol, and add the washings to the cylinder. Make the contents of the cylinder just acid with hydrochloric acid. Any sulfated oil and most or all of the saponifiable oil, a small amount of which may be with the mineral oil, separate together. If necessary, centrifuge, and read the volume. Experience proves that no hydrolysis of sulfated compounds occurs when this method is used.

The sulfated oil may be separated from the saponifiable oil by the method for total sulfated fatty matter on page 375.

Water-soluble. Evaporate the aqueous layer from the acid separation, including any partly precipitated gums, to near dryness. Take up with warm 95 per cent alcohol, and filter or centrifuge off any precipitated gums. Dry, weigh, and identify these. Evaporate the alcohol-soluble fraction to incipient fumes. This will normally be glycerin or one of the glycols. Weigh and identify as on page 466.

Small amounts of alkali chlorides may be present. For most purposes they can be ignored; if not, determine by the chloride method (page 128). Fixed alkalies are determinable by titration of ash from which the equivalent chloride may be deducted. However, remember that some salt may be precipitated by alcohol and some may remain in solution. The use of absolute alcohol will minimize this solubility.

Oil-gum Emulsions. Such emulsions are indifferent and cheap polishes but important cathartics. Extract the oil from the diluted sample with ether. Obtain the gum by evaporating the aqueous layer to dryness. Identify the oil according to Chap. 37.

Oil-in-water Emulsions Containing Suspended Solids. Automobile polishes, cleaners, and combinations of the two are illustrative of this type. In general, they contain less oil than emulsions free from suspended solids. Sulfated oils or sulfated synthetic compounds may be present in minor amounts. Water-dispersible gums, particularly karaya, act as dispersing agents. Alcohol is occasionally present. A trend toward the inclusion of a petroleum naphtha as a cleaning agent and colloidal clay, or bentonite, as suspending agent has been noted. Glycerin is usually in solution, and a small amount of wax is often incorporated. The abrasive, frequently diatomaceous earth but sometimes a soft silica, completes the essential constituents of the composition. Minor quantities of other substances, as indicated at the beginning of this chapter, may be present. Soap is rarely an ingredient. Small amounts of cover odor are usually added.

Sources of error arise when analysis is made on so-called total solids. There are also difficulties due to emulsion formation, especially when bentonite or much gum is present, if wet extraction is used. The principle of the following method is based on the fact that water in the sample is removed by distillation with benzene. Hence no emulsions can form. The low boiling point of benzene and the fact that the oils do not come in direct contact with air ensure that no decomposition or oxidation occurs. The whole analysis is performed on one sample except for minor ingredients such as small amounts of amines, particularly ammonia and triethanolamine.

SEPARATION. The first distillation step takes care of solvents volatile with steam and is omitted if such solvents are absent.

Weigh 100 grams of the sample into a 500-ml. round-bottomed flask, using an accurate trip scale. If the sample does not contain much water—such products usually do—add about 50 ml. of water. Ordinarily connect up to the type of receiver shown in Fig. 106, which in turn is connected to a water-cooled condenser; but for volatile solvents such as the chlorohydrocarbons having gravities higher than that of water use the receiver shown in Fig. 107. An assembly using all ground-glass joints is desirable. Put

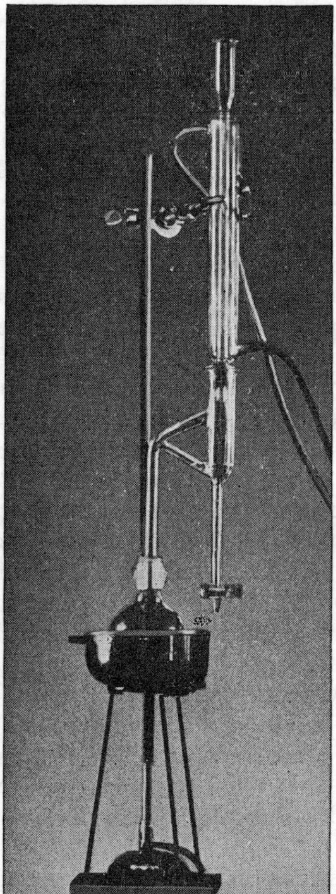

FIG. 106.—Apparatus for analysis of emulsions containing immiscible liquids lighter than water.

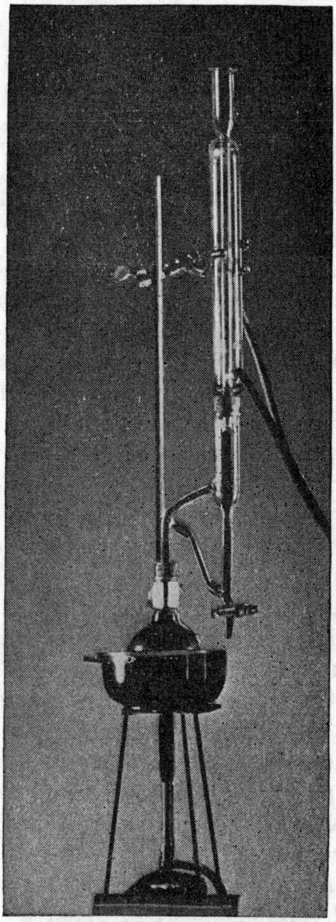

FIG. 107.—Apparatus for analysis of emulsions containing immiscible liquids heavier than water.

the flask in a cold oil bath, making sure that it does not touch the bottom of the bath and later overheat. Gradually heat the bath, and distill over any volatile solvent. Continue to distill, the water returning to the flask, until no more volatile solvent comes over. Read the volume of volatile solvent, which will usually be a petroleum naphtha, in the graduated tube. Allow the apparatus to cool, return the water in the receiver to the flask, and collect the volatile solvent. Identify in the usual manner by odor, specific gravity, and boiling point (Chap. 23).

Now add 100 ml. of benzene to the flask and contents, and connect to the receiver shown in Fig. 106. Using the same oil bath, reflux the benzene as fast as conveniently possible. Water distills over with the benzene and falls to the bottom of the receiver, while the benzene returns to the flask. At intervals draw off the water quantitatively into a 50- or 100-ml. volumetric flask. The process of distilling off all the water from the flask is somewhat long but may be greatly expedited by vigorously refluxing the benzene. Be careful not to overheat the contents of the flask. Although 2 hours or more may be necessary for this operation, little working time is consumed. The level of the water in the graduated receiver indicates when all the water has distilled over.

ALCOHOL. Dilute the collected water distillates to 50 or 100 ml., and obtain the specific gravity at a known temperature. If alcohol is present, it will be in this distillate and the amount can be calculated from specific-gravity tables. Confirm the presence of alcohol by odor or by the iodoform reaction. Isopropyl alcohol is often used. It is distinguished by its odor and immiscibility with saturated salt solution. For details, see Chap. 24.

OILS. The oils are now in benzene solution. Thoroughly shake the contents of the flask, and transfer to a 250-ml. centrifuge bottle. Rinse out the flask into the bottle with a little benzene. Centrifuge the mixture. No water being present, a sharp separation occurs. Decant off the benzene solution, add 50 ml. of warm benzene, and shake well after thoroughly breaking up any solids caked in the bottom of the bottle. Centrifuge again, and repeat the process until the benzene extract is perfectly colorless. Usually a total of three or four extractions is sufficient.

Combine the benzene extracts, and evaporate carefully. Drive off the final residues of benzene on a water bath, and dry the oils for a few minutes in an oven, avoiding overheating with consequent oxidation or loss by evaporation. Cool and weigh the oils. Then allow to stand overnight in a desiccator. If sulfated oil or castor oil is present, it will settle out as a viscous layer below the mineral oil.

Add chilled petroleum ether to the flask containing the benzene extract, and carefully swirl to dissolve the mineral oil. Decant, and repeat the operation as necessary. Cold, viscous, saponifiable oil will not dissolve appreciably in cold petroleum ether. Evaporate off the solvents to obtain, separately, the mineral and saponifiable oils. Analyze these according to the methods given in Chaps. 17 and 37.

WATER-SOLUBLE. Break up the residue in the centrifuge bottle, and shake well with 75 ml. of warm 95 per cent alcohol; SD3A will do. Cool to room temperature, and centrifuge. Decant the alcohol, and repeat the process twice more. Evaporate the combined alcohol extracts, finally heating in the oven to incipient fumes. Weigh and test for glycerin (page 466). In the improbable event that soap is present, it will be found here unless a volatile amine has been decomposed during the initial refluxing, and in this case the soap fatty acids will be extracted with the mineral oil. If soap is present, titrate in hot water solution to a methyl orange end point as on page 389, and then quantitatively extract the fatty acids with ether. Make such determinations on the fatty acids as appear necessary. Evaporate the water–hydrochloric acid layer to a small volume, and extract the glycerin from any chlorides present with warm 95 per cent alcohol (page 549).

Waxes. To the residue after the alcohol extraction add 50 ml. of hot trichloroethylene, break up the solids, and shake thoroughly. Centrifuge while still hot. Decant, and repeat the process twice more. Evaporate the combined trichloroethylene extracts. Finally heat to constant weight at 110°. Identify the extracted waxes by melting point, odor on burning, and other constants, as indicated in Chap. 22. Carnauba wax is the only fully satisfactory wax for use in polishing emulsions, but small amounts of other types may be found in other emulsions.

Water-dispersible Gums. After extracting the waxes, drive off any excess trichloroethylene from the centrifuge bottle, first on a water bath and then in the oven. Add boiling water to the hot centrifuge bottle, break up the solids, and shake thoroughly. Centrifuge hot, and decant the solution of gums. Repeat the process at least twice more, always using near-boiling water. Combine and evaporate the water extracts to obtain the water-dispersible gums. It is often convenient to dilute the extracts to a known volume and obtain total solids on an aliquot, testing the rest for the type of gum present.

Abrasive. The residue in the centrifuge bottle is now any abrasive or pigment present. Dry in the oven, conveniently overnight, and cool. Brush out, and weigh. Observe microscopically for the type of abrasive present (page 177), or analyze chemically according to Chaps. 10 or 27 to 29. A mesh analysis (page 70) is often informative.

Any bentonite used as a suspending agent will be in this abrasive. Usually 1 to 2 per cent is required to give a good emulsion. Its presence is indicated when little or no water-dispersible gum is found.

Amine. Occasionally an amine is present. If found qualitatively, determine separately by the Kjeldahl method (page 143), either directly on the sample or, better, on the filtrate after acidifying and heating the sample.

Water-in-oil Emulsions with and without Suspended Solids. Such products usually contain little water and are more properly called oil solutions and suspensions. The so-called oil may be a petroleum fraction, a coal-tar fraction, turpentine, or other solvents described in Chap. 23. Ammonia soap is commonly present.

Solvent. Weigh 50 or 100 grams of the sample into a distilling flask. Precipitate any soap by the addition of solid calcium chloride. This is essential, for otherwise foaming makes distillation impossible. Determine the distillation range in the usual manner. Exercise care to prevent decomposition and distillation of any free fatty acids present. An oil bath is convenient but not essential, for, with proper manipulation, it is easier to distill off the last fractions by using a bare flame. Any water and ammonia present distill over. Read the volume of distilled solvent. Wash with water, and obtain its specific gravity. Identify, or further fractionally distill, as in Chap. 25.

Soap Fatty Acids. Add hot 1:10 hydrochloric acid to the contents of the flask, and heat just to boiling, with vigorous shaking. Transfer the contents to a centrifuge bottle, and wash the flask with hot water, transferring the washings to the bottle. Cool the flask and bottle. Wash the flask several times with 10-ml. portions of ethyl ether, adding the ether to the centrifuge bottle. Shake the contents thoroughly, and centrifuge. Siphon off the ether layer into a separatory funnel, and repeat the extraction twice more, breaking up any solids each time. Wash the combined ether extracts free of acid, and evaporate. Drive off the last traces of ether, first on a water bath and finally for a few minutes in the oven. Cool, and weigh. Obtain an acid value and other constants as indicated in Chap. 17. Calculate to the appropriate soap. Small amounts of difficultly volatile solvent may be present; in this case the acid value will normally be less than 195. Correct the calculated soap accordingly.

Abrasive or Pigment. Drive off any residual ether in the centrifuge bottle by heating on top of the water bath, avoiding spattering. Add hot water, and shake. Centrifuge, decant, and discard the calcium chloride solution. Repeat the washing and centrifuging until free of chloride. Dry the abrasive or pigment in the over overnight. Cool, brush out, and weigh. Identify as indicated in the procedure for abrasives and pigments in oil-in-water emulsions (page 553).

Oil Emulsions Containing Large Amounts of Soap. Emulsions that do not strictly belong to any of the preceding types, especially those containing a considerable amount of soap, require different treatment. Kerosene-base creosote emulsions, such as are commonly used as garage floor cleaners, and dry cleaners' soaps fall in this category.

The following is a partial list of ingredients that may be present: ethyl, isopropyl, butyl and amyl alcohols, hexalin and methyl hexalin, all as blending agents; potassium, sodium, ammonium or triethanolamine soaps; pine oil, turpentine, and chlorinated hydrocarbons.

The method given is merely a suggestion; however, it has been found successful in a majority of cases. A few qualitative tests such as those for odor and the effect of dilution with water should precede a quantitative estimation.

Titration. Add a considerable amount of water, and titrate, first cold if a volatile amine is present, and finally in hot solution to a methyl orange or methyl red end point. This will give a total alkalinity and will be a check on the following work.

Total Ammonia. To a separate sample add water and calcium chloride. Distill according to the technic indicated on page 149.

Free Ammonia and Ammonium Soaps. If ammonia is present and other alkaline substances are absent, titrate directly as indicated on page 147. If other alkalies are present, an unusual case, follow the technic on page 148. Subtract the ammonia required for any fatty acids later found, and calculate the excess as free aqua ammonia.

Amines. These can often be determined by calculation from the two preceding determinations. A Kjeldahl determination (page 143) is helpful.

Distillation. Acidify 200 grams of the sample in a 500-ml. round-bottomed flask with just more than sufficient sirupy phosphoric acid to combine with all the soap and free alkali present. Distill carefully, using an oil bath. Be very careful not to heat-crack the fatty acids or high-boiling oils. In most cases, water, possibly containing alcohol, and volatile solvents such as kerosene and chlorinated hydrocarbons will distill over, sometimes forming three layers in the receiver. Read the volumes, and separate. Obtain specific gravities of the fractions, and if necessary redistill. Identify as indicated in Chap. 23. Partial distribution of one solvent in another may have occurred. Determine cresols on one or more of the fractions by extracting with dilute alkali. Acidify the extract, and take up the cresols with ether. The cresols may be part of the pine-tar or coal-tar creosote present. Depending on the boiling-point range of the cresols, all the cresol content may have distilled over, or part of it, or even none at all. This is also the case with regard to pine oil. Experience with the chemistry of such commercial ingredients is necessary to interpret the results successfully.

Residue in Flask. This will contain some or all of the following: soap fatty acids, blending agents of high boiling point, acid sulfated oils, part or all of the high-boiling pine-tar or coal-tar creosotes, and pine oil. Weigh the flask and contents. Subtract the weights of the flask and the added phosphoric acid, less the calculated amount of water formed by acidifying the soap present. The result is the weight of the mixture.

To separate this mixture requires ingenuity of a high order. Wash free of acid, centrifuging if sulfated oils are present, and obtain the acid value. Calculate to the kind of soap indicated to be present from prior qualitative tests. Sulfated oil, if present, complicates matters. If both soap and sulfated oil are present, neutralize the soap fatty acids with the minimum amount of 50 per cent sodium hydroxide solution and extract the oils from the sodium soap with ether, using a centrifuge for separation. Then acidify, extract, and identify the fatty acids in the soap by the usual methods, described in Chap. 17. If necessary, determine organic sulfate on the oil fraction according to page 373.

Water-soluble high-boiling blending agents may be extracted with water. A careful fractionation using a bare flame will sometimes help to separate other ingredients but is often complicated by decomposition.

SOAP AND ALKALIES. Ash a sample of total solids derived from the emulsion, and titrate as on page 390. Often the soap and/or alkalies can be titrated directly according to page 389. The two methods are sometimes useful cross-checks on each other. If no free alkali is present in the sample, calculate the titration to the appropriate soap. Otherwise, calculate the alkali due to the soap fatty acids found, and subtract from the total alkalinity as determined, to give added alkali.

Water-base Wax Emulsions. The so-called no-rubbing floor wax is the typical emulsion of this class, but others are used as shoe polishes, rubber renovators, etc. Wax is often a minor rather than a major ingredient in the solids. Owing to the complexity of these products and particularly owing to the extreme difficulty in separating added soap, soap formed from resins present, resins of many types, and carnauba and other waxes, only approximate, though useful, figures can be expected from an analysis of this type.

The following method will give reasonably accurate information, particularly in the hands of a worker of judgment and experience:

TOTAL SOLIDS. Rapidly weigh 2 to 3 grams of the sample into a weighed porcelain capsule crucible approximately 5 cm. in diameter and 1 cm. deep. Heat the capsule and sample on a steam bath for about 1 hour and then in an oven at 105 to 110° for 3 hours. Cool, and weigh. As the amines normally used in these products vary in rate of volatility, this size of sample, size of dish, and heating time are used as a compromise found in practice to give best results.

ASH. Ash the nonvolatile matter in the same capsule over a low flame, taking care to ignite off all the carbonaceous matter. Cool, and weigh. When the ash is to be analyzed, as when fixed alkalies are present, repeat with a 20-gram or larger sample. Ash, dissolve the ash in water, and titrate cold to phenolphthalein and methyl orange end points. Dilute the titrated solution to 100 ml. in a calibrated flask, and perform qualitative tests for borax, phosphate, and carbonate (Chap. 12). Determine any borax on an aliquot by titrating with standard alkali to a phenolphthalein end point in the presence of mannitol or neutral glycerin (page 271). Phosphate is best determined on another aliquot, using the volumetric molybdate method (page 203). Carbonate can be determined by the evolution method, using another ashed sample according to page 224, but is more commonly estimated from the total titration figure after subtracting the titrations equivalent to any borax or phosphate.

ACID-INSOLUBLE MATTER. Titrate a 100-gram sample, hot, to a methyl orange end point. This will give the total alkali in the sample. Add about 1 ml. of 0.5N acid in excess, and heat to clear the solution thoroughly. Allow to cool, when the waxes, resins, and fatty acids will separate as a waxy mass. Decant off the water layer, and wash the waxy mass. If necessary, filter the water layer and washings, and collect in a calibrated flask. Dilute to volume, and reserve for further work.

Waxes. Attempt to dissolve the waxy mass, including any on the filter paper, in 200 ml. of hot 95 per cent alcohol; SD3A will do. Carefully boil the mixture to ensure complete disintegration. Transfer to a 250-ml. centrifuge bottle, and cool to room temperature. Centrifuge until a clear upper layer is formed. With some resins a semigel structure may form; in this case add a large excess of alcohol, and use two centrifuge bottles. Decant off the clear liquid, add more alcohol to the bottle, and bring to a boil in a water bath. Shake thoroughly, cool, and centrifuge as before. Decant, and repeat the whole operation at least once more, or until the cold alcohol extract is colorless. Combine the alcohol extracts, and reserve.

Add 100 ml. of trichloroethylene or carbon tetrachloride to the residue in the centrifuge bottle, and bring to a boil in a water bath. Shake well; and if any insoluble matter is present, centrifuge while hot. Decant and wash the insoluble matter with small portions of hot solvent until the extract is colorless, centrifuging if necessary. Evaporate the combined extract and washings. It may be necessary to keep the extracted waxes in the oven at 105 to 110° for several hours before all the solvent is gone. After heating to constant weight, obtain constants such as melting point, acid value, saponification, and aniline point on the waxes (Chap. 22). Observe the odor on burning. In good-quality self-polishing waxes, carnauba wax is most commonly used. The cheaper grades use the minimum of the expensive carnauba wax, incorporating mineral waxes, cheaper vegetable waxes such as candelilla and ouricuri, coumarone-indene resins, and polymerized pine-type resins such as Piccolyte. These will all dissolve in hot trichloroethylene. If the character of this extract indicates a synthetic resin as well as waxes, boil it with 100 ml. of 95 per cent alcohol and centrifuge hot. Repeat at least twice more, being sure to keep the alcohol very hot. Coumarone resins and Piccolyte are insoluble in hot alcohol, whereas all vegetable waxes are soluble in hot alcohol on repeated extraction.

Small portions of some natural resins are soluble in trichloroethylene, and about 5 per cent of carnauba wax is soluble in cold alcohol. Such small sources of error tend to counteract each other.

Resins and Fatty Acids. These are in the alcohol-soluble portion. There are no simple methods for separating the resins and the soap fatty acids. Usually the soap in the original compound has been partly formed from the acidic portion of the resin and partly from added fatty acids or added as soap itself.

Dilute the alcohol extracts to 500 ml. On one aliquot obtain total solids content. On other aliquots determine the acid value and saponification value (Chap. 17). Test for rosin on the solids by the Liebermann-Storch reaction (page 395). Useful information as to colors obtained with various natural resins by using the Liebermann-Storch reaction has been published,[1] but space is lacking to give it here. In general, rosin-free Manila resins give brown or pink to old rose colors, which change to brown, instead of the characteristic violet of rosin. For constants on natural resins see Table 29, page 475.

AMINES. Soaps of ammonia, mono-, di-, and triethanolamine, morpholine, aminomethyl propanol, and other amines are all used in water-base waxes as emulsifying agents. Sometimes it is a matter of guesswork as to which is present. Experience indicates that some give off ammonia on heating with sodium hydroxide more readily than do others. With mixtures such as are at times employed, the difficulty increases.

Dilute the water-soluble portion after acidification and the washings to 250 ml., and obtain total solids on an aliquot. The rate at which the solids lose weight is somewhat indicative of the type of amine present. Keep one aliquot in the oven, after evaporating off the bulk of the water on the steam bath, for just sufficient time to be sure that all the water is removed. Cool, and weigh. Deduct from this weight the weight of chloride equivalent to any fixed alkali found in the ash or the weight of the resultant product after the titration of the alkali with acid. The residual weight will be that of the hydrochloride of the amine or mixture of amines.

On another aliquot determine total nitrogen by the Kjeldahl method (page 143). Calculate to the suspected amine, and further calculate to the hydrochloride of that amine. Thus, by trial-and-error calculations, a fair estimate of the type of amine and also its amount may usually be obtained.

As previously indicated, the interpretation of results so obtained in the shape of a formula requires considerable practical experience and knowledge of both the chemical and the physical properties of the waxes, resins, rosin, and amines and of the physical properties they impart to the finished product. Usually the results are modified and interpreted after making up the product formulated from the analysis.

[1] MANTELL, C. L., KOPF, C. W., CURTIS, J. L., and ROGERS, E. M., "The Technology of Natural Resins," John Wiley & Sons, Inc., New York, 1942.

CHAPTER 31

RUBBER

Rubber appears on the market as latex, smoked sheet, pale crepe, compounded with various fillers, and vulcanized with sulfur as hard or soft rubber. Then, as if this were not sufficiently confusing, there are various rubberlike materials, balata, gutta-percha, and synthetic elastomers, that are often described as synthetic rubbers. To confuse the issue further, mixtures of rubber and synthetic elastomers are used in some products.

This chapter will be confined to methods of analysis of rubber and rubber compounds. Associated tests, such as those for accelerators and the examination of rubberlike bodies, will be left for a subsequent chapter. Although these methods are not designed for application to synthetic elastomers, on which methods of analysis are very limited, they may nevertheless be applied to obtain specific data.

The methods presented are the standard procedures for chemical analysis of compounded and vulcanized rubber products of the American Society for Testing Materials (ASTM). By suitable modification they can be applied to crude rubber. The methods determine fractions of a complex mixture and therefore require close attention to detail.

No satisfactory simple method is known for determining directly the actual amount of rubber hydrocarbon in a vulcanized compound. It may be determined by calculation by the ASTM method and satisfactory results obtained except when decomposable compounding ingredients, such as carbonates that decompose at 550°, cellulose, clay, asbestos, asbestine, talc, factice, and high percentages of mineral rubber, are present. Most insulating compounds contain at least one of these materials. Therefore a modification of the methods of the ASTM is applied for rubber hydrocarbon in those compounds. This is known as the Joint Rubber Insulation Committee Procedure.

When carbonates decomposable at 550° are present, yet a third modification known as the rubber-solvent method is applied.

When clay, asbestos, asbestine, or talc are present, no known method is accurate unless correction is made by calculation for the loss of water of hydration on ashing these compounds. This correction can be made only if the nature and quantity of the fillers are known. The rubber-solvent method is preferred in the analysis of asbestos-rubber packings and of rubberized fabrics but is not applicable to hard rubber.

Preparation of Sample. Pass the sample, free from all foreign material such as adhering fabric, through a cold laboratory rubber mill, if available. Avoid the sticking together of the sample; if necessary, separate the layers from the mill by wrapping in Holland cloth. Cut the sample with scissors, either in pieces or strips, so that it will all pass through a 14-mesh sieve.

Evaporate rubber cements to dryness in a vacuum at a temperature not higher than 30°, and subdivide the resulting sheet.

Reduce hard rubber to powder by filing or grinding so that all will pass through a 30-mesh screen.

QUALITATIVE EXAMINATION

Carbonates. Drop a small piece of the rubber compound into a test tube containing concentrated hydrochloric acid saturated with bromine, A stream of bubbles indicates the presence of carbonates. In that case. use the rubber-solvent method for analysis.

Antimony. Ash a portion of the sample in a porcelain crucible, and heat the ash with a minimal volume of concentrated hydrochloric acid. Dilute to ten times its volume, filter, and pass in hydrogen sulfide. If an orange-red precipitate of antimony sulfide is formed, proceed as described for antimony (page 568) or use the rubber-solvent method.

Carbon Black. Heat a portion of the sample with concentrated nitric acid until frothing ceases. A black liquid indicates the presence of carbon black.

Glue. Extract a portion of the sample with a mixture of 32 per cent acetone and 68 per cent chloroform by volume for 8 hours. Dry the sample, and digest for 1 hour with hot water. Filter, cool, and add a few drops of a fresh 2 per cent tannic acid solution. Allow to stand for a few minutes. Turbidity indicates glue. In practice, it will usually be advisable to extract a weighed sample, weigh the dried, extracted sample, and test a portion of this qualitatively. If glue is found to be present, determine it quantitatively on another portion of the extracted sample.

Paraffin. The test will develop automatically in the standard method of analysis. If paraffin is present, it will solidify in the acetone extract as a white flocculent precipitate when it cools.

Barium Compounds. This test must be made during the progress of the analysis. Test qualitatively on the residue after filtration in the total sulfur determination, alternative method (page 565), by dissolving it in 10 per cent hydrochloric acid. Filter, and add a few drops of 10 per cent sulfuric acid. A white precipitate indicates barium.

QUANTITATIVE EXAMINATION

American Society for Testing Materials Methods

Specific Gravity. Determine the specific gravity of the sample by the use of a pycnometer (page 48) or by immersion in water or a suitable liquid.

Total Extract. This is defined as the extract obtained by extraction, for at least 8 hours, with a mixture consisting of 32 per cent acetone and

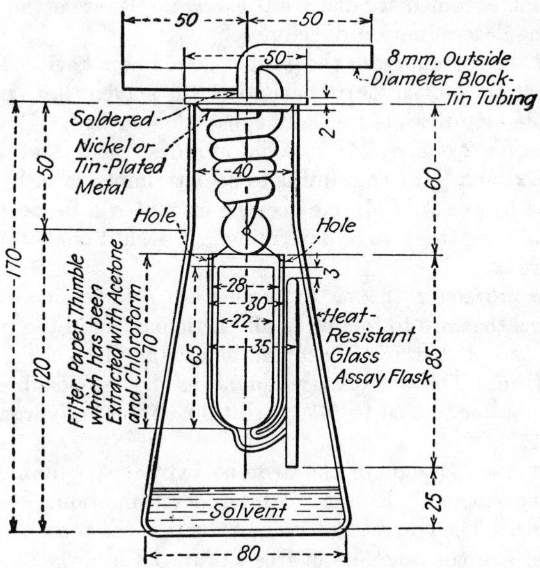

FIG. 108.—Rubber extraction unit according to ASTM standards. All dimensions in millimeters.

68 per cent chloroform by volume. The solvent selected is a constant-boiling mixture.

Weigh accurately about 2 grams of the sample, either comminuted or in thin strips, and fold in filter paper that has been previously extracted with the mixed solvents. Separate the pieces by paper if the sample is likely to become tacky during extraction. Transfer to an extraction apparatus of the type and dimensions shown in Fig. 108. Dry, cool, and weigh the flask in advance under the same conditions as are to be used after extraction. The feature of greatest importance in this type of equipment is that the sample being extracted is surrounded by the hot solvent vapors. The extraction is therefore like a Soxhlet procedure but at the temperature of boiling solvent instead of that of relatively cold condensed solvent.

Add 50 to 75 ml. of the mixed solvent to the flask. The minimum period of extraction is 8 hours. Heat the flask in such a way that the time required to fill and empty the siphon cup is about 3 minutes. Hard rubber and any soft rubber having a ratio of total sulfur to rubber hydrocarbon in excess of 1:10 must be extracted for 72 hours. Carefully note the characteristics of the extract both when hot and cold. A black color indicates a substantial pitch or bitumen content, which may not be completely removed by the mixed solvent. In that case follow the extraction with mixed solvent by one with chloroform, and add the value of this extract to that obtained for the total extract. Reserve the final rubber residue for the determination of cellulose.

Distill off the solvent from the extract on a steam bath. Do not allow the flasks to stand on the steam bath after the solvent has been removed, for appreciable quantities of free sulfur may thus be lost. Dry the extract in an air oven for 2 hours at 70°. Cool in a desiccator, and weigh.

Acetone Extract. If free sulfur is to be determined later by the method recommended by the ASTM, one acetone extract will be needed for that. If unsaponifiable matter is to be determined, two additional acetone extracts will be required.

Therefore proceed with one to three fresh samples, according to the determinations that are to follow, in all particulars as in the procedure for total extract except that acetone should be used in place of the chloroform-acetone mixture. The acetone should be as dry as possible, preferably the fraction obtained at 56 to 57° by distilling from anhydrous potassium carbonate.

CALCULATION. Designate the acetone extract so obtained as uncorrected acetone extract. Results of later determinations are needed for the corrections. The corrected acetone extract is the total acetone extract as determined less the amounts of free sulfur and hydrocarbons A and B. Another form of expression in less complete analyses is as organic acetone extract, which is the total acetone extract less the free sulfur. The latter value therefore counts hydrocarbons A and B as part of the acetone extract. Duplicates should check within 0.2 per cent.

INTERPRETATION. Rubber resins are extracted by acetone and, depending on the source and history of the rubber, may make up as much as 5 per cent of the rubber in the total compound. Usually they are 2 to 3 per cent in smoked sheet and around 5 per cent in guayule. Synthetic elastomers as manufactured contain substantially no rubber resins. Corrected acetone extracts may contain in addition to the true rubber resins any or all of the following: accelerators, antioxidants, cellulose esters, fatty oils, fatty-oil portions of sulfurized oils known as factice, resinous oils, mineral oils, acetone-soluble organic colors, naphthalene, and other organic cyclic compounds.

Chloroform Extract. This determination is not applicable to crude, unvulcanized, or reclaimed rubber. For all other rubbers, using the same sample on which total extract has been obtained, suspend the extraction cup in a second weighed flask, and extract with 50 to 75 ml. of chloroform for 4 hours. Hard rubber and any soft rubber sample having a ratio of total sulfur to rubber hydrocarbon in excess of 1:10 must be extracted for 24 hours. Complete as for total extract described on page 562, starting with "Distill off the solvent from the extract on a steam bath." The accuracy of the checking of duplicates depends on the amount of extract but is usually within 0.1 per cent.

INTERPRETATION. Substances insoluble in acetone but soluble in chloroform include some lower hydrocarbons of natural rubber, but these will normally not exceed 1 per cent of the rubber present in the compound. If higher than 1 per cent the extract is usually bitumen, pitch, or tarry material used as organic filler. The color of the extract frequently indicates the nature of such ingredients present.

Alcoholic Potash Extract. To purify the alcohol for this determination, dissolve 1.5 gram of silver nitrate in 3 ml. of water, and add it to 1 liter of pure alcohol or alcohol denatured with 10 per cent of methanol. Dissolve 3 grams of potassium hydroxide in the smallest possible amount of hot water, and cool. Add it to the alcoholic silver nitrate solution, and shake thoroughly. Allow to stand at least 24 hours, filter, and distill. For the alcoholic potash solution, dissolve 58 grams of potassium hydroxide in 1 liter of the alcohol so purified, the resulting solution being considered as *N*.

The sample is that remaining either from total extract or from the chloroform extraction if a supplementary extraction followed. In either case, remove the solvent from the rubber residue at about 70°, and transfer to a 200-ml. conical flask. Add 50 ml. of *N* alcoholic potash solution, and heat under a reflux condenser for 4 hours. In the case of hard rubber, continue the heating for at least 16 hours.

Filter into a 250-ml. beaker. Wash the filter twice with 25-ml. portions of boiling alcohol and then with three 25-ml. portions of boiling water. Save the residue on the filter paper if rubber hydrocarbon is to be determined later by the Joint Rubber Insulation Committee Procedure. Evaporate the total filtrate just to dryness. Transfer the dried extract to a separatory funnel with 75 ml. of water. Acidify with 1:3 hydrochloric acid, using Congo red paper as indicator. Extract the acidified residue with four 25-ml. portions of ether, or more if color is still extracted. Wash the combined ether extracts with 25-ml. portions of water until the washings are neutral to phenolphthalein. Let any suspended droplets of water settle out; then filter through a plug of absorbent cotton into a weighed flask. Wash this filter with ether. Evaporate the ether extract and washings, and dry to constant weight at 70°. Cool, and weigh. Duplicates should check within 0.1 per cent.

INTERPRETATION. This extract from rubber will normally not exceed 1 per cent of the rubber hydrocarbon compounded. If more than that amount is present, it usually consists of the soaps from saponification of sulfurized oils, known as factice, or similar soaps derived from chlorinated oils.

Total Sulfur. RECOMMENDED METHOD. By this method the total sulfur determined is the sulfur exclusive of that originally present as barium sulfate.

Prepare a zinc oxide–nitric acid mixture by adding 200 grams of zinc oxide to 1 liter of concentrated nitric acid. Transfer 0.5 gram of soft rubber or 0.2 gram of hard rubber to a 500-ml. Pyrex conical flask. Add 10 ml. of the zinc oxide–nitric acid solution, and moisten the sample thoroughly.

Let the sample stand at least 1 hour, preferably overnight. This will cause partial decomposition and remove any danger of ignition of the sample on the addition of fuming nitric acid. Add 15 ml. of fuming nitric acid, and whirl the flask rapidly to keep the sample immersed and thus prevent ignition. With some samples it may be necessary to cool the flask under running water. When solution of the rubber appears to be complete, add 5 ml. of saturated bromine water and slowly evaporate to a foamy sirup. For the determination of total sulfur in unvulcanized mixtures, use 3 ml. of liquid bromine in place of the 5 ml. of bromine water.

If organic matter or carbon remains at this point, add a few milliliters of fuming nitric acid and, very cautiously, a few crystals of potassium chlorate. Mix well, and boil. Repeat the procedure until all carbon is gone and the solution is clear and colorless or light yellow.

Place the flask on an asbestos gauze, and evaporate to dryness over a high-temperature burner. Bake at the highest temperature of the burner until all nitrates are decomposed and no more nitric oxide fumes can be detected. Anneal the flask by gradually decreasing the flame. Alternatively, evaporate the colorless or yellow solution, cool, add 10 ml. of concentrated hydrochloric acid, and evaporate to dryness, avoiding spattering. Repeat this alternative procedure until no more nitric oxide is evolved.

After the flask is cool, add 50 ml. of 1:6 hydrochloric acid and warm. Digest under these conditions until solution is as complete as possible. If the original mixture contained barium in any form, barium sulfate will be precipitated at this point. If the barium was originally present other than as the sulfate, such precipitation will represent a loss of sulfur from the sample for the barium will now combine with some of the sulfur trioxide formed by oxidation of the total sulfur. If lead oxide is present in moderate amounts, it will not interfere under the given conditions if all the filtrations and washings are made hot.

Filter out any barium sulfate precipitate, which may be contaminated with other mineral matter. Wash the filter well, and discard the paper and precipitate. Dilute the filtrate to 300 ml., and add 10 ml. of saturated aqueous solution of picric acid. Complete as on page 160, starting with "Bring to a boil, and add with constant stirring . . .".

The result, calculated as sulfur, must be corrected. The corrected total sulfur equals the total sulfur as so determined minus the percentage of sulfur present as antimony sulfide. When barium carbonate is present, a further correction is made by adding to the total sulfur as otherwise corrected the amount of sulfur necessary to convert the barium carbonate to barium sulfate. Duplicates should check within 0.1 per cent.

ALTERNATIVE METHOD. Add a considerable excess of bromine to concentrated nitric acid so that a layer of bromine is present in the bottom of the bottle. Shake thoroughly, and allow to stand 24 hours before using. Place 0.5 gram of soft rubber or 0.2 gram of hard rubber in a porcelain crucible of about 75 ml. capacity, and add 15 ml. of the clear nitric acid–bromine mixture.

Cover the crucible with a watch glass, and allow to stand for 1 hour in the cold. Heat for 1 hour on the steam bath. Remove, rinse the cover into the crucible, and evaporate the contents to dryness. Add 8 ml. of concentrated nitric acid, cover, and warm for a few minutes on the steam bath. Then let the crucible cool. Slightly raise the cover, and carefully add 4.5 to 5.5 grams of sodium carbonate in small portions down the side of the crucible. Avoid dropping this directly into the acid. Rinse the cover into the crucible with 2 to 3 ml. of hot water, and stir the mixture thoroughly with a glass rod. Digest for a few minutes; then spread the mixture halfway up the side of the crucible with a porcelain spatula or stirring rod to facilitate drying on a steam bath. Carefully fuse over a sulfur-free flame.

For the fusion, place the crucible in an inclined position on a wire triangle, and start the ignition over a low flame. The tendency for the organic matter to burn too briskly may be controlled by judicious use of the stirring rod with which the burning portion is scraped away from the rest. When part of the mass is burned white, work a fresh portion into it until all the organic matter is destroyed. It is necessary to hold the edge of the crucible with tongs during this operation. Toward the last half of the operation the flame should be increased, but at no time is it necessary to heat the crucible to redness. With care, a crucible can be used for at least 10 to 12 such fusions.

Cool the crucible, and place it in a 400-ml. beaker. Cover this with about 125 ml. of water, and digest on the steam bath overnight. Filter into a covered 400-ml. beaker containing 5 ml. of concentrated hydrochloric acid. Wash the residue on the filter thoroughly with hot 5 per cent sodium carbonate solution. All the barium present will be on the filter, largely in its original condition if sulfate or carbonate. Complete the acidification of the filtrate and washings, and add 2 ml. of concentrated hydrochloric acid in excess. Cover the beaker, and heat on a steam bath. The total volume should be about 300 ml. The solution should be acid to Congo red paper, that is, below pH 3, in order to ensure the complete decomposition of the soluble carbonates. Add 10 ml. of saturated aqueous picric acid solution, and complete as on page 160, starting with "Bring to a boil, and add with constant stirring . . .".

Free Sulfur. RECOMMENDED METHOD. To the flask containing the acetone extract, add 10 ml. of zinc oxide–nitric acid mixture, prepared as described for total sulfur, and 2 to 3 ml. of bromine. Cover with a watch glass, and allow to stand near the steam bath for 30 minutes. Then heat on the steam bath to a foamy sirup. Add 10 ml. of fuming nitric acid, and heat on the hot plate with the cover removed until all the bromine is expelled. Complete the determination as described on page 564 for total sulfur, recommended method, starting with "If organic matter or carbon remains at this point . . . ". As for total sulfur, duplicates should check within 0.1 per cent.

ALTERNATIVE METHOD. Place 2 grams of the sample, thinly sheeted to 0.05 to 0.07 cm., in a 400-ml. thin-walled Pyrex flask. Add 100 ml. of 5 per cent sodium sulfite solution, 5 ml. of a 0.1 per cent solution of sodium stearate, and approximately 1 gram of paraffin. Cover the flask with a small watch glass, and heat so as to boil gently for 2 hours. This converts the sulfur to thiosulfate. Remove the flask, and add 100 ml. of a 0.5 per cent solution of strontium chloride and 10 ml. of a 3 per cent solution of cadmium acetate. Separate the rubber and precipitates by filtration, using a Büchner funnel with suction. Prepare the funnel by forming a thin asbestos pad over a single sheet of qualitative filter paper. Filters so prepared can be used many times. Wash with two 75- to 100-ml. portions of a wash solution containing 40 ml. of a 3 per cent cadmium acetate solution per liter. To the filtrate add, with stirring, 5 ml. of 40 per cent formaldehyde solution, 10 ml. of glacial acetic acid, and 5 ml. of a 1 per cent starch solution. Reduce the temperature to below 15° by adding crushed ice, and titrate with standard iodine solution. Run a blank on reagents, and subtract this figure, usually 0.2 to 0.3 ml., from the titration of the sample.

CALCULATION.

$$\text{Ml. of iodine solution} \times \text{normality} \times 0.032 \times \frac{100}{\text{weight of sample}} = \text{per cent free sulfur}$$

The corrections that apply to the recommended method are not necessary here.

Ash. RAPID METHOD. Wrap a 1-gram sample of soft rubber in a filter paper, and extract with acetone for 4 hours. Transfer to a weighed crucible of about 50 ml. capacity. For hard rubber omit the extraction, and use a

2-gram sample. Ash the sample in a muffle furnace by heating at the following rate:

Time, minutes	5	10	15	70	75	80	85	145
Temperature, °C	100	200	300	300	400	500	550	550

Remove the crucible, cool, and weigh. If no muffle is available, distill off the rubber over a very small flame, not allowing it to catch fire. Then ignite quite gently until free of carbon, cool and weigh.

Calculation. The percentage of ash so obtained must be corrected. To do so, subtract the percentage of sulfur in the ash as subsequently determined, add the percentage of sulfur present as barium sulfate, and subtract the percentage of antimony in the ash, calculating the antimony as Sb_2O_4. Duplicates should check within 0.25 per cent.

Referee Method. This method is used only in case of controversy. Extract a 2-gram sample with acetone as described for the determination of acetone extract (page 562). Dry the extracted sample at 100°, cool, and weigh. Ignite two crucibles, each 41 mm. in diameter and 25 mm. in height, cool, and weigh. Divide the sample into two equal parts, and place them in the previously prepared crucibles. Place the two crucibles and contents together, with their tops touching centrally, in an electric furnace the bottom of which is covered with a sheet of asbestos 0.06 in. thick, cut to fit.

To measure the temperature use a calibrated thermocouple enclosed in a quartz tube closed at one end, 6 to 7 mm. in diameter, 0.5 to 1 mm. thick, and placed in a horizontal position in contact with the asbestos sheet on the bottom of the furnace, so that the junction of the thermocouple is on a line crossing the furnace halfway between its ends and halfway between the crucibles.

Regulate the temperature of the furnace by means of a rheostat to conform to the following rate of heating, with a maximum permissible variation of plus or minus 25°:

Time, minutes	12	50	100	150	200	250	300	350	390	450	457
Temperature, °C	100	200	250	275	290	300	310	400	425	490	540

Adjust the door of the furnace to conform to the following requirements:

First 400 minutes	0.5-inch opening
From 400 to 450 minutes	1.5-inch opening
Remaining 7 minutes	Wide open

At the end of 457 minutes, remove the crucibles, cool, and weigh. Correct as in the preceding method.

Sulfur in Ash. To the ash add 3 ml. of nitric acid–bromine mixture prepared for the determination of total sulfur, and cover with a watch glass. Heat for 1 hour on the steam bath. Remove, rinse the cover into the crucible, and evaporate the solution to dryness. Complete as for total sulfur by the alternative method (page 565), starting at "Add 8 ml. of concentrated nitric acid, cover, and warm . . . ".

Retain the insoluble residue, after filtering the solution of the fusion mixture in water, to use for the determination that follows.

Sulfur Present as Barium Sulfate. Wash the insoluble matter separated in the determination of sulfur in the ash back into the original beaker with hot water. Dissolve the residue in the beaker and any traces on the filter paper with hot 1:3 hydrochloric acid. Heat on the steam bath, filter, and wash thoroughly with hot water.

If lead was originally present in the sample, it will now be in this filtrate and must be removed. For this purpose, adjust the acidity to equal that of 0.2 to 0.3*N* hydrochloric acid. This corresponds to 1:20 to 1:30 dilution of the concentrated acid and is most conveniently accomplished by further dilution to about eight to ten times the volume of 1:3 hydrochloric acid used for solution of the insoluble matter. Cool, and saturate with hydrogen sulfide. When the lead sulfide has settled, filter into a 400-ml. beaker and wash thoroughly. The total volume of the solution should be not more than 200 ml. Precipitate the barium with hot 1:10 sulfuric acid. Digest, filter, ignite, and weigh in the usual manner (see page 189).

If barium was originally present only as barium sulfate, the value so obtained is a fair approximation. It will usually be less than the amount of commercial barium sulfate originally compounded. Conversely, because of sulfate and phosphate impurities often present in commercial barium sulfate, the value for total sulfur exclusive of that from barium sulfate will frequently be high when the calculation is made with data obtained from this procedure and interpreted as barium sulfate for comparison with that for total sulfur by the alternative method. If some of the barium was originally present as carbonate, it must be determined by a method given later, in order that a proper correction can be made. Thus part of the precipitate as barium sulfate in this determination may be the barium originally present as carbonate.

Total Antimony. Transfer a 0.5-gram sample to a 500-ml. Kjeldahl flask, and add 25 ml. of concentrated sulfuric acid and 10 to 12 grams of potassium sulfate. Place a funnel in the neck of the flask to condense sulfur trioxide vapors, and heat until the solution is colorless. Cool, wash the funnel, and dilute to 100 ml. with water.

If iron is present, the antimony must be separated from it. For this operation transfer the diluted contents of the Kjeldahl flask to a 400-ml. beaker, and further dilute to 250 ml. with hot water. Precipitate the antimony by saturating the solution with hydrogen sulfide. Filter, wash with hydrogen sulfide solution, and transfer the precipitate back to the Kjeldahl flask. Add 15 ml. of concentrated sulfuric acid and 10 to 12 grams of potassium sulfate, and heat as in the first digestion until the solution is colorless. Cool, wash the funnel, and dilute to 100 ml. with water as before. Transfer to a 400-ml. beaker. Add 1 to 2 grams of sodium sulfite, and boil until all the sulfur dioxide is driven off. This is shown when no blue color is obtained with starch-iodate paper held in the vapors. Add 25 ml. of concentrated hydrochloric acid, and dilute to 200 ml.

Regulate the temperature of the solution of sample to about 60°, and add 2 drops of 0.2 per cent methyl red solution. Titrate with the standard potassium bromate solution (page 727) until the solution is colorless. When the indicator starts to fade, add the bromate slowly, using another drop of indicator. As soon as all the antimony is oxidized, excess bromate will immediately decolorize the methyl red. The reaction is

$$KBrO_3 + 3SbCl_3 + 6HCl = 3SbCl_5 + KBr + 3H_2O$$

Provided the acidity of the solution is maintained within the specified limits, the titration is quantitative.

CALCULATION.

$$\text{Ml. titration} \times \text{normality of potassium bromate} \times 0.06088 \times \frac{100}{\text{weight of sample}} = \text{per cent antimony as such}$$

Other factors are antimony oxide, Sb_2O_4, 0.07688; antimony trisulfide, 0.08493; arsenious oxide, 0.04946.

Antimony in the Ash. Use the ash from a 1-gram sample. Transfer to a 600-ml. conical flask, and add 12 to 15 grams of potassium sulfate. Boil until solution is as complete as possible. It may be necessary to warm part of the sulfuric acid in the crucible to transfer any adhering particles to the flask. Rinse with the remaining portion of the acid. Cool and dilute to 100 ml. with water. Continue with the determination as under Total Antimony (page 568), starting with "If iron is present, the antimony must be separated from it."

CALCULATIONS. The percentage of sulfur present as antimony trisulfide equals the total percentage of antimony as such multiplied by 0.305. The percentage of antimony calculated to the form of trisulfide equals the percentage of antimony multiplied by 1.395.

Free Carbon. Extract a 0.5-gram sample for 8 hours with a mixture of 68 per cent chloroform and 32 per cent acetone by volume. Transfer the extracted sample to a 250-ml. beaker, and heat on a steam bath until the odor of chloroform is no longer detectable. Add a few milliliters of concentrated nitric acid, and allow to stand for about 10 minutes. Add 50 ml. more of the acid, taking care to wash down the sides of the beaker, and heat on the steam bath for at least 1 hour. No more bubbles or foam should remain on the surface. Pour the hot liquid into a Gooch crucible with an asbestos mat, the prepared crucible having been dried and then ignited at a dull red heat. Keeping as much as possible of the insoluble matter in the beaker, filter slowly with gentle suction, and wash well with hot concentrated nitric acid by decantation. Remove all the nitric acid from the filter flask because leaving it would result in a violent reaction at the next step. Wash the residue in the crucible with acetone and then with a mixture of equal parts of acetone and chloroform. Unless silicates are present as part of the mineral matter, digest the insoluble matter, which should still be in the beaker, for 30 minutes on the steam bath with 35 ml. of a 25 per cent solution of sodium hydroxide. Dilute to about 60 ml. with hot water, and again heat on the steam bath. Filter the alkaline solution and residue on the Gooch crucible, and wash well with hot 15 per cent sodium hydroxide solution. Remove the filtrate, and wash the residue four times with hot concentrated hydrochloric acid. Test the last washings for lead by neutralizing with sodium hydroxide solution, making just acid with glacial acetic acid, and adding potassium chromate solution. If lead is still present as shown by a precipitate of yellow lead chromate, continue to wash with hot concentrated hydrochloric acid until lead-free by this test. Finally wash with warm 1:20 hydrochloric acid. Remove the crucible from the filter funnel, clean the outside thoroughly, and dry in the air oven at 110° for 90 minutes. Cool, and weigh. Burn off the carbon by ignition at a dull red heat, and reweigh.

The difference in weight represents approximately 105 per cent of the carbon originally present in the form of lampblack or gas black.

CALCULATION.

$$\text{Loss in weight on ignition} \times \frac{100}{1.05 \times \text{weight of sample}} = \text{per cent free carbon}$$

Glue. Extract a 1-gram sample for 8 hours with acetone. Remove the solvent from the extracted sample, and transfer the residue to a 750-ml. Kjeldahl flask. Add 25 to 30 ml. of concentrated sulfuric acid, 10 to 12 grams of sodium sulfate, and 0.1 gram of copper sulfate or copper selenide. Complete the digestion and estimation of nitrogen by the Kjeldahl method as on page 144, starting at "Place the flask in an inclined position, resting on an asbestos board with a hole to fit the flask." Subtract a blank unless the amount of nitrogen found is substantial.

CALCULATION.

$$\text{Ml. acid consumed} \times \text{normality of acid} \times 0.014 \times 6.5 \times \frac{100}{\text{weight of sample}} = \text{per cent nitrogen as glue, subject to further correction}$$

The figure 6.5 is an empirical factor used to calculate nitrogen to the protein glue.

A correction should be made for the natural protein in rubber. The average content of nitrogen in rubber is 0.4 per cent. Multiply the percentage of rubber as compounded, as later obtained, by 0.004 and 6.5, and subtract this value from the percentage of nitrogen calculated as glue to obtain the corrected glue. This assumes that one is dealing with a rubber compound from conventional coagulated rubber. The protein content of latex is much higher than that of crepe, smoked sheet, etc.

Unsaponifiable Matter. Determine this on acetone extracts obtained in exactly the same way as described under Acetone Extract (page 562). It is well to extract three samples with acetone at the same time, using one for the acetone extract and free sulfur determination and the other two for the unsaponifiable matter.

To the acetone extracts from two 2-gram samples add 50 ml. of *N* alcoholic potassium hydroxide, and saponify by refluxing gently for 2 hours. Then evaporate off the alcohol. Transfer to a separatory funnel using 100 ml. of warm water, and cool. Rinse out the saponification flask with 25 ml. of ether, and add this to the funnel. Shake well but not sufficiently violently to cause emulsification, and allow to separate thoroughly. A small stream of alcohol on the surface of the ether will aid in separation if necessary. Draw off the lower layer, and repeat the extraction with at least two more 25-ml. portions of ether. If the last extract is not colorless, continue such extractions until it is. Wash the combined ether extracts with water until the aqueous extracts are neutral to phenolphthalein. Drain well, and evaporate the ether from a widemouthed flask. Finally dry to constant weight at 70°. Cool and weigh the unsaponifiable matter.

Hydrocarbons A. This is the cold alcohol-insoluble chloroform-soluble part of the unsaponifiable matter. It usually consists of any waxy material present. To the total unsaponifiable matter add 50 ml. of absolute alcohol, and heat on a steam bath for 30 minutes. Let the flask stand in a mixture of ice and salt for 1 hour. Surround a glass filter funnel with such a freezing mixture until it is thoroughly chilled. Filter off the waxy hydrocarbons on a coarse-textured qualitative filter paper, using gentle suction. Wash the filter with 90 per cent alcohol that has been chilled in a mixture of ice and salt.

Dissolve the precipitate from the filter paper with hot chloroform, and catch the solution in a widemouthed flask. Wash the original flask with hot chloroform, and add this to the widemouthed flask in order to include any wax not transferred to the funnel. Evaporate and recover the chloroform, taking care not to burn the waxes. Dry to constant weight at 70°, cool, and weigh.

Hydrocarbons B. This is the cold alcohol-soluble, carbon tetrachloride-soluble part of the unsaponifiable matter and is usually of a resinous nature. Evaporate the alcohol from the flask containing the alcohol-soluble unsaponifiable matter. Add 25 ml. of carbon tetrachloride, and transfer to a separatory funnel. Shake with 10 ml. of concentrated sulfuric acid, and drain off the discolored acid. Repeat with fresh 10-ml. portions of acid of the same strength until there is no longer any discoloration. Drain off the acid thoroughly, and wash the carbon tetrachloride solution with several 10-ml. portions of water until they react neutral to phenolphthalein. Transfer to a widemouthed flask and evaporate, taking care not to burn the extract. Dry to constant weight at 70°, cool, and weigh.

Cellulose. Treat 0.5 gram of rubber, previously extracted with the constant-boiling mixture of acetone and chloroform for total extract, with 25 ml. of rubber-solvent oil in a 250-ml. lipped assay flask (Fig. 109). The ASTM specification for this mineral oil is as follows:

Saybolt Universal viscosity at 68°F.	56 seconds
Saybolt Universal viscosity at 100°F.	45 seconds
Flash point	270°F.
Fire point	350°F.
Specific gravity	0.853
Color	Colorless

This oil consists of a fraction only a little higher than mineral seal oil or 300° burning oil, which is loosely describable as high-flash kerosene. A grade of these characteristics is supplied by most large oil companies.

Place the flask in an air bath at 145 to 150° for about 3 hours, after which small particles can usually be seen distinctly in the supernatant liquid. Cool somewhat, and dilute with 15 ml. of benzene and 200 ml. of petroleum ether. Filter through a Gooch crucible prepared with acid-washed asbestos. This crucible need not have been weighed. Wash the residue thoroughly with warm acetone. Treat repeatedly with hot 1:9 hydrochloric acid until free from acid-soluble fillers. Wash well with boiling water, and run small portions of acetone through it until the filtrate is colorless. Wash with alcohol, and dry at 105° to constant weight. Remove the pad from the crucible with the help of a pair of sharp-pointed tweezers, using the under portion of the pad to swab the sides of the crucible, and place in a tared weighing bottle. Replace in the oven for about 10 minutes, cool, and weigh.

Transfer the contents of the weighing bottle to a 50-ml. beaker, and pour over it 15 ml. of acetic anhydride and 0.5 ml. of concentrated sulfuric acid. Digest this mixture on the steam bath for at least 1 hour to acetylate the cellulose. Cool thoroughly, and dilute with 25 ml. of 90 per cent acetic acid. Filter through a weighed Gooch crucible. To guard against traces of the material being carried through, this filtration, as well as the ones that follow, should be very slow, with gentle suction applied. Wash the residue on the filter with hot 90 per cent acetic acid until the filtrate comes through absolutely colorless. Then wash about four times more. Wash the residue on the filter with acetone about five times. After having taken care that all the material has been washed into the funnel from the beaker in which the acetylation of the cellulose took place, remove the crucible from the funnel. Clean the outside, and dry for 2 hours at 150°. Cool, and weigh. The original weight of this crucible plus the weight of the pad, fillers, and cellulose used for acetylation after removal from a previous crucible, less the weight of this crucible and its contents, which now includes all those materials except the cellulose, gives the weight of cellulose. That cellulose has been acetylated and is now in solution in the acetic acid and acetone washings.

Barium Carbonate. Barium sulfate was determined by calculation to barium sulfate of all barium found in the sample. If barium carbonate is present, it must be determined in order that an undue correction will not be made for sulfur introduced as barium sulfate. This method must be used for hard rubber. A later method, known as the rubber-solvent method (page 576), is more rapid and equally accurate for barium carbonate in other rubbers.

Ash a 1-gram sample placed in a porcelain boat in a combustion tube through which passes a current of carbon dioxide. This inert atmosphere prevents oxidation of sulfur to sulfur trioxide, which would displace some of the carbonate, converting it to sulfate. Grind the residue in an agate mortar, and transfer to a 250-ml. beaker. Treat with 5 to 10 grams of ammonium carbonate, 15 to 20 ml. of concentrated ammonium hydroxide, and 50 ml. of water. Boil for 20 minutes, filter, and wash the precipitate thoroughly to remove all soluble sulfates. Wash the residue on the filter paper back into the original beaker. Add about 10 ml. of glacial acetic acid and sufficient water to make the total volume to 100 ml. Heat to boiling, and pass through the same paper as before. Pass hydrogen sulfide into the filtrate to precipitate lead if present. Filter the solution of the soluble barium, and add 10 ml. of saturated aqueous solution of picric acid. Complete as on page 568 starting with "Precipitate the barium with hot 1:10 sulfuric acid." Calculate as barium carbonate.

Analysis of Ash. Ash 1-gram samples for soft-rubber products and 2-gram samples for hard-rubber products by the rapid method.

Silica and Insoluble Matter. Dissolve the weighed ash, so far as possible, in 10 ml. of concentrated hydrochloric acid. Rinse the crucible thoroughly, dilute the solution to about 75 ml., and evaporate to dryness. Proceed as for silica in minerals on page 189, starting with "If it is necessary to expedite the operation, evaporate to separation of solids . . . ". One dehydration will usually be sufficient. When the silica and insoluble matter have been separated, continue as follows:

Examine the residue on the paper. If a large residue is present and appears to be other than silica, barium sulfate or titanium oxide is probably present. Lead should be in the filtrate rather than in the precipitate; but if faulty removal is suspected, ignite in a porcelain crucible. If lead is known to be absent, ignite in platinum. Cool, and weigh. If ignition was in a porcelain crucible, transfer to platinum. Add 2 to 3 ml. of 48 per cent hydrofluoric acid and a few drops of 1:4 sulfuric acid, and evaporate to dryness, finally igniting at a dull red heat. The loss in weight will be silica. If lead is present, be careful not to heat beyond a dull red. The barium sulfate if present will have been determined earlier. Barium carbonate will also have been determined but is now present as barium sulfate. If titanium is present, it can be determined on this residue by the method for paint pigments (page 507). Fuse any iron and aluminum oxides present, if only those oxides are present, with acid potassium sulfate, and add to the filtrate from the silica determination. Alternatively, if a lower accuracy is permissible, assume such a small residue to be iron and aluminum oxides.

LEAD. If present, determine this on the filtrate from silica as described for the analysis of paint pigments described on page 501, starting with "Dilute with water to about 200 ml., and filter if necessary."

OTHER ELEMENTS. Continue with the filtrate from separation of lead following the technic for paint pigments. For zinc use the method of precipitation as sulfide and subsequent titration.

Calculation of Rubber Hydrocarbon. If the Joint Rubber Insulation Committee Procedure that follows is not used for determination of rubber hydrocarbon, the amount may be approximated from the following calculation: Obtain the total of organic acetone extract, chloroform extract, alcoholic potash extract, corrected total sulfur, corrected ash, carbon, glue, and total antimony as trisulfide. Call this C in the formula

$$^{94}/_{97}(100 - C) = \text{per cent of rubber hydrocarbon}$$

Rubber Hydrocarbon. Joint Rubber Insulation Committee Procedure. Add sufficient water to the flask containing the rubber residue from the alcoholic potash extraction (page 563) to make the total 125 ml., then add 25 ml. of concentrated hydrochloric acid. Heat on a steam bath for 1 hour. Decant the supernatant liquid through a hardened filter paper on a 7-cm. Büchner funnel, using suction. An 11-cm. hardened paper in a conical funnel with a platinum cone may also be used. Wash the residue on the paper with 25 ml. of hot water.

Return the residue from the funnel to the flask, and repeat the entire treatment twice more, three times in all, using the same filter. At this stage, the rubber should be practically free from black specks of undissolved fillers. If it is not, repeat the treatment as many additional times as is necessary to cause the black specks to disappear. If carbon is present, all the particles of rubber will be grayish, bluish, or black, depending on the form and quantity of carbon used. Black specks in light particles of rubber usually indicate the presence of lead sulfide, which must be removed by continued treatment to prevent the formation of lead sulfate on subsequent ignition of the residue.

Add 150 ml. of water to the flask, and allow it to stand on a steam bath or hot plate for 30 minutes. Decant through the filter paper used previously, repeating the decantations until the washings are free from chlorides. Transfer all the rubber in the flask to the filter paper, and dry as much as possible by suction. Wash the residue on the paper with 95 per cent alcohol, using suction. Transfer the entire residue to a weighing bottle. Dry at 95 to 100° for 1 hour, cool in a vacuum desiccator under reduced pressure, and weigh. Dry for 30 minutes longer, cool, and weigh. Repeat this process either until constant weight is reached or until the weight starts to increase. Determine the ash on a weighed part of this residue of rubber hydrocarbon. Ignite at as low a temperature as possible, and consider the nature of the fillers present; those decomposable at the temperature of ignition will tend to cause low results for ash and high results for rubber hydrocarbon. Determine the sulfur in this ash (page 567) and in the unignited rubber hydrocarbon by the alternative method (page 565).

CALCULATIONS.

$$\text{Residue after treatment as described} \times \frac{100}{\text{weight originally used for acetone extract}} \times \left(1 - \frac{\text{weight of sulfur in residue}}{\text{weight of residue used to determine sulfur}} - \frac{\text{weight of ash of residue} - \text{weight of sulfur in the ash}}{\text{weight of residue used to determine ash}}\right) = \text{per cent rubber hydrocarbon}$$

Rubber as Compounded and Related Calculation. The percentage of rubber hydrocarbon plus 6 per cent of its weight is taken as rubber as compounded, except when the sum of the percentages of the corrected acetone extract–chloroform extract and alcoholic potash extract is less than the figure represented by the arbitrary 6 per cent, as in many high-grade products. Percentage of rubber as compounded is then equal to the percentage of rubber hydrocarbon plus the sum of the percentage of the corrected acetone extract, the chloroform extract, and the alcoholic potash extract.

$$\frac{\text{Rubber compounded} \times \text{specific gravity of the compound}}{0.911} = \text{per cent of rubber by volume}$$

$$\frac{\text{Organic extract}}{\text{Rubber compounded}} \times 100 = \text{ratio of organic acetone extract}$$

$$\frac{\text{Corrected total sulfur}}{\text{Rubber compounded}} \times 100 = \text{ratio of sulfur to rubber}$$

Report. A report of a complete rubber analysis includes the items that follow as determined by the methods given. The methods that make up the balance of this chapter are not standard methods, are not acceptable for use in accumulating those determinations, and in many cases are special types of determination not included in this list.

Acetone extract, corrected.
Waxy hydrocarbons (A and B).
Chloroform extract.
Alcoholic extract.
Free sulfur.
Total sulfur, corrected.
Ash, corrected.
Cellulose.
Sulfur as barium sulfate.
Total antimony.
Sulfur as antimony trisulfide.
Carbon.
Glue.
Rubber hydrocarbon.
Rubber, as compounded.
Rubber, by volume.
Ratio of acetone extract to rubber as compounded.
Ratio of sulfur to rubber as compounded.
Specific gravity.

Rubber-solvent Method. Solvents for vulcanized rubber have been suggested from time to time in attempts to isolate the compounding ingredients in rubber goods by methods that would prevent the thermal decomposition incidental to the ashing method. Aniline, kerosene, terebene, toluene, and cymene have been used with success. For the most part, filtration is slow with these solvents.

The rubber solvent used in this method consists of a petroleum fraction of the general composition of a 300° burning oil or a mineral seal oil. Before use, the oil is passed through a column of fuller's earth, a hydrous aluminum silicate, contained in a glass tube 3 feet in length and 1.5 inches in diameter. The filtered mixture should be practically colorless. The rubber hydrocarbon in vulcanized soft-rubber compounds is dissolved by this solvent in about 2 hours at 130 to 150°. The subsequent operations incident to isolating the compounding ingredients are conducted as readily as with an aqueous solution. This solvent is recommended because of the completeness of separation and the rapidity with which it may be filtered.

After the rubber is dissolved, the fillers remain in suspension in the solution for some time but settle out on continued heating. In mixtures of asbestos fibers and rubber compound, such as are present in compressed asbestos sheet packing, this difference in behavior can be used to separate the rubber-compound ingredients from the asbestos fibers. For this, while the fillers are still in suspension, the solution should be poured through a No. 80 mesh sieve, which retains the fibers. Subsequently, the rubber compounding ingredients should be separated. Asbestos present in the form of asbestine is separated with the compounding ingredients. In the case of rubber goods made from new rubber, a small amount of undissolved material is obtained for which a correction must be made. The method is not satisfactory for hard rubber.

Solution of Rubber. Weigh two equal portions of 0.5 to 1 gram each of the finely ground sample. Extract with a mixture of 32 per cent acetone and 68 per cent chloroform by volume for at least 8 hours. If necessary, continue the extraction for a longer time until no further color is removed by the solvents. Put each of the extracted samples into a separate 150-ml. lipped assay flask (Fig. 109) that has been weighed, and add 20 to 25 ml. of the oil. Cover with a watch glass, and heat in an air bath at 130 to 150° until the solution appears complete. Continue the heating for 15 to 30 minutes more. Solution may be considered complete when the rubber colloid has been broken down and the oil seems quite clear. Cool the flasks to about 100° and add in a small stream 10 to 25 ml. of benzene while mixing thoroughly. Cool further, and then dilute with sufficient petroleum ether to fill each flask to within about 2 cm. of the top. Mix thoroughly, cover to prevent evaporation, and allow to stand overnight. Set one sample aside at this point for later determination of sulfur in compounding ingredients.

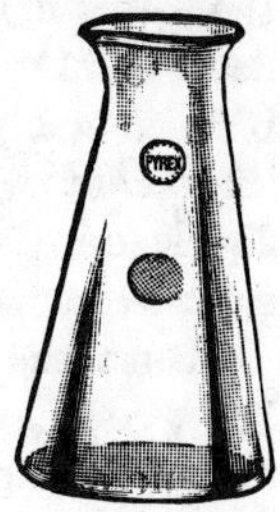

Fig. 109. Assay flask with lip.

Prepare a Gooch crucible with finely divided asbestos that has been previously treated with 20 per cent sodium hydroxide solution and then with concentrated hydrochloric acid and finally well washed with water. Ignite the crucible, cool, and weigh. Filter one sample by decantation through the crucible, using suction. Wash well with petroleum ether and then with warm acetone. If the filtrate is dark, follow this with a warm mixture of equal volumes of acetone and chloroform. Remove as much as possible of the organic residue by washing, and finally wash with hot alcohol. A small portion of the fillers may remain in the bottom of the flask. Dry the crucible and flask for 1 hour at 105 to 110°. Cool, and weigh. By difference from the weight of the crucible and flask, obtain the uncorrected weight of total compounding ingredients, and calculate to percentage of the original sample.

Removal of Acid-soluble Compounding Materials. To the flask and the crucible add a few ml. of boiling alcohol, and let soak for 2 to 3 minutes. Remove the alcohol in the crucible by suction, decant that from the flask into the crucible, and finally wash the flask and crucible two to three times with boiling water. Cool the flask, add about 10 ml. of concentrated hydrochloric acid, and swirl the flask to bring the acid in contact with the compounding ingredients. Pour the acid from the flask into the crucible, and let it stand until no more bubbles rise through the liquid. If carbonates are present, there is danger of loss by excessive frothing. Prevent this by first adding a few drops of the acid to the crucible and sucking it through. After the first violent action has ceased, add the remainder of the 10 ml. of acid.

When no further gas is evolved, suck the acid through the crucible and again wash with 20 ml. of acid, a little at a time. Wash well with hot water, and transfer the residue remaining in the flask to the asbestos pad. If previous qualitative tests have shown antimony to be present, save the filtrate and washings for further treatment. Dry the crucible and flask for 1 hour at 105 to 110°, cool, and weigh. The loss in weight of both crucible and flask represents the loss of acid-soluble compounding ingredients. This is often not calculated but serves as a valuable cross-check on the determination of such ingredients present as determined elsewhere.

Burn the organic residue from the crucible by igniting in a furnace at 700°, cool, and weigh. The percentage thus obtained is a measure of the organic compounding ingredients, possibly complicated by loss of moisture from hydrated inorganic ingredients. If the sample contains barium sulfate, save the contents of the crucible for further treatment.

SULFUR PRESENT AS ANTIMONY TRISULFIDE. Dilute the acid filtrate and washings to about 250 ml., and precipitate the antimony with hydrogen sulfide. Filter and wash the precipitate with water slightly acidified with hydrochloric acid and saturated with hydrogen sulfide. Transfer the filter paper containing the antimony sulfide to a Kjeldahl flask. Add 12 ml. of concentrated sulfuric acid and 5 grams of potassium sulfate. Heat gently until it is colorless. Cool, wash the funnel, and dilute with water to about 100 ml. Complete as directed for total antimony on page 569, starting with "Transfer to a 400-ml. beaker."

Calculate the antimony present to per cent of antimony trisulfide, and express the sulfur present in the antimony trisulfide as percentage of the sample.

BARIUM SULFATE. Transfer the mat and contents of the Gooch crucible, after acid washing, to a 50-ml. porcelain crucible. Fuse with 5 grams of a mixture of equal parts of sodium carbonate and sodium nitrate. Stir well during the fusion. Cool the crucible, put it into a 250-ml. beaker, and cover with distilled water. Digest on the steam bath or hot plate overnight. Filter the solution into a covered 400-ml. beaker containing 5 ml. of concentrated hydrochloric acid. Wash the residue thoroughly with hot 5 per cent sodium carbonate solution, receiving the washings in the same beaker. Adjust the acidity, by addition of 1:2 ammonium hydroxide or 1:1 hydrochloric acid, to between 0.2 and 0.3N with hydrochloric acid. This requires that roughly 0.7 to 1.1 per cent of hydrochloric acid shall be present and is conveniently accomplished by neutralizing to phenolphthalein with the ammonium hydroxide and adding 1 volume of 1:1 hydrochloric acid for each 20 volumes of solution. From this point proceed as directed for sulfur present as barium sulfate on page 568, starting with "Cool, and saturate with hydrogen sulfide." The precipitation of lead can be omitted if it is shown to be absent. Calculate the percentage of sulfur present as barium sulfate.

Sulfur in Compounding Ingredients. The sample for this determination was set aside after the rubber was in solution. Filter it on a Gooch crucible as described for the first sample, but omit the weighing of the crucible. Dry the crucible pad, and transfer it to a flask. Remove the last traces of the fillers from the sides of the crucible with wads of damp absorbent cotton, and put these in the flask. Add about 10 ml. of bromine water to the flask, and swirl the flask to moisten the contents. Add 20 ml. of concentrated nitric acid saturated with bromine. Allow to stand for 15 minutes, and then heat on the steam bath for about 1 hour. Transfer the contents of the flask to a 75-ml. crucible, and evaporate to dryness. Complete the determination of sulfur as described for total sulfur, alternative method, on page 565, starting with "Add 8 ml. of concentrated nitric acid, cover and warm for a few minutes on the steam bath." Calculate as the percentage of sulfur in the uncorrected weight of compounding ingredients.

Calculations. The calculation, in brief, is as follows:

Per cent of inorganic compounding ingredients
− per cent of sulfur other than that combined as $BaSO_4$ and Sb_2S_3
= corrected per cent of compounding ingredients

In more detail, this is as follows:

Uncorrected per cent of compounding ingredients
− per cent of organic compounding ingredients
+ per cent of sulfur as antimony trisulfide
+ per cent of sulfur in barium sulfate
− total per cent of sulfur in compounding ingredients
= corrected per cent of compounding ingredients

To determine the per cent rubber hydrocarbon, subtract from 100 the sum of the per cent acetone, chloroform, and alcoholic potash extracts, free carbon, glue, total sulfur corrected, and total inorganic compounding ingredients corrected.

Short Procedure. Use this procedure only for the analysis of simple rubber compounds consisting of rubber, sulfur, and undecomposable compounding ingredients as shown by the preliminary examination.

Determine acetone extract, free sulfur, alcoholic potash extract, total sulfur by the alternative method, and ash by the rapid method as described in the complete procedure.

Calculations. Consider the per cent rubber to be the difference between 100 and the sum of the total sulfur, ash, free carbon, and similar ingredients. If the alcoholic potash extract is over 2 per cent of the rubber as first calculated, subtract this excess also from the rubber. Obtain the organic acetone extract by taking the difference between the total acetone extract and the free sulfur. Calculate the organic acetone extract, free sulfur, total sulfur, and alcoholic potash extract on the amount of rubber as found by the above procedure.

CHAPTER 32

MISCELLANEOUS RUBBER ANALYSES AND SYNTHETIC ELASTOMERS

This chapter, essentially a continuation of the preceding, presents miscellaneous methods applied to rubber and rubberlike products, followed by a consideration of the properties of synthetic elastomers.

Artificial Aging Tests. Rubber deteriorates with age, particularly if exposed to light and if suitable antioxidants are not included in the compounds. Therefore, as a practical test, regardless of chemical composition, it is important to know the aging properties of a rubber compound. Natural aging is too prolonged a process to be practical; hence, accelerated aging processes have been devised to simulate natural aging. Such methods are discussed without going into detail as to the specialized equipment. Factors for conversion of results to natural aging are so unreliable that usually the results of comparison with a stock of known aging characteristics are reported.

GEER OVEN. Heat the specimen in an oven freely exposed to a slow stream of air, at 70 ± 1°. A special Geer oven is available, but for many comparative purposes any properly constructed and ventilated thermostatically controlled electric air oven will do. Observe for any visible deterioration. If necessary, at stated intervals make physical tests such as that for tensile strength on different specimens to determine the rate of deterioration. Normal exposure times are 2 to 14 days.

The aging effect in this test is mainly due to heat coupled with air oxidation. It has been found that, in a period of several days, deterioration in strength or bond of rubber to other materials may be produced comparable to the result of several years of natural shelf aging in darkness. Identical aging effects are not always obtained in duplicate tests, but a useful estimate of the life of a product may be obtained by this means.

BIERER-DAVIS OXYGEN BOMB. Heat the specimen in a special oxygen bomb containing oxygen at 300 ± 10 pounds per square inch pressure, at 70°. Special equipment is essential and must be located in a suitable bomb shelter, for explosions due to rapid oxidation have been reported. The effect in this test is largely due to oxidation. As the temperature employed is the same as in the Geer oven test the deterioration is usually much faster because of the presence of oxygen at a higher pressure. Normal exposure times are 1 to 8 days. Follow the rate of deterioration by observing the development of stickiness, brittleness, or loss of tensile strength, opening the bomb at suitable intervals for inspection of the samples.

Air Bomb. This modification of the oxygen bomb consists in heating the specimen in a bomb containing air maintained at 80 or 100 pounds per square inch pressure at 250 or 260°F. Precautions similar to those for the oxygen bomb are required. Specimens are often so tested at an elongation of 50 per cent. Under these conditions, the deterioration of the rubber is very rapid, the test being even more severe than the oxygen-bomb test. Exposures of 3 to 8 hours are frequently sufficient; they should not exceed 24 hours.

Latex. Latex is the form in which rubber is found in rubber-bearing trees. It is a milklike substance in which the rubber is present in very finely dispersed particles. These particles consist of a core of liquid rubber surrounded by a shell of rubber solid, or gel, which is itself surrounded by an outside protein layer. A typical analysis of natural latex is as follows:

	Per Cent
Rubber hydrocarbon	29–40
Proteins	2
Soaps, fat and fatty-acid esters, and sterols	1
Quebrachitol	1
Inorganic matter	0.4
Water	55–66

Small amounts of amino acids, lipins, sugars, enzymes, oxidases, etc., are present. Since rubber latex may coagulate on aging, generally less than 1.0 per cent of ammonia is added as a preservative.

Latex concentrates are commercial. They are prepared either by evaporation after the addition of sodium or potassium hydroxide or other permanent stabilizing agent such as potassium soap, or by partial separation of the solids from the serum. This separation may be accomplished by the use of a high-speed centrifuge or by a creaming agent such as ammonium alginate or gum karaya. The distinguishing characteristic of rubber prepared directly from latex rather than by the conventional coagulation and sheeting is that the protein content is much higher. This is due to the elimination of much of such protein in the coagulation of rubber but not in the deposition of rubber from latex.

Latex in which the rubber has been prevulcanized by the use of sulfur and ultra-accelerators is not uncommon. Its appearance is no different from that of an ordinary latex, but the film deposited on drying is far tougher than that of an unvulcanized latex.

Artificial latex is prepared by mechanically dispersing dry masticated rubber with soap. Such a dispersion has some advantages over natural latex since it contains no resinous substances and therefore on drying does not give such tacky solids as natural latex.

The tack of rubber latex solids finds use in its commercial application as an adhesive or cement, both compounded and uncompounded. Even without vulcanization the film is strong because the rubber has not been broken down by milling as is the case with solid rubber. These adhesives or cements find use on materials that are readily wetted with water since water is the external phase in the latex.

Oils, tars, natural resins, and other similar softeners that are soluble in rubber and cause it to swell produce a tacky soft rubber when added to latex. Materials such as alkalies, casein, gelatin, and silicates when mixed with latex produce dry films on evaporation. Thus in the various applications of rubber adhesives one must consider compounded latex. With and without compounding the latex, articles are made on forms by dipping. Rubber may be electrodeposited on forms from latex. Textiles may be impregnated with latex with or without fillers. Articles may even be molded after suitably bodying the latex. Latex is whipped up and vulcanized as a form of sponge rubber.

Analysis of Latex. Solid latex products and total solids deposited from latex may be analyzed by the conventional procedures for rubber. The methods that follow are for the liquid latex, compounded or otherwise.

Total Solids. Weigh out 2 to 3 grams of the latex, distributing it evenly in a dish with an area of approximately 32 sq. cm. Dry in an air oven at 70° for 16 hours. The weight so obtained includes all the natural resins and ingredients soluble in the serum as well as the rubber.

Dry Rubber Content. This is the actual rubber hydrocarbon present. Weigh rapidly, from a weighing bottle, 20 grams of normal latex or 10 grams of concentrated latex into a flat-bottomed dish containing a glass rod, and dilute to approximately 25 per cent solids. Coagulate the rubber by slowly adding 2 per cent acetic acid with constant stirring until the acid produces no further effect. Heat on a steam bath for 30 minutes, pour off the clear serum, and wash the clot of rubber at least six times with water while it is put through rollers to squeeze out the liquid. Dry the well-washed clot at 70° in an air oven until constant weight is attained. This usually takes 15 to 20 hours.

AMMONIA. Weigh about 10 grams of latex from a weighing bottle into a Kjeldahl flask. Add about 90 ml. of water. If ammonia soap is present, add 1 ml. of 10 per cent sodium hydroxide; otherwise, omit this addition. Such added alkali introduces error, for it tends to break down the proteins present, causing them to evolve additional ammonia. This effect is even more important if glue or casein is present.

Distill the ammonia as in the conventional Kjeldahl method, starting on page 144 at "Immediately close the flask with a rubber stopper through which a spray trap is inserted." Report the result as per cent ammonia based on the original sample.

PROTEINS. If the original sample contains ammonia, transfer the total solids previously determined to a 500-ml. Kjeldahl flask. Otherwise weigh 2 to 3 grams of latex directly into such a flask. Proceed by the Kjeldahl method described on page 143, starting at "Add 0.5 gram of copper sulfate crystals as a catalyst." Calculate the result as protein, using the factor 6.25 for protein nitrogen, after subtracting the nitrogen equivalent of any ammonia in the latex if an original sample is used.

This determination is not often required for natural latex. It may be used to determine with fair accuracy the amount of casein or glue in a compounded latex. This requires interpretation, to allow for the amount of nitrogen in the protein of the latex, which, in turn, depends on the method of separation of the rubber. If the sample is coagulated, usually about 0.4 per cent of nitrogen is present as protein. If dried down as latex or deposited as whole latex solids, the protein may be equivalent to nearly 1 per cent of nitrogen. No hard and fast rule can be stated for interpretation.

WATER-SOLUBLE SUBSTANCES. For plain latex this value is obtained with reasonable accuracy by subtraction of the dry rubber solids from the total solids. For compounded latices this involves also subtracting ash and perhaps casein, glue, etc.

For greater accuracy with latex not compounded with water-soluble compounding ingredients, use the following procedure: Weigh 5 grams accurately into a 400-ml. beaker, and immediately dilute with 200 ml. of water. Cover with a watch glass, and boil on a hot plate until the volume is reduced by one-half. This drives off ammonia. Transfer to a 200-ml. calibrated flask, which should have a wide neck to facilitate cleaning. Make up to within about 20 ml. of the mark, and add 1 ml. of 0.1 per cent methyl orange indicator solution. Add N sulfuric acid from a buret until the red-orange color indicating a pH of about 4.3 is obtained and the rubber is well coagulated. Too much acid causes bad coagulation. Dilute to the mark, shake thoroughly, and filter the solution. Obtain total solids on 100 ml. of the filtrate by drying in an air oven at 70° to constant weight. From twice the weight so obtained, subtract 0.0049 gram for every milliliter of *N* sulfuric acid used and 0.001 gram for the weight of indicator. Calculate the residual weight as water-soluble substance.

Ash. Make sure by qualitative tests that lead or mercury compounds and ingredients producing reducing gases such as hydrogen sulfide are not present before carrying this out in platinum. Then weigh 20 grams of latex into a platinum or porcelain dish, and evaporate to total solids on a water bath. Ash at a low temperature. Cool, and weigh. This is a measure of the inorganic matter present.

This determination is not often required for a natural latex but is a usual means of determining the inorganic filler of a compounded latex. If necessary, analyze the ash by the procedure for mineral analysis (Chap. 10), in order correctly to identify the added filler.

Copper and Manganese in Rubber Products. While many metals have the property of accelerating the deterioration of rubber, only two, copper and manganese, are important. This effect is present at the bond when rubber is joined to another material, as well as in the rubber itself. Experience indicates that the upper limit of 50 ppm. for copper and 10 ppm. for manganese used in the United States Army Specification 6-9A is a fair standard. Both should be determined and the equivalent total amount calculated. For example, 25 ppm. of copper and 5 ppm. of manganese are equivalent to either 50 ppm. of copper or 10 ppm. of manganese.

Rubber Compounds. Accurately weigh 5 grams of the sample, and transfer to a 500-ml. Kjeldahl flask. Treat with concentrated sulfuric acid and concentrated nitric acid as described with reference to the wet ashing of organic matter on page 206. If, after repeated treatments, the final solution is still yellow, cool, carefully add 5 ml. of 30 per cent hydrogen peroxide, and heat with a low flame to strong fumes of sulfur trioxide. Repeat the procedure until no decrease in color occurs. Cool, add 100 ml. of water, and boil down to fumes to destroy any hydrogen peroxide. Cool again, dilute with 100 ml. of water, filter off any insoluble matter, and dilute the filtrate and washings to 250 ml.

Determine copper on an aliquot of the solution by the method described on page 206. Determine manganese by one of the methods on pages 199 to 201.

Cloth for Rubberizing. The size of sample will depend on the amounts of metals present; in general, use as large a sample as possible. A weight of 25 to 50 grams is desirable. Add the cloth slowly in small pieces to a platinum or porcelain dish heated over a low flame, and char. Then place a clean triangle over the dish and a porcelain cover on top. All the carbon must be burned out, and this is more readily accomplished at a temperature low enough so that the ash does not fuse, rather than at a higher temperature.

Dissolve the ash as completely as possible in 10 ml. of 1:4 nitric acid. Filter into a 100-ml. flask, wash the filter thoroughly, and dilute to volume. Determine copper and manganese as recommended in the preceding method.

Gutta-percha and Balata. The hydrocarbons of gutta-percha and of balata are chemically similar to those of rubber but have important differences, the most outstanding of which is that they are thermoplastic but not resilient.

Gutta-percha, which is appreciably more expensive than balata, is used largely as submarine-cable insulation material, for molding, for surgical equipment, and as a thermoplastic cementing material. Balata finds considerable use in belting. Both can be vulcanized, but their value is not thus increased.

The following figures give typical compositions of the commercial materials:

	Gutta-percha, %	Balata, %	
		Before washing	After washing
Water	1–1.5	7	1
Gutta-percha hydrocarbon	30–83		
Balata hydrocarbon	......	43	52
Resins	......	39	47
Dirt	3–5	10	
Alban (alcohol-insoluble resins)	7–44		
Fluavil (alcohol-soluble resins)	3–21		

These materials, like rubber, oxidize readily and should be stored in the dark. Once oxidization sets in, deterioration is much more rapid than is the case with rubber.

Moisture. Samples as received are frequently much drier outside than inside. Make sure whether the average or the damper central part is required to be analyzed. Moisture determined at 100° in a vacuum oven on a 2- to 3-gram sample is unreliable. Preferably, therefore, weigh out about 25 grams, cut into lumps, and determine the moisture by xylene distillation (page 41).

DIRT AND FILLERS. Dissolve 1 gram of dried sample in 35 ml. of toluene. Filter through a dry filter paper or a Gooch crucible previously conditioned in a desiccator for 3 hours. Wash with three 15-ml. portions of hot toluene. Save the filtrate and washings. The process of filtration and washing is often slow. Evaporate the toluene from the paper or crucible at a low temperature. Condition the filter or crucible as before, and weigh.

Alternatively, dissolve the sample in a weighed centrifuge tube. Centrifuge until the toluene layer is clear. Decant and save the toluene. Wash three times with hot toluene, separating the clear layer by centrifuging. Dry the centrifuge tube and contents at 100°, and weigh. The contents may be brushed out for microscopic examination after weighing. This residue includes fungi, bark, sand, and twigs, as well as proteins and other serum products of the original latex. Compounds may also show filler such as clay, talc, chalk, or other pigments. The residue or its ash is analyzed if necessary by the usual methods for mineral analysis (Chap. 10).

RESINS. These are of a complex nature and consist largely of the esters of α- and β-amyrin, lupeol, and other phytosterols, associated with fats and organic acids. Cool the toluene filtrate and washings from the determination of dirt and filler to 15°, and slowly add three times the volume of acetone, stirring constantly. The hydrocarbon precipitates in clots, some usually remaining as a cloudy suspension. Let stand at least 2 to 3 hours, preferably overnight, when the solution should become clear by precipitation of the suspended hydrocarbons. Redissolve the hydrocarbons in 35 ml. of toluene, and reprecipitate with three volumes of acetone. Wash the hydrocarbons with cold acetone, squeezing the clot occasionally. Combine the solutions, evaporate off the solvents carefully, and heat to constant weight at 100°. The color and physical characteristics of the resin so obtained are significant to the worker experienced in handling gutta-percha and balata.

HYDROCARBONS. Dry the precipitated hydrocarbons obtained under Resins at 100°, and weigh.

Separation of Gutta-percha or Balata Hydrocarbon from Rubber. Roll the sample into a very thin sheet. Disperse 3 grams of this in 200 ml. of petroleum ether in a flask. Warm on a steam bath, with continued shaking, until a thorough suspension is formed. Allow to cool to room temperature, and then cool in a freezing mixture to −12°. The gutta-percha or balata hydrocarbon is precipitated. Filter the contents of the flask through a funnel surrounded by a freezing mixture, and wash with chilled petroleum ether. Dry the precipitate in a vacuum desiccator at 100° to obtain the gutta-percha or balata hydrocarbon. Either calculate the rubber and resins by difference or evaporate the petroleum ether and washings, and dry to constant weight.

Rubber Accelerators. Accelerators are used to increase the rate of vulcanization of rubber and also to decrease the amount of sulfur required to obtain the optimum physical properties of a vulcanizate. About 7 to 10 per cent of sulfur is required without an accelerator. If no accelerator had been used with a soft rubber, this would be indicated by the high total sulfur content. With the usual accelerators the sulfur used is 2.5 and 4.0 per cent. An extreme case is when an ultraaccelerator of the dithiocarbamate type is used. This carries its own sulfur, and frequently no additional sulfur is required. Improvement in physical properties is also obtained by the use of an accelerator. The more important inorganic accelerators are litharge and zinc oxide. Organic compounds are more powerful, and consequently less is required in a compound.

Typical organic accelerators may be divided into several classes. Typical organic bases so used are aniline, *p*-phenylenediamine, and *p*-aminodimethylaniline. Typical amine aldehydes are acetaldehydeaniline, butyraldehydeaniline, heptaldehydeaniline, and hexamethylenetetramine. Guanidines such as diphenylguanidine, di-*o*-tolylguanidine, and phenyl-*o*-tolylguanidine are used. A typical nitroso accelerator is *p*-nitrosodimethylaniline. These types of compound are now relatively little used commercially. Thiocarbamate-type compounds such as dithiocarbamates and dithio- and trithioacid derivatives, which are similar to the xanthates, are used. Others are thiocarbanilide, thiuram disulfides such as tetramethylthiuram disulfide, and thiuram monosulfides such as tetramethylthiuram monosulfide. Thiazole compounds used are typified by mercaptobenzothiozole. Zinc oxide is used as an activator with the thiocarbamate types. Not infrequently two or even three accelerators are used together to give the modified properties desired. It follows that only rarely will the analysis of a rubber compound indicate specifically the accelerator used, unless performed by a person who has spent years in such work. Owing to reaction and decomposition during vulcanization, only a very approximate indication of the amount originally present can be obtained, even if the specific accelerator is known.

Identification of Accelerators in Vulcanizates. Obtain the acetone extract from 2 grams of rubber,[1] using the conventional procedure described on page 562, with the modifications noted. If aldehyde ammonia is sus-

[1] Twiss, D. F., and Martin, G., *India-Rubber J.*, **61**, 1283–1284 (1921); *Rubber Age*, **9**, 379–380 (1921).

pected, add a few drops of 0.5*N* hydrochloric acid to the acetone to fix any ammonia that might otherwise volatilize. Determine the nitrogen content of the acetone extract by the Kjeldahl method (page 143). If the percentage of acetone-soluble nitrogen calculated on the original material exceeds 0.05 per cent, an organic accelerator is probably present. Although the nitrogen content of an accelerator is usually 12 to 25 per cent, hexamethylenetetramine exceeds these limits. To identify the accelerator it is best to examine the sample as soon as possible after vulcanization, for a freshly cut surface may have an odor characteristic of the accelerator used, such as of thiocarbanilide types, of aldehyde ammonia, and of hexamethylenetetramine. A typical surface darkening characterizes *p*-phenylenediamine. A rough qualitative scheme has been developed.

Nitroso-type Accelerators. These give a characteristic color to the acetone extract. The most commonly used substance of the type, *p*-nitrosodimethylaniline, gives a striking yellow to deep brown acetone extract, the proportion being roughly indicated by the depth of color. During the vulcanization *p*-nitrosodimethylaniline will have been converted in part to *p*-aminodimethylaniline. If a portion of the acetone extract is heated with a saturated aqueous solution of hydrogen sulfide and then with a ferric chloride solution, a blue to greenish blue is developed due to the formation of some methylene blue. For further confirmation, boil about 5 grams of a comminuted sample under a reflux condenser with 1:2 hydrochloric acid for several hours. Cool, and filter. This solution will be yellow but will change color when made alkaline. The yellow color is extractable with ether. As another test, dissolve a portion of the acetone extract in 1:5 hydrochloric acid, and add a small piece of metallic zinc. This reduces a nitroso compound to a primary amine. To show the presence of the primary amine, cool the reduced acid solution in cracked ice, and add a few drops of cold 1 per cent sodium nitrite solution. After mixing well, let this stand for 3 minutes, and pour into a solution of a few crystals of β-naphthol in aqueous sodium hydroxide. A colored compound results, the color of which varies. The reduced acid solution will give the methylene blue reaction previously discussed more readily than the acetone extract does.

Ammonia Types. These give ammonia in the aqueous extract obtained by soaking a finely divided sample for an hour. Determine the presence of ammonia with Nessler's reagent (page 88) as compared with a blank. A positive test usually is indicative of aldehyde ammonia or hexamethylenetetramine. It may occasionally result from furfuramide, hydrobenzamide, alkylamines, alkylammonium dithiocarbamates, and urea derivatives.

Amines and Amides. Those can be distinguished from ammonia types by their characteristic amine reactions. Extract about 5 grams of finely divided sample with 1:2 hydrochloric acid by boiling under a reflux condenser for several hours. When cool, filter. Make a portion of the extract alkaline with sodium hydroxide, and distill. Ammonia in the distillate results from nitroso compounds, ammonium compounds, amines, and amides. Warm another portion of the acid extract. Thiocarbanilide gives phenyl isothiocyanate—mustard oil. Others of this type give related products. Diazotize as described for nitroso compounds (page 589). The compound on coupling with β-naphthol is red if a primary amine is present. This may be aniline, toluidine, *p*-phenylenediamine, and many others.

Alkaloid Types. These should be suspected if the previous class reactions are negative. An aqueous solution of the extract will react with potassium iodide to show quinidine if quinoidine was used.

Thiocarbamate Class. This group is particularly difficult to detect, for they are used in much smaller proportion than the majority of aromatic accelerators and are likely to be lost by volatilization during vulcanization.

Mercaptobenzothiazole. This may be isolated as the pure compound. Extract 30 grams of rubber in a Soxhlet thimble with acetone for 6 hours. Concentrate the extract to about 50 ml., cool, and decant from free sulfur. Add 10 ml. of petroleum ether, and stir well. Filter the precipitate, and wash with 20 ml. of petroleum ether. Dry the paper and residue at 70°, and place in a 50-ml. beaker. Add 10 ml. of 5 per cent sodium carbonate solution. Boil gently, with stirring, to dissolve the compound. Filter, and wash the paper once. Acidify the filtrate with 1:5 hydrochloric acid to reprecipitate the compound. Filter, wash, and redissolve the precipitate in the minimum amount of 5 per cent sodium carbonate solution. Again precipitate with 1:5 hydrochloric acid. Wash the precipitate thoroughly, and dry at 105°. Determine the melting point of the product thus recovered. It should be 174 to 177° if mercaptobenzothiazole has been used.

Synthetic Elastomers. A synthetic elastomer is the type of product conventionally called a synthetic rubber. It may be defined as a substance that can be stretched to at least twice its original length and can then return to approximately its original length or position in a reasonable time. This definition is necessarily somewhat vague, for no national or international limits have been set for the amount of elongation or time of return of set for a substance that may be reasonably called a synthetic rubber.

Many substances such as those described in the preceding paragraph are available. In general, in some respects they are inferior to rubber, and in others they are superior. It is possible to synthesize a type of synthetic elastomer that will be superior to rubber for a particular kind of service.

Natural rubber is considered to be a polymer of isoprene, $CH_2{:}C(CH_3)CH{:}CH_2$. Some synthetic elastomers are similar, but not identical, in chemical composition. The chief types and more commonly known individuals will be briefly considered. There are no definite tests for synthetic elastomers as a class, and the series of chemical tests given for rubber compounds do not apply. Identification of a synthetic elastomer can be made only by class tests indicated by the structures and by physical properties.

TYPES. There are three general classes of synthetic elastomers, the simple polymers, the copolymers, and the elastothiomers. Tables 36 to 38 give the general chemical formulas in comparison with that of rubber.

The Bunas are German products; in the United States, GR-S is the term for the same products; SKA, SKB, and Sovprene are Russian. Ker was manufactured in Poland, while Mustone was made in Japan. The other products listed are of American manufacture. Such structural data as appear in the tables are given in the belief that they will permit class identification by qualitative tests, elastothiomers by testing for sulfur, for example, and will furnish a basis for the development of methods of examination not yet formulated.

TABLE 36.—CHEMICAL STRUCTURES OF SIMPLE POLYMERS USED AS SYNTHETIC ELASTOMERS

Type	Monomer	Unit of monomer	Unit of polymer
Rubber	Isoprene, 2-methyl-1,3-butadiene	$CH_2{:}C(CH_3)CH{:}CH_2$	—$CH_2C(CH_3){:}CHCH_2$—
GR-S, the numbered Bunas, SKA, SKB, or Ker	1,3-Butadiene	$CH_2{:}CHCH{:}CH_2$	—$CH_2CH{:}CHCH_2$—
GR-1, methyl rubber	2,3-Dimethyl-1,3-butadiene	$CH_2{:}C(CH_3)C(CH_3){:}CH_2$	—$CH_2C(CH_3){:}C(CH_3)CH_2$—
Neoprene, Sovprene, Mustone	Chloroprene, 2-chloro-1,3-butadiene	$CH_2{:}CClCH{:}CH_2$	—$CH_2CCl{:}CHCH_2$—
Vistanex, Oppanol	Isobutene, 2-methyl-1-propene	$C(CH_3)_2{:}CH_2$	—$C(CH_3)_2CH_2$—
Polyvinyl chloride, Koroseal, Korogel, Flamenol	Vinyl chloride	$HCCl{:}CH_2$	—$HCClCH_2$—

VULCANIZATION PROPERTIES. The butadiene copolymers are chemically nearest to natural rubber and can be vulcanized with sulfur, zinc oxide, and accelerators more or less as rubber is vulcanized. Butyl rubber is similarly vulcanized. Neoprene is vulcanized by heat alone, sulfur not being necessary. Sulfur lowers the heat and water resistance of neoprene compounds and tends to promote hardening or aging at elevated temperatures. On the other hand, sulfur improves the resilience and permanent set, which is advantageous, for neoprene is inferior to natural rubber in these properties.

TABLE 37.—CHEMICAL STRUCTURES OF COPOLYMERS USED AS SYNTHETIC ELASTOMERS

Type	Ingredients	Unit of polymer
GR-S, Buna S, or Buna SS	CH_2:CHCH:CH_2 + CH_2:CHC$_6H_5$ Butadiene Styrene (vinyl benzene)	—CH_2CH:CHCH$_2$CH$_2$CH— | C_6H_5
Perbunan or Perbunan extra	CH_2:CHCH:CH_2 + CH_2:CHC⋮N Butadiene Acrylonitrile vinyl cyanide	—CH_2CH:CHCH$_2$CH$_2$CH:C:N—
GR-1, butyl rubber, Chemigum, Hycar OR	Olefins + small amounts of diolefins Butadiene + other unnamed substances	

TABLE 38.—CHEMICAL STRUCTURES OF ELASTOTHIOMERS USED AS SYNTHETIC ELASTOMERS

Type	Ingredients	Unit of polymer
Thiokol A	H_2CClCH$_2$Cl + NaS$_2$—S$_2$Na Ethylene dichloride Sodium 1,2-dichloroethane tetrasulfide	—CH_2CH_2—S_2—S_2—
Thiokol D Perduren G	H_2CClCH$_2$OCH$_2$CH$_2$Cl + NaS$_2$—S$_2$Na Dichloroethyl ether Sodium tetrasulfide	—CH_2CH_2—O—CH_2CH_2—S_2—S_2—

The Thiokols require a metallic oxide like zinc oxide. In this case sulfur in the compound appears to be sufficient. Vistanex is not vulcanizable. It is thermoplastic and can be molded or mixed with rubber but remains thermoplastic. Koroseal is similar in these respects and cannot be vulcanized. It retains its shape and properties as long as it is not used above 75°.

When natural rubber is overvulcanized, a change occurs, known as reversion and consisting in a softening of the compound and a general lowering of desirable physical properties. The synthetic elastomers do not show reversion; even extra-long heating is seldom disadvantageous. The butadiene rubbers can be converted into hard rubbers by the use of a large amount of sulfur. The other types cannot be so converted but, when mixed with natural rubber and much sulfur, give flexible hard rubbers.

Other Properties. A summary of the properties of various synthetic elastomers may be useful in their identification. The limited available data on maximum tensile strength and elongation are shown in Table 39. Specific gravities of the uncompounded stock will often be helpful, particularly as the chloro and sulfo elastomers have high specific gravities. Some typical figures are as follows: natural rubber, 0.92; Buna 85, 0.92; Buna S, 0.91; Perbunan, 0.96; Perbunan extra, 0.97; Hycar OR, 1.00; butyl rubber, 0.91; Vistanex, 0.91; neoprene, 1.25; Thiokol A, 1.60; Thiokol D, 1.34; Koroseal with average amount of plasticizer, 1.30.

The general principles used in compounding rubber are modified for use in compounding synthetic elastomers. Softeners are more often used since the synthetics are tougher than natural rubber. As they age better, also, it is not always necessary to use antioxidants in compounding. However, the tendency of some synthetic elastomers to continue to polymerize in storage is often retarded by the addition of antioxidants. The butadiene copolymers and neoprene generally contain upward of 2 per cent of phenyl-β-naphthylamine for this purpose. More antioxidant is needed in heat-resistant compounds.

Both laboratory and service tests indicate that synthetic elastomers, as a class, show better resistance to aging than do natural-rubber compounds. This superiority is particularly noticeable in their resistance to sunlight, which causes checking of natural rubber compounds.

The permanent set of synthetic elastomer compounds is usually high. On long standing after stretching, the set may compare with that of natural rubber compounds. The elastoprenes, that is, the compounds related to butadiene including neoprene, are more like rubber, whereas Vistanex, Thiokol, and Koroseal show a high set.

The butadiene polymers compare favorably with and are sometimes better than rubber in abrasion resistance. Buna S, Buna N, Chemigum, and Hycar OR[1] tread stocks have shown better wear than natural rubber, with neoprene close to them in this

[1] Meaning "oil resistant."

respect. Koroseal is good in abrasion resistance if no heat is generated, but Thiokol is poor.

In general, synthetic-elastomer compounds are not so good as natural-rubber compounds in regard to tear resistance.

TABLE 39.—MAXIMUM TENSILE STRENGTHS AND CORRESPONDING ELONGATIONS

Compound	Unvulcanized		Vulcanized pure gum compound		Vulcanized carbon black compound	
	Tensile strength, p.s.i.	Elongation, %	Tensile strength, p.s.i.	Elongation, %	Tensile strength, p.s.i.	Elongation, %
Natural rubber	355	1,200	4,125	710	5,000	650
Buna 85	.....	.....	.....	...	2,490	600
Buna 115	.....	.....	.....	...	2,845	700
Buna S	.....	.....	.....	...	4,200	650
Perbunan	.....	.....	2,130	900	5,000	600
Chemigum	.....	.....	.....	...	4,425	630
Hycar OR	.....	.....	.....	...	3,500	450
Methyl rubber W	.....	.....	425	510	2,415	530
Neoprene	425	1,100	4,265	820	4,215	700
Thiokol A	.....	.....	.....	...	855	370
Thiokol D	100	570	500	750	1,705	610
Vistanex MM	285	1,000	Nonvulcanizable		Nonvulcanizable	
Vistanex HM	855	1,000	Nonvulcanizable		Nonvulcanizable	
Koroseal, unplasticized	5,690	2	Nonvulcanizable		Nonvulcanizable	
Koroseal, 30 per cent plasticizer	3,840	170	Nonvulcanizable		Nonvulcanizable	
Koroseal, 60 per cent plasticizer	995	500	Nonvulcanizable		Nonvulcanizable	

Resistance to swelling in various solvents, oils, fats, and greases is an outstanding property of most synthetic elastomers and their compounds. None is inferior in this respect to natural rubber. Buna S, Perbunan, neoprene, and butyl rubber swell considerably in benzene and carbon tetrachloride. Natural rubber is unaffected by acetone, but Perbunan is dissolved by it. Thiokol A either does not swell or in some cases swells only slightly in solvents and oils. Thiokol D is swelled somewhat by aromatic hydrocarbons. Koroseal is unaffected by the aromatic and aliphatic hydrocarbons, that is, by benzene and gasoline, but is affected by ketones, esters, and nitro and chloro compounds and amines, which remove the plasticizer. Hycar OR shows good resistance to swelling in gasoline, and certain varieties of Chemigum are very good.

Natural rubber and the butadiene polymers burn when a flame is applied and continue to burn when the flame is removed. Neoprene and Koroseal are flame resistant and usually do not continue to burn when the flame is removed. However, unless distinguished by sulfur or chlorine content, the synthetic elastomers present a field in which much must be left to guesswork.

CHAPTER 33

ANALYSIS OF COAL

The methods of analysis of coal given are selected from those of the United States Bureau of Mines and of the American Society for Testing Materials (ASTM). Many details required in referee work but not in normal use have been omitted.

The correct sampling of coal is of great importance. Mechanical devices are available for such sampling at the pit. In principle,

FIG. 110.—Mechanical pulverizer.

the cone and quartering method (page 38) is largely employed. The description given here applies to samples normally received at the chemical laboratory.

Preparation of Sample. For coarse coal in any amount too large to handle in a mortar, use a jaw crusher. Mix the crushed coal by forming a pile, and quarter the pile. Continue to pile and quarter until a 2-pound sample is obtained.

When crushed coal in amounts larger than 2 pounds is received, pile and quarter until a 2-pound sample is obtained. If this sample is wet with surface moisture so that it will stick to the mill or agglomerate in grinding, weigh, spread out to dry out excess moisture at room temperature and reweigh. Such loss is reported as moisture lost in air drying.

The 2-pound sample then goes directly to a mechanical pulverizer such as that shown in Fig. 110 or to a large mortar and pestle. Pulverize to pass 16 mesh. Mix by rolling on a large sheet of paper, work into a long narrow row, and lift out about 20 increments by using a spatula, so as to obtain about a 4-ounce sample. Reserve the rest for any retests. An alternative if many samples are being handled is to pass the sample several times through a riffle such as those shown in Fig. 111.

Remove and save any dust passing 40 mesh. Now further break down the sample, using a coffee-mill type of grinder with enclosed grinding plates

Fig. 111.—Riffles.

and a receptacle to catch the ground coal. Continue grinding until all will pass a 40-mesh sieve. Add to the reserved dust, mix well, and store a 2-ounce portion for analysis in a stoppered weighing bottle. Recommendations of the Bureau of Mines are for grinding to 60 mesh at this point. If the original sample was free from superficial moisture, there will be no appreciable loss of moisture during sampling and pulverizing.

PROXIMATE ANALYSIS

Moisture. Weigh 1 gram of coal into a porcelain capsule—1⅝ by ⅞ inch is a convenient size—and heat for 1 hour at 105° in an oven. If special capsules are unavailable, reasonably satisfactory results are obtained with low-form porcelain crucibles. Special ovens are available but are not essential. Cool, preferably over sulfuric acid, in a desiccator, weigh, and calculate the percentage loss in weight.

Volatile Matter. Weigh 1 gram of coal into a 10-ml. platinum crucible with a close-fitting capsule cover. Heat exactly 7 minutes in an electric furnace at 950°. Special vertical furnaces are available but are not essential. For routine work it is sufficient to use a Meker burner with artificial gas, placing the crucible 1 cm. from the top of the burner. The lid must fit snugly. Remove from the source of heat, cool, and weigh. Coals particularly high in volatile matter or moisture may cause the cover to pop out of the crucible when treated as above. In such cases, heat slowly over a low flame for not less than 5 minutes and not more than 10 minutes; then heat at 950° for 6 minutes. The percentage loss in weight minus the percentage of moisture is the volatile matter, often referred to as volatile combustible matter (vcm.).

Ash. Use the sample taken for the moisture determination. Place the porcelain capsule in a muffle furnace. Heat slowly until all volatile matter is gone, to prevent coking of the sample. Continue heating at 700 to 750°, at a red but not bright red glow, until all particles of carbon have disappeared. Occasionally stir the ash with a platinum wire. Cool in a desiccator, and weigh. Heat again for 30 minutes. Cool, and weigh. Repeat until a constant weight within 5 mg. is attained.

Fixed Carbon. Subtract the percentages of moisture, ash, and volatile matter from 100.

ULTIMATE ANALYSIS

Sulfur. ESCHKA METHOD. Follow the procedure given on page 162. Alternatively, as a method of heating, start with a cold electric muffle and gradually raise the temperature. Maintain a maximum temperature of 800 ± 25° for about 90 minutes. A blank is essential.

OXYGEN-BOMB METHOD. The most convenient and accurate method of determining sulfur is in the oxygen bomb as fully described on page 161. A convenient and timesaving alternative is to determine the sulfur in the washings from the oxygen-bomb calorimeter after the calorimetric determination, the same technic being followed.

Carbon and Hydrogen. This method, with suitable adjustment of the method of heating the sample depending on its composition, is useful in determining carbon and hydrogen in many organic compounds. Gas-heated furnaces may be used, but electrically heated ones are more easily manipulated. The latter method, as used by the Bureau of Mines, is described in the following paragraphs.

APPARATUS. The furnace of the Heraeus type consists of three independent heaters (Fig. 112, *a*, *b*, and *c*). It is substantially the same as all furnaces for determining hydrogen and carbon by combustion. Two of these heaters, *a* and *b*, connected to rheostats *f* and *g*, are movable along the tube. The third, *c*, which surrounds the lead chromate, is stationary. This form of furnace though not essential is highly convenient. The usual furnaces are not long enough, and the lead chromate has to be heated by a gas flame, with less flexibility and with frequent damage by melting to the tube.

As combustion tube, use a hard glass or fused silica with a transparent section. The total length is 100 to 110 cm., and the external diameter is about 21 mm. Insert a rubber stopper at one end, and draw out the other sufficiently to make a rubber tubing connection to the absorption train. To allow observation the 12-cm. transparent section should begin about 35 cm. from the large end. The first part of the tube up to the end of the transparent section is empty. Following this space is a roll of copper gauze about 2 cm. long; the next 33 cm. is filled loosely with copper oxide in short lengths. A second roll of copper gauze separates the wire from 18 cm. of fused lead chromate, which is held in place by a plug of glass wool extending to the constricted end of the tube.

Purify the oxygen or air entering the combustion tube by passing through a drying apparatus *H*, containing the following arranged in order of the passage of the gas through them: sulfuric acid *i* for removing traces of ammonia, 30 per cent potassium hydroxide solution *j*, granular soda lime *k* for removing traces of carbon dioxide, granular calcium chloride *l* for removing moisture. The absorption tube *i* is connected to an oxygen tank with a suitable reducing valve.

The absorption train consists of a 5-inch tube *m* filled with granular calcium chloride to absorb moisture. Saturate the calcium chloride, before use, with carbon dioxide to prevent possible absorption of the latter from the sample by any traces of calcium oxide that may be present. This is conveniently done by passing carbon dioxide through a tower filled with calcium chloride or by filling a large jar containing calcium

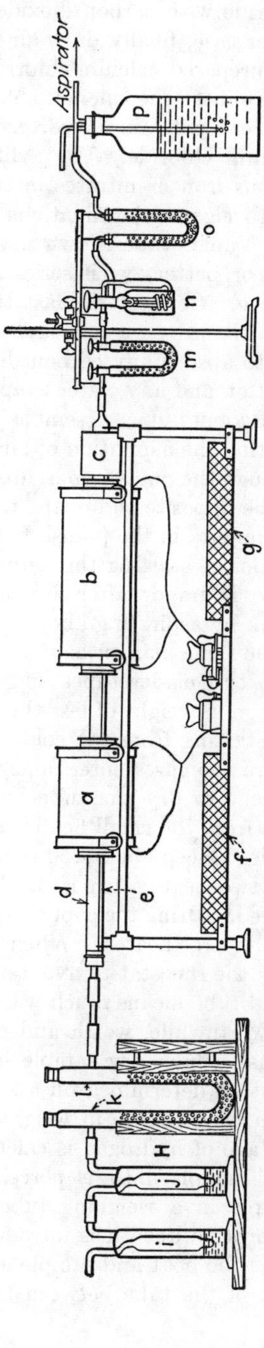

FIG. 112.—Furnace and other equipment for analysis of coal for carbon and hydrogen. [*Stanton, Fieldner, and Selvig, Methods of Analyzing Coal and Coke, U. S. Bur. Mines Tech. Paper* 8, 19 (1939).]

chloride with carbon dioxide and then allowing it to stand overnight. In either case, finally draw air through to remove any carbon dioxide. Keep the prepared calcium chloride in well-stoppered bottles.

Connect the calcium chloride tube to a potash bulb *n* containing a 30 per cent potassium hydroxide solution and some soda lime and granular calcium chloride. Use a little potassium permanganate to oxidize any ferrous iron or nitrates in the potassium hydroxide solution. Normally, six to eight determinations may be made with one charge in the bulb. The Vanier tube shown may be replaced by the more common Geissler tube or, better, two in series. Follow these by a soda lime–calcium chloride tube *o*. Connect the last tube to an aspirator, using a guard tube *p* to keep the pressure constant. The soda lime and granular calcium chloride are to absorb any carbon dioxide escaping from the potassium hydroxide solution and any water evaporating from it.

PROCEDURE. Assemble the apparatus, and test for tightness by starting the aspiration of air at the rate of about three bubbles per second through the potassium hydroxide bulb; then close the inlet for air or oxygen at the opposite end of the train. If there are no more than three bubbles per minute in the potash bulb, the apparatus is considered tight.

Before starting the determination, after the train has been out of use some hours or after any changes in chemicals and connections, run a blank by aspirating about a liter of air through the train, which is heated in the same manner as if a determination were being made. If the Vanier bulb or Geissler tubes and the tube containing calcium chloride show a change in weight of less than 0.5 mg. each, the apparatus is ready for use.

Starting from the cold, move the heaters toward the end of the tube where the gases enter, and warm the air enough by cutting in the electric current to dry the tube and its contents thoroughly. Cut off the current from the small heater *a*, and move the large heater *b* over the copper oxide. Keep about 250 mm. of that part of the combustion tube between the two heaters exposed where the boat containing the sample is to be placed. Bring the copper oxide to a red heat by turning on the full current on the large heater. When this temperature is reached, reduce the current with the rheostat to avoid overheating the tube. This is important, for a fused tube means much wasted work.

Meanwhile, weigh and connect the absorption train. It is advisable to use a bone-dry sample in this determination; otherwise, the moisture must be determined on a similar sample and the weight later subtracted from the increase in weight of the calcium chloride tube before the percentage of hydrogen is calculated. Weigh 0.2 gram of the finely meshed coal sample into a porcelain or platinum boat. Place the boat and sample in a weighing tube closed with a stopper to prevent moisture absorption until it is introduced into the combustion tube. When ready, place the boat and sample as rapidly as possible in the exposed and cooler part of the tube between the two heaters. The copper oxide should be red-hot and the lead chromate at dull red heat at this stage. As soon as the boat is in place near the asbestos plug at the beginning of the copper oxide, insert the stopper connecting with the purifying train and aspirate pure oxygen gas through the apparatus at the rate of about three bubbles per second.

Pass the current through the shorter heater. Control the rate of evolution of volatile matter by manipulating the rheostat and by gradually pushing this heater toward the boat. The heat must be increased gradually until all the volatile matter is driven off. Prevent too rapid an evolution of gas and tar, which would thus escape complete combustion or be driven back into the purifying train. After the sample is ignited, increase the heat more rapidly but take care not to overheat and soften the combustion tube.

Continue the aspiration with oxygen for 2 minutes after the sample ceases to glow. Then turn off the current, and aspirate about 1,200 ml. of air through the apparatus. Disconnect the absorption bulbs, cool to room temperature, wipe with a lintless cloth, and weigh. Weigh the ash in the boat, and carefully inspect for any unburnt carbon, which would vitiate the results. The percentage of ash found should check with that determined by proximate analysis.

CALCULATIONS.

$$\text{Increase in weight of calcium chloride tube} \times 0.1119 \times \frac{100}{\text{weight of sample}} = \text{per cent hydrogen}$$

$$\text{Increase in weight of potassium hydroxide bulbs} \times 0.2729 \times \frac{100}{\text{weight of sample}} = \text{per cent carbon}$$

Nitrogen. Using a 1-gram sample, determine the nitrogen by the Kjeldahl method (page 143). For coke, add about 0.1 gram of chromic oxide, Cr_2O_3, to the mixture before boiling. Bituminous samples require a total digestion period of 3 to 4 hours. Coke and anthracite coal require much longer.

Oxygen. In ultimate analysis the composition of coal is conventionally expressed in percentages of ash, sulfur, carbon, hydrogen, nitrogen, and oxygen. The sum of these constituents is taken as equal to 100 per cent. As there is no method for the determination of oxygen, it is estimated by subtracting the sum of the other five constituents from 100. This method throws the sum of all the errors incurred in the other determinations upon the oxygen.

Phosphorus. Ash a 5-gram sample at a temperature of 700 to 750° as in the determination of ash. Treat the ash so obtained in a 30-ml. platinum crucible with 10 ml. of concentrated nitric acid and 5 ml. of 48 per cent hydrofluoric acid. Evaporate to dryness. Make a second evaporation with 5 ml. of concentrated nitric acid. Heat the residue over a Meker burner to red heat until no more fumes are given off, and then fuse with 3 grams of sodium carbonate. Leach the melt thoroughly with hot water, and filter. Ignite the paper and residue, and again fuse with sodium carbonate. Leach and filter as before. Combine the two filtrates, and determine the phosphorus as on page 203, starting with "Add about 7 ml. of concentrated nitric acid, less an allowance for any present in the solution."

Analysis of Coal Ash. A substantial sample is required and may be obtained by separately ashing a large sample of coal. Generally the sample of ash should be 1 to 2 grams. Proceed as described on page 188, starting with "Add five to six times its weight of anhydrous C.P. sodium carbonate . . . ". The fusion melt will seldom be perfectly clear and transparent. A bluish green color of the cooled melt indicates manganese. If this is present, add 1 ml. of alcohol before acidifying with hydrochloric acid. This will reduce the manganate and prevent attack on the platinum by chlorine, which would be evolved on later addition of the hydrochloric acid. Proceed to the determination of the usual mineral elements as described in Chap. 10.

Calorimetric Determination. Various types of bomb are available. Essentially, the apparatus consists of a bomb made of or lined with acid-resisting metal and fitted with devices for holding and igniting the sample. The bomb is placed in a calorimeter jacket provided with a cover to protect the calorimeter from air currents, and with a stirrer. The whole is encased in a heat-insulating jacket. Full details for operation are furnished with each unit. Since the procedure consists largely in manipulating the specific equipment and calculating the data obtained, details will be omitted here.

CHAPTER 34

GAS ANALYSIS

Gas analysis, strictly speaking, is a branch of volumetric analysis. Gases are more conveniently measured by volume than by weight. Furthermore, at similar pressures and temperatures, 1 gram-molecule of any gas occupies the same volume as 1 gram-molecule of any other, 22.4 liters at 760 mm. pressure and 0°.

Ammonia, halogens, sulfur dioxide, and hydrogen sulfide are usually more conveniently estimated by their chemical properties in solution. They are therefore discussed in other chapters. The gases dealt with here are those which are substantially insoluble in water, an exception being carbon dioxide.

Such gases may be met with and determined singly or in mixtures. The best exemplification of the methods utilized in general gas analysis is the analysis of fuel gases. Normally, these contain carbon monoxide, carbon dioxide, oxygen, hydrogen, methane and ethane, nitrogen, and unsaturated hydrocarbons, which include ethylene, propylene, and benzene. The method of analysis of such gases will be outlined in this chapter. Typical analyses of some fuel gases are given in Table 40.

TABLE 40.—TYPICAL ANALYSES OF FUEL GASES

	CO_2, %	O_2, %	CO, %	H_2, %	Saturated hydrocarbons	Unsaturated hydrocarbons, %	N_2, %
Coal gas.......	2.8	0.5	9.3	44.1	33.2	3.7	6.4
Producer gas..	9.9	0.3	20.0	17.8	2.8	0.3	48.9
Water gas.....	4.8	0.4	37.9	49.2	3.0	...	4.7
Flue gas.......	11.3	7.0	1.0	0.8	....	...	79.9

Typical methods used in volumetric gas analysis are (1) successive absorption of the gases by suitable reagents each of which absorbs one gas only; (2) burning the gas with air or oxygen and measuring the change in volume and the amount of carbon dioxide produced; and (3) selective oxidation by passing the gas over a suitable reagent such as heated copper oxide.

Carbon dioxide, oxygen, carbon monoxide, and unsaturated hydrocarbons are estimated by method 1. Hydrogen, methane, and occasionally carbon monoxide are estimated by method 2 or 3. Nitrogen is usually obtained by difference. The various methods will be explained as the procedures are given.

Gas Collection. Great care must be taken that the gas sample is truly represented and is not contaminated with air. A convenient sampler is shown in Fig. 113. Thoroughly sweep out the air in the collection tube by allowing ten times the volume of the gas to pass through the tube if the gas being collected is under pressure. It will be necessary to connect the stopcock opposite the gas inlet to an aspirator if the gas is not under pressure. Such an aspirator may be a rubber bulb, a small hand pump, a water pump, or an aspirating bottle.

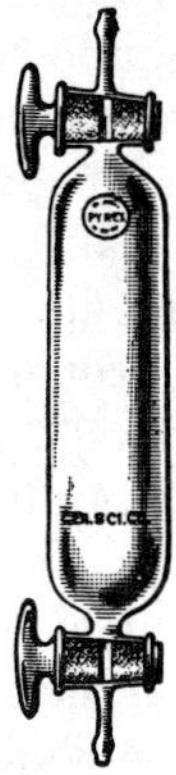

FIG. 113.—A typical form of sampling tube for collection of gas samples.

A convenient method in many cases, as with flue gases, is to fill the tube with a saturated sodium chloride solution, insert one end into the source of the gas, and open up the stopcock in the gas and then the farther stopcock in such a way that the salt solution slowly empties from one end of the tube, the gas itself taking its place by entry at the other end. The stopcocks must be well greased and well fitting. Analyze the sample as soon as possible after sampling. Slow air leakage always can occur on prolonged storage of the sample.

If one of the gases being collected is appreciably soluble in water, the use of a tube filled with water is not advisable. For many purposes, especially since the gas is not led through the water, this method is sufficiently accurate even though several per cent of carbon dioxide be present. Strictly speaking, the water or salt solution should be saturated with the gases being collected before the sample is taken. In a very few cases, gas samples are collected by displacement of a liquid other than water. Occasionally this is a light mineral oil. More often it is mercury, as in some of the methods of the U.S. Pharmacopoeia.

Simple Gas Buret, Leveling Tube, and Pipet. Hempel Apparatus. The simplest form of gas-measuring device is shown in Fig. 114. This consists essentially of a cylindrical glass measuring buret graduated to 100 ml. in 0.2-ml. divisions. It is surrounded by a glass water jacket so that a uniform temperature is maintained during the analysis. The measurements are taken at atmospheric pressure, which is assumed not to change during the determination.

Various modifications of the buret are obtainable. A stopcock at the bottom is convenient but not essential. A three-way stopcock at the top of the buret is a real convenience in that the buret can be connected at the same time both to the sample tube containing the gas to be analyzed and also to the buret containing the absorbent or reactant. A buret with a single one-way stopcock will suffice, but considerably more work is involved in continually changing connections.

The buret is connected at the bottom to a leveling tube or bottle by means of rubber tubing. As a leveling medium, slightly acid water or saturated salt solution may be used. Mercury, although expensive, does not dissolve gases, is far more efficient, and is used in work of high precision. Oxygen in high concentrations may tend to oxidize mercury if left in contact with it for any length of time. To allow for the weight of mercury when this is used, employ thick-walled rubber pressure tubing as connection.

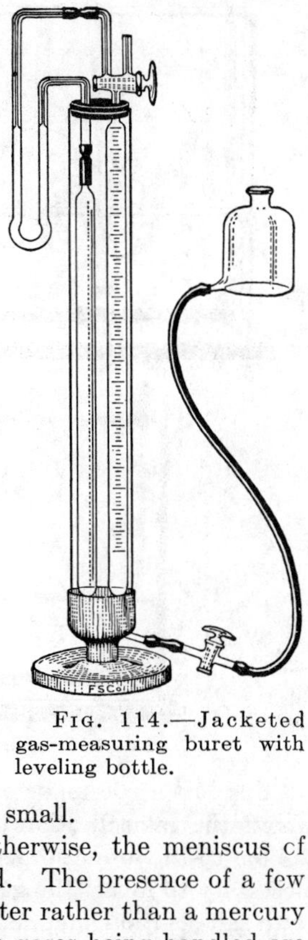

FIG. 114.—Jacketed gas-measuring buret with leveling bottle.

Figure 115 shows a group of typical Hempel bulbs used for absorption or oxidation of gases by technics to be described.

General Procedure. In the technics described here it will be assumed that high precision is required. Therefore, the Hempel gas buret, leveled with mercury, and a series of Hempel bulbs are the associated equipment. Accuracy to a few hundredths of a milliliter is obtainable with a 100-ml. sample, which is progressively absorbed to leave less than 1 ml. of residual gas. Such precision is not commonly attempted unless the actual determined volume of gas absorbed or remaining unabsorbed is very small.

The gas buret must be thoroughly clean; otherwise, the meniscus of the mercury will be irregular and difficult to read. The presence of a few drops of water on top of the mercury, giving a water rather than a mercury meniscus, is not objectionable, provided that the gases being handled are not highly soluble in water.

Before introducing the gas sample, fill the buret with mercury by opening the top stopcock and slowly raising the leveling tube. As soon as the buret stopcock and tubing including the capillary beyond the stopcock are full of mercury, close the cock and support the leveling tube so that the mercury in it is at approximately the same level as the top of the buret. If it is below this level there is a tendency to leakage of air through the greased stopcock.

Clamp the gas-sample pipet parallel and adjacent to the gas buret, with its upper stopcock parallel with that of the gas buret. Using thick-walled rubber pressure tubing and thick-walled glass capillary tubing, connect the top of the gas-sample pipet to the upper stopcock of the Hempel buret. Fill all the tubing with mercury before connecting, or with water or salt solution when one of them is used as the liquid in the Hempel buret.

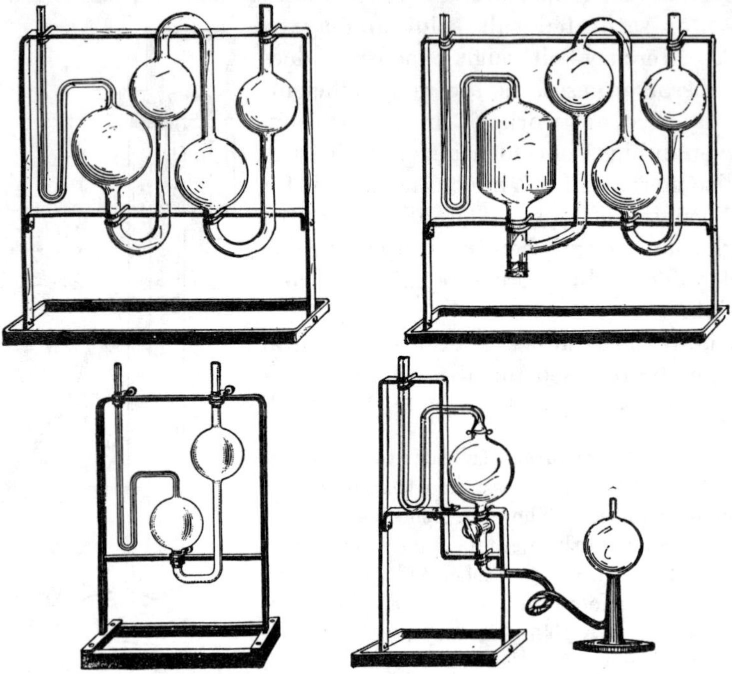

Fig. 115.—Various forms of Hempel absorption bulbs.

Support a 500-ml. separatory funnel containing saturated salt solution above the gas collection pipet, and connect its outlet by means of rubber tubing to the lower stopcock of the gas-sample pipet. To do so it will be necessary to invert the gas-sample pipet, fill the tube outside the stopcock with salt solution, and make the connection while holding it in this inverted position. This prevents contamination of the sample with the volume of air in this capillary. Fill the rubber tubing with the salt solution before connecting, and be certain that no air is entrapped in the connection. To ensure this, keep the gas collection pipet stopcocks closed, and open the separatory funnel stopcock. Gently shake or tap the rubber tubing so that any air in the tubing will rise through the salt solution. Reinvert the sample pipet.

Close the stopcock on the separatory funnel, and open all the others. At once lower the leveling tube gradually, and at the same time slowly open the separatory-funnel stopcock. The salt solution rises in the gas-collection tube, pushing out the sample into the buret in which the level of mercury is lowering as the leveling tube descends. The amount of salt solution that was in the capillaries goes over on top of the mercury but is minor in amount. In work of great precision, this salt solution is replaced by mercury.

Keep the top of the mercury meniscus in both the leveling tube and the buret level and gradually lower until exactly 100 ml. of sample has been drawn in. At once close the stopcock at the top of the buret and then all the other stopcocks. As the mercury levels are the same, the sample is now 100 ml. at atmospheric pressure. Unless the operator has a great deal of experience this exact adjustment to 100 ml. may require several attempts. Time spent in getting the accurate volume is well spent. Always keep the mercury levels approximately the same even when all stopcocks are closed, except while conducting the necessary transfers.

Hempel bulbs must be filled with absorbent solutions for the next operation. In its simplest form such a setup, as shown in Fig. 115, consists of two bulbs, the first being of about 150 ml. capacity and the second of about 100 ml. capacity. From the first issues a thick-walled capillary tube, which is connected to the buret. Other types of Hempel burets are shown in the same figure. They are filled with the appropriate absorbent liquid through the tube projected upward from the second bulb. It is convenient to connect a small funnel to this tube by a piece of rubber tubing and slowly pour in the liquid. Add approximately 150 ml. of absorbent so that the first bulb is full. There should be only a little liquid in the second bulb. For use manipulate this device so that not only the first bulb is filled but also the capillary inlet tube to the gas inlet.

Disconnect the gas collection pipet, and in its place connect the Hempel bulbs filled with the appropriate absorbent solution. Preferably use a short connection, and support the stand holding the Hempel pipet at a suitable height. The disconnection is not necessary in using a three-way stopcock. Exercise all the precautions in making this connection that were applied in the original connection of the sample pipet and the gas buret.

Now open the top buret stopcock, and gradually force the sample into the Hempel pipet by slowly raising the leveling tube. Drive the mercury through the capillary tube but not into the absorption bulb. At once close the buret stopcock. Gently shake the pipet so that the liquid comes in thorough contact with the gas. Open the buret stopcock, and suck back the sample into the buret by lowering the leveling bulb. Allow the absorbent to come as far as to fill the hole in the stopcock of the gas buret, but do not let the absorbents enter the gas buret. Level the mercury, and read the volume of gas. Again transfer the gas into the absorption pipet, and repeat the whole operation. Continue this procedure until the gas does not decrease in volume. The difference between the original and the final volume will be the amount absorbed. This procedure is then repeated for other absorbents until the operation is complete.

In the series that follows specific applications of this general procedure are given but without repetition of the details of manipulation.

Analysis of Fuel Gas. Such a gas must be analyzed for its various constituents in the order given below. Any other order will not selectively extract the different gases.

CARBON DIOXIDE. Use a Hempel pipet containing a 50 per cent solution of potassium hydroxide. Sodium hydroxide will serve, but it deposits carbonate more readily. Only carbon dioxide is removed by this reagent.

UNSATURATED HYDROCARBONS, OR ILLUMINANTS. After removing the carbon dioxide, connect the buret to another Hempel pipet filled with fuming sulfuric acid containing 15 to 20 per cent sulfur trioxide. In this case use as short a piece of capillary rubber tubing as possible for connection, and do not fill the exit capillary of the gas buret with salt solution. This will introduce an unavoidable error due to a few tenths of a milliliter of air introduced, but the error will be largely offset by leaving an equal volume of residual gas in the connections at the end of the manipulation. Bring the absorbent solution almost but not quite to the exit of the absorption bulb, and mark this level as the standard to be used. Pass the gas back and forth about four times. Be sure that the sulfuric acid does not reach the rubber connection. The gas in the buret is now contaminated with acid vapors. Remove these by passing it into the potassium hydroxide solution pipet, afterward returning it to the gas buret and making the final reading.

This procedure removes ethylene, propylene, acetylene, butylene, benzene, and other unsaturated hydrocarbons, often termed illuminants. In analyzing uncarbureted water gas and flue gas this procedure is normally unnecessary because no unsaturated hydrocarbons are present.

OXYGEN. As absorbent, use an alkaline pyrogallol solution made by mixing 1 volume of a solution containing 300 grams of pure pyrogallic acid per liter of solution with 4 to 6 volumes of 50 per cent potassium hydroxide solution.

A satisfactory alternative is a solution of sodium hydrosulfite. To prepare this reagent dissolve 50 grams of the salt in 250 ml. of water and 40 ml. of a 40 per cent solution of potassium hydroxide. The following reaction occurs: $2Na_2S_2O_4 + 2H_2O + O_2 = 4NaHSO_3$. The absorption is always complete after 5 minutes.

Stick phosphorus has also been used for this purpose. Aside from difficulties in manipulation it is not to be recommended because the concentration of oxygen must be less than 60 per cent and traces of heavy hydrocarbons, alcohol, or ammonia inhibit the absorption.

Carbon Monoxide. This gas may be absorbed by either an ammoniacal or an acid solution of cuprous chloride with which the carbon monoxide forms an addition compound. The ammoniacal solution is preferable, for the addition compound is less stable in hydrochloric acid solution. Prepare as follows: Mix 1 volume of concentrated ammonium hydroxide with 3 volumes of a solution containing 200 grams of technical cuprous chloride, 250 grams of technical ammonium chloride, and 750 grams of water. In order to maintain the cuprous chloride in a reduced condition keep a copper spiral in the first bulb of the pipet and also in the stock solution.

In practical work the first part of the absorption is carried out with an old, partly exhausted solution. For the final stages always replace the Hempel bulbs with a set containing relatively fresh solution. When no further absorption occurs, wash the residual gas, now containing some ammonia, with a 1:2 hydrochloric acid solution in another Hempel pipet. Read the final volume.

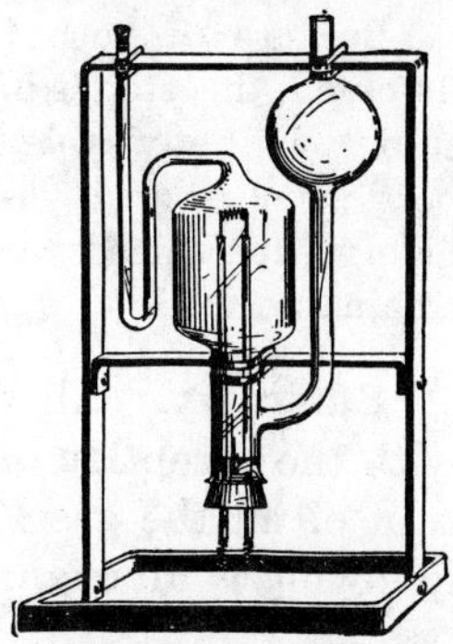

Fig. 116.—Winkler explosion bulbs, a modified Hempel unit.

Hydrogen and Methane. After the carbon monoxide estimation, return the gas to the hydrochloric acid pipet, and wash out the gas buret with a dilute hydrochloric acid solution in order to remove any trace of alkali.

Transfer 15 to 16 ml. of the residual gas, normally containing hydrogen, methane, and nitrogen, to the buret. Read its volume, and drive it over into a Winkler explosion bulb, as shown in Fig. 116, containing a saturated salt solution. This contains a platinum spiral connected to two rods that pass through a rubber stopper at the base of the first bulb. Be sure that the stopper is wedged in tightly enough to withstand a small explosion.

Now, from the buret, add exactly 100 ml. of air, which will contain 20.9 ml. of oxygen. Close the stopcock. While still connected to the buret, mix the contents of the pipet by gentle shaking and gradually heat the platinum spiral, which is connected to an electric current, reduced in voltage by suitable resistance. A slight, safe explosion takes place. The hydrogen, oxidized to water, condenses from the system. Methane is oxidized to carbon dioxide. Immediately after the explosion, return the residual gases to the buret. The solubility of carbon dioxide in the salt solution in this brief time has been shown to be negligible. Measure the volume, and note the contraction from the 115 to 116 ml. originally present. Now measure the carbon dioxide present by absorbing it in potassium hydroxide solution. By volume the simple reactions are

$$(1)\ 2H_2 + O_2 = 2H_2O \qquad \text{and} \qquad (2)\ CH_4 + 2O_2 = CO_2 + 2H_2O$$

From (1) it is clear that contraction due to the removal of the hydrogen is $1\frac{1}{2}$ times the volume of the hydrogen, x. Equation (2) shows that contraction due to the oxidation of methane is equal to twice the volume of the methane, y.

Hence the total contraction = $\frac{3}{2}x + 2y$. The volume of carbon dioxide formed is the same as that of the methane already present. Therefore it reads directly as the volume of methane. The hydrogen originally present is calculated by solving the above equations as follows:

$$(\text{Total contraction} - 2 \text{ carbon dioxide}) \times 0.67 = \text{hydrogen}$$

These are the volumes in the 15- to 16-ml. gas sample combusted. They must then be calculated to the total amount in the amount of residual gas remaining when absorption of carbon monoxide was completed.

Because of both the small size of samples and the manipulations involved, this step is inherently somewhat less accurate than the preceding steps. It is advisable therefore to carry it out with two or three successive portions of residual gas and average the results. Except in the hands of a highly experienced worker, accuracy to better than 1 per cent is hardly attainable.

Summary. All the common gases have now been determined with the exception of nitrogen, which is obtained by subtracting the sum of all the gases found from the original 100 ml. taken. The following is an example of an illuminating-gas analysis:

Original volume		= 100 ml.
Volume after absorption of CO_2	97.9 ml.	= 2.1 per cent CO_2
Volume after absorption of C_nH_{2n}	94.6 ml.	= 3.3 per cent C_nH_{2n}
Volume after absorption of O_2	94.4 ml.	= 0.2 per cent O_2
Volume after absorption of CO	86.5 ml.	= 7.9 per cent CO

Of the residual gas 15.5 ml. was taken.

Total volume, with air	= 115.5 ml.
Volume after explosion	= 92.5 ml.
Total contraction	= 23.0 ml.

Volume after absorption of CO_2 86.5 ml. = 6.0 per cent CH_4

Then methane in original $= 6.0 \times \frac{86.5}{15.5} = 33.5$ per cent CH_4

and $\frac{3}{2}$ volume of hydrogen $+ (2 \times 6) = 23.0$

Therefore, hydrogen $= 2\frac{2}{3}$

and hydrogen in original $= \frac{22}{3} \times \frac{86.5}{15.5} = 40.9$ per cent H_2

The final summary is therefore

CO_2	C_nH_{2n}	O_2	CO	CH_4	H_2	N_2 (by difference)
2.1	3.3	0.2	7.9	33.5	40.9	12.1

Orsat Apparatus for Flue Gases. The Orsat apparatus and its various modifications (Fig. 117) were devised to collect and analyze samples of flue gases outside the laboratory. These compact units also dispense with the necessity of connecting and disconnecting each pipet in turn as the different gases are absorbed. However, owing to greater length of capillary tubing, any error due to the volume of air thus introduced into the sample is increased. For most practical purposes, this error is negligible. For accurate work this volume can be measured and appropriate corrections made.

The simplest type of Orsat apparatus contains three absorption pipets. The first and third contain glass tubes for increasing the

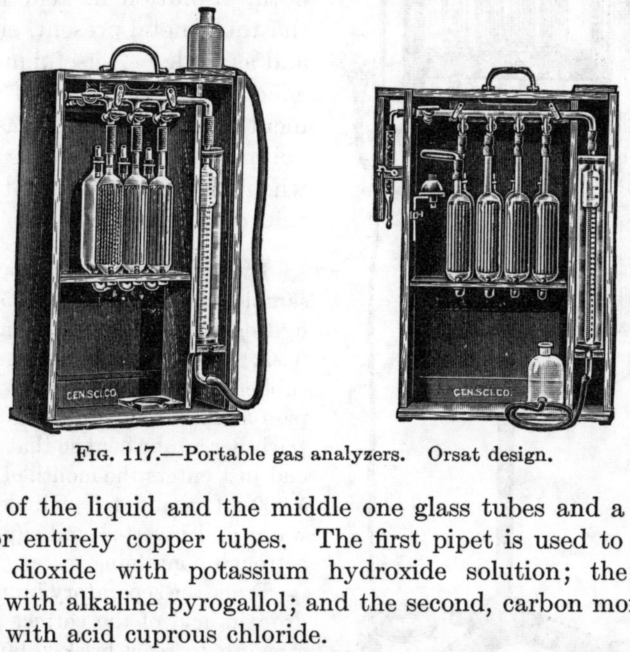

FIG. 117.—Portable gas analyzers. Orsat design.

surface of the liquid and the middle one glass tubes and a copper spiral or entirely copper tubes. The first pipet is used to absorb carbon dioxide with potassium hydroxide solution; the third, oxygen with alkaline pyrogallol; and the second, carbon monoxide, usually with acid cuprous chloride.

Therefore, in operation the carbon dioxide is first absorbed in the first tube, then oxygen in the third, and finally carbon monoxide in the second. Some types provide explosion wires in the gas pipet and are applicable to flue gases in much the manner of the Hempel bulbs.

In one of the units in Fig. 117 the fourth tube contains palladium, which is sometimes used to determine hydrogen. It is necessary to heat such a tube to promote this absorption. An alcohol lamp is provided for this. Still more complicated apparatus such as the Bureau of Mines gas analyzer for illuminating and combustible gases (Fig. 118) is available. Such gas analyzers are usually provided with full instructions as to how to operate the particular model. In essence the operation is the same as for the simple Hempel buret and pipet.

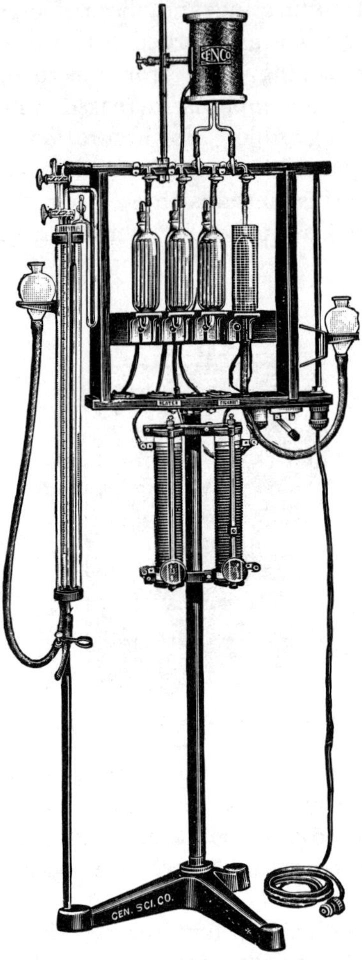

FIG. 118.—Laboratory-type gas analyzer. Bureau of Mines design.

Analysis of Metals by Hydrogen Evolution. Occasions arise, for example, if an active metal and its oxide are present, when it is desired to determine one or both. Solution in acid will give the total metal present, elemental and as oxide. A useful method to effect the estimation of the elemental metal is to measure the volume of hydrogen gas evolved when the metal is treated with acid or alkali.

PROCEDURE. Place sufficient sample to give at least 250 ml. of hydrogen in an empty 250-ml. conical flask. Fit the flask with a two-hole rubber stopper. Through one hole pass a separatory funnel and through the other a tube bent so that the other end just enters the mouth of a 500- or 1,000-ml. graduated cylinder full of water and inverted and supported in a trough containing water.

From the separatory funnel slowly drop in acid of the correct kind and strength to react briskly, but not too rapidly, on the metal in the flask. Always keep some acid in the funnel. If necessary, aid the evolution by heating the flask and contents. Hydrogen gas, which has a negligible solubility in water, 0.017 ml. dissolved in 1 ml. of water at normal temperature and pressure, and which rapidly passes through the water, leaving little opportunity for absorption, displaces the water in the cylinder.

Continue heating the flask and contents until the volume of gas in the cylinder does not increase. With the hydrogen is the air contained in the flask and the connecting tube. Now boil the contents of the flask vigorously until all the gases present are driven into the cylinder and the flask and tube are full of water vapor. The trough and cylinder must be of such capacity that this heating does not appreciably alter their temperature. At once read the total volume of gas in the cylinder, and observe the temperature at the side of the cylinder. Immediately disconnect the flask, or water will suck back on cooling. Read the barometric pressure at this time. Obtain the volume of air originally in the tube and in the flask up to the bottom of the stopper by pouring in measured volumes of water.

In the case of aluminum and zinc, it is often convenient to use a 10 per cent sodium hydroxide solution to dissolve the metal. This will differentiate these metals from any others present.

Calculation. An actual example is given. Using a 2-gram sample containing aluminum, the latter was dissolved using 10 per cent sodium hydroxide solution.

Total volume of gas in cylinder at 26° and 760 mm. = 520 ml.
Volume of air originally in cylinder, tubing, and flask = 255 ml.
Hence Volume of hydrogen evolved = 265 ml.
Vapor pressure of water at 26° = 25.2 mm.

Hence, 760 − 25.2 = 734.8, the effective pressure in mm.

Since $PV/T = P_1V_1/T_1$,

$$\text{Volume of hydrogen at N.T.P.} = \frac{265 \times 734.8}{(273 + 26)} \times \frac{273}{760} = 234.0 \text{ ml.}$$

Since $Al/3 = H_2/2$ and 1 gram-molecule of a gas at normal temperature and pressure (N.T.P.) measures 22,400 ml., therefore,

$$\frac{9 \times 234}{11{,}200} \times \frac{100}{2} = 9.40 \text{ per cent aluminum metal}$$

For greater accuracy, the negative pressure of the column of water in in the cylinder should be considered. Actually, outside pressure = pressure of (gas plus water vapor) + pressure of column of water in cylinder. This water pressure is normally small—for 6 inches it is equivalent to 10 mm. of mercury, which should be subtracted from 734.8 above.

Also, the aluminum itself may be, say, 99 per cent pure, or even 94%, and the calculation to the original aluminum alloy may thus be in error.

CHAPTER 35

PAPER, PAPER FIBERS, TEXTILE FIBERS, AND STARCHES

The testing methods applied to paper are mostly physical, although some chemical methods are used. These will be found in the publications of the Technical Association of the Pulp and Paper Industry, (TAPPI), from which the following methods have been drawn. The identification of paper fibers is taken up as a separate section in this chapter. Since textile fibers overlap in some cases, they follow. Lastly starches, widely used for textile sizing, are considered, mainly microscopically.

PAPER TESTING

Sample. This should consist of sheets not less than 11 by 11 inches in size and sufficient in number to complete the tests. They should be flat and must be protected from sunlight and from contamination.

Moisture. For large samples in a delivery of paper, use a sample not less than 50 grams in weight. Be careful not to lose moisture during handling. Place in a weighed, dry, airtight metal can, and reweigh. Place in an air oven at 100 to 105° for 2 hours, with the lid of the can removed. Replace the lid, cool in a desiccator, and weigh. Repeat the periodical drying and weighing until the difference in weight between successive weighings is not more than 0.1 per cent of the weight of the specimen.

For smaller samples used for chemical analysis, weigh 1 to 2 grams of the paper in a stoppered weighing bottle; heat at 100 to 105°, with the lid off the bottle, for 1 hour; cool; and weigh. Repeat as often as necessary, as before.

Ash. The ash of paper may be due to mineral matter in the cellulose, various residues from chemicals used in its manufacture, metallic matter from piping and machinery, and filling, coating, and pigmenting materials. Generally, though not always, if the ash content is less than 2 per cent, no filling, coating, or pigmenting material is present. The ash is usually a close approximation of any filling and coating materials in the paper unless ignition changes its character.

Allowing for the moisture previously determined, weigh not less than 1 gram of sample into a platinum, Alundum, silica, or porcelain crucible. Dry at 100 to 105°. Ignite, either in an electric muffle furnace or over a gas burner, first at a low temperature and with the lid on to avoid loss of small particles, and finally at 925° until no black particles remain. Cool in a desiccator, and weigh. Repeat the ignition, and weigh until the weight is constant to the nearest 0.1 mg. Compute the percentage of ash on the basis of the dried sample.

For accurate work take the average of two determinations. These should agree within 0.1 per cent for paper containing less than 5 per cent of ash, to within 0.15 per cent for paper containing 5 to 10 per cent of ash, and to within 0.2 per cent for paper containing over 10 per cent of ash.

Mineral Filler and Mineral Coating. Ash a substantial sample, varying according to the amount of ash present. Test a portion qualitatively, bearing in mind that carbonates will have been calcined to oxides and that other changes may have resulted from ignition. Usually the qualitative methods for pigments and fillers in Chaps. 25 to 27 or those in Chap. 5 are suitable. This should be followed by the corresponding quantitative methods.

Mineral Coating on Paper. This test differs from the preceding in being a determination of the amount of coating that can be separated, rather than of the ash. While suitable for the ordinary types, it is not effective for special types, as where lacquers or other materials are used to impart a high degree of water resistance.

Prepare an aqueous enzyme solution containing approximately 1.5 grams of enzyme and 25 ml. of $0.1N$ sodium hydroxide solution per liter. The enzyme may be trypsin or one of the commercial mixtures of enzymes used for desizing cotton and degumming silk. The latter are more stable, more rapid in action, and less expensive.

Weigh not less than 25 square inches of the properly conditioned paper, and place it in a suitable container in such a way that the enzyme solution has free access to the entire surface of the coating. To accomplish this, either use a flat tray or roll the sample in cylindrical form and place it in a buret.

Allow to stand for at least 1 hour at 50°; then carefully place the paper on a pane of glass, and brush off the coating with a camel's-hair brush, taking care not to dislodge the paper fibers. Additional heating, and in some cases additional enzyme solution, may be found necessary for very resistant coatings.

After the coating is entirely removed, stand the glass pane at a slight angle and wash the paper on both sides by means of a wash bottle, holding the paper on the glass by one corner. Condition the paper exactly as at the beginning, and weigh it. Make at least two determinations, and calculate the results as percentage coating on the decoated paper.

Owing to the variable nature of coating materials, the precision is variable. It normally should be correct to 0.5 to 2 per cent.

Casein. QUALITATIVE. This test is applicable only to papers containing relatively large amounts of casein, for example, mineral-coated papers on which casein is commonly used as a binder. It is not applicable to papers containing such small amounts of casein as may be used in the beater, as a constituent of the engine sizing, etc.

Cut about 0.5 gram of the paper into small pieces, and boil for 3 minutes in a test tube with 10 ml. of 1 per cent sodium hydroxide solution. This dissolves any casein hardened by formaldehyde or other agent. Filter, cool to 25°, and exactly neutralize with 1:5 nitric acid, using phenolphthalein as indicator. Now add 3 ml. of a special Millon's reagent made by dissolving 20 grams of mercury in 40 grams of concentrated nitric acid and diluting the solution to 180 ml. with water. On heating, the presence of casein is indicated by the development of a red coloration of the coagulated casein.

This reaction depends on the presence of tyrosin, present to the extent of about 5 per cent in casein but rarely present in animal glue and gelatin. Blood albumin and some other materials give a similar test but are not usually found in papers.

QUANTITATIVE. The next method is applicable.

Proteinaceous Nitrogenous Materials. QUALITATIVE. Prepare Schmidt's reagent by dissolving 3 grams of ammonium molybdate in 250 ml. of water and adding 25 ml. of 2:3 nitric acid. This reagent should be made fresh at frequent intervals. As in the casein test, boil 0.5 gram of paper with 10 ml. of a 1 per cent sodium hydroxide solution. Filter, cool, and neutralize with 1:1 hydrochloric acid to a phenolphthalein end point. Add 1 volume of Schmidt's reagent to 2 volumes of the filtrate. A white precipitate indicates the presence of protein material such as glue or casein. The test is very delicate. If no precipitate or only a slight one is obtained, there can be no appreciable amount of protein material present.

QUANTITATIVE. Use 3 to 5 grams of air-dried paper cut into squares of approximately 0.25 inch on a side. Determine nitrogen by the Kjeldahl method (page 143). Duplicate determinations should agree within 0.02 per cent of nitrogen. For accurate work, perform a blank determination on all reagents used. The factor to be used in later calculation varies with the grade and kind of nitrogenous material present. When possible, as in plant investigation of paper made with materials of known source, determine the nitrogen content of the nitrogenous material used and of the paper before addition of the glue or casein. Otherwise, the factor for calculation must be based on the prior qualitative examination.

CALCULATION.

$$\text{Per cent nitrogen} \times 5.6 = \text{per cent glue}$$
$$\text{Per cent nitrogen} \times 6.3 = \text{per cent casein}$$

Starch. QUALITATIVE. Boil about 0.5 gram of paper with 10 ml. of water for several minutes. Filter, cool the extract, and add a drop of approximately 0.001*N* iodine solution. A blue coloration indicates starch.

If only a faint violet coloration is obtained, this should be disregarded as due to nonstarch constituents.

QUANTITATIVE. *Preparation of Sample.* Weigh 5 grams of the paper to the nearest 5 mg. Thoroughly disintegrate soft paper in a small quantity of water and hard papers by tearing into small pieces and shaking in a stoppered flask with 75 ml. of water and some lead shot. Do not use a mechanical grinder, for considerable mineral filler and starch may thus be lost. Determine moisture on a sample at the same time the starch determination is made.

Remove the lead shot, if present. If the mixture is acidic make the suspension up to 100 ml., heat to boiling, and add 2 drops of 4:7 ammonium hydroxide solution. Adjust the pH to about 7, using a universal indicator, measuring pH in steps of one unit (page 114). If the mixture is alkaline, add 5 ml. of 5:4 hydrochloric acid solution and then neutralize with a slight excess of 4:7 ammonium hydroxide solution. Dilute to 100 ml., heat to boiling, and similarly adjust to pH 7.

To 100 ml. of water add 0.5 gram of sodium chloride. Wash out the mouth thoroughly with water, and collect 10 ml. of saliva; a convenient means is by chewing on clean paraffin wax. Add the saliva to the salt solution, and heat to 50°. Cool the water suspension of the sample to 50°, and add it to the salt-saliva solution. Store the mixture at 40 to 55° for 1 to 4 hours, then remove a few fibers, and test for starch with $0.02N$ iodine solution. If a blue coloration appears, filter the main solution and digest the pulp as before with another 10 ml. of saliva. Test the pulp again for unconverted starch. Continue the digestion with saliva-salt solution until the pulp gives no color with iodine solution. Salivas vary in activity. Some give a slight reducing reaction with Fehling's solution. Test for this effect, and run a blank if necessary. The second digestion with saliva is usually necessary where fairly large amounts of rosin size are used or the paper has been overdried. If the stock has been very thoroughly beaten or the paper overdried, the pulp may persistently give a faint blue color with iodine. This is not a positive reaction for unconverted starch but indicates that the cellulose has attained the same iodine-dispersing power as the starch. Few papers require more than two digestions with saliva.

Filter the dispersion through a Büchner funnel, being careful that no fibers reach the filtrate. After sucking all the water through the funnel, wash the pulp three times with boiling water, draining completely after each washing. Combine all filtrates, add a few drops of concentrated hydrochloric acid, and concentrate to 200 ml. by evaporation on the steam bath. Transfer to a conical flask, add 20 ml. of 5:4 hydrochloric acid solution, attach to a reflux condenser, and boil moderately for 2 hours. Cool the solution, transfer to a 250-ml. volumetric flask, dilute to the mark, and mix well. Pipet 25 ml. of the solution into a 400-ml. beaker; then, with stirring, slowly add dry sodium carbonate until effervescence ceases.

Gravimetric. Prepare a Gooch crucible with a 0.6-cm. pad of specially washed asbestos. The asbestos should be treated with 1:3 hydrochloric acid solution at room temperature for 2 to 3 days, washed free from acid, digested similarly with 10 per cent sodium hydroxide solution, and then treated for 3 to 4 hours with hot alkaline tartrate solution; old Fehling solution B will do. Wash free from alkali, digest with 1:3 nitric acid solution for 3 to 4 hours, wash free from acid, and shake with water into a fine pulp. Wash the prepared Gooch crucible with water to remove all fine particles, then with 10 to 15 ml. of alcohol, and finally with 10 to 15 ml. of ether. Dry at 105° for 30 minutes, cool, and weigh.

Prepare Fehling's solutions, Allihn modification, as follows: *Solution A.* Dissolve 69.3 grams of C.P. copper sulfate, $CuSO_4.5H_2O$, in 1 liter of water. *Solution B.* Dissolve 250 grams of C.P. potassium hydroxide and 34 grams of C.P. potassium sodium tartrate, Rochelle salts, in 1 liter of water.

Mix 30 ml. of solution A, 30 ml. of solution B, and 60 ml. of water in a 250-ml. beaker. Heat to boiling, making sure no visible reduction takes place. Add to the solution from the paper in the 400-ml. beaker, and boil the mixture for exactly 2 minutes, keeping the beaker covered. Filter at once through the prepared Gooch crucible without dilution. Decant the major portion of the solution containing the precipitated cuprous oxide through the crucible under suction, then transfer the precipitated oxide, and wash five times with hot water, three times with alcohol, and three times with ether. Dry, cool, and weigh as before. From the weight of cuprous oxide obtain from appropriate tables, found in the publications of the Association of Official Agricultural Chemists or handbooks and conventionally known as Allihn's tables, obtain the corresponding amount of dextrose. This amount multiplied by 0.9 gives the amount of starch in the aliquot part of the solution. The factor 0.9 is due to the fact that, on inversion, starch takes on water and the corresponding amount of dextrose is therefore greater. Care must be exercised to make certain that the Fehling solution used corresponds to the tables used. Thus the Munson and Walker modification might be applied here with appropriate tables, but the results would not then be by the official TAPPI method.

Volumetric. Proceed exactly as in the foregoing to the point at which the precipitated cuprous oxide is ready for filtration. Filter through a small Büchner funnel, and wash with about 200 ml. of hot water. Transfer the asbestos mat and the cuprous oxide back to the precipitation beaker, add 25 ml. of molybdophosphoric acid solution (page 621), and mix thoroughly with a glass stirring rod. Filter through the same Büchner funnel, and wash thoroughly with cold water until all trace of blue color is removed. Dilute the filtrate to 400 to 700 ml., and titrate with 0.0333*N* potassium permanganate solution to a faint pink color after the blue color has been discharged. The cuprous oxide reduces the molybdenum, which in turn is quantitatively oxidized by the permanganate.

If the titration requires less than 15 ml. of the 0.0333*N* potassium permanganate solution, repeat the analysis, using a 50-ml. aliquot of the prepared solution from the paper and 35 ml. instead of 60 ml. of added water.

Each ml. of 0.0333*N* potassium permanganate = 0.0011 gram dextrose
= 0.00099 gram starch

Rosin. QUALITATIVE. *Liebermann-Storch Method.* Heat 1 gram of paper, cut in small pieces, in a dry test tube. Add 5 ml. of acetic anhydride, and boil down to 1 ml., burning the vapors as they leave the test tube. Cool and filter if necessary. Transfer several drops to a spot plate, and carefully add one drop of concentrated sulfuric acid. A fugitive rose-violet color indicates rosin.

Raspail Method. Place the paper on a glass or porcelain plate, and apply a drop of a concentrated sugar solution. After a minute remove the excess sugar solution with filter paper. Add a drop of concentrated sulfuric acid to the sugar on the filter paper. A raspberry-red coloration indicates the presence of rosin.

Resin. The total resins in a finished paper consist of natural resins in the pulp from which the paper is made, plus any resins added in sizing. The method of quantitative determination depends on the conversion of insoluble resinates to alcohol-soluble resins by the action of a strong volatile acid of a definite composition for a definite length of time. Thus, a minimum of nonresinous material, such as cellulose hydrolysis products, is formed. Separation of the resinous from the nonresinous materials is accomplished by dissolving out the resin from the dried extract with anhydrous ether, in which nonresinous material, including starches and glues, is insoluble. If waxes and fats, which may dissolve in anhydrous ether, are present, they must be separated from the resinous material by saponification.

PROCEDURE. Weigh 5 to 7 grams of paper in strips to the nearest 10 mg. A moisture determination on this sample must be available. When mineral matter reacting with hydrochloric acid is present, immerse the strips in approximately normal hydrochloric acid solution for 5 minutes, drain off the acid, wash free from acid, and dry at room temperature. Omit this step if acid-soluble mineral matter is absent.

Place the strips lengthwise in a Soxhlet extractor, taking care not to pack them together tightly. Using as solvent 95 per cent alcohol containing 4 ml. of concentrated hydrochloric acid per liter, extract for 2 hours, heating so that 15 siphonings occur per hour. Coated or surface-sized papers require 2.5 hours' extraction. Do not extract much longer than specified.

Evaporate off the solvent on a steam bath until there is no odor of either alcohol or hydrochloric acid. Heat at 100° in an oven for 15 minutes, cool to room temperature, and add 20 ml. of anhydrous ether. The resin normally dissolves in 5 to 30 seconds. If the particles are covered with foreign matter, use a stirring rod to break up the mass. Let the solution stand for 20 minutes to further coagulation and settling out of the foreign matter. Filter the ether solution through a fine filter paper cut to fit a Gooch crucible, using suction. Refiltering through the same paper is often necessary. Rinse the flask with 20 ml. of ether, and wash this through the Gooch crucible. Evaporate ether and washings in a weighed flask, dry at 100° in an oven for 15 minutes, and weigh to the nearest milligram. Repeat drying and weighing until constant weight is attained. In the absence of paraffin or similar unsaponifiable wax, calculate the percentage directly.

If paraffin or similar unsaponifiable wax but no oil volatile at 100° is present, determine the unsaponifiable matter. Saponify the resin by adding 25 ml. of 0.5N alcoholic potassium hydroxide solution and heat to not over 60° for 15 minutes. Transfer to a 500-ml. separatory funnel with warm water, and dilute to about 300 ml. Add 6 grams of sodium chloride, and cool. Proceed as on page 349, starting at "Add 75 ml. of ethyl ether." The unsaponifiable material extracted will contain the wax and the unsaponifiable matter from the resin, which, for most purposes and in the absence of definite knowledge regarding the particular resin, may be assumed to be 5 per cent of the resin in the paper. Therefore,

$$\frac{\text{Weight of resin and wax} - \text{weight of total unsaponifiable matter}}{0.95} \times \frac{100}{\text{weight of sample}} = \text{per cent resin}$$

When oil volatile at 100° is present, with or without paraffin and other unsaponifiable waxes, determine the resin as follows: Saponify the resin by adding 25 ml. of 0.5N alcoholic potassium hydroxide solution, and heat to not over 60° for 15 minutes. Transfer to a 500-ml. separatory funnel with warm water, and dilute to about 300 ml. Add 6 grams of sodium chloride, and cool. Extract as on page 349, starting at "Add 75 ml. of ethyl ether," until the ether extracts have been combined. Wash the combined ether extractions with 2 per cent sodium chloride solution. and add the washings to the aqueous layer. Discard these ether extracts. Just acidify the aqueous solution of the resin soaps, and extract three times with ether. Wash these combined ether extracts with water containing 2 per cent of sodium chloride, drain off thoroughly, and wash the salt solution with 20 ml. of ether. Evaporate the combined ether extract and ether washings to dryness at 100° until constant in weight. This weight divided by 0.95 is that of the resin originally present.

When saponifiable fats or greases are present, this method is not accurate, for the fatty acids of these will also be found with the resin.

Paraffin. This method is for wax-impregnated papers but is not intended for paraffin-sized papers, because the extract from the latter generally contains soluble matter other than waxes, making the accuracy questionable for small amounts of paraffin.

PROCEDURE. Cut not less than 3 grams of the paper into strips ½ inch wide, weigh accurately, and fold into numerous small crosswise folds. Do not tear the paper into small pieces, or they will stick together. Place the paper in a Soxhlet tube, without a thimble, and extract with carbon tetrachloride for at least 6 hours. Take care that the strips are in the solvent and do not float when the extractor is filled.

Evaporate off the solvent, and take off the last traces on the steam bath. Add 25 ml. of 0.5*N* alcoholic potassium hydroxide, and again evaporate to dryness. Cool the dry residue, take it up in 25 ml. of petroleum ether and 150 ml. of water, and transfer to a separatory funnel. Wash out the flask into the funnel with a little petroleum ether. Shake the contents of the funnel well, and allow the two layers to separate. Draw off the water, and extract twice more with 25-ml. portions of petroleum ether. Wash the combined ether extracts with water until free from alkali. It may be necessary to add 2 per cent of sodium chloride to the water in the first extraction to inhibit emulsification during shaking with the petroleum ether. Drain the water thoroughly from the washed ether extracts, and evaporate the ether in a tared flask. Dry at 105° for 1 hour, cool, and weigh. This weight is considered as that of the paraffin in the sample. Occasionally, higher alcohols and other unsaponifiable matter that have been added to the paper may be present.

Copper Number. The copper number of paper is defined as the number of grams of metallic copper in the cuprous oxide resulting from the reduction of copper sulfate by 100 grams of the paper fibers. The paper is used to reduce the copper, which in turn serves to reduce a molybdophosphoric acid solution. The reduced solution so obtained is quantitatively titrated with standard potassium permanganate solution.

REAGENTS. As reagents prepare the following: *Copper sulfate solution.* Dissolve 100 grams of copper sulfate, $CuSO_4.5H_2O$, in water, and dilute to 1 liter.

Carbonate-bicarbonate Solution. Dissolve 129 grams of sodium carbonate, Na_2CO_3, and 50 grams of sodium bicarbonate, $NaHCO_3$, in water, and dilute to 1 liter.

Molybdophosphoric Acid. Dissolve 100 grams of sodium molybdate, $Na_2MoO_4.2H_2O$, and 75 ml. of 83 to 85 per cent phosphoric acid in a mixture of 275 ml. of concentrated sulfuric acid and 1,750 ml. of water.

Procedure. Completely disintegrate the sample, without heating or contaminating it, in a grinder of the Koerner type or its equivalent. Allow the ground sample to come to moisture equilibrium with the atmosphere of the balance. Weigh about 1.5 grams, to the nearest 10 mg. At the same time, if the determinations are not available at the same humidity, weigh a sample for the determination of moisture, ash, and other sizing, filling, or nonfibrous matter as may be necessary for correct calculation of the copper number.

Immediately before use, add 5 ml. of the copper sulfate reagent to 95 ml. of the carbonate-bicarbonate reagent. Bring the mixture to a boil in 2 minutes, and pour it over the weighed ground sample in a 125-ml. conical flask. Stir well with a glass rod in order to distribute the fibers and to remove air bubbles. Fit the flask with a loosely fitting glass bulb; or use a small watch glass, and submerge it as completely as feasible in a steam bath. Shake the flask from time to time to redistribute fibers that float to the surface. Remove the flask at the end of 3 hours, and filter the contents through an ashless filter paper in a 7.5-cm. Büchner funnel, using suction. Wash by flooding with 100 ml. of a 5 per cent sodium carbonate solution at about 20° and then by flooding with 250 ml. of water at about 95°. Discard the filtrates.

Transfer the fibers and the filter paper to a small beaker. Add 25 ml. of the molybdophosphoric acid solution, and macerate well with a flattened glass rod. Filter through the Büchner funnel again, and wash thoroughly with cold water until the blue color of the reduced molybdenum is removed from the fibers. Dilute the filtrate with water to approximately 700 ml., and titrate it with $0.05N$ potassium permanganate solution (page 728) to a faint pink coloration.

A copper number of 6 is seldom exceeded except in papers containing highly lignified fibers such as groundwood or in papers that have deteriorated considerably. In such cases, repeat the determination and increase the amount of the copper sulfate solution to 10 ml., and the amount of the molybdophosphoric acid solution to 50 ml. or as much more as may be necessary, retaining the ratio between the two solutions. This method is not suitable for papers containing zinc sulfide, which has a reducing action.

Calculation. Correct the weight of sample to that of the cellulose by subtracting moisture, ash, and other noncellulose materials present in significant amounts.

$$\text{Ml. potassium permanganate} \times \text{normality} \times 0.0636 \times \frac{100}{\text{corrected weight of sample}} = \text{copper number}$$

Report the results to the nearest 0.1 per cent. Duplicate determinations should agree within 0.2 per cent.

Water-soluble Acidity or Alkalinity. Disintegrate the sample without heating or contamination in a grinder of the Koerner type or its equivalent. Allow the ground sample to come to moisture equilibrium with the atmosphere, and weigh 5-gram portions to the nearest milligram for the acid extractions and moisture determinations. Determine moisture in duplicate on weighed portions. Transfer other weighed portions to 500-ml. conical flasks, and proceed with each as follows: Add 250 ml. of boiling water, first adding a little and thoroughly wetting the sample by shaking well. Attach an air or water condenser to the flask, and place it on a bath that will maintain the temperature of its contents at 98 to 100°. Heat for 1 hour with occasional shaking.

Now pour the contents of the flask on a Büchner funnel without filtering medium, and wash the fibers remaining in the flask with 10 ml. of water. Apply strong suction to the fibers. Cool the extract rapidly to room temperature, and at once add a few drops of phenolphthalein indicator solution. If the solution remains colorless, titrate at once with $0.01N$ sodium hydroxide solution to the appearance of a permanent light red coloration. If the solution becomes pink on adding phenolphthalein indicator, titrate at once with $0.01N$ hydrochloric or sulfuric acid solution until the color is just discharged. Make a blank titration on 250 ml. of the water heated for 1 hour in the same bath and with the same glassware used in the extractions.

CALCULATIONS.

$$\text{Ml. alkali corrected for blank} \times \text{normality} \times 0.040 \times \frac{100}{\text{weight of dried sample}} = \text{acidity as sulfuric anhydride, } SO_3$$

$$\text{Ml. acid corrected for blank} \times \text{normality} \times 0.031 \times \frac{100}{\text{weight of dried sample}} = \text{alkalinity as sodium oxide, } Na_2O$$

Report either to the nearest 0.01 per cent.

Hydrogen-ion Concentration. Disintegrate the paper in a Koerner or similar type of grinder. Weigh 5 grams of the air-dry sample to the nearest 0.1 gram. Transfer to an alkali- and acid-resistant 500-ml. flask, such as a Pyrex flask, that has been thoroughly boiled out with several changes of pure water. Add 250 ml. of boiling water, which should be thoroughly boiled before adding and is preferably double or triple distilled. This is very important in determining the pH of papers low in acidity, such as those which are unsized. In such cases the water should have a pH between 6.6 and 7.0, indicating a high degree of purity. If the fibers tend to float, wet with a small amount of the water, stirring well, and then add the remainder.

Attach a clean air condenser. Glass jointed apparatus is advantageous. Place the flask in a heating bath that will maintain the contents of the flask at 95 to 100° without boiling the water. Heat at this temperature for 1 hour, with occasional shaking. Pour the contents of the flask on a Büchner funnel, thoroughly cleaned and washed with the specially pure water, without filtering medium. A filter paper is likely to alter the pH of the liquid. Draw the solution through the fiber mat, using strong suction. The filter flask must be as scrupulously clean as the flask containing the sample. Disconnect the suction just before the filtration is complete. To filter out the few fibers in the filtrate, pass it again through the mat. Close the flask at the top and the side as a protection against carbon dioxide contamination, and allow to cool.

Determine the pH colorimetrically to the nearest 0.1 unit or electrometrically to the nearest 0.05 unit. Methods for this determination are given in Chap. 6 under Hydrogen-ion Determination.

A rough, but often useful, quick modification of this method for control work is to place 1 gram of the cut up paper in a clean beaker and add 100 ml. of pure boiling water. Heat on the steam bath for 1 hour, covering the beaker with a water glass. Remove, and allow to cool. Finally, determine the pH on a decanted portion of the extract.

Arsenic. Digest 5 grams of the air-dry paper for the Gutzeit method with nitric and sulfuric acids as described on page 208, starting at "*Organic by Wet Ashing.*" Report results as parts of arsenic trioxide per million parts of paper or grams of arsenic per square yard of paper.

Pentosans. Disintegrate the paper to a cottonlike form in a Koerner or similar type of disintegrator. Weigh out 1 gram after the sample has attained moisture equilibrium with the balance. Determinations of moisture, ash, and sizing materials under similar humidity conditions must be available.

Introduce the test specimen into a 500-ml. distilling flask in the neck of which is mounted a 300-ml. separatory funnel and which is connected to a long water-cooled condenser. The latter delivers through an adapter into a 1-liter reagent bottle, which is used as a receiver and later as a reaction vessel. Rubber stoppers may be used to make connections. If the temperature of the condensate in the receiver is over 30°, use a small U tube containing glass beads and 5 ml. of water to close off the system. Add the specimen through a clean, dry glass tube to prevent adherence of fibers to the neck of the flask.

Add 100 ml. of hydrochloric acid prepared by diluting 307 ml. of concentrated acid to 1 liter. Mark the flask to indicate this volume. To the separatory funnel now add 300 ml. of the diluted hydrochloric acid, and attach to the flask. Heat the contents of the flask; and as soon as distillation commences, let the acid drop from the funnel at such a rate that the level in the flask remains constant. Continue the distillation at a uniform rate for 100 minutes; during this time, 300 ml. of condensate should have collected. It is important to observe time and quantity.

Add 50 ml. of water to the condensate. If a U tube was used, combine its contents with the condensate, and add correspondingly less water. Then add 250 grams of crushed ice. After the temperature of the mixture has fallen to 0° or lower, add 20 ml. of 0.2*N* bromate-bromide solution made by dissolving 5.57 grams of potassium bromate and 50 grams of potassium bromide in water, adding 1 ml. of *N* sodium hydroxide solution and diluting to 1 liter. This solution should be added from a pipet with minimum agitation. At once close the bottle with a ground-glass stopper, shake well, and let stand for exactly 5 minutes. The temperature should still be 0° or lower.

Remove the stopper, add 10 ml. of 10 per cent potassium iodide solution from a small graduated cylinder, and replace the stopper as quickly as possible. Shake the mixture thoroughly to allow absorption of the bromine vapor liberated, and titrate with 0.1*N* sodium thiosulfate solution until colorless.

Determine a blank in precisely the same manner, using all reagents including the ice, except that the starting point is 270 ml. of the hydrochloric acid diluted to 350 ml., instead of 300 ml. of condensate plus 50 ml. of water.

CALCULATION.

$$\left[(\text{Ml. blank} - \text{ml. sample titration}) \times \text{normality of thiosulfate solution} \times 1.03 \times \frac{0.048}{0.727} \times \frac{100}{\text{corrected weight of sample}}\right] - 0.9 = \text{per cent pentosans}$$

The weight of specimen is that after correcting for moisture, ash, and any filling and sizing materials present. The factors used are as follows: 1.03 corrects for a 3 per cent loss during distillation; 0.048 is the weight of furfural in grams corresponding to 1 ml. of *N* sodium thiosulfate solution; 0.727 is the theoretical conversion factor of pentosans to furfural; and 0.9 is a correction for volatile material, probably hydroxymethylfurfural, arising from the action of hydrochloric acid upon the cellulose. This correction was found to be valid for the typical chemical wood fibers and papers made from them, including unbleached sulfate. For materials containing large quantities of lignin, such as newsprint, the correction is probably too low.

If Krober's tables, found in some handbooks, are used, the factor in the above calculation $\left(1.03 \times \frac{0.048}{0.727} \times 100 = 6.80\right)$ becomes 8.25 and the constant 0.9 is neglected. The conversion of pentosans to furfural is, in this case, arbitrarily taken as 80 per cent.

Determinations should be made in duplicate and should agree within at least 0.4 per cent. Report to the nearest 0.1 per cent.

Alpha, Beta, and Gamma Cellulose. Cellulose consists analytically of three fractions, alpha, beta, and gamma. In this method the alpha fraction is the cellulose that can be filtered out of a mixture consisting of the fibrous material and 7.3 per cent sodium hydroxide solution of maximum dissolving power, after the fibers have previously been swollen with 17.5 per cent sodium hydroxide solution. The beta cellulose is taken as that fraction which precipitates at 15 to 35° after the filtrate has been acidified, whereas the gamma fraction remains in solution.

After separation, the alpha cellulose is determined either by drying and weighing, or volumetrically by oxidation with dichromate. Both methods for alpha cellulose give practically the same values and are reproducible. In the volumetric method no moisture or ash determinations are made, and a much smaller sample is used, resulting in a shorter and more rapid procedure. For beta and gamma cellulose, only the volumetric method is practicable.

This procedure is intended primarily for papers from rags or chemical wood fibers, not for papers containing large amounts of lignin such as newsprint. All operations should be performed and all solutions stored at as near 20° as possible.

ALPHA CELLULOSE. *Volumetric.* Allow a 50 per cent solution of sodium hydroxide to stand about a week in a stoppered bottle to permit settling of sodium carbonate. Carefully draw off 2 ml. of the supernatant liquid with a pipet, and add about 50 ml. of water and 1 ml. of saturated barium chloride solution to precipitate residual carbonate. Titrate with N hydrochloric acid to a phenolphthalein end point. Knowing the approximate normality of the concentrated sodium hydroxide solution, dilute it with distilled water to $5.24 \pm 0.05N$, checking by titration. This is to be used as the 17.5 per cent solution.

If the sample is mineral coated, remove tne coating. Comminute the sample in a Koerner or similar type of disintegrator, and then mix thoroughly for a few seconds. Weigh 0.3 gram ± 10 mg. of the sample in a 100-ml. beaker. Add 20.0 ml. of 17.5 per cent sodium hydroxide solution, macerate until the fibers are uniformly wet and dispersed, and let stand 10 minutes from the time of addition of the alkali. Then add 33.0 ml. of water, stir the mixture thoroughly, and let stand 1 hour longer, stirring once during the interval. After stirring once more, pour about 5 ml. of the unsettled mixture on an 80-mesh copper or brass wire screen fitted into a Gooch crucible. Support the ring and crucible by a funnel fitted into the neck of a 100-ml. volumetric flask with a rubber stopper through which passes a glass tube for suction. Form a mat with gentle suction. Avoid excessive packing of the fibers, for this retards filtering. It may be necessary to refilter the first filtrate, but loss of small amounts of alpha cellulose to the filtrate does not appreciably affect the results. Pour the remainder of the mixture on the mat; and before the last of the liquid has run through, wash the beaker and the mat with 35 ml. of water. Dilute the filtrate to volume for later use in determination of beta and gamma cellulose.

Moisten the residue of alpha cellulose with water, and remove it from the crucible. Place the crucible upright in a 400-ml. beaker, and fill it at room temperature with 25 ml. of 3:2 sulfuric acid. Rinse it after a few minutes with 50 ml. more of this acid. Disintegrate the alpha cellulose pad in the acid, using a thermometer as a stirring rod. From a pipet add exactly 25 ml. of potassium bichromate solution containing 90 grams per liter, and heat to 140 to 150° for 10 minutes. Bubble air in a fine stream through the solution to prevent lumping, and keep the beaker covered with a watch glass notched to permit entrance of the thermometer and the bubbling tube. After the solution has cooled to 130°, add 50 ml. of water, rinse the watch glass and thermometer, and cool the solution to 60° or lower.

Dissolve 195 grams of ferrous ammonium sulfate, $Fe(NH_4)_2(SO_4)_2.6H_2O$, in water containing 10 ml. of concentrated sulfuric acid, and dilute to 1 liter. This solution needs daily standardizing by titration of a suitable aliquot with the bichromate.

As indicator, dissolve 0.3 gram of barium diphenylamine sulfonate and 0.5 gram of anhydrous sodium sulfate in 100 ml. of water. Use 5 to 10 drops in a solution sufficiently dilute for good observation of the end point, which is from red to green. As an alternative, use potassium ferricyanide solution as an outside indicator on a spot plate. In this case the end point is blue. Titrate the remaining bichromate with the ferrous ammonium sulfate solution.

Pipet exactly half of the filtrate from the alpha cellulose, after all fibers have settled, into a 400-ml. beaker containing exactly 5 ml. of the bichromate solution. If the paper contains oxidizable fillers such as zinc sulfide or calcium sulfite, filter the filtrate once through a thick pad of asbestos in a Gooch crucible before taking the portion for analysis. Any such fillers remaining with the alpha cellulose may cause some error, but this is usually slight. Cautiously and with constant stirring, pour 50 ml. of concentrated sulfuric acid down the side of the beaker containing the portion of the filtrate. Then heat and titrate as before.

If resin, starch, or glue is present, correct the bichromate volumes as follows: Take the amounts of sizing materials remaining with the alpha cellulose as 0.25 per cent glue, 0.2 per cent starch, and 0.2 per cent resin, based on the dry weight of the paper. These are average values, but actual amounts have been found to vary not more than 0.1 per cent from them, irrespective of the original content of glue, starch, or resin in the papers. For example, if a paper were found to contain 3.4 per cent glue, 0.7 per cent starch, and 1.1 per cent resin, then the sizing materials reaching the beta plus gamma portion would be 3.15, 0.5, and 0.9 per cent, respectively. After these values are converted into weights, convert them into milliliters of the bichromate solution by dividing each weight by the corresponding bichromate factors, which are 0.0154 gram per ml. for glue, 0.0129 gram per ml. for starch, and 0.0066 gram per ml. for resin. Subtract the resulting equivalent volumes in milliliters of the bichromate solution from the volumes of bichromate consumed by the alpha and by the beta plus gamma fractions, which are then substituted in the equation given for calculation.

CALCULATION.

25 − (volume of ferrous ammonium sulfate used
× its bichromate equivalent) = ml. bichromate solution used by alpha cellulose = A

2[5 − (volume of ferrous ammonium sulfate used
× its bichromate equivalent)] = ml. bichromate solution used by beta and gamma cellulose = B

Then

$$A \times \frac{100}{A + B} = \text{per cent alpha cellulose}$$

Base the percentage of alpha cellulose on the total cellulose including pentosans, but excluding moisture, ash, resin, or any sizing or other added nonfibrous materials. Determinations should be made in duplicate and should agree within 0.4 per cent. Express the average to the nearest 1 per cent, which is compatible with experience and is sufficient for the practical usefulness of the value. The same remarks apply to beta and gamma cellulose values.

Gravimetric. Weigh 1.5 grams of specimen prepared as already described for the volumetric method. Add 100 ml. of 17.5 per cent sodium hydroxide solution to the sample in a 400-ml. beaker. Macerate until uniformly wet and dispersed, and let stand 10 minutes from the time of addition of the alkali. Dilute with 165 ml. of water, stir the mixture thoroughly, and let stand 1 hour longer, stirring once during the interval. After stirring once more, pour the mixture upon a medium-weight cotton cloth cut to fit a 7.5-cm. Büchner funnel. The cloth should have been previously washed and the fitted piece dried to constant weight at 100 to 105° and weighed in a weighing bottle before use. Refilter if necessary. Before the last of the liquid has run through, add water, break up the pad well with a pointed stirring rod, and keep in a loose condition until 200 ml. of wash water has passed through. Then cover the alpha cellulose with 10 per cent acetic acid, and allow to soak for 5 minutes; then pass an additional 500 ml. of wash water through while breaking up the pad with the pointed stirring rod.

Dry the alpha cellulose on the cloth in the original weighing bottle overnight at 100 to 105°, cool, and weigh. Correct the weight for the ash of the alpha cellulose. Take the sizing materials, if any, remaining with the alpha cellulose as 0.25 per cent glue, 0 2 per cent starch, and 0.2 per cent resin, based on the dry sample weight, and subtract the corresponding weights from the total.

Beta and Gamma Cellulose. Use the remainder of the filtrate from the volumetric alpha cellulose method for this determination. Acidify with 15 to 16 ml. of 1:5 sulfuric acid. After cooling, dilute the mixture to 100 ml. Pour into a cylinder, cover the cylinder, and let stand at room temperature until the beta cellulose has settled, which is usually overnight. Pipet out 50 ml. of the supernatant liquid, oxidize, and titrate as before. Correct the bichromate volumes as previously described for any glue and starch, but not for resin.

Calculation.

$$\text{Ml. bichromate for gamma portion} \times \frac{200}{100} \times \frac{100}{50} \times \frac{100}{\text{ml. bichromate for alpha fraction plus filtrate}} = \text{per cent gamma cellulose}$$

$$100 - (\text{per cent alpha} + \text{gamma}) = \text{per cent beta cellulose}$$

PAPER FIBERS

Since paper fibers are a plant product that has been subjected to more or less drastic chemical action, it is natural to expect variations in different fibers from the same source. Careful examination from this viewpoint is essential. If the original sample is paper, sizing and coating materials must first be removed to free the original fibers and then these must be pulped. The identity of the fibers making up the pulp is then established in part by reactions to stains and in part by observation of structure under a microscope at a magnification of 100 diameters.

General Preparation. First the method of the TAPPI is given because it is a standard method. The short method that follows is recommended because it gives as uniform a pulp as is obtainable and requires much less time.

TAPPI Method. Tear a 0.2-gram sample into small pieces and cover with distilled water. Bring to a boil on a hot plate, decant the water, and roll the sample into small pellets with the fingers. Place these in a test tube with a little water and shake tube until the water is absorbed. Add a little more water and shake well. Continue until the paper is completely defibered. Dilute to a fiber content of about 0.05%.

If the paper cannot be disintegrated by water, cover the sample with 1% sodium hydroxide solution. Bring to a boil, decant the alkaline solution, and wash twice with water. Cover with 1:200 hydrochloric acid, decant after several minutes, and wash several times with water. Roll into pellets and proceed as previously described. For methods for specially treated papers, consult TAPPI T 401m-60. Thoroughly mix the fibers and water.

Keep slides and cover glasses in 50% alcohol. After a slide has been dried and polished, draw lines 1 inch from each end with a glass-marking pencil to keep the fiber suspension inside the 1-inch square at each end of the slide. Place the slide on a warm plate. Shake the test tube. Withdraw a portion of the fibers by inserting a dropper. Deposit 0.5 ml. of the fiber suspension on one of the squares. Repeat with another portion of the suspension on the other end. Allow the water to evaporate until there is just sufficient to float the fibers, gently tap the slide to distribute the fibers evenly, and leave them on the warm plate until completely dry. Let cool before staining.

Short Method. Sample as for the TAPPI method, using sodium hydroxide technic. Boil the representative sample of paper for about 1 minute in the alkali. Rinse in the beaker with water, avoiding loss of fibers in decantation. Roll the paper in a ball between the fingers sufficiently to loosen the fibers. Drop this ball into a small beaker half full of water, and stir mechanically for several minutes or as long as is necessary completely to separate the individual fibers. Remove a portion of the fibers, and place them on a slide, using a pipet with a rubber bulb at one end and with the discharge end unconstricted. Remove excess water with a hard blotting paper, and proceed to stain.

Qualitative Identification. There are many stains for identification of fibers. Complete discussion is far beyond the scope of this general procedure. Therefore, "C" stain is given with the suggestion that confirmatory stains are available in TAPPI T 401 m-60.

Graff "C" Stain is an iodine preparation purchasable from the Institute of Paper Chemistry. It is also preparable in the laboratory, as discussed below, with the caution that extreme accuracy is essential, that slight differences are significant, that specific gravities and volumes must be accurately measured, and that the finished product must be stored in dark-colored dropping bottles and renewed after 2–3 months. Only reagent-grade chemicals and distilled water may be used.

A. *Aluminum chloride* solution of 1.15 sp.gr. at 28°C., made by adding about 40 grams of $AlCl_3 \cdot 6H_2O$ to 100 ml. of water.

B. *Calcium chloride* solution of 1.36 sp.gr. at 28°C., made by adding about 100 grams of $CaCl_2$ to 150 ml. of water.

C. *Zinc chloride* solution of 1.80 sp.gr. at 28°C., made by adding approximately 25-ml. of water to 50 grams of dry $ZnCl_2$ (fused reagent-grade sticks in sealed bottles, or crystals). Do not use $ZnCl_2$ from a previously opened bottle.

D. *Iodide-iodine* solution, made by dissolving 0.90 gram of dry KI and 0.65 gram of dry iodine in 50 ml. of water. Dissolve the KI and iodine by first thoroughly intermixing and crushing together, then adding the required amount of water drop by drop with constant stirring.

Mix well together, 20 ml. of solution A, 10 ml. of solution B, and 10 ml. of solution C; add 12.5 ml. of solution D and again mix well. Pour the mixture into a tall, narrow vessel and place it in the dark. After 12 to 24 hours, when the precipitate has settled, pipet off the clear portion of the solution into a dark bottle and add a leaf of iodine. Keep this in the dark when not in use.

The reactions are as follows:

When lignin is present, a yellow color is developed with the "C" stain. Groundwood gives a very vivid yellow with a tendency toward orange; unbleached jute stains much the same color, but the two fibers can easily be distinguished by their structural appearance. Unbleached pulps of all kinds tend toward the yellow, with the depth of yellow determined by the degree of cooking and the type of cook. Thus, a raw unbleached sulfite will stain a vivid yellow, but as the degree of cooking increases, it tends toward a greenish yellow. Unbleached sulfate tends toward yellowish brown, whereas an unbleached alpha is more brown than yellow. The hardwood pulps have a tendency to appear bluish and greenish even in their unbleached state. Manila, straw, bamboo, sugar cane, flax hurds, and esparto also give yellow colors with raw, unbleached cooks.

When any pulp is bleached, it has a tendency to give a reddish hue with the "C" stain. In some cases this tendency is very slight, but any hint of red can generally be taken as an indication of some degree of bleaching. The shade of red usually indicates the type of bleached pulp. Thus, rag, which is the purest form of cellulose, gives the purest red, followed by bleached softwood alpha, bleached softwood sulfite, and bleached softwood sulfate in that order. The sulfite is weak enough in red so that it frequently appears purplish gray. Alkali cooking tends to give a bluish color to wood pulp, so that with bleached softwood kraft pulp the blue coloration nearly overshadows the red, and a bluish gray is seen. Hardwood pulps have a tendency to be bluer than softwood pulps; therefore hardwood alkaline pulps, even though bleached, show almost no red when stained. Unbleached hardwood alkaline pulps cannot be easily disdistinguished from hardwood soda pulp. Some special fibers lend their own colors to the system.

Quantitative Determination. With the mechanical stage, move the field so that the pointer is 2 or 3 mm. from a top corner of the cover glass, then slowly move it in a horizontal direction and count and record the fibers of each kind as they pass the pointer. A multiple tally counter, such as is frequently used for blood-counting, is most convenient. Alternatively, if care is taken and the slide is not moved vertically, repeated passes may be made for each type of fiber counted.

If part of a fiber passes the center of the pointer more than once, count it each time; but if it follows the center for some time, count it only once. With fiber bundles, such as are often present in groundwood, count every fiber in the bundle as it passes under the pointer. Ignore very fine fragments, but count the larger fragments as fractions so that when two or three of the same kind of fiber fractions are observed in the same field, mentally they can be added together to give a whole number.

When all the fibers in a line have been counted, move the stage 5 mm. vertically to a new line and count the fibers in the same way. Continue until the fibers in four separate lines, 5 mm. apart, have been examined. If the slide has been properly prepared, a total of 200 to 300 fibers will have been counted.

Multiply the total number of each kind of fiber by its respective weight factor (see following data) to obtain the equivalent weights, and calculate the percentages by weight of each kind of fiber in the total composition.

SUMMARY OF WEIGHT FACTORS. 1.55, Southern kraft; 1.50, Douglas fir; 1.30, Groundwood (variable); 1.25, Cotton linters; 1.20, Hemlock; 1.00, Rag, gum; 0.90, Unbleached bagasse (boards), unbleached, bleached softwood (sulfite, kraft) except hemlock, fir, southern; 0.80, Bleached, unbleached bagasse (paper); 0.70, Northern alpha; 0.65, Straw (board); 0.60, Sisal, hardwood (soda, sulfate, sulfite) except gum, alpha; 0.55, Manila, jute, alpha; 0.50, Bleached flax, ramie, esparto; 0.30, Bleached straw.

Examine both square-inch fields. If the results for the two fields vary for any type of fiber present by more than the amount stated in the next section under Precision, then prepare and examine one or more additional fields and include the results from all the fields in the reported average.

Precision. The precision depends upon the skill and experience of the operator and on the selection of the weight factors. Competent workers may be expected to be able to check the composition of a furnish that is not too complex within the following tolerances, provided that the weight factors employed are reliable.

Percentage of given fiber in total furnish	Tolerance	Percentage of given fiber in total furnish	Tolerance
Less than 20	±2	60 to 70	±4
20 to 30	±3	70 to 80	±3
30 to 40	±4	Over 80	±2
40 to 60	±5		

It is emphasized that to achieve this precision, authentic pulp mixtures should be examined from time to time to ensure that sound judgment is exercised when including or rejecting debris in the count. Under ideal conditions with weight factors determined on the pulps examined, it is possible for experienced analysts to check the composition of a furnish within half the stated limits.

References. For further details on the examination of paper fibers, consult the following: Federal Specification for Paper, General Specification UU-P-31b (1949); TAPPI Tentative and Official Testing Methods, T401m-60: Fiber Composition of Paper and Paperboard (1960).

TEXTILE FIBERS

General Examination of Samples. Either microscopically or with the hand glass, look for nonuniformity of color. On careful examination this is often a clue because the same dye "takes" differently on different types of fiber. For mixed colors or colored and white fabrics such an examination will show whether colors were printed on, dyed in the piece and discharge printed, or skein dyed. Separate the warp and filling to see whether they are observably different in structure. Separate pile, nap, or woven-in tufts, and examine them. Usually a low power of the microscope is desirable for this. Textures are produced in various ways as by warp and filling of different composition, yarns composed of strands of different composition, or even yarns spun with mixtures of fibers. This preliminary examination is a guide to future procedure; usually what are suspected to be fibers of different types will be so detected and this suspicion subsequently confirmed or invalidated by staining.

While no detailed quantitative method will be given, approximations as to the percentage composition of mixed fibers are obtainable microscopically, after staining by technics similar to those described for paper fibers.

More details than are given here can be obtained from the texts on microscopy already referred to, from the ASTM Standards on

Textile Fibers, from "Textiles and the Microscope" by E. R. Schwartz, and from "Elementary Textile Microscopy" by J. H. Skinkle.

Ignition. A valuable preliminary indication of the type of fiber can often be obtained by burning the individual fibers. Hold the fibers, or strands if mixed fibers are not suspected, over a low flame. Note the odor. Treatments such as casein finish, rubber impregnation, and pigmentation may affect the observation. Heavily weighted silk leaves a substantial residual ash in the shape of the original fiber. Asbestos fibers glow but do not fuse. Unless the flame is very low, glass fibers fuse without burning to leave a hard knob on the end. Animal and casein fibers burn slowly, with a characteristic odor best described as that of burning feathers, and leave a black knob as ash. Vegetable fibers and rayons, with the exception of cellulose acetate, burn rapidly with the odor of burning paper and leave little or no ash. Acetate rayon burns rapidly with an acid odor and leaves a black knob as ash. All these observations are much clarified by carrying them out with fibers of known origin.

Removal of Color. Except for white or lightly tinted fabrics the color is usually so deep as to interfere with staining reactions. Only general instructions can be given to apply experimentally. A few dyes can be removed by cold 1 per cent acid or alkali; usually acetic acid or ammonia is used. Heating nearly to boiling will often facilitate such removal The dyeing industry uses a strong solution of sodium hydrosulfite, which is correctly sodium hyposulfite, $Na_2S_2O_4$, for discharging many colors. Wool fibers often retain dye so tenaciously as to respond to none of these treatments. In that case warm the fibers with a solution consisting of 1 volume of 20 per cent titanous chloride, $TiCl_3$, and 2 volumes of concentrated hydrochloric acid. This usually achieves the desired result. Wash the fibers thoroughly with water to remove all traces of bleaching agents before applying stains.

Cross Sections. To obtain cross sections of fibers for examination, thread the fibers through a small cork, or embed them in parallel strands in collodion, cellulose acetate, or paraffin. After drying or chilling, as the case may be, cut with a sharp knife directly at right angles to the lengthwise direction of the fibers.

Differentiating Animal from Vegetable Fibers. Two methods are often used. Of these Millon's reagent is the simpler because the solution is also used as a mounting medium. Picric acid is often effective when the result is in doubt because of incomplete removal of dye.

PROCEDURE. Apply Millon's reagent (page 720) to the fibers, mount as a slide, and examine after about 2 minutes. Animal fibers assume a pink to red color that deepens on standing. Fibers of vegetable origin show no color.

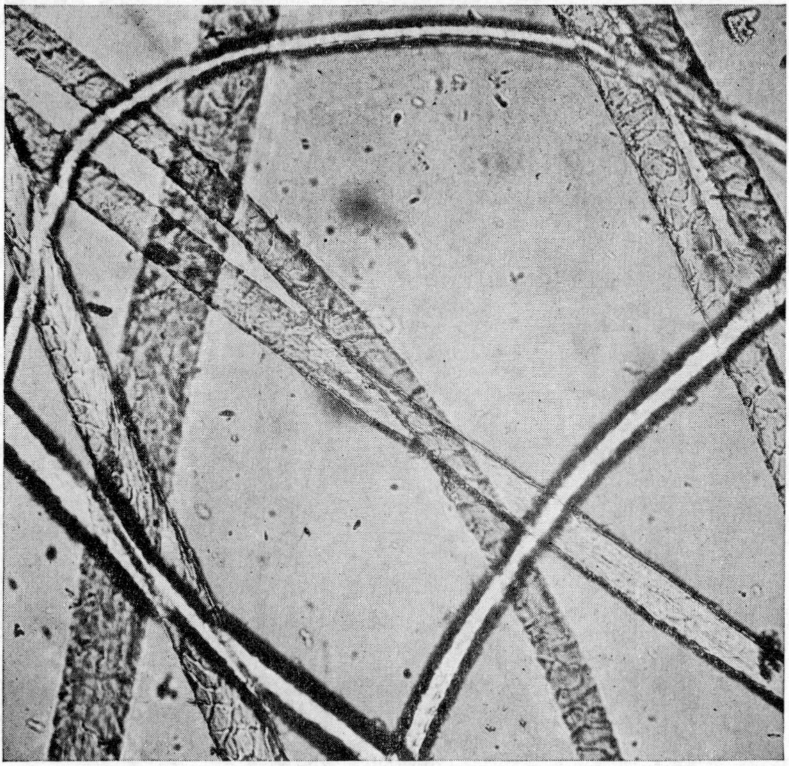

FIG. 119.—Wool. 250 ×.

To confirm, place other fibers in a few drops of 1 per cent aqueous picric acid on a slide for about 3 minutes, and wash with water. Animal fibers are stained yellow, but vegetable fibers are not.

Animal Fibers. The types of animal fiber available are limited largely to wool and silk. Each is characteristic, as shown by the accompanying photomicrographs.

WOOL. These fibers have a scaly appearance, as shown in Fig. 119, which is readily recognized in transmitted light. Fibers out of focus are deceptive unless carefully examined by moving the focus up and down. Even with this technic, scales are not readily observable on small woolen fibers, particularly by the worker who is not experienced in lighting adjustments. Different kinds vary in diameter from 0.06 to 0.004 mm., the finer diameters corresponding to the better wools. Hair has similar characteristics but

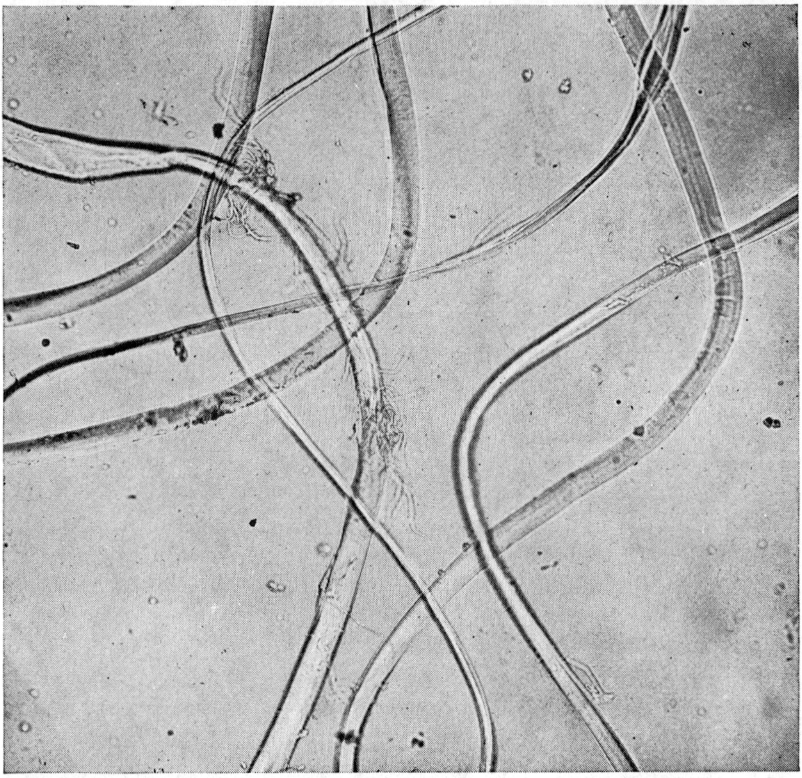

FIG. 120.—Silk. 250 ×.

is readily distinguished by its greater stiffness of fiber; in fact, there is little basis for distinction between wool and hair other than by stiffness.

While wool is very stable it may be injured by overexposure to light, by mechanical treatment, or by alkali. Such damage is shown by staining with a saturated solution of indigo carmine rendered just acid with sulfuric acid.

PROCEDURE. Place the fibers in this reagent for a few minutes, and wash. Mount in glycerin. Damaged wool shows a bright blue color varying with the degree of damage; undamaged wool is not affected. By additional treatment with 1 per cent picric acid solution before mounting, the damaged parts show green, the balance yellow. Casein fibers react like damaged wool.

SILK. Raw silk consists of two filaments of silk bound together by a coating of water-dispersible albumin known as sericin. This

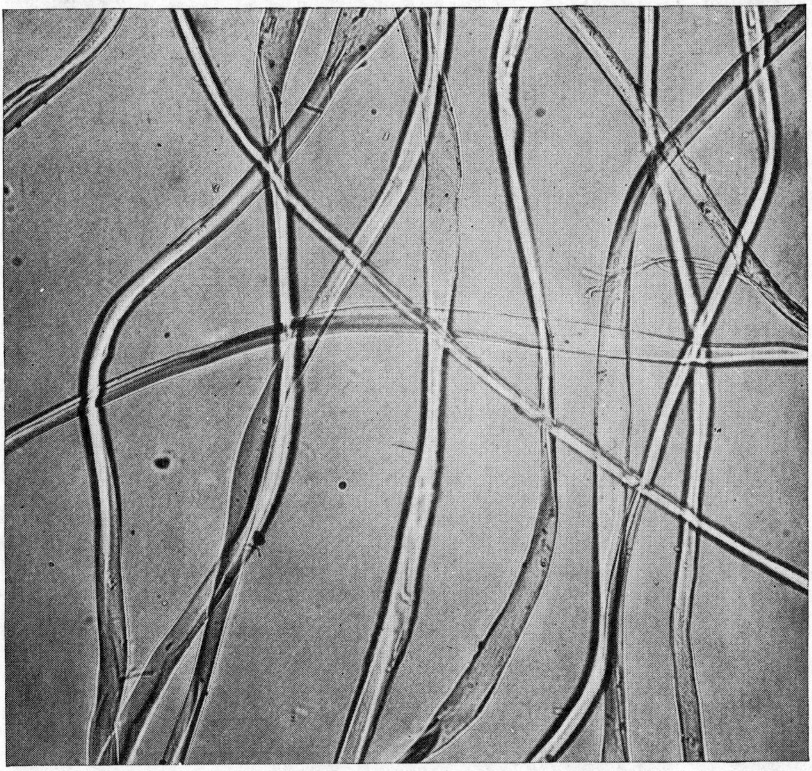

FIG. 121.—Wild silk, or tussah. 250×.

coating amounts to around 25 per cent of the total. The silk of commerce (Fig. 120) consists of the individual filaments after the sericin has been dissolved away. Silk filaments are smooth and rodlike. Under the polarized light of the Nicol prism they show the color effects that characterize doubly refractive substances. The diameter will be 0.009 to 0.024 mm. Silk is readily distinguishable from artificial fibers by its reaction with picric acid. The structure is hardly characteristic on direct examination.

PROCEDURE. Ordinary silk must sometimes be differentiated from what is known as wild silk or tussah (Fig. 121), a misnomer in the sense that this is in part a product of cultivation. For this purpose dissolve 25 grams of crystallized nickel sulfate, $NiSO_4.6H_2O$, in 500 ml. of water. Precipitate nickel hydroxide by addition of about 5 per cent sodium hydroxide solution, but do not add excess. Filter or centrifuge the precipitate, wash

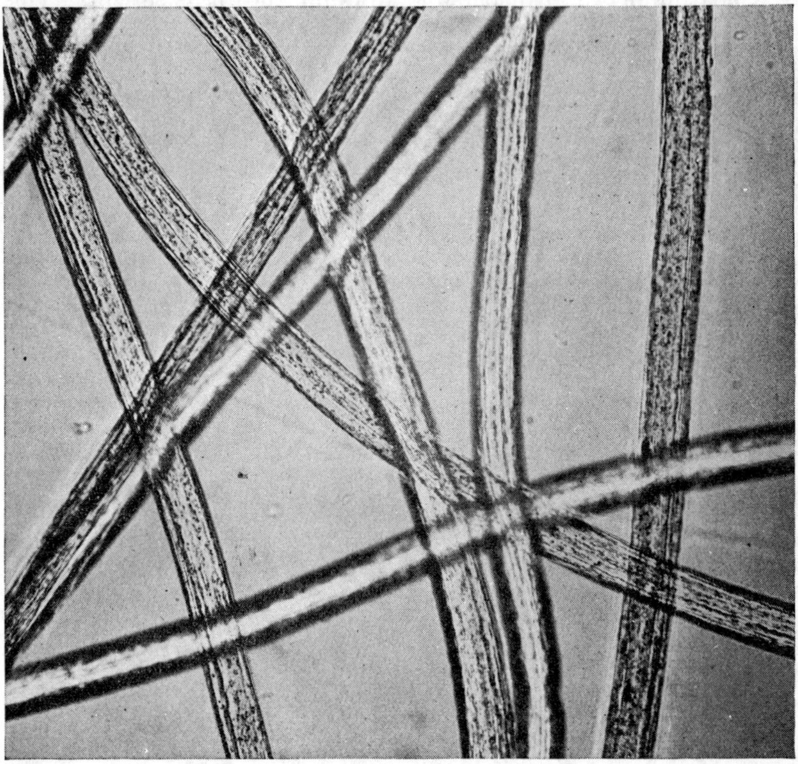

FIG. 122.—Artificial wool—Lanital. 250×.

with water, and dissolve in 250 ml. of 1:1 ammonium hydroxide solution. Place the fibers in a drop of this solution, and observe. Ordinary silk fibers swell and go into solution at room temperature; wild-silk fibers remain unaffected. The latter will also swell and dissolve on warming.

Artificial Animal Fiber. An artificial animal fiber is produced from casein. Commercially it is a wool replacement; chemically an animal fiber. Figure 122 shows a fair example. Casein fibers absorb the dye from a saturated aqueous solution of benzopurpurin in a few minutes and on washing will be found to be red. Similar exposure to saturated aqueous indigo carmine solution acidified with sulfuric acid produces a bright blue stain that does not wash out in water. Silk and damaged wool give the latter reaction, also, but are readily distinguished microscopically by the experienced worker.

Nylon and Other Synthetic Silk Replacents. These are readily identified by heating, for they fuse whereas silk chars on heating.

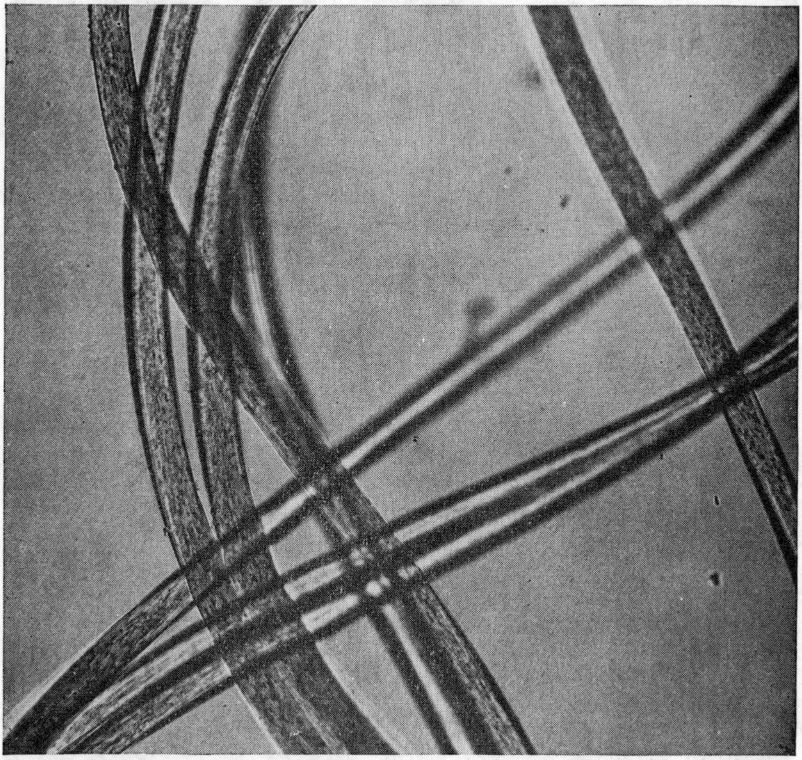

Fig. 123.—Nylon. 250×.

Different types of these synthetic fibers show characteristic cross sections. The appearance of nylon is shown in Fig. 123.

Modified Vegetable Fibers. Modifications of cellulose or its esters constitute a group of fibers similar to silk in appearance. The principal members of the group are viscose rayon (Fig. 124) and cellulose acetate rayon. Both are used as continuous filaments or cut into staple fiber and spun into yarn. Cuprammonium rayon is of lesser importance and so-called nitrocellulose rayon produced by denitration of nitrocellulose fibers of no commercial significance.

The methods that follow will identify the usual commercial rayons. Consideration should be given to the use to which the sample has been put. Special treatments such as delustering and "ironproofing" may cause deviation from the results indicated. Experimental rayons may differ radically in behavior from the usual commercial product, perhaps because of unusual purity.

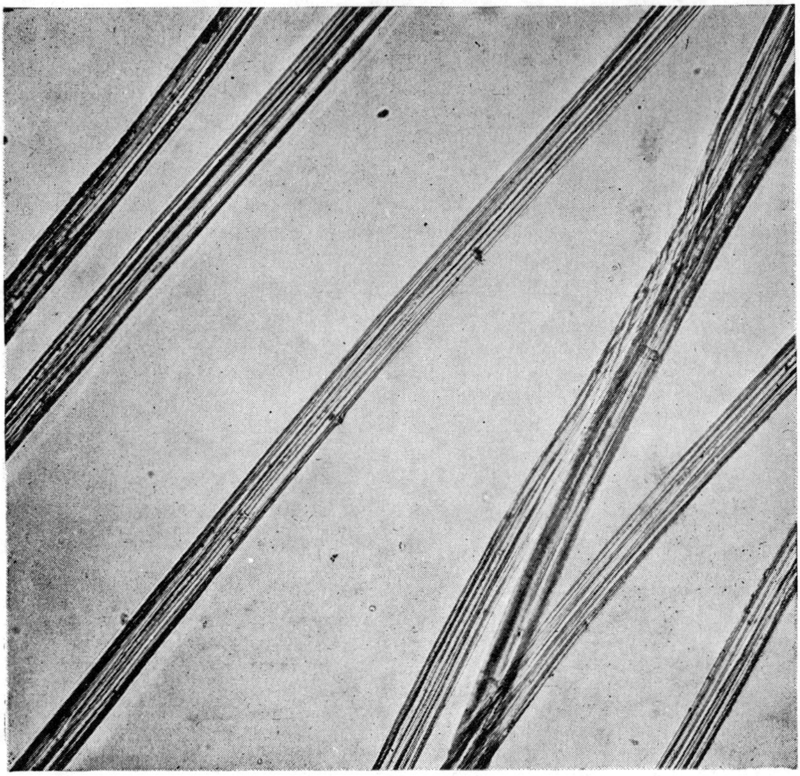

FIG. 124.—Viscose rayon. 250 ×.

For quantitative comparisons Wright's stain and Hahn's stain are found most satisfactory.

EXAMINATION UNDER POLARIZED LIGHT. Mount the fibers in water, and observe between crossed Nicol prisms, bearing in mind that variable thickness of fiber will affect the colors obtained. Viscose rayons show a white to green color and are streaked; cuprammonium rayons are a uniform yellow or green color; acetate rayons are silver gray; nitro rayons are light yellow to yellow and show a streaked appearance.

Examination of Cross Sections. When cross sections are examined as dry mounts, viscose rayon fibers have irregular serrated edges. Those of cuprammonium rayon fibers are smooth and almost circular. Cellulose acetate and nitro rayon fibers have relatively smooth circumferences, with a few indentations.

Identification of Viscose Rayon. Examine the fibers at a magnification of about 350 diameters by means of a dark field condenser. Viscose fibers reveal numerous white particles within the fiber, which give a cloudy effect. Fibers of other rayons are almost transparent and show few, if any, scattered particles. Larger and more brilliantly reflecting particles may be present on or in all rayons. They are usually particles of materials used as delustrants. These are not likely to be confused with the fine white particles that characterize viscose rayon.

A confirmatory test may be made with silver nitrate. Dissolve 1 gram of silver nitrate in 40 ml. of water and 4 grams of sodium thiosulfate in a similar amount of water. Add the first solution to the second. The cloudiness at first formed will finally disappear. Now add a solution of 4 grams of sodium hydroxide in 20 ml. of water, boil, make up to 100 ml., and filter.

Immerse the fibers in this solution, and bring to a boil. Remove, lightly rinse with water, and examine. Viscose fibers stain to a brown or reddish brown color. Cuprammonium and acetate fibers remain colorless. Nitrocellulose rayon acts somewhat like viscose rayon in this test.

Identification of Acetate Rayon. Observe the fibers under the microscope while adding acetone at the edge of the cover slip. Acetate fibers swell and dissolve readily. Other fibers are not affected.

Dissolve 0.015 gram of iodine in 1 liter of a 5 per cent solution of potassium iodide. Acidify with 1 or 2 drops of glacial acetic acid. Immerse the fibers in this reagent, and after a few minutes wash with water and observe. Acetate fibers become bright yellow. Other rayon fibers are either colorless or nearly so.

Identification of Cuprammonium Rayon. Cuprammonium rayon may be identified by a 0.2 per cent aqueous solution of Erie fast orange CG. Immerse the fibers for 3 minutes in the solution, wash well with water, and mount in water. Cuprammonium fibers stain deep orange. Other rayons are colorless or pale orange. The absorption of dye and resulting intensity of color can be increased by heating the solution to boiling with the fibers.

A modified Hahn's stain is also effective. Dissolve 0.2 gram of soluble blue 2B extra, 0.1 gram of eosin J, and 1 gram of tannic acid in 100 ml. of hot water. Cool, and add 0.2 ml. of 10 per cent hydrochloric acid. Dip the fibers in the cold solution for 3 minutes, wash well with cold water, and observe. Cuprammonium fibers are stained blue; other rayon fibers are lavender.

If the hydrochloric acid and eosin J are omitted, acetate fibers are yellow, cuprammonium blue, and others colorless.

Differentiating Viscose and Cuprammonium Rayon. This is best done with Wright's stain. Precipitate a water suspension of methylene blue 2B, with about one-half the quantity of eosin yellowish, stir thoroughly to complete the reaction, centrifuge, and decant or filter. Dissolve the precipitate in sufficient alcohol to give the correct coloration as determined on a known sample of fiber. Employ a white or light-colored sample of

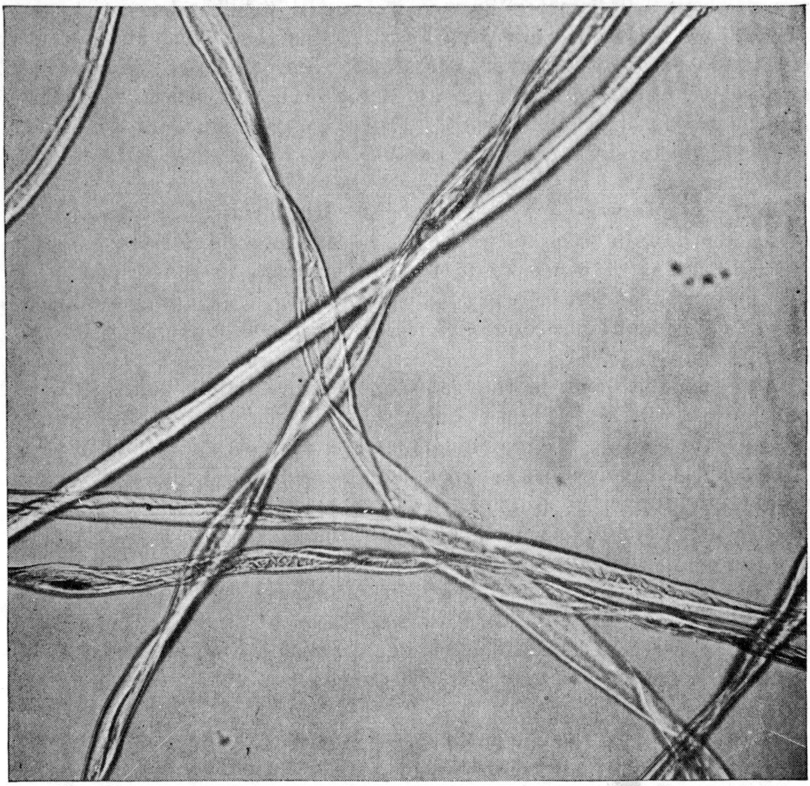

Fig. 125.—Cotton. 250×.

dry fiber for identification. Cover with the cold alcoholic solution, and bring just to a boil in a few seconds. Wash the stained sample thoroughly with water.

Viscose fibers are stained blue, cuprammonium fibers violet. Acetate rayon is also violet, but partly disintegrated. Nitrocellulose rayon fibers are stained much like viscose. Rayon fibers when wet with water before staining all give a violet coloration.

Vegetable Fibers. The outstanding vegetable fiber is cotton. Secondary types range from flax fiber or linen to the cordage fibers such as Manila and jute.

Cotton. The cotton fiber is flattened and twisted as shown in Fig. 125. This is in contrast to a bast fiber such as jute shown in Fig. 126. This structure of cotton is characteristic, but it is often necessary to differentiate between mercerized and unmercerized cotton. Hubner's reagent is satisfactory for qualitative differentiation provided that the fibers have previously been properly decolorized. For this reagent, dissolve 20 grams of iodine in 100 ml. of a saturated solution of potassium iodide in water.

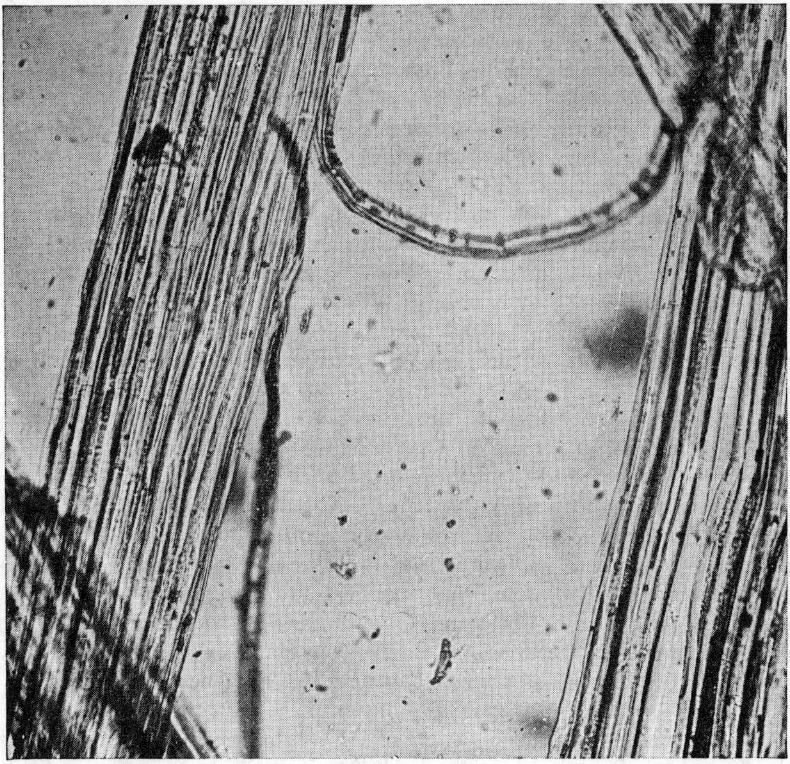

Fig. 126.—Jute. 250×.

Immerse the fibers in this solution for a few seconds; then wash several times. Unmercerized cotton is only slightly stained; mercerized cotton becomes deep indigo or appears black.

Differentiating Old and New Cotton. This is an examination which is often important because of laws which require that articles stuffed with old cotton shall be so labeled. A mercury quartz arc lamp with a glass filter passing rays of 4,047 to 2,900 angstrom units gives good results. Any convenient source of ultraviolet rays may be used provided that little of the visible spectrum is present.

New cotton fibers exhibit a definite violet fluorescence that must not be confused with any visible spectrum that may be present. As the cotton ages, the fluorescence turns from a violet to an ivory-white or brownish white. It is advisable to have known samples of new and old cotton for comparison.

DIFFERENTIATING FLAX AND HEMP. These two bast fibers are similar in structure and chemical composition. A mixture of the two fibers is rarely encountered. Several tests are given with the thought that several should be applied before results can be considered to be conclusive. Of these that based on the cuprammonium reagent is probably the best.

Cuprammonium Solution. Prepare the reagent by dissolving 2 grams of freshly precipitated cupric hydroxide, $Cu(OH)_2$, in 100 ml. of 1:3 ammonium hydroxide, and store in a dark bottle. Place the fibers in a few drops of the reagent, and immediately observe in order to follow the stages of swelling.

As the flax fibers swell, the protoplasmic remnants become prominent in the tortuous, narrow, central tube. The inner canal of hemp exhibits prominent horizontal striations. Swelling may take place at different rates, and hence the test is only qualitative. Highly bleached fibers do not react. The swelling process may be arrested by substituting a solution of equal parts of glycerin, alcohol, and water for the cuprammonium solution.

Ruthenium Red. Prepare cross sections of the bundles of fibers. Place on a slide, and moisten with a 1:10,000 solution of ammoniated basic ruthenium chloride, usually referred to as ruthenium red. Warm for 5 to 10 minutes, and wash the sections. The middle lamellae and the cell contents are stained red. In flax the cell contents are apt to be the more prominently stained; in hemp, the middle lamellae. Cross sections of hemp fibers tend to be somewhat flattened and have an elongated lumen with projecting points. Sections of flax fibers tend toward a more nearly circular lumen and cell wall. Conclusions must be drawn from the appearance of bundles as a whole and not from detached single fibers.

Cyanin. Prepare an aqueous solution of cyanin, nearly saturated at room temperature. Add one-third of its volume of glycerin. Macerate the fibers by boiling in a 1 per cent solution of sodium hydroxide, and then thoroughly wash out the alkali. Heat the fibers in a few drops of the reagent on a slide, and wash carefully with a solution of equal parts of glycerin, alcohol, and water. Finally mount in glycerin. Flax fibers remain colorless. Hemp fibers assume a greenish blue color because the middle lamella is slightly lignified. Distinguish carefully between the lignin in the middle lamellae of hemp fibers and the lignified shives in incompletely cleaned flax.

Moisture. Wet the fiber by drawing it between the moistened fingers, and holding the fiber in a vertical position view the tip of the fiber from above. Flax will be seen to move in a clockwise direction. Hemp may move in either a clockwise or a counterclockwise direction, but usually the latter. The motion is much weaker in hemp than in flax.

IDENTIFICATION OF MANILA FIBER. Manila, or abacá, fiber can be identified by Swett's reagent. To prepare, work bleaching powder containing 35 per cent available chlorine into a thick paste with water. Add about six times the resultant volume of water. Let this settle for 12 hours, and decant the clear strong solution of reagent into a dark-colored glass bottle. Before using, make just acid with acetic acid. Only a strong bleach solution should be used since the success of the test depends upon having an excess of available chlorine present for chlorination of the fibers. To ensure effectiveness of the reagent it is always desirable to run a blank on Manila and non-Manila fibers the same day the sample is tested.

For testing, separate the fibers, and wash out any natural oils with alcohol followed by ether. Dry by waving in the air. Immerse the fibers in the recently acidified bleach solution for 30 seconds and rinse in water. Shake, rinse in alcohol, and again shake until nearly dry. Finally expose to the fumes of strong ammonia solution for a few seconds. Manila fibers are colored brown and all adulterant fibers cherry red.

DIFFERENTIATING MANILA, COCONUT, AND SISAL FIBERS. The difference is best brought out by examination of the ash from the fiber. Ignite the sample in a small uncovered porcelain crucible over a low flame. Carefully transfer the ash to a slide, and mount in aniline or liquefied phenol. The silicified cells of Manila fibers, known as stegmata, appear as rows of rectangular hollow bodies. The ash of coconut fibers yields siliceous globules of various sizes. These globules have the appearance of possessing a foamy inner structure. In the ash of sisal fibers there are no siliceous skeletons, but some characteristic elongated aggregates of calcareous material are present.

Drop another portion of ash into a 20 per cent solution of hydrochloric acid. The siliceous stegmata of Manila and of coconut are unaffected and are more readily seen because of the destruction of other parts of the ash. The calcareous material in the ash obtained from sisal is dissolved by this treatment. When the moistened ash from sisal is treated with 1:1 sulfuric acid, the presence of calcium is shown by formation of needles of calcium sulfate.

To further differentiate Manila and coconut fibers, drop another portion of ash carefully into a 2 per cent solution of potassium ferrocyanide on a slide. Add a little 1:3 hydrochloric acid. The round stegmata of coconut fiber ash usually turn distinctly blue. In some samples of Manila the stegmata turn light blue.

IDENTIFICATION OF JUTE. Place the fibers in a drop of 1.25 per cent solution of phloroglucinol in alcohol. Add a drop of concentrated hydrochloric acid. Unbleached jute fibers turn dark violet-red because they are heavily lignified. Phloroglucinol also stains the lignified portions of other fibers such as phormium and groundwood. Jute, however, is readily identified by consideration of the structural characteristics in conjunction with its exceptionally heavy lignification.

IDENTIFICATION OF RAMIE. Ash the fibers in a porcelain crucible over a low flame. Mount the ash in aniline. Numerous characteristic spherical crystallites with spicules are prominent. Manila has crystallites that are smaller than those of ramie but also has characteristic rows of stegmata. Flax has neither crystallites nor stegmata.

Determination of Filling, Wool, Cotton, Rayon, and Silk in Mixtures. FILLING AND WOOL. Using the methods already given, determine what fibers are present. Obtain a representative 5-gram sample. Do not lose any fibers that may separate here or during later treatment. Oven-dry the sample at 105 to 110° to constant weight. Now extract in a Soxhlet extractor with carbon tetrachloride, siphoning over at least six times. Allow to dry in the air, and then wash by repeated immersion in hot distilled water, squeezing between each immersion. Place in a 3 to 5 per cent aqueous solution of a starch- and protein-solubilizing enzyme preparation at 50 to 60° or the optimum temperature range of the particular enzyme used. For detailed suggestions see pages 615 and 617.

To wet the fibers thoroughly, squeeze while immersed, remove, and squeeze again. Repeat the procedure at least three times. Keep immersed for 1 hour if the results are doubtful; some sizings are removed in 15 minutes. Immerse the sample for two 30-minute periods in baths of boiling water, rinsing after each bath. Thorough rinsing is necessary to remove China clay from heavily filled fabrics. Dry at 105 to 110° to constant weight. Then

$$100 \times \frac{\text{original oven-dry weight} - \text{weight after desizing}}{\text{original oven-dry weight}}$$
$$= \text{percentage of sizing, finishing, and other nonfibrous materials present}$$

Carefully unravel fabric or felt and partly untwist tightly twisted yarns to facilitate penetration of reagents. Immerse for 7 to 10 minutes in a boiling solution containing 1 per cent by weight of sulfuric acid, using 100 ml. of solution per gram of fibers. Filter through a Büchner funnel or Gooch crucible, and remove excess acid, using suction. Refilter if fibers come through. Allow the sample to cool, place in a 70 $\pm$ 1 per cent sulfuric acid solution, that is, 70 grams of sulfuric acid to 30 grams of water, using 100 ml. per gram of fibers. Allow to stand for 15 minutes, with frequent stirring. Filter again through the same filter, using suction. Refilter if necessary. Discard the clear filtrate, and wash the pad of fibers with water and then with a 2 per cent sodium bicarbonate solution and finally with water until free from sulfates. Dry the pad at 105 to 110° to constant weight. Then

$$100 \times \frac{\text{weight after acid treatment}}{\text{weight of desized sample}}$$
$$= \text{percentage of woolen fiber on the desized and filler-free sample}$$

Cotton, rayon, or silk are obtained by difference.

Correction for Reprocessed and Reused Wool. When such wool is present to the extent of 100 per cent, the woolen fiber as above determined may be lower than the actual amount by as much as 4 per cent. If such wool is claimed or known to be present, adjust results accordingly. For example, if it is claimed or known that a product contains 30 per cent of reprocessed or reused wool and a woolen fiber content of 78.7 per cent is obtained by the analysis, then there should be added to that percentage 30 × 0.04, or 1.2 per cent, making a corrected total of 79.9 per cent.

Acetate Rayon. *Acetone-soluble Types.* As in the wool determination use a 5-gram sample. Oven-dry, remove sizing and filling, and obtain the oven-dry weight. Now add about fifty times its weight of acetone, and agitate vigorously at room temperature for 15 minutes. Filter through a Büchner funnel or Gooch crucible, refiltering until the solution is clear. Rinse the pad of fibers by alternate squeezing and immersing in acetone, using two fresh portions. Allow the residue to dry, and then immerse in water at about 70°. Remove excess water by squeezing, and dry at 105 to 110° to constant weight. Then

$$100 \times \frac{\text{weight after desizing} - \text{weight after acetone treatment}}{\text{weight of desized sample}}$$
$$= \text{percentage of acetate rayon on the desized and filler-free sample}$$

Acetone-insoluble types. Treat as described in the preceding paragraph, using glacial acetic acid instead of acetone, but do not allow the fibers to dry before immersion in water at 70° and washing free of acetic acid. Dry at 105 to 110° to constant weight, and calculate as indicated.

Silk. This method is not suitable for tussah silk. Cut the residue from the acetate rayon determination into lengths of about 2 to 4 mm., and agitate vigorously for 1 hour in 200 ml. of a clear aqueous solution of calcium thiocyanate of specific gravity 1.20 to 1.21 at 70°, made just acid to litmus with acetic acid and maintained at 70 ± 2°. Take precautions to prevent evaporation and consequent concentration of the solution. Filter through a Büchner funnel or Gooch crucible, using suction. Refilter the hot solution until the filtrate is clear. Now agitate the fibers for 5 minutes in a fresh 200-ml. portion of the thiocyanate solution. Repeat the filtration, and wash the fibers with hot water until free from thiocyanate. Dry the residue at 105 to 110° to constant weight. Then

$$100 \times \frac{\text{weight after removal of acetate rayon} - \text{weight after removal of silk}}{\text{weight of desized sample}}$$
$$= \text{percentage of silk on the desized and filler-free sample}$$

Regenerated Cellulose Rayon. Treat the residue from the silk determination exactly as in that determination, but using an acid solution of calcium thiocyanate of specific gravity 1.35 to 1.36 at 70°, and weigh the dried residue. Then

$$\frac{100 \times \text{weight without silk} - \text{weight without regenerated cellulose}}{\text{weight of desized sample}}$$

= percentage of regenerated cellulose rayon on desized, filler-free sample

Cotton. The residue from the above determinations consists of cotton and wool. Treat exactly as described for wool, weigh the dried residue, and calculate the percentages of cotton and wool on the desized and filler-free sample, oven-dry basis. Obtain wool by difference.

Starches. As a general reagent for detection of starch, grind 2.0 grams of iodine in a mortar with 6.0 grams of potassium iodide, wash out with water, and stir until the iodine is completely dissolved. Dilute to 100 ml. A deep blue coloration is produced on the addition of a drop of this reagent to powdered starch, care being taken to wet the powder thoroughly, or to a starch solution. This is a very sensitive reaction and may indicate the presence of only a very small amount of starch in the presence of large amounts of nonstarch. Always test cold solutions, for the color is fugitive in hot, very dilute solution. The reaction is only a class test, for it is given by all starches. Partly hydrolyzed starch, which is dextrin, gives a reddish purple coloration.

The granules are characteristically different in each kind of starch. The shape, size, position, and character of the hilum, or nucleus, and any concentric markings and appearance under polarized light should be noted. Sometimes the particle size varies greatly in the same kind of starch. This is particularly marked in the case of potato starch.

With polarized light, all starches appear to show a black cross, but some so much more definitely than others as to make it a characteristic marking. Some starches show much more color than others under the polarizing microscope. It is often desirable to use a highly refracting mounting medium such as glycerin for such examination.

It is not possible to describe here minutely the differences shown by all different starches under the microscope. Careful comparison with known specimens under exactly similar conditions of magnification, illumination, and medium is the best method of identification. For more comprehensive descriptions of technic and characteristics, see "Commercial Organic Analysis" by A. H. Allen and editors, which though now out of print is available in the majority of libraries.

For estimation of particle size, place a pin point of the starch powder or a drop of the solution on a miscroscope slide. Add a drop of diluted iodine reagent. Too much iodine will make the

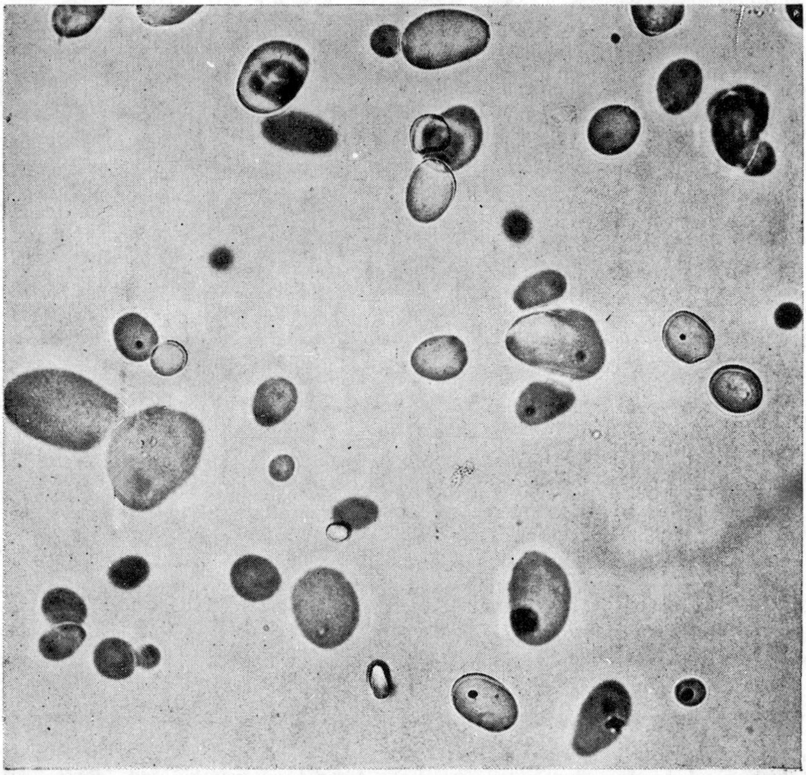

FIG. 127.—Potato starch. 250 ×.

granules too dark to observe satisfactorily; too little iodine will not show up the granules sufficiently. Apply a cover glass, and observe, using a somewhat oblique light. Usually a magnification of 100 times is satisfactory although a higher magnification is occasionally necessary. Use a micrometer scale, if available, to measure the particle size. Focus slowly up and down to observe the whole surface of the individual granule. In such an examination the extent to which the granules are broken up is a useful indication of the amount of boiling which a starch solution has undergone. Such boiling affects the particle size, of course.

In mixtures the amount of starch present may often be roughly estimated by observing the amount of blue-colored substance under the microscope after the addition of iodine reagent.

Potato Starch. This starch varies considerably in particle size, generally from 0.005 to 0.06 mm. The shape is oval or

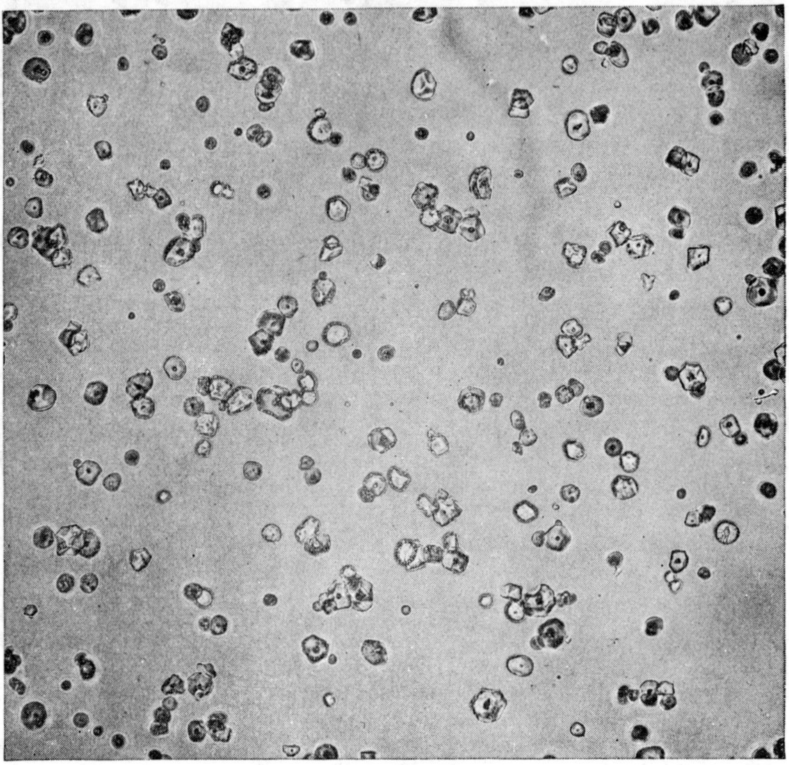

Fig. 128.—Cornstarch. 250×.

oysterlike. The hilum is a spot near the smaller end. Concentric rings are clearly observable (Fig. 127). Polarized light produces brilliant colors, and the smaller end shows a distinct cross.

Cornstarch. The particle size is fairly uniform at 0.007 to 0.02 mm. The general shape is circular to polyhedral, usually with rounded angles. The hilum is in a central position and in shape is stellate or like a crack. Rings are not readily visible (Fig. 128). Polarized light shows a distinct cross and faint colors.

BEAN STARCH. This is nearly uniform in particle size, about 0.034 mm. in diameter. The general shape is oval or kidney-shaped. The hilum is stellate, sometimes showing a longitudinal furrow. Rings are practically invisible. Polarized light produces no characteristic effects.

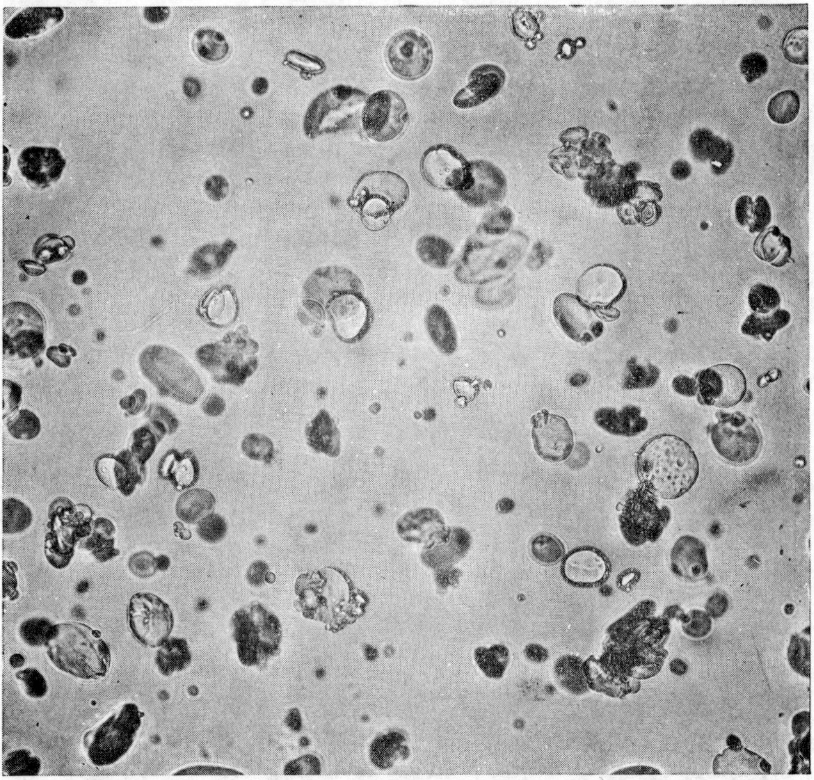

FIG. 129.—Wheat starch. 250 ×.

PEA STARCH. This varies in the range 0.018 to 0.028 mm. The general shape is oval or kidney-shaped. The hilum is elongated. Polarized light produces no characteristic effects.

WHEAT STARCH. This varies very considerably in size from 0.002 to 0.05 mm. Granules are usually uniform in size in some part of this range. The general shape is flat and circular, or nearly so. Hilum rings are not visible (Fig. 129). Polarized light with glycerin as medium shows a cross and faint rings and colors.

RYE STARCH. This is very similar in size, appearance, and behavior under polarized light to wheat starch. Some granules show a three- or four-armed fissure extending over most of the visible surface.

BARLEY STARCH. This starch is nearly uniform in size, the variation being 0.013 to 0.039 mm. It is similar to wheat starch in appearance and behavior under polarized light. Some granules are slightly angular or oval-shaped.

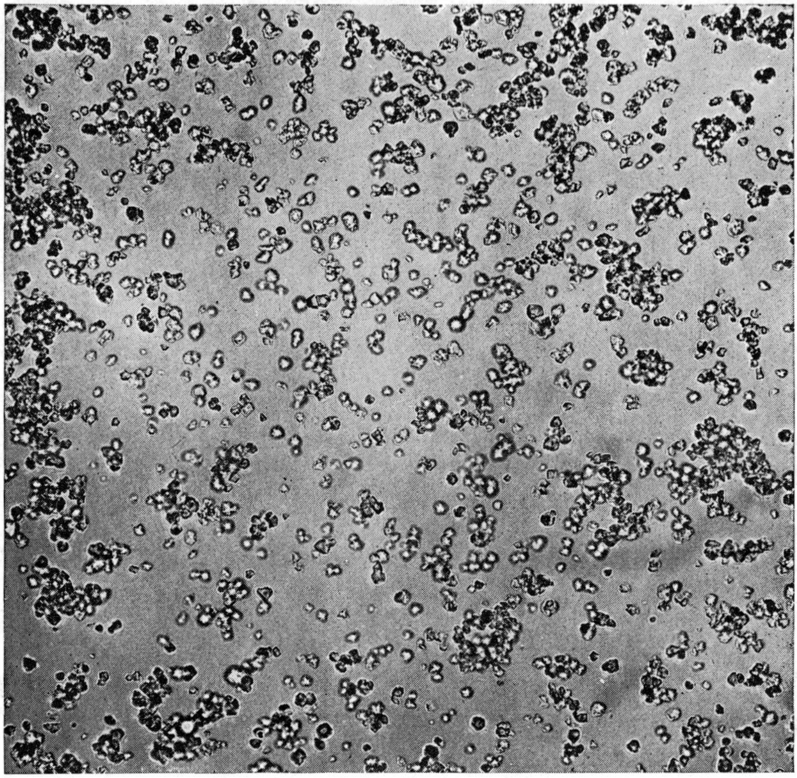

FIG. 130.—Rice starch. 250 ×.

OAT STARCH. This varies in size from 0.004 to 0.03 mm. The general shape is oval, with triangular to hexagonal divisions. The hilum and rings are not visible. Polarized light produces only a faint cross.

RICE STARCH. The particles are nearly uniform in size, from 0.005 to 0.008 mm. The general shape is pentagonal or hexagonal and occasionally triangular (Fig. 130). The angles are sharply defined. Very high magnifications show a distinct hilum. Polarized light shows a cross in the larger granules.

BUCKWHEAT STARCH. This starch varies in size in the range 0.005 to 0.02 mm., depending on the variety. The general shape is similar to that of oat and rice starches, but the angles are more rounded. The hilum is a spot or star and is centrally placed. No rings are visible. Polarized light shows a distinct cross.

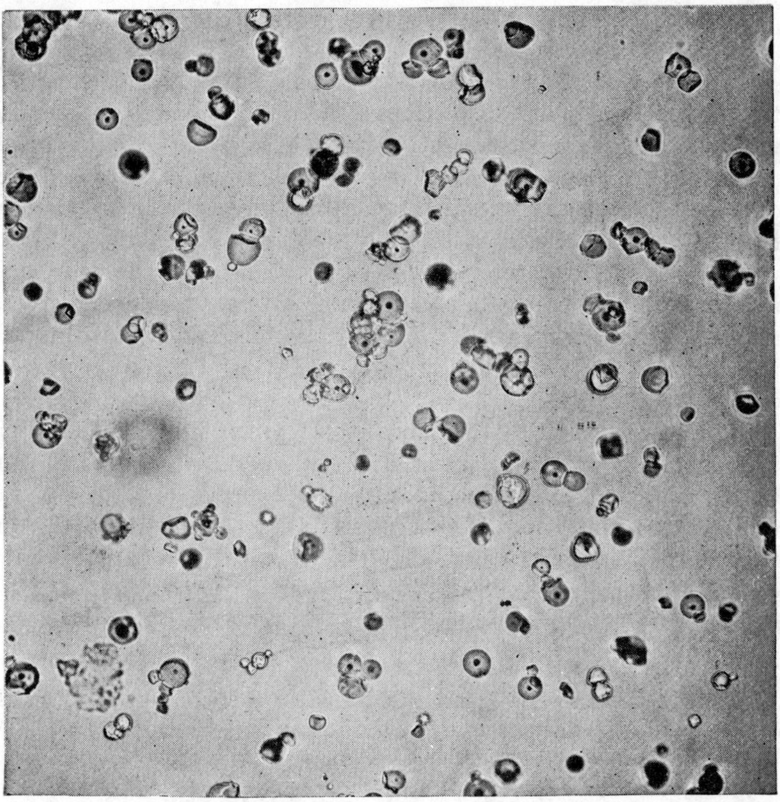

FIG. 131.—Tapioca starch. 250 ×.

TAPIOCA STARCH. The particles vary in size from 0.008 to 0.022 mm. The general shape is circular or hemispherical. The hilum is a dot or short slit and is nearly central (Fig. 131). Polarized light produces colors and a distinct cross.

CHAPTER 36

SUGARS AND SUGAR PRODUCTS

Sugar, or sucrose, $C_{12}H_{22}O_{11}$, may be accurately determined by two general methods. The Clerget method consists in observing the difference in polarimetric rotation before and after inversion or breaking down by hydrolysis to invert sugar, $C_6H_{12}O_6$, which contains 1 mole of glucose, often called dextrose, and 1 mole of fructose, often called levulose.

The copper reduction method uses the fact that a solution of invert sugar will reduce an alkaline copper tartrate solution, giving, under specific conditions, a weight of cuprous oxide approximately proportional to the amount of invert sugar present and accurately estimated by tables.

Various methods are available for carrying out these determinations. The presence of other sugars such as raffinose, maltose, and lactose interferes, and modified methods are necessary. A full description of such methods is given in the chapter on Sugar and Sugar Products in the Methods of the Association of Official Agricultural Chemists. Only the simpler and more common methods are described here, including those using the Brix hydrometer, the pycnometer, and the refractometer.

Sucrose and Invert Sugar. QUALITATIVE. If necessary, clear the solution with basic lead acetate as later described (page 657). To 5 ml. of the filtrate add 1 ml. of combined Fehling's solution (page 659), and boil. A yellow or red precipitate indicates the presence of invert sugar or of any reducing hexose.

When no precipitate forms, to another 5-ml. portion of the filtrate add 0.5 ml. of concentrated hydrochloric acid and boil for a few minutes. Cool, and neutralize with 10 per cent sodium hydroxide solution. Add 1 ml. of combined Fehling's solution, and continue to boil. A yellow or red precipitate indicates the presence of sucrose or any polyose that hydrolyzes with acid. The nature of the product analyzed will generally give a clue as to what is actually present. Sucrose is by far the most common polyose.

Both invert sugar and sucrose may be together in solution. The presence of a larger precipitate after hydrolysis than before will indicate this. When in doubt, quantitative methods must be used.

Sucrose. Quantitative. *Hydrometric and Pycnometer Methods.* These methods are not accurate when applied to low-grade sugar products, molasses, and other materials containing large quantities of nonsugar solids. They are extensively used for approximate results with juices, sirups, etc.

Obtain the reading on a Brix hydrometer (page 47) at 20°; on a Baumé hydrometer, modulus 145 (see page 46); or with a pycnometer (page 43), obtaining the specific gravity at 20°/4° or at 20°/20°. Take care that all air bubbles have escaped and all fatty or waxy matter has come to the top and been skimmed off. Tables are given in most handbooks correlating readings of Brix and Baumé, modulus 145, hydrometers and specific gravities at 20°/4° and 20°/20°, of sugar solutions. Degrees Brix correspond to per cent by weight of sucrose.

If the sample is too dense to determine the density directly, (1) dilute a weighed portion with a weighed quantity of water, or (2) dissolve a weighed portion and dilute to a known volume with water.

Calculation.

(1) $$\frac{W \times P}{S} = U$$

where W = weight of diluted sample, P = percentage solids in diluted material, S = weight of sample taken for dilution, and U = percentage solids in undiluted sample.

(2) $$\frac{V \times G \times P}{S} = U$$

where V = volume of diluted solution at a given temperature, G = specific gravity of diluted solution at same temperature, P = percentage of solids in diluted sample at same temperature, and S and U are the same as in Eq. (1).

Refractometer Method. Determine the refractometer reading of the solution, which must contain no undissolved solids, at 20°, and obtain the corresponding percentage of dry substance from either the direct reading, if a sugar refractometer is used, or from the appropriate table found in most handbooks, if the instrument gives readings in terms of refractive index (see page 57).

If the determination is made at a temperature other than 20° or if the humidity causes condensation of moisture on the exposed faces of the prisms, make the readings at room temperature and correct the readings to standard temperature. If the solution is too dark to be read in the instrument, dilute with a concentrated sugar solution; never use water for this purpose. Mix weighed quantities of the solution under examination and a solution of pure sugar of about the same strength, and obtain the quantity of dried substance in the former as follows:

$$\frac{(A + B)C - BD}{A} = \text{percentage of dry substance}$$

where A = weight in grams of the sample mixed with B, B = weight in grams of pure sugar solution used in the dilution, C = percentage of dry substance in the mixture $A + B$ obtained from the refractive index, and D = percentage of dry substance in the pure sugar solution obtained from its refractive index.

Polarimetric Method. Verify the saccharimeter (Fig. 132) and make all polarizations at 20°. For instruments using the Ventzke scale, one normal weight or 26 grams of pure dry sucrose dissolved in water so that 100 ml. of solution is obtained at 20° and polarized in a room or cabinet also at 20° must give a saccharimeter reading of exactly 100.00, the temperature of the sugar solution remaining constant at 20°.

FIG. 132.—Saccharimeter.

The saccharimeter used may have either a single or a double wedge and should be a half-shadow instrument with either double or triple field (Fig. 132). Keep the instrument so far removed from the source of light that the polarizing Nicol prism is not warmed. As sources of light, employ lamps giving strong illumination, such as a triple gas burner with metallic cylinder, lens, and reflector, an electric lamp, a petroleum duplex lamp, or sodium light. With an electric light, place a thin ground-glass plate between the source of the light and the polariscope.

Whenever a white light is used in polarimetric determinations, it must be filtered through a solution of potassium bichromate of such a concentration that the percentage content of the solution multiplied by the length of the solution in centimeters is equal to 9. This concentration must be doubled in polarizing carbohydrate materials of high rotation dispersion, such as commercial glucose.

In determining the polarization of a sample, use the whole normal weight of 26 grams for 100 ml. or a multiple or fraction thereof for any corresponding value. As clarifying and decolorizing agents use basic lead acetate, neutral lead acetate, basic lead nitrate, or alumina cream. Bone black and decolorizing powders must not be used, for they sorb sugar. Whenever reducing sugars are determined in the solution for polarizing, use only neutral lead acetate for clarification, for basic lead acetate causes precipitation of some of the reducing sugars. More specific details are given later in this chapter for treatment of specified types of samples.

After bringing the solution exactly to the mark at the proper temperature, pour all the well-shaken clarified solution on a rapid-acting, dry filter. Cover the funnel during filtration, and reject the first 25 ml. that runs through. In no case should the whole solution or any part be returned to the filter. If cloudy after the 25 ml. has been rejected, begin a new determination.

Clarification. Many sugar solutions have to be clarified. This usually consists in adding a suitable clarifying agent to a normal weight of 26 grams partly but not completely diluted to 100 ml. Always avoid excess.

Basic lead acetate is used primarily for clarifying dark-colored cane, sorghum, and beet products when sucrose is determined polarimetrically. This may be the solution prepared as indicated on page 718 or it may be the dry powdered salt. The salt should contain 72.8 per cent of lead, corresponding to a composition of $3Pb(C_2H_3O_2)_2.2PbO$. Of this salt about 1 gram equals 3 ml. of the basic lead acetate solution.

Neutral lead acetate is used in saturated aqueous solution for clarifying light-colored sugar products when sugars are determined polarimetrically. Its use is imperative when reducing sugars are determined in the solution used for polarization.

Alumina cream (page 718) is suitable as a clarifying agent and as an adjunct to other agents when sugars are determined by polariscopic or reducing sugar methods.

Mutarotation. Products, such as honey and commercial glucose, that contain dextrose, or other reducing sugars in crystalline form or in solution at high density may exhibit mutarotation. When freshly dissolved, such substances have a rotatory power different from that after the lapse of a comparatively short period. The explanation of this phenomenon lies in partial conversion of the dextrose or other sugar into another modification of different rotatory power. When the rotation has become constant, an equilibrium has been set up. Only this constant rotation must be used.

PROCEDURE. To obtain the constant rotation, allow the solution to stand overnight before taking a reading. To expedite the conversion, heat the neutral solution of approximate pH 7.0 to boiling for a few minutes, or add a few drops of concentrated ammonium hydroxide before diluting to volume. If the solution has been made to volume, add dry sodium carbonate until distinctly alkaline to litmus paper. Do not allow the slightly alkaline solutions to stand at such high temperatures or for such lengths of time as to cause destruction of fructose. Determine the completion of mutarotation by making readings at 15- to 30-minute intervals until these become constant.

Sucrose. POLARIZATION BEFORE AND AFTER INVERSION WITH HYDROCHLORIC ACID. Raffinose must be absent. In the presence of much levulose, as in honeys, fruit products, sorghum sirup, cane sirup, and molasses, this method gives erroneous results. In such cases use invertase to invert.

DIRECT READING. Dissolve the double normal weight of the substance, 52 grams, or a fraction thereof, in water in a 200-ml. volumetric flask. Add the necessary clarifying agent, avoiding excess. Shake, dilute to the mark with water, mix well, and filter, keeping the funnel covered. Reject the first 25 ml. of the filtrate. If a lead clarifying agent was used, remove the excess lead from the solution when sufficient filtrate has collected by adding anhydrous sodium carbonate a little at a time, avoiding excess. Mix well, and refilter, rejecting the first 25 ml. Pipet a 50-ml. portion of the lead-free filtrate into a 100-ml. flask, dilute to the mark, mix well, and polarize in a 200-mm. tube. The result, multiplied by 2, is the direct reading or polarization before inversion.

INVERT READING. Pipet a 50-ml. portion of the lead-free filtrate into a 100-ml. flask, and add 25 ml. of water. Add, little by little, while rotating the flask, 10 ml. of 1:3 hydrochloric acid. Heat a water bath to 70°, and maintain at about that temperature. Place the flask in the bath, insert a thermometer, and heat with constant agitation until the thermometer in the flask indicates 67°. This should require 2.5 to 2.75 minutes. From this moment, leave the flask in the bath for exactly 5 minutes longer. During this time the temperature should rise to about 69.5°. Plunge the flask at once into water at 20°. Cool to 35°, remove and rinse the thermometer, and fill the flask almost to the mark. Leave the flask in the bath at 20° for at least 30 minutes longer, and finally make up exactly to volume. Mix well, and polarize the solution at 20° in a 200-mm. tube. Multiply by 2 to obtain the invert reading. If it is necessary to work at a temperature other than 20°, which is permissible within narrow limits, the volumes must be completed, and both direct and indirect polarizations must be made at exactly the same temperature.

CALCULATION.

$$\frac{100(\text{direct reading, normal solution} - \text{invert reading, normal solution})}{143 + 0.0676(m - 13) - t/2} = \text{per cent sucrose}$$

where m = grams of total solids in 100 ml. of the invert sugar solution in the polariscope and t = temperature at which readings are made.

Determine the total solids as percentage by weight as found by the refractometer, and multiply this figure by the density at 20° as obtained from corresponding tables. The correction due to solids content is quite small.

INVERSION AT ROOM TEMPERATURE. To 50 ml. of the clarified lead-free solution, add 10 ml. of 1:3 hydrochloric acid, and set aside for 24 hours at a temperature not below 20°. If the temperature is above 25°, set aside for 10 hours. Make up to 100 ml., and polarize as in the preceding method. Under these conditions replace 143 by 143.2 in the formula.

Sucrose. REDUCING SUGARS BEFORE AND AFTER INVERSION. Determine the reducing sugars, clarification having been effected with neutral, never with basic, lead acetate, as described for the Munson and Walker method, in the following section. Calculate to invert sugar from the invert sugar and sucrose columns. Invert another aliquot of the solution as directed for polarization after acid hydrolysis, and similarly determine reducing sugars, but calculate them to invert sugar from the table, using the invert sugar column alone. Deduct the percentage of invert sugar obtained before inversion from that obtained after inversion, and multiply the difference by 0.95 to obtain the percentage of sucrose. One molecule of sucrose and one of water give two molecules of invert sugar.

The solutions must be diluted in both determinations so that not more than 240 mg. of invert sugar is present in the quantity taken for reduction. It is essential that all lead be removed with anhydrous sodium carbonate or powdered potassium oxalate before reduction and that tables used have been prepared for the specific form of Fehling solution applied.

Invert Sugar. MUNSON AND WALKER METHOD. Prepare the Soxhlet modification of Fehling's solution used in this method as follows: (1) Dissolve 34.639 grams of copper sulfate, $CuSO_4.5H_2O$, in water, dilute to 500 ml., and filter through prepared asbestos. (2) Dissolve 173 grams of sodium potassium tartrate, $NaKC_4H_4O_6.4H_2O$, Rochelle salt, and 50 grams of sodium hydroxide in water. Dilute to 500 ml., allow to stand for 2 days, and filter through prepared asbestos. Mix equal volumes of (1) and (2) immediately before using.

Digest amphibole asbestos with 1:3 hydrochloric acid for 2 to 3 days. Wash free from acid, digest for a similar period with 10 per cent sodium hydroxide solution, and then treat for a few hours with hot alkaline tartrate solution of the same strength as (2) above. An old solution of (2) will do. Wash the asbestos free from alkali, and digest for several hours with 1:3 nitric acid. After washing free from acid, shake into a fine pulp with water.

In preparing the Gooch crucible make a film of asbestos 0.25 inch thick, and wash thoroughly with water to remove fine particles of asbestos. Wash with 10 ml. of alcohol and then with 10 ml. of ether, dry for 30 minutes at 100°, cool in a desiccator, and weigh.

PROCEDURE. Transfer 50 ml. of the mixed Fehling solution to a 400-ml. Pyrex beaker, and add 50 ml. of the reducing sugar solution, which should contain not more than 0.25 gram of reducing sugars. If a smaller volume is used, add water to make the final volume 100 ml. Heat the beaker on an asbestos gauze over a Bunsen burner, regulate the flame so that boiling begins in 4 minutes, and continue the boiling for exactly 2 minutes. It is important that these directions shall be strictly observed. To regulate the burner for this purpose it is advisable to make preliminary tests, using 50 ml. of the reagent and 50 ml. of water before proceeding with the actual determinations. Keep the beaker covered with a watch glass during the heating.

Filter the hot solution at once through an asbestos mat in a prepared Gooch crucible, using suction. Wash the precipitate of cuprous oxide, Cu_2O, thoroughly, from the beaker and then in the crucible with water at about 60°. Now wash with 10 ml. of alcohol and finally with 10 ml. of ether. Dry the precipitate for 30 minutes at 100°, cool, and weigh. Obtain from Munson and Walker tables, found in most handbooks, the corresponding weight of invert sugar. For very accurate work conduct a blank determination, using 50 ml. of the reagent and 50 ml. of water; and if the weight of cuprous oxide obtained exceeds 0.5 mg., correct the result of the reducing sugar determination accordingly. The alkaline tartrate solution deteriorates on standing, and the quantity of cuprous oxide obtained in the blank increases.

The number of milligrams of copper reduced by a given quantity of reducing sugar varies, depending on whether or not sucrose is present. In the tables the absence of sucrose is assumed except in the entries under invert sugar, where, in addition to the column for invert sugar alone, one column is given for mixtures of invert sugar and sucrose containing 0.4 gram of total sugar in 50 ml. of solution and one column for invert sugar and sucrose when the 50 ml. of solution contains 2 grams of total sugar. Two entries are also given under lactose and sucrose mixtures.

Commercial Glucose. APPROXIMATE POLARIMETRIC METHODS FOR SUBSTANCES CONTAINING LITTLE OR NO INVERT SUGAR. Commercial glucose cannot be determined accurately, owing to the varying quantities of dextrin, maltose, and dextrose present in the product. However, in sirups in which the quantity of invert sugar is so small as not to affect the result appreciably, calculate as follows:

$$\frac{\text{Direct polarization, normal solution} - \text{per cent of cane sugar}}{211} = \text{per cent of commercial glucose solids}$$

Express the results in terms of commercial glucose solids polarizing at +211° Ventzke. This result may be recalculated in terms of commercial glucose of any Baumé reading desired.

SUBSTANCES CONTAINING INVERT SUGAR. Prepare an inverted 0.5*N* solution of the substance as directed for polarization by acid hydrolysis, except to cool the solution after inversion. Make neutral to phenolphthalein indicator with 10 per cent sodium hydroxide solution, slightly acidify with 1:5 hydrochloric acid, and treat with 5 to 10 ml. of alumina cream before making up to the mark. Filter, and polarize at 87° in a 200-mm. jacketed metal tube, preferably silver. Multiply the reading by 200, and divide by the factor 196 to obtain the quantity of commercial glucose solids polarizing +211° Ventzke. This result may be recalculated in terms of commercial glucose of any Baumé reading desired

CHAPTER 37

PETROLEUM PRODUCTS

Petroleum varies in composition according to its source. The main portion of all crudes consists of saturated hydrocarbons of the formula C_nH_{2n+2}, called paraffins. Petroleum from some sources contains substantial amounts of naphthenes. These are of the general formula C_nH_{2n} but show characteristics very similar to those of the paraffins. They are not unsaturated compounds but have a closed carbon chain, of which cyclohexane is a simple example, 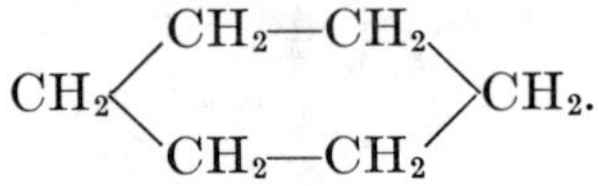 Lesser ingredients are unsaturated chain compounds called olefins, containing one double bond conforming to the formula C_nH_{2n}, or more double bonds such as C_nH_{2n-2}, etc. Sulfur is present to an appreciable amount in most petroleum products.

The analyses most frequently conducted are those on the commercial fractions produced by distillation; petroleum naphtha, gasoline, kerosene, and lubricating oils. To a lesser extent these are applied to petrolatum, fuel oils, paraffin waxes, etc. These various fractions are so much further processed by cracking, hydrogenation, alkylation, hydroforming, etc., that they possess only a remote resemblance to the crude.

The tests applicable to such hydrocarbon products are generally physical rather than chemical and consist in determining boiling range, specific gravity, viscosity, flash point, etc. Since the possible number of fractions obtainable from petroleum is almost infinite, definite standards have been proposed and accepted for specified fractions and products that have specified uses. The American Petroleum Institute (API) and the American Society for Testing Materials (ASTM) have issued many such specifications. In general, the specifications and requirements of the two are very similar. The tests that follow are largely based on these specifications. They are given roughly in the order of their application to gasoline, lubricating oils, fuels, and miscellaneous products, but necessarily with much overlapping. Only the main tests for the main products can be covered; when other tests are required, the specifications of the ASTM or API should be consulted. To permit duplication by different laboratories the tests are necessarily given in highly specific detail, and usually these conditions must be scrupulously followed. Generally speaking, the petroleum industry operates on Fahrenheit temperatures, although a few of the more recent methods, that for aniline point, for example, are specified in centigrade. All temperatures in this chapter are therefore designated as either Fahrenheit or centigrade.

Terms Relating to Petroleum. The following are standard definitions of terms connected with petroleum. They are useful in distinguishing the different products as referred to in commerce.

Benzine is a term formerly applied to a petroleum naphtha. It is archaic and should not be used.

Crude petroleum is the naturally occurring mixture, consisting predominantly of hydrocarbons and/or sulfur, nitrogen, and/or oxygen derivatives of hydrocarbons, that is removed from the earth in a liquid state or is capable of being so removed. It is commonly accompanied by varying quantities of extraneous substances such as water, inorganic matter, and gas. The removal of such substances alone does not change the status of the mixture as crude petroleum. If such removal appreciably affects the composition of the oil mixture, then the resulting product is no longer crude petroleum.

Crude shale oil is the oil obtained as a distillate by the destructive distillation of oil shale.

Engine distillate is a refined or unrefined petroleum distillate similar to naphtha, but often of higher distillation range.

Fuel oil is any liquid or liquefiable petroleum product burned for the generation of heat in a furnace or firebox or for the generation of power in an engine, exclusive of those with a flash point below 100°F. (38°C.) in a closed tester and oils burned in cotton- or wool-wick burners. Fuel oils in common use fall into four classes:

1. Residual fuel oils, which are topped crude petroleums or viscous residua obtained in refinery operations.

2. Distillate fuel oils, which are distillates derived directly or indirectly from crude petroleum.

3. Crude petroleums and weathered crude petroleums of relatively low commercial value.

4. Blended fuels, which are mixtures of two or more of the three preceding classes.

Thus, by definition, fuel oils include a range from just above kerosene, the domestic fuel oils, to some that are semisolid at ordinary temperature and must be softened with heat in order to feed them, as some of the bunker oils.

Gas oil is a liquid petroleum distillate having a viscosity intermediate between that of kerosene and lubricating oil. The oils carbureted in the manufacture of gas are not limited to gas oils but range from gasoline to the heaviest fuel oils.

Gasoline is a refined petroleum naphtha that by its composition is suitable for use as a carburant in internal-combustion engines.

Kerosene is a refined petroleum distillate having a flash point not below 73°F. as determined by the Abel tester, which is approximately equivalent to 73°F. as determined by the Tag closed tester, and suitable for use as an illuminant when burned in a wick lamp. According to city ordinances or insurance regulations this flash point is usually raised to well above 100°F., often to 130 to 135°F. This is economically justified. for the demand for kerosene is relatively small.

Lubricating oil is a term applied to any distilled fraction of petroleum having sufficient body and other properties to serve as a lubricant. The properties are often modified by minor additions, as of oxidized or chlorinated hydrocarbons.

Oil shale is a compact rock of sedimentary origin, with an ash content of more than 33 per cent and containing organic matter that yields oils when destructively distilled but does not yield them to an appreciable extent when extracted with the ordinary solvents for petroleum.

Petroleum grease is a semisolid or solid combination of a liquid petroleum lubricant and a soap or a mixture of soaps, with or without fillers, suitable for certain types of lubrication. It is discussed in more detail in Chap. 21.

Petroleum naphtha is a generic term applied to refined, partly refined, or unrefined petroleum products and liquid products condensed from natural gas, not less than 10 per cent of which distills below 347°F. and not less than 95 per cent of which distills below 464°F. when subjected to distillation in accordance with the standard method of test for distillation of gasoline, naphtha, etc., of the ASTM. There are numerous subclasses. For example, rubber solvent is a naphtha of much greater volatility used for cutting rubber to produce rubber cements. As another example, dry cleaners' naphtha, often called Stoddard solvent from the originator of the specification, is a cut designed for safe use in dry cleaning and therefore has a flash point not less than 100°F. by the Tag closed tester.

Petroleum spirits, also often called white spirits, is a refined petroleum distillate with a minimum flash point of 70°F., determined by the Tag closed tester or by the Abel tester, with volatility and other properties making it suitable as a thinner and solvent in paints, varnishes, and similar products. It is often referred to as VM&P naphtha, meaning varnish makers' and painters' naphtha. The obsolete term turpentine substitute is sometimes applied to this and to other fractions but has no specific meaning. The term mineral spirits is sometimes used in the paint and varnish industry for this and other hydrocarbon mixtures, such as coal-tar distillates. In Great Britain the term petroleum spirits is applied to a very light hydrocarbon mixture having a flash point below 32°F.

Pitch and tar are nonspecific terms that may designate asphaltic residues from the petroleum industry, naturally occurring, or from the stills of the fatty-acid industry.

Topped crude petroleum is a residual product remaining after the removal, by distillation or other artificial means, of an appreciable quantity of the more volatile components of crude petroleum.

Tops is the unrefined distillate obtained in topping a crude petroleum.

Weathered crude petroleum is the product resulting from crude petroleum through loss due to natural causes, during storage and handling, of an appreciable quantity of the more volatile constituents.

Distillation of Gasoline, Naphtha, Kerosene, and Similar Petroleum Products. The operation of this standard ASTM and API test depends on the use of a specific piece of commercial equipment shown with dimensions in Figs. 133 and 134. Results will be only roughly approximated by setting up the usual glass laboratory equipment, even if this is of similar dimensions.

The equipment in brief consists of the following: A 100-ml. ASTM Engler distillation flask rests on two asbestos boards with holes, placed on a ring stand and heated by a Bunsen burner or electric heater. The whole is surrounded by a draft shield containing ventilation holes. An accurate thermometer is inserted in the flask. The side arm of the flask leads to a metal condenser tube surrounded by a metal bath. The exit end of the tube dips into a standard 100-ml. graduated cylinder.

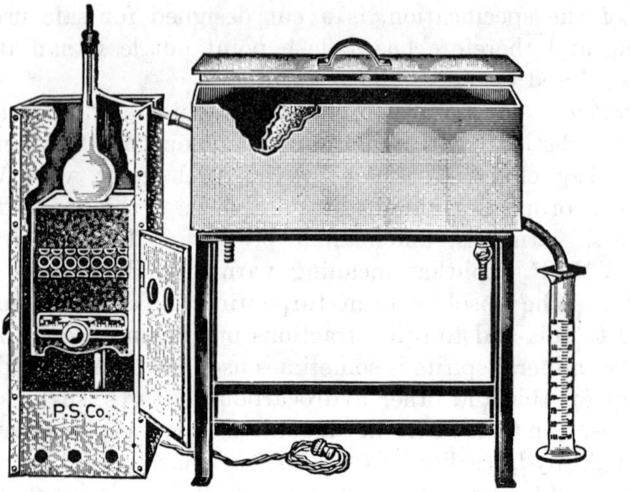

FIG. 133.—Equipment for distillation of petroleum products. ASTM D 86.

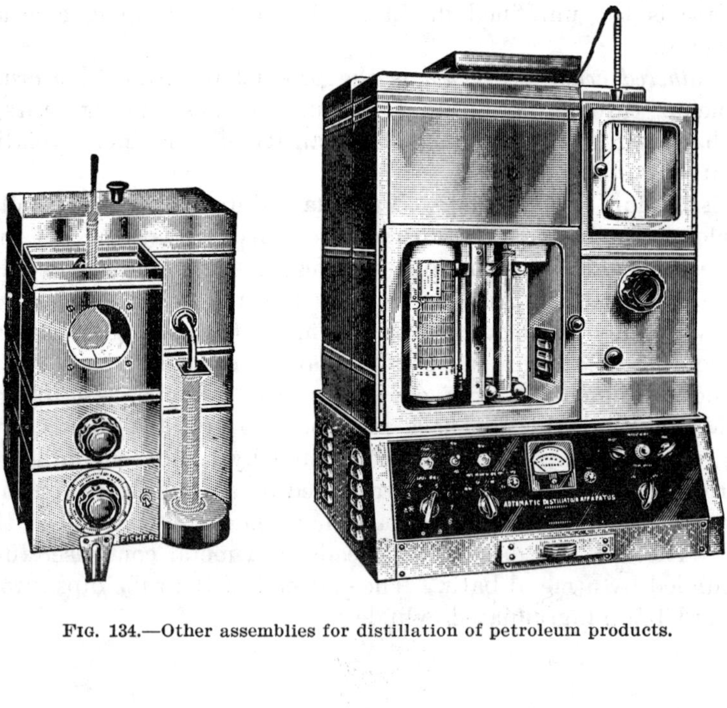

FIG. 134.—Other assemblies for distillation of petroleum products.

The ASTM low-distillation total-immersion thermometer has a range of 0 to 300°C. or 30 to 580°F. The ASTM high-distillation total-immersion thermometer has a range of 0 to 400°C. or 30 to 760°F. The ring stand supports a 4-inch or larger diameter ring on which is placed a hard asbestos board 6 by 6 by 0.25 inches with a 4-inch-diameter opening. On this latter is placed a second asbestos board 6 by 6 by 0.25 inches with a 1.25-inch-diameter hole. For petroleum products having an end point above 470°F., this hole should be 1.5 inches in diameter. In the use of a gas flame this should never be so large that it spreads over a circle of diameter greater than 3.5 inches on the undersurface of the asbestos board.

PROCEDURE. Fill the condenser bath with cracked ice and enough water to cover the condenser tube. Maintain the bath temperature at 32 to 40°F. Swab the condenser tube with a piece of soft lint-free cloth attached to a cord or copper wire to remove any liquid remaining from a previous test.

Measure a 100-ml. sample in an ASTM 100-ml. graduated cylinder at 55 to 65°F., and transfer directly to the distillation flask, taking care that none of the sample flows into the side arm.

Place a thermometer in the neck of the flask, using a tightly fitting cork, and adjust so that the low end of the capillary tube is on a level with the inside of the bottom of the side arm at its junction with the neck of the flask. For samples having end points not higher than 428°F. or initial boiling points not higher than 212°F. use the low-distillation thermometer. For samples having end points higher than 428°F. or initial boiling points higher than 212°, use the high-distillation thermometer.

Place the flask and contents on the hole of the asbestos board, and insert the side arm into the condenser tube so that 1 to 2 inches extend into the tube. Use a tight cork stopper to make the connection. Without drying, place the 100-ml. graduated cylinder used to measure the sample, under the outlet of the condenser, so that the tube extends at least 1 inch into the cylinder but not below the 100-ml. mark. If the surrounding temperature is above 55 to 65°F., immerse the cylinder up to the 100-ml. mark in a transparent bath maintained between these temperatures. Cover the top of the graduate during the distillation with a piece of stiff paper cut so as to fit the condenser tube tightly.

Apply heat at a uniform rate, and regulate so that the first drop of condensate falls from the condenser in not less than 5 or more than 10 minutes. Read the distillation temperature 2 minutes after heat is applied, and record this reading as the "correction temperature." This figure is of significance only in cases in which there is a question as to the accuracy of the initial boiling point as subsequently determined. When the first drop falls from the end of the condenser, the reading of the distillation thermometer is recorded as the *initial boiling point.* This temperature is indicative of the quality of gasoline. It is not easy to check this point owing to variations due to operating conditions. A correction may be made as follows: If the correction temperature is below 70°F. or above 80°F., consider the observed initial boiling point, if it is below 150°F., incorrect by an amount not less than one-third the difference between the correction temperature and 75°F.

CALCULATION.

$$\text{Revised initial boiling point} = \text{observed initial boiling point} - \frac{\text{correction temperature in °F.} - 75}{3}$$

Move the receiving cylinder so that the end of the condenser tube touches the side of the cylinder. Regulate the heat so that the distillation proceeds at a uniform rate of not less than 4 ml. per minute or more than 5 ml. per minute. Record the volume of the distillate to 0.5 ml. when the mercury temperature reaches each point that is a multiple of 10°C., that is, 30, 40, 50, 60°C., etc. These are usually expressed in Fahrenheit. Alternatively, record the temperature when the distillate reaches each 10-ml. mark on the graduate.

Do not adjust the heat after the liquid residue in the flask is approximately 5 ml. unless the time required to bring over the last 5 ml. of distillate and reach the end point exceeds 5 minutes. The *end point* is the maximum temperature observed on the distillation thermometer and is usually reached after the bottom of the flask has become dry. If the bottom of the flask is not dry, record this fact. If the time required to bring over the last 5 ml. of distillate and reach the end point exceeds 5 minutes, repeat the test and adjust the heat when the liquid residue reaches 5 ml. so that the time required to vaporize the last 5 ml. of distillate and reach the end point is 3 to 5 minutes.

The total volume of the distillate collected in the receiving graduate is called the *recovery.* Pour the cooled residue into a small cylinder, and measure to 0.1 ml. The difference between 100 ml. and the sum of the recovery and the residue is called the *distillation loss.*

The sum of the volume collected in the cylinder at any specified temperature and the distillation loss may be recorded as the percentage evaporated at the temperature in question. Make a clear distinction between the inclusion or exclusion of the distillation-loss figure in the report of the volume of distillate obtained at any prescribed temperature. In the first case the figure is reported as the *volume evaporated,* in the second as the *volume recovered.*

With proper care and strict attention to details, duplicate results obtained for initial boiling point and end point, respectively, should differ from each other by not more than 6°F. Duplicate readings of the volume of the distillate collected in the cylinder when each of the prescribed temperature points is reached should differ from each other by not more than 2 ml. If observations were made on the basis of prescribed percentage points, the differences in temperature readings should not exceed the amounts equivalent to 2 ml. of distillate at each point in question.

Make a correction for barometric pressure only in case of dispute. By using the Sydney Young equation, the corrections to be added to the observed temperature t_f are $0.00012(760 - P)(460 + t_f)$ where P is the actual barometric pressure in millimeters of mercury.

For all specification work on petroleum products, strict attention to all the details and requirements of this test must be given. However, in cases in which during the course of analysis a petroleum fraction is separated, for example, a petroleum naphtha residue after sulfonating a mixed thinner, it would be useless to observe such exact precautions. An exact initial boiling point cannot be obtained. On the other hand, the general trend of the distillation is often sufficient to identify the type of petroleum product recovered in such cases.

Flash Point by the Tag Closed Tester. The flash point of a product is the temperature at which it gives off just enough vapors to produce a combustible mixture which will flash over the entire surface of the sample. Except for fuel oils the Tag closed tester (Fig. 135) is applicable to all mobile liquids flashing below 175°F. The instrument as sold conforms to ASTM requirements, which it is therefore unnecessary to repeat in detail. Two thermometers are required, one for the bath and the other for the sample. The two ASTM Pensky-Marten and Tag thermometers have a range of −7 to +110°C. or 20 to 230°F.

Procedure. Perform the test in a partly darkened place for better detection of the flash. Keep the apparatus steady, and protect from drafts by a shield. Fill the water bath with water to overflowing, at a temperature at least 20°F. below the probable flash point of the oil. Place the oil cup in the water bath, and add 50 ml. of the sample from a graduate. The sample must be at a temperature at least 20°F. below its flash point. Remove any air bubbles, and properly attach the cover and the flash-point thermometer to the cup.

Light the test flame, and adjust to the size of the small white bead on the cover. If gas is not available, insert a wick of cotton waste in the oil chamber to which the burner tip is attached and fill the chamber with signal, sperm, or lard oil. Heat the water bath, using a gas flame or an alcohol lamp placed centrally under the bath. Adjust the flame so that the temperature of the oil in the cup rises at as nearly as possible the rate of 1.8°F. per minute, but in no case faster than 2°F. or slower than 1.6°F. per minute. Record the initial temperature of the oil.

When the temperature of the oil is 9°F. below its probable flash point, turn the knob on the cover so as to introduce the test flame into the vapor space of the cup and immediately turn it back again. This operation should take about 1 second or the time required to pronounce "one thousand and one." Record the time and oil temperature at which the first introduction of the test flame is made. Repeat this application of the flame after each 1°F. rise in temperature of the oil until a distinct flash in the interior of the cup is observed. This flash must not be confused with the bluish halo that sometimes surrounds the test flame immediately

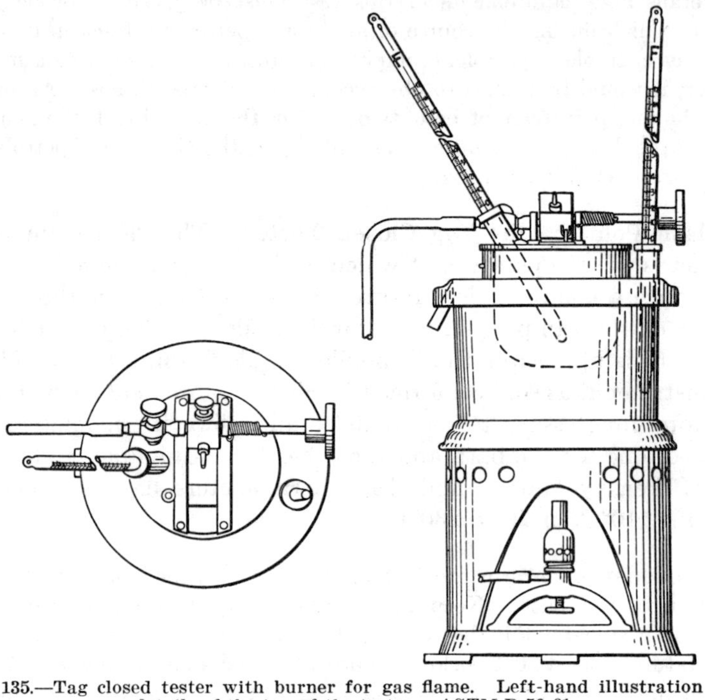

FIG. 135.—Tag closed tester with burner for gas flame. Left-hand illustration is of details of the top of the tester. ASTM D 56-61.

preceding the actual flash. Record the time and oil temperature at which the flash occurs.

Repeat the whole procedure if the rise in temperature of the oil was greater or less than that prescribed or if the flash occurred less than 20°F. above the initial temperature. Empty and wipe the oil cup, and dry the thermometer. Empty the water jacket, and refill with cold water at 15°F. Add a fresh sample of 50 ml. of the oil to the cup, and again carry out the test as described above.

If two or more tests agree within 1°F., consider the average as the correct flash point. Otherwise, make a further test, and if the maximum variation of all tests is not greater than 2°F. consider the average of all as the correct flash point. A correction for barometric pressure is made only in disputes or when the barometer reading at the time of test varies more than 13 mm. from the standard pressure of 760 mm. When below this, add to the thermometer reading 1.6°F. for each 25 mm. of barometer pressure difference to obtain the true flash point. When the barometric reading is more than 13 mm. above 760 mm., deduct 1.6°F. for each 25 mm.

In determining flash points below 40°F. the bath must be filled with prechilled brine or other low-freezing liquid instead of water. The range of the thermometers used should then be to 120°F.

Flash points of solvents that have been separated from mixtures are not significant, owing to the process through which the solvent has gone. The most volatile fraction is most likely to be lost during separation. Also, small residual amounts of other and more volatile solvents may be introduced in the process of separation.

Flash and Fire Points by the Open Cup. After a flash point has been obtained, if the temperature is further raised a point is reached at which the emission of inflammable vapors occurs rapidly enough to support a flame. This is the fire point. Since excess air is needed to support combustion, the fire point cannot be obtained with the closed type of tester.

The standard unit for the purpose is the Cleveland open-cup tester shown in Fig. 136. This instrument is an ASTM standard unit specified for all petroleum products other than fuel oils and those flashing below 175°F. Practically speaking, it is the only instrument for determining fire point. The cup used in this test is supported in a heating plate that is covered with a sheet of hard asbestos.

Procedure. Use a gas burner, electric heater, or even an alcohol lamp as heat source, but do not permit products of combustion or free flame to come up around the cup. Protect a free flame from drafts by a shield that does not project above the level of the upper surfaces of the asbestos board. The ASTM thermometer employed has a range of −6 to +400°C. or 20 to 760°F.

Suspend the thermometer so that the bottom of the bulb is 6 mm. from the bottom of the cup and above a point halfway between the center and back of the cup. Fill the cup with the oil to be tested in such a manner that the top of the meniscus, at room temperature, is exactly at the filling line engraved on the inner side of the cup. Asphalt or other solid bituminous material must first be heated to not less than 300°F. or more than 350°F. to render it sufficiently fluid so that the cup can be filled at this temperature in the same manner as when oil is used. The surface of the sample must be free from bubbles, and there must be no oil above the filling line or on the outside of the apparatus.

The test flame should be approximately 4 mm. in diameter. A convenient device is a small glass stopcock with one end drawn out as in a buret, the other end being attached through rubber tubing to the gas supply. Heat the oil at a rate not exceeding 30°F. temperature rise per minute until a point is reached approximately 100°F. below the probable flash point of the oil. The rate of heating is then decreased and for at least the last 50°F. before the flash point is reached must be not less than 9°F. or more than 11°F. per minute.

Apply the test flame as the temperature reaches each successive 5°F. by passing the flame in a straight line or on the circumference of a circle having a radius of at least 6 inches across the center of the cup or at right angles to the diameter passing through the thermometer. The test flame should be in the plane of the upper edge of the cup. The time for the passage of the test flame across the cup should be approximately 1 second.

The flash point is the temperature registered on the thermometer when a flash appears at any point on the surface of the oil. It must not be confused with a bluish halo that sometimes surrounds the test flame shortly before reaching the flash point.

To obtain the fire point continue to heat at the specified rate and apply the test flame at the specified intervals until the oil ignites and continues to burn for a period of at least 5 seconds. The temperature must be read at the time of the flame application that causes burning for at least 5 seconds. The burning oil thereafter rapidly raises the reading of the thermometer.

Make both determinations in a room or compartment free from drafts, and avoid breathing over the surface of the oil. The flash may be more readily discerned by working in a darkened room or compartment.

Free Sulfur and Corrosive Sulfur Compounds. Petroleum products always contain sulfur. The following method is accepted for setting an upper limit to the sulfur content of gasoline:

Procedure. Place a clean strip of mechanically polished pure sheet copper, about 0.5 inch in width and 3 inches in length, in a clean tube or sample bottle. Add the sample under test so that the copper strip is completely immersed. Close with a loosely fitting cork, and place in a suitable bath at 122°F. At the end of 3 hours, remove the strip, and compare with a similar strip of freshly polished copper. Gasoline should impart not more than an extremely slight discoloration as compared with the fresh copper strip. Otherwise, it must be reported as not passing the test.

A similar method of testing for naphthas flashing over 100°F. and for lubricating oils is as follows: Place a clean strip of mechanically polished pure sheet copper, about 0.5 inch in width and 3 inches in length, in a clean test tube. Add enough of the sample to be tested to cover the strip completely. Close the tube with a vented stopper, and maintain for 3 hours at 212°F. Rinse the copper strip with sulfur-free acetone, and compare it with a similar strip of freshly polished copper. Discoloration or pitting indicates corrosion.

Doctor Test. The test with doctor solution, which is sodium plumbite, indicates whether many kinds of organic sulfides are present. It does not show whether any form of sulfur is present.

Preparation of Sodium Plumbite Solution. Dissolve approximately 125 grams of sodium hydroxide in 1 liter of water. Add 60 grams of litharge; and shake vigorously for 15 minutes, or let stand with occasional shakings for at least a day. Allow to settle, and decant or siphon off the clear liquid. Filtration through a mat of asbestos may be employed if the solution does not settle clear. Keep the solution in a tightly stoppered bottle, and refilter if necessary before use.

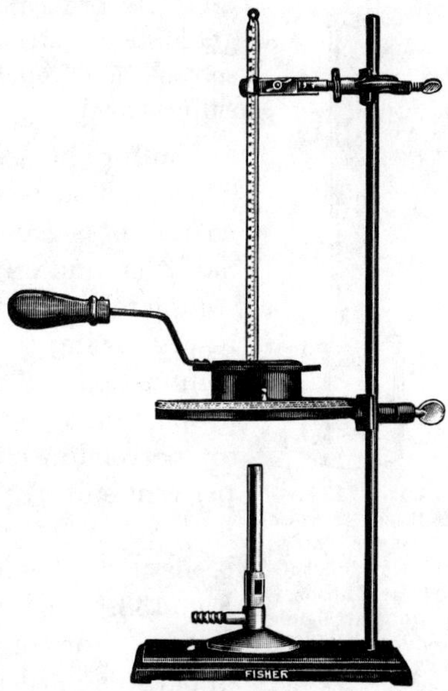

Fig. 136.—Cleveland open-cup tester for determination of flash and fire points. ASTM D 92-57.

Procedure. Shake vigorously together in a test tube 10 ml. of the sample to be tested and 5 ml. of sodium plumbite solution for about 15 seconds. Add a small pinch of pure, dry flowers of sulfur, again shake for 15 seconds, and allow to settle. The quantity of sulfur used should be such that practically all of it floats on the interface between the sample and the sodium plumbite solution.

If the sample is discolored or if the yellow color of the sulfur film is noticeably masked, report the test as positive and condemn the sample as "sour." If the sample remains unchanged in color and if the sulfur film is bright yellow or only slightly discolored with gray or flecked with black, report the test as negative and consider the sample "sweet."

Sulfuric Acid Absorption. This is a measure of the amount of unsaturated material present in the sample.

Procedure. Use a modified Babcock bottle, as shown in Fig. 137. Measure into the dry bottle with a standard pipet 10 ml. of the sample at 20 ± 1°C., and cool in ice water for 5 minutes. Add from a graduate 20 ml. of concentrated sulfuric acid, pouring down the side of the bottle to prevent splashing. Again cool for 10 minutes in ice water, the level of which is above the level of the sample in the bottle. Remove the bottle from the bath, place the glass stopper, previously wet with sulfuric acid, in the bottle, and shake violently for 1 minute. Carefully add to the bottle sufficient sulfuric acid to bring the liquid level almost to the top graduation, and allow the stoppered bottle to stand at least 12 hours. Place in a water bath at 20 ± 1°C. for 15 minutes. Add sulfuric acid at 20°C., to bring the liquid level exactly to the top of the graduation. Read the scale at the lower surface of the solvent, and report the loss as percentage absorbed in sulfuric acid.

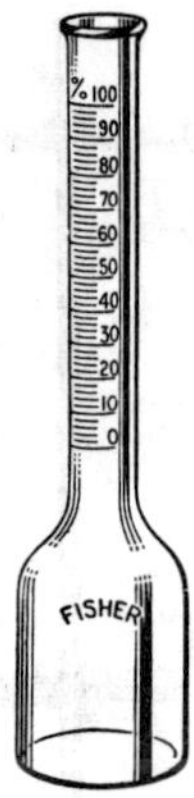

FIG. 137.—Modified Babcock bottle for determination of sulfuric acid absorption of petroleum products. ASTM D 483-61T.

Sulfur in Petroleum Oils by Lamp Method. This method is applicable to quantitative determination of sulfur in gasoline, kerosene, petroleum naphtha, and any petroleum oil that can be burned completely in a wick lamp. It gives slightly high results owing to the formation of nitric acid during combustion. The amount of this acid is nearly proportional to the amount of sample burned, and for petroleum oils the error is usually around 0.01 per cent sulfur.

The combustion and absorption apparatus, which is made of chemically resistant glass, is depicted in Fig. 138. The complete apparatus is available as a unit. For comparative results, the official ASTM apparatus must be used.

The lamp is a 25-ml. conical flask. The burner consists of two concentric glass tubes, the external one having an arm, provided with tapered glass joints for connection with the flask and the chimney. The upper ends of both tubes are polished and have plane surfaces in the same horizontal plane. A small opening near the base allows equalization of pressure between the chimney and the flask. When connected with the chimney, the lamp is held in position by rubber bands stretched between glass hooks in the flask and chimney. The latter connects through standard tapered glass joints with the burner and absorber. A sintered glass plate with an average pore diameter of 150 to 200 microns is sealed in the larger of the two bulbs of the absorber. The spray trap is also provided with a standard taper for connection with the absorber. A clean, unused uniform two-strand twisted cotton wicking of good quality, weighing 5 to 6 mg. per cm. of strand, is used in the burner. A convenient suction system is depicted in Fig. 139. Under normal conditions a pressure of 12 inches of water below atmospheric

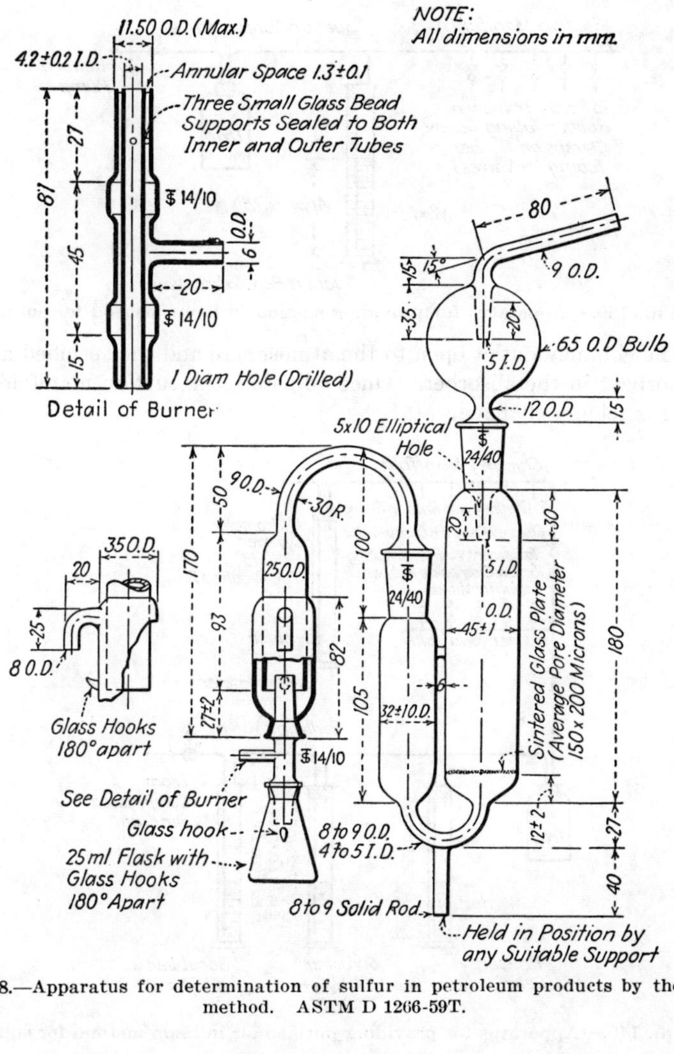

FIG. 138.—Apparatus for determination of sulfur in petroleum products by the lamp method. ASTM D 1266-59T.

pressure is required. In Fig. 140 is shown an air-purifying system to supply purified air to the burner manifold at a pressure equivalent to about 8 inches of water and to the chimney manifold at a pressure equivalent to 0.5 to 0.75 inch of water.

Procedure. This procedure is for a series of complete sets of apparatus all using the same suction and air-purifying systems. Adjust the valves on the suction manifold leading to the individual spray traps so that 2.5 to 3 liters of air per minute will be drawn through each absorber

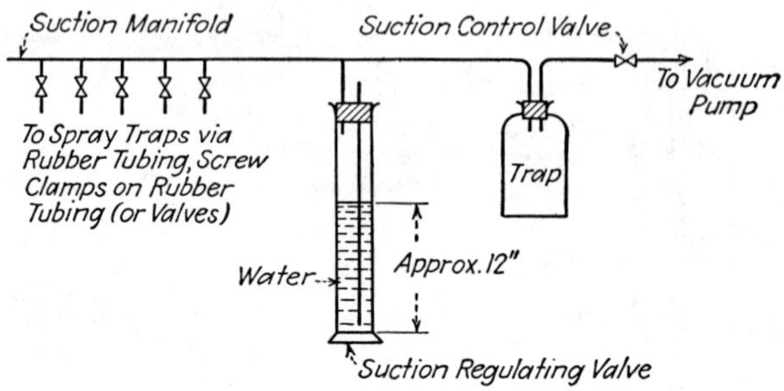

Fig. 139.—Apparatus for providing suction in lamp method for sulfur.

with the chimney outlet open to the atmosphere and the specified amount of absorbent in the absorber. Once adjusted, the suction manifold valve setting is seldom changed.

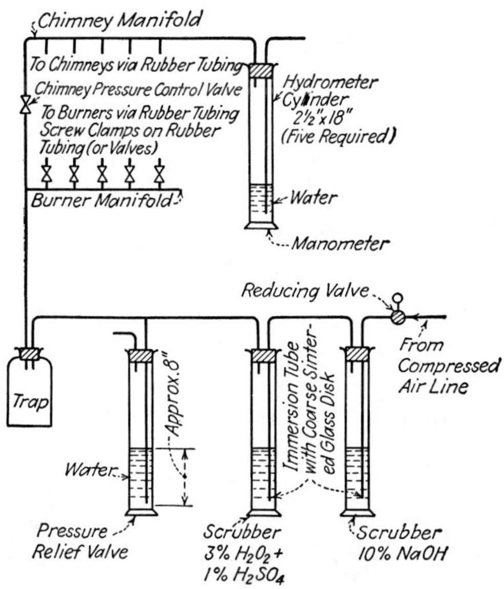

Fig. 140.—Apparatus for providing purified air in lamp method for sulfur.

Thread the burner with four 15-cm. lengths of the two-strand cotton wicking obtained by quadrupling a 60-cm. length, inserting the double-looped end at the top of the burner, and drawing through by means of a crochet hook. Carefully trim the wick as close as possible to the top of the burner, and draw the trimmed end down so that it is flush with the top edge of the burner. Remove any protruding cotton fibers.

Add 40 ml. of 3 per cent hydrogen peroxide, neutralized to the methyl orange end point, to the larger bulb of the absorber. Attach the spray trap and chimney to the absorber, and connect the two outlets to their respective manifolds by means of rubber tubing. Stopper the burner opening of the chimney with a cork.

Before starting the apparatus, close the burner manifold valves. To start the flow of air, adjust the reducing valve on the compressed air line, the suction control valve, and the chimney-pressure control valve so that, at the required flow through the absorbers, a small stream of air escapes at the pressure-relief valve, a small stream of air enters at the suction-regulating valve, and the pressure on the chimney manifold is at 0.5 to 0.75 inch of water. Minor adjustments may also be necessary on the individual valves connecting the suction manifold to the spray traps.

The amount of sample to be burned depends on the anticipated sulfur content. If this is below 0.25 per cent, add slightly in excess of 10 grams of sample to the conical flask. If above 0.25 per cent, add proportionately less. If unknown, use 10 grams and, when necessary, because of very high sulfur content, repeat with less. Connect the burner to the flask, and weigh the assembled lamp to the nearest 10 mg. Connect the lamp to the burner manifold by means of rubber tubing.

As soon as the oil rises to the top of the wick, light the burner with a sulfur-free alcohol lamp and insert it in the chimney. Do not let the sample evaporate from the wick before being lighted. Adjust the flow of air so that smoking does not occur. Continue the burning until the flame becomes small and approaches extinction; but do not let it go out, or incomplete combustion of the sulfur will result. At the beginning and end of the run, minor burner manifold valve adjustments are necessary. Remove the lamp, extinguish the flame, and at once weigh to the nearest 10 mg. Stopper the chimney with a cork, and shut off the supply of air.

Always supply sufficient air, but not so much as to cause jets from the inner cone of the flame, with resultant incomplete combustion. A light, highly volatile oil, such as aviation gasoline, requires a moderately snugly fitting wick, while a heavier oil like kerosene burns best with a fairly loose wick. It may therefore be necessary to vary the number of threads in the wick. With highly volatile samples the flame may be controlled by immersing the conical flask in a cooling bath during burning or by drawing the wick down into the burner. Occasionally, it may be necessary to add about 50 per cent by weight of the sample of a sulfur-free paraffinic diluent, such as normal heptane. Burn as described, and toward the end extinguish the flame and add 2 ml. of the diluent. Resume the burning until the flame becomes small. Repeat the addition of diluent and burning 2 more times, when the original sample will have all been burned. Make a blank run on a similar amount of diluent.

Carefully rinse the chimney spray trap with about 35 ml. of water, and collect in the absorber. Add 5 drops of a methyl red indicator made by dissolving 0.1 gram of methyl red in 37 ml. of 0.0624*N* sodium hydroxide solution and diluting to 100 ml. with water. Titrate the solution with standard sodium hydroxide solution made 0.0624*N* by dissolving 2.5 grams of C.P. sodium hydroxide in 1 liter of water and adjusting to this normality with respect to total alkalinity, using methyl orange indicator. Add the alkali to the smaller of the 2 bulbs; and, during the titration, carefully agitate the contents by repeatedly drawing air through the absorber by means of a rubber tube, one end of which is connected to the larger bulb and the other held between the lips. Incomplete oxidation is readily detected by a characteristic taste and odor while aspirating. Repeat the test if this has occurred.

It will be noticed that the sulfurous acid formed in the solution is titrated to a methyl red end point, the final product being sodium bisulfite. This could be further titrated to sodium sulfite, phenolphthalein being used as indicator. However, the methyl red end point is far sharper and is preferable even though the actual titration is half the theoretical titration to phenolphthalein. Also, the sulfite present tends to decolorize the phenolphthalein indicator.

CALCULATION:

$$\frac{\text{Ml. of standard alkali}}{\text{Weight of sample burned} \times 10} = \text{per cent sulfur}$$

If the sodium hydroxide used is not of the exact normality, calculate as follows:

$$\text{Ml. of alkali} \times \text{normality} \times 0.03206 \times \frac{100}{\text{weight of sample}} = \text{per cent sulfur}$$

As a check on proper operation of the apparatus, make periodic tests with solutions of ethyl disulfide of known sulfur content, dissolved in normal heptane or alcohol. Such solutions are made to contain 0.1, 0.2, and 0.5 per cent sulfur. Run a blank on the diluent to correct for nitric acid formed during the combustion. Results should differ from the true value by not more than 0.002 plus 0.01 times the percentage of sulfur. This also applies to the actual samples themselves.

Sulfur in Petroleum Oils by the Bomb Method. This method is applied to petroleum oils that cannot be burned completely in a wick lamp. It is fully described on page 161. As sample, use 0.6 to 0.8 gram of the oil weighed to the nearest 2 mg. Report the result as per cent by weight of sulfur.

Aniline Point of Petroleum Products. The aniline point is defined as the minimum equilibrium solution temperature for equal volumes of aniline and petroleum product. This method of test is for use with petroleum products not darker than No. 8 ASTM color, provided that the aniline point lies between the initial boiling point and the solidification point of the aniline-sample mixture.

The following apparatus is required for the test: A Pyrex test tube approximately 25 mm. in diameter and 150 mm. in length and another Pyrex tube approximately 38 mm. in diameter and 175 mm. in length. A variable-speed mechanical stirrer, fitted with an auger-type glass tip, as shown in Fig. 141. ASTM aniline-point

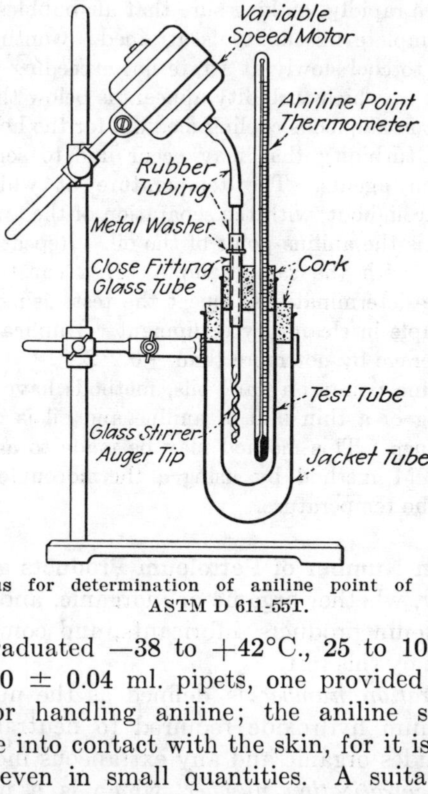

FIG. 141.—Apparatus for determination of aniline point of petroleum products. ASTM D 611-55T.

thermometer graduated −38 to +42°C., 25 to 105°C., and 90 to 170°C. Two 10 ± 0.04 ml. pipets, one provided with a suitable rubber bulb for handling aniline; the aniline should never be allowed to come into contact with the skin, for it is extremely toxic by absorption even in small quantities. A suitable nonaqueous, nonvolatile, transparent liquid bath; an open flame may be substituted, but, as aniline is hygroscopic, hot water should not be used.

PROCEDURE. Dry the sample by shaking for at least 2 minutes with anhydrous sodium sulfate. In the case of viscous samples, it may be necessary to remove the drying agent by filtering through a fritted glass filter of medium porosity.

Clean and dry the apparatus. Pipet 10 ml. of aniline and 10 ml. of the sample into the jacketed test tube fitted with stirrer and thermometer.

Do not use direct suction with the mouth. The aniline, C.P. grade, should be dried over potassium hydroxide pellets, decanted, and freshly distilled on the day of use, the first and last 10 per cent being discarded. If desired, it may be checked against suitable secondary standards, such as normal heptane or paraffinic fractions of known aniline point.

If the material is too viscous to pipet, weigh a quantity of the sample corresponding to 10 ml. at room temperature in the test tube to the nearest 0.01 gram. Place the thermometer in the test tube at the proper level of immersion, making sure that the bulb does not touch the side of the tube.

Stir the mixture rapidly, making sure that air bubbles are not formed, and heat until complete miscibility is attained. Continue stirring, and allow the mixture to cool slowly at a rate not exceeding 1°C. per minute. If the temperature at which turbidity appears is below that of the atmosphere, substitute a nonaqueous cooling medium for the heating medium.

Disregard any turbidity that may occur due to separation of non-petroleum addition agents. The temperature at which the mixture becomes cloudy throughout, with the separation of the aniline and oil into discernible layers, is the aniline point of the oil. Repeat the heating and cooling operations with the same sample. If a constant value is not obtained after five determinations, repeat the test, using fresh quantities of aniline and sample in clean, dry equipment. Duplicate results should differ from the average by not more than ±0.2°C.

To obtain aniline points on dark oils, methods have been devised in which the clouding of a thin film of aniline and oil is observed to give approximate readings. This method may be made to agree closely with the standard ASTM method by using a thermocouple and a potentiometer to record the temperature.

Neutralization Number of Petroleum Products and Lubricants. The total acidity, whether organic or inorganic, and any alkalinity present in petroleum products, lubricants, and compounded products, is measured by this test.

The *neutralization number* is defined as the number of milligrams of potassium hydroxide required to neutralize 1 gram of sample. It includes organic and any extraneous inorganic acidity.

The *alkali neutralization number*, which is a measure of any extraneous water-soluble alkali present, is defined as the weight of acid required to neutralize 1 gram of sample, expressed in equivalent milligrams of potassium hydroxide.

The *strong-acid neutralization number*, which is a measure of any extraneous inorganic or water-soluble acid present, is defined as the number of milligrams of potassium hydroxide required to neutralize the strong acid present in 1 gram of sample.

Reagents. To prepare the standard alkali for strict conformity to this method, dissolve 5.1 grams of C.P. potassium hydroxide in 1 liter of freshly boiled distilled water. Add sufficient barium hydroxide to precipitate any carbonate present, and filter. Standardize this solution against a solution of National Bureau of Standards benzoic acid in alcohol, either SD30 or SD3A, that has been boiled and neutralized to phenolphthalein with the potassium hydroxide solution. Titrate without heating to a phenolphthalein end point. Adjust the alkali solution so that 1 ml. contains 5 mg. of potassium hydroxide, equivalent to 10.88 mg. of benzoic acid. Adjust C.P. sulfuric acid solution so that 1 ml. of the acid is required to neutralize 1 ml. of the standard alkali, using phenolphthalein as indicator and titrating at the boiling point.

The alkali solution containing 5 mg. per liter of potassium hydroxide is rather less than $0.1N$. It is permissible and more convenient to use accurately standardized $0.1N$ alkali and acid solutions in place of the special solutions.

Neutralization Number. Weigh 10 $\pm$ 0.1 gram of a representative sample into a 250-ml. conical flask. Prepare an alcohol solution by adding 10 per cent by volume of methanol to 95 per cent ethanol, or by adding 5 per cent of methanol to SD3A, and dilute the mixture with an equal volume of distilled water. Add 100 ml. of this solvent to the sample, and put on a hot plate. Agitate by swirling during the heating process, as frequently as may be required to prevent bumping. When boiling begins, remove from the hot plate, add 1 ml. of 1 per cent alcoholic phenolphthalein solution and agitate by vigorous swirling for not more than 5 seconds. Titrate immediately. Add the standard alkali solution in increments of 0.1 to 1.0 ml., depending on the rate at which the color fades in the alcohol layer, with gentle swirling. After sufficient alkali has been added so that the color does not fade instantly on gentle swirling, add 50 ml. of distilled water at 175 to 190°F. Agitate by vigorous shaking without using a stopper, and tilt the flask at a 45-degree angle to observe the color of the separated alcohol layer. Without waiting for a complete separation, proceed with the final titration, adding the alkali solution in increments of 0.2 ml. When a definite pink coloration persists, again agitate by vigorous shaking. If the pink coloration persists after shaking, the end point has been reached. If too much alkali has been added, the sample may be back-titrated with standard acid. The entire titration should not consume more than 1 to 1.5 minute for a new or used oil. Even with oils containing much insoluble matter the total titration time must not exceed 3 minutes. Subtract a blank obtained by heating 100 ml. of the alcohol solution to boiling, adding 1 ml. of the phenolphthalein solution and titrating to a definite pink coloration.

The foregoing method must be used in referee work. Generally, when it is necessary merely to make a comparison of total acidity a method from the saponifiable oil industry is satisfactory and much simpler. For this, dissolve the sample in a neutralized mixture of equal volumes of benzene and alcohol, and warm and titrate to a phenolphthalein end point. Normally, with $0.1N$ alkali, 1 drop excess will give a distinct coloration unless the oil is extremely dark. With very dark oils the ASTM method must be employed in order to read the end point. Titration with a glass electrode to determine the end point has been applied to highly colored oils where the usual procedure is applicable only with extreme difficulty.

CALCULATION.

$$\frac{\text{Ml. of potassium hydroxide solution} \times 5}{\text{Weight of sample}} = \text{neutralization number}$$

If $0.1N$ alkali has been substituted for the special solution specified, this becomes

$$\frac{\text{Ml of standard alkali} \times \text{normality} \times 56.1}{\text{Weight of sample}} = \text{neutralization number}$$

ALKALI OR STRONG-ACID NEUTRALIZATION NUMBER. To 25 to 50 ± 0.1 gram of a representative sample in a warm separatory funnel, add 100 ml. of boiling water. Shake vigorously, let settle, and drain off the water into a 500-ml. casserole. Extract twice more with 50-ml. portions of boiling water, separate, and drain into the casserole. Add 1 ml. of 1 per cent alcoholic phenolphthalein solution to the water extractions, and bring just to the boil. If pink, titrate the extract with standard sulfuric acid solution. If not pink, add 0.1 ml. of a 0.1 per cent water solution of methyl orange, and titrate with standard alkali solution. For very accurate work, run a blank on an equal volume of water, and subtract from the sample titration.

CALCULATION. If the special alkali or acid has been used, the calculation is as follows:

$$\frac{\text{Ml. of alkali} \times 5}{\text{Weight of sample}} = \text{strong-acid neutralization number}$$

$$\frac{\text{Ml. of special acid} \times 5}{\text{Weight of sample}} = \text{alkali neutralization number}$$

If laboratory stocks of alkali or acid have been used, this becomes

$$\frac{\text{Ml. of alkali} \times \text{normality} \times 56.1}{\text{Weight of sample}} = \text{strong-acid neutralization number}$$

$$\frac{\text{Ml. of acid} \times \text{normality} \times 56.1}{\text{Weight of sample}} = \text{alkali neutralization number}$$

If either hydrochloric acid or sulfuric acid is known to be present, it may be desirable to calculate the percentage as follows:

$$\text{Ml. of alkali} \times \text{normality} \times 0.03654 \times \frac{100}{\text{weight of sample}} = \text{per cent hydrochloric acid}$$

The factor for sulfuric acid is 0.04901.

Saponification Number of Petroleum Products and Lubricants. The determination of saponification number or saponification value of petroleum products for nonreferee purposes can be conducted exactly as for saponifiable fats and oils (page 343). The ASTM method calls for dissolving a 2- to 3-gram sample, or more for samples with low saponification number, in 25 ml. of technical grade methylethyl ketone, adding $0.5N$ alcoholic potassium hydroxide and refluxing for 30 minutes. The procedure and calculation are otherwise identical with those for saponifiable oils.

This procedure is not applicable to oils containing free sulfur, sulfurized oils, or oils heated with chlorine or phosphorus compounds, for alkali will be consumed by the sulfur, chlorine, or phosphorus during the saponification. For such samples, the only method that will give an indication as to the type of oil present is to saponify the sample, acidify its aqueous solution, extract and weigh the liberated fatty acids and then determine their acid value.

Cloud and Pour Points. The *cloud point* of a petroleum oil is defined as the temperature at which paraffin wax or other solid substances begin to crystallize out or separate from solution when the oil is chilled under definite prescribed conditions. The test is applicable only to those oils which are transparent in layers 1.5 inches in thickness.

The *pour point* of a petroleum oil is defined as the lowest temperature at which the oil will pour or flow when it is chilled without disturbance under definite prescribed conditions. This test may be applied to any petroleum oil.

The apparatus used in these tests is shown in Fig. 142. The cylindrical glass test jar *a* is approximately 1.5 inches wide internally and 5 inches high. A 4-ounce oil sample bottle is suitable for this purpose and is most commonly used. The thermometer used is either an ASTM cloud and pour test thermometer graduated -38 to $+50$°C. or -36 to $+120$°F. or an ASTM low cloud and pour thermometer graduated -60 to $+20$°C. or -70 to $+70$°F. The jacket *d* is of either glass or metal. It is cylindrical, about 4.5 inches in depth, with inside diameter 0.37 to 0.5 inch greater than the outside diameter of the test jar, and watertight. A disk of cork or felt *e* rests on the bottom of the jacket. The test jar rests on this disk and is centrally located in the jacket by a ring gasket *f*, which may be of any suitable material. Practically, the tests for both cloud and pour point are often done on the same sample in a 4-ounce oil bottle with a thermometer fitted through the cork. The precautions are approximated.

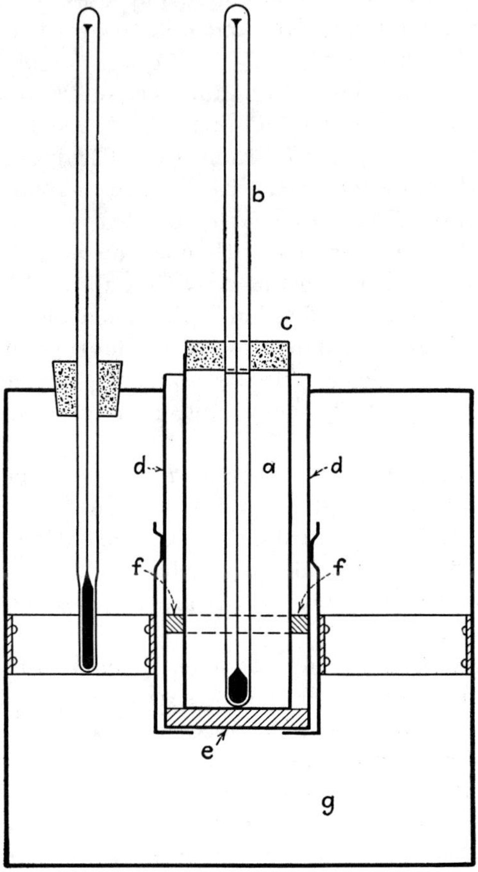

FIG. 142.—Apparatus for cloud and pour point. ASTM D 97-57.

PROCEDURE. Support the assembled apparatus in a suitable cooling bath g, and as a cooling medium use one of the following mixtures.

	For Temperatures Down to
Ice and water	+50°F.
Crushed ice and sodium chloride	+10°F.
Crushed ice and calcium chloride	−15°F.
Solid carbon dioxide and acetone, gasoline, or alcohol	−70°F.

CLOUD POINT. The oil must be at least 25°F. above the cloud point. If necessary, remove moisture by filtration through a dry paper. Pour the clear oil into the test jar to a depth of 2.0 to 2.25 inches. Insert the thermometer through a cork *c*, fit the cork tightly in the test jar, and allow the thermometer bulb to rest centrally on the bottom of the jar. The disk and jacket must be clean and dry.

Maintain the cooling bath at 30 to 35°F., and fill the bath to within 1 inch of the top of the test-jar jacket. At each test thermometer reading that is a multiple of 2°F., remove the test jar quickly from the jacket, without disturbing the oil, inspect for cloud, and at once return to the jacket. This should take about 3 seconds. If the oil does not cloud at 50°F., replace the cooling medium by one with a temperature of 0 to 5°F. Observe as before, and, if necessary, further cool the bath to between −30 and −25°F. When a distinct cloudiness or haze appears in the oil at the bottom of the test jar, read the thermometer. This is the cloud point.

Pour Point. Pour the oil, heating it on a water bath if necessary to allow it to flow, into the test jar *a* to a height of 2.0 to 2.25 inches. Tightly stopper with a cork carrying the test thermometer in a vertical position in the center of the jar, with the thermometer bulb immersed so that the beginning of the capillary is ⅛ inch below the surface of the oil.

Heat the oil, without stirring, to a temperature of 115°F. in a bath maintained at not higher than 118°F. Cool to 90°F. in air or in a water bath at approximately 77°F. Oils on which a pour point below −30°F. is expected should be heated, as already described, with the high cloud and pour test thermometer in position; they should then be cooled to 60°F. and the low cloud and pour test thermometer used as replacement before the assembly is placed in the jacket. Oils on which a pour point of above 90°F. is expected should be heated to a temperature 15°F. above the expected pour point, with the high cloud and pour test thermometer in position, and the test jar immediately introduced into the jacket. The apparatus is assembled as in the cloud point test.

After the oil has cooled enough to allow the formation of paraffin wax crystals, great care must be taken not to disturb the mass of the oil or to allow the thermometer to shift in the oil. Any disturbance of the spongy network of wax crystals will cause low results.

Maintain the cooling-bath temperature at 30 to 35°F. Beginning at a temperature 20°F. before the expected pour point, at each test thermometer reading that is a multiple of 5°F., carefully remove the test jar from the jacket and tilt it just enough to ascertain if there is a movement of the oil. The complete operation should take not more than 3 seconds. If the oil has not ceased to flow at 50°F., replace the test jar in a second bath maintained at 0 to 5°F. If the oil has not ceased to flow at 20°F., use a third bath maintained at −30 to −25°F. For very low pour points still colder baths may be necessary; these are changed at temperature differentials of 30°F. In each case transfer the test jar when the temperature of the oil reaches a point 50°F. above the temperature of the new bath. Never place the cold test jar directly in the cooling medium. As soon as the oil in the test jar does not flow when the jar is tilted, hold it in a horizontal position for exactly 5 seconds and observe carefully. If any movement ensues, immediately replace in the jacket and observe for flow at the next 5°F. lower.

Continue the test under these conditions until the oil does not flow when held in a horizontal position for exactly 5 seconds. Some lubricating oils tend to move as a whole and should be very carefully observed. The temperature at this solid point plus 5°F. is taken as the pour point.

For black oils, cylinder stocks, nondistillate fuel oils, samples which have been subjected to a temperature higher than 115°F. during the preceding 24 hours or samples of which the history is not known, the sample should be held 24 hours before testing unless three consecutive tests of the same sample in the same jar give check results. This point is called the upper, or maximum, pour point.

To obtain the lower, or minimum, pour point, heat the sample, while stirring, to 220°F. Pour into the test jar, cool to 90°F., and determine the pour point as described with reference to the main method. Report both upper and lower pour points. Individual results on the same oil in one laboratory should differ by not more than 5°F.

Carbon Residue of Petroleum Products. Conradson Carbon Residue. This test is a means of determining the amount of carbon residue left after evaporating an oil under specified conditions and is intended to throw some light on the relative carbon-forming tendency of an oil. The results of the test must be considered in connection with those of other tests and the use for which the oil is intended. The test is applied to lubricants for internal-combustion engines, domestic fuel oils, and oils used in the manufacture of gas.

The apparatus used may be purchased as a unit and is depicted in Fig. 143. The wide-form porcelain or silica crucible *a* is of 30 ml. capacity and 46 to 49 mm. in rim diameter. The Skidmore iron crucible *b* is flanged and ringed, is of 65 to 82 ml. capacity, has 53 to 57 mm. inside and 60 to 67 mm. outside diameter of flange, is 37 to 39 mm. in height, and is supplied with a cover without delivery tubes and having the vertical opening closed. The horizontal opening of about 6.5 mm. must be kept open and clean. The outside diameter of the flat bottom is 30 to 32 mm.

The spun sheet-iron crucible *c*, with cover, is 78 to 82 mm. in outside diameter at the top, 58 to 60 mm. in height, and approximately 0.8 mm. in thickness. Place at the bottom of this crucible, and level before each test, a layer of about 25 ml. of dry sand or enough to bring the Skidmore crucible, with cover on, nearly to the top of the sheet-iron crucible.

The bare Nichrome triangle *d*, of approximately 13 Brown and Sharpe gauge, has an opening small enough to support the bottom of the sheet-iron crucible at the same level as the bottom of the asbestos block or hollow sheet-metal box. The asbestos-block, refractory ring, or hollow sheet-metal box *f*, is 150 to 175 mm. in

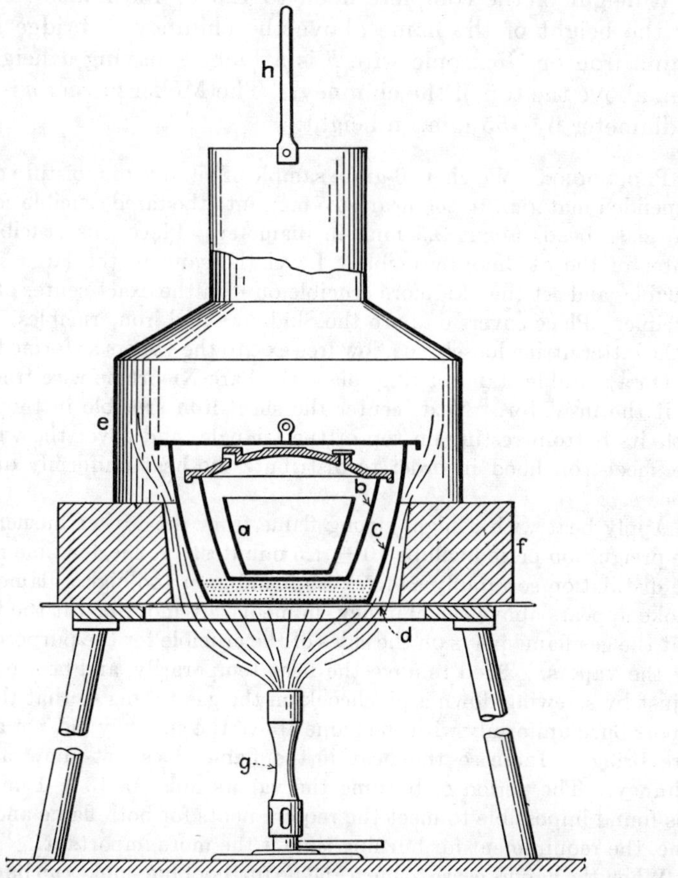

FIG. 143.—Conradson apparatus for determination of carbon residue. ASTM D 189-61.

diameter if round or on a side if square, to 32 to 38 mm. in thickness, and is provided with a metal-lined, inverted cone-shaped opening through the center, 83 mm. in diameter at the bottom and 89 mm. in diameter at the top. If a refractory ring is used, no metal lining is necessary, provided that the ring is of hard, heat-resistant material.

The circular sheet-iron hood *e* is 120 to 130 mm. in diameter, and the height of the lower perpendicular side is 50 to 53 mm. It is provided at the top with a chimney 50 to 60 mm. in height, of 50 to 56 mm. inside diameter, and attached to the lower part having the perpendicular sides by a cone-shaped member, bringing the total height of the complete hood to 125 to 130 mm. As a guide for the height of the flame above the chimney, a bridge made of 3-mm. iron or Nichrome wire *h* is attached, having a height of 50 mm. above the top of the chimney. The Meker burner *g* is 24 mm. in diameter by 155 mm. in height.

PROCEDURE. Weigh a 10-gram sample of oil, free of moisture and other suspended material, to the nearest 5 mg., into the tared crucible containing two glass beads about 2.5 mm. in diameter. Place this crucible in the center of the Skidmore crucible. Level the sand in the large sheet-iron crucible, and set the Skidmore crucible on it in the exact center of the iron crucible. Place covers on both the Skidmore and iron crucibles, the cover of the latter fitting loosely to allow free exit to the vapors as formed.

On a suitable stand or ring, place the bare Nichrome wire triangle and on it the insulator. Next, center the sheet-iron crucible in the insulator with its bottom resting on top of the triangle, and cover the whole with the sheet-iron hood in order to distribute the heat uniformly during the process.

Apply heat with a high, strong flame from the Meker burner, so that the preignition period will be 10 ± 1.5 minutes. A shorter time may start the distillation so rapidly as to cause foaming or too high a flame. When smoke appears above the chimney, immediately move or tilt the burner so that the gas flame plays on the sides of the crucible for the purpose of igniting the vapors. Then remove the heat temporarily, and before replacing adjust by screwing down a pinchcock on the gas tubing so that the ignited vapors burn uniformly with the flame above the chimney but not above the wire bridge. Increase the heat if the flame does not show above the chimney. The period of burning the vapors must be 13 ± 1 minute. If it is found impossible to meet the requirements for both flame and burning time, the requirement for burning time is the more important.

When the vapors cease to burn and no further blue smoke can be observed, readjust the burner as at the beginning of the process so as to make the bottom and lower part of the sheet-iron crucible a cherry red, and maintain for exactly 7 minutes. The total time of heating should be 30 ± 2 minutes, which constitutes an additional limitation on the tolerances for the preignition and burning periods. There should be no difficulty in carrying out the test exactly as directed, using city gas of 550 B.t.u. and a Meker burner, with the top of the burner about 2 inches below the bottom of the crucible.

Remove the burner, and allow the apparatus to cool until no smoke appears. Then remove the cover of the Skidmore crucible, usually after about 15 minutes. Remove the porcelain or silica crucible with heated tongs, place in the desiccator, cool, and weigh. Calculate the percentage of carbon residue on the original sample.

When the carbon residue, with a 10-gram sample, is in excess of 5 per cent, as may be the case with heavy crude oils, residuum, heavy fuel oils, and heavy gas-enrichment oils, difficulties may be experienced due to boiling over of the sample. Trouble may also be encountered with samples of heavy products that are difficult to dehydrate. In such cases, repeat the test in duplicate, using a 5-gram sample weighed to the nearest 5 mg.

Light distillate oils having carbon residues less than 0.05 per cent, such as No. 1 and No. 2 fuel oils, should be tested by using as a sample a 10 per cent residuum obtained by an adaptation of the distillation procedure in the standard method (page 665). To receive the distillate from a 200-ml. sample of the oil, use the 200-ml. graduate employed to measure the initial sample. Maintain the condenser outlet at 32 to 40°F. throughout the distillation in the case of products having distillation end points below 600°F., and use a condenser temperature of 90 to 100°F. for products having end points above 600°F. or in those cases in which waxy distillates are obtained. The 200-ml. sample and a condenser outlet temperature of 90 to 100°F. are standard for the distillation of gas, oil, and similar distillate fuel oils.

Regulate the heat so that condensate falls from the condenser in 10 to 15 minutes, this being the initial boiling point, and then so that the distillation will proceed at a uniform rate of 8 to 10 ml. per minute.

Distill exactly 178 ml. into the graduate, discontinue the heating, and allow the condenser to drain until 180 ml. has been collected in the graduate. Replace the graduate with a small conical flask, and catch any final drainage in this flask. To the conical flask add, while still warm, the residue left in the distillation flask, the whole being 10 per cent of the original sample. While warm enough to flow freely, pour approximately 10 grams of this residuum into the weighed crucible to be used in the carbon residue test. Cool, accurately weigh, and proceed with the determination as described. Report the percentage of carbon residue so obtained as "carbon residue on 10 per cent residuum."

Run the tests in duplicate, and repeat if necessary until the percentages of carbon residue differ by not more than 10 per cent from the average. In the case of a 10 per cent residuum, duplicate determinations should differ by not more than 20 per cent from an average if the value is 0.10 per cent or higher. For lower values the deviations of individual determinations will be higher, increasing as the carbon residue decreases.

Precipitation Number of Lubricating Oils. The ASTM precipitation number is the number of milliliters of precipitate formed when 10 ml. of lubricating oil is mixed with 90 ml. of petroleum naphtha as specified and centrifuged under prescribed conditions. This test is employed with steam-cylinder stocks and black oils and may be used for other lubricating oils.

A certified ASTM conical centrifuge tube, graduated every 0.05 to 0.5 ml., every 0.1 to 2 ml., etc., as shown in Fig. 144, is used in this test. The centrifuge should preferably have a diameter of swing, tip to tip of whirling tubes, of 15 to 17 inches and a speed of at least 1,500 r.p.m. or the equivalent. The thermometer used is an ASTM aniline-point thermometer having a range of 100 to 180°F.

The following are the requirements for the petroleum naphtha.

Specific gravity, API at 60°F.	70–73
Aniline point	58–60°C.
Initial boiling point	Not less than 122°F.
50 per cent point	158–176°F.
End point	Not more than 266°F.

Procedure. Measure 10 ml. of the sample into each of two clean, dry centrifuge tubes. Fill each tube to the 100-ml. mark with the specified petroleum naphtha, and close tightly with a softened cork, not a rubber stopper. Thoroughly mix by inverting each tube at least twenty times, allowing the liquid to drain from the tapered end of the tube each time. Place the tube in a water bath at 90 to 95°F. for 5 minutes. Momentarily remove the corks to relieve any pressure, and again invert each tube at least twenty times as before. The success of this method depends to a large degree upon having a homogeneous mixture that will drain quickly and completely from the tapered tip when the tube is inverted.

Place the tubes in carriers so designed that the glass centrifuge tubes may be cushioned with water, rubber, or other suitable material. Should sand be used, be careful that none gets inside the tubes. Use particular care to cushion the narrow end of the tube. With tubes and carriers on opposite sides, centrifuge at a rate of 1,400 to 1,500 r.p.m. or the equivalent for 10 minutes. Allow the centrifuge to come to a standstill without applying any braking action, which is likely to stir up the precipitate from the bottom of the tube. Read the volume of sediment to the nearest 0.05 ml. if possible. Replace the tubes, and centrifuge again for 10 minutes. Repeat this procedure until the volume of sediment in each tube remains constant for three consecutive readings. In general, not more than 4 cycles will be required for oils having a low precipitation number.

If the readings of the tubes differ by not more than 0.1 ml., report the mean of the two as the ASTM precipitation number. If the readings differ by more than 0.1 ml., make two more determinations and report the average of the four. It is not normally necessary to make more than the original two determinations if proper attention is given to details.

Viscosity by the Saybolt Viscosimeter. The oil industry is almost entirely standardized with respect to the measurement of viscosity by time of flow from a standardized tube, the instrument being called the Saybolt Universal viscosimeter. There are other methods of measurement of viscosity in absolute units, and such measurements are convertible into Saybolt

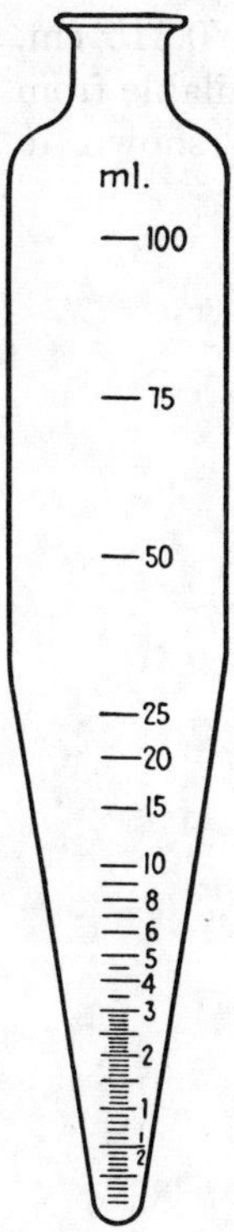

FIG. 144.—Conical centrifuge tube for precipitation number of lubricating oils. ASTM D 91-61.

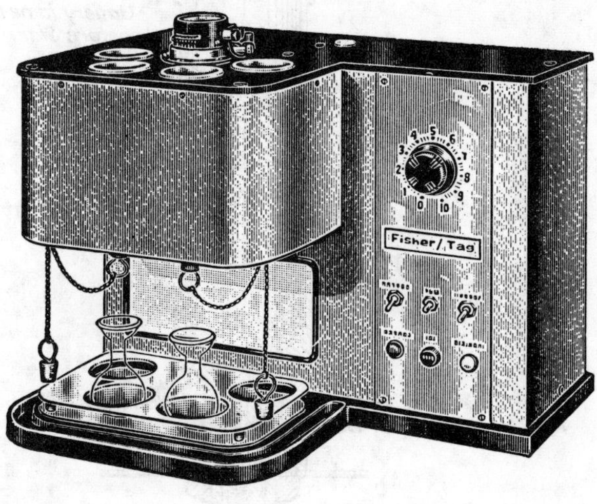

FIG. 145.—Saybolt Universal viscosimeter. ASTM D 88-56.

values by a method given later (page 694). The Saybolt instrument if often used with oils other than those of petroleum origin, but values are relative only.

The lower limit of applicability is to oils with times of flow under 32 seconds. There is no maximum limit to viscosity to be measured. In general, liquids having an outflow time of the order of 1,000 seconds and higher, Saybolt Universal, are tested more conveniently by a modified instrument designed for fuel and road oils. This is known by an abbreviated name as the Saybolt Furol viscosimeter.

Use of the Furol instrument is restricted to oils with a flow time greater than 25 seconds. In general, the time of flow with the Furol instrument is approximately one-tenth that with the Universal. The apparatus is the same in either case except that the size of the outlet tube, or orifice, differs. In the Universal orifice the inside diameter is 0.1765 cm.; in the Furol orifice, 0.315 cm. The whole apparatus and parts are standardized and available from apparatus supply houses. A Saybolt viscosimeter is shown in

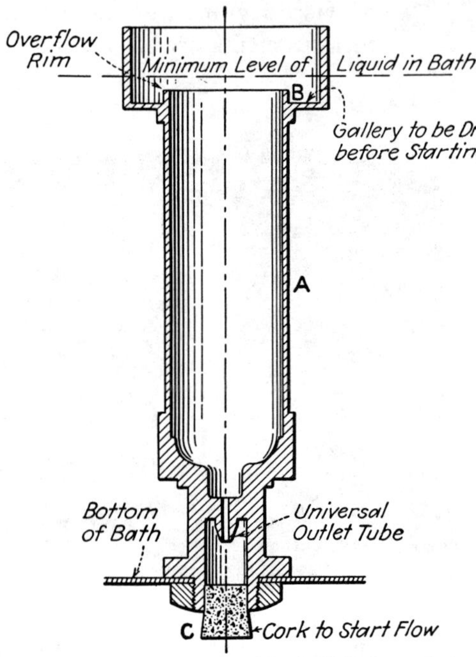

FIG. 146.—Details of viscosity tube of Saybolt Universal viscosimeter. ASTM D 88-56.

Fig. 145. Units contain up to six viscosity tubes to permit a corresponding number of simultaneous determinations. They are heated by either gas or electricity.

The corrosion-resistant oil tube, shown in Fig. 146, is provided with a nut for locking it in place in the bath, and with a cork to prevent flow until the test is started. The ASTM requires that the tube shall be standardized and any correction in excess of 0.2 per cent applied. The time of flow must be within ±1 per cent of the time obtained with the National Bureau of Standards Master Saybolt oil tube.

The bath is equipped with a stirring device and a means for heating or cooling. If an external heater such as a flame is used, it must be more than 2 inches from the oil tube. The bath temperature necessary to maintain thermal equil brium, while the oil in the oil tube is well stirred by the oil tube thermometer, must be within ±0.1°F. of the standard temperatures of 70 or 77°F. or should not exceed 100.25, 122.35, 130.5 141.0, 181.5, or 212°F., respectively, for the corresponding standard temperatures of 100, 122, 130, 140, 180, or 210°F. at which Saybolt Universal readings are made. With the Saybolt Furol viscosimeter, determinations are made at 77, 100, 122, or 210°F. In tests on road and paving materials, determinations are sometimes made at 140 and 180°F. When oil is used as a bath medium, it may be necessary to maintain the bath temperatures slightly higher than those mentioned above. The level of the bath liquid should be no lower than 5 mm. from the overflow rim of the oil tube. A standard glass receiver is shown in Fig. 147. This has a capacity up to the graduation mark on its neck of 60 ± 0.05 ml. at 68°F.

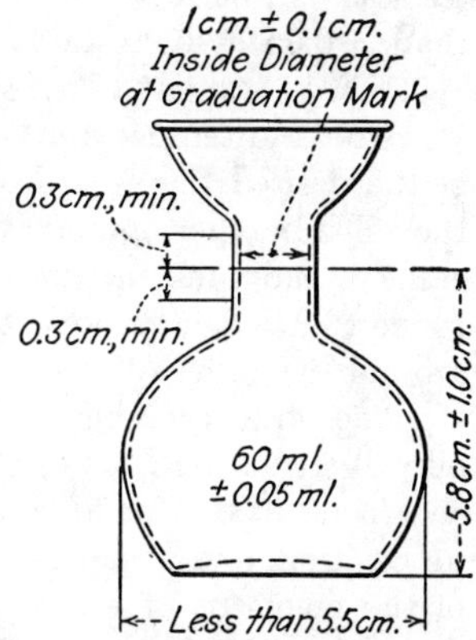

FIG. 147.—Receiving flask for use with Saybolt viscosimeter. ASTM D 88-56.

The ASTM Saybolt viscosity thermometers are sold in sets of six thermometers, each reading in either Fahrenheit or centigrade degrees.

PROCEDURE. Make the determinations in a room free from drafts and subject to no rapid changes in temperature. For referee tests the room must be at 68 to 86°F. For routine tests, temperatures up to 100°F. may be employed without introducing errors in excess of 1 per cent. Determinations must not be made at temperatures below the dew point of the atmosphere.

Clean and thoroughly dry the oil tube, using a solvent such as benzene. Pass the oil sample through a 100-mesh wire strainer before placing it in the oil tube. Wet the inside of the oil tube with the sample, and allow to drain. Do not use the plunger supplied with the viscosimeter if the apparatus is used as a standard. If the flash point of the oil is less than 50°F. above the temperature of the test, omit the wetting operation in order to prevent an increase in viscosity due to evaporation of a portion of the light fraction.

Insert the cork stopper tightly not less than 6 or more than 9 mm. into the lower end of the air chamber at the bottom of the oil tube.

If the test temperature is above that of the room, heat the strained oil to not more than 3°F. above the test temperature; if below, cool correspondingly. Never preheat the oil above a temperature 50°F. below the flash point. Pour the oil into the oil tube until it ceases to overflow into the gallery. Keep the oil well stirred with the oil-tube thermometer, being careful to avoid hitting the overflow tube. Adjust the bath temperature until the oil temperature is constant, and maintain at this level. The tests should be repeated if the indicated bath temperature varies by more than ±0.05°F. in tests at 70, 77, 100, 122, and 130°F. or by more than ±0.1°F. in tests at 140, 180, and 210°F.

After the temperature of the oil has remained constant within 0.02°F. of the desired temperature for 1 minute with constant stirring, withdraw the oil-tube thermometer, insert the withdrawal tube at a point in the gallery, and suck up the oil until the level is below that in the oil tube proper. Do not remove the excess oil by rotating the withdrawal tube around the gallery.

Place the receiving flask under the outlet tube so that the stream of oil will strike the neck of the flask. The graduation on the receiving flask should be 10 to 13 cm. from the bottom of the bath. Snap the cork from its position, and at once start the timer. Stop the timer when the bottom of the meniscus of the oil reaches the mark on the neck of the receiving flask when this is viewed horizontally.

The time in seconds as so measured, with any calibration correction that may be necessary, is the Saybolt Universal or Saybolt Furol viscosity of the oil at the temperature at which the test is made. Report results below 200 seconds to the nearest 0.1 second; to the nearest whole second for results above 200 seconds. Results should differ among different operators by not more than 0.5 per cent.

Conversion of Kinematic Viscosity to Saybolt Universal Viscosity. The Saybolt Universal viscosity equivalent to a given kinematic viscosity varies with the temperature at which the determination is made. The basic conversion tables are those given in Table 41 for 100°F. Equivalent values at temperatures of 130 and 210°F. are also given.

The conversion equivalent may be read directly from the tables. For kinematic viscosities between 2 and 70 centistokes, which are not listed in the table, obtain the equivalent Saybolt Universal viscosities by linear interpolation. For kinematic viscosities over 70 centistokes, obtain the equivalent Saybolt Universal viscosities by use of the multiplication constant given in the table corresponding to the proper temperature. The Saybolt Universal viscosity

TABLE 41.—CONVERSION OF KINEMATIC VISCOSITY TO SAYBOLT UNIVERSAL VISCOSITY*

Kinematic viscosity, centistokes	Equivalent Saybolt Universal viscosity, sec.			Kinematic viscosity, centistokes	Equivalent Saybolt Universal viscosity, sec.		
	At 100°F.†	At 130°F.	At 210°F.		At 100°F.†	At 130°F.	At 210°F.
2	32.6	32.7	32.8	31	145.3	145.6	146.3
2.5	34.4	34.5	34.6	32	149.7	150.0	150.7
3	36.0	36.1	36.3	33	154.2	154.5	155.3
3.5	37.6	37.7	37.9	34	158.7	159.0	159.8
4	39.1	39.2	39.4	35	163.2	163.5	164.3
4.5	40.7	40.8	41.0				
5	42.3	42.4	42.6				
6	45.5	45.6	45.8	36	167.7	168.0	168.9
7	48.7	48.8	49.0	37	172.2	172.5	173.4
8	52.0	52.1	52.4	38	176.7	177.0	177.9
9	55.4	55.5	55.8	39	181.2	181.5	182.5
10	58.8	58.9	59.2	40	185.7	186.1	187.0
11	62.3	62.4	62.7	41	190.2	190.6	191.5
12	65.9	66.0	66.4	42	194.7	195.1	196.1
13	69.6	69.7	70.1	43	199.2	199.6	200.6
14	73.4	73.5	73.9	44	203.8	204.2	205.2
15	77.2	77.3	77.7	45	208.4	208.8	209.9
16	81.1	81.3	81.7	46	213.0	213.4	214.5
17	85.1	85.3	85.7	47	217.6	218.0	219.1
18	89.2	89.4	89.8	48	222.2	222.6	223.8
19	93.3	93.5	94.0	49	226.8	227.2	228.4
20	97.5	97.7	98.2	50	231.4	231.8	233.0
21	101.7	101.9	102.4	55	254.4	254.9	256.2
22	106.0	106.2	106.7	60	277.4	277.9	279.3
23	110.3	110.5	111.1	65	300.4	301.0	302.5
24	114.6	114.8	115.4	70	323.4	324.0	325.7
25	118.9	119.1	119.7				
26	123.3	123.5	124.2	Over 70	Saybolt seconds = centistokes × 4.620	Saybolt seconds = centistokes × 4.629	Saybolt seconds = centistokes × 4.652
27	127.7	127.9	128.6				
28	132.1	132.4	133.0				
29	136.5	136.8	137.5				
30	140.9	141.2	141.9				

* ASTM D 446-53.

† To obtain the Saybolt Universal viscosity equivalent to a kinematic viscosity determined at t°F. multiply the equivalent Saybolt Universal viscosity at 100°F. by $1 + (t - 100)\ 0.000064$; for example, 10 centistokes at 210°F. are equivalent to 58.8×1.0070 or 59.2 Saybolt Universal seconds at 210°F.

equivalent to a kinematic viscosity determined at t°F. may be calculated by multiplying the equivalent Saybolt Universal viscosity at 100°F. by $1 + (t - 100)0.000064$. For example, 10 centistokes at 210°F. is equivalent to 58.8×1.0070, or 59.2 Saybolt Universal seconds at 210°F.

CONVERSION EQUIVALENTS

Poise = c.g.s. unit of absolute viscosity
Centipoise = 0.01 poise
Centistoke = 0.01 stoke
Centipoises = centistokes × density at temperature under consideration
Reyn (1 pound-second per square inch) = 69×10^5 centipoises

Viscosity Index. The viscosity index is an empirical number indicating the effect of change of temperature on the viscosity of an oil. A low viscosity index signifies relatively large change of viscosity with temperature. The calculation is made from the viscosity of the oil at 100 and at 210°F.

CALCULATION.

$$\text{Viscosity index} = \frac{L - U}{L - H} \times 100$$

where U = viscosity at 100°F. of the oil whose viscosity index is required, L = viscosity at 100°F. of an oil of 0 viscosity index that has the same viscosity at 210°F. as does the oil whose viscosity index is required, and H = viscosity at 100°F. of an oil of 100 viscosity index having the same viscosity at 210°F. as the oil whose viscosity index is required.

Table 42 contains values of L and H in Saybolt Universal seconds corresponding to various viscosities at 210°F. Table 43 contains similar values of kinematic viscosity in centistokes. For convenience, $L - H$, designated as D values, are given in both tables.

In the appropriate table find the value that represents the viscosity of the oil at 210°F., and read off the values of L and D. Subtract the viscosity of the oil at 100°F. from L, divide by D, and multiply by 100 to obtain the viscosity index of the oil.

For example, if the oil has a Saybolt Universal viscosity of 40 seconds at 210°F. and 10 seconds at 100°F., the viscosity index is as follows: $100(137.9 - 110)/30.8 = 90.6$.

If the viscosity of the oil at 210°F. is not listed in Table 42 or 43, for routine purposes obtain the values of L and H by linear interpolation. Viscosity indices obtained in this manner may be in error by one-half unit.

Since this value is purely empirical and the figures in the tables were obtained from actual determinations on what were considered to be representative samples of oil, it is not unlikely that with some oils values below 0 and above 100 may result.

TABLE 42.—VALUES OF H, L, AND D FOR SAYBOLT UNIVERSAL VISCOSITY*

H = viscosity at 100°F. of an oil of 100 viscosity index

L = viscosity at 100°F. of an oil of 0 viscosity index

Saybolt Universal viscosity at 210°F., seconds	H	L	D $(L-H)$	Saybolt Universal viscosity at 210°F., seconds	H	L	D $(L-H)$	Saybolt Universal viscosity at 210°F., seconds	H	L	D $(L-H)$
40.0	107.1	137.9	30.8	62.5	469.5	876.9	407.4	85.0	888	1,865	977
40.5	113.1	147.3	34.2	63.0	478.4	896.5	418.1	85.5	898	1,890	992
41.0	119.3	157.9	38.6	63.5	487.2	916.2	429.0	86.0	907	1,914	1,007
41.5	125.9	169.4	43.5	64.0	496.1	936.2	440.1	86.5	917	1,939	1,022
42.0	132.7	181.7	49.0	64.5	505.0	955.9	450.9	87.0	927	1,964	1,037
42.5	139.7	194.7	55.0	65.0	514.0	976.1	462.1	87.5	937	1,989	1,052
43.0	146.9	208.4	61.5	65.5	522.9	996.2	473.3	88.0	947	2,014	1,067
43.5	154.2	222.4	68.2	66.0	531.8	1,016.3	484.5	88.5	957	2,039	1,082
44.0	161.5	236.5	75.0	66.5	540.8	1,036.7	495.9	89.0	966	2,064	1,098
44.5	168.9	250.9	82.0	67.0	549.8	1,057.1	507.3	89.5	976	2,090	1,114
45.0	176.3	265.1	88.8	67.5	558.9	1,077.7	518.8	90.0	986	2,115	1,129
45.5	183.8	279.4	95.6	68.0	567.8	1,098.3	530.5	90.5	996	2,141	1,145
46.0	191.3	293.9	102.6	68.5	577.0	1,119.2	542.2	91.0	1,006	2,166	1,160
46.5	198.7	308.2	109.5	69.0	586.1	1,140.1	554.0	91.5	1,016	2,192	1,176
47.0	206.2	322.4	116.2	69.5	595.2	1,161.1	565.9	92.0	1,026	2,217	1,191
47.5	214.0	337.0	123.0	70.0	604	1,182	578	92.5	1,036	2,244	1,208
48.0	222.0	352.3	130.3	70.5	614	1,204	590	93.0	1,046	2,270	1,224
48.5	230.3	368.5	138.2	71.0	623	1,225	602	93.5	1,056	2,296	1,240
49.0	239.0	386.2	147.2	71.5	632	1,247	615	94.0	1,066	2,322	1,256
49.5	247.2	404.2	157.0	72.0	641	1,268	627	94.5	1,077	2,349	1,272
50.0	255.1	422.0	166.9	72.5	651	1,290	639	95.0	1,087	2,375	1,288
50.5	263.4	439.2	175.8	73.0	660	1,311	651	95.5	1,097	2,402	1,305
51.0	271.7	456.1	184.4	73.5	669	1,333	664	96.0	1,107	2,428	1,321
51.5	280.0	473.1	193.1	74.0	678	1,355	677	96.5	1,118	2,455	1,337
52.0	288.4	490.5	202.1	74.5	688	1,377	689	97.0	1,128	2,481	1,353
52.5	296.8	507.7	210.9	75.0	697	1,399	702	97.5	1,138	2,509	1,371
53.0	305.3	525.2	219.9	75.5	707	1,422	715	98.0	1,148	2,536	1,388
53.5	313.8	543.0	229.2	76.0	716	1,444	728	98.5	1,158	2,564	1,406
54.0	322.2	560.5	238.3	76.5	725	1,468	743	99.0	1,168	2,591	1,423
54.5	330.7	578.1	247.4	77.0	734	1,489	755	99.5	1,179	2,619	1,440
55.0	339.2	596.0	256.8	77.5	744	1,512	768	100.0	1,189	2,646	1,457
55.5	347.8	614.1	266.3	78.0	753	1,534	781	101.0	1,210	2,701	1,491
56.0	356.4	632.3	275.9	78.5	763	1,557	794	102.0	1,231	2,757	1,526
56.5	364.9	650.2	285.3	79.0	772	1,580	808	103.0	1,252	2,814	1,562
57.0	373.5	668.6	295.1	79.5	782	1,604	822	104.0	1,273	2,870	1,597
57.5	382.2	687.0	304.8	80.0	791	1,627	836	105.0	1,294	2,928	1,634
58.0	390.8	705.5	314.7	80.5	801	1,651	850	106.0	1,315	2,985	1,670
58.5	399.4	723.9	324.5	81.0	810	1,674	864	107.0	1,337	3,043	1,706
59.0	408.2	742.8	334.6	81.5	820	1,698	878	108.0	1,358	3,102	1,744
59.5	416.9	761.7	344.8	82.0	829	1,721	892	109.0	1,379	3,161	1,782
60.0	425.6	780.6	355.0	82.5	839	1,745	906	110.0	1,401	3,220	1,819
60.5	434.3	799.7	365.4	83.0	849	1,769	920	111.0	1,422	3,280	1,858
61.0	443.1	818.9	375.8	83.5	859	1,793	934	112.0	1,444	3,340	1,896
61.5	451.9	838.1	386.2	84.0	868	1,817	949	113.0	1,466	3,400	1,934
62.0	460.6	857.4	396.8	84.5	878	1,841	963	114.0	1,488	3,462	1,974

* Method for Calculating Viscosity Index ASTM D 567-53.

TABLE 42.—VALUES OF H, L, AND D FOR SAYBOLT UNIVERSAL VISCOSITY.*—(*Continued*)

Saybolt Universal viscosity at 210°F., seconds	H	L	D $(L-H)$	Saybolt Universal viscosity at 210°F., seconds	H	L	D $(L-H)$	Saybolt Universal viscosity at 210°F., seconds	H	L	D $(L-H)$
115	1,510	3,523	2,013	160	2,580	6,740	4,160	205	3,816	10,831	7,015
116	1,532	3,585	2,053	161	2,606	6,821	4,215	206	3,845	10,931	7,086
117	1,554	3,648	2,094	162	2,631	6,903	4,272	207	3,874	11,033	7,159
118	1,576	3,711	2,135	163	2,657	6,985	4,328	208	3,904	11,134	7,230
119	1,598	3,774	2,176	164	2,683	7,068	4,385	209	3,933	11,237	7,304
120	1,620	3,838	2,218	165	2,709	7,151	4,442	210	3,963	11,339	7,376
121	1,643	3,902	2,259	166	2,735	7,234	4,499	211	3,993	11,442	7,449
122	1,665	3,966	2,301	167	2,761	7,318	4,557	212	4,023	11,546	7,523
123	1,688	4,031	2,343	168	2,788	7,403	4,615	213	4,053	11,649	7,596
124	1,710	4,097	2,387	169	2,814	7,488	4,674	214	4,083	11,754	7,671
125	1,733	4,163	2,430	170	2,840	7,573	4,733	215	4,113	11,858	7,745
126	1,756	4,229	2,473	171	2,867	7,659	4,792	216	4,143	11,964	7,821
127	1,779	4,296	2,517	172	2,893	7,745	4,852	217	4,173	12,069	7,896
128	1,802	4,363	2,561	173	2,920	7,832	4,912	218	4,203	12,175	7,972
129	1,825	4,430	2,605	174	2,947	7,919	4,972	219	4,234	12,282	8,048
130	1,848	4,498	2,650	175	2,974	8,006	5,032	220	4,264	12,389	8,125
131	1,871	4,567	2,696	176	3,000	8,094	5,094	221	4,295	12,496	8,201
132	1,894	4,636	2,742	177	3,027	8,182	5,155	222	4,325	12,604	8,279
133	1,918	4,705	2,787	178	3,054	8,271	5,217	223	4,356	12,712	8,356
134	1,941	4,775	2,834	179	3,082	8,360	5,278	224	4,387	12,820	8,433
135	1,965	4,845	2,880	180	3,109	8,450	5,341	225	4,418	12,930	8,512
136	1,988	4,915	2,927	181	3,136	8,540	5,404	226	4,449	13,039	8,590
137	2,012	4,986	2,974	182	3,163	8,630	5,467	227	4,480	13,149	8,669
138	2,036	5,058	3,022	183	3,191	8,721	5,530	228	4,511	13,259	8,748
139	2,060	5,130	3,070	184	3,218	8,813	5,595	229	4,542	13,370	8,828
140	2,084	5,202	3,118	185	3,246	8,904	5,658	230	4,574	13,481	8,907
141	2,108	5,275	3,167	186	3,274	8,997	5,723	231	4,605	13,593	8,988
142	2,132	5,348	3,216	187	3,302	9,089	5,787	232	4,636	13,705	9,069
143	2,156	5,422	3,266	188	3,329	9,182	5,853	233	4,668	13,818	9,150
144	2,180	5,496	3,316	189	3,357	9,276	5,919	234	4,700	13,930	9,230
145	2,205	5,570	3,365	190	3,385	9,370	5,985	235	4,731	14,044	9,313
146	2,229	5,645	3,416	191	3,414	9,464	6,050	236	4,753	14,158	9,395
147	2,254	5,721	3,467	192	3,442	9,559	6,117	237	4,795	14,272	9,477
148	2,278	5,796	3,518	193	3,470	9,654	6,184	238	4,827	14,387	9,560
149	2,303	5,873	3,570	194	3,498	9,750	6,252	239	4,859	14,502	9,643
150	2,328	5,949	3,621	195	3,527	9,846	6,319	240	4,891	14,617	9,726
151	2,353	6,026	3,673	196	3,555	9,942	6,387	241	4,923	14,733	9,810
152	2,378	6,104	3,726	197	3,584	10,039	6,455	242	4,955	14,850	9,895
153	2,403	6,182	3,779	198	3,613	10,137	6,524	243	4,988	14,966	9,978
154	2,428	6,260	3,832	199	3,641	10,235	6,594	244	5,020	15,084	10,064
155	2,453	6,339	3,886	200	3,670	10,333	6,663	245	5,053	15,201	10,148
156	2,478	6,418	3,940	201	3,699	10,431	6,732	246	5,085	15,319	10,234
157	2,503	6,498	3,995	202	3,728	10,531	6,803	247	5,118	15,438	10,320
158	2,529	6,578	4,049	203	3,757	10,630	6,873	248	5,151	15,557	10,406
159	2,554	6,659	4,105	204	3,786	10,730	6,944	249	5,184	15,676	10,492

* Method for Calculating Viscosity Index ASTM D 567-53.

TABLE 42.—VALUES OF *H*, *L*, AND *D* FOR SAYBOLT UNIVERSAL VISCOSITY.*—(*Continued*)

Saybolt Universal viscosity at 210°F., seconds	*H*	*L*	*D* (*L*−*H*)	Saybolt Universal viscosity at 210°F., seconds	*H*	*L*	*D* (*L*−*H*)	Saybolt Universal viscosity at 210°F., seconds	*H*	*L*	*D* (*L*−*H*)
250	5,217	15,796	10,579	285	6,420	20,263	13,843	320	7,724	25,260	17,536
251	5,250	15,917	10,667	286	6,456	20,399	13,943	321	7,763	25,410	17,647
252	5,283	16,037	10,754	287	6,492	20,535	14,043	322	7,802	25,561	17,759
253	5,316	16,158	10,842	288	6,528	20,671	14,143	323	7,841	25,712	17,871
254	5,349	16,280	10,931	289	6,564	20,808	14,244	324	7,880	25,864	17,984
255	5,382	16,402	11,020	290	6,601	20,945	14,344	325	7,919	26,017	18,098
256	5,416	16,524	11,108	291	6,637	21,082	14,445	326	7,958	26,169	18,211
257	5,449	16,647	11,198	292	6,673	21,220	14,547	327	7,997	26,322	18,325
258	5,483	16,771	11,288	293	6,710	21,359	14,649	328	8,036	26,476	18,440
259	5,517	16,894	11,377	294	6,746	21,498	14,752	329	8,076	26,630	18,554
260	5,550	17,019	11,469	295	6,783	21,637	14,854	330	8,115	26,784	18,669
261	5,584	17,143	11,559	296	6,819	21,777	14,958	331	8,155	26,939	18,784
262	5,618	17,268	11,650	297	6,856	21,917	15,061	332	8,194	27,094	18,900
263	5,652	17,394	11,742	298	6,893	22,057	15,164	333	8,234	27,250	19,016
264	5,686	17,520	11,834	299	6,930	22,198	15,268	334	8,274	27,406	19,132
265	5,720	17,646	11,926	300	6,967	22,340	15,373	335	8,314	27,563	19,249
266	5,755	17,773	12,018	301	7,004	22,482	15,478	336	8,354	27,720	19,366
267	5,789	17,900	12,111	302	7,041	22,624	15,583	337	8,394	27,877	19,483
268	5,823	18,028	12,205	303	7,079	22,767	15,688	338	8,434	28,035	19,601
269	5,858	18,156	12,298	304	7,116	22,910	15,794	339	8,474	28,193	19,719
270	5,892	18,284	12,392	305	7,153	23,054	15,901	340	8,514	28,352	19,838
271	5,927	18,413	12,486	306	7,191	23,198	16,007	341	8,555	28,511	19,956
272	5,962	18,542	12,580	307	7,228	23,342	16,114	342	8,595	28,671	20,076
273	5,996	18,672	12,676	308	7,266	23,487	16,221	343	8,636	28,831	20,195
274	6,031	18,802	12,771	309	7,304	23,632	16,328	344	8,676	28,991	20,315
275	6,066	18,933	12,867	310	7,342	23,778	16,436	345	8,717	29,152	20,435
276	6,101	19,064	12,963	311	7,379	23,924	16,545	346	8,758	29,314	20,556
277	6,136	19,196	13,060	312	7,417	24,071	16,654	347	8,798	29,475	20,677
278	6,172	19,328	13,156	313	7,456	24,218	16,762	348	8,839	29,638	20,799
279	6,207	19,460	13,253	314	7,494	24,366	16,872	349	8,880	29,800	20,920
280	6,242	19,593	13,351	315	7,532	24,513	16,981	350	8,921	29,963	21,042
281	6,278	19,726	13,448	316	7,570	24,662	17,092				
282	6,313	19,860	13,547	317	7,609	24,808	17,199				
283	6,349	19,994	13,645	318	7,647	24,960	17,313				
284	6,385	20,128	13,743	319	7,686	25,110	17,424				

* **Method for Calculating Viscosity Index ASTM D 567-53.**

TABLE 43.—VALUES OF H, L, AND D FOR KINEMATIC VISCOSITY

H = viscosity at 100°F. of an oil of 100 viscosity index

L = viscosity at 100°F. of an oil of 0 viscosity index

Kinematic viscosity at 210°F., centistokes	H	L	D $(L-H)$	Kinematic viscosity at 210°F., centistokes	H	L	D $(L-H)$	Kinematic viscosity at 210°F., centistokes	H	L	D $(L-H)$
2.00	6.620	8.360	1.740	7.50	57.669	96.528	38.859	13.00	131.153	256.690	125.537
2.10	7.143	9.043	1.900	7.60	58.873	98.958	40.085	13.10	132.658	260.180	127.522
2.20	7.684	9.752	2.068	7.70	60.063	101.398	41.335	13.20	134.166	263.684	129.518
2.30	8.243	10.485	2.242	7.80	61.305	103.925	42.620	13.30	135.716	267.293	131.577
2.40	8.821	11.244	2.423	7.90	62.513	106.388	43.875	13.40	137.230	280.824	133.594
2.50	9.417	12.028	2.611	8.00	63.723	108.859	45.136	13.50	138.745	274.369	135.624
2.60	10.031	12.838	2.807	8.10	64.969	111.419	46.450	13.60	140.270	277.964	137.694
2.70	10.664	13.672	3.008	8.20	66.251	114.067	47.816	13.70	141.784	281.498	139.714
2.80	11.315	14.532	3.217	8.30	67.501	116.650	49.149	13.80	143.348	285.177	141.829
2.90	11.984	15.417	3.433	8.40	68.753	119.306	50.553	13.90	144.874	288.776	143.902
3.00	12.671	16.328	3.657	8.50	70.041	121.926	51.885	14.00	146.402	292.388	145.986
3.10	13.377	17.263	3.886	8.60	71.296	124.528	53.232	14.10	147.933	296.014	148.081
3.20	14.101	18.224	4.123	8.70	72.542	127.153	54.611	14.20	149.507	299.749	150.242
3.30	14.843	19.210	4.367	8.80	73.793	129.786	55.993	14.30	151.043	303.402	152.359
3.40	15.603	20.222	4.619	8.90	75.089	132.515	57.426	14.40	152.582	307.069	154.487
3.50	16.382	21.258	4.876	9.00	76.352	135.176	58.824	14.50	154.124	310.749	156.625
3.60	17.179	22.320	5.141	9.10	77.617	137.841	60.224	14.60	155.708	314.540	158.832
3.70	17.994	23.407	5.413	9.20	78.880	140.517	61.637	14.70	157.255	318.247	160.992
3.80	18.828	24.520	5.692	9.30	80.184	143.284	63.100	14.80	158.804	321.968	163.164
3.90	19.680	25.657	5.977	9.40	81.448	145.989	64.541	14.90	160.396	325.801	165.405
4.00	20.550	26.820	6.270	9.50	82.714	148.695	65.981	15.00	161.950	329.549	167.599
4.10	21.400	28.060	6.660	9.60	83.986	151.411	67.425	15.10	163.548	333.410	169.862
4.20	22.280	29.360	7.080	9.70	85.262	154.147	68.885	15.20	165.190	337.385	172.195
4.30	23.180	30.730	7.550	9.80	86.575	156.982	70.407	15.30	166.793	341.275	174.482
4.40	24.100	32.180	8.080	9.90	87.856	159.722	71.866	15.40	168.399	345.179	176.780
4.50	25.040	33.720	8.680	10.00	89.178	162.494	73.316	15.50	170.007	348.881	178.874
4.60	26.000	35.350	9.350	10.10	90.458	165.361	74.903	15.60	171.660	353.131	181.471
4.70	26.980	37.060	10.080	10.20	91.814	168.303	76.489	15.70	173.274	357.078	183.804
4.80	27.980	38.840	10.860	10.30	93.128	171.194	78.066	15.80	174.891	361.039	186.148
4.90	29.000	40.680	11.680	10.40	94.461	174.075	79.614	15.90	176.552	365.117	188.565
5.00	30.040	42.570	12.530	10.50	95.825	177.068	81.243	16.00	178.174	369.107	190.933
5.10	31.090	44.500	13.410	10.60	97.152	179.980	82.828	16.10	179.841	373.214	193.373
5.20	32.150	46.460	14.310	10.70	98.492	182.907	84.415	16.20	181.552	377.439	195.887
5.30	33.210	48.440	15.230	10.80	99.818	185.849	86.031	16.30	183.224	381.577	198.353
5.40	34.270	50.430	16.160	10.90	101.206	188.873	87.667	16.40	184.942	385.729	200.787
5.50	35.330	52.430	17.100	11.00	102.537	191.848	89.311	16.50	186.577	389.897	203.320
5.60	36.390	54.430	18.040	11.10	103.909	194.899	90.990	16.60	188.300	394.184	205.884
5.70	37.450	56.430	18.980	11.20	105.305	197.966	92.661	16.70	189.983	398.381	208.398
5.80	38.510	58.430	19.920	11.30	106.721	201.150	94.429	16.80	191.670	402.594	210.924
5.90	39.570	60.430	20.860	11.40	108.100	204.238	96.138	16.90	193.401	406.929	213.528
6.00	40.630	62.430	21.800	11.50	109.493	207.339	97.846	17.00	195.094	411.172	216.078
6.10	41.690	64.430	22.740	11.60	110.887	210.468	99.581	17.10	196.831	415.537	218.706
6.20	42.750	66.430	23.680	11.70	112.273	213.601	101.328	17.20	198.571	419.917	221.346
6.30	43.810	68.430	24.620	11.80	113.711	216.829	103.118	17.30	200.357	424.421	224.064
6.40	44.880	70.430	25.550	11.90	115.114	219.981	104.867	17.40	202.104	428.832	226.728
6.50	45.970	72.460	26.490	12.00	116.507	223.145	106.638	17.50	203.853	433.260	229.407
6.60	47.080	74.550	27.470	12.10	117.948	226.412	108.464	17.60	205.605	437.704	232.099
6.70	48.220	76.740	28.520	12.20	119.438	229.784	110.346	17.70	207.361	442.161	234.800
6.80	49.390	79.040	29.650	12.30	120.883	233.078	112.195	17.80	209.162	446.745	237.583
6.90	50.590	81.440	30.850	12.40	122.330	236.392	114.062	17.90	210.923	451.237	240.314
7.00	51.820	83.920	32.100	12.50	123.781	239.713	115.932	18.00	212.687	455.743	243.056
7.10	52.980	86.460	33.480	12.60	125.274	243.136	117.862	18.10	214.350	460.266	245.916
7.20	54.150	89.040	34.890	12.70	126.730	246.482	119.752	18.20	216.268	464.915	248.647
7.30	55.273	91.660	36.387	12.80	128.189	249.842	121.653	18.30	218.042	469.469	251.427
7.40	56.473	94.095	37.622	12.90	129.689	253.305	123.616	18.40	219.818	474.039	254.221

TABLE 43.—VALUES OF *H*, *L*, AND *D* FOR KINEMATIC VISCOSITY.—(*Continued*)

Kinematic viscosity at 210°F., centistokes	*H*	*L*	*D* ($L-H$)	Kinematic viscosity at 210°F., centistokes	*H*	*L*	*D* ($L-H$)	Kinematic viscosity at 210°F., centistokes	*H*	*L*	*D* ($L-H$)
18.50	221.597	478.625	257.028	28.00	415.1	1,018.4	603.3	52.5	1,090.9	3,280.4	2,189.5
18.60	223.423	483.339	259.916	28.20	419.7	1,032.0	612.3	53.0	1,107.1	3,339.0	2,231.9
18.70	225.208	487.956	262.748	28.40	424.3	1,045.6	621.3	53.5	1,123.9	3,399.9	2,276.0
18.80	226.996	492.590	265.594	28.60	428.9	1,059.4	630.5	54.0	1,140.6	3,460.8	2,320.2
18.90	228.831	497.352	268.521	28.80	433.5	1,073.2	639.7	54.5	1,157.2	3,521.4	2,364.2
19.00	230.625	502.017	271.392	29.00	438.1	1,087.0	648.9	55.0	1,173.7	3,582.0	2,408.3
19.10	232.466	506.812	274.346	29.20	442.8	1,101.0	658.2	55.5	1,190.5	3,643.6	2,453.1
19.20	234.354	511.739	277.385	29.40	446.9	1,113.5	666.6	56.0	1,207.2	3,705.2	2,498.0
19.30	236.201	516.568	280.367	29.60	451.6	1,127.6	676.0	56.5	1,224.2	3,767.8	2,543.6
19.40	238.052	521.413	283.361	29.80	456.2	1,141.8	685.6	57.0	1,241.1	3,830.4	2,589.3
19.50	239.906	526.274	286.368	30.0	460.9	1,156.0	695.1	57.5	1,258.6	3,895.4	2,636.8
19.60	241.806	531.247	289.441	30.5	472.8	1,192.0	719.2	58.0	1,276.1	3,960.4	2,684.3
19.70	243.666	536.164	292.498	31.0	484.1	1,226.8	742.7	58.5	1,293.5	4,025.0	2,731.5
19.80	245.529	541.075	295.546	31.5	496.1	1,263.7	767.6	59.0	1,310.8	4,089.6	2,778.8
19.90	247.440	546.120	298.680	32.0	508.2	1,301.1	792.9	59.5	1,328.3	4,155.2	2,826.9
20.00	249.31	551.07	301.76	32.5	520.4	1,338.9	818.5	60.0	1,345.8	4,220.7	2,874.9
20.20	253.10	561.12	308.02	33.0	532.6	1,377.2	844.6	60.5	1,363.9	4,288.8	2,924.9
20.40	256.86	571.13	314.27	33.5	544.9	1,416.0	871.1	61.0	1,382.0	4,356.8	2,974.8
20.60	260.59	581.08	320.49	34.0	557.3	1,455.3	898.0	61.5	1,399.9	4,424.4	3,024.5
20.80	264.64	591.94	327.30	34.5	569.9	1,495.0	925.1	62.0	1,417.7	4,492.0	3,074.3
21.00	268.26	601.66	333.40	35.0	582.4	1,535.2	952.8	62.5	1,435.8	4,560.6	3,124.8
21.20	272.35	612.67	340.32	35.5	595.8	1,577.7	981.9	63.0	1,453.9	4,629.1	3,175.2
21.40	275.99	622.52	346.53	36.0	608.5	1,618.9	1,010.4	63.5	1,472.6	4,700.2	3,227.6
21.60	280.10	633.67	353.57	36.5	621.4	1,660.6	1,039.2	64.0	1,491.2	4,771.3	3,280.1
21.80	284.22	644.89	360.67	37.0	634.3	1,702.7	1,068.4	64.5	1,509.7	4,841.9	3,332.2
22.00	287.90	654.94	367.04	37.5	647.4	1,745.3	1,097.9	65.0	1,528.1	4,912.4	3,384.3
22.20	292.05	666.30	374.25	38.0	660.5	1,788.3	1,127.8	65.5	1,546.8	4,984.0	3,437.2
22.40	296.22	677.75	381.53	38.5	674.4	1,833.9	1,159.5	66.0	1,565.4	5,055.5	3,490.1
22.60	299.96	687.98	388.02	39.0	687.7	1,877.9	1,190.2	66.5	1,584.7	5,129.7	3,545.0
22.80	304.13	699.57	395.44	39.5	701.1	1,922.4	1,221.3	67.0	1,603.9	5,203.8	3,599.9
23.00	308.34	711.24	402.90	40.0	714.6	1,967.4	1,252.8	67.5	1,622.9	5,277.4	3,654.5
23.20	312.09	721.67	409.58	40.5	728.3	2,013.1	1,284.8	68.0	1,641.9	5,350.9	3,709.0
23.40	316.32	733.47	417.15	41.0	741.9	2,058.7	1,316.8	68.5	1,661.1	5,425.5	3,764.4
23.60	320.57	745.35	424.78	41.5	756.1	2,106.4	1,350.3	69.0	1,680.3	5,500.0	3,819.7
23.80	324.36	755.98	431.62	42.0	770.2	2,154.1	1,383.9	69.5	1,700.1	5,577.2	3,877.1
24.00	328.63	768.00	439.37	42.5	784.2	2,201.7	1,417.5	70.0	1,719.9	5,654.4	3,934.5
24.20	332.45	778.76	446.31	43.0	798.2	2,249.3	1,451.1	70.5	1,739.5	5,731.0	3,991.5
24.40	336.75	790.92	454.17	43.5	812.7	2,299.0	1,486.3	71.0	1,759.1	5,807.5	4,048.4
24.60	341.05	803.17	462.12	44.0	827.2	2,348.6	1,521.4	71.5	1,779.3	5,886.8	4,107.5
24.80	345.40	815.49	470.09	44.5	842.2	2,400.4	1,558.2	72.0	1,799.5	5,966.0	4,166.5
25.00	349.3	826.5	477.2	45.0	857.2	2,452.1	1,594.9	72.5	1,819.5	6,044.6	4,225.1
25.20	353.6	839.0	485.4	45.5	872.1	2,503.8	1,631.7	73.0	1,839.4	6,123.1	4,283.7
25.40	358.0	851.5	493.5	46.0	886.9	2,555.4	1,668.5	73.5	1,860.0	6,204.4	4,344.4
25.60	362.4	864.1	501.7	46.5	902.0	2,608.1	1,706.1	74.0	1,880.6	6,285.6	4,405.0
25.80	366.8	876.8	510.0	47.0	917.1	2,660.7	1,743.6	74.5	1,901.0	6,366.2	4,465.2
26.00	371.2	889.6	518.4	47.5	932.7	2,715.5	1,782.8	75.0	1,921.3	6,446.7	4,525.4
26.20	375.1	901.0	525.9	48.0	948.2	2,770.3	1,822.1				
26.40	379.6	913.9	534.3	48.5	963.7	2,825.0	1,861.3				
26.60	384.0	926.9	542.9	49.0	979.1	2,879.6	1,900.5				
26.80	388.5	939.9	551.4	49.5	994.8	2,935.2	1,940.4				
27.00	393.0	953.1	560.1	50.0	1,010.4	2,990.8	1,980.4				
27.20	397.5	966.3	568.8	50.5	1,026.2	3,047.4	2,021.2				
27.40	402.0	979.6	577.6	51.0	1,042.0	3,104.0	2,062.0				
27.60	406.0	991.4	585.4	51.5	1,058.4	3,162.9	2,104.5				
27.80	410.6	1,004.9	594.3	52.0	1,074.7	3,221.8	2,147.1				

Steam Emulsion Test on Lubricating Oils. This method is applied not only to lubricating oils but also to turbine and other oils when an emulsion, demulsibility, or emulsification test is required. The *steam emulsion number* is defined as the number of seconds required for an oil to separate when emulsified and separated under definitely prescribed conditions.

The apparatus employed is shown in Fig. 148. The steam generator, of not less than 1 liter capacity, is made of metal or glass and is fitted with three outlets as shown. The emulsifying and separating baths must be of glass and have a capacity of 3.0 to 3.5 liters and a depth of 19 to 23 cm. Beakers or good-quality battery jars are suitable. Wooden or metal covers are provided to hold the oil container and thermometers. Floating-type thermometers are useful but not essential.

The oil container *H* is a test tube 200 mm. long and of about 23 mm. internal diameter with a wall thickness of 0.8 to 1.4 mm. This test tube is graduated from 0 to 50 ml. or 10 to 50 ml. in milliliters, each 5-ml. line encircling the tube, and the distance from the 10-ml. line to the 50-ml. line must be 93 to 103 mm.

The steam pipe *C* is a glass tube of inside diameter 2.3 to 2.7 mm., wall thickness 0.8 to 1.1 mm., and 140 mm. length. It is bent at right angles 60 mm. from the discharge end. The steam delivery tube *I* is a glass tube 305 mm. long, of the same dimensions as the steam pipe. At the discharge end it is cut diagonally at an angle of 30 degrees with the axis of the tube and is bent at right angles 250 to 260 mm. from the discharge end. The steam pipe is connected to the steam delivery tube by a rubber tube 65 mm. long, with a distance of about 40 mm. between the ends of the glass tubes.

PROCEDURE. Assemble the apparatus as shown in Fig. 148. Fill the steam generator half full with water, and heat. Fill the baths with 3 liters ± 60 ml. of water. Raise the temperature in the separatory bath to 200 to 203°F., and maintain it at that point. When battery jars are used, do not heat by a bare flame but use an immersion heater or pass in live steam through the line *G*. The emulsifying bath should be at a temperature of 67 to 78°F. at the start of the test but is not controlled thereafter.

Place 20 ml. of the sample in the oil container at room temperature, and place the container in the holder of the emulsifying bath. Stopper the test tube with a two-hole cork, one hole of which contains a thermometer so adjusted that the bottom of the bulb is 20 to 25 mm. from the bottom of the oil container. Through the other hole passes the steam delivery tube, which is connected to the steam pipe with rubber tubing and screw pinchcocks placed as shown. The apparatus, particularly the oil container, the thermometer in the latter, and the steam pipe and steam delivery tube must be chemically clean; the water in the steam generator must also be clean.

Steam out the delivery tube until all condensation disappears; then disconnect it from the rubber tubing, and insert it through the second opening in the cork until the end of the tube touches the center of the

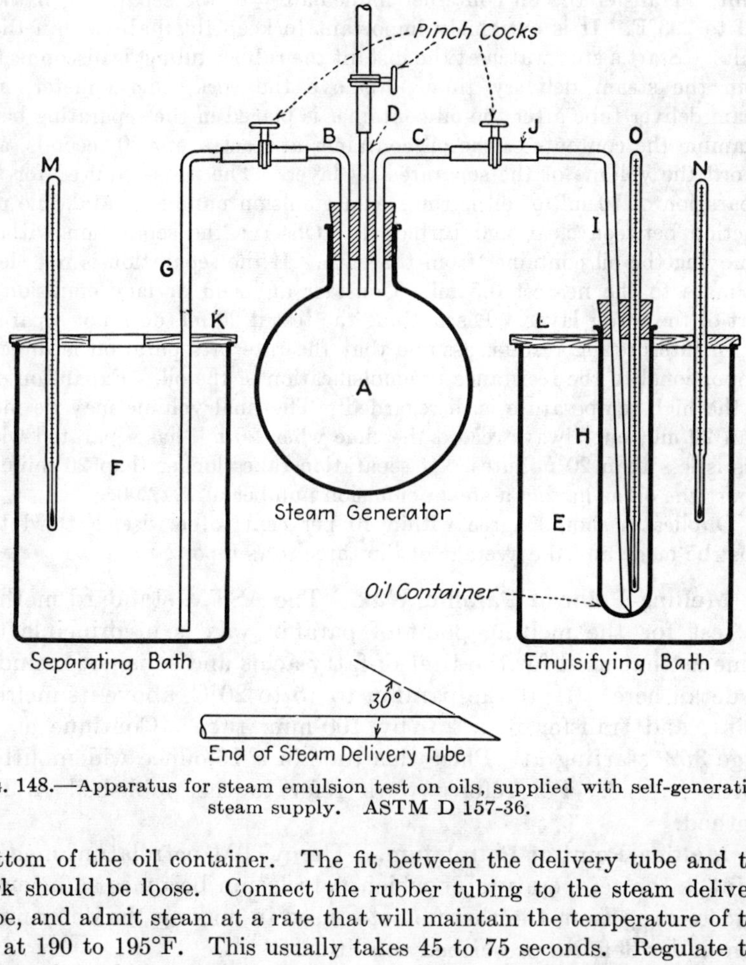

FIG. 148.—Apparatus for steam emulsion test on oils, supplied with self-generating steam supply. ASTM D 157-36.

bottom of the oil container. The fit between the delivery tube and the cork should be loose. Connect the rubber tubing to the steam delivery tube, and admit steam at a rate that will maintain the temperature of the oil at 190 to 195°F. This usually takes 45 to 75 seconds. Regulate the steam through manipulation of the pinchcocks on the steam pipe and steam exhaust. The steam supply must be sufficient at all times to cause a generous discharge from the exhaust. Continue steaming until the volume of the condensed water and oil is 40 ± 3 ml. The apparent volume in the tube near the end of the steaming operation is approximately 12 to 15 ml. greater than the actual volume, owing to displacement caused by steam, the thermometer, and the steam delivery tube. The time required for this operation should be 4 to 6.5 minutes, depending on the quality of the oil, altitude, etc. If the steam is wet and the condensed water is more than 20 ml. in less than 4 minutes, rerun the test.

Disconnect the rubber tubing from the steam delivery tube at this point. Transfer the oil container immediately to the separating bath at 200 to 203°F. It is extremely important to keep this bath within these limits. Start a stop watch at the instant the rubber tubing is disconnected from the steam delivery tube. Remove the cork, thermometer, and steam deliver tube after the oil container is placed in the separating bath. Examine the contents of the oil container at least every 30 seconds, and record the volume of the separated oil layer. The time required for the separation of 20 ml. of oil is the steam emulsion number. Make no distinction between clear and turbid oil. Observe the separation without removing the oil container from the bath. If the separation is not clear, estimate to the nearest 0.5 ml. Consider any cuff or lacy emulsion as part of the water layer. Discontinue the test if 20 ml. does not separate.

In interpreting results, assume that the rate of separation is directly proportional to the resistance to emulsification of the oil. Expansion due to the high temperature is disregarded. The final volume may be more than 20 ml., but always record the time when 20 ml. has separated when this is less than 20 minutes. If separation takes longer than 20 minutes report the oil as having a steam emulsion number of +1,200.

Duplicates should agree within 10 per cent; otherwise, a third test must be made and the average of the three tests reported.

Melting Point of Paraffin Wax. The ASTM standard method of test for the melting point of paraffin wax is in principle the same as that used for the titer of fatty acids and so is not included in detail here. Heat the paraffin to 15 to 20°C. above its melting point, and transfer to a 25- by 100-mm. tube. Continue as on page 352, starting at "Place this tube in a 16-ounce widemouthed bottle . . . ". For referee work refer to the published ASTM method.

Melting Point of Petrolatum. The ASTM petrolatum melting point is the temperature at which petrolatum becomes sufficiently fluid to drop from the thermometer used in making the determination under definite prescribed conditions.

Standard test tubes 25 mm. in outside diameter and 100 mm. in length, supplied with corks grooved at the sides to permit air circulation and bored in the center for the thermometer, are used. The ASTM paraffin-wax melting-point thermometer used is graduated either 38 to 82°C. or 100 to 180°F. The bath thermometer may be of any suitable type, accurate to 2°F. The beaker of at least 500 ml. capacity should permit the immersion of the test tube to a depth of at least 75 mm. and still leave a depth of 15 mm. of water below the bottom of the test tube.

Procedure. Slowly melt a representative sample in a casserole or suitable container until the temperature reaches 200°F. Use a fresh portion of the sample for each determination. Allow the sample to cool to 15°F. above the expected melting point. Chill the thermometer bulb to 40°F., wipe dry, and while still cold thrust into the melted sample so that approximately the lower half is submerged. Withdraw immediately, hold vertically away from the heat until the surface dulls, and then place for 5 minutes in a water bath having a temperature not over 60°F.

Securely fix the thermometer in the test tube by means of a cork so that the lowest point is 15 mm. above the bottom of the test tube. Surround the latter with a water bath at a temperature of 60°F. Raise the temperature at a rate of 3°F. per minute until the first drop of petrolatum leaves the thermometer.

If the variation of three such determinations does not exceed 2°F., report the average of the three determinations as the melting point of the sample. Should the variation be greater than 2°F., make two more determinations and report the average of the five as the melting point.

Water in Petroleum Products and Other Bituminous Material. This test may be used for a variety of materials but is especially applicable to petroleum, fuel oil, road oil, coal tar, water-gas tar, coke-oven tar, and other petroleum products or bituminous materials.

The method, essentially, and the apparatus employed are those of the general distillation method described on page 665. For road materials and tars a metal still, preferably of copper and with a clamped-on head, is generally used because it can be more readily cleaned than glass.

The solvent employed in this test is gasoline, free from water. The ASTM requirements are that 5 per cent shall distill at a temperature of 194 to 212°F. and 90 per cent shall distill below 410°F. This solvent is for use in testing petroleum or bituminous materials derived from petroleum. For testing bituminous materials derived from coal tar, water gas, etc., coal-tar naphtha or a light oil is used of which 98 per cent distills at 284 to 482°F.

Procedure. Measure 100 ml. of sample in a 100-ml. graduate, and transfer to the distillation flask. Rinse out the graduate with 50 ml. of the solvent and then with two 25-ml. portions of the solvent, draining well each time. If the sample contains less than 10 per cent of water, use a trap graduated from 0 to 10 ml. in 0.1-ml. divisions. With samples containing more than 10 per cent of water use a 10-ml. trap and proportionately less of the sample; in special cases in which the water content is more than 10 per cent and an accurate sample cannot be obtained with less than 100 ml., use a distilling receiver graduated from 0 to 25 ml.

Connect the apparatus, and distill at the rate of 2 to 5 drops per second. Continue the distillation until no further water comes over. This operation usually requires less than an hour. Remove a persistent ring of condensed water in the condenser tube by increasing the rate of distillation for a few minutes. Duplicate determinations by this method should differ from each other by not more than one division on the trap.

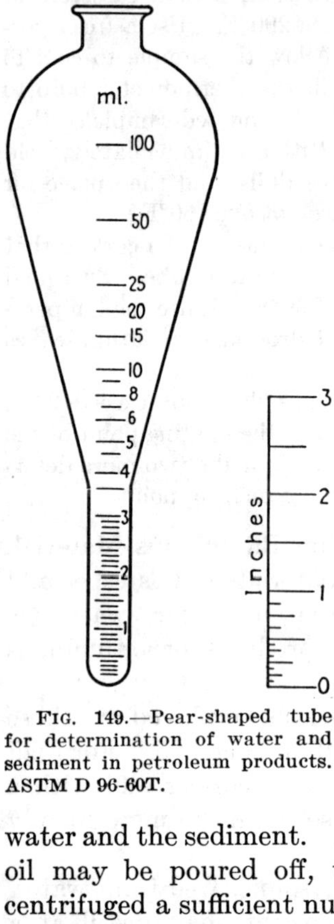

FIG. 149.—Pear-shaped tube for determination of water and sediment in petroleum products. ASTM D 96-60T.

Water and Sediment in Petroleum Products by the Centrifuge. This test is used to determine water and sediment in crude mineral oils and fuel oils. It is not entirely satisfactory, for the amount of water obtained is nearly always lower than the actual water content. Because of its simplicity it is widely used. Accurate water estimation may be made by the Standard Test immediately preceding. It must be remembered that the total volume read in the centrifuge tube is that of the water and the sediment. If desired, the benzene used to dissolve the oil may be poured off, the sediment washed with benzene and centrifuged a sufficient number of times to get rid of all the oil, and the sediment dried at 110°C. and weighed.

The centrifuge and one form of centrifuge tube used are similar to those in determining the precipitation number of lubricating oils. A pear-shaped tube used as an alternative is shown in Fig. 149.

PROCEDURE. The difficulties in obtaining representative samples for this test are great; hence, it is important to observe the utmost care in sampling. Measure 50 ml. of benzene into each of two centrifuge tubes, and then add exactly 50 ml. of the oil to each. Tightly stopper the tubes, and thoroughly mix the contents. Loosen the stoppers and immerse the tubes for 10 minutes in a bath maintained at 120°F. Centrifuge the two tubes exactly as described for the precipitation number. Read the volume of water and sediment at the bottom of the tubes to the nearest 0.1 ml. Replace the tubes, and centrifuge again for 10 minutes. Repeat this process, if necessary, until the volume of water and sediment remains constant. The sum of the two readings, equivalent to the value on a 100-ml. sample, is the percentage of water and sediment. Duplicate determinations should differ by not more than 0.2 ml. when accurately calibrated tubes are used.

Ash. Weigh approximately 20 grams of the sample into a previously heated and weighed platinum, porcelain, or silica wide-form crucible. Avoid platinum if lead, zinc, or other similar metals are present. To prevent foaming due to the presence of water, add 1 to 2 ml. of absolute alcohol before heating. Gently heat until the oil can be ignited at the surface. Remove the burner, and allow the oil to burn completely. Rid the crucible of all carbonaceous matter, using a hot flame or a muffle furnace. Cool, weigh, and report as percentage ash. Duplicate determinations should agree within 10 per cent.

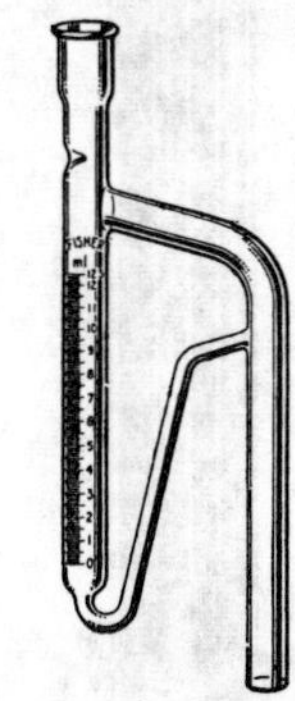

FIG. 150.—Apparatus for measuring dilution of crankcase oils. ASTM D 322-58T.

Gravity. HYDROMETER METHOD. This method is for use with crude petroleum and petroleum products normally handled as liquids, but not for mixtures with other substances or for road oils, tars, asphalt cements, and soft-tar pitches.

Using a specially graduated API hydrometer, determine the gravity at 60°F. as described on page 46. No officially recognized tables are available to convert gravities from one temperature to another; hence, determination at 60°F. is imperative. The API gravity can be calculated from the specific gravity at 60°F. according to the formula on page 47.

Duplicate determinations should agree within 0.1° API or among separate laboratories using different hydrometers within 0.5° API.

Dilution of Crankcase Oils. This method is used to determine the amount of dilution in crankcase oils of engines when gasoline has been used as fuel.

The apparatus consists of a 1-liter short-necked, round-bottomed flask connected to a trap, shown in Fig. 150, designed to collect and measure the diluent, returning the water to the flask. The trap is in turn connected to a reflux condenser.

PROCEDURE. Mix the sample thoroughly, and measure 25 ml. in a 25-ml. graduate. Transfer this to the flask, and wash out the graduate into the flask with hot water. Add 500 ml. of water to the flask, and assemble the apparatus. Apply heat, using the full flame of a Meker-type burner. Adjust so that refluxing starts within 7 to 10 minutes after the heat is applied, the water initially being at room temperature. Boiling must be continuous and vigorous enough to prevent the oil from forming, even momentarily, a continuous film on top of the water. To reduce bumping with dirty oils, add to the flask broken glass, steel wool, or 5 ml. of concentrated hydrochloric acid.

Make readings of the amount of diluent at 5, 15, and 30 minutes from the time refluxing starts and each 15 minutes thereafter until the test is complete. The test is considered complete either when the volume of diluent increases by not more than 0.1 ml. in any 15-minute period during the test or when the volume of diluent obtained in a given time is no greater than the following: 5 minutes—0.1 ml.; 30 minutes—2.0 ml.; 60 minutes—4.0 ml.; 90 minutes—7.0 ml. If only 0.1 ml. of diluent collects, report as "no dilution."

When the test is complete by any of the above criteria, remove the heat. Allow the cylinder to stand for at least 15 minutes, and read the volume of diluent. Multiply this by 4 to obtain the percentage dilution.

Determinations by independent operators should agree within 20 per cent for dilutions under 5 per cent and within 10 per cent for dilutions over 5 per cent. Duplicate determinations by the same operator should agree more closely than this.

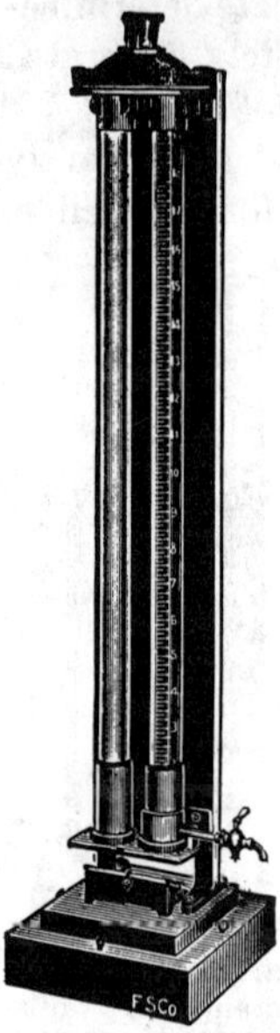

FIG. 151.—Saybolt chromometer for reading colors of refined petroleum oils. ASTM D 156-53T.

Color of Petroleum Products. For determining the color of refined oils such as naphthas and kerosene, the Saybolt chromometer shown in Fig. 151 is used. The color standards consist of calibrated glass disks.

The color of lubricating oils and petrolatum is read with the ASTM Union colorimeter (Fig. 152) which employs Lovibond red, yellow, and blue glasses. The resulting NPA numerical designation of color is often referred to as ASTM color.

IODINE METHOD. Although not official with the ASTM, this method is useful when colorimeters and glass color standards are not available.

Dissolve 10 grams of iodine and 20 grams of potassium iodide in a small amount of water, and dilute to 1 liter. Make up frequently as the color deteriorates. Dilute oils with colorless benzene; a convenient means is a glass-stoppered comparison tube. In a similar tube, having the same cross section, accurately measure 1 ml. of the iodine solution, and dilute with water until the colors in the two tubes match when these are viewed transversely.

For lubricating oils a dilution of 1:100 with benzene, or higher if the oil is dark, is usually convenient. Naphthas and kerosene may normally be matched without dilution.

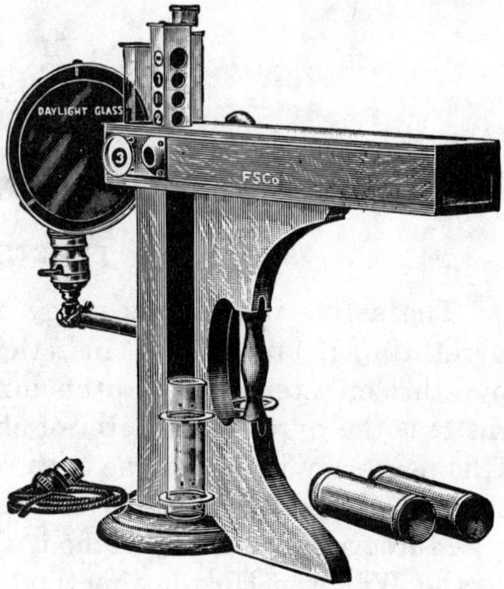

Fig. 152.—Union oil colorimeter for color of lubricating oils. ASTM D 1500-58T.

CALCULATION.

Color = milligrams iodine per 100 ml. of water in standard tube × (ml. of benzene to 1 ml. of oil + 1)

Read the corresponding National Petroleum Association (NPA) color from Table 44, which shows the NPA standards for lubricating oils and the equivalent iodine colors, expressed in milligrams of iodine per 100 ml. of solution.

TABLE 44.—NPA COLOR STANDARDS FOR PETROLEUM OILS

	NPA standard	NPA color	Iodine color
A	Cylinder, extra light filtered	...	50
D	Cylinder, light filtered	...	100
E	Cylinder, medium filtered	...	500
G	Lily white	1	2.8
H	Cream white	1½	5.7
I	Extra pale	2	10.8
J	Extra lemon pale	...	20.1
K	Lemon pale	3	32.1
L	Extra orange pale	...	38.4
M	Orange pale	4	70.7
N	Pale	...	112.0
O	Light red	5	195.0
P	Dark red	6	300.0
Q	Claret red	...	460.0

CHAPTER 38

PYRETHRINS AND ROTENONE

PYRETHRINS

The active ingredients of pyrethrum flowers, pyrethrin I and pyrethrin II, are efficient insecticidal agents. Pyrethrin I is the pyrethrolon ester of chrysanthemum monocarboxylic acid. Pyrethrin II is the pyrethrolon ester of chrysanthemum dicarboxylic acid. It is necessary to determine each separately, for pyrethrin I is considerably more toxic than pyrethrin II.

Mercury Reduction Method. This method, sometimes known as the Wilcoxen-Holaday method, depends on the fact that chrysanthemum monocarboxylic acid reduces the mercury sulfate in Denige's reagent. The pyrethrin I is isolated and then so reacted. A series of colors form, beginning with phenolphthalein red, which gradually changes to purple, then blue, and finally bluish green. Quantities as low as 1 mg. may usually be detected. If the color reaction is negative, pyrethrin I is absent. The extent of reduction is quantitatively determined.

Denige's Reagent. Mix 5 grams of yellow mercuric oxide—red will not do—with 40 ml. of water, stir, and add 20 ml. of concentrated sulfuric acid. Carefully add 40 ml. more water, and stir until solution is complete. Test for the absence of mercurous mercury by adding a few drops of iodine monochloride reagent, and titrate as described in the procedure on page 711.

Iodine Monochloride Reagent. Dissolve 10 grams of potassium iodide and 6.44 grams of potassium iodate in 75 ml. of water in a glass-stoppered bottle. Add 75 ml. of concentrated hydrochloric acid and 5 ml. of chloroform. Shake well, and adjust to a faint iodine color in the chloroform by addition of dilute potassium iodide solution or potassium iodate solution as necessary. If much iodine is set free, use a stronger solution of potassium iodate than $0.01M$, making the final adjustment with $0.01M$ solution. Keep in a dark cupboard, and readjust when necessary.

Potassium Iodate Reagent. $0.01M$. Dissolve in water 2.14 grams of C.P. potassium iodate, dried at 105°, and dilute to 1 liter. One milliliter of this solution is equivalent to 0.0044 gram of pyrethrin I.

Procedure. Pyrethrum Powder. *Pyrethrin* I. Using a Soxhlet apparatus, extract a finely ground sample containing 20 to 75 mg. of pyrethrin

I, usually 12 to 20 grams, for 7 hours with petroleum ether. Evaporate the solvent at as low a temperature as possible, and heat for the minimum length of time. Do not pass a current of air through the flask during the evaporation.

Add 15 to 20 ml. of 0.5*N* alcoholic sodium hydroxide (page 725) to the extract, connect to a reflux condenser, and heat carefully for 1 to 1.5 hours. Transfer to a 600-ml. beaker, and add enough water to bring the volume to 200 ml. Add a few beads, and boil down to 150 ml. Pour into a 250-ml. calibrated flask, and add 1 gram of filter medium and 10 ml. of 10 per cent barium chloride solution. The filter medium used is a good grade of calcined diatomaceous earth such as Filter-Cel. The barium chloride causes precipitation of the fats, resins, and inert materials present, the pyrethrins remaining in solution. Do not shake before making to volume, for foaming would occur. Make to volume, mix well, and filter. Measure 200 ml., exactly neutralize with 1:4 sulfuric acid to a phenolphthalein end point, and add 1 ml. of excess acid. If necessary to leave the solution over night, it must be alkaline. Filter through a sprinkled layer of filter medium on a 7-cm. filter paper in a Büchner funnel, and wash several times with water.

Transfer the filtrate and washings to a 500-ml. separatory funnel, and extract with two 50-ml. portions of petroleum ether. Wash the combined extracts with two or three 10-ml. portions of water. Discard the washings, and filter the petroleum ether extract through a plug of cotton into a clean 250-ml. separatory funnel. Wash the cotton with 5 ml. of the solvent. Add 5 ml. of approximately 0.4 per cent sodium hydroxide solution, shake vigorously, and draw off the aqueous layer into a 100-ml. beaker. Wash the petroleum ether with 5 ml. more of the 0.4 per cent sodium hydroxide solution, and add this to the beaker.

Add 10 ml. of the Denige's reagent to the beaker, note the changing colors for an indication of the amount of pyrethrin I present, and let stand for 1 hour. Add 20 ml. of alcohol, and precipitate the reduced mercury sulfate with 3 ml. of saturated sodium hydroxide solution. Warm to about 60°, and filter through a small filter paper. Wash with 10 ml. or more of hot alcohol, transferring all the precipitate to the filter paper. Wash with two or more 10-ml. portions of hot chloroform, and place the filter paper and mercurous sulfate precipitate in a 250-ml. glass-stoppered flask. Add 30 ml. of concentrated hydrochloric acid and 20 ml. of water, and cool. Now add 6 ml. of chloroform or carbon tetrachloride and 1 ml. of iodine monochloride solution, and titrate with the iodate solution, shaking vigorously after each addition, until there is no iodine color in the chloroform layer. The end point is not permanent, and therefore the titration should be completed rapidly with vigorous shaking after each addition of iodate. The added iodine monochloride aids in determining the end point in the titration of small amounts of mercury.

Potassium iodate oxidizes the mercurous to the mercuric salt with the liberation of iodine, which is oxidized by the further addition of potassium iodate to iodine monochloride. The reaction is

$$4HgCl + KIO_3 + 6HCl = 4HgCl_2 + ICl + KCl + 3H_2O$$

CALCULATION.

$$\text{Ml. iodate solution} \times 0.0044 \times \frac{250}{200} \times \frac{100}{\text{weight of sample}} = \text{per cent pyrethrin I}$$

Pyrethrin II. The ester that constitutes pyrethrin II is separated by solvent extraction and titrated with alkali. If necessary, filter the aqueous residue from the petroleum ether extraction in the determination of pyrethrin I through a Gooch crucible. Concentrate the filtrate to about 50 ml., and transfer to a separatory funnel. Make alkaline with sodium bicarbonate. Extract twice with chloroform, and wash the chloroform extract with about 15 ml. of water in each of two separatory funnels. Combine the aqueous solutions and washings, and make strongly acid with 8 ml. of concentrated hydrochloric acid, added cautiously to prevent too rapid evolution of carbon dioxide. Saturate the solution with sodium chloride, and extract with 50 ml. of ethyl ether.

Draw off the aqueous layer into a second separatory funnel, and reextract with 50 ml. of ethyl ether. Continue this extraction and drawing off of the aqueous layer, using 35 ml. for the third and fourth extractions. Wash the four ether extracts successively with 10 ml. of water, and repeat with a further 10 ml. of water. Combine the ether solutions, draw off any more water that separates, and filter through a plug of cotton into a 500-ml. conical flask. Evaporate the ether on a water bath, and dry the residue at 100° for 10 minutes. Add 2 ml. of neutral alcohol and 20 ml. of water. Heat to dissolve the chrysanthemum acid. Cool, filter through a Gooch crucible, and wash with neutral alcohol. Add 2 drops of phenolphthalein indicator solution, and titrate with 0.02*N* sodium hydroxide solution.

CALCULATION.

$$\text{Ml. sodium hydroxide} \times 0.00374 \times \frac{250}{200} \times \frac{100}{\text{weight of sample}} = \text{per cent pyrethrin II}$$

PYRETHRUM EXTRACT IN MINERAL OIL. Weigh or measure a quantity of sample that will contain 20 to 75 mg. of pyrethrin I, and transfer to a 300-ml. conical flask. Add 20 ml. or more, if necessary, of *N* alcoholic sodium hydroxide solution, and reflux for 1.5 hours. Transfer to a 600-ml. beaker, and add sufficient water to make the aqueous layer to 200 ml. If more than 20 ml. of alcoholic hydroxide solution has been used, add sufficient water so that all the alcohol will be removed when the volume has been reduced to 150 ml.

Add a few glass beads or a boiling tube, and boil down the aqueous layer to 150 ml. Cool somewhat and transfer the contents of the beaker to a 500-ml. separatory funnel, and draw off the aqueous layer into a 250-ml. volumetric flask. Wash the oil layer with 25 ml. of water, and add the wash water to the aqueous solution. If, after the drawing off of the aqueous layer and the washings, a slight emulsion still persists, break it by the addition of 2 to 3 ml. of a 10 per cent barium chloride solution.

Do not shake vigorously after adding the barium chloride solution, for a reversed emulsion that is difficult to separate may thus be formed. To the solution in the flask add 1 gram of filter medium and 10 ml. or more of 10 per cent barium chloride solution. Make to volume without shaking, mix thoroughly, and filter 200 ml. Test the filtrate with barium chloride to see if sufficient has been added to obtain a clear solution. Neutralize with 1:4 sulfuric acid, and add 1 ml. in excess, using a drop of phenolphthalein solution as indicator. From this point, proceed as directed for pyrethrum powder in the determinations of pyrethrins I and II (page 711), starting at "If necessary to leave the solution overnight, it must be alkaline."

Presence of Perfume Oils. To 100 ml. of an ordinary household extract, or less of a concentrate, in a 500-ml. flask, add 50 ml. of water, and steam-distill until the distillate coming over has no perfume odor. Cool the flask and contents, and transfer the latter to a separatory funnel. Allow to separate, and draw off the aqueous layer into a second separatory funnel. Wash this with 25 ml. of kerosene. Transfer the oil from the first separatory funnel to a 250-ml. conical flask, and wash the funnel with the 25 ml. of wash kerosene, which is then added to the flask. Now proceed in the usual manner to determine the pyrethrins in mineral oil (page 712), starting with "Add 20 ml. or more, if necessary, of *N* alcoholic sodium hydroxide solution . . . ".

Rotenone. Rotenone is obtained in particular from derris and cubé root. It is a powerful insecticide but is harmless to man. If the sample consists of roots or twigs, take great care to obtain a correct sample with proportionate amounts of small and large roots or twigs. Grind to about 60-mesh, but do not decompose the rotenone by causing the sample to heat excessively during grinding. The method of analysis involves extraction of the rotenone with chloroform, evaporation of the solvent, and finally forming its solvate with carbon tetrachloride and weighing, after purification.

Procedure. Weigh 30 grams of the finely powdered sample, or less if it contains more than 7 per cent of rotenone, into a 500-ml. glass-stoppered conical flask. A clean cork fitted tightly into a flask or bottle with metal foil will do as a substitute. Add 10 grams of a good grade of decolorizing carbon and 300 ml. of chloroform, making a record of the room temperature. Agitate mechanically for not less than 4 hours, preferably interrupting the shaking to let stand overnight.

Remove the flask or bottle from the shaking apparatus, and allow to cool in a refrigerator for at least an hour. Rapidly filter the mixture into a flask without suction, using a fluted paper and keeping the funnel covered to prevent evaporation loss. Stopper the flask, and adjust the temperature of the filtrate to that at which the original chloroform was measured.

Transfer exactly 200 ml. of this solution to a 500-ml. conical flask, and distill until only about 25 ml. remains in the flask. Transfer the extract to a 125-ml. conical flask, using a few milliliters of carbon tetrachloride to rinse out the 500-ml. flask. Evaporate almost to dryness on a steam bath in a current of air. Then remove the remainder of the solvent under reduced pressure, applying suction directly to the flask. When it is necessary to hasten evaporation, heat cautiously on the steam bath.

Dissolve the extract in 15 ml. of hot carbon tetrachloride, and again, in a similar manner, remove all the solvent. Repeat with another 10- to 15-ml. portion of hot carbon tetrachloride. This treatment removes all the chloroform from the resins. The chloroform extract is usually completely soluble in carbon tetrachloride. If small quantities of insoluble material are present, the purification procedure described later will eliminate them. However, if large quantities of insoluble residue remain when the extract is dissolved in the first portion of carbon tetrachloride, filter and wash thoroughly with further portions of hot solvent. Then treat the filtered solution plus washings as directed for the removal of chloroform.

Add exactly 25 ml. of carbon tetrachloride, and heat gently to dissolve the solid extract completely. Cool the flask in an ice bath for several minutes, and seed with a few small crystals of rotenone–carbon tetrachloride solvate if necessary. Stopper the flask, and swirl until crystallization is apparent. If at this stage only a small quantity of crystalline material appears, add an accurately weighed quantity of pure rotenone, estimated to be sufficient to ensure that the final weight, expressed as pure rotenone, is at least 1 gram. Warm to dissolve, and again induce crystallization. At the same time prepare a saturated solution of rotenone in carbon tetrachloride for washing, by adding excess of rotenone, warming to effect solution, and then keeping at 0° in an ice bath overnight. Filter immediately before using. Alternatively, dissolve 0.27 gram of pure rotenone in 100 ml. of carbon tetrachloride, and then cool to 0° for use.

Place the flasks containing the extract and the wash solution in an ice bath capable of maintaining a temperature of 0°, and allow to remain overnight. After 17 to 18 hours, rapidly filter the extract through a weighed Gooch crucible fitted with a disk of filter paper, removing the flask from the ice bath only long enough to pour each fraction of extract into the crucible. Rinse the residue of crystalline material from the flask, and wash under suction with sufficient ice-cold carbon tetrachloride saturated with rotenone, usually 10 to 12 ml., to remove the excess mother liquor. Allow the crucible to remain under suction for 5 minutes, and then dry to constant weight at 40°. This usually requires 1 hour. The weight so obtained is called the "crude rotenone–carbon tetrachloride solvate."

Break up the contents of the crucible with a spatula, mix thoroughly, and weigh 1 gram into a 50-ml. conical flask. Add 10 ml. of alcohol that has been saturated with rotenone at room temperature. Swirl the flask a few minutes, stopper tightly, and set aside at least 4 hours, preferably overnight, at the same temperature. Filter on a weighed Gooch crucible fitted with a disk of filter paper. Rinse the crystals from the flask, and wash under suction with alcohol saturated with rotenone at room temperature. Usually 10 ml. will suffice. Allow the crucible to remain under suction for 3 to 5 minutes, and then dry at 105° to constant weight. This should require 1 hour, and during this time the carbon tetrachloride is removed from the solvate.

In hot weather it is advisable to weigh flasks and contents at the beginning of the extraction and again before filtering. If there is appreciable loss, the original weight should be restored by the addition of chloroform before the contents of the flask are filtered. It is also advisable to cool crucibles in the refrigerator immediately before use in hot weather.

A suitable crystallizing bath may be made by enclosing a covered pan in a carton or box and insulating with excelsior or other packing material. In use, flasks are placed in the pan and surrounded with cracked ice.

CALCULATION.

$$\left[\left(\text{Weight of pure rotenone in crucible} \times \frac{\text{weight of crude rotenone–carbon tetrachloride solvate}}{\text{weight of purified rotenone–carbon tetrachloride solvate}}\right) + 0.07 - \text{weight of pure rotenone added}\right] \times \frac{300}{200} \times \frac{100}{\text{weight of sample}} = \text{per cent rotenone}$$

The 0.07-gram correction represents the rotenone held in solution in the 25 ml. of carbon tetrachloride used in crystallizing.

ALTERNATIVE EXTRACTION PROCEDURE. If the sample is one in which the ratio of rotenone to total ether extract is greater than 40 per cent, use a quantity sufficient to contain 1.0 to 1.5 grams of rotenone and successively extract four times with chloroform using 200 ml. each for the second to fourth extractions. Filter after each extraction, and return the marc to the flask for extraction with fresh solvent. Finally combine the extracts, evaporate almost to dryness, and continue from this point as in the preceding directions.

Total Ether Extract in Derris or Cubé Root. Extract 5 grams of the finely powdered root in a Soxhlet apparatus with ethyl ether for 48 hours. After extraction, concentrate the extract, and filter off any insoluble material that may be present. Receive the filtrate in a tared beaker, evaporate the ether on a steam bath, and dry in the oven at 105° to constant weight.

CHAPTER 39

INDICATOR, REAGENT, AND VOLUMETRIC SOLUTIONS

Many indicators, reagents, and standard volumetric solutions have already been referred to. In some cases, for greater convenience, their preparation has been described in connection with the methods. The preparation of others is discussed in this chapter.

INDICATOR SOLUTIONS

Bromothymol Blue. Dissolve 1 gram of bromothymol blue in about 75 ml. of 50 per cent alcohol by addition of $0.1N$ sodium hydroxide until faintly blue. Dilute to 100 ml. with 50 per cent alcohol. This solution is 1 per cent.

Ferric Nitrate. Dissolve 10 grams of purest grade ferric nitrate, $Fe(NO_3)_3.9H_2O$, in water, add 1 ml. of 1:4 nitric acid in order to prevent hydrolysis, and dilute to 100 ml. As an alternative, 25 grams of ferric alum, $Fe_2(SO_4)_3.(NH_4)_2SO_4.24H_2O$, in 100 ml. of solution may be used in place of ferric nitrate. Test either solution with silver nitrate to ensure that no chloride is present.

Methyl Orange. Dissolve 1 gram of purest grade methyl orange in 1 liter of water, and filter if necessary. Keep in a glass-stoppered bottle. This is 0.1 per cent. It is sometimes diluted to other desired concentrations.

Methyl Red. Dissolve 0.1 gram of purest grade methyl red in 1000 ml. of alcohol—SD3A will do—and filter if necessary. Keep in a glass-stoppered bottle.

Phenolphthalein. Dissolve 1 gram of purest grade phenolphthalein in 50 ml. of alcohol; SD3A will do. Dilute with water to 100 ml., and add $0.1N$ sodium hydroxide solution until a faint pink color appears. Remove the color with $0.1N$ acid. Keep in a glass-stoppered bottle. This is 1 per cent. Occasionally other concentrations or dilutions are required.

Potassium or Sodium Chromate. Dissolve 25 grams of potassium chromate, K_2CrO_4, or of sodium chromate, Na_2CrO_4, in 50 ml. of water. Add sufficient silver nitrate solution to give a slightly brick red precipitate. Filter, preferably after standing overnight, and dilute the filtrate and washings to 250 ml.

Starch. To 2.5 grams of arrowroot or potato starch add a few milliliters of cold water, and rub up to a paste. Add this, with stirring, to 1 liter of boiling water, so slowly that boiling never ceases. Boil for an additional minute or two, and allow to cool. Preserve by the addition of 1 ml. of chloroform. Keep in a glass-stoppered bottle. Discard and make up a new solution when mold appears or if a good strong color is not given with very dilute iodine solution.

Prepared soluble starch is satisfactory when dispersed in water, provided only that an intense blue color is given by a trace of iodine.

REAGENTS

Aluminum Hydroxide Suspension. Dissolve 125 grams of potassium alum, $K_2SO_4.Al_2(SO_4)_3.24H_2O$, in 3 liters of distilled water. Add 1:1 ammonium hydroxide solution slowly, with constant stirring, until the solution is alkaline and smells faintly of ammonia. Allow the precipitate to settle, and decant and discard the supernatant liquid. Wash the precipitate with distilled water by decantation until free from ammonia and sulfate. This is a tedious task. For smaller quantities, centrifuging and decanting will greatly expedite the process. Dilute the final suspension to 1 liter with distilled water, and preserve for use. One milliliter equals approximately 0.02 gram of aluminum hydroxide, $Al(OH)_3$.

Ammonium Acetate Reagent. Mix 120 ml. of glacial acetic acid and 130 ml. of water. Add, with stirring, 95 ml. of concentrated ammonium hydroxide. This is a reagent for solution of lead compounds.

Ammonium Molybdate. Dissolve 100 grams of commercial 85 per cent molybdic oxide in a mixture of 144 ml. of concentrated ammonium hydroxide and 271 ml. of water. When solution is complete, pour this slowly, with constant stirring, into a mixture of 489 ml. of concentrated nitric acid and 1,148 ml. of water. Store in a warm place for several days until precipitation of ammonium phosphomolybdate due to phosphate impurities has ceased. No yellow precipitate should then be obtained on heating the reagent to boiling. Decant the clear solution, and keep well stoppered.

Ammonium Polysulfide. Cool 200 ml. of concentrated ammonium hydroxide in running water, and saturate with hydrogen sulfide. Add 200 ml. more of concentrated ammonium hydroxide, and dilute to 1 liter. Add 25 grams of flowers of sulfur, warm for several hours, and filter.

Basic Lead Acetate. Boil 430 grams of neutral lead acetate and 130 grams of litharge, in 1 liter of water for 30 minutes. Allow the mixture to cool and settle; then dilute the supernatant liquid to a specific gravity of 1.25 with recently boiled water. Solid basic lead acetate may be substituted for the normal salt and litharge in the preparation of the solution. The solid salt is dissolved in water and diluted to the proper specific gravity.

Dimethylglyoxime. Dissolve 10 grams of the oxime in 1 liter of ethyl or methyl alcohol or in 13:7 ammonium hydroxide. Ten milliliters of this solution allows sufficient excess to precipitate 0.015 gram of nickel. The alcoholic solution keeps almost indefinitely. The ammoniacal solution should be kept for not more than a week. Occasionally, ammonium hydroxide contains impurities that cause incomplete precipitation of nickel.

Hanus Solution. IODINE MONOBROMIDE. This halogen reagent is to be prepared slightly more than $0.1N$ with respect to iodine and to bromine, corresponding to a solution of the compound IBr.

Add 13.615 grams of C.P. iodine crystals to 825 ml. of glacial acetic acid, and dissolve by warming and stirring. Cool and pipet out 25 ml. of solution. Dilute with water to about 200 ml., and titrate with $0.1N$ sodium thiosulfate solution. It is desirable to accomplish such pipetting of the solutions used in preparation of the reagent and the reagent itself with a water pump or a suction bulb, for the vapors are corrosive and poisonous.

Add 3 ml. of C.P. bromine from a buret to 200 ml. of glacial acetic acid. Mix well, and pipet out 5 ml. Dilute this to about 150 ml. with water, and add 10 ml. of 15 per cent potassium iodide solution. Titrate with $0.1N$ sodium thiosulfate solution. This titration of 5 ml. of bromine solution should be approximately 80 per cent of the titration of 25 ml. of iodine solution.

Calculate the amount of bromine solution to add to the remaining approximately 800 ml. of iodine solution as follows:

$$800 \times \frac{\dfrac{\text{titration of iodine solution}}{25}}{\dfrac{\text{titration of bromine solution}}{5}} = \text{ml. of bromine solution to be used}$$

After mixing, dilute to 1 liter with acetic acid, and store in an amber glass-stoppered bottle. This solution remains of reasonably constant strength. A blank must be run with each determination, but if such values are being obtained frequently the approximate concentration is known in advance.

Hydrogen Peroxide. This cannot be a standard solution because of its instability. Often the concentration of a solution must be checked to indicate the approximate concentration if used immediately. This will be illustrated by the usual 3 per cent commercial solution. The reaction is $5H_2O_2 + 2KMnO_4 + 4H_2SO_4 = 2MnSO_4 + 2KHSO_4 + 5O_2 + 8H_2O$.

Pipet or weigh 2 ml. of sample into a conical flask containing 250 ml. of cold 10 per cent sulfuric acid. Mix well, and titrate with $0.1N$ potassium permanganate.

CALCULATION.

$$\text{Ml. } KMnO_4 \times N \times 0.017 \times \frac{100}{\text{weight of sample}} = \text{per cent } H_2O_2$$

If checking the concentration of 30 per cent hydrogen peroxide, dilute 10 ml. to 100 ml. with distilled water and titrate as just described.

Magnesia Mixture. Dissolve 55 grams of magnesium chloride hexahydrate, $MgCl_2.6H_2O$, and 140 grams of ammonium chloride in 500 ml. of water. Add 130.5 ml. of concentrated ammonium hydroxide, mix well, and dilute to 1 liter. Filter if necessary.

Magnesium Nitrate. Dissolve 150 grams of magnesium oxide in 1:1 nitric acid, avoiding excess of acid. Add a little excess magnesium oxide, heat to boiling for about 2 minutes, and filter to separate iron and aluminum hydroxides and other contaminants.

Millon's Reagent. Dissolve metallic mercury in an equal weight of concentrated nitric acid, using a reflux condenser, and dilute the solution with an equal volume of water.

Nessler Reagent. Dissolve 50 grams of potassium iodide in 35 ml. of cold ammonia-free distilled water. Add a saturated solution of mercuric chloride made with ammonia-free water (page 146) until a slight precipitate exists. This will require 415 to 420 ml. Add 400 ml. of a solution containing 300 grams of sodium hydroxide per liter prepared in ammonia-free water. Dilute the reagent solution to 1 liter, allow it to clarify by standing a few days, and decant. This solution will remain unimpaired for months.

Schiff's Reagent. Dissolve 0.2 gram of fuchsin, rosaniline hydrochloride, in 120 ml. of hot water, and cool. Add 2 grams of anhydrous sodium sulfite dissolved in 20 ml. of water and 2 ml. of concentrated hydrochloric acid. Make up to 200 ml., and place in a refrigerator at least 24 hours before use. Store in a well-filled amber bottle.

Stokes' Reagent. Dissolve metallic mercury in twice its weight of concentrated nitric acid, and dilute this solution to twenty-five times its volume with water.

Thiocyanogen Solution. It is first necessary to prepare lead thiocyanate and then decompose it. Dissolve 331 grams of C.P. lead nitrate in 700 ml. of water, and filter. Dissolve 194 grams of C.P. potassium thiocyanate in 500 ml. of water, and filter. Slowly add the first solution to the second with stirring, continue the stirring for 30 minutes, and allow the precipitate to settle. Decant the supernatant liquid through a filter paper in a Büchner funnel, using slight suction. Wash the precipitate several times with water by decantation. Transfer the precipitate with water and a nonmetallic spatula to the Büchner funnel. The precipitate will react with a metallic instrument. Wash until the washings give no test for nitrates. Suck as dry as possible, transfer to a watch glass, and dry to constant weight in a vacuum desiccator over sulfuric acid. This takes about 7 days. The lead thiocyanate should be white in color and weigh about 260 grams. Keep in an airtight amber bottle and in the dark.

Prepare anhydrous acetic acid by gently refluxing 500 ml. of 99.5 per cent acetic acid, or better, with 40 ml. of acetic anhydride. This should be in a liter flask with a ground-glass connection to the condenser and should take about 1 hour of heating. Attach a calcium chloride tube to the end of the condenser, and allow the acid to cool to room temperature.

From a buret, weigh 4.2 grams of dry bromine into a 250-ml. calibrated flask, add 100 ml. of dry carbon tetrachloride, and fill to the mark with anhydrous acetic acid. Weigh 12.5 grams of the dry lead thiocyanate into a dry, glass-stoppered liter bottle. Add the bromine solution to this in small quantities, shaking vigorously after each addition and taking care that decoloration of the previous increment takes place before each addition of the bromine solution. After all has been added, allow the precipitated lead bromide and excess lead thiocyanate to settle. Filter the solution through a dry paper into a glass-stoppered amber bottle, and store in the dark. The solution should be clear and colorless or only slightly yellow. In about a week a yellow color and turbidity develop, followed shortly by a fine yellow precipitate. When color and turbidity develop, the solution should be discarded.

Wijs Solution. Iodine Monochloride. This halogen reagent is prepared to be slightly more than 0.1*N* with respect to iodine and chlorine, corresponding to a solution of the compound ICl.

To prepare from the elements dissolve 13 grams of C.P. iodine crystals in 1 liter of glacial acetic acid, warming to promote solution but cooling before proceeding further. This solution is slightly more than 0.1*N*. Generate chlorine gas by dropping 1:1 hydrochloric acid on potassium permanganate crystals and leading the gas through a wash bottle of concentrated sulfuric acid. Pass this gas through the solution until the titration with sodium thiosulfate has not quite doubled. Excess chlorine must not be present. If a slight excess is added inadvertently, correct by addition of the calculated amount of iodine and dilute with the corresponding amount of acetic acid. Store in an amber bottle having a glass stopper, and protect from heat or light. Standard methods call for replacement of the solution when it is 30 days old. Experience indicates that for routine work it can be used for several months.

To prepare from iodine monochloride, which has a molecular weight of 162.38, dissolve about 16.3 grams of the liquid in glacial acetic acid and check by titration with sodium thiosulfate that it is approximately 0.2*N*. The iodine monochloride is highly corrosive and must be handled with the same care as bromine. For routine work, preparation in this way is satisfactory and much simpler than production from the elements.

VOLUMETRIC SOLUTIONS

In making up standard volumetric solutions it is essential that the glassware used, especially flasks and burets, shall be accurately calibrated by the worker or that National Bureau of Standards calibrated apparatus be employed. Pipets, if used, should also be accurate. It is better practice to use burets for measuring out volumes throughout the standardization procedures, for thus conditions of drainage, time of flow, and actual volume are more nearly comparable. An error in one piece of apparatus may impair the accuracy of a number of standard solutions when one solution is tested against another solution of supposed known normality.

Highest quality chemicals should always be used in making up standard solutions. Some of especially high quality may be purchased from the National Bureau of Standards, Washington, D.C., for use as primary standards. Examples are potassium acid phthalate, sodium oxalate, arsenious acid, and potassium dichromate.

Once made up, standard solutions, except those which are alkaline, should be kept in glass-stoppered bottles in a cool dark place. Iodine and potassium permanganate solutions must be kept in amber or dark-colored bottles. Various standard solutions should be recalibrated at intervals, depending on their stability.

Since the usual laboratory buret has a capacity of 50 ml., it follows that the highest degree of accuracy would be obtained by a titration slightly under 50 ml. As a precaution, standardizations should be based on an expectation of a 40-ml. titration. The probable degree of accuracy can then be roughly estimated as follows. Since 1 ml. is about 20 to 25 drops, the 40 ml. would be 800 to 1,000 drops. A proper titration is accurate to 1 drop, and it is therefore accurate to about 1 part per 800 to 1,000, or around 0.1 per cent.

Sodium Hydroxide. It is always best to use standard sodium hydroxide solution in preference to potassium hydroxide solution. Not only is the chemical less expensive, but it is easier to prepare solutions free of carbonate. Fifty per cent solutions of sodium hydroxide precipitate the carbonate present on standing. This is not the case with potassium hydroxide solutions.

PROCEDURE. Use carbonate-free water prepared by boiling it for 20 minutes, cooling with soda-lime protection; or by bubbling air, purified from carbon dioxide by passing through a soda-lime tower, through distilled water for 12 hours. Prepare 1:1 alkali by adding an equal weight of water to C.P. sodium hydroxide and swirling until dissolved. Store in a rubber-stoppered flask or bottle for 10 days until all the carbonate has settled out.

The following tabulation shows the approximate amount of 1:1 alkali necessary to make 10 liters of standard solution.

Approximate Normality	1:1 Alkali, Milliliters
0.01	5.4
0.05	27.0
0.10	54.0
0.50	270.0
1.0	540.0

STANDARDIZATION. Dry potassium acid phthalate, National Bureau of Standards Standard for Acidimetry, molecular weight 204.22, for 2 hours at 120°, and cool in a desiccator over sulfuric acid. Weight out sufficient to give a titration of approximately 40 ml., which will vary according to the normality being prepared. Transfer the potassium acid phthalate to a 250-ml. conical flask swept free from carbon dioxide, and dissolve in 50 ml. of cool carbonate-free water. Add 3 drops of 1 per cent alcoholic solution of phenolphthalein, and titrate with the alkali to a faint pink color.

Calculate the normality of the stock solution as follows:

$$\frac{\text{Grams potassium acid phthalate}}{\text{Ml. alkali} \times 0.20422} = \text{normality}$$

The solution should be slightly overstrength. For occasional use it is not necessary to adjust to the exact normality. For commercial use, where relatively large volumes of such standard solutions are consumed, there is sufficient economy of time in the use of an exact normality to dictate such adjustment. For obtaining the desired normality, use the following formula:

$$\frac{\text{Normality} \times \text{volume of stock solution}}{\text{Desired normality}} = \text{volume after dilution}$$

Check the exact final strength of the alkali solution by further titrating against potassium acid phthalate. Store the standard alkali in alkali-resistant glass. Automatic pipets should have all exits to the air protected from carbon dioxide contamination by suitable guard tubes containing soda lime.

The normality so determined is correct only when phenolphthalein is used as an indicator. To prepare more dilute standards from this, dilute accurately, according to the formula, with carbon dioxide-free distilled water.

ALTERNATIVE STANDARDIZATION. The alkali may be standardized against a standard succinic acid solution or against hydrochloric acid that has been standardized by precipitation and weighing of silver chloride.

Hydrochloric Acid. Always use carbon dioxide-free distilled water to prepare these solutions. The following tabulation shows the approximate amounts of C.P. concentrated hydrochloric acid, 34 to 35 per cent hydrogen chloride, needed to make 10 liters of standard solution.

Approximate Normality	Concentrated Acid, Milliliters
0.01	8.9
0.05	44.5
0.10	89.0
0.50	445.0
1.0	890.0

STANDARDIZATION. Measure 40 ml. of the stock solution from an accurate buret into a 250-ml. conical flask, and titrate with standard alkali solution of approximately the same strength as the acid being tested. Use phenolphthalein indicator, and titrate to a faint pink color.

Calculate the normality of the stock solution as follows:

$$\frac{\text{Ml. standard alkali} \times \text{normality of alkali}}{\text{Ml. of stock solution of hydrochloric acid}} = \text{normality}$$

The solution as prepared should be over strength. If desired to adjust to an exact normality, use the following formula:

$$\frac{\text{Normality} \times \text{volume of stock solution}}{\text{Desired normality}} = \text{volume after dilution}$$

Check the final strength of the acid solution as before. The normality so determined is correct only when phenolphthalein is used as an indicator.

If the standard acid solution is to be used with methyl orange as an indicator, determine a correction for the volume of acid required to pass from the end point of phenolphthalein to that of methyl orange. Add 3 drops of a 1 per cent alcoholic solution of phenolphthalein to 100 ml. of carbon dioxide-free water, and then add sufficient alkali solution to give a faint pink with phenolphthalein. Disregard this amount used, and take buret readings from this point. Add 3 drops of a 0.02 per cent solution of methyl orange and sufficient 0.1*N* acid to produce the pink color of methyl orange. Titrate back with 0.1*N* alkali solution to the same end point taken in the usual methyl orange titration, preferably about pH 4.2. Buffered solutions of 3.8, 4.0, and 4.2 pH are useful in accurately determining the methyl orange end point. Prepare as follows:

pH 3.8; 2.041 grams potassium acid phthalate + 5.26 ml. 0.1*N* HCl. Dilute to 200 ml.
pH 4.0; 2.041 grams potassium acid phthalate + 0.80 ml. 0.1*N* NaOH. Dilute to 200 ml.
pH 4.2; 2.041 grams potassium acid phthalate + 7.40 ml. 0.1*N* NaOH. Dilute to 200 ml.

If the acid and alkali solutions are equivalent,

Ml. of acid − ml. of alkali solution = ml. of acid required to pass from the phenolphthalein to the methyl orange end point

ALTERNATIVE STANDARDIZATION. When standard alkali is not available, standardize a suitable volume of the stock acid solution by precipitation of silver chloride. First just neutralize the acid with chloride-free alkali. Then following the procedure on page 129.

CALCULATION.

$$\frac{\text{AgCl} \times 0.2544 \times 1{,}000/\text{aliquot}}{36.47} = \text{normality}$$

Alcoholic Potassium Hydroxide. For routine preparation of 0.5*N* solution, heat about 28 grams of C.P. or U.S.P. potassium hydroxide in 500 ml. of alcohol, with constant stirring; SD3A, which contains 5 per cent of synthetic methanol, is entirely satisfactory for the purpose. Practically speaking, the only serious interference occurs with alcohols containing sufficient aldehydes as impurities to darken the solution to such a degree that the end point will be obscured. Allow the solution to stand overnight. The potassium carbonate should precipitate to leave a clear solution. Decant, centrifuge or filter through asbestos, and dilute to 1 liter with alcohol. It is preferable to make the solution not more than 0.5*N*, for the usual blank in use for saponification of oils would otherwise require more than 50 ml. of 0.5*N* acid to neutralize it. For purposes requiring *N* alcoholic potassium hydroxide, double the concentration.

Umpire work according to standard methods calls for greater precautions. To conform to the method of the Association of Official Agricultural Chemists, reflux 1,200 ml. of 95 per cent alcohol for 30 minutes in a distilling flask with 10 grams of potassium hydroxide and 6 grams of aluminum foil or granular aluminum. This destroys aldehydes. Distill, discarding the first 50 ml. and collecting the next 1,000 ml. Dissolve 40 grams of potassium hydroxide in this, and keep in a glass-stoppered bottle.

This reagent is never standardized in the usual way. Procedures usually require it to be standardized by titration of a blank at the time of making a determination with it.

Alcoholic Sodium Hydroxide. The preparation is exactly analogous to that for potassium hydroxide except that 20 grams is used per liter to give a 0.5*N* solution.

Ammonium Thiocyanate. The usual concentration is 0.1*N*. For this dissolve 7.6118 grams of purest ammonium thiocyanate, NH_4CNS, molecular weight, 76.118, in water. Dilute to exactly 1 liter, and mix well. To standardize, measure 40 ml. of 0.1*N* silver nitrate solution into a 250-ml. conical flask or a porcelain casserole, and dilute to about 150 ml. Add 5 ml. of 1:4 nitric acid and 5 ml. of chloride-free ferric nitrate indicator solution. Titrate with the thiocyanate solution until a permanent red coloration in the liquid appears. This is best seen in artificial light.

CALCULATION.

$$\frac{\text{Ml. silver nitrate used} \times \text{normality}}{\text{Ml. of thiocyanate used}} = \text{normality of thiocyanate solution}$$

Barium Hydroxide. This reagent is occasionally required. It shows any carbonate contamination by precipitation, which is an advantage. Available grades are pure enough to permit its use as a primary standard. Dry the barium hydroxide in the oven at 110°, but protect it from exposure to carbon dioxide. Cool in a desiccator, and then weigh out 8.569 grams. Dissolve this in freshly boiled and cooled water, and dilute to 1 liter.

If the solution is clear it is $0.1N$. If the solid has become carbonated, it will not dissolve completely; in this case he solution may be either discarded or filtered and standardized by titration with hydrochloric acid.

Ferrous Ammonium Sulfate. This is a reagent usually prepared and used at about $0.1N$. Unless the solution is strongly acid, it deteriorates because of air oxidation. While frequent standardization is advisable, if it is properly protected the change with time is only minor. Freshly prepared, it can be used as a primary standard.

Weigh out 39.214 grams of ferrous ammonium sulfate of the highest purity, $FeSO_4(NH_4)_2SO_4.6H_2O$, molecular weight 392.14. Dissolve this in 1:9 sulfuric acid that has been boiled to drive out dissolved air and cooled. Dilute to 1 liter with the same acid.

Potassium Permanganate Standardization. Measure 40 ml. of the ferrous ammonium sulfate solution into a conical flask. Dilute to about 150 ml. with recently boiled and cooled 1:9 sulfuric acid. Titrate to a faint pink end point with $0.1N$ potassium permanganate.

Calculation.

$$\frac{\text{Ml. of permanganate used} \times \text{normality}}{\text{Ml. of ferrous solution used}} = \text{normality of ferrous solution}$$

Iodine. This oxidizing solution is normally made up at $0.1N$ but occasionally is further diluted. Weigh out 12.692 grams of pure iodine, I_2, molecular weight 126.92. Dissolve 20 to 25 grams of potassium iodide in the smallest quantity of water possible, and add the iodine. Stir well until fully dissolved. Dilute the solution to 1 liter, and preserve in a dark, glass-stoppered bottle. When freshly prepared, this is a primary standard. The solution is affected by light and should be standardized about every 2 weeks.

To standardize, measure 40 ml. of $0.1N$ arsenious oxide solution into a conical flask, and titrate with the iodine solution, adding 1 ml. of starch solution as indicator near the end of the titration. Titrate until the blue color just appears. The reaction is

$$Na_2HAsO_3 + H_2O + I_2 = Na_2HAsO_4 + 2HI$$

The standardization may be made in exactly the same manner, 40 ml. of $0.1N$ sodium thiosulfate solution being used in place of the arsenious oxide solution. The reaction is then $2Na_2S_2O_3 + I_2 = Na_2S_4O_6 + 2NaI$.

Calculation.

$$\frac{\text{Ml. of standard} \times \text{normality}}{\text{Ml. of iodine}} = \text{normality of iodine solution}$$

Nitric Acid. While it is used much less frequently than hydrochloric acid, there are occasions when standard nitric acid is required. To prepare, measure out 7.0 ml. of high-quality reagent acid of specific gravity about 1.42. Dilute to about 100 ml. with distilled water. If any yellow color is detectable, boil until it is driven off and cool. Dilute to 1 liter, and standardize by titration with standard alkali as already described for hydrochloric acid. The solution as so prepared is approximately 0.1*N*.

Oxalic Acid. This is commonly prepared at 0.1*N*. It must be standardized, if only because of the variability of composition of the hydrate. To prepare, dissolve 6.3034 grams of purest oxalic acid, $(COOH)_2.2H_2O$, molecular weight 126.07, in boiled, distilled water. Cool, make up to 1 liter, and mix well.

Sodium Hydroxide Standardization. To check the accuracy of preparation of the solution, measure out 40 ml. from a buret into a 250-ml. conical flask. Dilute to 100 ml. with boiled and cooled water, add a few drops of phenolphthalein indicator solution, and titrate to a faint pink color with 0.1*N* sodium hydroxide solution from a similar buret or, preferably, the same one.

Calculation.

$$\frac{\text{Ml. sodium hydroxide} \times \text{normality}}{\text{Ml. oxalic acid}} = \text{normality of oxalic acid}$$

Potassium Permanganate Standardization. Measure out 40 ml. of the oxalic acid solution from a buret into a 250-ml. conical flask, dilute to 100 ml. with boiled 1:19 sulfuric acid solution, and titrate as in the standardization of potassium permanganate against sodium oxalate.

Calculation.

$$\frac{\text{Ml. potassium permanganate} \times \text{normality}}{\text{Ml. oxalic acid}} = \text{normality of oxalic acid}$$

Potassium Bromate. This is a primary standard. Dry the salt in a 120° oven, and cool in a desiccator. For 0.1*N* solution weigh out 2.7835 grams. Dissolve this in water, and dilute to 1 liter. If it is of doubtful purity, standardize it. For the latter weigh 0.1 to 0.2 gram of arsenious oxide, and dissolve it in potassium hydroxide solution. Neutralize with concentrated hydrochloric acid, and add 15 ml. in excess. Dilute to 100 ml., warm to 60°, and titrate with the potassium bromate solution, using 2 drops of 0.2 per cent methyl red solution as indicator. When the indicator fades, add the potassium bromate solution slowly, using more indicator if desired. At the end point the solution turns colorless, and an added drop of indicator should be decolorized.

Calculation.

$$\frac{\text{Weight of arsenious oxide}/0.04946}{\text{Ml. potassium bromate solution}} = \text{normality of bromate solution}$$

Potassium Dichromate. This reagent is very stable and is used as a primary standard. A good grade of analyzed dichromate is usually satisfactory. To use greater precautions, fuse the reagent at the lowest possible temperature, grind, and dry at 110° before using. It is usually prepared at 0.1*N*, though occasionally at lower strengths.

Dissolve 4.9037 grams of purest potassium dichromate, $K_2Cr_2O_7$, molecular weight 294.21, in boiled, distilled water. Cool, and make up to exactly 1 liter for 0.1*N*.

Potassium Iodate. This is also a primary standard. Dry the salt in a 120° oven, and cool in a desiccator. Weigh out 3.567 grams to make 1 liter of 0.1*N* solution. Dissolve in water with 15 grams of potassium iodide, and dilute to volume.

Potassium Permanganate. This is commonly prepared at 0.1*N*. For special purposes 0.03*N*, 0.02*N*, and other strengths are prepared by dilution with carbon dioxide-free water. That the equivalent weight is one-fifth the molecular weight follows from the equation of the reaction with sodium oxalate,

$$2KMnO_4 + 5Na_2C_2O_4 + 8H_2SO_4 = K_2SO_4 + 2MnSO_4 + 5Na_2SO_4 + 8H_2O + 10CO_2$$

Therefore, the equivalent weight is 158.03/5 = 31.606.

To prepare 0.1*N* solution dissolve slightly over 3.160 grams of potassium permanganate in hot water, and dilute to approximately 1 liter. Allow to stand for a few hours; then filter through asbestos in order to free it from manganese dioxide, which aids in decomposing the reagent. Dilute the filtrate to 1 liter, and store in a dark bottle.

Adjustment of potassium permanganate to exactly 0.1*N* is economical with respect to the time spent only if many daily determinations are made with it. Otherwise it changes too rapidly to make the expenditure of time worth while.

Sodium Oxalate Standardization. Weigh 0.3000 gram of purest sodium oxalate, dried at 105°, and transfer to a 600-ml. beaker. Add 250 ml. of 1:19 sulfuric acid that has been boiled for 10 minutes to drive off dissolved gases and cooled. Stir to dissolve the oxalate, and titrate with the permanganate solution until the color begins to disappear. Heat to near boiling, and finish the titration, adding the permanganate slowly until a faint pink color persists for 30 seconds.

For very accurate work, determine the amount of permanganate required to give the faint pink color by titrating a similarly treated sulfuric acid solution without the added sodium oxalate. This blank is usually less than 0.05 ml. and should be subtracted from the previous titration.

Calculation.

$$\frac{\text{Weight of sodium oxalate}}{0.067007 \times \text{ml. of permanganate}} = \text{normality of potassium permanganate}$$

Ferrous Ammonium Sulfate Standardization. This method may be used in place of sodium oxalate standardization and is appropriate when the potassium permanganate solution is to be used to titrate ferrous iron. As an absolute standard, ferrous ammonium sulfate is not so exact as sodium oxalate, for the end point is not so sharp and the salt cannot be obtained in so high a degree of purity.

The reaction is

$$2KMnO_4 + 10Fe(NH_4)_2(SO_4)_2 + 8H_2SO_4$$
$$= 5Fe_2(SO_1)_3 + K_2SO_4 + 2MnSO_4 + 10(NH_4)_2SO_4 + 8H_2O$$

The milliequivalent of ferrous ammonium sulfate is therefore 0.39216.

Using boiled and cooled 1:19 sulfuric acid as in the previous method, dissolve in it 2 grams of purest ferrous ammonium sulfate, $Fe(NH_4)_2(SO_4)_2 \cdot 6H_2O$, and titrate at room temperature with the potassium permanganate solution to a permanent faint pink color.

Calculation.

$$\frac{\text{Weight of ferrous ammonium sulfate taken}}{0.39216 \times \text{ml. of permanganate used}} = \text{normality of the potassium permanganate solution}$$

Potassium Thiocyanate. Prepare exactly as described for ammonium thiocyanate, but use 9.7174 grams of potassium thiocyanate, KCNS, molecular weight 97.174. Standardize in exactly the same way.

Silver Nitrate. This is ordinarily used at 0.1*N* concentration, though occasionally a more dilute reagent is required. Weigh out 16.989 grams of purest silver nitrate, $AgNO_3$, molecular weight 169.89, dried at 120°. Dissolve in water, and dilute to exactly 1 liter. For the majority of purposes, this solution is sufficiently accurate. When necessary it is standardized either by precipitating and weighing silver chloride or by titrating against pure sodium chloride.

Gravimetric. Measure 25 ml. of the solution from a buret into a 250-ml. conical flask. Dilute to about 200 ml. with water, and add a slight excess of 1:10 hydrochloric acid, drop by drop, with stirring. Let stand until clear, or coagulate by bringing just to a boil. Allow to cool, and filter through a weighed and ignited Gooch crucible. Wash well with hot water until free from chlorides. Dry at 110°, and then heat to incipient melting at the edges of the precipitate. Cool, and weigh.

Calculation.

$$\frac{[(\text{Weight of AgCl} \times 169.89/143.34)] \times 1{,}000/25}{169.89} = \text{normality}$$

Volumetric. Measure out 40 ml. of 0.1*N* sodium chloride solution from a buret into a 250-ml. conical flask. Dilute to 150 ml., add 3 drops of sodium chromate indicator solution, and titrate to a slight but definite darkening of the yellow color.

CALCULATION.

$$\frac{\text{Ml. sodium chloride used} \times \text{normality}}{\text{Ml. of silver nitrate used}} = \text{normality of silver nitrate solution}$$

Sodium Arsenite. This solution is stable and is a primary standard. It is nearly always used at 0.1N for titration by iodine. Dissolve about 10 grams of purest sodium hydroxide in about 40 ml. of distilled water. Dissolve 4.9455 grams of purest arsenious oxide, As_2O_3, molecular weight 197.82, in the caustic solution by gently warming the mixture. Dilute to 250 ml., and saturate with carbon dioxide. Remove, and wash the tube into the flask. Make up the solution to 1 liter, and mix well.

A somewhat less stable solution can be made by dissolving the arsenious oxide in water containing 2.5 grams of sodium hydroxide. Neutralize the excess alkali with dilute sulfuric acid, using phenolphthalein as indicator. Then add 25 grams of sodium bicarbonate in 500 ml. of water, and again neutralize if necessary. Neutralization at this point will be due to some sodium carbonate in the bicarbonate. Make up the solution to 1 liter, and mix well.

Sodium Chloride. This is suitable for use as a primary standard. A convenient concentration is 0.1N. Dry the highest quality sodium chloride at 110°, cool in a desiccator, and weigh out 5.8454 grams. Dissolve in water, and dilute to 1 liter.

Sodium Thiosulfate. This is commonly prepared at 0.1N. Occasionally more dilute solutions, such as 0.005N or 0.002N, are required; these are conveniently made by accurate dilution. The solution deteriorates on standing and should be standardized about every 2 weeks.

For the 0.1N solution, dissolve 24.82 grams of purest sodium thiosulfate, $Na_2S_2O_3.5H_2O$, molecular weight 248.19, in boiled and cooled distilled water. Dilute to 1 liter, and add as a preservative 0.1 gram of sodium carbonate. The solution is now ready for use. If the carbon dioxide had not been removed by boiling the distilled water, it would be necessary to allow the solution to stand about 10 days for the sulfur precipitated by reaction to settle out.

POTASSIUM DICHROMATE STANDARDIZATION. From a buret measure 40 ml. of 0.1N potassium dichromate solution into a 350-ml. glass-stoppered flask. Add 100 ml. of boiled and cooled distilled water and 25 ml. of a 10 per cent solution of potassium iodide. Finally, add 7 ml. of concentrated hydrochloric acid, stopper, shake, and allow to stand for 2 to 3 minutes. Remove the stopper carefully, and wash it down into the flask with distilled water. The reaction that has occurred is

$$K_2Cr_2O_7 + 6KI + 14HCl = 8KCl + 2CrCl_3 + 7H_2O + 3I_2$$

Titrate with the sodium thiosulfate solution until the yellow color has nearly gone. Add 1 ml. of starch solution, and continue to titrate until the deep blue color changes to a sea green. The reaction is

$$I_2 + 2\ Na_2S_2O_3 = Na_2S_4O_6 + 2\ NaI$$

This standardization can also be performed against weighed 0.2-gram portions of potassium dichromate.

CALCULATION.

$$\frac{\text{Ml. of potassium dichromate} \times \text{normality}}{\text{Ml. of sodium thiosulfate}} = \text{normality}$$

$$\frac{\text{Weight of potassium dichromate}}{0.04904 \times \text{ml. of sodium thiosulfate}} = \text{normality}$$

IODINE STANDARDIZATION. Weigh out 0.5-gram portions of resublimed iodine as the primary standard. Dissolve each in a few milliliters of water with 3 grams of potassium iodide, and after solution is complete dilute to about 200 ml. Titrate with the thiosulfate solution to the usual starch end point. Calculate as follows:

CALCULATION.

$$\frac{\text{Weight of iodine}}{0.1269 \times \text{ml. of thiosulfate solution}} = \text{normality}$$

Succinic Acid. This is a primary standard for standardization of alkali solutions. A suitable strength is 0.1*N*. Dry the highest quality succinic acid in the oven at 105°, and cool in a desiccator. Weigh out 5.9044 grams, dissolve in water, and dilute to 1 liter.

Sulfuric Acid. Though not so suitable as hydrochloric acid or even nitric acid for titration, there are occasions when sulfuric acid should be used. This is usually when a standard acid is to be added in excess and a solution boiled vigorously before titration. The end point given by sulfuric acid tends to be less distinct because of the lesser ionization of the second hydrogen.

For 0.1*N* solution, carefully add 3 ml. of C.P. concentrated acid to about 25 ml. of water. Mix well, cool, and dilute to 1 liter. Standardize as described for hydrochloric acid.

Other Standard Solutions. The standard solutions outlined are representative. Other concentrations, such a 0.5*N*, 0.05*N*, etc., are simple modifications. Every chemical handbook gives the composition of many other volumetric reagent solutions. Such a source should be consulted when solutions are needed other than those given here.

INDEX

A

Abbé refractometer, 54
Abrasives in emulsions, 546, 553–554
Absorption bulbs, Hempel, 606
Accelerators, rubber, 588
  ammonia-type, 589
  nitroso-type, 589
Accuracy, 2
Acetate rayon, 647
  identification of, 641
Acetone, detection of, 436
Acetone extract from rubber, 562
Acetyl value of fats and oils, 351
Acid, effect of, 79
Acid-soluble compounding materials in rubber, 577
Acid sulfites, 172
Acid titration of sodium bicarbonate, 261
Acid value, of fats and oils, 341
  of fatty acids of soap, 396
  of plasticizers and oils of lacquers, 480
  of waxes, 429
Acidity, of paper, 623
  of water, 233
Acrolein test for glycerol, 466
Adurco wax, 428
Air bomb test on rubber, 582
Albumen test for phosphates, 258
Albuminoid nitrogen in water, 240
Alcohol, 16
  denatured, 457–458
  in emulsions, 552
Alcoholic potassium hydroxide, 725
Alcoholic sodium hydroxide, 725
Alcohols, 457
  hydroxyl value of, 439
  monohydric, detection of, 438
    estimation of, 439
  sulfated, 367
Alcoholic potash extract from rubber, 563
Aliphatic amines, 141
Alizarin as test for aluminum, 86
Alkali, effect of, 79
  metal perchlorates, 135
  neutralization number, 680, 682
Alkalies, in cement, 277
  fixed, and alkaline salts, 254
Alkaline builders in soap, 401
Alkaline permanganate solution, 241
Alkaline salts, 254
Alkaline silicates in soaps, 402
Alkaline titration of sodium bicarbonate, 261
Alkalinity, of paper, 623
  of water, 231
  of water-soluble matter of soap, 386
Alkaloids as rubber accelerators, 590
Alloy steel, copper in, 294
  manganese in, 286
  silicon in, 292
  sulfur in, 290
Alloys, aluminum, 311
  copper-base, 323
Alpha cellulose in paper, 625
Alumina cream, 718
Aluminon, 87
Aluminum, 311
  by alizarin, 86
  by aluminon, 87
  bismuth in, 313
  in bismuth, 314
  chromium in, 319
  copper in, 313
  in copper and lead, 314
  by difference, 205
  iron in, 312
  lead in, 313
  manganese in, 316
  micro test for, 84
  nickel in, 320
  as oxide, 191–194
  as phosphate, 193
  silicon in, 311
  tin in, 321
  titanium in, 313
  zinc in, 313, 315
Aluminum alloys, 311
Aluminum hydroxide suspension, 718
Aluminum metal, analysis by hydrogen evolution, 612
Aluminum oxide, phosphorus absent, 191
Alundum crucible, 23
Amides as rubber accelerators, 589
Amine soaps, analysis of, 403
Amines, aliphatic, 141
  aromatic, 142
  in emulsions, 553, 555, 558
  nitrous oxide test for, 141
  primary, 141
  primary aromatic, 142
  as rubber accelerators, 589
  secondary, 141
  tertiary, 141
  by titration, 147
Ammonia, and amines by titration, 147
  combined, by distillation, 149

Ammonia, free, 148
in latex, 584
by Nessler reagent, 88, 145
in the presence of organic amines, 149
by silver nitrate and formaldehyde, 88
soaps, analysis of, 403
test for, 86
by titration, 147
and volatile amines, 140
in water, 240
Ammonia- and magnesium-free water, 146
Ammonia-type rubber accelerators, 589
Ammonium acetate reagent, 718
Ammonium ferrocyanide, 92
Ammonium mercuric thiocyanate method for zinc, 315
Ammonium molybdate reagent, 718
Ammonium molybdate test for phosphates, 258
Ammonium polysulfide reagent, 718
Ammonium thiocyanate, standard, 725
Amorphous paraffins, 427
Amyl alcohols, Pettenhoffer furfural reaction, 465
Analysis, of ammonia and amine soaps, 403
of ash of water, 228
of heavy-metal soaps, 407
of rubber, 573
of dissolved solids, 228
of fibers, 632
of light-colored greases, 411
purpose of, 2
of total solids of water, 228
Analyst, tools of, 7
Analytical balance, 27
Analytical weights, 30
Analyzer, gas, Bureau of Mines design, 612
portable gas, 611
Aniline point, of petroleum products, 678
of waxes, 431
Animal fats, phytosterol in, 362
Animal fiber, 634, 635
artificial, 638
Anions, qualitative, 106
Antimony, and arsenic by distillation-volumetric method, 333
in the ash of rubber, 569
in copper-base alloys, 333
electrode, 113
micro test for, 83
oxide, 513
by rhodamine B, 88
in rubber, 568
by 2-methylbenzothiazole, 89
API hydrometer, 47
Apparatus, Hempel, 604
micro, 82
Approximate melting point, 49
Aromatic amines, benzenesulfonyl chloride test for, 142
secondary, 142
Aromatic hydrocarbons, 443
Arsenic, and antimony by distillation-volumetric method, 333
as arsenate by silver nitrate, 89
in copper-base alloys, 333
Gutzeit method, 208
by mercuric bromide, 89
micro test, 83
in paper, 624
Artificial animal fiber, 638
Artificial wool, 638
Asbestos, microscopic, 186
Ash, of coal, 598
analysis of, 602
on dissolved solids, 228
and driers in paint, 470
of greases, 408
of lacquers, 478
of latex, 585
in paper, 614
of rubber, 566
Assay flask, 577
ASTM distillation equipment, 666
Aurin tricarboxylic acid, aluminon, 87
Austenitic steel, phosphorus in, 288
Automobile cleaner, 550
Automobile polish, 550
Available chlorine by sodium arsenite, 136
Azo compounds, 115

B

Babcock bottle, 674
Balance, analytical, 27
chainomatic, 28
rough, 27
Westphal, 48
Balata, 586
Balata hydrocarbon, separation from rubber, 587
Ball and ring apparatus, 52
melting point, 52
softening point, 52
Barium carbonate in rubber, 573
Barium hydroxide, standard, 725
Barium, micro test, 85
in minerals as sulfate, 189
by rhodizonic acid, 89
by sodium rhodizonate, 90
Barium sulfate in rubber, 578
Barley starch, 652
Basic lead acetate reagent, 718
Basic lead carbonate, lead as lead chromate, 501
lead as lead sulfate, 501
Basic lead sulfate, 503
Baumé hydrometer, 46
Bayberry wax, 424
Beaker, Griffin, 8
Beakers, 7
Bean starch, 651
Becké test, for refractive index, 60
Beeswax, 424
Bentonite, in emulsions, 553
microscopic, 185

Benzine, 663
α-Benzoinoxime, 94
Benzyl alcohol, 467
Beryllium, by quinalizarin, 90
Beta cellulose in paper, 625
Bichromate method for glycerol, 397
Bierer-Davis oxygen bomb test for rubber, 581
Bismuth, aluminum in, 314
in aluminum, 313
by 8-hydroxyquinoline, 91
micro test, 83
by oxychloride method, 314
by thiourea, 91
Bismuthate method for manganese, 284
Bituminous material, 705
Bleach, sodium hypochlorite, 136
Bleached fibers, 631
Blowdown, 226
Blue, Prussian, 533
Boiler contents, 226
Boiler feed water, 226
Boiling range, of solvents, 446
of thinners, 446
Bomb method for sulfur in petroleum, 678
Borate, by curcumin, 107
methanol flame test for, 258
by methyl alcohol, 106
Borax buffer, 119
Boric acid buffer, 119
Bottles, weighing, 38
Bright's stain, 631
Brix hydrometer, 47
Bromate, by manganous sulfate, sulfuric acid, and benzidine, 107
Bromate method for tin, 321
Bromothymol blue indicator, 717
Bromine value of fats and oils, 346
Buckwheat starch, 653
Buffer, borax, 119
boric acid, 119
citric acid, 120
disodium phosphate, 120
hydrochloric acid, 118–120
monopotassium phosphate, 119
potassium acid phthalate, 118–120
potassium chloride, 118–120
sodium carbonate, 119
sodium hydroxide, 118–120
Buffer mixtures, 118–120
Buffer solutions, 117
Builders in soap, alkaline, 401
Bulkiness of heavy-metal soaps, 407
Burets, 10
Burner, Argand, 18
blast, 17
Bunsen, 17
gas, 17
Meker, 17
Butanol, 463

C

Cadmium, by dinitrodiphenylcarbazide, 91
micro test for, 83
by p-nitrobenzenediazoaminoazobenzene, 91
Cadmium sulfide, 533
Calcium, by ammonium ferrocyanide, 92
in cement, 276
and magnesium in aluminum, 317
micro test for, 85
as the osazone of sodium dihydroxytartrate, 92
as oxalate, phosphorus absent, 216
phosphorus present, 217
Calcium chloride solution, standard, 235
Calculation, of iron pigments, 529
of mixed thinners, 450
of results, 5
of results on mixed white pigments, 518
of rubber hydrocarbon, 574
Calorimetric determination on coal, 602
Candelilla wax, 426
Carbon in coal, 598
Carbon dioxide in gas, 608
Carbon monoxide, determination of, 609
Carbon residue of petroleum, 686
Carbon steel, copper in, 294
manganese in, 284
molybdenum in, 302
nickel in, 296
phosphorus in, 288
silicon in, 291
sulfur in, 290
titanium in, 309
total carbon in, 282, 283
vanadium in, 301
zirconium in, 310
Carbonate and noncarbonate hardness of water, 234
Carbonates, by evolution, 224
in soaps, 403
test by acid, 256
Carnauba wax, 426
Casein on paper, 615
Cast iron, chromium in, 299
copper in, 294
graphite in, 284
high-nickel chromium-alloy, 300
manganese in, 286, 288
molybdenum in, 303
nickel in, 296
phosphorus in, 289
silicon in, 292
sulfur in, 290
titanium in, 309
vanadium in, 301
zirconium in, 310
Catechol and sulfuric acid, colors developed with, 466
Cations, tests for, 86
Cellulose in rubber, 572

Cement, 275
alkalies in, 277
aluminum in, 276
calcium in, 276
calculation of original composition, 279
free lime in, 277
insoluble residue of, 276
iron in, 276
iron oxide in, 276
loss on ignition of, 275
magnesium in, 276
phosphorus in, 277
soluble silica in, 276
sulfur trioxide in, 276
Centrifuge, 33
Ceresin, 426
C_H, 112
Chainomatic balance, 28
Chemicals, 14
chemically pure, 15
technical-grade, 15
U.S.P., 15
Chinese insect wax, 426
Chloramine in water, 250
Chlorates, chlorides and perchlorates, 136
and perchlorates, 135
quantitative, 134
reduction with ferrous sulfate, 135
reduction with sulfurous acid, 135
reduction with zinc, 135
spot test for, 107
by sulfuric acid, 127
Chloride, gravimetrically, 128
nephelometrically, 229
in soap, 400
volumetric with sodium chromate indicator, 130
volumetric with dichlorofluorescein indicator, 130
volumetric with thiocyanate indicator, 131
in water, 228
Chlorides, chlorates, and perchlorates, 136
Chlorinated hydrocarbons, bomb method, 442
detection of, 440
fractionation of, 452
quantitative methods, 441
Chlorine, available, by sodium arsenite, 136
Carius method, 131
demand of water, 248
by orthotolidine, 246
sodium peroxide fusion method, 134
total, 137
Chlorite, qualitative, 128
Chloroform extract from rubber, 563
Chloroplatinate method for potassium, 221
Cholesterol in vegetable fats, 362
Chromate, indicator, 717
lead, 529
zinc, 532
Chrome green, 537
Chrome yellow, 529
Chromic oxide, 539
Chromium, in aluminum, 319
in cast iron, 299
by diphenylcarbazide, 93
by ferrous ammonium sulfate, 319
by ferrous sulfate titration, 530
in high-chromium high-nickel steels, 298
in high-nickel chromium-alloy cast iron, 300
in high-nickel steels, 298
in low chromium steels and irons, 299
micro test for, 84–85
as silver or lead chromate, 93
Chromium-alloy cast iron, chromium in, 300
Chromium-nickel steel, manganese in, 287
phosphorus in, 288
Chromium steel, 297
Chromium-tungsten steel, manganese in 287
Chromium-tungsten-vanadium steel, vanadium in, 301
Chromium-vanadium, stainless and similar steels, manganese in, 285
Chromium-vanadium-molybdenum steel, manganese in, 287
Chromium-vanadium steel, manganese in, 287
phosphorus in, 289
vanadium in, 300
Chromometer, Saybolt, 708
Citric acid buffer, 120
Clarification of sugar solutions, 657
Clay, microscopic, 185
Cleaner, automobile, 550
Cleaning solution, 10
Cleveland open-cup tester, 671
Cloth for rubberizing, copper and manganese in, 585
Cloud point, 683–684
of petroleum products, 684
Cloud and pour point, apparatus for, 684
of petroleum products, 683
Coal, analysis of, 596
ash of, 598
calorimetric determination on, 602
carbon in, 598
fixed carbon in, 598
hydrogen in, 598
moisture in, 597
nitrogen in, 601
phosphorus in, 601
preparation of sample, 596
proximate analysis of, 597
sulfur in, 598
ultimate analysis of, 598
volatile matter in, 598
Coal ash, analysis of, 602
Cobalt, by ammonium thiocyanate, 93
in cobalt steel, 306
micro test for, 85
by zinc oxide-α-nitroso-β-naphthol method, 306
Cobalt oxide, 536
Cobalt steel, cobalt in, 306
Coconut fibers, differentiating, 645
Cold alcohol, solubility of waxes in, 432

Cold titration of sodium carbonate, 260
Collection of samples, water, 227
Color, of petroleum products, 708
 of water, 243
Color changes, 116
Color standards, for free chlorine, 247
 for petroleum oils, 709
Colored pigments, 522
Colorimeter, 61
 Duboscq, 63
 Wesson, 66
Colorimetric pH, 113, 121–122
Colorimetry, balancing method, 63
 dilution method, 61
 duplication method, 62
Colors, mineral, 522
Combined alkali in soap, 387
Combined ammonia by distillation, 149
Combined nitrogen, 140
Commercial glucose, approximate polarimetric methods, 660
Commercial gums, qualitative reactions of, 545
Commercial metallic soaps, constants of, 409
Commercial silicas, silica in, 191
Commercial solvents, constants on, 455
Common fats and oils, fatty-acid distribution in some, 338
Common indicators, 116
Comparison of pH with buffers, 115
Compounding ingredients of rubber, sulfur in, 579
Compounds, rubber, 559
 sulfur, 157
Concentration, hydrogen-ion, 112
Concrete, 278
Coniferous fibers, 631
Conradson apparatus, 687
Conradson carbon residue, 686
Consistency of lubricating greases and petrolatum, 418
Constants, on commercial metallic soaps, 409
 on commercial solvents, 455
 on waxes, 425
Copolymers, as synthetic elastomers, 592
Copper, in alloy steel, 294
 in aluminum, 313
 by α-benzoinoxime, 94
 in carbon steel, 294
 in cast iron, 294
 colorimetrically, 206
 electrolytic method, 323
 micro test, 83
 in open-hearth iron, 294
 in pigments, 528
 by salicylaldoxime, 94
 in steel, 293
 in wrought iron, 294
Copper number of paper, 621
Copper phthalocyanine, 537
Copper steel, titanium in, 308
Copper and lead, aluminum in, 314
 electrolytic method, 326
Copper and manganese, in cloth for rubberizing, 585
 in rubber, 585
Copper-base alloys, 323, 327
 antimony in, 333
 arsenic in, 333
 iron in, 330
 nickel in, 333
 phosphorus in, 331
 tin in, 329
Corn starch, 650
Corrosive sulfur compounds, 672
Cotton, 642–643
 determination of, in mixtures, 646
 differentiating new and old, 643
Cover odors in emulsions, 547
Crankcase oils, dilution of, 707
Cross sections of fibers, examination of, 641
Cross sections of textile fibers, 634
Crucible, alundum, 23
 filter, 22
 glass, filter, 23
 Gooch, 22
 plain, 24
 porcelain, 24
Crude petroleum, 663
 topped, 665
 weathered, 665
Crude shale oil, 663
Crystalline silica, microscopic, 177
Cubé root, rotenone in, 713
 total ether extract in, 716
Cupferron-phosphate method, for zirconium, 310
Cuprammonium rayon, differentiating viscose, 642
 identification of, 641
Cuprammonium solution, 644
Cyanide, Liebig method, 155
 Volhard method, 154
Cyanin, 644
Cylinder method for unsaponifiable matter, 348

D

Dean-Stark tube, 41
Deciduous fibers, 631
Decomposed silica, microscopic, 179
Denatured alcohol, 457–458
Denige's reagent, 710
Density of solids, 48
Derris root, rotenone in, 713
 total ether extract in, 716
Desiccants, 26
Desiccators, 26
Diatomaceous earth, 180
Diatomite, 180
Diethylene glycol, 467
Differentiating animal from vegetable fibers, 634

Dilution of crankcase oils, 707
p-Dimethylaminobenzylidenerhodanine, 96, 98, 101
Dimethylglyoxime, method for nickel, 320, 333
  reagent, 718
Dinitrodiphenylcarbazide, 91
Diphenylcarbazide, 93
α,α'-Dipyridyl, 95
Dirt and fillers in gutta percha, 587
Disodium phosphate, 266
  buffer, 120
Dissolved oxygen in water, 241
Dissolved solids, ash on, 228
  in water, 227
Distillate, engine, 663
Distillation, of emulsions, 555
  fractional, 72
  volumetric method for arsenic and antimony, 333
Distillation equipment, ASTM, 666
Distilled water, 16
Distilling head, Hempel, 72
Dithizone method for lead, 211
Doctor test on oils, 673
Dropping point of grease, 421
Dry extraction of unsaponifiable matter, 350
Drying ovens, 30
Duboscq colorimeter, 63

E

Ebullators, 11
Elastomers, synthetic, 581, 590–592
Elastothiomers, as synthetic elastomers, 592
Electric hot plates, 32
Electrode, glass, 123
  hydrogen, 113

Electrodeposition apparatus, 34
Electrolytic method, for copper, 323
  for copper and lead, 326
  for lead, 327
Electrometric titration, 112, 125
Emulsifying agents, 542
Emulsions, abrasive in, 546, 553–554
  alcohol in, 552
  amines in, 553, 555, 558
  bentonite in, 553
  containing suspended solids, oil-in-water, 550
  cover odors in, 547
  distillation of, 555
  emulsifying agents for, 542
  free ammonia and ammonium soaps in, 555
  free from suspended solids, oil-in-water, 547
  glycerine in, 553
  miscellaneous ingredients in, 547
  modified separation method for, 549
  oil-gum, 550
  oils in, 552
  petroleum naphtha in, 546
Emulsions, pigments in, 546
  residue in flask from, 555
  resins and fatty acids in, 557
  saponifiable oils in, 546
  separation of, 548
  separation by extraction of, 549
  soap fatty acids in, 554
  solvent in, 553
  titration of soap in, 555
  total ammonia in, 555
  types of, 541
  unsaponifiable oils in, 546
  water-dispersible gums in, 553
  water-soluble liquids in, 547
  water-soluble matter in, 549
  wax, 540
  waxes in, 553, 557
  with and without suspended solids, water-in-oil, 553
Engine distillate, 663
Ester value of fats and oils, 344
Esters, detection of, 435
  estimation of, 435
Ethanol, 457, 459
  hydrometer method, 460
  immersion refractometer method, 460
  methanol in the presence of, 462
  volumetric method, 458
  Zeiss immersion refractometer readings, 464
Ether alcohols, detection of, 440
  estimation of, 440
  factors for calculation of, 441
  specific gravities of, 441
Ethyl xanthogenate, potassium, 99
Ethylene glycol, 467
Evolution method for carbonates, 224
Explosion bulbs, Winkler, 609
Extenders, 496, 509
  bulking value of, 498
  specific gravity of, 498
Extraction, Soxhlet, 67
Extractor, rubber, 69

F

Fats and oils, acetyl value of, 351
  acid value of, 341
  bromine value of, 346
  ester value of, 344
  insoluble acids in, 356
  iodine value of, 345
  Maumené number of, 364
  melting point of, 353
  peroxide value of, 354
  physical and chemical constants of, 365
  ratio of saturated and unsaturated fatty acids in, 357
  refractive index of, 341
  saponification value of, 343
  soluble and insoluble acids in, 355
  specific gravity of, 341
  thiocyanogen number of, 361

Fats and oils, titer value of, 352
unsaponifiable matter in, 347
volatile matter in, 340
water-insoluble volatile acids in, 357
water-soluble volatile acids in, 356
Fatty-acid distribution in some common fats and oils, 338
Fatty acids, from plasticizers, 482
of soap, acid value of, 396
iodine value of, 396
Feldspar, microscopic, 184
Ferric iron, by potassium ferrocyanide, 94
by potassium thiocyanate, 95
Ferric nitrate indicator, 717
Ferricyanide, 156
by benzidine, spot test for, 108
Ferrocyanide, 155
by uranyl acetate, spot test for, 108
Ferrocyanide method for zinc, 504
Ferrous ammonium sulfate, standard, 726
Ferrous iron, by α,α'-dipyridyl, 95
by potassium ferricyanide, 95
Ferrous metals, sampling, 282
total carbon, 282
Fibers, 632, 645
analysis of, 632
animal, 634–635
artificial animal, 638
bleached, 631
coniferous, 631
cross sections of textile, 634
deciduous, 631
examination of cross sections of, 641
identification of Manila, 645
ignition of textile, 634
modified vegetable, 639
paper, 614, 629
preparation of paper, 629
removal of color of textile, 634
sulfate, unbleached, 632
sulfite and unbleached sulfate, 632
textile, 614, 633
unbleached, 631
unbleached sulfate, 632
vegetable, 634, 642
Fillers, bulking value of, 498
mineral, 496
specific gravity of, 498
Filling, determination of in mixtures, 646
Filter cone, 22
Filter crucibles, 22
Filter paper, 19
Carl Schleicher and Schull, 21
Munktells, 21
qualitative, 20
quantitative, 20
semiquantitative, 20
Whatman, 20
Filter pulp, 21
Fineness of heavy-metal soaps, 407
Fire point, 671
Fixed alkalies and alkaline salts, 254
Fixed carbon in coal, 598
Flame reactions, sample, 77
Flash point, 669, 671
Flask, Erlenmeyer, 8
Flasks, 7
Flax, differentiating flax and hemp, 644
Floor wax, no-rub, 556
Flue gases, Orsat apparatus, 611
Fluorescence of new cotton, 643
Fluoride, in water, by zirconium-alizarin, 250
by zirconium-alizarin, spot test for, 108
Fluoride standard for water, 251
Fractional distillation, of solvents, 446, 448
of thinners, 446
Fractionation, 72
of chlorinated hydrocarbons, 452
of hydrocarbons, 452
of hydrocarbon fraction, 450
of solvents soluble in 80 per cent sulfuric acid, 451
Free acid, in greases, 416
in soap, 384
Free alkali, in greases, 416
in soap, 384
Free ammonia and ammonium soaps in emulsions, 555
Free carbon in rubber, 569
Free carbon dioxide in water, 234
Free chlorine in water, 246
Free fatty acid in sulfated oil, 373
Free lime in cement, 277
Free mineral acids in water, 233
Free stearic acid in stearate soaps, 407
Free sulfur, 672
in rubber, 566
Fuel gas, analysis of, 608
typical analyses of, 603
Fuel oil, 663
Funnels, 22
glass, 22
separatory, 12
Furfuryl alcohol, 467
Furol viscosimeter, Saybolt, 691

G

Gamma cellulose in paper, 625
Gas, carbon dioxide in, 608
collection of, 604
illuminants in, 608
oxygen in, 608
supplies of, 18
unsaturated hydrocarbons in, 608
Gas analysis, 34
general procedure, 605
sampling tube for, 604
Gas analyzer, Bureau of Mines design, 612
portable, 611
Gas buret, 604
Gas burners, 17
Gas oil, 664

Gases, calculation of combustible, 610
  moisture in, 43
Gasoline, 664
  distillation of, 665
Gauzes, wire, 19
Geer oven test on rubber, 581
Glass electrode, 113, 123
Glass filter crucibles, 23
Glass joints, 10
Glasses, Lovibond, 67
Glassware, volumetric, 8
Glucose, commercial, approximate polarimetric methods, 660
Glue in rubber, 570
Glycerin, acrolein test for, 466
  by bichromate method, 397
  in emulsions, 553
  in greases, 415
  in soap, 396
Glycols, detection of, 440
  estimation of, 440
  factors for calculation of, 441
  specific gravities of, 441
Gold, by benzidine, 96
  by p-dimethylaminobenzylidenerhodamine, 96
Gooch crucibles, 22
Graphite in cast iron, 284
Grease, 418
  ash of, 408
  dropping point of, 421
  free acid in, 416
  free alkali in, 416
  glycerin in, 415
  gypsum in, 412
  lubricating, 408
  neutral fat in, 415
  petroleum, 664
  petroleum oil in, 415
  soap in, 413
  unworked consistency of, 419
  water in, 417
  worked consistency of, 420
Green, chrome, 537
Gums, commercial, qualitative reactions of, 545
  water-dispersible, 543
Gutta-percha, 586
  dirt in, 587
  fillers in, 587
  moisture in, 586
  resins in, 587
  separation of, 587
Gutzeit method for arsenic, 208
Gypsum in greases, 412

H

Halogens, by potassium permanganate, 127
  by silver nitrate, 127
Halowax, 428
Hanus method, iodine value, 345
Hanus solution, 719
Hardness of water by soap method, 234
Heavy-metal soaps, 405
  analysis of ash of, 407
  bulkiness of, 407
  fineness of, 407
  insoluble ash of, 406
  lead soaps in, 406
  moisture in, 406
  soluble ash of, 407
  total ash of, 406
Hehner number, 356
Hemp, differentiating from flax, 644
Hempel absorption bulbs, 606
Hempel apparatus, 604
Hempel distilling head, 72
Herzberg's stain, 630
Hexahydrophenol, 467
High-chromium high-nickel steels, chromium in, 298
  nickel in, 295
  vanadium in, 301
High-nickel chromium-alloy cast irons, nickel in, 296
High-nickel steel, chromium in, 298
  nickel in, 295
  vanadium in, 301
High-speed steels, phosphorus in, 289
Hot plates, electric, 32
Hot titration of sodium carbonate, 260
Hydrocarbons, 452
  aromatic, 443
  chlorinated, 452
    detection of, 440
  fractionation of, 450–452
  nitration of, 443
  separation from rubber, 587
  sulfonation of, 444
  and turpentine, 453
Hydrocarbons A, in rubber, 571
Hydrocarbons B, in rubber, 571
Hydrochloric acid, buffer, 118–120
  standard, 723
Hydrogen, in coal, 598
  determination of, 609
  electrode, 113
Hydrogenated oils, identification of, 364
Hydrogen-ion concentration, 112
  of paper, 623
Hydrogen peroxide, standardization, 719
Hydrometer, API, 47
  Baumé, 46
  Brix, 47
  specific-gravity, 45
  Twaddell 47
Hydrosulfide and sulfide, by iodine, 171
Hydroxyl value of alcohols, 439
8-Hydroxyquinoline, 91
Hypochlorite, qualitative, 128

I

Identification, of hydrogenated oils, 364
  of rubber accelerators, 588
  of thinner in paint, 469
Ignition of textile fibers, 634
I.G. wax, 428
Illuminants in gas, 608
Immersion refractometer, 57
Index, viscosity, 696
Indicator, bromothymol blue, 717
  chromate, 717
  ferric nitrate, 717
  methyl orange, 717
  methyl red, 717
  phenolphthalein, 717
  starch, 717
Indicator papers, 115
Indicator solutions, 115–116, 717
Indicators, 78
  for alkalies, 256
  common, 116
  universal, 114
Inorganic salts in soap, 386
  in sulfated oil, 371
Inorganic sulfate by benzidine, 167
Inorganic sulfur trioxide in sulfated oil, 376
Insoluble acids in fats and oils, 356
Insoluble ash of heavy-metal soaps, 406
Insoluble residue of cement, 276
Insoluble sulfates, 157
Insoluble sulfides, 158
Invert sugar, 654
  Munson and Walker method, 659
  substances containing, 661
Iodate, by hypophosphorous acid, spot test for, 109
Iodine monobromide, 719
Iodine monochloride, 710, 721
Iodine, standard solution, 726
Iodine value, of fats and oils, 345
  of fatty acids of soap, 396
  Hanus method, 345
  of plasticizers, 480
  of waxes, 431
  Wijs method, 346
Ions, tests for, 86
Iron, in aluminum, 312
  in cement, 276
  colorimetrically by thiocyanate, 195
  in copper-base alloys, 330
  by dichromate, 195
  ferric, 94–95
  ferrous, 95
  micro test for, 84
  by permanganate method, 194
  sulfide separation of, 312
Iron and aluminum, as phosphates, 193
Iron and aluminum oxides, phosphorus absent from, 191
  in cement, 276
Iron pigments, 526
  calculation of, 529
Isopropanol, iodoform test for, 463

J

Jackson turbidimeter, 243
Japan wax, 426
Joint rubber insulation committee procedure for rubber hydrocarbon, 574
Joints, glass, 10
Jones reductor, 13
Jute, identification of, 645

K

Kappelmeier method for organic phthalates, 483
Kerosene, 664
  distillation of, 665
Ketones, detection of, 436
  estimation of, 437
Kieselguhr, 181
Kinematic viscosity, conversion to Saybolt Universal viscosity, 694–695
Kjeldahl-Gunning-Arnold method, nitrates absent, 143
  nitrates present, 145
Kjeldahl method, nitrogen by, 142

L

Lacquer solvents, 447
Lacquers, 468, 478
  ash of, 478
  pigments in, 478
  plasticizers and oils in, 479
  solvents and thinners in, 479
  total solids in, 478
Lakes, red dyes used as, 525
Lamp method for sulfur, 674
Lanette wax, 428
Lanital, 638
Latex, ammonia in, 584
  ash of, 585
  proteins in, 584
  rubber, 582
  rubber content of, 583
  total solids in, 583
  water-soluble substances in, 584
Lead, in aluminum, 313
  by benzidine or by tetramethyldiaminodiphenylmethane, 96
  and copper, aluminum in, 314
    electrolytic method, 326
  by diphenylthiocarbazone, 211
  dithizone method for, 211
  electrolytic method for, 327
  by excess dithizone, 213
  as lead chromate, 501
  micro test for, 83
  by rhodizonic acid, 96
  in rubber, 574
  as sulfate, 503

Lead chromate, 529
volumetrically, 501
Lead soaps in heavy metal soaps, 406
Lead sulfate, basic, 503
Leveling tube for gas analysis, 604
Liebermann-Storch method, rosin on paper, 619
for rosin, 395
on synthetic resins, 493
Light-colored greases, analysis of, 411
Linoleic acid, 339
Linolenic acid, 339
Lithopone, barium sulfate in, 506
total zinc in, 507
zinc sulfide in, 507
Lovibond glasses, 67
Low-alkali glassware, 7
Low-chromium steels and irons, chromium in, 299
Lubricating greases, 408
and petrolatum, 418
Lubricating oil, 664
precipitation number of, 689

M

Magnesia mixture, 719
Magnesium, and calcium in aluminum, 317
in cement, 276
micro test for, 85
as the phosphate, 218
by quinalizarin, 97
by thiazole yellow, 97
Magnesium nitrate reagent, 720
Magnesium silicate, microscopic, 183
Magnetic damp, 30
Malachite green stain, 631
Manganese, 198
in the absence of cobalt, bismuthate method, 284
in alloy steels, 286
in aluminum, 316
by bismuthate, 199
in carbon steel, 284, 286
in cast iron, 286, 288
by catalytic oxidation with ammonium persulfate, 98
in chromium-nickel steels, 287
in chromium-tungsten steels, 287
in chromium-vanadium-molybdenum steels, 287
in chromium-vanadium, stainless and similar steels, 285
in chromium-vanadium steels, 287
in cloth for rubberizing, 585
micro test for, 84–85
in nickel steels, 287
in open-hearth iron, 286–287
by periodate, 200
by persulfate, 199
in the presence of cobalt, persulfate method for, 286
Manganese, by the pyrophosphate method, 316
in rubber, 585
by tetramethyldiaminodiphenylmethane, 97
Volhard method for, 527
Williams method for, 528
in wrought iron, 286, 287
Manila fiber, differentiating, 645
identification of, 645
Matter insoluble, in alcohol, in soap, 383
in water, in soap, 385
Matter volatile, at 105° in soap, 382
from soap with steam, 399
Maumené number of fats and oils, 364
McNicoll method for rosin, 395
Melting point, approximate, 49
ball and ring, 52
closed-end capillary, 50
of fats and oils, 353
mercury-cup method, 53
open-end capillary, 51
of waxes, 430
Wiley method for, 353
Mercaptobenzothiazole as rubber accelerator, 590
Mercuric bromide paper, 209
Mercuric mercury, microtest for, 83
Mercuric sulfide, 524
Mercury, by aniline and stannous chloride, 98
by p-dimethylaminobenzylidenerhodamine, 98
mercuric, micro test for, 83
mercurous, micro test for, 83
as sulfide, 524
Mercurous mercury, micro test, 83
Metallic soaps, constants of commercial, 409
Metals, analysis by hydrogen evolution, 612
Metaphosphate, sodium, 266
Methane, determination of, 609
Methanol, in the presence of ethanol, 462
quantitative, 461
Zeiss immersion refractometer readings, 464
Method, choice of, 4
2-Methylbenzothiazole, 89
Methyl cyclohexanol, 467
Methyl ethyl ketone, detection of, 436
Methyl orange indicator, 717
Methyl red indicator, 717
Microscope, 57
Micro qualitative analysis, 81
Micro qualitative sample, 83
Micro qualitative scheme, 83
Millon's reagent, 720
Mineral acids in water, 233
Mineral coating on paper, 615
Mineral colors, 522
Mineral fillers, 496, 509
in paper, 615
Minerals, insoluble in hydrochloric acid, 187
microscopic, 177
quantitative analysis, 186–189

Minerals, sample of, 176
soluble in hydrochloric acid, 187
Miscellaneous ingredients in emulsions, 547
Mixed solvents, 445
Mixed thinners, 445
calculation of, 450
Mixed white pigments, calculation of results on, 518
Mixing equipment, 35
Mixtures, buffer, 118–120
Modified vegetable fibers, 639
Moisture, in alkalies, 259
in coal, 597
in gutta-percha, 586
in heavy-metal soaps, 406
in paper, 614
by solvent distillation, 41
in sulfated oil, 370
Molybdenum, in carbon steel, 302
in cast iron, 303
colorimetrically by thiocyanate, 304
in molybdenum steel, 302
in open hearth iron, 302
as oxide, 302
by potassium ethyl xanthogenate, 99
by potassium thiocyanate, 99
as sulfide, 302
in tungsten steel, 303
in wrought iron, 302
Molybdenum steel, manganese in, 287
molybdenum in, 302
Molybdophosphoric acid, 621
Monohydric alcohols, detection of, 438
estimation of, 439
Monopotassium phosphate buffers, 119
Montan wax, 427
Mortar, 275–281
Municipal water supplies, 227
Munson and Walker method for invert sugar, 659
Mutarotation, 657

N

Naphtha, distillation of, 665
petroleum, 664
in emulsions, 546
Naphthylamine acetate solution, 236
Natural resin glycerides, constants on, 475
Natural resin varnishes, 473
Natural resins, constants on, 475
Negative radicals, tests for, 79
Nessler reagent, 720
Nessler tubes, 61
Neutral fat in greases, 415
Neutralization number, 680–682
of petroleum products, 680
Nickel, aluminum in, 320
in carbon steels, 296
in cast iron, 296
in copper-base alloys, 333
Nickel, by dimethylglyoxime, 99
dimethylglyoxime method for, 320, 333
in high-chromium high-nickel steel, 295
in high-nickel chromium-alloy cast irons 296
micro test for, 85
in open-hearth iron, 297
by rubianic acid, 100
in wrought iron, 297
Nickel steel, 294
manganese in, 287
phosphorus in, 288
Nitrate, by brucine, 109
by diphenylamine, 140
by ferrous sulfate, 140
Kjeldahl-Gunning-Arnold method for, 145
by nitron, 152
by reduction to ammonia, 150
by reduction with ferrous sulfate, 152
by sulfanilic acid and α-naphthylamine, 109
in water, 237
Nitration of hydrocarbons, 443
Nitric acid, standard, 726
Nitrite, 140
by potassium permanganate oxidation, 153
by sulfanilic acid, 236
by sulfanilic acid and α-naphthylamine, 109
in water, 236
p-Nitrobenzenediazoaminoazobenzene, 91
Nitrocellulose, separation from resins, 485
Nitrogen, in coal, 601
combined, 140
by Kjeldahl method, 142
Nitrogenous materials on paper, 616
Nitrophenols, 115
Nitroso-type rubber accelerators, 589
Nitrous oxide test for amines, 141
Nonvolatile vehicle of paint, 469
NPA color standard, 709
Nylon, 639

O

Oat starch, 652
Odor, of solvents, 446
of thinners, 446
Oil emulsions containing large amounts of soap, 554
Oil-gum emulsions, 550
Oil-in-water emulsions, containing suspended solids, 550
free from suspended solids, 547
Oils, 540
doctor test on, 673
in emulsions, 552
fuel, 663
gas, 664
iodine value of, 345
in lacquers, 479
lubricating, 664
(*See also* Fats and Oils)

Oils, and resins, Boughton method, 474
separation of, 474
saponifiable, in emulsions, 546
separation of, 474
steam emulsion test on, 702
sulfonated, 367
sulfuric acid absorption of, 673
unsaponifiable, in emulsions, 546
in water, 253
Oil shale, 664
Oil-type vehicle of paint, 471
Oleic acid, 339
Opal wax, 428
Open cup tester, 671
Open-hearth iron, copper in, 294
manganese in, 286–287
molybdenum in, 302
nickel in, 297
phosphorus in, 288
silicon in, 292
titanium in, 309
vanadium in, 301
zirconium in, 310
Open-hearth steel, sulfur in, 290
Organic compounds, chlorine in, 131
Organic ester sulfur trioxide of sulfated oil, 373
Organic materials, sulfur in, 161
Organic phosphates, 482
Organic phthalates, Kappelmeier method, 483
test with resorcinol, 483
Organic sulfate, by benzidine, 167
qualitative tests, 368
Orsat apparatus for flue gases, 611
Orthophosphate in soaps, 403
Orthotolidine reagent, 246
Ouricuri wax, 427
Ovens, drying, 30
vacuum, 31
Oxalic acid, standard, 727
Oxide, aluminum, phosphorus absent, 191
cobalt, 536
iron, phosphorus absent, 191
Oxychloride method for bismuth, 314
Oxygen bomb test, Bierer-Davis, on rubber, 581
Oxygen in gas, 608
Ozocerite, 427

P

Paint, 468
ash and driers in, 470
identification of thinner in, 469
nonvolatile vehicle of, 469
oil-type vehicle of, 471
separation of pigment of, 469
volatile thinner in, 469
water in, 468
Paint vehicle, interpretation of constants on, 473
Palmitic acid, 339
Paper, 614
acidity of, 623
alkalinity of, 623
alpha cellulose in, 625
arsenic in, 624
ash of, 614
beta cellulose in, 625
casein on, 615
copper number of, 621
fibers, 614, 629
preparation of, 629
filter, 19
gamma cellulose in, 625
hydrogen-ion concentration of, 623
indicator, 115
mercuric bromide, 209
mineral coating on, 615
mineral filler in, 615
moisture in, 614
paraffin on, 621
pentosans in, 624
proteinaceous nitrogenous materials on, 616
resin on, 619
rosin on, Liebermann Storch method, 619
Raspail method, 619
starch on, 616
Paper stains for identification of fibers, 630
Paper testing, 614
Paraffin, 427
amorphous, 427
on paper, 621
Paraffin wax, melting point of, 704
Particle size, 69
by the microscope, 71
by special equipment, 72
Pea starch, 651
Pentosans in paper, 624
Perchlorate method for potassium, 222
Perchlorates, alkali-metal, 135
and chlorates, 135
chlorides and chlorates, 136
qualitative, 128
Periodate, by manganous sulfate and phosphoric acid, spot test for, 109
Peroxide value of fats and oils, 354
Persulfate, by benzidine, 109
ferrous sulfate method, 175
method for manganese in the presence of cobalt, 286
oxalic acid method, 175
Petrolatum, consistency of, 418
melting point of, 704
Petroleum, carbon residue of, 686
crude, 663
saponification number of, 683
terms relating to, 663
topped crude, 665
weathered crude, 665
Petroleum grease, 664
Petroleum naphtha, 664
in emulsions, 546

Petroleum oils, color standards for, 709
  in greases, 415
Petroleum products, 662
  aniline point of, 678
  cloud and pour point of, 683
  color of, 708
  neutralization number of, 680
  sediment in, 706
  water in, 705–706
Petroleum solvents, 442
Petroleum spirits, 665
Phenolphthalein indicator, 717
pH, colorimetric, 113
  by colorimetry, 114
  methods of measuring, 113
pH scale, 112
pH values for 0.1N solutions, 113
pH value of water, 240
Phenol, 467
Phenols, qualitative, 484
  quantitative, indophenol method, 484
Phenoldisulfonic acid, 238
Phosphate, by ammonium molybdate and benzidine, 110
  disodium, 266
  iron and aluminum as, 193
  organic, 482
  in soaps, 403
  trisodium, 265
Phosphate test by ammonium molybdate, 258
Phosphorus, in austenitic steel, 288
  in carbon steel, 288
  in cast iron, 289
  in cement, 277
  in chromium-nickel steel, 288
  in chromium-vanadium steel, 289
  in coal, 601
  colorimetrically, 204
  in copper-base alloys, 331
  in high-speed steel, 289
  magnesium phosphate precipitation of, 201
  in nickel steel, 288
  in open-hearth iron, 288
  phosphomolybdate-pyrophosphate method, 202
  as phosphomolybdate volumetrically, 203
  in stainless steel, 288
  in steel, 288
  in wrought iron, 289
Phthalates, organic, Kappelmeier method, 483
  test with resorcinol for, 482
Phthaleins, 115
Phthalocyanine, copper, 537
Phytosterol in animal fats, 362
Pigments, acid-soluble, aluminum in, 515
    barium in, 514
    calcium in, 517
    magnesium in, 517
    sulfate in, 518
  antimony oxide in, 513
  barium sulfate in, 510
  bulking value of, 498
Pigments, calcium in, 515
  carbon dioxide in, 517
  in emulsions, 546, 553–554
  in lacquers, 478
  lead in, 511–513
  silica in, 510
  specific gravity of, 498
  sulfide sulfur in, 518
  total acid- and water-soluble sulfur in, 517
  total antimony in, 514
  total zinc in, 516
  white, 496
    calculation of results on mixed, 518
    qualitative, 497
  zinc oxide in, 517
Pipet for gas analysis, 604
Pipets, 10
Pitch, 665
Plastizicers, fatty acids from 482
  iodine value of, 480
  and oils in lacquers, 479
    acid value of, 480
  saponification value of, 482
  synthetic, 481
  unsaponifiable matter in, 482
Platinumware, cleaning of, 25
  molds, 25
  precautions for, 24
Point, cloud, 684
  pour, 685
Polenske number, 357
Policeman, 14
Polish, automobile, 550
Polyhydric alcohols, 466
  acrolein test for glycerol, 466
  reaction with catechol and sulfuric acid, 466
Polysulfide sulfur and sulfide, 165
Portable gas analyzers, 611
Potassium, by calculation, 220
  as the chloroplatinate, 221
  micro test for, 85
  as the perchlorate, 222
  and sodium, by calculation, 220
    as chlorides, 219
    in insoluble silicates, 219
    as sulfates, 218
  by sodium cobaltinitrite and silver nitrite, 100
  by sodium dipicrylamine, 100
  as sulfate, 218
Potassium acid phthalate buffer, 118–120
Potassium bromate, standard, 727
Potassium chloride buffer, 119
Potassium dichromate, standard, 728
Potassium ferricyanide, 95
Potassium ferrocyanide method for zinc, 214
Potassium hydroxide, alcoholic, 725
Potassium iodate, reagent, 710
  standard, 728
Potassium nitrate solution, standard, 238
Potassium permanganate, standard, 728
Potassium thiocyanate, standard, 729

Potato starch, 649–650
Pour point, 683–685
  of petroleum products, 684
Powder, pyrethrum, 710
Powders, specific gravity of, 49
Precipitates, washing of, 74
Precipitation number of lubricating oils, 689
Precipitation values for waxes, 432
Preparation of paper fibers, 629
Primary amines, 141
Primary aromatic amines, 142
Propanol, iodoform method, 463
  iodoform test for, 463
Proteinaceous nitrogenous materials on paper, 616
Proteins in latex, 584
Proximate analysis of coal, 597
Prussian blue, 533
Pulverizer for coal, 596
Pumice, 181
Pumicite, 181
Pycnometer, 43
Pyrethrin I, 710
Pyrethrin II, 712
Pyrethrins, 710
  mercury reduction method, 710
Pyrethrum extract in mineral oil, 712
  perfume oils present, 713
Pyrethrum powder, 710
Pyrex, low-alkali, 7
Pyrex glassware, 7
Pyrophosphate method for manganese, 316
Pyrophosphate, in soaps, 403
  sodium, 266
Pyrophyllite, microscopic, 183

Q

Qualitative analysis, 81
Quantitative methods, selection of, 80
Quinalizarin, 90, 97

R

Ramie, identification of, 646
Raspail method, rosin on paper, 619
Rayon, acetate, 647
  identification of, 641
  regenerated cellulose, 648
Reagent, Denige's, 710
  iodine monochloride, 710
  Nessler, 720
  potassium iodate, 710
  Schiff's, 720
  Stoke's, 720
Reagent solutions, 717
Red, ruthenium, 644
Red dyes used as lakes, 525
Red lakes, 524
Red lead, lead peroxide in, 523
  total lead in, 522
Reductor, Jones, 13
Refractive index, 53
  Becke test for, 60
  of fats and oils, 341
  of solids, 60
  of synthetic resins, 491
Refractometer, Abbé, 54
  immersion, 57
Reichert-Meissl value, 356
Removal of color of textile fibers, 634
Replacements, synthetic silk, 639
Report of rubber analysis, 575
Reporting results of water analysis, 239
Residual chlorine in water, 249
Resins, and fatty acids in emulsions, 557
  in gutta-percha, 587
  and oils, Boughton Method, 474
    separation of, 474
  on paper, 619
  separation of, 474
  separation from nitrocellulose, 485
  synthetic, 487
Results, calculation of, 5
Rhodamine B, 88
Rhodizonic acid, 89, 96
Rice starch, 652
Ricinoleic acid, 339
Riffles, 597
Rilan wax, 428
Rosin, 471
  Liebermann-Storch test for, 395
  McNicoll method for, 395
  on paper, 619
    Liebermann-Storch method, 619
    Raspail method, 619
  in soap, 394
Rotenone, 710, 713
Rottenstone, 178
Rough balance, 27
Rubber, 559
  acetone extract from, 562
  acid-soluble compounding materials in, 577
  air bomb test on, 582
  alcoholic potash extract from, 563
  analysis of ash of, 573
  analysis, report of, 575
  antimony in, 568
  antimony in the ash of, 569
  artificial aging tests on, 581
  ash of, 566
  barium carbonate in, 573
  barium sulfate in, 578
  Bierer-Davis oxygen bomb test on, 581
  cellulose in, 572
  chloroform extract from, 563
  as compounded, 575
  copper in, 585
  free carbon in, 569
  free sulfur in, 566
  Geer oven test on, 581
  glue in, 570

Rubber, hydrocarbon, calculation of, 574
hydrocarbons A in, 571
hydrocarbons B in, 571
joint rubber insulation committee procedure, 574
identification of accelerators in, 588
latex, 582
lead in, 574
manganese in, 585
preparation of sample of, 560
qualitative examination of, 560
rubber-solvent method for, 576
solution of, 577
specific gravity of, 561
sulfur in, 564
sulfur in ash of, 567
sulfur as antimony trisulfide in, 578
sulfur in compounding ingredients of, 579
sulfur present as barium sulfate in, 568
synthetic, 581
total extract from, 561
unsaponifiable matter in, 571
Rubber accelerators, 588
amines and amides as, 589
ammonia type, 589
alkaloids as, 590
mercaptobenzothiazole as, 590
nitroso-type, 589
thiocarbamate, 590
Rubber compounds, 559
Rubber content of latex, 583
Rubber extractor, 69
Rubber-solvent method for rubber, 576
Rubianic acid, 100
Ruthenium red, 644
Rye starch, 651

S

Saccharimeter, 656
Salicylaldoxime, 94
Sample, of alkalies, 259
appearance of, 75
coal, preparation of, 596
effect of acid on, 79
effect of alkali on, 79
flame reactions of, 77
inflammability of, 76
micro qualitative, 83
odor of, 75
size of, 3
solubility of, 77
Sampling, 37
coarse solids in, 38
fine solids in, 38
of gases, 39
of liquids, 38
of nonhomogeneous liquids, 39
Sampling tube for gas analysis, 604
Saponifiable fats and oils, 337
Saponifiable oils in emulsions, 546
Saponification number of petroleum, 683
Saponification value, of fats and oils, 343
of plasticizers, 482
of waxes, 429
Saybolt chromometer, 708
Saybolt Furol viscosimeter, 691
Saybolt Universal viscosimeter, 691
Scale, pH, 112
Schiff's reagent, 720
Secondary amines, 141
Secondary aromatic amines, 142
Secondary propyl alcohol, iodoform method, 463
iodoform test for, 463
Sediment in petroleum products, 706
Selection of quantitative methods, 80
Selenium, by hydriodic acid, 101
by thiourea, 100
Semidecomposed silica, microscopic, 178
Separation of oils and resins, 474
Separation of pigment of paint, 469
Separatory funnel method for unsaponifiable matter, 349
Separatory funnels, 12
Shaking machine, 36
Shale oil, 664
crude, 663
Shellac, 486
Silica, hydrochloric acid method, 189
perchloric acid method for, 190
present as alkaline silicates in soaps, 402
microscopic, 177
Silicate, by ammonium molybdate and benzidine, 110
magnesium, microscopic, 183
qualitative test for, 256
sodium, 268
Silicon, in alloy steel, 292
in aluminum, 311
in carbon steel, 291
in cast iron, 292
in open-hearth iron, 292
in tungsten steels, 292
in wrought iron, 292
Silk, 636–637, 647
determination of, in mixtures, 646
wild, 637
Silver, by p-dimethylaminobenzylidenerhodamine, 101
gravimetric method for, as chloride, 128
micro test for, 83
by potassium chromate, 101
Silver nitrate, standard, 729
Silver sulfate solution, standard, 238
Simple polymers as synthetic elastomers, 591
Sisal fibers, differentiating, 645
Soap, 381
acid value of fatty acids of, 396
alkaline builders in, 401
alkaline silicates in, 402

Soap, alkalinity of water-soluble matter of, 386
  analysis of ammonia and amine, 403
  by ashing of soap and soap mixtures, 390
  carbonates in, 403
  chlorides in, 400
  combined alkali in, 387
  fatty acids, in emulsions, 554
    titer value on, 396
  free acid in, 384
  free alkali in, 384
  glycerol in, 396
  in greases, 413
  heavy metal, 405
  inorganic salts in, 386
  iodine value of fatty acids of, 396
  matter insoluble in alcohol in, 383
  matter insoluble in water in, 385
  matter volatile from, with steam, 399
  matter volatile at 105° in, 382
  orthophosphate in, 403
  phosphates in, 403
  pyrophosphate in, 403
  qualitative tests on, 381
  rosin in, 394
  starch in, 396
  sugars in, 396
  sulfates in, 401
  by titration, 389
    of fatty acids, 388
  total anhydrous, 387
  total fatty matter in, 391
  unsaponifiable matter in, 391, 393
  unsaponified matter in, 394
Soap method for hardness, 234
Soap products, 381
Soap solution, stock, 235
Sodium, by calculation, 220
  micro test for, 85
  and potassium, by calculation, 220
    as chlorides, 219
    in insoluble silicates, 219
    as sulfates, 218
  by zinc uranyl acetate, 222
    spot test for, 102
Sodium bicarbonate, acid titration of, 261
  alkaline titration of, 261
  and sodium carbonate, 264
  in sodium hypochlorite, 138
Sodium carbonate, buffer, 119
  cold titration of, 260
  hot titration of, 260
  and sodium bicarbonate, 264
  and sodium hydroxide, 262
  and sodium tetraborate, 272
  trisodium phosphate, 267
    and sodium metasilicate, 270
    and sodium tetraborate, 273
Sodium chloride, standard, 730
Sodium cobaltinitrite, 100
Sodium dipicrylamine, 100
Sodium hydroxide, 260
  alcoholic, 725
  and sodium carbonate, 262
  in sodium hypochlorite, 138
  standard, 722
Sodium hydroxide buffer, 118, 119, 120
Sodium hypochlorite, sodium bicarbonate in, 138
  sodium hydroxide in, 138
  total alkalinity in, 138
  total chlorine in, 137
Sodium hypochlorite bleach, 136
Sodium-ion correction for glass electrode, 124
Sodium metaphosphate, 266
Sodium metasilicate, sodium carbonate, and trisodium phosphate, 270
Sodium nitrite, standard, 237
Sodium oxide, total alkalinity as, 138
Sodium peroxide fusion method, for chlorine, 134
Sodium plumbite solution, 673
Sodium pyrophosphate, 266
Sodium rhodizonate, 90
Sodium silicates, 268
Sodium sulfite, 174
  and sodium thiosulfate, 174
Sodium tetraborate, 270
  and sodium carbonate, 272
  sodium carbonate, and trisodium phosphate, 273
Sodium thiosulfate, 174
  and sodium sulfite, 174
  standard, 730
Softening point, ball and ring method, 52
Solids, density of, 48
  refractive index of, 60
  in water, 227
Solubility, of samples, 77
  of solvents, 446
  of thinners, 446
  of waxes, 433
    in cold alcohol, 432
Soluble ash of heavy-metal soaps, 407
Soluble and insoluble acids of fats and oils, 355
Soluble silica in cement, 276
Soluble sulfates, 157
Soluble sulfides, 158
Solutions, buffer, 117
  indicator, 115, 717
  reagent, 717
  of rubber, 577
  standard indicator, 116
  volumetric, 717
Solvent distillation, volatile matter by, 41
Solvents, 434
  boiling range of, 446
  emulsions of, 541
  in emulsions, 553
  fractional distillation of, 446, 448
  fractionation of, salt-solution soluble fraction, 449
    soluble in 80 per cent sulfuric acid, 451

Solvents, lacquer, 447
in lacquers, 479
mixed, 445
odor of, 446
petroleum, 442
saponification value of, 449
solubility of, 446
in 20 per cent sodium chloride solution of, 448
in 80 per cent sulfuric acid of, 448
specific gravity of, 446
and thinners in lacquers, 479
Soxhlet extraction, 67
Specific gravity, 43
definition of, 45
determination of, 45
of ether alcohols, 441
of glycols, 441
of fats and oils, 341
of powders, 49
of rubber, 561
of solids, 48
of solvents, 446
of synthetic resins, 491
of thinners, 446
of waxes, 429
Specific-gravity bottle, 43
calibration of, 44
Specific-gravity hydrometer, 45
Spermaceti, 427
Spirit varnish, 468
Spirits, petroleum, 665
Spot tests, 81
Standard indicator solutions, 116
Stain, Bright's, 631
cyanin, 644
Herzberg's, 630
malachite green, 631
paper, for identification of fibers, 630
ruthenium red, 644
Sutermeister's, 631
Stainless and similar steels, manganese in, 285
Stainless steel, phosphorus in, 288
Standard calcium chloride solution, 235
Standard glasses, color with, 65
Standard solutions, 721
Starch, barley, 652
bean, 651
buckwheat, 652
corn, 650
indicator, 717
oat, 652
on paper, 616
pea, 651
potato. 649–650
rice, 652
rye, 651
in soap, 396
tapioca, 653
Starches, 614, 648
Steam bath, 31
Steam emulsion number, 702
Steam emulsion test, apparatus for, 703
on oils, 702
Stearate soaps, free stearic acid in, 407
Stearic acid, 339
in stearate soaps, 407
Steel, chromium, 297
chromium-vanadium-molybdenum, manganese in, 287
copper in, 293
and iron, low-chromium, chromium in, 299
manganese in, 285
nickel in, 294
phosphorus in chromium-nickel, 288
phosphorus in nickel, 288
phosphorus in stainless, 288
Still, Barnstead, 16
Precision, 16
Stokes, 16
Stirrer, flexible shaft, 36
Stirring rods, 11
Stock soap solution, 235
Stoke's reagent, 720
Stopcocks, 35
Stoppers, 35
Strong-acid neutralization number, 680, 682
Strontium, micro test, 85
by sodium rhodizonate, 102
Succinic acid, standard, 731
Sucrose, 654
hydrometric and pycnometer methods, 655
inversion at room temperature, 659
polarimetric method, 656
polarization before and after inversion, 658
reducing sugars in, before and after inversion, 659
refractometer method, 655
Sugar, invert, 654
Munson and Walker method, 659
substances containing, 661
Sugar products, 654
Sugar solutions, clarification of, 657
Sugars, 654
in soap, 396
Sulfate, benzidine method, 166
by benzidine, 167
by barium rhodizonate, 111
insoluble, 157
organic, qualitative tests, 368
in soap, 401
soluble, 157
titrimetrically by tetrahydroxyquinone, 236
volumetrically, 166
in water, 230
Sulfate fibers, unbleached, 632
Sulfate and sulfide, 166
Sulfate and sulfite, 165
Sulfated alcohols, 367
Sulfated oil, 367
free fatty acid in, 373
inorganic salts in, 371

Sulfated oil, inorganic sulfur trioxide in, 376
  moisture in, 370
  organic ester sulfur trioxide of, 373
  total fatty matter in, 377
  total alkalinity of, 372
  total sulfated fatty matter in, 378
  unsaponifiable matter in, 379
Sulfide, as barium sulfate, 164
  cadmium, 533
  as cadmium sulfide, 164
  and hydrosulfide by iodine, 171
  insoluble, 158
  by iodine, 171
  as lead sulfate, 163
  method for zinc, 327
  and polysulfide sulfur, 163, 165
  separation of iron, 312
  by sodium nitroprusside, 111
  soluble, 158
  and sulfate, 166
Sulfites, 158, 172
  and acid sulfites, 172
  by sodium nitroprusside, 111
  and sulfate, 165
  in water, 252
Sulfite fibers, 632
Sulfonated oil (*see* sulfated oil)
Sulfonates, 367
Sulfonation of hydrocarbons, 444
Sulfonephthaleins, 115
Sulfur, 290
  in alloy steel, 290
  as antimony trisulfide in rubber, 578
  in ash of rubber, 567
  in carbon steel, 290
  Carius method for, 163
  in cast iron, 290
  in coal, 598
  in compounding ingredients of rubber, 579
  Eschka method for, 162
  free, 672
  by lamp method, 674
  in open-hearth steel, 290
  in organic materials, 161
    Carius method, 163
    Eschka method, 162
    oxygen bomb method, 161
    sodium peroxide method, 163
  in petroleum by the bomb method, 678
  polysulfide and sulfide, 165
  present as barium sulfate in rubber, 568
  in rubber, 564
  as sulfate, 158
  sulfide, 163
  in tungsten steel, 290
  in wrought iron, 290
Sulfur compounds, 157
  organic, 158
Sulfur dioxide, Monier-Williams method, 168
Sulfur trioxide, in cement, 276
  organic, 373
Sulfuric acid absorption of oils, 673
Sulfuric acid, standard, 731
Suspended solids in water, 227
Sutermeister's stain, 631
Swimming pools, 227
Synthetic elastomers, 581, 590
  chemical structures of, 591–592
  copolymers as, 592
  elastothiomers as, 592
  maximum tensile strengths and corresponding elongations of, 594
  properties of, 593
  simple polymers as, 591
  vulcanization properties of, 591
Synthetic plasticizers, 481
Synthetic resin varnishes, 477
Synthetic resins, 487
  distillates from, 492
  halogens in, 492
  Lieberman-Storch test on, 493
  nitrogen in, 492
  odors on heating of, 492
  petroleum ether–insoluble matter of, 487
  physical constants of, 491
  quantitative examination of, 493
  refractive index of, 491
Synthetic rubber, 581
Synthetic silk replacements, 639
Synthetic waxes, 428
  specific gravity of, 491
  sulfur in, 492
  test for phenols in, 493
  test for phthalates in, 493
  typical classes of, 488–490

T

Tag closed tester, 669–670
Talc, microscopic, 183
Tapioca starch, 653
Tar, 665
Terms relating to petroleum, 663
Terpenes, 445
Tertiary amines, 141
Testing, paper, 614
Tests, spot, 81
Tetrahydrofurfuryl alcohol, 467
Tetramethyldiaminodiphenylmethane, 96–97
Textile fibers, 614, 633
  cross sections of, 634
  ignition of, 634
  removal of color of, 634
Thallium, by benzidine, 102
  by potassium iodide, 102
Thinners, 434, 447
  boiling range of, 446
  fractional distillation of, 446
  in lacquers, 479
  mixed, 445
    calculation of, 450
  odor of, 446

Thinners, solubility of, 446
specific gravity of, 446
Thiocarbamate rubber accelerators, 590
Thiocyanogen number of fats and oils, 361
Thiocyanogen solution, 720
Thiosulfates, 158
Thiourea, 91
Tin, in aluminum, 321
by the bromate method, 321
by cacotheline, 103
in copper-base alloys, 329
by mercuric chloride and aniline, 103
micro test, 83
volumetric method for, 329
Titanium, in aluminum, colorimetric method, 313
in carbon steel, 309
in cast iron, 309
by chromotropic acid, 103
colorimetrically, 197
after cupferron separation, 308
in copper steel, 308
by hydrogen peroxide, 103
in open-hearth iron, 309
titrimetrically, 197
in wrought iron, 309
Titanium dioxide, barium sulfate in, 507
calcium sulfate in, 508
iron in, 508
Titanium steel, titanium in, 308
Titanium-vanadium steel, 309
Titration, electrometric, 125
of soap in emulsions, 555
Titer value, of fats and oils, 352
on soap fatty acids, 396
Tools of the analyst, 7
Topped crude petroleum, 665
Tops, 665
Total acidity of water, 233
Total alkalinity, in sodium hypochlorite, 138
as sodium oxide, 138
of sulfated oil, 372
Total ammonia in emulsions, 555
Total anhydrous soap, 387
Total ash of heavy-metal soaps, 406
Total carbon in ferrous metals, 282
Total chlorine, in sodium hypochlorite, 137
Total extract from rubber, 561
Total fatty matter, in soap, 391
Total hardness of water, 234
in sulfated oils, 377
Total solids, in lacquers, 478
in latex, 583
in water, 227
Total sulfated fatty matter in sulfated oils, 378
Triangles, 19
Tripoli, 178
Trisodium phosphate, 265
and sodium carbonate, 267
Trisodium phosphate, and sodium metasilicate, 270
and sodium tetraborate, 273
Tube, Dean-Stark, 41
Tubes, Nessler, 61
Tungsten, by potassium thiocyanate, 104
in tungsten steel, 305
Tungsten steel, molybdenum in, 303
silicon in, 292
sulfur in, 290
tungsten in, 305
Turbidity of water, 243
Turpentine, 445
and hydrocarbons, 453
Tussah, 637
Twaddell hydrometer, 47

U

Ultimate analysis of coal, 598
Ultramarine, 535
Unbleached fibers, 631
Unbleached sulfate fibers, 632
Union oil colorimeter, 709
Universal indicators, 114
Universal viscosimeter, Saybolt, 691
Unsaponifiable matter, cylinder method for, 348
dry extraction for, 350
in emulsions, 546
in fats and oils, 347
in plasticizers, 482
in rubber, 571
separatory funnel method for, 349
in soap, 393
in sulfated oils, 379
in waxes, 430
Unsaponified, matter in soap, 394
and unsaponifiable matter of soap, 391
Unsaturated hydrocarbons in gas, 608
Unworked consistency of grease, 419
Uranium, by potassium ferrocyanide, 104

V

Vacuum drying of volatile matter, 41
Vanadium, in carbon steel, 301
in cast iron, 301
in chromium-tungsten-vanadium steel, 301
in chromium-vanadium steel, 300
in high-chromium high-nickel steel, 301
by 8-hydroxyquinoline, 105
by hydrogen peroxide, 104
in open-hearth iron, 301
in wrought iron, 301
Vanadium steel, titanium in, 309
Varnish, 468
natural resin, 473
spirit, 468
synthetic resin, 477
volatile thinner in, 473
Vegetable fats. cholesterol in, 362

Vegetable fibers, 634, 642
  modified, 639
Vermilion, 523
Viscose and cuprammonium rayon, differentiating, 642
Viscose rayon, 640
  identification of, 641
Viscosity index, 696
Viscosity by Saybolt viscosimeter, 691
Viscosity tube, 692
Volatile amines and ammonia, 140
Volatile matter, in coal, 598
  conductimetric method, 43
  determination of, 40
  by drying in a desiccator, 43
  in fats and oils, 340
  by ignition, 43
  by oven drying, 40
  by solvent distillation, 41
  by vacuum drying, 41
Volatile thinner, in paint, 469
  in varnish, 473
Volhard method for manganese, 527
Volumetric glassware, 8
Volumetric solutions, 717, 721
Vulcanization properties of synthetic elastomers, 591
Vycor glassware, 7

W

Wash bottles, 12
Washing of precipitates, 74
Watch glasses, 11
Water, acidity of, 233
  albuminoid nitrogen in, 240
  alkalinity of, 231
  ammonia in, 240
  ammonia- and magnesium-free, 146
  analysis, 226
  in bituminous material, 705
  carbon dioxide in, 234
  carbonate and noncarbonate hardness of, 234
  chloramine in, 250
  chloride in, 228
  chlorine demand of, 248
  chlorine in, 246
  color of, 243
  dissolved solids in, 227
  distilled, 16
  fluoride in, 250
  in greases, 417
  mineral acids in, 233
  nitrate in, 237
  nitrite in, 236
  oil in, 253
  in paint, 468
  in petroleum products, 705–706
  pH value of, 240
Water, reporting results of analysis of, 239
  residual chlorine in, 249
  samples, collection of, 227
  sulfate in, 230
  sulfite in, 252
  suspended solids in, 227
  turbidity of, 243
Water-base wax emulsions, 556
Water bath, 31
Water-dispersible gums, 543
  in emulsions, 553
Water-in-oil emulsions with and without suspended solids, 553
Water-insoluble volatile acids in fats and oils, 357
Water-soluble liquids in emulsions, 547
Water-soluble substances in latex, 584
Water-soluble volatile acids in fats and oils, 356
Water supplies, municipal, 227
Wax, 423
  acid value of, 429
  Adurco, 428
  aniline point of, 431
  bayberry, 424
  bees, 424
  candelilla, 426
  carnauba, 426
  Chinese insect, 426
  constants of, 425
  emulsions, 540, 553, 557
    water-base, 556
  general properties of, 423
  I.G., 428
  iodine value of, 431
  Japan, 426
  Lanette, 428
  melting point of, 430
  montan, 427
  Opal, 428
  ouricuri, 427
  precipitation value for, 432
  qualitative solubility of, 433
  Rilan, 428
  saponification value of, 429
  solubility in cold alcohol, 432
  specific gravity of, 429
  synthetic, 428
  unsaponifiable matter in, 430
Waxy substances, 423
Weathered crude petroleum, 665
Weighing, 39
  accuracy of, 39
Weights, analytical, 30
Wesson colorimeter, 66
Westphal balance, 48
White lead, lead as lead sulfate, 501
White pigments, 496, 509
  calculation of results on mixed, 518
  qualitative tests on, 497

Wijs method, iodine value by, 346
Wijs solution, 721
Wild silk, 637
Wiley method for melting point, 353
Williams method for manganese, 528
Winkler explosion bulbs, 609
Winkler method for dissolved oxygen, 241
Wire gauzes, 19
Wool, 635
  artificial, 638
  determination of, in mixtures, 646
  reprocessed and reused, 647
Worked consistency of greases, 420
Wrought iron, copper in, 294
  manganese in, 286–287
  molybdenum in, 302
  nickel in, 297
  phosphorus in, 289
  silicon in, 292
  sulfur in, 290
  titanium in, 309
  vanadium in, 301
  zirconium in, 310

Y

Yellow, chrome, 529
  zinc, 532

Z

Zinc, in aluminum, 313, 315
  by ammonium mercuric thiocyanate method, 315
  chromate, 532
  by cobalt-mercury thiocyanate, 105
  in copper-base alloys, 327
  by diethylaniline and ferricyanide, 105
  micro test for, 85
  by potassium ferrocyanide, 214
  as the sulfide, 327
  volumetrically by ferrocyanide, 504
  yellow, 532
  by zinc oxide method, 315
Zinc oxide, 504
Zinc oxide method for zinc, 315
Zinc oxide-α-nitroso-β-naphthol method, for cobalt, 306
Zinc uranyl acetate method for sodium, 222
Zirconium, by alizarin, 106
  in carbon steel, 310
  in cast iron, 310
  by cupferron-phosphate, 310
  by β-nitroso-α-naphthol, 106
  in open-hearth iron, 310
  in wrought iron, 310
  in zirconium steel, 310